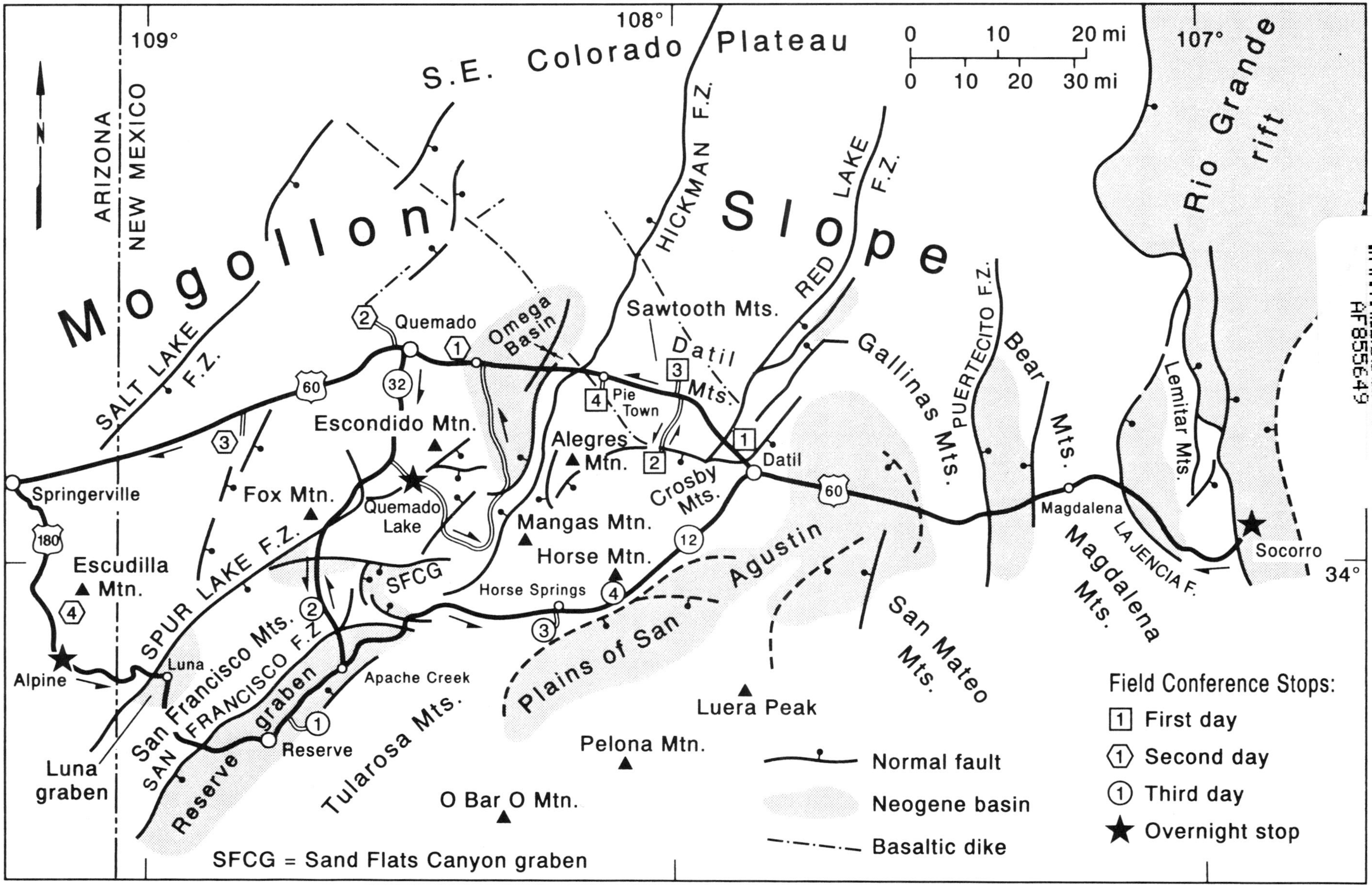
S.E. Colorado Plateau
Mogollon Slope
Rio Grande rift
109°
108°
107°
34°
0 10 20 mi
0 10 20 30 mi
ARIZONA
NEW MEXICO
N
SALT LAKE F.Z.
HICKMAN F.Z.
RED LAKE F.Z.
PUERTECITO F.Z.
SPUR LAKE F.Z.
SAN FRANCISCO F.Z.
LA JENCIA F.
Springerville
Escudilla Mtn.
Alpine
Luna
Luna graben
San Francisco Mts.
Reserve graben
Reserve
Apache Creek
Tularosa Mts.
SFCG
Fox Mtn.
Escondido Mtn.
Quemado
Quemado Lake
Omega Basin
Pie Town
Sawtooth Mts.
Datil Mts.
Alegres Mtn.
Mangas Mtn.
Horse Mtn.
Horse Springs
Crosby Mts.
Datil
Gallinas Mts.
Bear Mts.
Magdalena
Magdalena Mts.
Lemitar Mts.
Socorro
Plains of San Agustin
San Mateo Mts.
Luera Peak
Pelona Mtn.
O Bar O Mtn.
60
32
12
180
Normal fault
Neogene basin
Basaltic dike
Field Conference Stops:
1 First day
1 Second day
1 Third day
Overnight stop
SFCG = Sand Flats Canyon graben

Cowboys ready to ride from the Navajo Lodge in Datil, New Mexico, ca. 1920. Photograph courtesy of Buz and Beverly Easterling, Quemado Lake Estates.

Mogollon Slope, West-Central New Mexico and East-Central Arizona

Editors

RICHARD M. CHAMBERLIN
BARRY S. KUES
STEVEN M. CATHER
JAMES M. BARKER
WILLIAM C. MCINTOSH

New Mexico Geological Society
Forty-Fifth Annual Field Conference
September 28–October 1, 1994

This spectacular south-facing cliff (Peak "8919") in the eastern Sawtooth Mountains provides a cross-section view of soft-sediment deformation and liquefaction-related structures in the upper Eocene Dog Springs Formation. Regional soft-sediment deformation of these argillaceous andesitic sandstones and overlying andesitic debris-flow deposits may have been triggered by a major seismic event in late Eocene time. An ESE-dipping low-angle fault, rising to the left, juxtaposes steeply ENE-dipping debris-flow beds on upper right with underlying strongly folded sandstones. Note thick sandstone bed, doubled up in an east-plunging isoclinal fold on lower right. This cliff is approximately 150 m high on its west face, to the left. Photograph by Richard M. Chamberlin, photographic enhancement by William DeMarco, DeMar Co., Photographics.

CONTENTS

Volcanic Geology, Geochemistry and Geochronology

Stratigraphy, Sedimentology and Paleontology

Economic Geology

Quaternary Geology, Hydrogeology and Environmental Geology

DEDICATION

The 1994 NMGS Guidebook is dedicated to James C. Ratté, a USGS geologist who has mapped in southwestern New Mexico for nearly 30 years. He is widely known not only to fellow geologists, but also to ranchers, packers, foresters, miners, and other local residents throughout the area.

Originally from Vermont, Jim joined the U.S. Navy in 1943. After the war ended, he completed an undergraduate degree in Geology at Michigan State University in 1950, marrying Fran Brut in his Junior year. (He and Fran have raised five daughters and celebrate their 45th wedding anniversary this year.) Jim earned his Master's degree from Dartmouth in 1952, and followed this with a year of study at Cal Tech. He began his 40-year career with the U.S. Geological Survey in 1953, and immediately began work in the San Juan volcanic field in Colorado. Jim, together with Tom Steven, delineated calderas and ignimbrites in the Creede and Summitville areas, laying the foundations of understanding for what is now probably the best studied large silicic volcanic field in the world. Following his work in the San Juan Mountains, Jim spent several years evaluating mineral and water resources in South Dakota, then turned his attention to the Mogollon-Datil volcanic field in southwestern New Mexico. From 1967 to 1970 Jim and his family spent summers in the area between Alpine and Glenwood, mapping wilderness and primitive areas as part of the U.S.G.S. resource evaluation program. Since 1970 Jim has returned nearly every spring and fall to continue field mapping in southwest New Mexico. Jim received the Meritorious Service Award from the Department of the Interior in 1988 and officially retired in 1993, but is continuing his work for the U.S.G.S. as scientist emeritus.

Of the 89 publications attributed to Jim Ratté by GEOREF, 58 are concerned with New Mexico geology. The remainder pertain to the geology of widely ranging areas including Colorado, South Dakota, Alaska, Saudi Arabia, and Bolivia. Jim's New Mexico publications include more than 10 quadrangle maps, numerous smaller scale maps, several journal articles, and five N.M.G.S. guidebook contributions. Although he has concentrated on mapping, stratigraphy, and mineral potential of Cenozoic volcanic rocks, Jim's work has included volcanology, geochemistry, and geochronology. His fundamental contributions to the geology of the Mogollon-Datil volcanic field include development of a regional stratigraphic framework, determining the location, relative age, and structure of calderas, mapping the distribution of ignimbrite sheets and eruptive vents of lavas, characterizing mineralization, and deciphering regional structure.

Above all, Jim Ratté is known for the quality, quantity, accuracy, and thoroughness of his geologic mapping. Even now, as black box geology continues to become more fashionable and fundable than mapping, Jim continues to go into the field and produce first quality maps in geologically and topographically challenging volcanic terrain.

In addition to his geological talents, Jim Ratté is a kind, polite, and generous human being. He always has abundant time to help and advise other geologists, especially students. Also, as evidenced by the photos above, Jim does not age at the same rate as the rest of us. Although officially a "scientist emeritus" he is continuing his work in New Mexico and can still walk the legs off most younger geologists.

W. C. McIntosh

PRESIDENT'S MESSAGE

Welcome to this year's New Mexico Geological Society Field Conference—our forty-fifth! The Society is more than half-way through its fifth decade and quickly approaching its fiftieth anniversary. The long history of the Society reflects the hard work and dedication of the many individuals who have contributed their time, expertise and efforts over the years to the success of the organization.

This conference is the collective brainchild of general chairpersons Richard Chamberlin and Jim Ratté with additional insights from Steve Cather. Northeastern Catron County was last visited as part of the tenth Society Field Conference in 1959, so a new look is welcome and overdue.

Organizing this trip and publishing the guidebook required tremendous efforts by numerous people, most of whom responded "above and beyond the call of duty." Such people are necessary for the success of the trip and the publication of the guidebook, and are many times taken for granted. Many of the volunteers are listed by committee in the front matter of this guidebook. Please take the opportunity to thank these individuals because without their support, organizing future trips and guidebooks would be more challenging than it already is.

The Society is and has been basically self-supporting, in the sense that no outside financial support is necessary to keep its activities and publications progressing from year to year. The economic health of the Society seems to have stabilized in recent years, and appears to be quite good, particularly when we reflect on the status and future of the oil and gas, and mining industries in this country. The principal source of revenue for the Society is publication sales, which were down nearly 11 percent in 1993. Additionally, the cost of publishing new guidebooks and operating costs are increasing. Therefore, the Society has taken some steps to lower the cost of publishing the guidebooks and reduce the number of guidebooks in storage. This has also involved streamlining the process of putting the guidebook together, using less-costly materials, while maintaining the quality we're accustomed to.

The ability to award annual scholarships and fellowships, totaling several thousand dollars, is a point of pride for the Society. The Society has benefitted from recent donations by Ms. Lucille Pipkin and the Wellnitz family. The Society's current scholarships include the Beverly A. Wellnitz Memorial Scholarship, the Kottlowski Fellowship, the Wellnitz Fellowship, the Pipkin Book Scholarships and the Grants-in-Aid. These funds are awarded annually to deserving geology students at New Mexico universities and colleges. The Society also benefits greatly from the continued support of Chuck Chapin and the New Mexico Bureau of Mines and Mineral Resources.

This year's conference will provide an opportunity to see and appreciate an area of New Mexico and Arizona which hasn't been visited by the Society in many years. I hope you enjoy the trip and have a great time!

Bob Newcomer, President

EDITORS' MESSAGE

The 1994 New Mexico Geological Society Fall Field Conference journeys through the serene outback of northeastern Catron County, New Mexico and part of adjacent Apache County, Arizona. We refer to this scenic terrane of volcanic capped ranges on the northern periphery of the Mogollon-Datil volcanic field and the downwarped southern margin of the Colorado Plateau as the Mogollon slope. This area was last visited by the Society in 1959 as numerous reconnaissance maps of west-central New Mexico were being completed in preparation for the now familiar geologic map of New Mexico by C. H. Dane and G. O. Bachman (1965). For the last 35 years our geologic understanding of west-central New Mexico has developed on a foundation laid by Frank Kottlowski, John Shilling, Robert Weber, Max Willard, Roy Foster, Clay Smith, Gus Armstrong, Sherm Wengerd, Charlie Stearns, Carl Dane and George Bachman, just some of the contributors to the 1959 guidebook. Robert Weber, now retired, here again contributes to the Society's collective knowledge with his description of pluvial lakes of the Plains of San Agustin. Also Clay Smith and his colleagues of the Socorro Lions Club will cater a barbecue and breakfast among the pines at the old Quemado Lake campground; the only place we could find to put 100 or more "rock-nockers" overnight in northern Catron County.

New regional geologic maps of the Tularosa Mountains 30' x 60' quadrangle by Jim Ratté and the late Tommy Finnell; and the Quemado 30' x 60' quadrangle by Richard Chamberlin, Steve Cather, Orin Anderson and Glen Jones provide the geologic picture of the field conference area. As with the previous generation of maps in this area, the impetus for new regional maps is the upcoming geologic map of New Mexico (Anderson and Jones, in preparation) which will supplant the map of Dane and Bachman. Bill McIntosh and Matt Heizler at the New Mexico Geochronology Research Laboratory provided high-precision $^{40}Ar/^{39}Ar$ ages for regional ignimbrite sheets and lavas that represent critical time lines on these maps.

The 26 articles, 23 minipapers and 4 road logs that comprise this guidebook represent a new geological understanding of west-central New Mexico, adjacent parts of the Mogollon-Datil volcanic field and the southern Colorado Plateau.

In addition to the many volunteers listed in the credits and committees, we here express our appreciation to all individual contributors to the guidebook and field conference as listed in the table of contents. Our special thanks to Chuck Chapin, State Geologist and Director of the New Mexico Bureau of Mines and Mineral Resources, who generously provided personnel, word processing, cartographic drafting, secretarial help, photographic work, vehicles and helpful encouragement. Lynne Hemenway and Terry Telles quickly and pleasantly typed most of the guidebook. Many quality line drawings and illustrations were cheerfully and expertly made by Rebecca Titus, Kathryn Campbell and Jesse Dengate. Tasty lunches will be provided by Xi Theta Chapter of Beta Sigma Phi, Socorro.

Finally, our thanks to all landowners and citizens of Catron County who have graciously permitted access through their property to examine the rocks that make up the Mogollon slope.

Richard M. Chamberlin, Barry S. Kues, Steven M. Cather, James M. Barker, William C. McIntosh

COMMITTEES

EXECUTIVE COMMITTEE

Robert W. Newcomer, Jr., *President*	John W. Shomaker, Inc.
Richard M. Chamberlin, *Vice President*	New Mexico Bureau of Mines and Mineral Resources
David A. Schoderbek, *Treasurer*	Meridian Oil
James C. Witcher, *Secretary*	New Mexico State University
Ronald F. Broadhead, *Past President*	New Mexico Bureau of Mines and Mineral Resources

FIELD CONFERENCE

Richard M. Chamberlin, *General Co-chairman*	New Mexico Bureau of Mines and Mineral Resources
James C. Ratté, *General Co-chairman*	U.S. Geological Survey, Retired

GUIDEBOOK

Barry S. Kues, *Managing Editor*	University of New Mexico
Richard M. Chamberlin	New Mexico Bureau of Mines and Mineral Resources
Steven M. Cather	New Mexico Bureau of Mines and Mineral Resources
James M. Barker	New Mexico Bureau of Mines and Mineral Resources
William C. McIntosh	New Mexico Bureau of Mines and Mineral Resources

ROAD LOGS

Richard M. Chamberlin	New Mexico Bureau of Mines and Mineral Resources
James C. Ratté	U.S. Geological Survey, Retired
Steven M. Cather	New Mexico Bureau of Mines and Mineral Resources
William C. McIntosh	New Mexico Bureau of Mines and Mineral Resources
James C. Witcher	New Mexico State University
Orin J. Anderson	New Mexico Bureau of Mines and Mineral Resources
Dana J. Bove	U.S. Geological Survey
Steven G. Crews	ARCO Exploration and Production Technology
Spencer G. Lucas	New Mexico Museum of Natural History and Science

REGISTRATION

Orin J. Anderson	New Mexico Bureau of Mines and Mineral Resources

LOGISTICS

Gretchen K. Hoffman	New Mexico Bureau of Mines and Mineral Resources
James C. Ratté	U.S. Geological Survey, Retired
George S. Austin	New Mexico Bureau of Mines and Mineral Resources
Norma J. Meeks	New Mexico Bureau of Mines and Mineral Resources
James M. Barker	New Mexico Bureau of Mines and Mineral Resources
James A. Olsen	U.S. Bureau of Land Management
Orin J. Anderson	New Mexico Bureau of Mines and Mineral Resources

TECHNICAL ASSISTANCE

Halliburton Logging Services	Mobile Sound Equipment
Salt River Project (SRP)	Beverages

PUBLICATION SALES

Norma J. Meeks	New Mexico Bureau of Mines and Mineral Resources
James M. Barker	New Mexico Bureau of Mines and Mineral Resources
Gretchen K. Hoffman	New Mexico Bureau of Mines and Mineral Resources

1994
FIELD CONFERENCE SCHEDULE

Wednesday, September 28—Registration Day

4:00–8:00 p.m.	Registration: Lower lobby of Macey Center, New Mexico Tech, Socorro, New Mexico.
6:00–9:00 p.m.	Icebreaker: North patio of Macey Center, Socorro, New Mexico.

Thursday, September 29—First Day

7:30 a.m.	Assembly point, frontage road at State Forestry Division Office; approximately 1 mile west of Socorro on U.S. 60. Registration for late arrivals at first stop.
7:45 a.m.	Caravan departs headed west on U.S. 60 for tour of Cenozoic structure and stratigraphy of the Datil-Pie Town region. (Lunch provided)
5:00 p.m.	Arrive at campgrounds of Quemado Lake Recreation Area. Camp at older facility southeast of Quemado Lake. No potable water, but modern looking outhouses.
6:00–7:30 p.m.	Catered outdoor barbecue at Quemado Lake campground.

Friday, September 30—Second Day

6:30–7:30 a.m.	Catered outdoor breakfast at Quemado Lake campground.
7:45 a.m.	Assembly Point, southeast end of Quemado Lake campground on FR 13.
8:00 a.m.	Caravan departs for tour of Mangas Mountains and the Quemado-Red Hill-Alpine region. (Lunch provided)
*5:00 p.m.	Arrive Alpine, Arizona. (local time 4:00 p.m.)
*6:30–8:00 p.m.	Banquet at Alpine Country Club, Alpine, Arizona
*8:00 p.m.	Speaker: Dr. Richard V. Fisher, Prof. Emeritus, University of California Santa Barbara, "Catastrophic volcaniclastic sedimentation."

Saturday, October 1—Third Day

*7:00–7:50 a.m.	Breakfast at Alpine Country Club, Alpine, Arizona
*7:50 a.m.	Assembly point, parking area at Alpine Country Club.
*8:00 a.m.	Caravan departs for tour of Reserve graben, Apache Creek and Horse Springs areas. (Lunch provided)
4:00 p.m.	Conference ends at Horse Springs, ~20 miles south of Datil.

* Schedule in Arizona will be on New Mexico time (Mountain *daylight* time).

CREDITS

Front Cover: Eastern Sawtooth Mountains, color separation by Starline Printing Co., from a slide by Richard M. Chamberlin, New Mexico Bureau of Mines and Mineral Resources.

Front End Sheets: Satellite image, Jane C. Love, New Mexico Bureau of Mines and Mineral Resources; Trip diagram, Rebecca Titus, New Mexico Bureau of Mines and Mineral Resources.

Photography: Steve M. Cather and Richard M. Chamberlin, New Mexico Bureau of Mines and Mineral Resources; William DeMarco, De Mar Co. Photographics; and James C. Ratté, U.S. Geological Survey, retired.

Historical Photographs: Kenneth and Carol Coker, Eagle Guest Ranch; Buz and Beverly Easterling, Quemado Lake Estates; and Jimmy and Irene Jaramillo, El Sarape Cafe.

General Interest Photographs: Richard M. Chamberlin, New Mexico Bureau of Mines and Mineral Resources; Kelly D. Gatlin, La Luz Photography; and Wayne Lambert, West Texas A & M University, assisted by Kirby Hoffman and supported by a grant from Killgore Research Center, West Texas A & M University.

Drawings: Louann C. Jordan, Santa Fe.

Word Processing: Lynne Hemenway, Terry Telles, and Theresa Lopez, New Mexico Bureau of Mines and Mineral Resources.

Drafting: Rebecca Titus, Kathryn Campbell and Jesse Dengate, New Mexico Bureau of Mines and Mineral Resources.

Printer and Graphic Design: Starline Printing Company

Cliffs of upper Eocene Dog Springs Formation in the eastern Sawtooth Mountains approximately 25 km northwest of Datil, New Mexico. view is S26°E. Location of Stop 3, first day, is below and right of the tooth-like pillar. Pillar and upper half of cliff face to left consist of nearly vertical dacitic debris-flow beds above a subhorizontal decollment or detachment fault. Highly contorted and folded dacitic sandstone beds below the subhorizontal fault grade downward into gently dipping sandstones conformable with the underlying Baca Formation. Piñon-juniper woodlands on lower slopes generally mask the underlying Baca Formation; ponderosa pine is present on the higher slopes. Camera station is in NE¼ sec. 4, T1N, R11W. Wayne Lambert photograph No. 93L48. August 11, 1993, 5:33 pm MDT.

FIRST-DAY ROAD LOG FROM SOCORRO TO MAGDALENA, DATIL, WESTERN CROSBY MOUNTAINS, SAWTOOTH MOUNTAINS, PIE TOWN, QUEMADO AND QUEMADO LAKE

RICHARD M. CHAMBERLIN, STEVEN M. CATHER, WILLIAM C. MCINTOSH, ORIN J. ANDERSON, and JAMES C. RATTÉ

THURSDAY, SEPTEMBER 29, 1994

Assembly point: Frontage road next to New Mexico State Forestry Division office approximately 1 mile southwest of Socorro on US-60.
Departure time: 7:45 a.m.
Distance: 161.6 mi.
Stops: 4

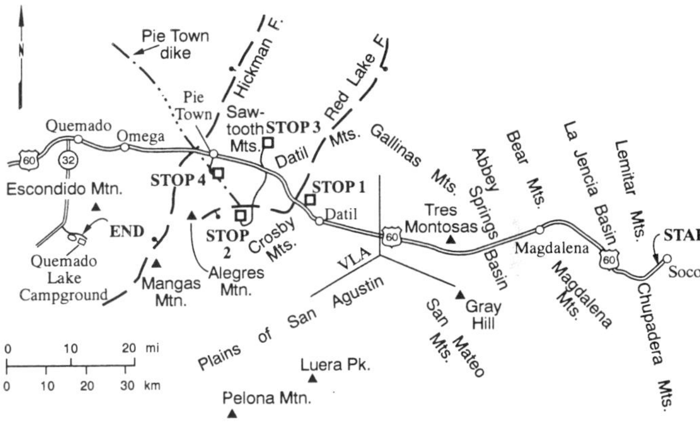

SUMMARY

The First Day tour emphasizes the Cenozoic stratigraphy and structure of the Mogollon slope in northeastern Catron County. Gently south-dipping Eocene and Oligocene orogenic to volcanogenic strata as much as 1.3 mi thick mark the downwarped (loaded) southeastern margin of the stable Colorado Plateau microplate, referred to here as the Mogollon slope. US-60 west from Socorro generally follows strongly to moderately extended topographically lower terrain of the San Agustin arm of the Rio Grande rift until it crosses onto the Mogollon slope at the Red Lake fault zone west of Datil. The San Agustin arm marks a complex transition zone from rift to plateau; thus US-60 crosses several north-trending extensional basins and distended horsts between Socorro and Datil.

Stop 1 west of Datil provides an overlook of the Red Lake fault zone, a narrow NNE- trending belt of Neogene extension that locally cuts across the Mogollon slope. This mildly distended belt was probably guided by weak lithosphere below a zone of earlier Laramide transpression. The massive south-sloping core of the Datil Mountains visible to the west of Stop 1 locally defines an edge of the gently depressed Mogollon slope.

Stop 2 permits a hands-on look at upper Eocene ignimbrites of the Datil Group and intercalated volcaniclastic sedimentary apron deposits of the middle Spears Group that are well exposed on the west flank of the Crosby Mountains. Both epiclastic braided-stream deposits and syneruptive tuffaceous sandstones deposited by an overloaded stream system are locally well exposed here at Saulsberry Ranch.

Today's third and lunch stop is near Monument Rock in the eastern Sawtooth Mountains. Bold cliffs to the northeast of Monument Rock provide spectacular views of intense soft sediment deformation within the upper Eocene Dog Springs Formation. Intensely folded andesitic sandstones and upturned andesitic debris-flow deposits visible in these cliffs may reflect a seismically triggered regional liquefaction event of late Eocene age.

Stop 4 is along the late Oligocene Pie Town dike, where it forms a large wall-like outcrop at Bright Lake south of Pie Town. Country rocks of Eocene andesitic sandstone and the chilled margin of this basaltic-andesite dike are well exposed here. This great dike, approximately 45 mi long, indicates regional extension in late Oligocene time was to the WSW.

From Pie Town the route follows US-60 west to Quemado and then south along NM-32 to the Apache National Forest and the Quemado Lake Recreation Area. Following a catered barbecue we will camp out below the tall ponderosa pines at the old Quemado Lake campground in the valley southeast of the reservoir.

The first 43.8 mi of this road log is modified and updated after the road log of Chapin et al. (1978), which emphasized then-newly-discovered calderas and their relationship to well-known mining districts in the Mogollon-Datil volcanic field. This log and following road logs adopt the Tertiary stratigraphic nomenclature proposed by Cather et al. in this guidebook. Road logs also use 33.4 Ma as the Eocene-Oligocene boundary following the lead of McIntosh et al. (1992); this boundary was previously placed at 36.6 Ma (Palmer, 1983). Ages of Tertiary ignimbrites and other volcanogenic units are from McIntosh et al. (1991, 1992) and papers in this guidebook unless otherwise specified.

Mileage

0.0 Stop sign at US-60 and road from the Dicaperl perlite mine. **Turn right, and head west on US-60. 0.2**

0.2 Bridge over Socorro flood-control channel. Remnants of the Pliocene basalt of Socorro Canyon (Osburn and Chapin, 1983) cap ridge from 1:00 to 2:00. Pale red piedmont-slope conglomerates and buff axial-river sands of the Sierra Ladrones Formation (upper Santa Fe Group) are conformably overlain by this basalt flow near the white water tank at the east point of the ridge. **0.2**

0.4 Milepost 136. Thick cauldron-facies Hells Mesa Tuff is exposed in low rounded peaks of the central Chupadera Mountains at 12:30. This structurally high, resurgent-like block lies near the eastern center of the Socorro cauldron, from which the Hells Mesa Tuff was erupted at 32.06 Ma (Osburn and Chapin, 1983; McIntosh et al., 1991). **0.3**

0.7 Approaching fault scarp on right that displaces inset terrace gravels of late Pleistocene age (~ 0.1 Ma) as much as 10 ft. Note gravel pile, from trench across fault scarp, near the fence line. This is the recurrent Socorro Canyon fault, which may have moved as recently as the early Holocene (Machette and McGimsey, 1982; Machette, 1982). At 9:00 the middle Pleistocene piedmont slope and remnants of a 3-ft-thick laminar calcrete are offset 100 ft by the Socorro Canyon fault. Pliocene to lower Pleistocene strata of the Sierra Ladrones Formation may be as much as 1150 ft thick on the downthrown eastern side of the Socorro Canyon fault (Chamberlin, 1981a). Based on gravity data, Sanford (1978) estimated that the total structural relief along the west side of the Socorro Basin is as much as 2.2 mi. From 9:15 to 12:15 a thin veneer of middle Pleistocene piedmont gravels unconformably overlies discontinuously and poorly exposed strata of the upper Popotosa Formation (lower Santa Fe Group). Where better exposed in the next canyon to the south these upper Miocene Popotosa beds dip 30° to the east. **0.2**

0.9 At 2:30, bulldozer cuts of the Dicaperl perlite mine expose red mudstones of the upper Popotosa Formation overlapping light gray hills of perlite that define the buried crest of a 7.85-Ma, high-silica, rhyolite lava dome (Chamberlin et al., 1994a). Chemical weathering (hydration) of this small, 0.02 mi^3, glassy lava dome has produced a world-class perlite deposit (see following minipaper). **0.4**

GEOLOGY AND MINING OF THE SOCORRO PERLITE DEPOSIT

James M. Barker[1], Timothy L. Hall[2], and Richard M. Chamberlin[1]

[1]New Mexico Bureau of Mines & Mineral Resources, Socorro, NM 87801;
[2]Dicaperl, Socorro, NM, 87801

The Socorro perlite deposit, mined by Dicaperl (a unit of Grefco), is about 4 mi southwest of Socorro in central New Mexico. The Dicaperl Socorro mine and processing plant is in sec. 27 T3S R1W and occupies 70 to 90 acres on patented claims. The mine is adjacent to the plant and employs about 35 people. Socorro was one of the first perlite mines in the United States when it opened in 1949. Closed from 1959 to 1975, large reserves and high quality have led to continual operation since then.

The small volume, high-yield Socorro perlite deposit lies along a Quaternary fault-line escarpment that defines the eastern flank of the west-tilted Socorro Peak uplift (Weber, 1963, Chamberlin, 1980; Bobrow et al 1983). The Socorro perlite deposit is a zone of weathering (hydration) on a small glassy lava dome of high-silica rhyolite with an $^{40}Ar/^{39}Ar$ age of 7.85 ± 0.04 Ma, from sanidine phenocrysts (Chamberlin et al., 1994), that agrees with an earlier K-Ar whole-rock age of 7.4 ± 0.4 Ma (Chamberlin, 1980).

The following geology is largely from Chamberlin, Barker and Jenkins (1994). The lava dome has a maximum lateral extent of 2800 ft, and is exposed over a topographic interval of 450 ft. The partially buried crest of the dome is flat over a distance of 1200 ft. The high-silica rhyolite lava (76-77% SiO_2, anhydrous), was strongly enriched in lithophile trace elements Rb, Mn, Nb, Ta, Pb, Th, U and heavy rare earths and is relatively depleted in Sr and Ba (Bobrow et al., 1983). The steep flanks of the original dome are largely buried by younger sandstone and mudstone of the lower Santa Fe Group. The southeast margin of the dome is a fault contact; another fault bisects the main perlite outcrop into east and west halves. The bisecting fault is down-to-the-east normal with an approximate throw of 200 ft. Topographic relief in the western block and recent drilling at the south end of the eastern block both require a minimum thickness of 300 ft for the original unfaulted lava dome.

An early phase of cogenetic lithic-rich ash-flow tuff and bedded tuff, as much as 60 ft thick, is discontinuously exposed around the flanks of the lava dome. Large mudstone clasts (>2 ft), in an ash flow cropping out about 0.5 mi north of the dome, indicate proximity to the source vent. Greater than 95% of the exposed lava dome consists of glassy, microvesicular, flow-banded rhyolite. Sparse (<1%), fine-grained phenocrysts of oligoclase, sanidine, quartz and rare biotite are evenly distributed throughout the dome. A large lens of crystalline felsite (approximately 200 ft x 600 ft) at the northeast margin of the dome is interpreted as a syneruptive "crystalline mush." Flow-banding is dominantly defined by flattened microvesicles and locally by stringers of syneruptive felsite. Open tension fractures in the felsite bands pinch out in adjacent glass, thereby demonstrating a ductility contrast. Flow banding is generally steep, irregular and is roughly concentric within the dome. Density and microvesicularity are inversely proportional; density increases downward. Small lenses of dense nonvesicular perlite locally occur near the center of the dome. Some samples of perlite from the core area contain microscopic nodules of obsidian that range in diameter from 0.0125 to 0.025 in.

The eruptive history (Chamberlin et al., 1994) began at about 7.85 Ma with throat-clearing, small-volume, ash-flow tuffs followed by ash falls. A highly viscous, high-silica magma rose slowly in the open conduit. Chilling of the magma cap, at or near the vent walls (at moderate vapor pressure?), locally formed a granophyric crystallite mush that was part of the initial surface extrusion. Finally, a viscous, foamy, high-silica melt was very slowly extruded through a confined(?) orifice, forming a small, circular, steep-sided and flat-topped dome with a volume of about 0.02 mi^3. Cooling by gas expansion (vesiculation) and by gas escape (dehydration), extreme viscosity, and relatively large ratio of surface area to thermal mass, allowed this small body to cool quickly so that the crystalline core, commonly observed in other rhyolite lava domes, did not form here. Flow fractures, cooling cracks, microfractures and flattened microvesicles created a highly permeable obsidian that facilitated relatively complete and homogeneous alteration (hydration) by meteoric waters to form a commercial perlite.

The mining and milling of Socorro perlite described below are largely from Austin and Barker (1994). A mining crew of three works half a shift to replenish the surge pile and then finishes with development and reclamation work. Perlite is ripped with a D9 Caterpillar, and a scraper (Caterpillar 631D with a screw auger) moves it to the 14 in. grizzly over the primary jaw crusher (-2 in.). A 0.25-mi covered belt conveys the perlite to a seven-day surge pile. A vibratory feeder moves ore into the plant where it is screened to -3/8 in. The +3/8 in. is recycled to a vertical impact crusher. The crushed perlite is dried in a 400°F rotary dryer fired with recycled crankcase oil, to 0.25% free moisture at a discharged temperature of about 250°F. The dry -3/8 in. perlite is rescreened over one of two banks of screens and stored as eight grades, in 12 storage tanks, for blending to customer specifications.

About 150,000 st of crude perlite of all sizes were sold in 1993. Addition of regrind circuits has added capability to better match surges in demand for a particular grade. The capacity of the plant is 45 stph, dropping to 35 stph if intense regrind is underway (maximum of 650 short tons per day [stpd] of crushed and sized perlite). A Santa Fe Railroad spur extends about 3.5 mi from the center of Socorro to the Dicaperl mill.

About 95% of processed perlite is shipped in 100 st bottom-dump railcars or in pressure differential railcars for pneumatic unloading of finer sizes. A Dresser 510 frontend loader switches railcars and moves product around the plant. About 5% of crude perlite is trucked to customers perlite in 1 yd^3 supersacks. Dicaperl expansion plants in Lafayette, LA, and Jackson, MS, take about 10-15% of the total Socorro production. About 10% of production is exported, principally to Canada, with small amounts to Mexico. About 60% of production goes into ceiling tile with most of the remainder used in horticulture. A minor amount goes into filter aids, wallboard, and pipe insulation.

1.3 Road to site of old Great Lakes Carbon perlite mill on right. At 3:30 through small saddle is a view of northerly aligned necks and vent on "M" Mountain, all part of the upper Miocene Socorro Peak Rhyolite (Osburn and Chapin, 1983). **0.3**

1.6 At 3:00 in middle slope, reddish brown craggy outcrops of densely-welded upper Lemitar Tuff (28.0 Ma) define a west-tilted fault block that is unconformably overlain by east-dipping beds of lower and upper Popotosa Formation. High mesa at 3:00 is formed by a flat-topped dacite lava flow of late Miocene age. Vertical columnar jointing visible on west flank of the dacite flow is perpendicular to the flat upper cooling surface. Black Mountain (Mesa) at 2:00 is capped by a 4.1-Ma basalt flow, (Bachman and Mehnert, 1979) designated by Osburn and Chapin (1983) as the basalt of Socorro Canyon. On Black Mountain the basalt of Socorro Canyon unconformably overlies gypsiferous playa deposits of the upper Popotosa Formation. Quaternary landslide blocks of basalt, sliding on incompetent claystones, form the hummocky slope below the mesa. A high-velocity test-track facility of New Mexico Tech (EMRTC) is visible at the east end of the mesa. **1.0**

2.6 Gully at 9:00 exposes varved Popotosa mudstones and siltstones unconformably overlain by middle Pleistocene piedmont gravels. Red Popotosa claystones and siltstones are locally well exposed at 2:45 on north bank of Socorro Canyon. Bowie and McLemore (1987) reported that these gypsiferous claystones consist predominantly of sodium smectite. For the next 0.9 mi, road cuts on the left provide discontinuous exposures of colluvial gravels and underlying red mudstones of the upper Popotosa Formation. **0.9**

3.5 Crossing range-bounding fault on the northeast flank of the Chupadera Mountains. Roadcuts expose west-tilted red andesite porphyry lava and light gray rhyolitic ash beds collectively assigned to the Luis Lopez Formation, that represent the locally erupted post-collapse fill of the Socorro cauldron. The rhyolitic ash beds have been zeolitically altered and contain potassic clinoptilolite (G.S. Austin, oral commun., 1993). A thin ignimbrite below a rhyolite lava just south of this cut has yielded a K-Ar age of 28.6 ± 1.1 Ma from biotite phenocrysts (Chamberlin, 1980). This rhyolite lava member caps the Luis Lopez Formation and is presumably only slightly older than the 28.85 Ma La Jencia Tuff that overlies the Luis Lopez Formation in the southern Chupadera Mountains (Eggleston et al., 1983; McIntosh et al., 1991). The Oligocene andesite porphyry lava is potassium metasomatized; coarse-grained plagioclase phenocrysts are replaced by a chalky white mixture of adularia, clay minerals and quartz (see Dunbar et al., this volume). Preliminary $^{40}Ar/^{39}Ar$ ages of metasomatic adularia in the Socorro region (7.4 and 8.9 Ma) suggest a long period of metasomatism, contemporaneous with playa sedimentation (ca. 15-7 Ma) in the early rift Popotosa basin. Widespread potassium metasomatism in the Socorro-Magdalena region has been attributed to deep circulation of alkaline saline brines associated with playa deposits of the Popotosa basin (Chapin and Lindley, 1986). **0.1**

3.6 Entering north-trending strike valley cut in soft claystones of the upper Popotosa Formation that are downfaulted between resistant rocks of the underlying Luis Lopez Formation. **0.2**

3.8 Roadcut on left in thin basalt flow interbedded in red and green claystones of the upper Popotosa Formation. This thin flow is locally folded into an open northwest-trending syncline apparently sympathetic to a local zone of northeast-trending scissors faults that separate west-tilted Luis Lopez strata on the north from east-tilted Luis Lopez strata on the south. This zone of scissors faults crosses the north end of the Chupadera Mountains about 1.2 mi south of US-60, and is presumably part of the Socorro accommodation zone (SAZ), described in detail by Chapin (1989). The SAZ is a regional domain boundary that trends west-southwest from Socorro and separates strongly extended domains of imbricate fault blocks. Imbricate, domino-style, fault blocks and half grabens north of the SAZ are predominantly tilted to the west, compared to strongly east-tilted domino blocks south of the SAZ. Accommodation zones and transfer faults (strike-slip boundaries of extended domains) are common features in the Basin and Range extensional orogen (Wernicke, 1992) and in the Rio Grande rift (Chapin and Cather, in press). **0.4**

4.2 Cliffs of Luis Lopez Formation tuffs and lavas define the footwall of a fault-line scarp from 9:30 to 1:00; Popotosa claystones are locally exposed on the downthrown block just east of the scarp. **0.2**

4.4 Bridge over Box Canyon; recent attempts to mine placer gold from gravels on the floor of Box Canyon have been unsuccessful. **0.1**

4.5 Large road cuts on right and left expose locally erupted lithic-rich ignimbrites, andesite lavas and a feeder dike collectively assigned to the Luis Lopez Formation. Bouldery red debris flows and pebbly conglomerates of the lower Popotosa Formation disconformably overlie the Luis Lopez ignimbrite and andesite dike toward the west end of the cut. All of the

Luis Lopez members and the lower Popotosa beds are potassium metasomatized.

The light gray, poorly welded, pumiceous, lithic-rich ignimbrite is locally 900 ft thick and characterized by large rounded to angular clasts of densely welded Hells Mesa Tuff and less common clasts of andesitic lavas (Chamberlin, 1980). Another poorly welded, pumiceous, lithic-rich ignimbrite near the base of the Luis Lopez Formation contains mostly andesitic lithic fragments.

The lower Luis Lopez ignimbrite is about 200 ft thick where well exposed about 1.2 mi southeast of here in Black Canyon. Sanidine from the potassium metasomatized lower Luis Lopez ignimbrite at Black Canyon has yielded a relatively low precision $^{40}Ar/^{39}Ar$ age of 29.85 ± 0.31 Ma (M. Heizler, written commun., 1993); approximately 2 Ma younger than the underlying Hells Mesa Tuff. However, potassium metasomatism may have slightly reset this sanidine age. If the apparent 2 Ma hiatus in volcanism is real, then the Luis Lopez Formation may represent a series of small precursor eruptions that culminated in eruption of the large-volume La Jencia Tuff (28.8 Ma) from the Sawmill-Magdalena cauldron (Osburn and Chapin, 1983). The Sawmill-Magdalena cauldron is nested within the older Socorro cauldron. The eastern margin of the Sawmill Canyon depression is locally well defined south of Box Canyon, where the 28.0-Ma Lemitar Tuff laps unconformably onto Luis Lopez Formation at the topographic rim of the caldera. **0.2**

4.7 Road to Luis Lopez manganese district on left. Vent of small andesitic volcano in Luis Lopez Formation is defined by sheeted andesite plug at 2:30, across Socorro Canyon. **0.1**

4.8 Red and green upper Popotosa claystones unconformably overlie red fanglomerates of the lower Popotosa Formation in small pit at 10:00. Bowie and McLemore (1987) summarized published and unpublished clay mineralogy data for this pit. Illite is the dominant clay mineral in the lower Popotosa fanglomerates, and the illite content of the upper Popotosa claystones decreases upward from the unconformity until dioctehedral sodium smectite becomes the dominant species several feet above the contact. They attributed this trend to illitization of smectite by downward percolating potassium metasomatizing fluids. **0.1**

4.9 Road cuts on right and left in Luis Lopez lithic-rich tuff and andesite dikes, unconformably overlain by lower Popotosa fanglomerates. Route crosses buried northeast margin of Sawmill-Magdalena cauldron just west of this road cut. **0.2**

5.1 Road to telephone cable relay station on right. Vent area for the late Miocene (7.1 ± 0.2 Ma; C. E. Chapin, unpubl. data) basalt of Sedillo Hill at 12:30. The upper Miocene basalt of Sedillo Hill was erroneously correlated with the Pliocene basalt that caps Black Mountain at 2:30 (Chamberlin, 1980). Both of these basalt flows overlie claystones of the uppermost Popotosa Formation and are in turn overlain by piedmont-slope gravels of the Sierra Ladrones Formation. **0.3**

5.4 Milepost 131. Range crest west of Socorro Peak, from 3:00 to 1:00, is underlain by thick rhyolite flows and domes of late Miocene age. Below the cliff-forming rhyolite lavas is a hummocky Quaternary landslide terrane consisting of lava blocks that have slid downslope on upper Popotosa claystones. The claystones are only locally exposed in deeper ravines that cut across the landslide terrane. **0.9**

6.3 Road cuts on right in red claystones of the upper Popotosa Formation, largely masked by Quaternary colluvial gravels. Beginning of steep grade up Sedillo Hill. **0.5**

6.8 Break in slope at 2:30 defines poorly exposed contact of upper Santa Fe Group piedmont-slope deposits (Sierra Ladrones Formation) disconformably overlying playa claystones of the lower Santa Fe Group (upper Popotosa Formation). Map patterns at Sedillo Hill (Chamberlin, 1980) suggest that the Sierra Ladrones piedmont gravels locally fill an east-trending paleovalley cut through the basalt of Sedillo Hill and about 200 ft down into the underlying Popotosa claystones. The piedmont gravels thin abruptly to the north, to less than 100 ft, where they overlap the basalt of Sedillo Hill at the apparent north rim of the paleovalley. Pale red, poorly-indurated, conglomeratic sandstones and gravels near the bottom of the paleovalley are locally well exposed in the arroyo wall at 9:00. These beds dip 4° to the west. The thickness of Sierra Ladrones piedmont gravel west of Sedillo Hill is at least 800 ft (Chamberlin, 1980) and may be as much as 1475 ft (C. E. Chapin, unpubl. drill hole log). **1.2**

8.0 Curve right at top of Sedillo Hill. Road cut on left exposes red paleargid soil of early middle Pleistocene age (McGrath and Hawley, 1987). Colorado Plateau (Lucero uplift) is on the distant horizon from 12:00 to 12:30. Hells Mesa, the small butte at 11:50, is capped by gently southwest-dipping Oligocene ignimbrites and marks the northeast end of the Mogollon slope.

Apatite fission-track data from the Ladron Mountains at 12:30 indicate that this triangular west-tilted block, bound by faults on three sides, was rapidly uplifted during the period from 14 to 9 Ma (Kelley et al., 1992). Wernicke and Axen (1988) interpreted the Sierra Ladrones as the isostatically uplifted footwall of a low-angle (25°), east-dipping detachment fault, previously mapped by Black (1964) as the Jeter fault. Fission track data also indicate that temperatures in the Ladron block did not exceed 240°C (Kelley et al., 1992), well below the minimum 300°C required for ductile deformation of quartz as observed in Cordilleran metamorphic core complexes below detachment faults (Wernicke, 1992). **0.4**

8.4 Milepost 128. For the next 11.7 miles, US-60 traverses the western margin of the La Jencia Basin (11:00 to 1:00), a late Miocene to Recent half graben of the central Rio Grande rift (Cather et al., in press). The late Pleistocene to early Holocene scarp of the La Jencia fault (Machette, 1982, 1988), which cuts across the piedmont slope of the Magdalena range from 9:00 to 11:30, delineates the western edge of the La Jencia half graben. The Lemitar Mountains, from 1:30 to 2:00, and Socorro Peak at 3:30, mark the upturned eastern flank of the west-tilted La Jencia half graben. An active fault zone along the west side of the Socorro Basin, the Socorro Canyon-Loma Pelada fault zone (Machette,

1982) bounds the east side of the Lemitar Mountains and presumably interacts with the structurally lower La Jencia fault to facilitate block rotation and spreading within the active domain. A mid-crustal, sill-like, magma body first defined by microearthquake seismograms (Sanford et al., 1977) and later corroborated as a mid-crustal seismic reflection bright spot (Brown et al., 1980) underlies the active domain of extension from Socorro northward to Bernardo. Recently, Lipman (1992) suggested that the Socorro mid-crustal bright spot may represent hydrothermal fluids concentrated along the brittle-ductile transition zone or at the solidification interface of a tabular granitic intrusion. Lipman's interpretation implies that the combined La Jencia and Socorro Canyon-Loma Pelada fault systems may represent an active domain of detachment faulting balanced above a rising or expanding mid-crustal granitic intrusion or hydrothermally softened zone. In this sense, the La Jencia fault would represent a breakaway fault, structurally lower than all other active faults in the domain (Wernicke, 1992). **1.0**

9.4 Milepost 127. Crest of the South Canyon alluvial fan. From 1:45 to 2:30 west-tilted hogbacks of Oligocene ignimbrites in the Lemitar Mountains are the most obvious expression of a late Oligocene episode of domino-style crustal extension (Chamberlin, 1983). High precision $^{40}Ar/^{39}Ar$ ages of Oligocene ignimbrites and of Miocene ash-beds within the middle to upper Popotosa Formation have been combined with stratal tilt data to demonstrate two episodes of relatively rapid extension and block rotation in the Lemitar Mountains, from 28.6 to 27.4 Ma and from 16 to 10 Ma (Cather et al., in press). In the northern Lemitar Mountains a stack of thin basaltic andesite lava flows, each about 23-33 ft thick, forms a wedge-shaped prism as much as 1675 ft thick between the 28.6-Ma Vicks Peak Tuff and the 27.4-Ma South Canyon Tuff. If this lava wedge represents 51 to 73 flows erupted over a span of 1.2 Ma, then on average a basaltic lava flow was erupted every 16,000 to 24,000 yrs during this period. The middle to late Miocene episode of relatively rapid extension began with eruption of the Silver Creek basaltic andesite at 16.2 ± 1.5 Ma (Cather et al., in press). This middle Miocene basaltic andesite is now tilted about 35° to the west. The overlying main body of Popotosa beds, ~2000 ft thick, and intercalated ash beds show a fan-shaped array with dips decreasing to about 10-20° in 12-Ma beds. Total late Cenozoic crustal extension in the Lemitar Mountains is estimated to be about 175% (Chamberlin, 1983). **1.0**

10.4 Milepost 126. **0.9**

11.3 Water Canyon Lodge on right. Route crosses buried northern margin of 32.1-Ma Socorro cauldron and the nested 28.8-Ma Sawmill-Magdalena cauldron near here (Osburn and Chapin, 1983). These volcanic collapse structures locally merge to the southwest where they pass beneath Water Canyon Mesa at 9:00. Cliff-forming debris flows of the lower Popotosa Formation, at canyon mouth on left side, unconformably buried these cauldron margins in Miocene time. Farther south, the cauldron subsidence faults separate and then swing west over the crest of the Magdalena range. An exposure of white, partially welded Hells Mesa Tuff onlaps the northern topographic rim of the Socorro cauldron near North Baldy at 10:00. Palinspastically restored cross sections and elongation of the Socorro cauldron, now 15 mi N-S by 23 mi E-W, indicate an average of 50% crustal extension across the Socorro-Magdalena caldera complex (Chamberlin and Osburn, 1986; Chapin, 1989). **1.6**

12.9 Road to Water Canyon on left. The south-central portion of La Jencia Basin is hydrographically closed and drains into the small playa at 1:00, in middle distance below Ladron Peak. Aggradation and northeastward progradation of the upper Pleistocene Water Canyon alluvial fan, which dominates the southern third of the basin, may have been a factor in the formation of this playa. **0.5**

13.4 Milepost 123. Good view of the crest of the northern Bear Mountains at 11:45 and Hells Mesa at 12:00. The southern Bear Mountains on the lower horizon, at 11:30-11:45, include west-tilted, imbricate fault blocks that represent 25% of west-southwest directed extension (Chamberlin and Osburn, 1986) in comparison to about 5% extension in the northern Bear Mountains (Massingill, 1979). A northeast-trending oblique-slip fault separates the weakly extended northern domain, (part of the Mogollon slope), from the more strongly extended southern domain, which is assigned to the San Agustin arm of the Rio Grande rift (Chapin, 1971). This oblique-slip fault, or transfer fault (Wernicke, 1992), locally marks the southeast margin of the Mogollon slope as defined in the following minipaper by Chamberlin and Cather. **0.9**

DEFINITION OF THE MOGOLLON SLOPE, WEST-CENTRAL NEW MEXICO

Richard M. Chamberlin and Steven M. Cather

New Mexico Bureau of Mines and Mineral Resources, Socorro, NM 87801

Although previously mentioned by Kelley (1957), the Mogollon slope was first defined by Fitzsimmons (1959, p. 114):

> "The Mogollon slope is the southern part of the Colorado Plateau and, in many ways, is a rather nondescript structural unit. In general, the sedimentary rocks in this unit dip gently to the south. The slope is dotted by accumulations of volcanic rocks of many types. Along the south end only volcanic rocks, or sedimentary deposits derived from them, are to be found. At the north end, older sedimentary rocks may be observed in places and the volcanic rocks form local ranges or individual mountains. The precise southern boundary of this unit is not easy to delineate everywhere. Along the Mogollon Rim, where faulting has produced a prominent escarpment the margin is drawn closely. But in much of New Mexico the volcanic accumulations are so thick that, although faults are present, a sharp line cannot be perceived."

The purpose of this minipaper is to present a more explicit definition of the Mogollon slope and to delineate its boundaries relative to other more recently recognized tectonic elements of western New Mexico and eastern Arizona. A generalized tectonic map (Fig. 1.1) shows a first approximation of Cenozoic extensional strain relative to the Mogollon slope and the New Mexico area. Regional topographic trends (Godson, 1981) and physiographic patterns (NMGS, 1982, physiographic map) are used as a guide, but known or inferred extensional strain is the ultimate criteria (e.g., Wernicke, 1992). For example, positive topographic elements such as the Magdalena Mountains and eastern San Mateo Mountains that show strong domino-style crustal extension (e.g.,

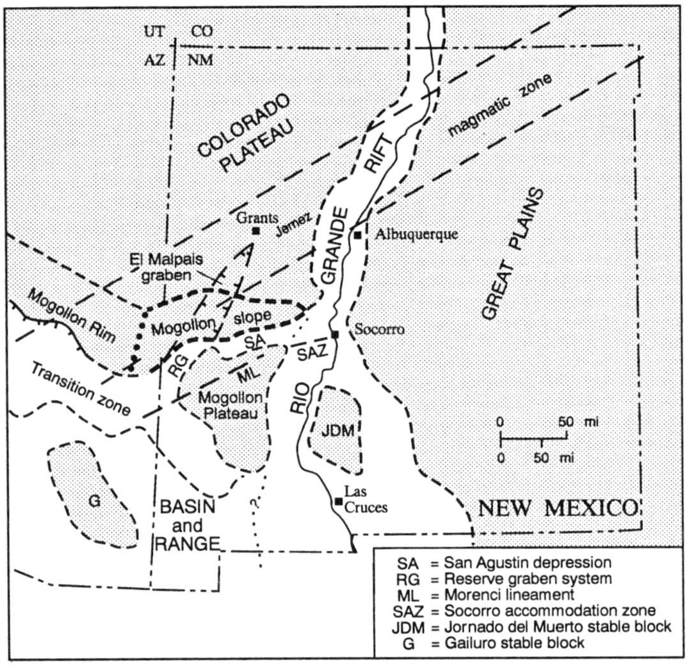

FIGURE 1.1. Generalized tectonic map of New Mexico and eastern Arizona showing the relationship of the Mogollon slope to regional structural elements. Provinces and domains of moderate to strong Cenozoic crustal extension are shown in white. Crustal provinces and blocks that have generally resisted Cenozoic extension are shaded gray. Compiled from Aldrich and Laughlin (1984), Chapin (1971, 1989), Godson (1981), Pierce (1985), Ratté (1989) and Wernicke (1992).

Ferguson, 1991) are mapped as part of the Rio Grande rift. More commonly the rift is defined by the distribution of deep Neogene half grabens. Areas of moderate to strong extension in the Rio Grande rift, Basin and Range, and Arizona transition zone have been lengthened (generally to the west) by about 10 to 100% or moré (Wernicke, 1992). Belts of weak extension show about 1 to 10% lengthening in east-west cross sections.

The Mogollon slope is primarily defined by gently south-dipping (2–10°) Eocene to Oligocene orogenic and volcanogenic strata approximately 1.3 to 2.1 km thick that generally parallel the classic (physiographic) southeastern margin of the Colorado Plateau. Eocene continental red-bed deposits were derived from Laramide transpressional welts along the Morenci lineament (Chapin and Cather, 1981; Garmezy, 1990). Overlying volcanic-apron deposits were derived from andesitic volcanoes that buried these welts (Elston et al., 1976) and from subsequent rhyolitic caldera complexes (McIntosh et al., 1991) that developed within the andesitic terrane to form the batholithic nucleus of the Mogollon Plateau. Thick piles of upper Oligocene basaltic andesite lavas that erupted contemporaneously with early strong extension along the axis of the Rio Grande rift locally cap highlands of the Mogollon slope. From east to west, landmarks formed by middle Tertiary volcanic strata along the Mogollon slope include: northern Bear Mountains, Gallinas Mountains, Datil Mountains, Crosby Mountains, Alegres Mountain, Mangas Mountains, Escondido Mountain, Gallo Mountains, Fox Mountain and Escudilla Mountain. The latter is located in eastern Arizona, where the south-dipping Mogollon slope grades into north-dipping strata on the Mogollon Rim.

The northern margin of the Mogollon slope generally parallels and lies a few kilometers north of the outcrop belt of the Eocene Baca Formation (Cather and Johnson, 1984). Here, the underlying Upper Cretaceous strata typically flatten out to a gentle north-northeast dip that is common throughout the Colorado Plateau. The southern margin of the Mogollon slope is defined by moderate to strongly extended domains along the San Agustin arm of the Rio Grande rift (Chapin, 1971). This weaker arm of the rift forms a regional west-southwest-trending topographic low (Godson, 1981) referred to here as the San Agustin depression. North-northeast-trending zones of weak extension along the Puerticito, Red Lake and Hickman faults locally cut across the Mogollon slope and southern margin of the Colorado Plateau (Cather, 1989; Chamberlin and Anderson, 1989b). For simplicity, only the El Malpais graben that parallels the Hickman fault zone is shown here (Fig. 1.1).

The Mogollon Rim section of the southwestern Colorado Plateau margin (between the Mogollon escarpment and the Little Colorado River) probably formed in Neogene time by isostatic uplift during extensional unloading of its footwall in the Arizona transition zone (Pierce, 1985). The northwest-trending extensional structures of the Arizona transition zone appear to give way to the northeast-trending structures of the San Agustin arm of the rift near Alma, New Mexico (Houser, this volume). A narrow, topographically low, distended zone west of Horse Springs (Godson, 1981; Ratté et al., 1991) appears to link the Reserve graben extended domain with the western San Agustin depression (see Third-Day road log). The en echelon pattern of the Reserve graben system and San Agustin depression may be attributed to sinistral transtension associated with Neogene clockwise rotation of the Colorado Plateau as it pulled away from the batholithically rooted Mogollon Plateau block (Chapin and Cather, in press; Fig. 1.1).

The Mogollon slope most likely represents downwarping of the southeastern margin of the cratonic Colorado Plateau microplate as it was loaded by as much as 2.1 km of orogenic and volcanogenic strata in Eocene to Oligocene time. Middle to upper Miocene Fence Lake Formation conglomerate derived from these south-tilted volcanic highlands near Quemado and Mangas show gentle primary dips to the northwest. Thus, the southerly tilt is pre-middle Miocene in the Quemado-Mangas region of the Mogollon slope.

Middle Miocene to Quaternary extension along the El Malpais graben was apparently penecontemporaneous with volcanism along the Jemez magmatic zone, or Jemez lineament of Aldrich and Laughlin (1984). The El Malpais graben is a narrow finger of weak crustal extension that penetrates deeply into the Colorado Plateau as it follows a pre-existing zone of Laramide strike-slip deformation (e.g. Hickman fault zone; Cather, 1989; Chamberlin and Anderson, 1989a). This should not be surprising, since the axis of the Rio Grande rift follows a major dextral strike-slip zone with as much as 120 km of lateral displacement in Laramide time (Chapin, 1983).

14.3 US-60 bends to left near milepost 122. The northern Magdalena Range from 9:00 to 11:30 is formed by a west-tilted hogback of Mississipian Kelly Limestone unconformably overlying a variety of Proterozoic metamorphic rocks and plutonic rocks. Late Oligocene monzonitic stocks and dikes that intrude the Proterozoic rocks in this structurally high block, north of the Socorro cauldron, may be regarded as an expression of the plutonic roots of the caldera complex. **4.6**

18.9 At 10:00, the La Jencia fault scarp displaces a middle Pleistocene fan surface as much as 23 ft (Machette, 1982, 1988). Total structural relief, presumably Neogene, in the La Jencia Basin is about 1.85 mi (Sanford, 1978). As much as half of this displacement may occur on the La Jencia fault. A southeast-trending seismic reflection profile across the La Jencia basin (COCORP line 3, Brown et al, 1980) shows several buried faults below Santa Fe Group basin fill, which is as much as 4500 ft thick. Seismic refraction data of Roberts et al. (this volume) also support the presence of a thick low-velocity basin fill in the La Jencia basin. **1.0**

19.9 Highway climbs over the La Jencia fault scarp. For the next 20 mi, from Magdalena to Tres Montosas, US-60 traverses a strongly extended domain that was stretched rapidly in late Oligocene time (Chapin and Seager, 1975; Ferguson, 1991), moderately extended in early Miocene time (Abbey Springs basin, Chapin and Cather, in press) and apparently inactive since middle Miocene time (Chamberlin and Osburn, 1986). East-northeast-trending transfer faults separate this domain from gently dipping Oligocene ignimbrites farther north on the Mogollon slope (Fig. 1.1). **0.2**

20.1 Knobby peak at 2:30 is capped by 34.17-Ma Rock House Canyon Tuff dipping about 35° to the west. The Rock House overlies and is interbedded with volcaniclastic sandstone and conglomerate of the Spears Group (see Cather et al., this volume). The Rock House Canyon Tuff has been assigned to the redefined Datil Group of Cather et al. (this volume). **0.3**

20.4 Milepost 116. **0.1**

20.5 Highway curves to left. Microwave tower at 1:00 is on crest of Granite Mountain, where 600 ft of granitic-looking, hornfelsed, crystal-rich Hells Mesa Tuff (32.1 Ma) is well exposed as a resistant west-dipping hogback. Volcanclastic sedimentary units of the lower Spears Group and intercalated Datil Group tuffs and lavas occupy the mid-slope. The basal Spears Group here rests in angular unconformity on the Permian San Andres Formation, exposed in the low hill at the east foot of Granite Mountain. This early Tertiary unconformity indicates that the west-tilted domino block of Granite Mountain is superimposed on the north flank of an east-northeast-trending Laramide uplift named the Morenci uplift by Cather and Johnson (1984). Similar exposures of the basal Tertiary unconformity delineate remnants of Laramide uplifts in the Lemitar Mountains, southern Chupadera Mountains, northern Magdalena Mountains, southern San Mateo Mountains, southern Gallinas Mountains near Tres Montosas, and on the southeast flank of Horse Mountain. These are also areas of strong crustal extension related to the Rio Grande rift. Thus it appears that domains of Neogene crustal extension and subsidence are superimposed on the same weak lithosphere that was shortened and uplifted during the Laramide orogeny. Correspondence of Neogene extended domains with Mesozoic and early Tertiary contractional domains is widely recognized in the Basin and Range (Wernicke, 1992) and the Rio Grande rift (Chapin and Cather, in press). **2.2**

22.7 At 3:00 is good view to north-northwest along the structural axis of the Bear Mountains. Northward decrease in west tilt marks southern margin of Mogollon slope within this range. **0.1**

22.8 Road sign on right proclaims Magdalena as the "Trails end". From 1885 to 1971 hundreds of thousands of cattle and sheep were driven eastward from Springerville, AZ and Horse Springs, NM to the rail head at Magdalena, making Magdalena the "trails end" for cows and cowboys. The tracks of the AT & SF Magdalena spur were pulled up in 1973, leaving trucking as the only alternative method for shipping livestock out of west-central New Mexico. The Magdalena spur was originally built in 1884 to haul lead-silver ores from the Kelly Mining district, south of Magdalena, to the Billing smelter in Socorro (Eveleth, 1983). **0.2**

23.0 Beyond the red brick house on the hill is Magdalena Peak, at 10:00 on the skyline. Near the middle of the slope, talus and vegetation form the face (side view, looking east) and bunned-up hair of "Mary Magdalene" for whom the Magdalena Mountains and the town are named. Magdalena Peak is capped by a thick rhyolite lava flow fed by an intrusive plug (dated at 13.1 Ma; Chapin et al., 1978) visible at its eastern end. This late Miocene rhyolite flow unconformably overlies Miocene Popotosa Formation and 28-Ma monzonite; the latter is part of a composite pluton related to the distended northern margin of the Magdalena-Sawmill cauldron (Chapin and Seager, 1975, Osburn and Chapin, 1983). **0.3**

23.3 Magdalena village limit. A major mining center at the turn of the century (silver, lead, zinc) and a key shipping point for cattle after WW I, Magdalena is now a quiet ranching community and a gathering place for the annual Old-timers Festival. **1.1**

24.4 Highway junctions at west end of Magdalena. **Continue straight ahead on US-60.** NM-169 on right leads to Alamo Chapter of the Navajo Indian Reservation; NM-107 on left goes to San Marcial at I-25. **0.1**

24.5 Road sign indicates Datil 34 mi, Springerville, 142 mi. Landmarks on the skyline ahead include the northern San Mateo Mountains at 11:00, Cat Mountain at 11:30, Gray Hill, with microwave tower, at 12:00, and Tres Montosas at 1:00. The Mt. Withington cauldron, source of the 27.4-Ma South Canyon Tuff, occupies most of the northern San Mateo Range and is also largely coincident with a strongly extended domain of the central Rio Grande rift (Ferguson, 1991). Tres Montosas, Gray Hill and Cat Mountain lie in a strongly extended and hydrothermally mineralized domain associated with a monzonite-granophyre stock (north of Tres Montosas; Wilkinson, 1976) and the western margin of the Magdalena-Sawmill cauldron (Osburn and Chapin, 1983). All these landmarks lie south of the relatively unextended and unmineralized Mogollon slope. See McLemore (this volume) for distribution of volcanic epithermal mining districts in west-central New Mexico. **0.5**

25.0 Roadcuts for next 2.6 mi expose gravelly to sandy alluvium of probable middle to late Pleistocene age. Red argillic zones and thin calcretes locally cap these gravels. Pebble imbrications suggest northeasterly paleoflow subparallel to Arroyo Gato, which trends east-northeast on the north side of the highway. These may be axial-stream deposits of ancestral Arroyo Gato or in some cases younger inset terrace gravels. Near here US-60 crosses the projected trace of a gently south-plunging synclinal axis defined by Oligocene ignimbrites a few miles to the north (Brown, 1972). To the west, Oligocene ignimbrites are strongly tilted to the east, as at Silver Hill (1:30) and east of the Tres Montosas (12:30). Another antiformal domain boundary separates west-tilted ignimbrites at Tres Montosas and Gray Hill from this east-tilted domain. Similar synformal and antiformal tilt-block domain boundaries are common in moderately- to strongly-extended terranes of the Basin and Range (Stewart, 1980; Wernicke, 1992) and in the central Rio Grande rift (Chamberlin and Osburn, 1986; Chapin, 1989). **0.9**

25.9 Gravelly sand in roadcut on right shows pebble imbrications indicative of northeasterly paleoflow. **0.5**

26.4 Milepost 110. Silver Hill at 2:30 is comprised by Oligocene La Jara Peak Basaltic Andesite of the Mogollon Group. **1.2**

27.6 Enter large roadcut, mostly in hydrothermally altered, strongly sheared, and flow-banded La Jencia Tuff (28.8 Ma). Except near the west end of the cut, flow banding dips approximately 60° to the east-southeast. The light gray, phenocryst-poor tuff with sparse andesitic lithic fragments is locally bleached, iron stained and partially silicified (at southwest end of cut). Cubic limonites, pseudomorphs after pyrite, locally form 2 to 3% of the tuff. At the southeast end of the cut medium gray La Jara Peak Basaltic Andesite appears to conformably overlie the steeply-dipping La Jencia Tuff. Several north-trending normal faults, dipping steeply to east and west, cut the La Jencia Tuff. Fault gouges and breccias are as much as 6 ft thick, and dip-slip slickensides are prevalent. At the northwest end of the cut is a nearly vertical east-northeast-striking shear surface that exhibits subhorizontal slickensides (plunging 20° to S60°W) superimposed on older dip-slip slickensides that plunge 70° to S30°E. The sense of lateral slip on this late-stage shear surface is not readily apparent. If left lateral, it would tend to support Chapin and Cather (in press), who interpret the Colorado Plateau as having rotated 1.0° to 1.5° clockwise in Neogene time, in association with the formation of a zone of sinistral transtension along the San Agustin arm of the Rio Grande rift. **0.2**

27.8 Highway crosses projected trace of the down-to-the-west Hells Mesa fault and enters the Abbey Springs Basin of late Oligocene to middle Miocene age (Chapin and Cather, in press). Chapin et al (1979) estimated the maximum thickness of basin fill to be 2450 ft and assigned this fill to the Popotosa Formation. J. Abbott (oral commun., 1993) suggested that this basin-fill unit has a southerly trending axial-stream facies. Chapin and Cather (in press) refer to the Abbey Springs Basin fill as lower Santa Fe Group. Fanglomerate deposition within the Abbey Springs Basin is well constrained in time. Approximately 2 mi north of here, at Arroyo Montosa, 26-Ma dacite flows are interbedded with basal fanglomerates of the lower Santa Fe Group (Chapin and Cather, in press). Also, 5 mi to the west, subhorizontal lavas of the 16.3-Ma basalt of Council Rock (Bachman and Mehnert, 1978), rest conformably on top of the basin-fill fanglomerates near the western margin of the Abbey Springs Basin. **1.1**

28.9 Lower Santa Fe conglomerate in roadcut on right contains abundant basaltic andesite clasts. Pebble imbrications indicate westerly or southwesterly paleoflow. **0.3**

29.2 Roadcuts in gently west-tilted (3-4°), light brown conglomerate and conglomeratic sandstone of lower Santa Fe Group. Basaltic andesite clasts dominate; minor clasts of South Canyon-like ignimbrite and glassy flow-banded rhyolite lavas are present. Pebble imbrications indicate southwesterly or westerly paleocurrents. **0.2**

29.4 Flat-lying 16.3-Ma basalt of Council Rock forms rounded hills at 10:00. **0.9**

30.3 Roadcuts in non-indurated gravel and sand probably represent inset terrace deposits of middle to late Pleistocene age. **1.4**

31.7 Climbing hill of lower Santa Fe fanglomerate, roadcut on left in moderately indurated brown to buff conglomerate containing subequal concentrations of hydrothermally altered ignimbrite clasts and andesitic clasts. Pebble imbrication indicates easterly paleoflow. **0.2**

31.9 Note horizontal ledges of basalt of Council rock capping hill at 9:00. **0.5**

32.4 Milepost 104. Roadcuts in bouldery lower Santa Fe fanglomerate. Hydrothermally altered boulders of Hells Mesa Tuff, Rock House Canyon Tuff and andesitic conglomerate (lower Spears Group-type) are present and reflect increasing proximity to the western margin of the Abbey Springs Basin. **0.3**

32.7 Road cuts in subhorizontal 16.3-Ma basalt of Council Rock that caps the Santa Fe Group here. Approximately 22 mi to the northeast, the 16.2-Ma andesite of Silver Creek is tilted 35° to the west and *lies near the base* of the main body of the lower Santa Fe Group (Popotosa Formation). This relationship implies that the Magdalena-Abbey Springs-Tres Montosas extensional domain is relatively old and inactive in comparison to the Socorro-La Jencia Basin domain, which is distinctly active. Feeder dikes for the basalt of Council Rock are locally well exposed 4 mi north of here (Chamberlin, 1974). **0.4**

33.1 Road side picnic tables on right with USBLM-NMSHD sign commemorating the Magdalena Livestock Driveway. This sign reads as follows:

> "South across the road lies one of the west's historic "hoof highways" which was used annually from 1885 until 1971. Sheep and cattle were driven to and from the railroad at Magdalena, NM. The 125 mile driveway extended west to Datil, NM then forked leading to Horse Springs, NM or to Springerville, AZ. The driveway was 5-10 miles wide and covered 200 square miles. The peak trailing year 1919 saw 150,000 sheep and 21,677 cattle pass this point. In the 1930's the Civilian Conservation Corps fenced the driveway and drilled water wells every 10 miles. During the drives cowboys and herders moved cattle about 10 miles a day and sheep about 5 miles, allowing them to graze as they went with chuckwagons and remudas of horses following behind. Trailing gave way to trucking and the last portion of the driveway was officially closed in November, 1971." **0.1**

33.2 Crossing projected trace of the Puerticito fault zone, which bounds the western flank of the Abbey Springs Basin. **0.1**

33.3 Highway rises onto the Tres Montosas section of the

southern Gallinas Mountains uplift. Entering long roadcut in hydrothermally altered and structurally complex andesitic debris-flow deposits and sandstones of the middle Spears Group that are interbedded with light gray, 34.2-Ma Rock House Canyon Tuff of the Datil Group (McIntosh and Chamberlin, this volume). Compaction foliation in the silicified ignimbrite is locally apparent on the north side of the cut (0.15 mi from east end) and dips 40° to the east. Spears Group outcrops near the northeast end of the cut also contain a low-angle normal fault, dipping 28° west, and conjugate quartz veins with both vein sets dipping west, one steep and one shallow. These structural observations indicate this cut is part of a strongly extended domain on the east flank of Tres Montosas. Geologic maps of this area (Wilkinson, 1976; Chamberlin, 1974) show numerous down-to-the-west normal faults repeating strongly east-tilted Datil Group ignimbrites and interbedded Spears Group strata. Exposures of Permian Abo Formation a mile north of here are unconformably overlain by Spears Group conglomerate (Wilkinson, 1976); thus the Tres Montosas section is also on a Laramide uplift. Approximately 4 mi north of here the strongly east-tilted extensional domain terminates against a northeast-striking zone of scissors faults. North of this scissors zone the Hells Mesa Tuff is tilted about 5° to the west (Chamberlin, 1974). This scissors zone accommodates a dip reversal and apparently a greater degree of extension on the south side. It may be classified as an accommodation zone or a transfer fault. This accommodation zone locally defines the south margin of the Mogollon slope where it crosses the Gallinas uplift west-southwest of Council Rock. **0.4**

33.7 Three wooded hills of Tres Montosas at 11:00. An anticlinal tilt-block domain boundary trends north along the east foot of Tres Montosas. Tres Montosas is capped by moderately west-dipping (20°) Hells Mesa Tuff; the Hells Mesa is repeated by normal faults downthrown to the east (Wilkinson, 1976). To the north, this anticlinal axis intersects the Oligocene (?) monzonitic Tres Montosas stock. Domed and hornfelsed volcanic strata form a resistant annular rim around the stock. The accommodation zone west-southwest of Council Rock passes north of the stock thereby separating it from the Mogollon slope. **1.7**

35.4 Road to Montosa Cattle Company on right. Microwave tower marks the crest of Gray Hill at 9:30. Excellent exposures of beautifully flow-banded La Jencia Tuff are found along the road to the microwave tower (Fig. 1.2). Thc 28.8-Ma La Jencia Tuff fills a deep northeast-trending paleovalley here (Wilkinson, 1974) that is cut in the underlying 32.1-Ma Hells Mesa Tuff. The west rim of the Sawmill-Magdalena cauldron, source of the La Jencia Tuff, lies about 3 mi east of Gray Hill (Osburn and Chapin, 1983). **1.8**

37.2 Roadcuts in Hells Mesa Tuff within the west-tilted moderately extended domain that underlies Tres Montosas (Wilkinson, 1976). **0.6**

37.8 Road to North Lake and Dog Springs Canyon on right. Approximately 25 mi to the northwest, Dog Springs Canyon is the type area for the Dog Springs Member of the Spears Formation of Osburn and Chapin (1983), or the Dog Springs Formation of the Spears Group (Cather et al., this volume). Andesitic to dacitic (58-64% SiO_2; Cather et al., 1987) debris-flow deposits and breccias of the Dog Springs comprise much of the northern Gallinas Mountains, the Datil-Sawtooth Mountains, and the low hills along J-L Draw (west of Crosby Mountains). Highly contorted and chaotic bedding characterize the Dog Springs Formation, compared to gentle uniform dips in underlying and overlying strata (Osburn and Chapin, 1983). This deformation is attributed to widespread soft-sediment deformation, or liquefaction, possibly triggered by one or more seismic events. Rapid loading of saturated basin-floor sediments (early andesitic sandstones and mudstones) by thick debris flows may have also triggered soft-sediment deformation. Spectacular soft-sediment deformation in the Dog Springs Formation will be viewed at Stop 3 today, in the Sawtooth Mountains west of Datil. **0.2**

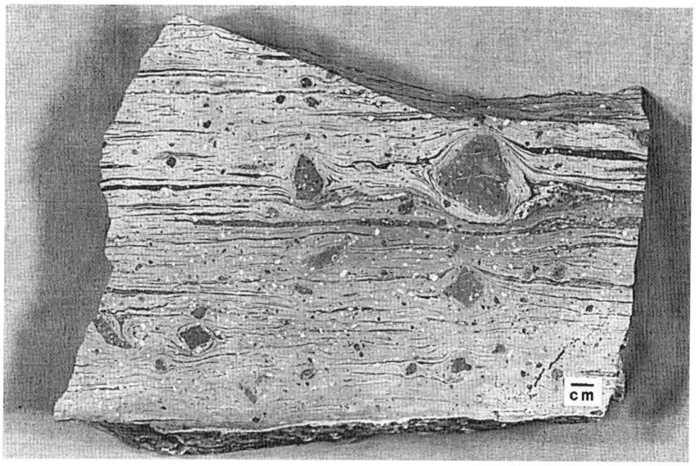

FIGURE 1.2. Flow-banded La Jencia Tuff showing effect of differential compaction and minor flowage around incompressible lithic fragments; sample from Gray Hill.

38.0 Entering northern section of the Plains of San Agustin, site of several late Pleistocene pluvial lakes, as described in the following minipaper by R. H. Weber. Highway crosses the projected high stand of these lakes near here, but shoreline features are locally masked by Holocene dune fields. Giant antennas of the VLA (Very Large Array) radio telescope dot the basin floor ahead. **2.4**

PLUVIAL LAKES OF THE PLAINS OF SAN AGUSTÍN

Robert H. Weber

New Mexico Bureau of Mines and Mineral Resources, Socorro, NM 87801

The 1994 Field Conference will cross the northern section of the Plains of San Agustín on the first day, and skirt the southwestern section on the third day. It is accordingly appropriate to summarize briefly some of the salient features of this structural and topographic intermontane basin that was inundated by an extensive lake system during at least the latest pluvial stages of the Pleistocene. Mapping on a scale of

1:24,000 of relict shorelines and surficial deposits of the floor and lower slopes of the basin by the present author has delineated a series of lake stages and associated littoral and lacustrine deposits of Pleistocene Lake San Agustín and subsequent Holocene playas.

Earlier investigations by Bryan (1926), Powers (1939, 1941), Hurt and McKnight (1949), Clisby and Sears (1956), Foreman (1956), Potter (1957), Foreman, Clisby, and Sears (1959), Stearns (1962), and Dick (1965) focused on various aspects of the geology, archaeology and palynology of the southwestern section of the area, herein designated as the Horse Springs (topographic) basin. Powers (1939) mapped a series of shorelines and associated littoral features in the Horse Springs basin extending to a maximum elevation of 6940 ft, recording the former presence of pluvial Lake San Agustin.

Later field investigations by Weber (1980, 1982) recognized a sequence of 16 principal lake levels ranging from 6800 to 7056 ft (later corrected to 7050 ft), the higher levels extending throughout three integrated topographic basins. These consist, from northeast to southwest, of the White Lake, C-N, and Horse Springs basins. North Lake, a very small, anomalous topographic basin, is linked to the northern end of White Lake basin. What here is designated as the Horse Springs basin formerly was shown as the limiting host for Lake San Agustín (Weber, 1980, fig. A-1). Later mapping of that basin indicated that the term Lake San Agustín should be extended throughout the conjoined topographic basins. At the 7050-ft maximum shoreline, the lake was approximately 57 mi long, with a maximum width of approximately 19 mi.

Absolute ages and chronology of the various lake stages remain essentially unknown. The degree of preservation of shorelines below the 7050-ft level (Fig. 1.3) and the presence of contiguous lake beds to these levels indicate a late classical Wisconsinan age, with a maximum stand at about 18,000 years B.P. Shorelines at 7005 ft, 7040 ft, and 7050 ft are highly fragmentary and lack contiguous lake beds. It therefore is suspected that these highest levels may reflect earlier pluvial stages; possibly early Wisconsinan (Bull Lake equivalent) or older. It should be noted that typical lacustrine deposits occur at 7060 ft in Nester Draw, indicating a higher lake stage than that recorded by the 7050-ft shoreline.

The present floors of the three contiguous basins are progressively higher in elevation from southwest to northeast: 6775 ft in Horse Springs basin, 6894 ft in C-N basin, and 6952 ft in White Lake basin. As a consequence, climatic changes toward the terminus of the Pleistocene resulted in desiccation and retreat of lake levels from northeast to southwest. On the basis of archaeological evidence, White Lake basin was reduced to a series of shallow ponds and marshes by ca 12,000 B.P. At 6960 ft, White Lake was separated from Lake San Agustín in C-N basin by a low sill, although a connection through a narrow channel, now covered by later deposits, is possible. The modern playa in White Lake basin is bounded by the 6955-ft level.

In the C-N basin, the reduced lake at the 6920-ft level connected with Lake San Agustín in Horse Springs basin through a narrow, steep-walled channel only 160 ft wide. The modern playa lies within the northern lobe of the area bounded by the 6895-ft shoreline (shown as 6900 ft in Weber, 1980, fig. A 2, as plotted on an uncorrected advance map sheet). A small lake may have persisted in the western part of Horse Springs basin well into the Holocene, possibly represented by a shoreline at 6810 ft, but this remains highly speculative. A probable playa shoreline is weakly expressed at approximately 6790 ft (Fig 1.3).

Drastically lowered lake levels resulted in vigorous eolian activity. Effective winds from the southwest drifted sands into sand sheets and

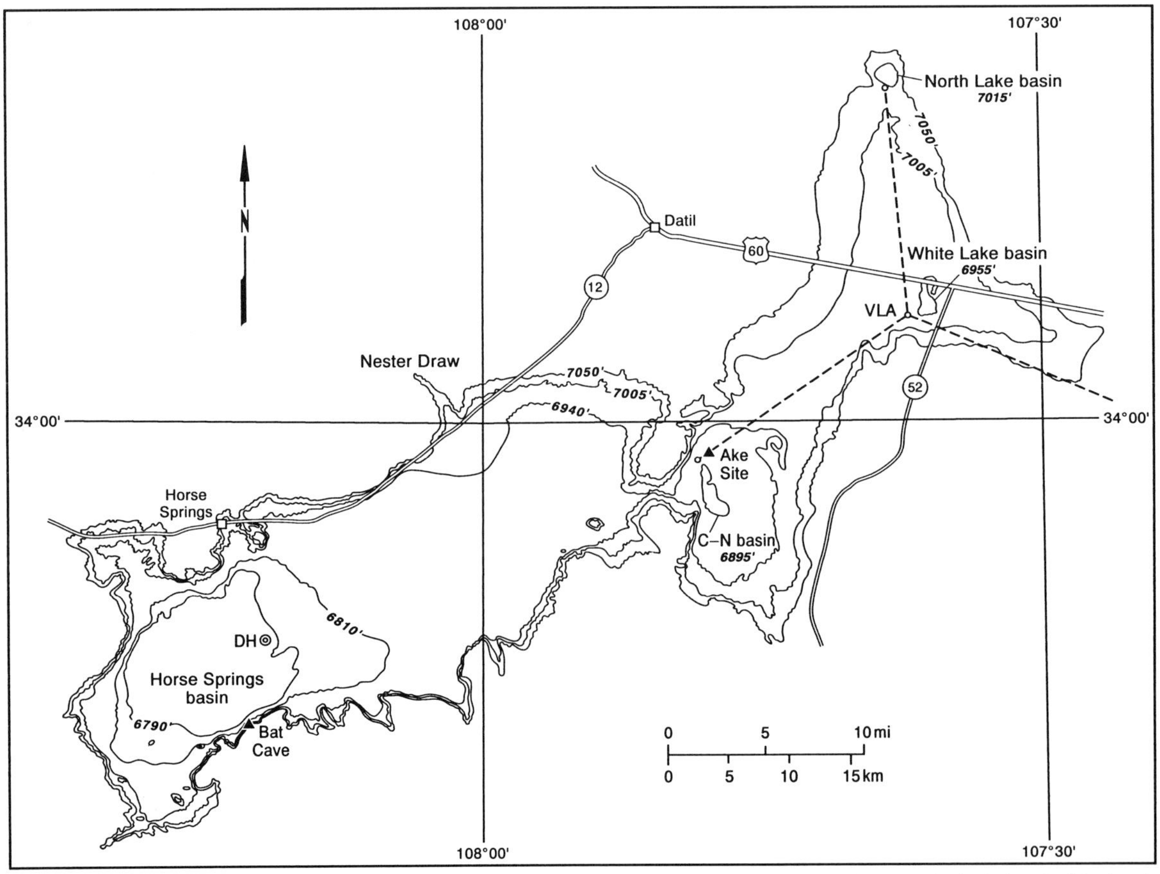

FIGURE 1.3. Map showing four stages of Pleistocene Lake San Agustin, Holocene playas (basins), and excavated archaeological sites (solid triangles). Oberlin core hole (Clisby and Sears, 1956) indicated by symbol labeled DH.

small dune fields on the northeastern side of Horse Springs basin, the northwestern edge of the C-N basin, and a much more extensive sand sheet and dune field on the northeastern side of White Lake basin (Osburn, et al., 1993). Sands driven over the crest of Lion Mountain cut pronounced northeasterly grooves in the Andesite of Lion Mountain (Osburn, et al., 1993, fig. 4). These dune fields are now largely stabilized by vegetation, although they were still active during the drought years of the 1950s. Several episodes of active dune formation have been recognized, extending back to ca 11,000 B.P.

The deeper stratigraphy of the basin fill is poorly known, except in the 2000-ft cored Oberlin hole on the basin floor south of Horse Springs (DH of Fig. 1.3). For details on this core see Clisby and Sears, 1956; Foreman, 1956; and Foreman et al., 1959. The lack of deep dissection that results from the closed internal drainage and low relief of the basin floor limit stratigraphic exposures to the uppermost few feet. Excavations for the Very Large Array (VLA of Fig. 1.3) radio-telescope facilities wave-guide trenches and antenna foundations exposed deeper strata that permitted documentation of the latest features of the stratigraphy. Alternating sands, silts, and clays indicate a complex sequence of rising and falling lake levels. Shrink-and-swell activity in deep-water clays results in unstable foundation conditions for highways and other engineering structural loads. In the central part of the Horse Springs basin this process is manifested in the hummocks and hollows of well-developed gilgai. Giant desiccation polygons are prominent in both the Horse Springs and White Lake basins, described and illustrated by Neal (1965) and Neal and Motts (1975). That these are actively opening and healing episodically was observed during mapping in the White Lake basin. Open fissures and associated conical soil pipes are effective in draining large volumes of surface water following thunderstorms.

Despite the complex faulted structure surrounding and probably underlying the lake basins, the undeformed, consistent elevation of shorelines throughout the area indicate structural stability during the latest Pleistocene and Holocene. Only one late fault was found to cut alluvial fan deposits of the western bajada of the San Mateo Mountains (Bruton Ranch fault of Weber; VLA faults of Menges et al., 1984). The fault does not truncate the oldest shoreline of the White Lake basin. None of the other late faults reported for the surrounding area could be confirmed on the ground. At least some of these were found to be based on linear air-photo images of gullied stock and ranch trails.

Further evidence reflecting past climatic and environmental conditions on the Plains of San Agustín derive from fossil faunal and floral assemblages. The megafauna of the surficial deposits includes extinct bison, horse, mammoth, mastodon(?), camel and short-faced bear of Rancholabrean aspect. Smaller faunal components include muskrat, pigmy rabbit, ground squirrel, vole, horned owl, water fowl, neotenic salamander, frog and fish. The molluscan fauna comprises more than 30 species of which only 6 are still living in the surrounding region. Abundant ostracodes and diatoms are proving useful in environmental reconstructions, as noted below.

Re-evaluation of the upper part of the original 2000-ft Oberlin core has yielded additional information on the pollen, ostracodes, diatoms, algae, magnetic polarity and isotopic ages, as reported by Markgraf et al. (1983, 1984). Additional analyses from a pit and shallow auger sample 4 mi northwest of the Oberlin core hole are from an area of gilgai and giant desiccation cracks that may have led to mixing of stratigraphic relationships.

Phillips and Campbell (1992) reviewed the geologic and hydrologic features of the Plains of San Agustín and reported analyses of two cores taken in the southwestern part of the Horse Springs basin. A chronology of paleoenvironmental reconstructions based on ostracode stratigraphy and oxygen isotope variations in ostracode tests was presented.

A recent geologic map of the Horse Springs West quadrangle illustrates some of the littoral features of the Horse Springs section of lake San Agustín (Ratté et al., 1991). In view southward across the basin toward Bat Cave from this area, in the vicinity of Horse Springs are a stacked series of relict shorelines, locally marked by prominent wave-cut cliffs. Some of the best examples of beaches, bay-mouth bars and spits are preserved along this section of the lake margin. Many of these are illustrated by Powers (1939).

Further information on the archaeology of Bat Cave, resulting from recent excavations by the University of Michigan, are found in Wills et al., (1984) and Wills (1988).

40.4 Milepost 96. Entrance to road-metal quarry ahead on right. This quarry provided ballast for trackways used to move VLA antennas into various configurations. The Oligocene andesite of Lion Mountain provided this ballast. A geologic map of the Lion Mountain and northern Arrowhead Well quadrangles (Osburn et al., 1993) provides detailed coverage of the Cenozoic geology from here north to Lion Mountain at 2:30. Structure sections of the Lion Mountain area illustrate a weakly extended domain (<3% extension) defined by gently folded and faulted Oligocene ignimbrites; thus, Lion Mountain is here placed on the Mogollon slope. A possible vent area for flow-banded rhyolite lavas (the rhyolite of Piñon well, Osburn et al., 1993), located about 3 mi south of Lion Mountain, lies along the projected accommodation zone that trends west-southwest from Council Rock. Osburn et al. (1993) also documented the intense scouring ability of sands blown across the crest of Lion Mountain with photographs of striated and grooved lava blocks. **1.0**

41.4 Entering Holocene dune field approximately 2 mi wide. Good view of Lion Mountain at 2:45. Other landmarks on the Mogollon slope skyline include the flat-topped Crosby Mountains from 11:30-12:30, the Datil Mountains from 12:15 to 2:00 and the Gallinas Mountains from 2:00 to 2:45. To the left and south of the San Agustin plains, high peaks mark late Oligocene to early Miocene volcanic centers aligned east-northeast along the Morenci lineament (Chapin et al., 1978; Ratté, 1989). Mt. Withington at 8:45 is at the apex of the Mt. Withington caldera, source of the 27.4-Ma South Canyon Tuff (Ferguson, 1991). Upper Oligocene to lower Miocene shield volcanoes composed of Bearwallow Mountain Andesite underlie Luera Peak at 10:00, Pelona Mountain at 10:15 and O Bar O Mountain at 10:20. Garmezy (1990) recognized the trend of the Morenci lineament as a major right lateral shear zone of Laramide age. Horse Mountain at 11:00 is a dissected upper Miocene dacitic to rhyolitic volcano (Ratté et al., 1991), summarized in the Third-Day road log of this guidebook. **2.4**

43.8 Junction with NM-52, to VLA visitor center and Dusty on left. **Continue straight ahead on US-60.** At 8:45 (looking along NM-52) the piedmont slope descending from Mt. Withington is offset approximately 100 ft down to the west, by the VLA fault of middle to late Pleistocene age. McFadden et al. (see following mini-paper) estimate the most recent offset on the VLA fault took place more than 100,000 years ago. The VLA fault and another Pleistocene fault scarp near North Lake (at 2:30, scarp not visible from here) are the most easterly indication of a late Miocene to Quaternary domain of extension in the Datil-Quemado-Springerville region that is temporally and, in part, spatially associated with the Jemez lineament (Aldrich and Laughlin, 1984). Tomographic mapping of the mantle across the southwestern U.S. indicates an anomalously low velocity upper mantle under west-central New Mexico, generally coincident with the southwestern part of the Jemez lineament (Spence and Gross, 1990). This upper-mantle anomaly may represent a zone of extensional decompression and partial melting that has been feeding basaltic magmas to the Quemado-Red Hill volcanic field (McIntosh and Cather, this volume) and adjacent volcanic fields along the Jemez lineament for the past several million years. **0.6**

SOIL, TECTONIC AND CLIMATIC GEOMORPHOLOGIC INVESTIGATIONS IN THE SAN AGUSTIN PLAINS AREA, NM

Leslie D. McFadden[1], Richard P. Lozinsky[2], Christopher M. Menges[3], Jerry R. Miller[4], John Ritter[5]

[1]Dept. of Earth & Planetary Sciences, University of New Mexico, Albuquerque, NM 87131-1116; [2]Division of Natural Sciences, Fullerton College, Fullerton, CA 92632-2905; [3]U.S. Geological Survey, 101 Convention Center Dr., Suite 860, Las Vegas, NV 89109; [4]Desert Research Institute, University of Nevada, Reno, NV, 89506; [5]Department of Geology, Wittenberg University, Springfield, OH 45501

The San Agustin Plains area has been the site of numerous geologic investigations. Published Quaternary studies of the San Agustin Plains have focused primarily on the stratigraphy and paleoenviromental analysis of the deposits of pluvial lakes in the ancestral San Agustin Basin that episodically formed during the Pleistocene (Bryan, 1926; Powers, 1939; Clisby and Sears, 1956; Foreman et al., 1959; Weber, 1980; Margraf et al., 1983, 1984; Phillips et al., 1992; Johnson, 1992), which form a major part of the 1.6 Ma stratigraphic record. This mini-paper describes the results of studies of Quaternary alluvial deposits (particularly the piedmont deposits in the northeastern part of the basin derived from the San Mateo Mountains) geomorphic surfaces, soils and the VLA fault. We estimate age ranges for the geomorphic surfaces by comparing observed soil development with independently dated soils (soil chronosequences) formed in similar soil-forming circumstances, following methods of Jenny (1980) and modified and elaborated by Birkeland (1984).

Quaternary alluvial deposits in the study area consist primarily of lenticular to weakly stratified, poorly sorted, gravelly sandy alluvium derived from Oligocene to Miocene, basaltic to rhyolitic, volcanic and volcaniclastic rocks in the San Mateo and nearby mountains. The deposits, depicted as Qp by Osburn et al. (1993) on their map of the area just north of this study area, are alluvial fans or terraces inset into older Quaternary fan remnants. The piedmont deposits range in altitude from 2140 m (approximate height of pluvial shoreline features) to over 2300 m near the mountain front. Presently, the drainages are ephemeral, and graded to the San Agustin plains, where there are numerous shoreline features and locally extensive eolian dunes and sand sheets. Because the alluvial deposits are high-energy gravels, datable materials were found only at one site.

On the basis of field and photogeologic studies, four major geomorphic surfaces associated with Quaternary surficial fluvial deposits are recognized. The oldest such surface in the area is Q1. Although we did not study this surface in detail, field and photogeologic observations indicate that it is a deeply dissected constructional surface remnant associated with relatively thick alluvial-fan deposits composed of coarse, poorly-sorted bouldery alluvium. Fragments of petrocalcic horizons, some of which exhibit laminar character, occur on the surface and lower hillslopes. These are interpreted as remnants of a very strongly developed (stage IV, after Gile et al., 1966) calcic soil. This interpretation is supported by a massive K horizon (> 1 m thick) in deposits that underlie the eroded Q1 surface in a railroad construction-related exposure associated with the VLA facilities. The age of the deposits of Q1, given the very strong degree of soil development and the strong degree of soil development on the younger Q2 surface, is estimated to be middle Pleistocene. The preservation of constructional surface remnants suggests that it is not a pre-Quaternary deposit. This estimated age is also based on a correlation with the oldest Quaternary surface (altitude 1865 m) in the nearby La Jencia Basin, described by McGrath and Hawley (1987), who recognized the associated fluvial deposits as the final stage of early to middle Pleistocene basin aggradation. In other studies of Quaternary deposits and soils at similar elevations and climate (1800- 2150 m; ppt = 35 cm) in the Española Basin in northern New Mexico (Dethier and Dempsey, 1984; Dethier et al., 1988), the oldest Quaternary surface possessing soils with maximal stage IV carbonate morphology is estimated to be between 0.35 and at least 1.1 Ma. The stage IV soil of the Lower La Mesa surface in the Las Cruces region (Gile et al., 1981) is at least 0.5 Ma, but soil development in this altitudinally lower site has occurred in a more arid, warmer climate.

The next youngest surface, Q2, occurs primarily as a fluvial terrace inset as much as 25 m below the Q1 surface. The soil on this surface exhibits a moderately developed, moderately to strongly reddened (5YR 4/6) argillic (Bt) horizon and a calcic horizon with stage III morphology and a content of up to 34% $CaCO_3$ (Table 1). The ochric A epipedon is 8.5 cm thick, and some pedogenic carbonate occurs in the argillic horizon (Btk) at a depth of 27 cm. Morphological attributes of the soil indicate that at least part of the lower argillic horizon has been massively engulfed by the calcic horizon. The degree of development indicates that the soil has formed in alluvium no younger than about 75 to 100 kyrs, based on comparison with soils on the late Pleistocene deposits of the Española Basin and Las Cruces soil chronosequences.

A variety of surfaces, recognized as Q3 and Q4, are either inset in or bury deposits of the Q2 surface. The Q3 surface may be latest Pleistocene to early Holocene in age; whereas Q4 is clearly Holocene. A radiocarbon age of 3350 ± 110 yrs B.P. (DIC-2928) was obtained from charcoal recovered from alluvium associated with a Q4-associated deposit in which a soil that exhibited A-AB-Bw-Bk-2Bk horizon sequence had formed. The calcic horizon exhibits only stage I morphology and the Bw horizon only slight reddening and weak structure development. Other soils on Q4 surfaces in the study area possess even weaker soil development, although the ochric epipedon may have an organic carbon content as high as 2.7% (Table 1). Weakly to moderately developed cumulic soils, also inferred to be Holocene in age, have formed in colluvial aprons derived from erosion of older Quaternary deposits. These aprons locally bury Q2 or Q3 surfaces.

Soils have formed on prominent beach ridges associated with the most recent high stands of the pluvial lake of the San Agustin Basin. These beach ridges probably formed no earlier than the last high stand (about 18 ka) and no later than the early Holocene (8 ka; Phillips et al., 1992; Johnson, 1992), after which paleoenviromental data indicate increasingly stronger lake salinity, a more xeric regional vegetation, and lake desiccation. Soils developed in the littoral sand and high energy, gravelly shoreline deposits exhibit organic-matter rich epipedons, a moderately developed, reddened argillic horizon, and stage I to locally II calcic horizon morphology. Soils on surface Q3 exhibit relatively similar characteristics (Table 1), supporting the latest Pleistocene to early Holocene age estimate for the associated fluvial parent materials.

Deeply incised and recently formed continuous and locally discontinuous arroyos locally expose the stratigraphy of the uppermost 3 to 4 m of late Quaternary valley-fill deposits. Complex stratigraphic relations among Quaternary units, some of which are not expressed as geomorphic surfaces, are revealed in these exposures. For example, the Q2 surface is buried by a stratigraphic unit in which a soil with stage III to locally stage IV carbonate morphology is present. This unit clearly is older than the Q3-associated deposits. In addition, Bt lamellar bands are present in the soil associated with deposits of Q2, reflecting deep translocation of illuvial clay, which has accumulated in distinct, separate bands, rather than as a continuous, at least several meter-thick zone that typifies argillic horizons. The processes by which these lamellae form are not completely understood, but their origin is presumed to be pedogenic and is usually favored in sandy parent materials. In detailed studies of such deposits, the formation of the lamellae precedes

TABLE 1.1. Summary of morphological, textural and chemical data for soils on Quaternary deposits, San Agustin Plains, New Mexico.

TABLE 1. SUMMARY OF MORPHOLOGICAL, TEXTURAL AND CHEMICAL DATA FOR SOILS ON QUATERNARY DEPOSITS, SAN AGUSTIN PLAINS, NM

Profile Horizon	Thickness (cm)	Color (dry)	Structure^	Consistence# (wet)	CaCO3 Stage	Clay Films***	Sand %	Silt %	Clay%	CaCO3 %	Organic Carbon (%)	pH
Q4 (Torriorthent)												
A1	0 to 4	10YR 4/3	sg	no, po	NO *	NO	62.9	30.4	6.7	nd	2.7	5.7
A2	4 to 56	10YR 3/2	1 gr	ss,sp	NO	NO	62.3	26.5	11.2	nd	1.2	6.9
A3	56 to 72	10YR 3/3	m	ss, sp	NO	NO	49.9	38.1	12	nd	1.1	7.3
Bk1	72 to 89	10YR 3/3	1fsbk	no,po	I	NO	51	37.1	11.9	nd	0.6	7.4
Bk2	89 to 139	10YR 3/2	1fsbk	no,po	I	NO	57.9	32.5	9.6	nd	0.5	7.5
Bk3	139 to 151	10YR 4/3	m	no,po	I	NO	59.7	30.9	9.4	nd	0.3	7.6
Q3 (Haplustoll)												
A	0 to10	10YR 4/3	sg	ss,sp	NO	NO	50.4	39.5	10.1	nd	1.7	6.1
Bt	10 to 46	10YR 3/3	1fsbk	ss,p	NO	cs	30.2	50.1	19.7	nd	0.8	6.9
Bw	46 to 60	10YR 4/3	1fsbk	ss,sp	NO	NO	60.4	28.4	11.2	nd	0.7	7.4
Bwk	60 to 100	10YR 3/3	2vc to fsbk	ss,sp	I+	NO	39.2	54.6	6.2	nd	0.6	7.4
Q3 (Calciustoll)												
A	0 to 4	10YR 5/3	1f&msbk	s,sp	NO	NO	46.1	26	23.7	0.2	nd**	nd
Bw1	4 to 19	7.5YR 5/4	1f&msbk	s,p	NO	NO	27.9	20.9	44	0.2	nd	nd
Bw2	19 to 40	7.5YR 5/6	2m&csbk	nd	NO	NO	20.5	30.4	31	0.7	nd	nd
Bwk	40 to 64+	7.5YR 5/4	1f&msbk	nd	II	NO	26.8	52.9	8.3	2.2	nd	nd
							21.9	23.6	8.5			
Q2 (Argiustoll)												
A	0 to 8.5	7.5YR 3/4	1m&csbk	s,p	NO	NO	46.4	26	23.7	nd	nd	nd
Bt	8.5 to 27	5YR 4/6	2m&csbk	vs,vp	NO	1vt&tgr;4tped	27.9	20.9	44	nd	nd	nd
Btk	27 to 37	7.5YR 4/6	1&2msbk	vs,p	I	1tped	20.5	30.4	31	nd	nd	nd
Bwk	37 to 46	7.5YR 5/6	2&3msbk	vs, p	II	NO	26.8	52.9	8.3	nd	nd	nd
K	46 to 111+	7.5YR 8/4	m	ss,po	III	NO	21.9	23.6	8.5	nd	nd	nd
Q2 (Calciustoll) ~'												
A1	0 to 4	7.5YR 4/2	1msbk	s,p	NO	NO	48.1	27.3	8.4	0.9	nd	nd
A2	4 to 11	7.5YR 4/2	1f&msbk	ss,sp	NO	NO	38.7	25.8	10.4	0.9	nd	nd
AB	11 to 19	7.5YR 5/4	1fsbk	ss,sp	NO	NO	37.2	21.3	14.1	1.6	nd	nd
Bwk	19 to 30	7.5YR 5/4	1fsbk	ss, sp	NO	NO	24	14.7	8.5	26.3	nd	nd
Bk1	30 to 70	7.5YR 7/4	m	s, sp	II	NO	27.5	14.6	12.1	34.2	nd	nd
Bk2	70 to 78+	7.5YR 6/4	m	ss,sp	II	NO	25.2	17.7	16.1	23.1	nd	nd

^ sg-single grain; m-massive; 1-weak; 2-mediu m; 3-strong; f-fine; m-moderate; s-strong; sbk-subangulary blocky

#no-nonsticky; ss-slighty sticky; s-sticky; vs-very sticky; po-non plastic; sp - slightly plastic; p-plastic; vp-very plastic;

* none observed

** not determined

*** cs-colloidal stains on grains;f-films; t-thin; mk-moderately thick; k-thick; 1-few; 2-common; 3 many; 4 -continuous; ped-ped face

~ Described on scarp slope formed in Q2-associated alluvium

coalescence and development of the argillic horizon. Such lamellae, however, are not observed in soils of younger deposits in the study area. This suggests that a strongly developed argillic horizon must be present before genesis of the lamellae is favored.

Very strongly reddened (3.75YR 6/6) clayey matrix occurs in gravelly lenses several decimeters below the slightly reddened (7YR 6/6) argillic horizon of the Q2 soil. In contrast to the Bt lamellae, accumulation of this clay is not attributed to pedogenesis, as the clay's color shows that it could not have been derived from the overlying argillic horizon. Also, this deposit does not exhibit other features typically associated with soil profiles; therefore, it does not represent buried soils. Instead, we attribute its formation to introduction of clay to the unit from an external source, perhaps via vadose-zone transport of suspended clay through hydrologically very transmissive gravels. The only possible source of such reddened, clay-rich material is a strongly developed argillic horizon once present on surface Q1, presumably associated with the well developed calcic horizon still partly preserved. The presence of the Q2-associated deposits topographically below the Q1 surface, and the presence of reddish clay colluvial deposits of Q1 slopes favor this process. Finally, the presence of a very strongly developed and reddened argillic horizon in the soil of the much less extensively dissected and oldest deposit of the La Jencia Basin described by McGrath and Hawley (1987) shows that such a horizon may well have been present on the Q1 deposit prior to its deep dissection and erosion during the late Quaternary.

Major horizon characteristics of soils in Holocene and latest Pleistocene deposits that were not subject to the effectively wetter, late Pleistocene (full glacial) climate reflect the currently semiarid (annual precipitation = approximately 35 cm) and moderately warm (annual average temperature = 8.5° C)) climate of area. Available soil moisture is sufficient to sustain locally abundant grasses, including Galleta and Blue Grama grasses. Biomass production rates are sufficiently large to provide the soil organic matter to enable Mollic epipedon development in many Holocene soils (Table 1.1) (Ritter et al., 1984). Within 3 ka of development, Bw horizons also have formed as iron oxyhydroxide accumulation and concomitant soil structural development progress. The weak argillic horizons in older Holocene and latest Pleistocene soils form via translocation of suspended silicate clay. Micromorphologic studies of such horizons show development of cutanic agglomeroplasmic or porphyroskelic fabrics, which reflects formation of oriented clay and clay bridging, accompanied by concomitant conversion of simple packing (interstitial) voids to channel and plain soil voids. Formation of the overlying Mollic horizon promotes substantial soil respiration, which favors development of a noncalcareous B horizon environment that, in turn, is conducive to clay translocation and the observed Bt horizon development. However, in all but a few soils, carbonate is leached to depths typically no deeper than 40 cm, despite the relatively high permeability of the parent materials (Table 1.1). This is partly because a significant fraction of annual precipitation falls during the summer when the seasonally maximal soil evapotranspiration limits soil infiltration depth. The ultimate development of strongly developed calcic horizons, however, must also reflect a high influx and subsequent incorporation of calcareous dust. This conclusion is indicated by the relatively low calcium content of the parent materials and the minimal evidence of strong chemical alteration of the parent material. This process has apparently strongly influenced Quaternary soil development elsewhere in New Mexico (Gile et al., 1981; Machette, 1985) , where strong calcic horizons form in Pleistocene soils, even at elevations that exceed 2600 m. This is an elevation at which soils are typically noncalcic elsewhere in much of the southwestern United States. Certainly, during relatively arid periods of the Quaternary, the plains of San Agustin provide one obvious local source for at least some of the dust.

Tectonic geomorphic studies in the San Agustin Basin were concentrated in the area of a large fault scarp located in northernmost subbasin. This scarp, the VLA fault, had been recognized previously by R. Weber (personal commun., 1983) and Machette and McGimsey (1983), but had not been studied. The topographic scarp can be discerned over a distance of 10 km, and in some places it is as high as 30 m. The VLA fault thus reflects substantial Quaternary movement. In addition it is somewhat anomalous, in that it is separated by several kilometers from the mountain fronts and projects northward into the center of northern subbasin, rather than following the basin margin.

Evaluation of the VLA Fault integrated studies of the Quaternary alluvial deposits reported above, their relation to the fault, and morphological studies of the scarp, largely based on the scarp degradation model of Wallace (1977). Topographic profiles across the VLA scarp at 15 locations and geomorphological evidence show that surface Q1 has been offset 20-30 m along a composite scarp and Q2 by 2-5 m along a single-rupture scarp (Menges et al., 1984). Q3 and younger surfaces are unruptured by the fault, constraining the age of most recent rupture to late Pleistocene. Following the method of Bucknam and Anderson (1979), linear regression analysis of log scarp height versus maximum scarp-slope angle on single-event scarps (Fig. 1.4) also indicates an older late Pleistocene age (10 ka) for the most recent rupture. This estimate is supported by comparison of scarp-height—scarp-slope trends observed on reference scarps locally and regionally. For example, topographic profiles of a scarp (La Jencia fault) located north of the Magdalena Mountains by Machette and McGimsey (1983); Machette (1986, 1988) and Menges et al. (1984) indicate a Holocene age for most recent rupture (Fig. 1.4). In addition, the soil on the VLA fault scarp slope is better developed than Q3 soils, which are primarily Holocene in age, strongly suggesting that the late Pleistocene last-rupture age estimate is reasonable. Comparisons between the heights of composite and single-rupture VLA fault scarps indicate 5 to 10 events on the fault since formation of the Q1 surface, which suggests a 10,000 yr recurrence interval. This rate is consistent with those reported for Quaternary faults in southern New Mexico and Arizona (Menges and Pearthree, 1989).

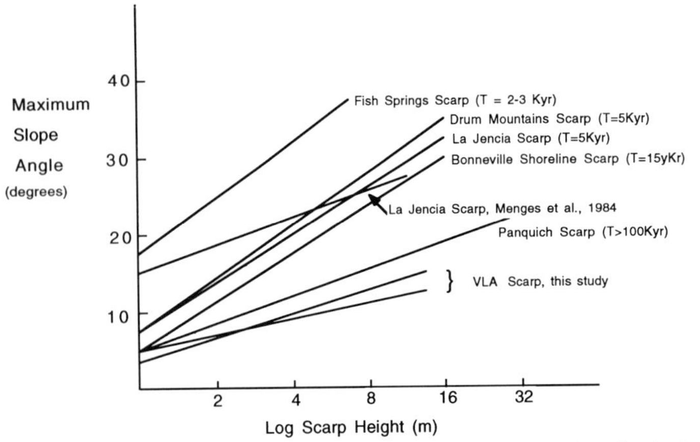

FIGURE 1.4. Composite "best-fit" lines for the VLA scarp and lines for the La Jencia scarp and other reference scarps described in studies noted in the text. T refers to the estimated age of the formation of the scarp. Best-fit line determined using simple linear regression analysis.

Margraf et al. (1983) suggested that generally constant basin subsidence rates in the San Agustin Basin, calculated on the basis of paleomagnetic, radiometric and amino acid-racemization ages of basin sediments deposited during the last 0.75 Ma, imply a strong tectonic-subsidence control on sedimentation rates. Documentation of sustained Quaternary movement along the VLA fault through soil and geomorphic studies also indicates that tectonism has continued to influence San Agustin Basin evolution significantly. Formation of the 30-m-high scarp has also clearly influenced geomorphic evolution of the piedmont. Uplift of the Q1 deposits along the scarp has confined more recent deposition to valleys formed subsequent to their deposition. Uplift has also precluded burial of the older Quaternary deposits despite continuous base-level rise in the basin, where aggradation during the last 1.6 Ma might otherwise be expected to cause regional base-level rise and burial of older Quaternary units. Many valleys above the scarp associated with the largest drainages are strongly asymmetrical, with consistently steep south-facing slopes and gentle north-facing slopes. Given evidence for continued tectonic movement on the VLA fault in this area during the Quaternary, such valley asymmetry may reflect local tectonic tilting. Similarly asymmetric valleys also are present on the Taos Plateau and on the Pajarito Plateau in northern New Mexico. Such valley asymmetry also may be related to microclimately driven, aspect-related geomorphic processes. Exposures of stratigraphy of deposits of Q1 at one location, however, show that beds have strikes of N22° to 60°W, and tilts that range from 5 to 9° to the north. Such attitudes differ significantly from beds of younger Quaternary deposits, are steeper than those expected for medial piedmont channel deposits, and are inconsistent with the expected major northeast direction of channel orientation on the piedmont. This evidence is consistent with the observed pattern of valley asymmetry; thus, although microclimatic influences may favor maintenance of valley asymmetry, tectonically driven processes are more likely the fundamental cause.

The lack of movement on the VLA fault during the last 10,000 yrs implies that the major latest Pleistocene and Holocene aggradational-degradational cycles in this area are not tectonically controlled. Thus their origin might be explained instead by climate, another important factor that influences fluvial systems. In the San Agustin Plains, Pleistocene-to-Holocene climate changes not only caused the complete desiccation of a once-extensive pluvial lake, but also were associated with major changes in regional vegetation (Margraf et al., 1983; 1984). Many geomorphologic studies have demonstrated that such climate changes significantly impact soil formation, vegetation, and fluvial processes over large areas and in diverse climates. Throughout the southwestern United States, these changes induced a cycle of rapid and thick valley aggradation and channel incision, caused primarily by dramatic changes in drainage basin sediment supply and hillslope/channel discharge relations (Bull, 1991). The precise nature of the interrelationships among climate changes, long-term Neogene tectonism, lithology, local relief, and other factors that have influenced soil and fluvial system dynamics in this area, however, is complex and remains an important research problem.

We thank several students in the Department of Earth and Planetary Sciences Quaternary Studies Program who also helped to describe soils during studies in the field area: Mary Ann Joca, Glen Kawaguchi, Larry Smith, and Keith Kelson. Funding for part of this research was provided by a University of New Mexico Research Allocations Grant to the senior author.

44.4 Milepost 92. Entering playa of White Lake; US-60 crosses projected southeastern margin of the Mogollon slope near here. The Bouger gravity map of Keller and Cordell (1983) suggests that the Plains of San Agustin include three linked structural basins, two deep half-graben basins coincident with the Horse Springs Basin and the C-N Basin (Fig. 1.3), and a third shallow basin north of US-60, between the White Lake Basin and the North Lake Basin. Local unconformities between Datil Group and Mogollon Group ignimbrites and thickness trends in the Dog Springs area at 2:00 (Harrison, 1980), suggest that the northern basin is a shallow, south-plunging, synclinal sag of late Oligocene age. This downwarp formed at about the same time (29-27 Ma) as initial extension in the Socorro segment of the Rio Grande rift (Chapin and Cather, in press), and presumably has been inactive during Neogene time. Uranium test wells drilled by Mobil Alternative Energy Inc. in the early 1980s indicate that basin-fill deposits in the area north of US-60 (HH Ranch area) are about 500-650 ft thick (G.R. Osburn, unpubl. report to Mobil Alternative Energy Inc., 1984). **0.4**

44.8 Low roadcuts in gray lacustrine muds. **1.3**

46.1 Trackway and tourist stop (on left) for one of the world's largest radio telescopes (Fig. 1.5). The Very Large Array, established in 1980, is now a sister instrument to the Very Long Baseline Array (established in 1993), which has its outermost antenna at Mauna Kea, Hawaii. The 10 VLBA antennas are electronically connected to and controlled by the National Radio Astronomy Observatory Array Operations Center located on the New Mexico Tech campus in Socorro. Each antenna weighs approximately 230 tons and has a reflecting dish 82 ft in diameter. The most distant stars, galaxies, quasars and other objects recorded by the VLA are 6×10^{21} mi away and are seen as they were more than 10 billion years back in time. The VLBA can measure image detail to better than one thousandth of a second of arc, or equivalent to being able to see a football on the surface of the moon. VLBA antenna stations are also able to measure their exact location in the universe to within one centimeter and can detect tiny variations in the earth's rotation. Self guided tours of the visitors center (near the ten-story building at 9:30) can be made from 8:30 a.m. to sunset every day. **3.1**

49.2 Road to old Augustine Well, (now B. and B. Gallaher Ranch) on left; road to HH Ranch on right. Augustine Well was a stop on the Overland Stage route of the 1880s and later a watering stop along the western New Mexico livestock driveway. Another important well 10 mi southwest of here (low hills at 2:00) is the 1966 Sun Oil Company No.1 Well, which was drilled to Precambrian gneiss and a total depth of 12,284 ft. This well intersected a nearly complete Tertiary stratigraphic section (Table 1.2), which is generally representative of the Datil-Crosby Mountains area (Osburn and Chapin, 1983). Only 590 ft of the basaltic andesite of Crosby Mountains (Lopez and Bornhorst, 1979) and 65 ft of the South Canyon Tuff (McIntosh and Chamberlin, this volume) are missing; presumably they were removed by Neogene erosion. Based on the Sun No. 1 Well and the above missing units, the total thickness of the Tertiary volcanic pile (Mogollon, Datil, and Spears Groups) in the Datil region was as much as 5000 ft. In addition, a 2095 ft intercept of Eocene Baca Formation indicates that the Sun No. 1 Well is in the Laramide Baca Basin (Cather and Johnson, 1984), probably near its southern structural margin. The total thickness of Eocene-Oligocene strata on the Mogollon slope near Datil is 6800 ft. A thick aplite intrusion of probable Miocene age was intersected within Permian rocks near the bottom of the Sun No. 1 well (Ratté, written commun., 1994). **1.2**

50.4 Begin ascent of low-gradient alluvial fan that emanates from White House Canyon, a major source of sediment to the northwestern San Agustin Plains. **1.3**

51.7 Entering Catron County (Fig. 1.6), named for New Mexico's first U.S. Senator Thomas B. Catron. It is the largest and least populated county in New Mexico. Cattle ranching is the primary economic activity. A

FIGURE 1.5. View to south along north arm of the VLA radio telescope. Quaternary scarp of the VLA fault offsets piedmont slope below lowest saddle on skyline (left center).

TABLE 1.2. Generalized Cenozoic stratigraphy of the Sun Oil Company No. 1 Well, San Agustin Plains (sec. 29, T3S, R9W; T.D. 12,284 ft). Based on an unpublished lithologic log by Roy Foster (New Mexico Bureau of Mines and Mineral Resources, Library of Subsurface Data) and a preliminary visual inspection of cuttings from Tertiary units by R. M. Chamberlin. All correlations are tentative; the more tentative correlations of Tertiary units are queried. Unit contacts and thicknesses are primarily based on Foster's log. Nomenclature of Tertiary units is after Osburn and Chapin (1983), Ratté (1989), Ratté et al., (1991) and Cather et al. (this volume). Ignimbrite ages from McIntosh et al., (1991); age of Dog Springs Formation from Cather et al., (1987) and McIntosh and Chamberlin (this volume).

Thickness (ft)	Unit
230	Quaternary alluvium
150	Vicks Peak Tuff (28.6 Ma), Mogollon Gr.
90	La Jencia Tuff (28.9 Ma), Mogollon Gr.
375	South Crosby Peak Fm., upper Spears Gr.
285	Hells Mesa Tuff (32.1 Ma), Datil Gr.
470	andesite of Dry Leggett Canyon (?), Datil Gr.
320	Rincon Windmill Fm. (?), middle Spears Gr.
220	tuff of Lebya Well (?), Datil Gr.
240	Kneeling Nun Tuff (?) (34.9 Ma), Datil Gr.
360	Chavez Canyon Fm., middle Spears Gr.
1760	Dog Springs Fm. (ca. 37-40 Ma), lower Spears Gr.
2095	Baca Formation (Bridgerian to Uintan)
6595 ft = (2005 m)	Total Cenozoic section

FIGURE 1.6. View west along US-60, the serene corridor to northeastern Catron County. Flat-topped Crosby Mountains on horizon to left. Gently south-dipping cuesta immediately right of road sign is underlain by 34.98-Ma Datil Well Tuff on the Datil Mountains block of the Mogollon slope.

generalized geologic map of the Quemado-Datil area (Fig. 1.7) provides the regional setting for today's stops. **1.7**

53.4 Milepost 83. View of flat-topped peaks of Crosby Mountains from 11:00 to 12:15 (Fig. 1.8). Twin Peaks in foreground are capped by the basaltic andesite of Deep Well directly overlying the La Jencia Tuff (Osburn and Chapin, 1983). These same units were originally mapped by Lopez and Bornhorst (1979) as the basaltic andesite of Twin Peaks overlying the A-L Peak Rhyolite. Lopez and Bornhorst's 1979 geologic map of the Datil area provides the best available data on outcrop distribution and structure in the Datil area, but most of their ignimbrite names have been abandoned because they are synonymous with previously named and mapped ignimbrites in the Socorro-Magdalena region (Osburn and Chapin, 1983; McIntosh et al., 1991; McIntosh and Chamberlin, this volume). Osburn and Chapin (1983) formally defined two ignimbrites named by Lopez and Bornhorst; the 34.98-Ma Datil Well Tuff and the 33.66 Ma-Blue Canyon Tuff. Lopez and Bornhorst's Tuff of Main Canyon is synonymous with the 34.17-Ma Rock House Canyon Tuff (Osburn and Chapin, 1983; McIntosh and Chamberlin, this volume). Lopez and Bornhorst's Tuff of Rock Tank and Tuff of Ary Ranch are both synonymous with the 32.1-Ma Hells Mesa Tuff; also their A-L Peak Rhyolite (as mapped) is now known to include the 28.8-Ma La Jencia Tuff, the 28.6-Ma Vicks Peak Tuff and the 27.4-Ma South Canyon Tuff (McIntosh et al., 1991; McIntosh and Chamberlin, this volume). All of the following identifications of stratigraphic units in the Datil area follow the nomenclature of Osburn and Chapin (1983) as redefined by Cather et al. (this guidebook). **0.3**

53.7 Route enters northeast portion of the geologic map of the Datil area by Lopez and Bornhorst (1979). **0.7**

54.4 Milepost 82. East-sloping cuestas at 3:00 in northeastern Datil Mountains include about 40-100 ft of South Crosby Peak Formation (conglomeratic sandstones and thin pumiceous ignimbrites) sandwiched between the La Jencia Tuff and the Hells Mesa Tuff (Harrison, 1980). Approximately 3 mi further west, within a structurally higher part of the Red Lake fault zone, the Hells Mesa and La Jencia tuffs fill west-trending paleovalleys and the South Crosby Peak Formation is absent (Harrison, 1980). In contrast, in the Crosby Mountains (11:00), the South Crosby Peak Formation ranges from 600 to 800 ft in thickness (Lopez and Bornhorst, 1979; Osburn and Chapin, 1983). In the Crosby Mountain area the South Crosby Peak Formation comprises all of the upper Spears Group volcaniclastic sedimentary rocks shown on the regional geologic map (Fig. 1.7) and contains less than 10% non-welded pumiceous ignimbrites by volume. The significant southward increase in thickness of the South Crosby Peak Formation is apparently associated with an east-trending zone of down-to-the-south normal faults that crosses the Datil-Crosby Mountains at the latitude of Datil (Lopez and Bornhorst, 1979). Based on thesis studies in the Datil-Horse Springs area (Lopez, 1975; Bornhorst, 1976; Jones, 1980), Elston (1978, 1984, 1989) interpreted the South Crosby Peak Formation as the cogenetic fill (intracaldera- or moat-fill) of the Crosby Mountains caldera or volcano-tectonic depression. As summarized in the following minipaper, the "Crosby Mountains caldera" might alternatively be described as the "Crosby Mountains volcano-tectonic collage", a seemingly disparate collection of temporally unrelated structural and volcanic elements. **2.0**

THE CROSBY MOUNTAIN "CALDERA": A VOLCANO-TECTONIC COLLAGE OF DISPARATE ORIGIN

Richard M. Chamberlin

New Mexico Bureau of Mines and Mineral Resources, Socorro, NM 87801

Elston (1989) described the Crosby Mountains caldera as one of the best documented ash-flow tuff calderas in the Mogollon-Datil volcanic field. As outlined by Elston, however, the Crosby Mountain caldera clearly lacks the thick, densely welded, intracaldera tuff, silicic intrusions and extrusions, and hydrothermal alteration or mineralization that are characteristic of large ash-flow calderas (Lipman, 1984). This minipaper summarizes current observations that indicate the Crosby Mountains caldera (or volcano-tectonic depression) as proposed by Bornhorst (1976), Jones (1980) and Elston (1978, 1984, 1989) represents a temporally unrelated collection of volcanic and tectonic elements (Fig. 1.9).

Except for block and ash deposits of the 33.7-Ma Horse Springs dacite (Ratté et al., this volume), all ignimbrites within and adjacent to the proposed Crosby Mountains caldera form thin sheets (0-130 ft) typical of distal ash-flow deposits (Chamberlin et al., 1994b; Osburn and Chapin, 1983a). Block and ash deposits of the Horse Springs dacite form a north-trending series of exposures between Horse Springs and the west flank of Alegres Mountain. Bornhorst (1976) originally interpreted these exposures to represent a vent zone (inside dotted line of Fig. 1.9) along the west flank of the "Crosby Mountains depression," later renamed the Crosby Mountains caldera (Elston, 1989). Exposures of the Horse Springs dacite become progressively finer grained and thinner from south to north (Ratté, 1989; Ratté et al., this volume). The 33.7-Ma Horse Springs dacite must have erupted from a source under the San Agustin Plains near Horse Springs and has no apparent relationship to the proposed Crosby Mountains caldera of Elston (1989). Two petroleum exploration wells to Precambrian basement (Fig. 1.9), a regional seismic reflection profile (Armstrong and Chamberlin, this volume), and regional mapping (Chamberlin et al., 1994b), demonstrate that a normal volcanic apron sequence (ca. 40-26 Ma) overlaps a reverse-fault bounded (AMF, DF, Fig. 1.9) Laramide synclinal basin within the proposed caldera. Several arcuate normal faults, such as the Alegres Mountain fault (AMF), in the Datil-Quemado-Reserve region apparently represent Laramide transpressional structures that have collapsed under later regional extension.

Field observations (Ratté, oral commun., 1993) and regional seismic reflection profiles (Armstrong and Chamberlin, this volume; L. Garmezy, oral commun., 1993) imply that Permian rocks exposed on the south flank of Horse Mountain represent a fragment of a Laramide uplift that was buried by Spears Group volcaniclastic conglomerates. Limestone and granite clasts in these conglomerates (Ratté, oral commun. 1993) imply the presence of a nearby erosional unconformity developed on pre-Tertiary rocks about 37-35 Ma, prior to development of the putative caldera. Bornhorst (1976) suggested these Permian rocks represented a resurgent dome *within* the Crosby Mountains volcano-tectonic depression.

Elston (1989) indicated that the proposed Crosby Mountains caldera

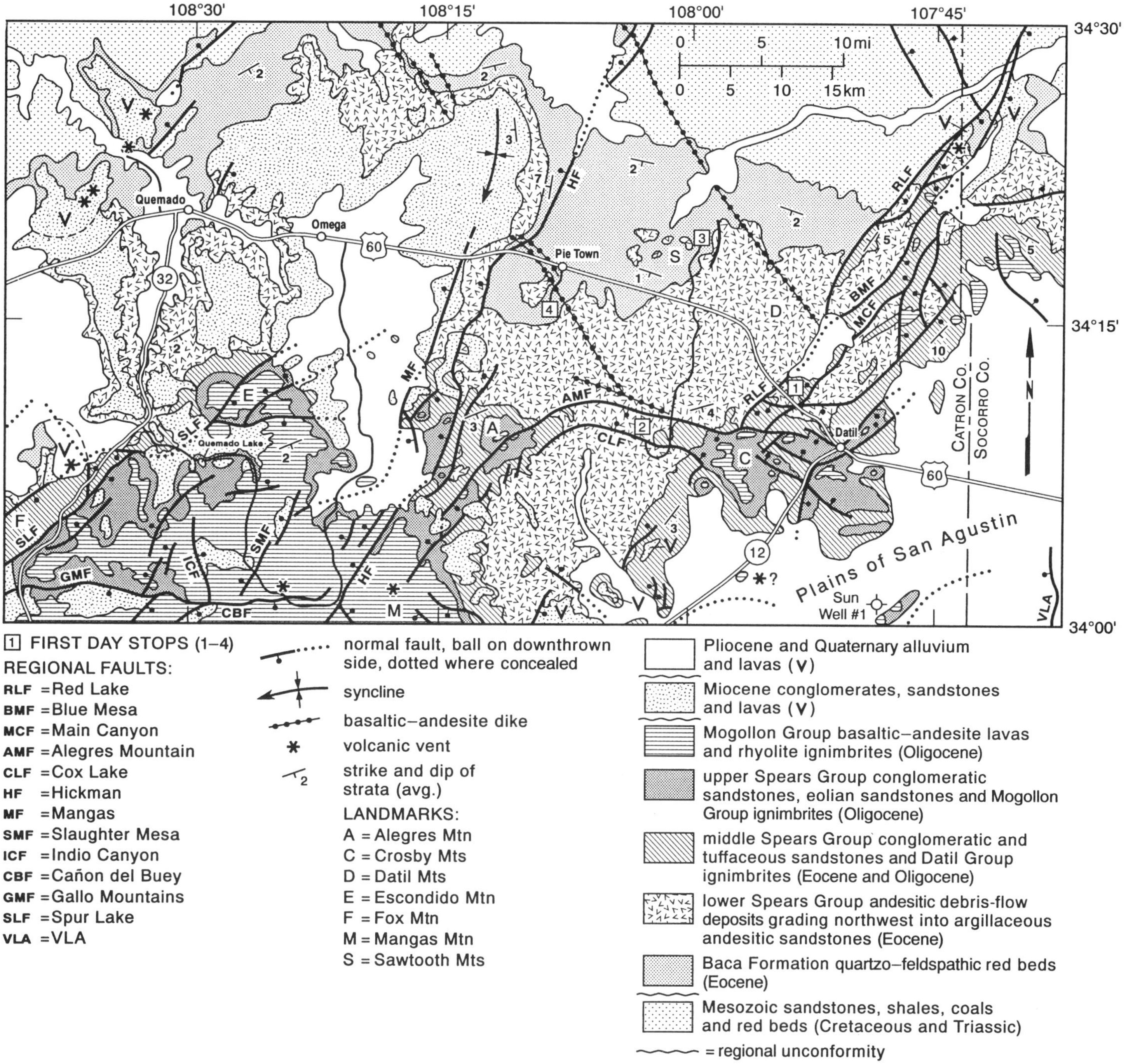

FIGURE 1.7. Generalized geologic map of the Quemado-Datil region, northeastern Catron County (modified after Chamberlin et al., 1994b; and unpublished digital data of the state geologic map project by O. J. Anderson and G. E. Jones, in prep.). As mapped here, the upper Spears Group locally contains thin ignimbrites of the Mogollon Group (McIntosh and Chamberlin, this volume). Eocene-Oligocene units generally intertongue, but unconformities are also apparent locally at several horizons within the volcanic pile.

FIGURE 1.8. The northern twin of Twin Peaks (left center) is underlain by light gray pumiceous ignimbrites (ca 29.7 to 28.9 Ma) intercalated with epiclastic conglomerates of the South Crosby Peak Formation of the upper Spears Group. Twin Peaks are capped by basaltic andesite of Deep Well and the La Jencia Tuff. Note absence of resistant ignimbrites on south twin peak, an indication that the pumiceous ignimbrites probably filled local paleovalleys.

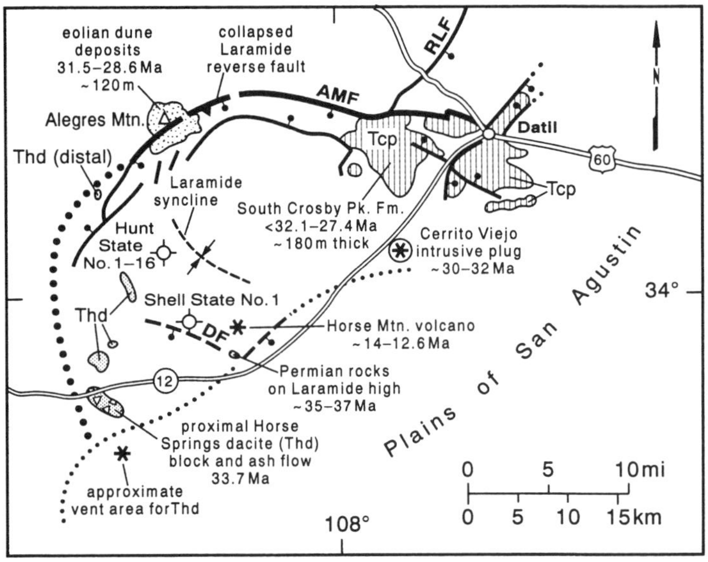

FIGURE 1.9. Sketch map of the Crosby Mountains caldera or volcano-tectonic depression (heavy line and dots), as proposed by Bornhorst (1976), Jones (1980) and Elston (1978, 1984, 1989). The distribution of volcanic and structural elements of their proposed caldera and currently available age constraints are summarized on the map. Data from McIntosh et al. (1991), McIntosh and Chamberlin (this volume), Ratté et al. (1991; this volume), third-day road log (this volume, mile 112.3), Lopez and Bornhorst (1979) and Chamberlin et al. (1994b). AMF = Alegres Mountain fault, RLF = Red Lake fault, and DF = Dutch Oven Pass fault zone.

formed between eruption of the 32.1-Ma Hells Mesa Tuff from the Socorro caldera and the 28.9-Ma La Jencia Tuff from the Sawmill Canyon-Magdalena caldera. A series of three thin (0-65 ft), nonwelded, pumiceous ignimbrites intercalated with epiclastic conglomeratic sandstone of the South Crosby Peak Formation are the only volcanic rocks of this age range within the proposed caldera (Lopez and Bornhorst, 1979; Osburn and Chapin, 1983; McIntosh and Chamberlin, this volume). The small volume (<4 km^3) of these ignimbrites in the South Crosby Peak Formation cannot account for collapse of the large diameter (25-30 mi) Crosby Mountains caldera proposed by Elston (1989). I propose that the dioritic hypabyssal porphyry (volcanic neck?) at Cerrito Viejo (Fig. 1.9), imprecisely dated at 30.56 ± 1.42 Ma, could represent a potential source of the lowest pumiceous ignimbrite in the South Crosby Peak Formation, which is imprecisely dated at 29.67 ± 0.86 Ma (McIntosh and Chamberlin, this volume). Regardless of possible genetic relationships, this small-volume pyroclastic unit cannot reasonably be linked to a large Crosby Mountains caldera as proposed by Elston (1989).

Detailed mapping and sedimentological studies of the South Crosby Peak Formation (32.1-27.4 Ma) will be necessary to evaluate why this unit thickens so rapidly to the south in the vicinity of Datil, New Mexico (cf. Lopez and Bornhorst, 1979; Harrison, 1980). Eolian sandstones in the Alegres Mountain area (Fig. 1.9), which are correlative to the South Crosby Peak Formation, span the proposed caldera margin fault (AMF) and show no apparent change in thickness (Jones, 1980; Chamberlin et al., 1994b).

At present, available stratigraphic relations, isotopic age data and regional mapping are contrary to the notion of a Crosby Mountains caldera, or a Crosby Mountains volcano-tectonic depression as proposed by Elston (1989, 1984, respectively). It is recommended that future maps of ash-flow calderas in New Mexico omit this apparent volcano-tectonic collage.

56.4 Milepost 80. Reconnaissance mapping by R. M. Chamberlin indicates that the east ridge of Twin Peaks, at 11:00, consists of three nonwelded pumice-rich (10-20%) and lithic rich (5-10%) ignimbrites, each about 30-50 ft thick, which are interbedded with andesitic conglomerates of the South Crosby Peak Formation. The lowest ignimbrite contains vitrophyric and andesitic rock fragments. The upper two ignimbrites contain a mixture of andesitic and welded-tuff fragments. Fragments of densely welded, crystal-rich, quartz-rich Hells Mesa Tuff are largest (~5 cm) and most abundant in the uppermost pumiceous ignimbrite. A similar sequence of three thin lenticular pumiceous ignimbrites on Anderson Mountain (peak at 11:45) has yielded $^{40}Ar/^{39}Ar$ ages confirming the presence of xenocrystic Hells Mesa sanidines in the upper two ignimbrites (McIntosh and Chamberlin, this volume). Also, at Anderson Mountain the lowest pumiceous ignimbrite has yielded an apparent eruptive age of 29.67 ± 0.86 Ma, as determined from plagioclase phenocrysts hand picked from cognate pumice clasts. Lopez and Bornhorst (1979) did not differentiate or map the distribution of these pumiceous ignimbrites within the South Crosby Peak Formation. Their age is bracketed by the underlying 32.1-Ma Hells Mesa Tuff and the overlying 28.8-Ma La Jencia Tuff. Similar pumice-rich ignimbrites, generally not associated with interbedded conglomerates, are commonly observed throughout the Socorro-Datil region (Osburn and Chapin, 1983). Where confined between the Hells Mesa and La Jencia tuffs, the thin pumiceous tuffs have been informally referred to as the tuff of South Crosby Peak (C. E. Chapin, oral commun., 1993). **0.7**

57.1 The northern twin of Twin Peaks at 9:00 is capped by 100 ft of basaltic andesite of Deep Well overlying 65 ft of densely welded La Jencia Tuff. Approximately 500 ft of conglomeratic sandstones and pumiceous ignimbrites of the South Crosby Peak Formation are exposed in the slopes below the La Jencia Tuff. **0.3**

57.4 Milepost 79. Highway descends onto flat floor of White House Canyon. North wall of canyon at 1:30 is cut in Hells Mesa Tuff. **1.0**

58.4 Milepost 78. New Mexico State Highway Department maintenance yard on right. **0.3**

58.7 Roadcuts on left in nonwelded pumice-rich ignimbrite within the South Crosby Peak Formation. The presence of large Hells Mesa-type lithic fragments (Fig. 1-10) indicates this is probably the uppermost pumice-rich ignimbrite in the Crosby Peak Formation. Zeolitization of the pumiceous ignimbrites and the presence of numerous lithic fragments have thwarted attempts to determine their paleomagnetic orientation. **0.3**

59.0 Roadcuts in conglomeratic sandstones of the South Crosby Peak Formation. **0.2**

59.2 Entering downtown Datil (Spanish, "date", as in fruit of the date palm), settled in 1884 (Pearce, 1965). Junction with NM-12 on left. **Continue straight ahead on US-60.** Third-Day road log will bring us back to Datil. Early travelers along the Ocean-to Ocean Highway (ca. 1912-1920) could have stayed at the luxurious Navajo Lodge near here (Fig. 1-11). Since 1933, the Eagle Guest Ranch has provided rustic and peaceful lodging, good food, and gasoline to tourists and local ranchers (Fig. 1.12). **0.2**

59.4 Milepost 77. For the next 9.3 mi, US-60 trends northwest along White House Canyon, a major drainage that generally cuts across the regional structural grain. **0.6**

60.0 Road to Datil Well campground on left. Highway crosses alluvium-covered trace of major down-to-south fault zone near here (Lopez and Bornhorst, 1979; Fig. 1.7). Thick South Crosby Peak Formation and Mogollon Group volcanic units are preserved on the south side of this fault zone. This arcuate, concave-to-the-south, normal-fault zone can be traced for 25 mi westward to Alegres Mountain, where it is referred to as the Alegres Mountain fault (Alegros fault of Jones, 1980; spelling and name changed to match that of Alegres Mountain 7.5' quadrangle). A regional north-trending seismic reflection profile (Armstrong and Chamberlin, this road log) shows that this fault was previously a zone of Laramide transpressional faulting, upthrown on the south, and then buried by the Eocene Baca Formation.

Near Datil, the Alegres Mountain fault zone merges in a complex pattern with the north-northeast-trending Red Lake fault zone, which also has indications of a Laramide ancestry (Cather, 1989). **0.2**

60.2 Dark ledges of coarsely porphyritic andesite of White House Canyon (Fig. 1.13) form southwest wall of canyon from 9:00 to 11:00. The andesite of White House Canyon and the immediately overlying Datil Well Tuff are interbedded in the Chavez Canyon Formation of the middle Spears Group (Osburn and Chapin, 1983; Cather el al., this volume). This gently southwest-dipping cuesta (9:00-11:00) locally marks the upthrown footwall of the Alegres Mountain fault. **0.2**

60.4 Milepost 76. Narrow ledge of 35.0-Ma Datil Well Tuff caps lower mesa at 2:00; Osburn and Chapin (1983) measured 31 ft of Datil Well Tuff at the type section approximately 200 ft east of the mesa corner. Upper mesa at 3:00 is capped by the 34.2-Ma Rock House Canyon Tuff. Volcaniclastic conglomerates and sandstones of the Chavez Canyon Formation typically underlie rubble-covered slopes below the Datil Group ignimbrites. Today's second stop provides an unusually good exposure of these widespread volcaniclastic sedimentary apron deposits assigned to the middle Spears Group. **0.7**

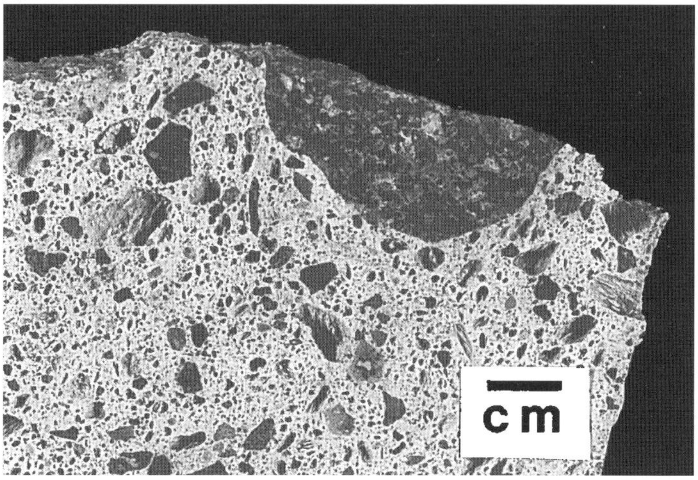

FIGURE 1.10. Slab of non-welded pumiceous and lithic-rich ignimbrite from the South Crosby Peak Formation; fragile zeolitic pumice forms striated pockets in rock face. Note large fragment of crystal-rich Hells Mesa Tuff (upper right).

FIGURE 1.11. Cowboys ready to ride, Navajo Lodge in Datil, ca. 1920. Photograph courtesy of Buz and Beverly Easterling, Quemado Lake Estates.

FIGURE 1.12. Local rancher George Farr and family at the Eagle Guest Ranch Trading Post in Datil, ca. 1935. Photograph courtesy of Kenneth and Carol Coker, Eagle Guest Ranch.

FIGURE 1.13. Vesicular turkey-track basaltic-andesite (55.3% SiO_2, anhydrous) lava from the andesite of White House Canyon.

FIGURE 1.14. Ledge-forming andesite of White House Canyon caps knobby hill at Vega Well overlook. Vantage point of Stop 1 is at hill crest on right. A small east-northeast-trending normal fault passes through saddle and drops lavas down to north. View is to northwest.

61.1 Junction with Forest Road 100 on right. FR-100 follows Main Canyon to Blue Canyon and the type section of the 33.7-Ma Blue Canyon Tuff (Osburn and Chapin, 1983) about 7 mi north of here. **0.1**

61.2 Dead Horse Mesa at 12:00 is capped by Datil Well Tuff and andesite of White House Canyon (Lopez and Bornhorst, 1979). **0.1**

61.3 Road cuts and slopes on left expose light gray argillaceous feldspathic sandstones and andesitic conglomerates of the Chavez Canyon Formation. Lopez and Bornhorst (1979) recognized and described the mostly fluviatile unit exposed here, but did not differentiate it from the coarse andesitic debris-flow deposits and breccias of the underlying Dog Springs Formation (Osburn and Chapin, 1983). Several mapping projects (Harrison, 1980; Coffin, 1981; Brouillard, 1984; Chamberlin et al., 1994b) have differentiated the fluvial conglomerates of the Chavez Canyon Formation from the underlying Dog Springs debris-flow deposits and noted that the contact is often an angular unconformity. Angular unconformities are associated with soft-sediment deformation structures that are widespread within the underlying Dog Springs Formation (Osburn and Chapin, 1983). The Chavez Canyon Formation is about 400 ft thick in the Datil region. **0.9**

62.2 Road sign announces entering Cibola National Forest. Most of the woodland along White House Canyon, however, is private property, originally acquired as sections and quarter sections under the Homestead Act. **0.5**

62.7 Baldwin Cabin, named for 1880s rancher Levy Baldwin, on hill top at 2:00. **0.4**

63.1 **Slow down, prepare for Stop 1.** Southeast crest of knobby hill at 1:00 (Fig. 1.14) is site of Stop 1. **0.1**

63.2 **STOP 1. Park along right shoulder of US-60**, as directed by flagpersons. Walk northwest along highway to private road to Vega Well (at cattle guard). S. Taubman is the local landowner at Vega Well and permission to use the overlook (Fig. 1.15) is required. The high point west of the Vega Well overlook is an alternative view point within the Cibola National Forest. Enter Vega Well property through gate at cattle guard and walk to southeast crest of the low hill immediately north of the road. This hill is capped by ledge-forming coarsely porphyritic andesite of White House Canyon that conformably overlies andesitic sandstones and conglomerates of the Chavez Canyon Formation of late Eocene age.

At the Vega Well overlook, use the index map (Fig. 1.15) and the panoramic photo (Fig. 1-16) as your guide to locating landmarks and geologic features. To the north, the almost treeless hill in the left middle foreground of the panorama is reference point "A" on the index map. Winchester (1920) was the first to map the north-northeast-trending Red Lake fault where it cuts Mesozoic strata in the Acoma section of the Colorado Plateau. Later studies by Wengerd (1959) and Maxwell (1976) in the Acoma region showed that this normal fault zone is associated with northwest-trending en echelon folds and domes in Cretaceous strata; this suggests it had an earlier history of Laramide right-lateral transpression. Robinson (1981) estimated the stratigraphic throw of Cretaceous strata along the main strand of the Red Lake fault at D-Cross Mountain to be about 1180 ft. Observations of striated fault surfaces in Cretaceous rocks near Alamocito Creek imply a late-stage, dip-slip, normal displacement for the Red Lake fault zone.

The main purpose of this stop is to present an introduction to the Cenozoic stratigraphy and structure of the Mogollon slope in northeastern Catron County. Major Cenozoic stratigraphic units and regional structural patterns are illustrated on Figure 1.7. The most significant map pattern is one of gently (2-5°) south-dipping Eocene and Oligocene strata locally offset and distended by north-northeast-trending normal fault zones. Miocene conglomerates are preferentially preserved on the downthrown side of the Neogene normal fault zones. These epiclastic conglomerates were mostly derived from erosional remnants of Oligocene volcanic highlands (e.g., Mangas Mountains, Gallo Mountains, Alegres Mountain and Crosby Mountains). Eocene andesitic debris flows in the Datil Mountains

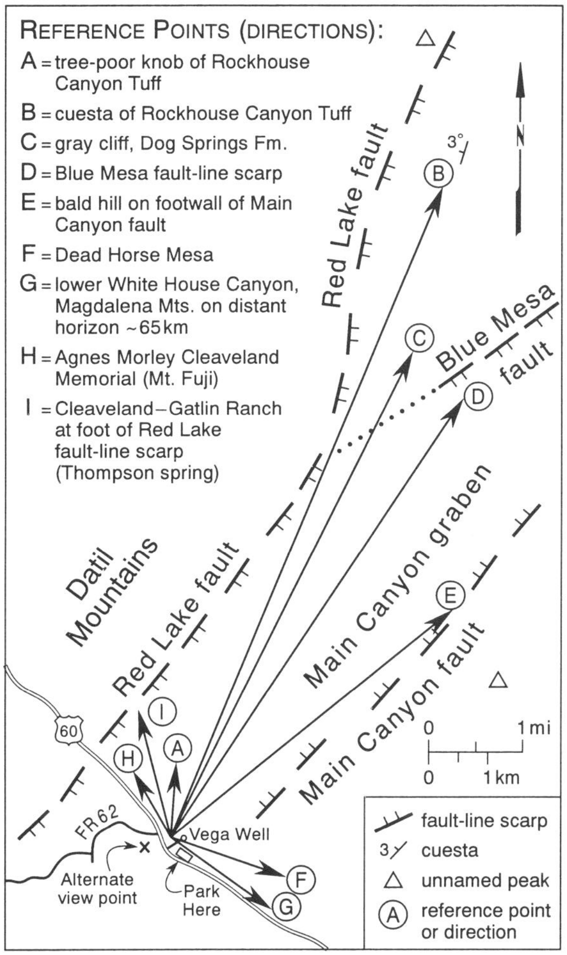

FIGURE 1.15. Panoramic index map of the Main Canyon graben area from the Vega Well overlook at Stop 1.

(lower Spears Group), which lie on a broad horst block between the Red Lake fault and the Hickman fault, were also a source of Miocene conglomerates in the Pie Town area. The estimated maximum Neogene normal displacement on the Hickman and Red Lake fault zones is about 2000 ft. Miocene units include the Fence Lake Formation, west of the Hickman fault zone; the Gila Group, south of the Spur Lake and Cañon del Buey faults; Santa Fe Group, south of the Alegres Mountain fault; and the conglomerate of Rock Tank Canyon (Robinson, 1981) east of the Blue Mesa fault.

Regional thickness trends of the middle to upper Eocene Baca Formation (nonvolcanic fluvial red beds) and upper Eocene andesitic debris flows of the Dog Springs Formation (Cather and Johnson, 1984; Cather, 1989) indicate that the Neogene Datil Mountains horst was a depressed synclinal trough between reverse faults along the Red Lake and Hickman fault zones during late Laramide transpressional deformation. Increasing evidence indicates that Laramide transpressional welts (uplifts) in west-central New Mexico have collapsed to form Neogene extensional or transtensional basins approximately contemporaneous with similar inversion of Laramide uplifts along the axis of the Rio Grande rift (Cather, 1989; Garmezy, 1990; Chapin and Cather, in press). However, the possibility of *local* Laramide trantensional basins at releasing bends in lateral or oblique slip regimes (e.g., Cather, 1992) should not be overlooked, because they may have preserved Paleozoic and Mesozoic sedimentary strata favorable for hydrocarbon exploration.

Andesitic debris-flow deposits of the Dog Springs Formation (ca. 40-37 Ma, Osburn and Chapin, 1983; Cather et al., 1987; McIntosh and Chamberlin, this volume) are widespread in the Gallinas and Datil Mountains and along the west flank of the Crosby Mountains (J-L Draw area). Debris flows of the Datil Mountains area appear to intertongue with and grade

FIGURE 1.16. Panorama of the Red Lake fault zone and topographic low of the Main Canyon graben; view to north from Vega Well overlook. Arrows indicate where major fault traces cross the horizon. Compare with index map, Fig. 1-15. Abbreviations are: RLF= Red Lake fault, BMF= Blue Mesa fault, and MCF= Main Canyon fault.

northwestward into andesitic sandstones and mudstones of the volcaniclastic unit of Largo Creek (new name, Chamberlin and Harris, this volume). Both of these upper Eocene andesitic units are informally assigned to the lower Spears Group. Regional distribution patterns of the lower Spears Group in the Datil-Quemado region suggest an andesitic source area located under the southern San Agustin Plains in the vicinity of the C-N Basin.

Volcaniclastic alluvial apron deposits of the middle Spears Group are mostly epiclastic conglomerates deposited by north-flowing braided stream systems emanating from andesitic to rhyolitic eruptive centers to the south. Outflow sheets of Datil Group ignimbrites derived from calderas to the south and east (McIntosh et al., 1991) are interbedded with the volcaniclastic sedimentary units (e.g. Chavez Canyon and Rincon Windmill Formations) to form a multi-story sediment and ignimbrite layer cake collectively lumped together here as the middle Spears Group and Datil Group. Porphyritic basaltic andesite to andesite lava flows (e.g. andesite of White House Canyon and andesite of Dry Leggett Canyon) are also interbedded in this upper Eocene to lower Oligocene interval (ca. 35 to 32 Ma). A major eruptive center for the andesite of Dry Leggett Canyon has been documented in the San Francisco Mountains area southwest of Reserve by Ratté (1989). As will be observed at Stop 2, syneruptive tuffaceous sandstones and pumiceous mudstones of rhyolitic composition are another significant lithology in the middle Spears Group. Syneruptive tuffaceous sandstones (terminology after Smith, 1991) largely define the middle Spears Group in the Quemado-Escondido Mountain area where Datil Group ignimbrites are locally absent (volcaniclastic unit of Cañon del Leon, new name, Chamberlin and Harris, this volume).

Epiclastic conglomerates and thin, non-welded, pumiceous ignimbrites in the South Crosby Peak Formation disconformably overlie the Hells Mesa Tuff (32.1 Ma) in the Crosby Mountains area (Osburn and Chapin, 1983). These non-welded lithic-rich ignimbrites must be older than the overlying 28.9-Ma La Jencia Tuff. However, additional epiclastic conglomerates that occur above the La Jencia Tuff, to the base of the South Canyon Tuff (27.4 Ma) have been mapped with the South Crosby Peak Formation in the Crosby Mountains (Lopez and Bornhorst, 1979). This dominantly sedimentary interval (ca 32-27.4 Ma) in the Crosby Mountains is here informally assigned to the upper Spears Group. A thick interval (400-800 ft) of massive and commonly high-angle cross-bedded sandstones occupies the same stratigraphic interval in the Mangas-Alegres Mountains region. Distal ignimbrite sheets, including the Caballo Blanco Tuff (31.7 Ma), Vicks Peak Tuff (28.6 Ma) and Bloodgood Canyon Tuff (28.0 Ma) locally fill paleovalleys in the massive fine- to medium-grained sandstones, and demonstrate that this unit is contemporaneous with the South Crosby Peak Formation. These massive cross-bedded to planar-bedded sandstones of apparent eolian origin (cross beds indicate prevailing westerly winds) have been informally assigned to the sandstone of Escondido Mountain (new name, Chamberlin and Harris, this volume). Presumably these sands once intertongued with distal alluvial apron deposits of the South Crosby Peak Formation in the erosional gap between the Crosby Mountains and Alegres Mountain. The Mogollon Group ignimbrites also erupted from calderas to the south and southeast of the Datil-Quemado area (McIntosh et al., 1991).

Upper Oligocene to lower Miocene(?) basaltic andesite lavas of the Bearwallow Mountain Andesite and the basaltic andesite of Crosby Mountains generally cap the middle Tertiary volcanic pile in the Datil-Quemado region. Erosional remanents of stacked basaltic andesite flows in the Mangas, Alegres and Crosby Mountains are commonly about 600 ft thick. Individual flows are typically about 30 ft thick. A much thicker pile of basaltic andesite lavas under Mangas Mountain (1200 ft) implies that this was a major eruptive center for the Bearwallow Mountain lavas.

Large basaltic-andesite dikes at Pie Town (Stop 4) and Lehew (aka Hickman) are lithologically similar to the Bearwallow Mountain lavas and yield K-Ar ages of 27.7 ± 0.6 and 27.3 ± 1.3 Ma, respectively (Laughlin et al., 1979) . Basal Bearwallow flows in the Mangas Mountains area have been dated at 27.8 ± 1.7 Ma (K-Ar, Baldridge et al., 1989) and at 26.0 ± 0.1 Ma ($^{40}Ar/^{39}Ar$ total gas isochron, McIntosh and Chamberlin, this volume). Considering analytical error of the K-Ar method these basaltic andesite dikes might be equivalent to the Bearwallow. These long (30-45 mi) dikes mark the onset of regional southwest-directed extension across the southeastern Colorado Plateau (Laughlin et al., 1983). Incipient regional extension, however, may have begun in the Datil area as early as 35-36 Ma (Cather, 1989). Relatively few regional normal faults in the Datil-Quemado area parallel the initial basaltic andesite dikes. Ancient colluvial breccias derived from Bloodgood Canyon Tuff adjacent to the Indio Canyon Fault (Fig. 1.7) suggest that it may have been active as early as 28 to 26 Ma.

On returning to vehicles note that the depositional contact of the andesite of White House Canyon on conglomerates of the Chavez Canyon Formation is well exposed on the southwest side of the Vega Well overlook (Fig. 1.17). Andesitic cross- bedded sandstones of the Chavez Canyon Formation are also well exposed in the roadcuts of US-60 west of the overlook.

Return to vehicles and continue west on US-60. 0.3

63.5 Roadcut on right in low-angle cross-bedded sandstone and andesitic conglomeratic sandstones of the Chavez Canyon Formation. A thin debris flow is also exposed near the northwest end of the cut, but this formation is dominantly fluviatile. Well-developed pebble imbrication in Chavez Canyon conglomerate at 11:00 (immediately below lava cap) indicates northwesterly paleoflow. **0.2**

63.7 Cuts on right and left in light gray conglomeratic sandstones of the Chavez Canyon conglomerate. Well-rounded clasts of medium-gray plagioclase-hornblende-biotite dacite porphyry, typical lithology of the underlying Dog Springs Formation, are ubiquitous in the lower Chavez Canyon Formation of the Datil area. **0.3**

FIGURE 1.17. Road logger (and honorary NMGS piano player) points to andesite of White House Canyon resting conformably on andesitic conglomeratic sandstones of the lower Chavez Canyon Formation. View to south on southwest side of Vega Well overlook.

FIGURE 1.18. Massive outcrop of andesitic debris-flow-breccia deposit in the upper Dog Springs Formation in upper White House Canyon.

64.0 Marble column at 10:00 marks memorial site for William Ramond Morley (1848-1884), pioneer railroad construction engineer for the Santa Fe Railroad, famous for designing and building the rail line through Raton Pass in northeastern New Mexico. He was accidentally shot and killed while working for the Mexican Central Railroad in old Mexico. His widow, Ada, and three young children came to the Datil area in 1886, later to become well-known successful cattle ranchers. Morley's tomboy daughter, Anne Morley Cleaveland, tells many tales of pioneer ranching experiences in the Datil region in her quasi-biographical book *No Life for a Lady*. Although rare, Ms. Cleaveland's brief geological notes are accurate and insightful. For example, she states "Datil is not a mineralized country. Not an ounce of valuable ore has ever been dug from its igneous mountains" (Cleaveland, 1941, p. 325). A brass plaque in memory of Anne Morley Cleaveland (1874-1958) is present on the hill at 1:00. Considering the view east down White House Canyon from the top of this hill, locally known as "Mt. Fuji", it is no wonder that "she loved this country". **0.6**

64.6 Road to Cleaveland-Gatlin ranch on right. Highway crosses alluvium-covered trace of the Red Lake fault near here. Roadcut on left in massive dacitic breccia (autoclastic?) of the Dog Springs Formation. For the next 5 mi, the relatively narrow upper reach of White House Canyon cuts across resistant dacitic debris flows and breccias of the upper Eocene Dog Springs Formation. **0.5**

65.1 Roadside table on right. Low cliffs of massive Dog Springs breccia at 1:30 (Fig. 1-18) have yielded an $^{40}Ar/^{39}Ar$ age of 36.94 ± 0.07 Ma from a single sanidine crystal and a less precise age of 37.57 ± 1.29 Ma from plagioclase crystals (McIntosh and Chamberlin, this volume). Some aspects of this dacitic breccia (Fig. 1-19) suggest a pyroclastic origin. The angular clasts of plagioclase-hornblende-biotite dacite porphyry float in a nonsorted massive matrix of similar composition. The microcrystalline matrix contains abundant broken slivers of phenocrysts and sparse angular fragments of granular sandstone and mudstone. The sedimentary fragments are soft and tend to weather out, leaving angular pock marks, which can easily be mistaken for pumice. However, pumice and glass shards are not evident in thin section. These massive breccias are commonly interbedded with thick debris-flow deposits containing well-rounded cobbles of dacite porphyry. Attempts to determine emplacement temperatures of Dog Springs volcaniclastic breccias using paleomagnetic properties have been unsuccessful. By association with sedimentary-looking debris-flow deposits, these dacitic breccias are generally regarded to be of volcaniclastic sedimentary origin as opposed to hot pyroclastic deposits (Cather, 1986). **0.8**

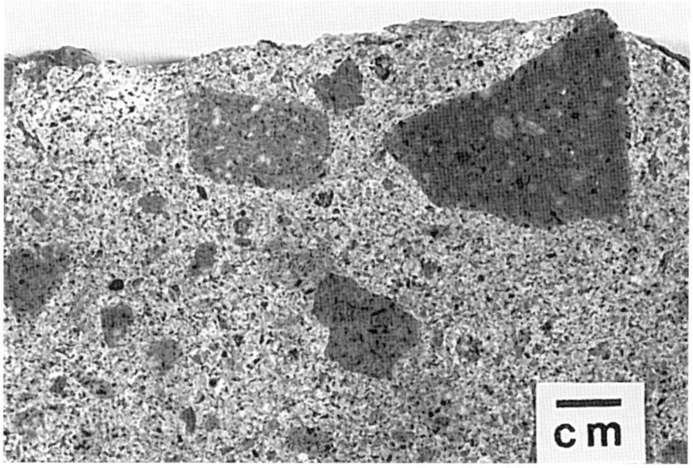

FIGURE 1.19. Typical plagioclase-hornblende-biotite porphyry clasts in Dog Springs debris-flow breccia deposit.

65.9 Rock fin at 11:00 is formed by a resistant debris-flow bed that is standing on end. Although hard to find, bedding in the adjacent hillside, like the rock fin, is nearly vertical and strikes N45°W. Lopez and Bornhorst (1979) erroneously mapped this fin and other similar outcrops as pyroclastic dikes. **0.3**

66.2 Massive dacite breccia of Dog Springs Formation in roadcut on left lacks discernible bedding. **0.7**

66.9 Roadcut on left in Quaternary valley fill. Little is known about the thickness and age of valley fill here. Approximately 130 ft of Pleistocene to Pliocene (?) valley-fill deposits are exposed near Monument Saddle, at the head of Davenport Canyon, about 5 mi north of here. **1.3**

68.2 Bedded dacitic breccias in roadcut on left show attitude of N22°W, 84°NE at south end of cut and are subhorizontal at north end of cut. FR-6 to Davenport Canyon and Monument Saddle on right, just beyond. **0.2**

68.4 Milepost 68. White House Spring, near valley wall at 12:15, was original site of Ada Morley's 10 room white log house built in 1886. **0.2**

68.6 Highway begins ascent through narrow reach of upper White House Canyon. **0.3**

68.9 Remnants of old roadcut and road bed of Ocean-to-Ocean Highway parallel powerline on north side of US-60. This old road bed, in use in the 1920s (Fig. 1.20) was abandoned in the mid-1950s. **0.2**

69.1 Roadcut of US-60 on right (Fig. 1-21) exposes thick-bedded, dacitic, debris-flow deposits of the Dog Springs Formation dipping 30° to east-southeast. Well-rounded cobbles and boulders of dacite porphyry (Fig. 1-22) were presumably derived from mountain-stream

FIGURE 1.20. Buttress and wooden beams of old Ocean-to-Ocean Highway bridge over drainage of upper White House Canyon. Carbide-lamp graffiti of W. H. Creamer Sep. 4, 1922 shows little or no fading over the last 71 years. Note medium to large pines growing from road bed.

FIGURE 1.21. Andesitic debris-flow deposits in Dog Springs Formation at Mile 69.1.

FIGURE 1.22. Close-up view of rounded stream cobbles and boulders of plagioclase-hornblende porphyry in massive debris-flow deposit of Dog Springs Formation.

bottoms and imply a relatively low-temperature environment of deposition for these deposits. **0.2**

69.3 Cut on right in interbedded argillaceous dacitic sandstones and dacitic debris-flow deposits dipping 30° to east-southeast. Small displacement conjugate normal faults exposed here strike N70°E and imply a local north-northwest oriented tensile stress axis. About 4 mi north of here, a 0.6-mi-long, 30-ft-wide, red mudstone dike cuts vertically across Dog Springs debris-flow deposits. This clastic injection dike, presumably derived from liquefied upper Baca Formation, strikes N50°E and also implies a northwest oriented tensile stress axis. All of these features are attributed to regional soft-sediment deformation and liquefaction in late Eocene time, shortly after or during deposition of the Dog Springs Formation. **0.1**

69.4 Milepost 67. Roadcut on left in Dog Springs debris-flow deposits and sandstones. The next 11-mi segment of US-60 was widened and "modernized" (roadcuts converted to rubble and then seeded) in a construction project completed in July 1993. **0.6**

70.0 Highway crests at West Pass; approximate elevation is 8100 ft. Headwaters of White House Canyon at 3:00. Entering drainage area of Z Slash Draw, which becomes Z Slash Canyon about 5 mi to the south. Dog Springs beds at 2:00 look horizontal from here. **1.3**

71.3 **Turn left onto FR-63 and head south. 0.1**

71.4 Alegres Mountain on skyline at 1:00 is the highest point on the Mogollon slope of west-central New Mexico. It rises to an elevation of 10,220 ft and is capped by the Mangas Mountain member of the Bearwallow Mountain Andesite of the Mogollon Group. This erosional remnant of late Oligocene basaltic andesite lavas is approximately 650 ft thick.

For the next 5 mi FR-63 follows the drainage of Z Slash Draw, which generally parallels and defines an upper Eocene facies transition from dominantly Dog Springs debris-flow deposits on the east (higher hills to left) to dominantly argillaceous fluviatile sandstones on the west (lower hills on right). These poorly-sorted sandstones, compositionally equivalent to the coarser Dog Springs deposits, are newly assigned to the upper Eocene volcaniclastic unit of Largo Creek (Chamberlin and Harris, this volume). Most of the poorly-sorted, planar-bedded Largo Creek sandstones probably represent distal hyper-concentrated flow deposits derived from downstream transformation of the medial Dog Springs debris-flow deposits. Similar transitions (transformations) are commonly observed in the overall facies lineage of continental volcanic terranes (Fisher and Smith, 1991). The Dog Springs Formation and the volcaniclastic unit of Largo Creek are both assigned to the lower Spears Group (Fig. 1.7).

Willard's (1959, fig. 1) reconnaissance geologic map of northern Catron County shows that his "latite facies (Tdl) of the Datil Formation" is essentially equivalent to the Dog Springs Formation as presently defined. Willard's "volcanic sediment facies (Tds)" is also generally equivalent to the volcaniclastic unit of Largo Creek. However, the Miocene Fence Lake Formation was not discriminated on Willard's map where it unconformably overlies the volcaniclastic unit of Largo Creek in the Quemado-Pie Town area. **0.8**

72.2 Flat-topped Crosby Mountains at 12:00 and gently west-tilted top of Sugarloaf Mountain at 12:15. Both are capped by undated late Oligocene(?) basaltic andesite lavas assigned to the Mogollon Group. These basaltic andesites locally overlie the 27.4-Ma South Canyon Tuff (McIntosh and Chamberlin, this volume) and are probably correlative with the Bearwallow Mountain Andesite. They are, however, presently assigned to the informal unit of the basaltic andesite of Crosby Mountains (Lopez and Bornhorst, 1979). **0.4**

72.6 Cliffs and pinnacle of Dog Springs Formation in wooded hillside at 11:00. **0.4**

73.0 Cox Peak on middle skyline at 1:00 is capped by 250 ft of gently south-dipping (undeformed) Dog Springs debris-flow deposits overlying at least 300 ft of andesitic sandstones of the volcaniclastic unit of Largo Creek (Chamberlin et al., 1994b). **0.7**

73.7 Profile of upper Miocene Horse Mountain volcano is visible on skyline at 12:15. Cuesta with white stripe in middle slope just below and left of Horse Mountain is near site of Stop 2. Cuesta is capped by 33.6-Ma Blue Canyon Tuff and the white stripe is formed by the 34.2 Ma Rock House Canyon Tuff. **1.0**

74.7 Bumpy road crosses outcrop of light gray andesitic sandstones of the volcaniclastic unit of Largo Creek. **0.7**

75.4 **Bear right at fork in road**. Primitive Forest Road 62 on left leads to Flying V Draw. **0.3**

75.7 Wooded hills in foreground from 11:00 to 1:00 are underlain by Dog Springs Formation. **0.2**

75.9 Cattle guard. Entering private land; road curves to right. **0.4**

76.3 Road enters Quemado 30′ x 60′ quadrangle, area of the reconnaissance geologic map of Chamberlin et al. (1994b), the source of much of the geologic information presented from here on. **0.2**

76.5 Road begins sweeping S curve, first right then left, around windmill at Black Well. **0.2**

76.7 Road enters narrows of upper Z Slash Canyon. Wooded hills to east and west are mostly capped by Dog Springs Formation debris-flow deposits that are locally interbedded with and overlie andesitic sandstone of Largo Creek lithology. On a reconnaissance level, these hills are mapped as Dog Springs Formation. **0.7**

77.4 Ledge-forming Dog Springs beds at 10:00. **0.6**

78.0 Butte at 2:30 is capped by apparently horizontal debris-flow-breccia deposit overlying Largo Creek-type sandstones; this horizontal-looking breccia bed actually lies on the crest of a complex isoclinal fold system that is well exposed on the south flank of the butte (Fig. 1-23). On the west limb of this fold system, the breccia bed dips 60° to the west-northwest; and sub-horizontal axes of isoclinal folds trend N20°E. **0.2**

78.2 Cattle guard. Enter T. L. Mannering ranch. **0.4**

78.6 Rounded peak on skyline at 10:30 is capped by 33.6-Ma Blue Canyon Tuff, just slightly older than the Eocene-Oligocene boundary, which has recently been shifted from 36.6 Ma (Palmer, 1983) to 33.4 Ma (McIntosh et al., 1992). This peak and underlying section of Datil Group ignimbrites and middle Spears Group conglomeratic sandstones are preserved on the downthrown side of the Alegres Mountain fault in a narrow east-west trending graben approximately 2 mi wide. **0.3**

78.9 Axes of tightly-folded Dog Springs beds dip about 50° to south-southwest at 9:00. Just beyond, the road crosses projected trace of Alegres Mountain fault (AMF, Fig. 1.7) with an estimated local stratigraphic throw of 600 ft, downthrown to the south. Displacement on the AMF apparently increases toward the east, where it may be as much as 1200 ft in the northern Crosby Mountains. The AMF forms the north margin of what is here termed the Cox Lake graben. The east end of the Cox Lake graben is downwarped to the east-southeast at an angle of about 4°; this accounts for the progressive easterly increase of stratigraphic throw on the AMF. To the west of Alegres Mountain the AMF displaces the Bearwallow Mountain Andesite about 400 ft (Jones, 1980). **0.1**

79.0 Red ledges of Chavez Canyon Formation conglomerates at 9:00 to 9:30. Datil Well Tuff is locally absent here. **0.2**

79.2 Road to Mannering ranch house on right. Blue Canyon Tuff of the Datil Group caps cuestas at 12:30 and 1:00, and Rock House Canyon Tuff of the Datil Group forms

FIGURE 1.23. Isoclinal fold in the Dog Springs Formation, NW¼ sec. 31, T1S, R11W; view to northeast. Note fold noses above brushy piñon on lower left. Small sausage-like nose, just above brush is approximately 2 ft wide. Subhorizontal debris-flow breccia bed on skyline is at crest of broader anticlinal fold.

white band in middle slope. Slope-forming interval between Blue Canyon and Rock House is the Rincon Windmill Formation of the middle Spears Group. **1.0**

80.2 White ledge forming Rock House Canyon Tuff is at road level on the left. **0.3**

80.5 Beds of Rincon Windmill Formation exposed at road level on left. Blue Canyon Tuff caps ridge on left, but descends to road level ahead. **0.4**

80.9 Cattle guard. Road crosses projected trace of the Cox Lake fault. This down-to-the-north normal fault has an estimated stratigraphic throw of 800 ft and forms the south margin of the Cox Lake graben. **0.2**

81.1 Cattle guard. Low hills at 2:45 to 3:00 are underlain by Dog Springs Formation in the footwall of the Cox Lake fault. **0.5**

81.6 Entrance to Jay Saulsberry Ranch on right. **0.4**

82.0 Cattle guard. **Prepare for hard right turn 0.2 mi ahead. 0.2**

82.2 **Make hard right turn,** onto ranch road that heads north along Nester Draw. **0.1**

82.3 Road forks at side-by-side gates. Take right fork; the right gate is usually open. Now on private land of the Fred Saulsberry ranch; permission to enter this area may be obtained at the ranch headquarters 1.2 mi to the south. **0.1**

82.4 Dog Springs Formation exposed on hillside to left. **0.3**

82.7 Datil Group ignimbrites form resistant ledges and caprock in hills at 12:00 (Fig. 1.24). **0.1**

82.8 Gate across ranch road, normally closed. Continue straight ahead. Folded Dog Springs beds at 2:00. **0.2**

83.0 Lower Chavez Canyon Formation is faulted against Dog Springs Formation at 3:00. **0.1**

83.1 Gate across ranch road, usually open. **0.1**

83.2 Cross projected trace of Cox Lake fault. Note gentle synclinal trough at 2:00. **0.3**

83.5 Ravine above small alluvial fan at 3:00 is site of Stop 2. **Continue straight ahead to turn around at triangular intersection. 0.2**

83.7 **Take right fork at Y intersection, make counterclockwise U turn, and head south to Stop 2. 0.4**

84.1 **STOP 2. Park at side of road, remain headed south,** for return to US-60. The purpose of this stop is to provide a hands-on look at three distal ignimbrite sheets of the Datil Group and interbedded volcaniclastic sedimentary rocks of the middle Spears Group that are unusually well exposed here at Saulsberry Ranch.

These upper Eocene rocks on the west flank of the Crosby Mountains are locally preserved in the Cox Lake graben, between the Alegres Mountain fault and the Cox Lake fault. For orientation, see photograph and measured section through these upper Eocene rocks (Figs. 1.25, 1.26). Correlation of the ledge-forming ignimbrites has been verified by ^{40}Ar/^{39}Ar geochronology (McIntosh and Chamberlin, this volume) and by determination of thermal remanent magnetic directions. These new ^{40}Ar/^{39}Ar ages, based on multiple analyses of individual sanidine crystals, are superior to previous bulk-sample sanidine analyses of the Datil Well Tuff and the Rock House Canyon Tuff, which were apparently contaminated by older

FIGURE 1.24. View to north of ledge-forming Datil Group ignimbrites at Saulsberry Ranch. Site of Stop 2 is at light-colored area on far left. Blue Canyon Tuff caps mesas, white ledges in mid-slope are formed by Rock House Canyon Tuff and narrow ledge in lower third of slope is Datil Well Tuff. Volcaniclastic sedimentary rocks of middle Spears Group between ignimbrites are poorly exposed, except for low cliff of andesitic conglomerate at base of section.

xenocrysts of sanidine. These new isotopic age data strongly suggest that the Datil Well Tuff is correlative with the Bell Top Tuff No. 4 (McIntosh et al., 1991), but more data are needed to confirm this possibility. If this tentative correlation is correct, the Datil Well Tuff (34.98 ± 0.19 Ma) may represent an initial (premonitory) phase of eruption from the Emory caldera area that culminated in eruption of the Kneeling Nun Tuff at 34.89 ± 0.05 Ma (McIntosh et al., 1991).

With the correlation of the ignimbrites at Saulsberry Ranch firmly established by isotopic ages and paleomagnetic directions, these formations may be mapped regionally on the basis of their physical characteristics. The Datil Well Tuff is a light gray, moderately phenocryst-rich (~15%) partially to densely welded, pumice-poor rhyolite ignimbrite. Medium-grained phenocrysts of sanidine are dominant and minor green pyroxene is distinctive; plagioclase and quartz are rare to absent. The light gray to white, phenocryst-poor (~2%), pumice-rich Rock House Canyon Tuff, is poorly

FIGURE 1.25. View showing location of measured stratigraphic section at Saulsberry Ranch. Distal ignimbrites of the Datil Well Tuff (Tdw), Rock House Canyon Tuff (Trh) and Blue Canyon Tuff (Tbc) are interbedded with and cap volcaniclastic sedimentary rocks of the Chavez Canyon Fm. (Tch) and the Rincon Windmill Fm. (Trw), of the middle Spears Group. Finer lines show approximate path of measured section (Fig. 1.26).

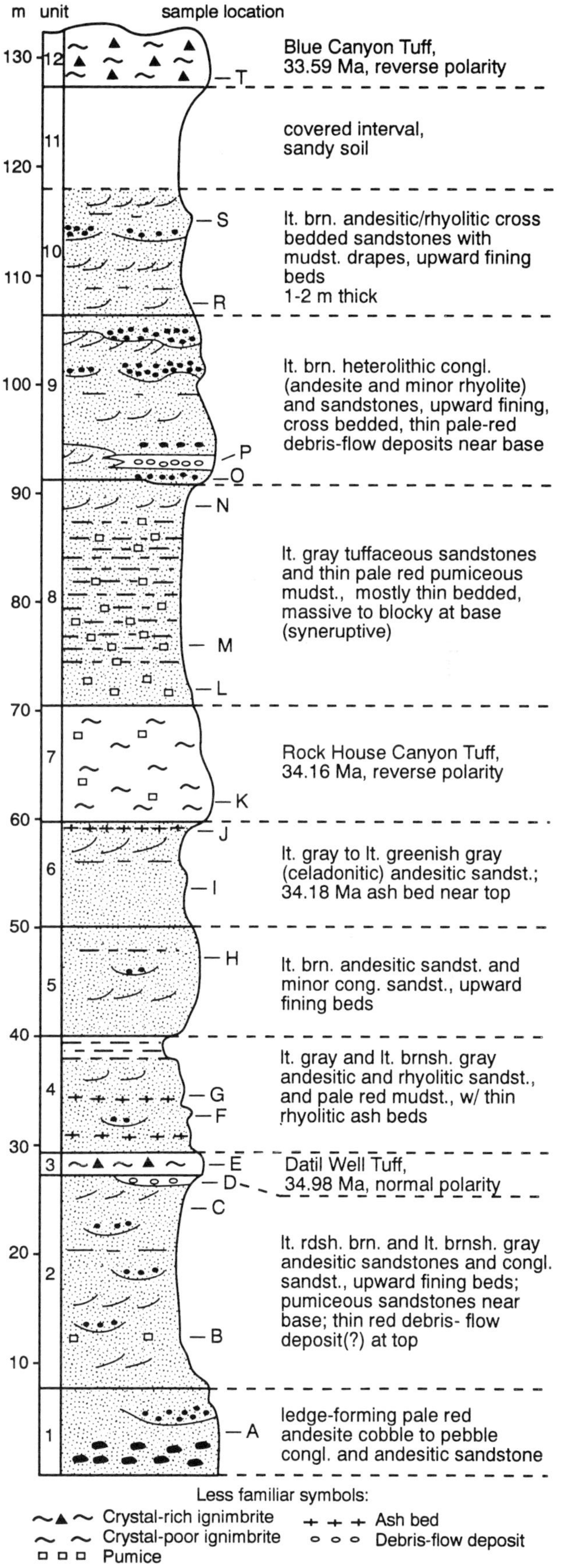

FIGURE 1.26. Measured section of Datil Group ignimbrites and middle Spears Group at Saulsberry Ranch; section measured by R. M. Chamberlin and O. J. Anderson. Rock House Canyon Tuff separates underlying Chavez Canyon Formation from overlying Rincon Windmill Formation (Osburn and Chapin, 1983).

welded to nonwelded. Small sparse phenocrysts of sanidine and plagioclase with minor quartz and biotite are typical. Unusually large phenocrysts of plagioclase (5-6 mm) and smaller sanidine phenocrysts comprise about 20% of the Blue Canyon Tuff. Biotite and sparse quartz are also present. Plagioclase in the Blue Canyon at Saulsberry Ranch is mostly altered to clay minerals; probably a reflection of deuteric or vapor-phase alteration. The partially to densely welded, light to medium gray Blue Canyon Tuff generally contains 2-4% small andesite lithic fragments; small lithic fragments also occur in the Rock House and Datil Well Tuffs but tend to be more sparse. All of the ignimbrites are ledge-forming units and locally form caprocks.

Thickness variations in these distal ignimbrites (typically 0-100 ft) can be an important indicator of paleotopography. The 6-ft-thick Datil Well Tuff locally pinches out to the northeast of the measured section; also the Rock House Canyon Tuff, about 30 ft thick in this region, is abruptly truncated at a local normal fault-controlled unconformity approximately 1.2 mi east of this site. The field relationships there (SW¼ sec. 32, T1S, R11W) indicate down-to-the-south normal faulting along the Alegres Mountain fault zone as early as 34.2 Ma. A similar normal fault appears to truncate the Hells Mesa Tuff 1.8 mi farther east (Lopez and Bornhorst, 1979). Abrupt thickness changes of ignimbrites at ancient fault scarps are key indicators of contemporaneous displacement because ignimbrites are emplaced essentially instantaneously and produce a mold of the underlying paleotopography (Chamberlin, 1983). Distal ignimbrites also fill paleovalleys in the Quemado-Datil region; paleovalley margins should not be confused with local fault-controlled unconformities.

The general lithology and compositional character of Spears Group volcaniclastic sedimentary rocks at Saulsberry Ranch are summarized in Fig. 1.26. Upward fining beds of volcaniclastic conglomerate to sandstone, low-angle trough crossbedding, and thin mudstone drape deposits indicate that the Chavez Canyon and Rincon Windmill Formations were primarily deposited by braided channel systems on a broad alluvial apron peripheral to volcanic highlands. Limited observations of pebble imbrications at widespread sites suggest that paleoflow was dominantly northerly. This is consistent with the general recognition that major late Eocene volcanic centers were primarily south of the Datil region, although specific source cauldrons for the Datil Well, Rock House Canyon, and Blue Canyon Tuffs have not yet been located. Both the Chavez Canyon and the Rincon Windmill formations consist of intercalated beds of andesitic to rhyolitic composition and variable mixtures of heterolithic conglomerates containing both andesitic and rhyolitic clasts.

The Rock House Canyon Tuff is locally overlain by 70 ft of mostly thin-bedded, light gray, tuffaceous (shard-rich) sandstone and thin, pale, red, pumiceous mudstones (unit 8, Fig. 1.26). Preferential concentration of pumice in the mudstone drapes (Fig. 1.27) provides convincing evidence that the pumice was floating and therefore freshly introduced into the fluvial system at the time of deposition. In the terminology of Smith (1991), this pumiceous unit is a syneruptive deposit, presumably derived from the non-welded top of the Rock House Canyon Tuff. Similar late Eocene syneruptive units are newly assigned to the volcaniclastic unit of Cañon del Leon in the Quemado area where the Datil Well, Rock House and Blue Canyon ignimbrites are locally absent. Tuffaceous Cañon del Leon beds are intercalated with andesitic sandstones and conglomerates that must be laterally equivalent to the Chavez Canyon beds observed here.

Another noteworthy lithology in the Saulsberry Ranch section is a rhyolitic sandstone bed (sample location F, Fig. 1.26), that probably represents reworked, nonwelded, crystal-rich Kneeling Nun Tuff. This interpretation is supported by its stratigraphic position and its mineralogy (sanidine-and plagioclase-rich with about 10% quartz and 5% biotite). Occurrences of heterolithic conglomerates (Fig. 1.28) and thin muddy debris-flow beds (Fig. 1.29) in the Rincon Windmill Formation are also notable.

The above observations and regional mapping suggest little difference in the general lithology, provenance or depositional environment of the Chavez Canyon and Rincon Windmill formations. They would be essentially indistinguishable where the Rock House Canyon Tuff is not present. Some revision in the nomenclature of these middle Spears Group formations may be appropriate when the Datil-Quemado region is mapped in detail. Considerably more sedimentological data is needed to evaluate the possible influence of contemporaneous normal faulting on upper Eocene and Oligocene depositional systems in the Datil region. For example, Mack et al. (this volume) have used detailed sedimentological data from the Bell Top Formation to define a large half-graben type basin of late Eocene to Oligocene age in the southern Rio Grande rift. Sedimentary units of the Bell Top Formation are similar in age and lithology to those observed here at Saulsberry Ranch.

Field conference attendees should use proper caution when climbing through the section at Saulsberry Ranch. Be careful not to dislodge loose rocks and watch your footing. Quick climbers may enjoy the view of Horse Mountain (to the south) and Alegres

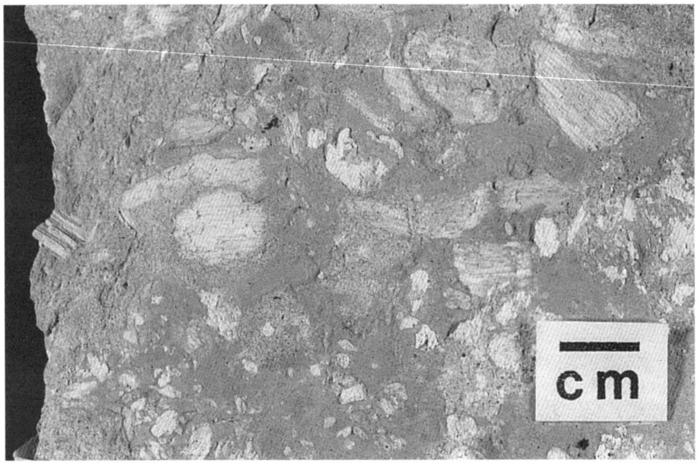

FIGURE 1.27. Pumiceous mudstone (Fig. 1-26, sample location M) as exposed on a bedding plane.

FIGURE 1.28. Heterolithic conglomerate (Fig. 1.26, sample location O) showing mixture of mostly darker andesitic clasts, and lighter rhyolitic clasts. Some rhyolitic clasts are densely welded ignimbrites.

FIGURE 1.29. Contact between muddy debris-flow deposit and underlying coarse-grained sandstone (Fig. 1.26, sample location P). Note diffuse lighter-colored mixing zone at base of debris flow. Total thickness of debris-flow bed approximately 10 in.

Mountain (to the west) from the top of this mesa.

After stop return to vehicles and retrace route to US-60. 12.2

96.3 Stop sign at junction of FR-63 and US-60. **Turn left onto US-60. 0.2**

96.5 **Turn right onto Forest Road 6A.** Next stop is the lunch stop, below spectacular cliffs of the Sawtooth Mountains. For the next 1.9 mi route traverses andesitic sandstones of the volcaniclastic unit of Largo Creek of the lower Spears Group. **0.7**

97.2 Road crests hill of andesitic sandstones locally exposed to west. **1.2**

98.4 Road crests ridge line of basal Spears Group sandstones; view of central Sawtooth Mountains at 12:00 to 12:30. **0.1**

98.5 At north base of ridge line; bottom of hill. This slope break marks the approximate contact of the upper Eocene Baca Formation and the conformably overlying unit of Largo Creek. For purposes of reconnaissance mapping, where cliff-forming debris-flow deposits overlie Largo Creek type sandstones in the Sawtooth Mountains, this same contact with the Baca is mapped as the base of the Dog Springs Formation (Fig. 1.7). At this point we are on the Madre Mountain 7.5´ quadrangle just east of the Quemado 30´ x 60´ quadrangle mapped by Chamberlin et al., (1994b). The Madre Mountain quadrangle was mapped in reconnaissance by Chamberlin (1981b). **0.2**

98.7 Low roadcut on right in red sandstone of the Baca Formation, which contains abundant quartz in comparison to the nearly quartz-free beds of the overlying lower Spears Group. **0.6**

99.3 Good view of pinnacles and cliffs of Dog Springs Formation overlying moderately folded argillaceous andesitic sandstones in central Sawtooth Mountains at 12:00. Baca Formation underlies gentle wooded slopes at base of cliffs. **0.6**

99.9 Large pedestal, called Monument Rock, at 2:00 is capped by undeformed Dog Springs Formation with Baca locally exposed at the break in slope. Cliffs of Dog Springs Formation from 12:00 to 1:30 are focal point of Stop 3. From this vantage point these cliffs look deceptively like a simple layer cake. **0.3**

100.2 Road curves to left; road cuts and natural exposures of Baca Formation. **0.2**

100.4 Road sign indicates Sawtooth Mountain looking north and Sawtooth Mountains when looking south (Fig. 1.30). Area of Stop 3 is at 2:30. **0.2**

100.6 **Turn right on Forest Road 325.** Monument Rock to southeast at 1:00. **0.8**

101.4 **STOP 3. Make counter clockwise U-turn, as directed by flagpersons,** in open meadow. Park headed west for return to US-60. This is our lunch stop, timed for slightly after noon when these west-facing cliffs are out of the morning shadows. Spectacular soft-sediment structures well exposed within these cliffs of Dog Springs Formation are the focus of this stop.

The snaggle-toothed peak to the north-northeast attains a height of 8919 ft and the notched cliff to the

FIGURE 1.30. View of unnamed peaks and cliffs in the eastern Sawtooth Mountains from mile 100.4. Intensely folded andesitic sandstone beds, a low-angle fault, and overlying nearly vertical debris-flow deposits are all locally well exposed in these cliffs. This low-angle fault forms a distinct line (rising to left) approximately two-thirds of the way up the southern cliff, and projects across the canyon to just under the tooth-like spire. Gently dipping red beds of the Eocene Baca Formation. underlie the wooded slopes at the base of the cliffs.

east of it reaches 8761 ft. These unnamed cliffs and peaks are simply referred to here as peak 8919 and peak 8761, respectively. These bold cliffs of internally deformed Dog Springs Formation have not been mapped in detail. The following discussion is derived largely from Chapin and Cather (1989).

The early Ecocene (36.9 to 39.6 Ma) Dog Springs Formation consists primarily of andesitic to dacitic debris flows that commonly overlie and intertongue with argillaceous andesitic sandstones near its base (Cather, 1986; Cather and Chapin, 1989). Plastically folded sandstone beds, chaotic and steep dips in thick-bedded debris-flow deposits and numerous clastic injection dikes occur throughout the Dog Springs outcrop belt (Cather and Chapin, 1989). This outcrop belt forms a south-facing arc about 40 mi long and about 6-9 mi wide from the central Gallinas Mountains, west to Greens Gap (15 mi west of Datil). Estimated thicknesses of the total debris-flow section range from 1000 to 2000 ft (Cather, 1986), and the lower sandstones may be 100 to 200 ft thick where undeformed (e.g., Monument Rock). Regional mapping demonstrates that the underlying Baca Formation and overlying Chavez Canyon Formation and Datil Group ignimbrites dip gently to the south; thus regional deformation in the Dog Springs must be intraformational and of late Eocene age. Cather and Chapin (1989) emphasized that the Dog Springs debris-flows were deposited in an actively subsiding area between the Hickman and the Puerticito fault zones near the hydrographic center of the Baca Basin. The anomalously thick debris-flow sequence (>1000 ft) may have enabled fluidization of the sediments due to the presence of shallow groundwater. Rapid loading and overpressurization of unconsolidated sediments during accelerated debris-flow sedimentation, and basal shear by overriding debris flows, are possible mechanisms to trigger widespread soft-sediment deformation.

Similar soft-sediment deformation and liquefication structures of regional extent have been described in the Eocene Absoraka Volcanic Supergroup of Wyoming by Decker (1990). Decker aptly noted that historic regional soft-sediment deformation and liquefaction events have been triggered by major seismic shocks such as the 1811 New Madrid earthquake and the 1964 south-central Alaskan earthquake. Cyclic loading associated with earthquake vibrations was cited by Decker as the most likely cause for widespread liquefaction. Sudden deposition of lava flows, debris flows and explosive volcanic eruption are also possible triggers for liquefaction events. Thick intervals of rapidly deposited, low-permeability volcaniclastic sediments are particularly susceptible to liquefaction-related deformation. Following liquefaction of a sedimentary unit, the associated chaotic deformation patterns are primarily caused by gravitational collapse and diapirism (Decker, 1990).

As viewed from the southwest, the soft-sediment or liquefaction-related deformation visible in peak 8919 and peak 8761 can be divided into four structural zones or elements. From the base of the cliffs to the top, these elements include (1) a lower zone of undeformed Baca beds and basal Dog Springs fluvial sandstones (Fig. 1.31), (2) a medial zone of intensely folded and sheared fluvial sandstones (Fig. 1.32) (3) a gently southeast-dipping low-angle detachment fault or decollement (Fig. 1.31) and (4) an upper zone (or plate) of steeply-dipping debris-flow deposits, of which only the massive-looking base of a single bed is visible in the upper cliff of peak 8761. The 60-70° east-northeast dip of these debris-flow deposits and truncation by the low-angle fault is much more impressive when viewed looking north from peak 8761 to the south face of peak 8919 (Fig. 1.33). This cross-section view shows all of the above-mentioned structural elements in one single frame. However, the crudely bedded and massive character of the debris-flow deposits is more safely observed and accessible in the notch of peak 8761 (Figs. 1.34, 1.35). The primary evidence of liquefaction at this stop is the zone of nearly recumbent and isoclinal folds in the medial zone below the low-angle fault. Clastic injection dikelets in flat lying dacitic-breccia beds about 1.2 mi northeast of here (Fig. 1.36) provide additional evidence of liquefaction in this area.

FIGURE 1.31. View to east-northeast of cliff-face below peak 8761 from Stop 3. Gently-dipping Baca Formation and light gray basal Dog Springs sandstones are poorly exposed in the lower slope on the right. Note intensely folded bedding in middle slope below gently south dipping fault surface (line between folded beds and massive looking top of cliff).

FIGURE 1.32. Close-up view of intensely folded fluvial sandstone beds of the lower Dog Springs Formation below low-angle fault in west-face of peak 8761. Noses of at least 8 nearly isoclinal folds can be recognized. Photograph from Cather and Chapin, 1989.

After lunch and a discussion of these spectacular cliffs, field conference attendees may hike to the notch in peak 8761 and get a close view of the folded beds, the low-angle fault and the nearly vertical debris-flow deposits above the fault. Alternatively, the south face of peak 8919 can be viewed by walking up the canyon between these cliffs. **After stop, return to vehicles and retrace route to US-60. 4.9**

106.3 Stop sign at junction of FR-6A and US-60. **Turn right and head west on US-60. 0.1**

106.4 Sandstones of the volcaniclastic unit of Largo Creek are locally exposed in low hills, in part the Airplane Hills, for next 2.8 mi. Recently reconstructed road cuts for next 8 mi consist of pulverized bedrock and smoothed soils that have been manicured and seeded so as to blend(?) into the environment. **2.1**

108.5 Leaving Cibola National Forest. Highway begins descent from the Airplane Hills into broad valley of Red Flats. **0.3**

108.8 Stream cut on right exposes light gray, medium- to coarse-grained argillaceous sandstone beds about 1-3 ft thick, separated by thin, pale red, mudstone drapes. Petrographic analysis of a medium-grained sandstone from this outcrop (Chamberlin and Harris, this volume; sample MP-62.4) shows that in decreasing order of abundance it consists of plagioclase, andesitic rock fragments, ferromagensian minerals (mostly hornblende and biotite with minor opaques) and very sparse quartz. Volume percentages of the above framework grains are 41.4, 40.8, 15.2 and 2.6 respectively. Low-birefringent clay minerals or zeolites (?) comprise about 10-15% of the matrix of this sandstone. **0.3**

FIGURE 1.33. View to north of peak 8919 from northwest flank of peak 8761. This cliff is approximately 500 ft high along the west face (left side). This spectacular cliff provides a cross-section view of a low-angle fault (rising to left) that juxtaposes steeply-dipping debris-flow beds (upper right) against strongly folded fluvial sandstones (note fold nose on lower right). Folded sandstones grade downward into essentially undeformed sandstone beds (lower left). Bedding in the upper plate appears to become more massive or obscure from right to left and the true character of the tooth-like spire is presently unknown.

FIGURE 1.34. Crudely bedded debris-flow deposits well exposed in notch of peak 8761; view to south. Beds dip approximately 70° and presumably are not overturned. Height of visible outcrop is about 50 ft.

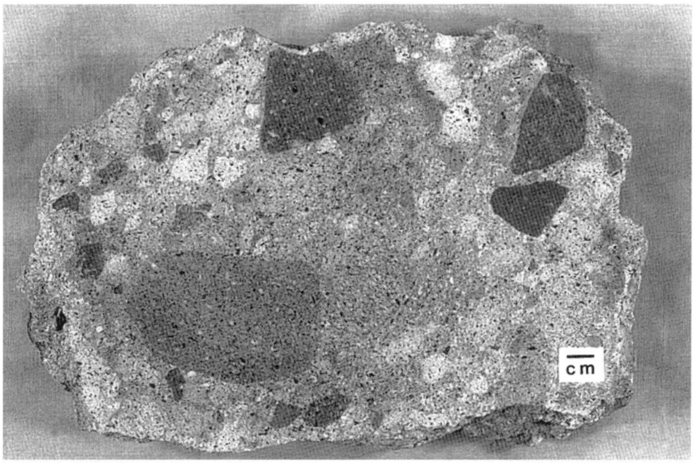

FIGURE 1.35. Slabbed andesitic debris-flow deposit from notch area of peak 8761. Note mixture of rounded and angular plagioclase-hornblende porphyry clasts floating in matrix of similar composition.

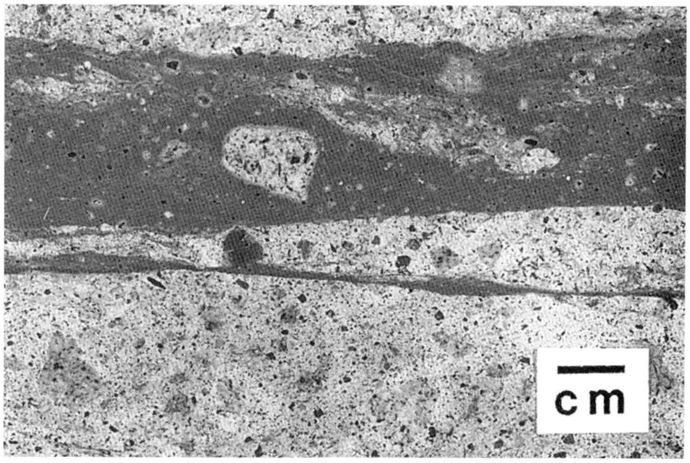

FIGURE 1.36. Clastic dikes within andesitic debris-flow breccia of the lower Dog Springs Formation. Sample collected near the head of Hay Canyon about 1.2 mi northeast of Monument Rock. Red muddy matrix of fluidal looking veinlets may have been partly derived from liquefaction of red mudstones in the uppermost Baca Formation, 200 ft stratigraphically below this site. Clastic dikes are subparallel to local bedding here, which is subhorizontal.

109.1 Rubble pile and small exposure on right used to be a revealing roadcut (Fig. 1.37). **0.2**

109.3 Good view of serrate Sawtooth Mountains to north. Highway descends across lower Spears Group/Baca Formation contact, which is locally concealed by Quaternary alluvium. Red Baca sandstones are poorly exposed on flanks of low hill at 1:00. **0.7**

110.0 Red soil in smoothed and seeded cut on right represents the Baca Formation. **0.2**

110.2 Milepost 61. Rounded hill at 1:00 provides good exposure of the base of the lower Spears Group overlying Baca Fm red beds. Local soft sediment deformation, expressed as low amplitude folds (~30 ft) in the Largo Creek sandstones might suggest that this is an angular unconformity, but only the *younger* volcaniclastic sandstone beds are visibly deformed. **0.3**

110.5 Manicured cuts in Baca Formation as indicated by red soil. **0.5**

111.0 Capitol Dome at 3:00 is the westernmost erosional monolith of the Sawtooth Mountains; it is formed by the Dog Springs Formation. **0.7**

FIGURE 1.37. Roadcut on north side of US-60 in volcaniclastic unit of Largo Creek at west end of the Airplane Hills. Upper photograph shows old fashioned pre-1993 roadcut. Lower photograph shows new manicured roadcut at completion of construction project in July 1993.

111.7 Road to West Pie Town Tank on right. A regional seismic reflection profile, the ARMA Baca Basin Line No. 1, crosses US-60 here (see following minipaper). **0.3**

PRELIMINARY GEOLOGIC INTERPRETATION OF THE ARMA BACA BASIN SEISMIC REFLECTION PROFILE, NORTHEASTERN CATRON COUNTY, NEW MEXICO

Rodney M. Armstrong[1] and Richard M. Chamberlin[2]

[1]ARMA Geophysical Co. Inc., 907 N. Poplar St. Suite 230, Casper, WY 82601
[2]New Mexico Bureau of Mines and Mineral Resources, Socorro, NM 87801

Data of the ARMA Baca Basin seismic reflection profile No. 1 were acquired in 1987 as a speculative survey underwritten by Hunt Oil Company. A cooperative petroleum exploration venture of Shell (SWEPI), Elf Aquitane and BP was active in the region at that time (Kopacz, 1989; Garmezy, 1990). The location of the ARMA Baca Basin seismic line No. 1 and its relationship to regional structures are shown on Figure 1.38. The profile is approximately 47.1 mi long, mostly followed existing roads, and thus contains several moderate to tight bends. The seismic line is roughly parallel to the south-southwest regional dip in the Datil Mountains block.

The ARMA Baca Basin data were acquired using the Vibroseis source and recorded 30-fold. The recording system had 120 channels which recorded 24 geophones per group at a 110 ft group interval with a symmetric split spread. The four vibrators took 8 8-second sweeps each per vibrator point. The total array length was 220 ft and the vibrator point interval was 220 ft.

The data were processed by ARMA using a standard flow-through migration. For the purpose of this preliminary interpretation the original seismic sections were reduced by plotting every fourth trace and then filtered and phase rotated 90°. The generalized interpretation of the filtered data set has been reduced by an additional factor of about 5.5 times in order to produce a book-sized cross section (Fig. 1.39). Here, we describe the general construction of this preliminary pick section and discuss the regional structure and stratigraphy of northeastern Catron County as revealed by this line. Cenozoic extensional collapse (structural inversion) of a Laramide compressional zone near Alegres Mountain (AMF; Figs. 1.38, 1.39) and Laramide basement uplift southwest of Horse Mountain (DF, Figs. 1.38, 1.39) are significant aspects of this preliminary interpretation.

The major seismic reflection events picked on the seismic profiles have been identified by synthetic seismograms derived from the Shell (SWEPI et al.) No. 1 State, the Hunt Oil Co. No. 1-16 State and the Tenneco Oil No. 1 Federal petroleum exploration wells. For a description of the stratigraphy in the Hunt Oil Co. No. 1-16 well, see Broadhead (this volume). In addition to this well control, the top of the Baca and top of the Cretaceous (Crevasse Canyon Formation) were correlated to mapped outcrop locations. Angular unconformities at the top of the Cretaceous and the top of the Triassic were also identified by event terminations below the unconformities. In both cases, reflectors appear to terminate toward the south under these regional unconformities.

Strata of the lower Spears Group (upper Eocene) appear to be massive and produce no internal reflections. The top of the Baca Formation (middle to upper Eocene) is the shallowest continuous reflective horizon. Although there are numerous reflectors in the Baca, none of them are continuous across the section. Presumably this reflects the lenticular nature of fluvial sandstones and mudstones in the Baca Formation. In general, the top of the Baca looks parallel to the top of the Cretaceous section, but the top of the Cretaceous seems to have been a topographically rough surface that is filled in by the lower Baca. Assuming a uniform seismic velocity of about 10,000 ft/sec for the Baca Formation yields a thickness of approximately 2000 ft for this formation along US-60 near Pie Town. The calculated dip of the Baca Formation here is approximately 100 ft/mi to the south-southwest.

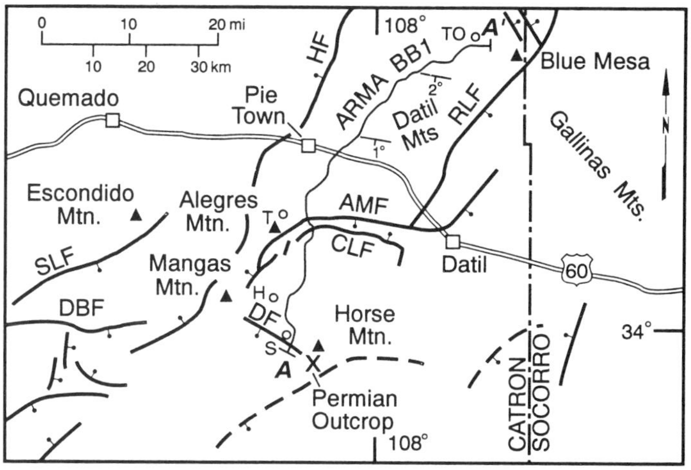

FIGURE 1.38. Location map of ARMA Baca Basin seismic reflection line No. 1 (A-A') and its relationship to regional structures in northeastern Catron County. RLF = Red Lake fault, HF = Hickman fault, AMF = Alegres Mountain fault, CLF = Cox Lake fault, DF = Dutch Oven Pass fault zone, DBF = Cañon del Buey fault, SLF = Spur Lake fault. S, H, T, and To indicate the approximate locations of the Shell (SWEPI et al.) No. 1 State, Hunt Oil Co. No. 1-16 State, Tenneco Oil No. 1 Federal, and Transocean Oil No. 1 Santa Fe RR petroleum exploration wells, respectively.

The unconformity at the top of the Cretaceous is tightly constrained in the Shell (SWEPI et al.) No. 1 State and the Hunt Oil Co. No. 1-16 State wells. It is clearly a major erosional surface associated with the Cretaceous section thinning fairly rapidly to the south. The exact event on the seismic section is difficult to identify but the angular unconformity between the Cretaceous and Baca is obvious. Numerous event terminations below the unconformity are obvious on the southern third of the line. The Baca reflectors and top of the Cretaceous dip rather steeply adjacent to the Alegres Mountain (AMF) and Cox Lake (CLF) normal faults. Rapid thinning of the Cretaceous section to the south across this fault zone demonstrates that the southern blocks were uplifted in a Laramide reverse-fault zone, prior to Baca deposition in middle Eocene time (Cather and Johnson, 1984). Lateral-slip may have occurred along this Laramide reverse-fault zone but cannot be demonstrated here.

The top of the Triassic (Chinle Formation) appears to be an erosional unconformity with the total Triassic section thinning to the south. It seems to be a rough surface that was overlain and filled in by the lower Dakota Formation, much like the basal Tertiary unconformity.

The San Andres event (top of Permian San Andres and Glorieta Formations) is not as obvious as it should normally be for a carbonate unit. It may be that the top of the San Andres is an erosional surface that is irregular and karsted. The San Andres event does appear to be more continuous north of US-60.

The Yeso event is a fairly easy pick and is closely tied to the Shell State No. 1 and Hunt well. The Tenneco well is also in reasonably good agreement, but the San Andres to Yeso interval is thicker on the seismic than indicated in the Tenneco well. This interval also thickens to the north. Several basement structures and faults appear to terminate in the San Andres-Yeso interval, suggesting a period of tectonic deformation in the lower to upper Permian (ancestral Rocky Mountains deformation?). Comparison of the Trans Ocean Well data and the seismic profile suggests the presence of a late Permian normal fault near the northeast end of the line. A strong lenticular reflection in the Yeso interval near the Hunt well is most likely equivalent to the diorite sill that was intersected in this well (Broadhead, this volume).

The Abo event is nearly conformable to the unconformity at the top of the Precambrian. Numerous changes in the seismic character of the whole Yeso–Abo interval occur along the section between the basement and the top of the Yeso. These changes might reflect variations in depositional environment and the influence of late Paleozoic structures.

Although tentative, the basement reflection at the southern end of the line appears to rise considerably to the south across the west-northwest-trending normal fault zone mapped by Ratté et al. (1991) near Dutch Oven Pass (Dutch Oven fault, DF, on Figs. 1.38, 1.39). If this is correct, the Dutch Oven fault zone may also have been a Laramide reverse fault zone that has partly collapsed under extension in Neogene time. Local Laramide uplift to the south would help explain the general rise in the basement south of the Hunt Oil Co. No. 1-16 well, and exposures of Permian rocks on the southeast flank of Horse Mountain.

112.0 Resistant outcrops of Largo Creek sandstones form low bluffs at 10:00 and 11:00; wooded slopes at base of bluffs are underlain by poorly exposed Baca Formation. Bare knob of Baca Formation at 1:30. **1.3**

113.3 Road sign, (well below crest of hill) declares that this is the Continental Divide, elevation 7796 ft. Apparently for next 650 ft water runs up hill to west, a local phenomenon presumably limited to the Pie Town area. **0.1**

113.4 Road crests at Continental Divide. **0.8**

114.2 Milepost 57. Forested hillside at 1:00 hides the Pie Town antenna of the VLBA. **0.1**

114.3 Road to Pie Town VLBA antenna on right. Visitors are welcome to tour this site, locally supervised by Kelly Gatlin and associates. **0.7**

115.0 Crest of hill in downtown Pie Town. Candelaria's Trading Post ahead on right carries on tradition of serving pie, mainly to tourists. Sign at entrance to park on left shows mountain-sized pies of the good old days. Pie Town was established by World War I veteran, Clyde Norman, who homesteaded here (ca. 1918) and began baking pies and serving coffee to local ranchers. Thus Norman's settlement was to become known as Pie Town. Hard times of the great depression and futile attempts at pinto bean farming in the Pie Town area were recorded by Farm Security Administration photographer Russell Lee (Smith, 1983). Slow, prepare for left turn ahead. **0.6**

115.6 **Turn left and head south on unpaved county road to Greens Gap.** NM-603 on right goes to Tres Lagunas and on to Grants. **0.2**

115.8 Four-way stop sign in west Pie Town. Continue straight ahead on county road. **0.1**

115.9 Hilltop to right is held up by small basaltic-andesite dike, which terminates just south of here as road curves to right. This dike is simply referred to here as the "small" Pie Town dike. It lies east of the larger dike referred to as "the Pie Town" dike. **0.2**

116.1 Late Oligocene (27.7 Ma) Pie Town dike of basaltic andesite composition forms ridge line from 1:00 to 2:30 (Laughlin et al., 1979). Volcaniclastic sandstones of the unit of Largo Creek are intermittently preserved here along the walls of the dike; Baca Formation underlies the gentle wooded slopes. A broad open syncline is partly visible in the Largo Creek beds from 3:00 to 4:00; north-northwest-dipping beds at 3:00 define the south limb of this syncline. Regional dips of Largo Creek beds in the Pie Town-Quemado area are about 2° to the south. Gentle open folds and minor faults, however, occur locally in the Largo Creek beds between Pie Town and

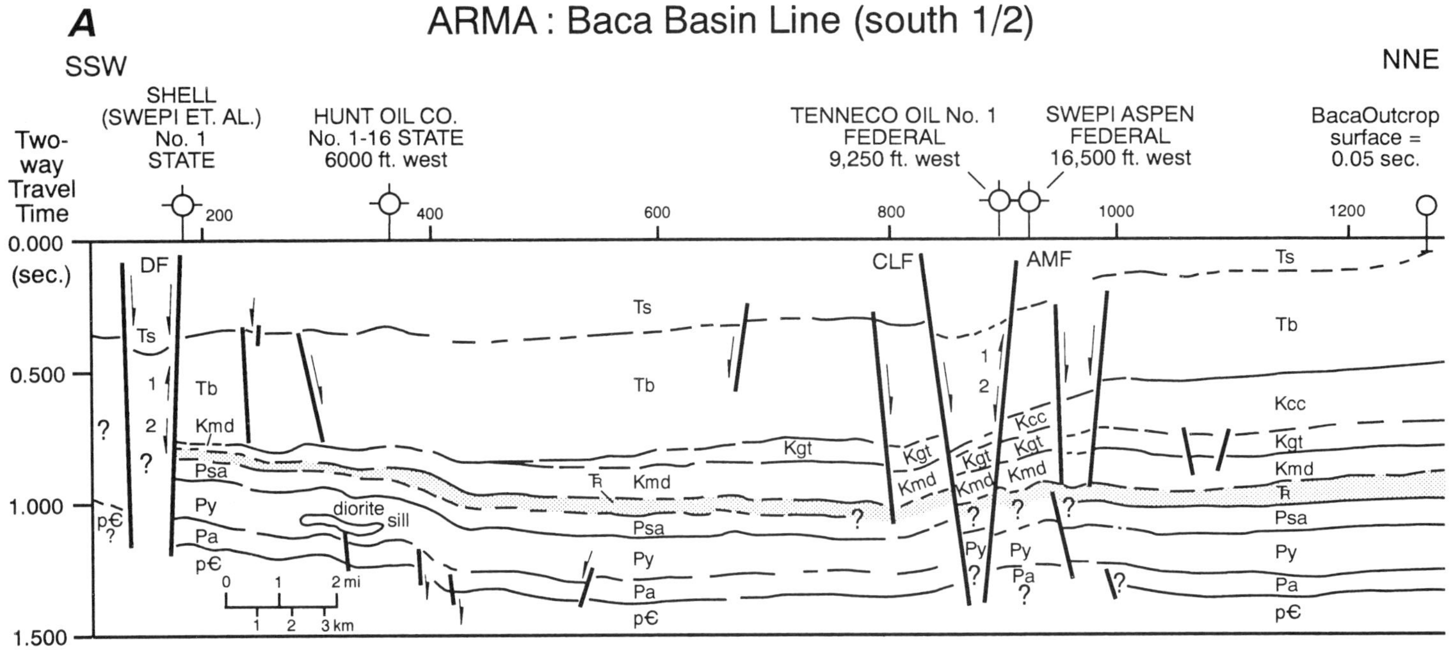

FIGURE 1.39. Generalized geologic section along the ARMA Baca Basin seismic reflection profile (A-A'; Fig. 1.38). Approximate vertical exaggeration in the Tertiary section is 5.5; in Paleozoic section vertical exaggeration decreases to about 3.7. Formations identified are: Ts = lower Spears Group, Tb = Baca Formation, Kcc = Crevasse Canyon Formation, Kgt = Gallup and Tres Hermanos Formations, Kmd = Mancos and Dakota Formations, TR = Chinle Formation, Psa = San Andres

Quemado Lake. The tectonic significance of these minor isolated(?) structures is presently uncertain; they may be contemporaneous with soft-sediment structures in the Dog Springs Formation, or they may reflect late Laramide deformation. **0.1**

116.2 Road to Pie Town landfill ahead on right. **0.1**

116.3 Cattle guard; entering W-Bar Ranch. **0.4**

116.7 Depression east of dike at 11:00 is Bright Lake (cattle tank) near the site of Stop 4. The Pie Town dike forms a great wall-like outcrop for at least 6 mi to the south of Pie Town (Fig. 1.40). North of Pie Town the late Oligocene dike disappears under the Miocene Fence Lake Formation in the Omega Basin syncline (Fig. 1.7). **0.2**

116.9 Note gently south-dipping Largo Creek beds across valley at 9:00 to 10:00. **0.3**

117.2 Road crosses through gap in ridgeline of dike. Remnants of unit of Largo Creek are present on the

FIGURE 1.40. Aerial view along the Pie Town dike, looking north-northwest. Site of Stop 4 at Bright Lake in open area on east side of dike. Pie Town at right center. White calcrete ledge above end of dike on left marks top of the Fence Lake Formation on the downthrown northwest side of the Hickman fault.

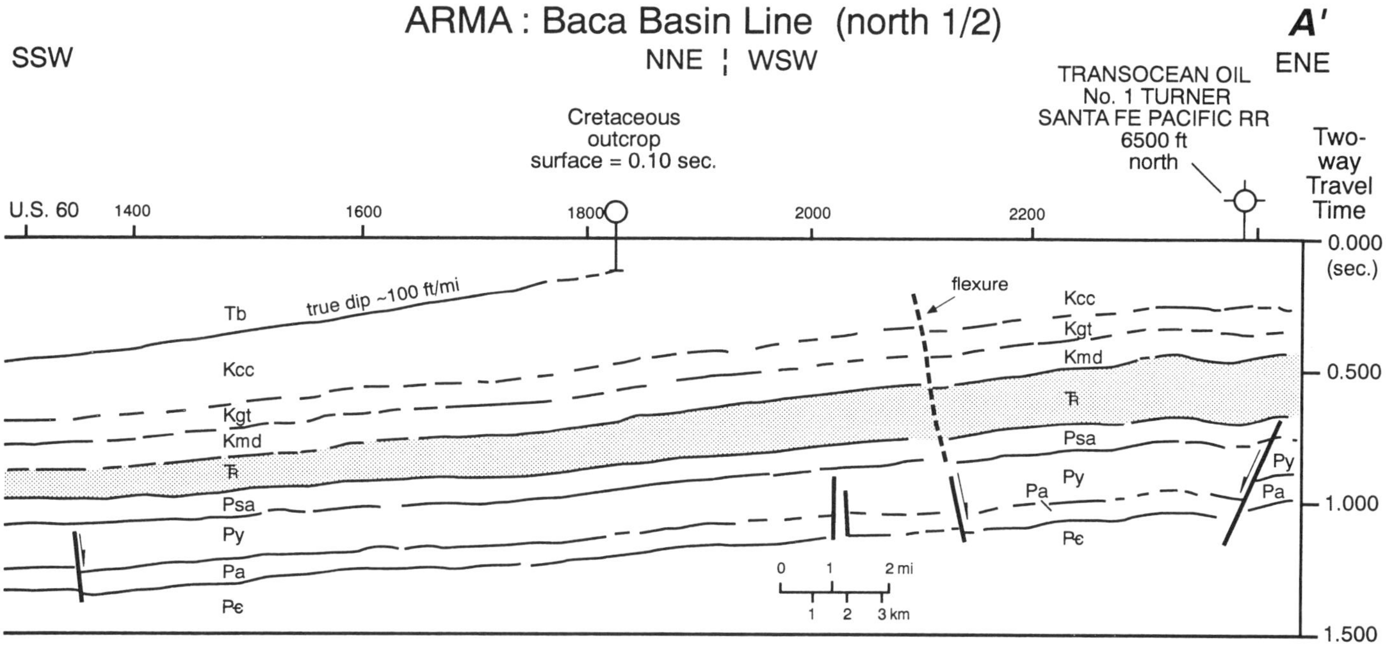

Limestone and Glorieta Sandstone, Py = Yeso Formation, Pa = Abo Formation and pC = Precambrian basement rocks. Arrows along faults indicate inferred Laramide component of reverse movement (1), followed by Neogene component of normal movement (2). Abbreviated designations of faults are explained in caption to Figure 1.38.

east side of dike; Baca Formation is wall rock (host rock) on west side. The contact between these units is near road level in the gap. **0.1**

117.3 Volcaniclastic unit of Largo Creek forms ledges on skyline at 2:30. **0.1**

117.4 Sandstone of Baca Formation exposed in bar ditch on left. Red Baca-derived soils on hillside to left are generally shown as outcrop on regional maps (Fig. 1-7). **0.2**

117.6 **Turn left** and head east on road to Bright Lake Dam to view wall-like exposures of the Pie Town dike and adjacent country rock. This is private land of the W-Bar Ranch. Permission to visit this locality may be obtained at the W-Bar headquarters 2 mi south of here on the west side of the Greens Gap road. **0.2**

117.8 **STOP 4. Follow directions of flagpersons, make U-turn in flat to be headed west for return to US-60.** Walk to Bright Lake Dam on east side of the Pie Town dike. The purpose of this stop is to discuss the petrology and tectonic significance of the Pie Town dike and its relationship to other regional structures such as the Hickman fault zone. Conference participants may then climb the east flank of the dike to examine the chilled dike contacts and hornfelsed blocks of Baca Formation and Largo Creek-type sandstone within the dike. **All observers should be cautious and not get too close to the 30-ft-high wall on west side of dike.**

The Pie Town dike at Bright Lake forms a large resistant wall (Fig. 1.41) where it cuts poorly exposed Baca Formation and overlying andesitic sandstones of the volcaniclastic unit of Largo Creek (upper Eocene). The dominant strike of the 15-20-ft-wide basaltic andesite dike is N32°W and it dips approximately 85°NE. Locally, the dike makes a sharp bend to the right just north of the Bright Lake Dam. Wall rocks consisting of baked (locally reddened) Largo Creek sandstones are well exposed on the east side of the dike near the bend. This bend represents a minor deviation in this magma-filled tension fracture that is approximately 45 mi long; and presumably just as deep.

A sample of the Pie Town basaltic-andesite dike collected about 3 mi to the northwest provided a K-Ar whole rock age of 27.67 ± 0.59 Ma (Laughlin et al., 1979). This aphyric basaltic andesite, lacking in olivine microphenocrysts, contains about 53.5% SiO_2 and is distinctly higher in TiO_2 (2.26%) and P_2O_5 (1.6%) than

FIGURE 1.41. View along the Pie Town dike at Bright Lake. The chilled eastern margin of the nearly vertical basaltic andesite dike is locally well exposed where it bends sharply to the right adjacent to a small light-colored knob of baked andesitic sandstones in the volcaniclastic unit of Largo Creek. The dike contains a 3x3 m block of hornfelsed quartzo-feldspathic Baca sandstone just left of the Largo Creek knob. Ledge-forming Largo Creek sandstones on horizon; wooded slopes are underlain by Baca Formation. Largo Creek-Baca depositional contact is discontinuously exposed near mid-slope on flanks of wall-forming dike.

most other late Cenozoic basaltic andesites in the region (Laughlin et al., 1983). Xenocrystic quartz (largely resorbed) is a significant component (~ 0.5%) of the dike at Bright Lake.

The Pie Town dike is one of a group of three long north-northwest-striking late Oligocene basaltic andesite dikes that cut across the southeast margin of the Colorado Plateau. The great length, similar composition (high TiO_2; 51-53% SiO_2) and analytically equivalent K-Ar ages of these subparallel dikes has led to the conclusion that they formed orthagonal to the regional least principal horizontal stress (LPHS=S55-60°W) in late Oligocene time (Laughlin et al., 1983). More recent investigations using isotopically-dated dikes to determine secular variations in LPHS have documented a shift from west-southwest directed extension to west-northwest extension during a regional magma lull in early Miocene time (~ 22-17 Ma; Aldrich et al., 1986; Baldridge, et al., 1989).

Based on apparent similarity in age and chemical composition Laughlin et al (1983) suggested that basaltic andesite lavas of the Squirrel Springs Canyon Andesite (Rhodes and Smith, 1976) might be genetically linked to the Pie Town dike. More recent observations show that the suggested Pie Town-Squirrel Springs link is unlikely and that the basaltic andesite lavas of the Bearwallow Mountain Andesite are a more likely candidate to be the extrusive equivalent of the Pie Town dike, if any exists.

Lavas of the Squirrel Springs Canyon Andesite are texturally distinct turkey-track flows that contain abundant tabular plagioclase phenocrysts as much as 3 cm long. Regional mapping has also shown that the source area of this unit must lie somewhere to the southwest of Mangas Mountain and Escondido Mountain. Since the Pie Town dike lies to the northeast of both mountain ranges, it seems physically impossible for the Pie Town dike to be a source of the Squirrel Springs Canyon Andesite.

Aphyric to phenocryst-poor basaltic andesite lavas of the Bearwallow Mountain Andesite in the Mangas Mountain area *could* be an eruptive equivalent to the Pie Town dike. However, the Bearwallow lavas appear to be somewhat younger (26.1 ± 0.1 Ma) and lower in TiO_2 (1.14%, Ratté, 1989) than the Pie Town dike. A similar, but much shorter 0.6 mi basaltic-andesite dike is present on Escondido Mountain, where it cuts an ash bed equivalent to the 28.6-Ma Vicks Peak Tuff (Fig. 1-7). This dike could have been a feeder for the adjacent Bearwallow flows.

Regional mapping has shown that the Hickman fault zone offsets and downwarps the Pie Town dike. The dike is buried by the Fence Lake Formation where it is preserved in the Omega basin syncline between Pie Town and Adams Diggings. Near Adams Diggings the Pie Town dike appears to transect the west limb of a gently north plunging Laramide anticline. Likewise, the Fence Lake Formation unconformably overlaps the west flank of this Laramide anticline. The Pie Town dike terminates to the north near Fence Lake, which is also near the south end of the southwest facing Laramide Atarque monocline (Anderson, 1987). As interpreted by Laughlin et al. (1983), southwesterly directed dilation of the Pie Town dike probably represents a significant relaxation of northwest-directed Laramide compressional stress in late Oligocene time.

Return to vehicles and retrace route to US-60. 2.2

120.0 Stop sign at junction of Greens Gap road and US-60 in west Pie Town. **Turn left and head west on US-60. 0.8**

120.8 Roadcuts on right expose the "small" Pie Town dike. Red sandstones of the Baca Formation are locally bleached adjacent to this 6-ft-wide dike that dips about 80° to the southwest. The "small" Pie Town dike has been sampled for paleomagnetic analysis and $^{40}Ar/^{39}Ar$ age dating by W. C. McIntosh, R. M. Appelt and R. M. Chamberlin. An isotopic age is not available, but the paleomagnetic data indicate that the "small" Pie Town dike was downwarped and rotated (about a horizonal axis) with the adjacent Baca beds in post-late Oligocene time. **0.5**

121.3 Highway passes through water gap in the Pie Town dike at Wyche Draw. The dike is relatively well exposed south of the highway, where it cuts locally bleached and reduced (greenish) Baca Formation sandstones. It is about 15 ft wide here and dips approximately 85° to the northeast. **0.6**

121.9 Baca Formation in cut on left. Ridge on skyline from 12:00 to 1:00 is designated as the "Top of the World" on the Pie Town 7.5′ quadrangle. Although not quite as impressive as the Himalayas, this ridge line marks the top of the Fence Lake Formation on the downthrown northwestern side of the Hickman fault (Fig. 1-7). A subhorizontal laminar calcrete about 6 ft thick locally caps this ridge and has been quarried for road metal. **0.5**

122.4 Highway descends onto alluvial valley floor of Chavez Draw and crosses projected trace of the Hickman fault zone that locally strikes northeast as opposed to north-northeast for the overall fault trend. Northeast of here the Hickman fault zone offsets of the Pie Town dikes (100 to 200 ft), in both a right lateral and a left lateral sense. Stratigraphic displacement and northwesterly down-warping of Miocene Fence Lake beds here at Pie Town is estimated to be about 1000 ft, well in excess of the observed offsets. Similar apparent horizontal offsets of the late Oligocene dike at Lehew (aka Hickman) occur along the Hickman fault zone approximately 12 mi northeast of here. Some researchers (Aldrich and Laughlin, 1984; Baldridge et al., 1983) have interpreted the apparent horizontal offsets of the 27.3-Ma basaltic andesite dike at Lehew (Laughlin et al., 1979) as the expression of left-lateral strike slip. Regional mapping (Chamberlin et al., 1994b), however, has shown that the Hickman fault zone has a post-28-Ma-stratigraphic throw of as much as 2000 ft (down to the west) near Mangas. Rough estimates of the down-to-west stratigraphic throw at Lehew range from 800 to 1600 ft depending on the regional dip used in cross sections to determine the offset of the base of the Baca Formation. It should be noted that *pure* normal dip slip of Neogene age could have produced the apparent 500 ft of left offset (total of three offsets) of the nearly vertical dike (N28°W, 85°NE) at Lehew, because the Hickman fault zone intersects the dike at an acute angle of about 53°. The Hickman fault zone here strikes N25°E; thus a pure dip-slip normal fault vector would plunge to N65°W and lie to the left of the dike trend (N28°W). Assuming a pure dip-slip normal displacement for the Hickman fault zone and a dip of 60° to the west-northwest; the observed 500 ft of apparent left offset at Lehew would require a dip-slip displacement of 1600 ft. A few observations of essentially dip-slip slickensides on post-Oligocene faults near Mangas and the above constraints near Lehew suggest that the Neogene displacement of the Hickman fault zone is essentially dip-slip; however, a minor component of left-oblique slip remains a possibility. A previous estimate of Neogene net slip on the Hickman zone by Chamberlin (1993) was incorrect.

Laramide reverse displacement on the Hickman fault zone, upthrown on the west, has been previously recognized (Cather and Johnson, 1984, Cather, 1989) by differences in thickness of the Baca Formation between Quemado (600 ft; Guilinger, 1981) and the Datil Mountains (1875 ft, Chamberlin, 1981b). More recent thickness data for the Baca Formation at Adams Diggings (0-400 ft; Chamberlin et al., 1994b) and near Pie Town (2000 ft; Armstrong and Chamberlin, this road log), both of which lie closer to Hickman fault zone, substantiate the previous interpretation.

Laramide deformation patterns (contractional strain) in the Zuni Mountains region (Chamberlin and Anderson, 1989; Chamberlin, 1991) also suggest that a significant component (>0.6 mi) of Laramide right-lateral slip occurred on the Hickman zone. **0.5**

122.9 Highway crosses Chavez Draw on upper Quaternary alluvial valley fill. **0.5**

123.4 Flat-topped Alegres Mountain on skyline at 9:00. Low mesa, with light gray landslide scar on its side at 10:00, is capped by basaltic boulder conglomerates of the middle to upper Miocene Fence Lake Formation unconformably overlying upper Eocene volcaniclastic unit of Largo Creek, (light gray beds in scar). Main trace of Hickman fault offsets the Fence Lake caprock to right of the landslide scar. The total post-Fence Lake displacement across two down-to-the-west faults of the Hickman zone in this area is about 500 ft. **0.5**

123.9 Volcaniclastic unit of Largo Creek forms small hill just north of highway. Petrographic analysis of framework grains in the medium-grained volcaniclastic sandstone yields 74.4% plagioclase, 14.4% andesitic rock fragments, 8.8% Fe-Mg minerals (hornblende, biotite and opaques), and 2.8% quartz (Chamberlin and Harris, this volume, sample MP-51.8). Clay minerals comprise about 10-15% of the matrix of this argillaceous sandstone. **0.5**

124.4 Low cuesta in foreground at 10:00 is composed of pebbly conglomeratic sandstones of the lower Fence Lake Formation dipping about 5-10°. At least 300 ft of lower Fence Lake beds are exposed here. An abrupt facies change occurs at Thaeger Draw on the south end of this cuesta. North of Thaeger Draw the lower Fence Lake beds consist of trough-cross-bedded, pebbly conglomerates and sandstones. Pebbles are mostly hornblende plagioclase porphyry dacites (Dog Springs type) mixed with less abundant basaltic andesite and rare ignimbrite clasts. South of Thaeger Draw the Fence Lake Formation is a subhorizontal basalt boulder to cobble conglomerate about 65 to 100 ft thick. Both facies unconformably overlie the unit of Largo Creek in this area. These facies relationships reflect different stream systems and different provenance for the Fence Lake Formation. Pebbly lower Fence Lake sediments were derived from the Datil Mountains area, and bouldery beds south of Thaeger Draw were derived from Alegres Mountain. A preliminary structure section in the Pie Town area (Chamberlin et al., 1994b) suggests that the lower Fence Lake pebble conglomerates may be locally 600-900 ft thick adjacent to the downwarped western flank of the Hickman fault zone. These moderately tilted lower Fence Lake beds near Pie Town may be somewhat older than the middle to upper Miocene Fence Lake conglomerate described by Lucas and Anderson (this volume) in the area southwest of Quemado. **0.2**

124.6 Roadcut on right in pale red muddy siltstones in unit of Largo Creek. **0.4**

125.0 Highway crosses approximate basal contact of the Fence Lake Formation. This regional unconformity is well exposed along the road to "Under the Mesa Well" in the Tres Lagunas 7.5′ quadrangle about 7 mi to the north. **0.1**

125.1 Lower Fence Lake Formation poorly exposed in roadcut on right. **0.1**

125.2 Lower Fence Lake pebble conglomerates and conglomeratic sandstones are well exposed in the upper bench of the road cut on the right. Note 25° dip. Pebble imbrications indicate westerly to northwesterly paleoflow. Hornblende-dacite clasts are mixed with basaltic andesites and rare ignimbrite clasts. **0.3**

125.5 Projected trace of north-northwest-trending, down-to-the-west Mangas fault (Fig. 1.7) passes under alluvial valley fill of Tres Lagunas Draw here. **0.4**

125.9 Roadcut on right is in poorly consolidated pebbly sandstone of the Fence Lake Formation; bedding attitude not discernible here. **0.9**

126.8 Milepost 49. Escondido Mountain at 10:30 on the skyline is capped by upper Oligocene Bearwallow Mountain Andesite of the Mogollon Group. **1.0**

127.8 Low roadcuts for next 1.2 mi are in light gray, poorly- to moderately-consolidated, pebbly sandstones of the upper Fence Lake Formation; bedding is approximately horizontal. **1.1**

128.9 Piñon-juniper covered hills on skyline from 1:00 to 2:00 are underlain by Fence Lake Formation. **0.1**

129.0 Highway begins descent onto Pliocene to early Pleistocene alluvial deposits of the Quemado Formation (Cather and McIntosh, and McIntosh and Cather, this volume). Within this area of the topographic Omega Basin (not to be confused with the Omega basin syncline), Quemado Formation deposits are derived from and inset against the Miocene Fence Lake Formation. Quemado deposits are generally poorly consolidated, often buff colored, and occupy lower levels in the landscape. Clast compositions are essentially the same as in the Fence Lake Formation, which is their primary source. **0.6**

129.6 Low roadcut on right in sandy Quemado Formation alluvium. **0.2**

129.8 Milepost 46. Cuestas that slope west and east at 2:30-3:00 respectively are capped by Fence Lake Formation. Knob in middle is undated late Miocene(?) basaltic volcanic neck of Techado Mountain. Cuestas may define a Pliocene(?) anticlinal axis near Toms Rock, 2.5 mi west of Adams Diggings. Upper Eocene volcaniclastic sandstones of the unit of Largo Creek are discontinuously exposed on the flanks of these mesas between Adams Diggings and Mariano Springs. One resistant andesitic debris flow is present within the unit of Largo Creek, 1.2 mi west of Toms Rock. It forms a distinctive elongated pedestal resting on typical Largo Creek sandstone beds (Fig. 1.42). Judging from its elongation, a distant volcanic source to the south or southeast seems likely. **0.4**

130.2 Grassy plains along Mangas Creek from 9:00 to 10:00 are underlain by Quemado Formation. Mangas Mountain, on the horizon at 9:00 was probably an eruptive center for the Bearwallow Mountain Andesite. **0.4**

130.6 Low cuts on right and left in sandy Quemado Formation. **0.5**

131.1 Windmill for well on right produces fresh water probably from saturated Fence Lake Formation conglomerates that project under the Quemado Formation here (see minipaper by Newcomer in Second-Day road log). **0.2**

131.3 Low roadcuts in medium-grained loamy sand. Landscape position suggests that this is an upper Pleistocene terrace deposit of San Ignacio Creek, which locally parallels the south side of US-60. **1.4**

132.7 County road to Mangas on left. First-Day road log overlaps with miles 38.2 to 47.5 of Second-Day road log for next 9.3 miles. The more detailed discussions of features along this stretch are in the Second-Day log. **1.1**

133.8 Milepost 42 at Omega Village limit. **1.1**

134.9 Roadcuts in sandy Quemado Formation alluvium. **0.3**

135.2 Road rises onto Fence Lake Formation. **0.4**

135.6 Site of Stop 1 on second day is at top of hill behind borrow pit at 1:00. **0.2**

135.8 Milepost 40. Begin descent of steep grade into valley of Rito Creek and Escondido Creek. Roadcut ahead on right exposes Fence Lake Formation unconformably overlying sandstones and mudstones of the volcaniclastic unit of Largo Creek. **1.2**

137.0 Roadcut on left exposes the Quemado Formation unconformably on volcaniclastic unit of Largo Creek. **0.7**

137.7 Cuts in the volcaniclastic unit of Largo Creek. **0.3**

138.0 Roadcut on left in pebbly stream sand of early Pleistocene to late Pliocene age. **0.2**

138.2 Round knob on right is the volcaniclastic unit of Largo Creek (upper Eocene). **1.0**

139.2 Cut on left in Quemado Formation. **0.5**

139.7 Cliffs of Fence Lake Formation capping lower Spears Group, 9:00 to 10:00. **1.2**

140.9 Old town of Quemado on left. Well-exposed Fence Lake Formation over slope-forming lower Spears Group on horizon (Fig. 1.43). **0.2**

141.1 Roadcuts on right in upper Baca Formation. **0.2**

141.3 Junction, NM-36 to Fence Lake on right. **Continue straight ahead on US-60.** Boundary stratotype for the base of the volcaniclastic unit of Largo Creek (upper Eocene) conformably overlying the Baca Formation (middle to upper Eocene) is well exposed on a low peak above the Quemado Community Cemetery 1.3 mi north-northeast of here. **0.1**

141.4 Quemado Post office on left. Downtown Quemado has changed some in the last 60 years (Fig. 1.44) but still has good food, comfortable lodging, gas and friendly Quemadans. **0.6**

142.0 Junction with NM-32 on left. **Turn left and head south on NM-32.** Road follows valley of Largo Creek for next 12 mi. Mesas on skyline at 12:00 and 1:00-4:00 are capped by commonly cliff-forming basaltic conglomerates of the Fence Lake Formation. The base of the Fence Lake Formation is a regional angular unconformity. The underlying Baca Formation and

FIGURE 1.42. Singular debris-flow deposit forms pedestal in the volcaniclastic unit of Largo Creek west of Adams Diggings near Tom's rock, in SE¼, sec. 25, T3N, R15W.

FIGURE 1.43. Cliff-forming conglomerate of Miocene Fence Lake Formation unconformably overlies upper Eocene slope-forming beds of the volcaniclastic unit of Largo Creek to create a geologic backdrop to old east Quemado in foreground.

FIGURE 1.44. A great depression era scene of downtown Quemado in February 1933; looking east-northeast. Building of Engle Mercantile Co. and Tavern remain standing. Photograph courtesy of Jimmy and Irene Jaramillo, El Sarape Cafe.

Spears Group strata near Quemado generally dip about 2° to the south-southeast; thus the Fence Lake rests on progressively older strata as it extends northward from source areas near Escondido Mountain and Agua Fria Mountain. The Fence Lake Formation is commonly 60-80 ft thick where well exposed in landslide scars on these mesa flanks. **0.3**

142.3 Low hill at 10:30 exposes red Baca Formation on its flanks, capped by Quemado Formation gravels. Landslide scar on mesa point exposes light gray andesitic sandstones of the volcaniclastic unit of Largo Creek. The depositional contact of the unit of Largo Creek on Baca Formation is approximately located at the foot of the mesa. **1.3**

143.6 Road curves to left and crosses concealed basal contact of the volcaniclastic unit of Largo Creek near here. Axis of Largo Creek valley ahead at 12:15; route on middle(?) to late Pleistocene valley fill for next 5 mi. **0.4**

144.0 Dissected graded surfaces, at 10:00 and 2:00, about 65-130 ft above the valley floor, are underlain by Quemado Formation piedmont gravels mostly derived from the Fence Lake Formation and lower Spears Group. Similar remnants of Quemado Formation can be observed in the middle landscape position along Largo Creek valley for the next 9.5 mi. **0.5**

144.5 Crest of Escondido Mountain (elev. 9854 ft,) on skyline at 10:30 consists of aphanitic basaltic-andesite lavas of the Bearwallow Mountain Andesite (26.1 Ma), disconformably overlying 28.05-Ma Bloodgood Canyon Tuff (not visible from here) and porphyritic Squirrel Springs Canyon Andesite, which locally forms 200 ft-high cliffs on the shoulders of Escondido Mountain. Vicks Peak Tuff (28.56 Ma) locally underlies the Squirrel Springs Canyon Andesite. All of these Oligocene volcanic units are assigned to the Mogollon Group (Cather et al., this volume).

Two newly named informal volcaniclastic units are locally well exposed along the upper reaches of Cañon del Leon, the deep cut below the west shoulder of Escondido Mountain. Upper Cañon del Leon is designated as the type section for approximately 600 ft of pumiceous and tuffaceous sandstones and intercalated andesitic conglomeratic sandstones of the volcaniclastic unit of Canyon del Leon (new name, Chamberlin and Harris, this volume). Sanidine phenocrysts from pumice in basal and uppermost pumiceous siltstones (presumably fresh and floating when deposited) have yielded $^{40}Ar/^{39}Ar$ ages of 35.33 to 34.17 Ma and imply this new map unit is of late Eocene age. The unit of Cañon del Leon conformably overlies andesitic sandstones and minor pumiceous sandstones of a 200-ft-thick transition zone at the top of the Largo Creek. As much as 800 ft of buff, massive planar-bedded to high-angle cross-bedded sandstones of the sandstone of Escondido Mountain (new name, Chamberlin and Harris, this volume) appear to disconformably(?) overlie the volcaniclastic unit of Cañon del Leon. Uniform grain size (fine- to medium-grained sandstone) and high-angle cross beds, generally facing east, indicate that the sandstone of Escondido Mountain is largely an eolian deposit. Ignimbrites of the Vicks Peak Tuff and Bloodgood Canyon Tuff locally occur as paleovalley fills within the sandstone of Escondido Mountain. Near Mangas, the 31.65-Ma Caballo Blanco Tuff occurs just 60 ft above the base of the sandstone of Escondido Mountain. These data indicate that sandstone of Escondido Mountain is Oligocene in age. As shown on the geologic map of the Datil-Quemado region (Fig. 1.7), the volcaniclastic unit of Largo Creek, the volcaniclastic unit of Cañon del Leon, and the sandstone of Escondido Mountain are informally assigned to the lower, middle and upper Spears group, respectively. Datil Group volcanic units, namely the tuff of Bishop Peak, the andesite of Dry Leggett Canyon and the tuff of Luna are locally intercalated with the unit of Cañon del Leon in the Fox Mountain area west of Escondido Mountain. **1.0**

145.5 Volcaniclastic unit of Largo Creek well exposed in deep scar at 8:45. These volcaniclastic sandstones consist primarily of andesitic rock fragments and oscillatory zoned (volcanic) plagioclase, with minor hornblende and biotite, and usually little or no quartz. **0.5**

146.0 Headquarters of Jim Williams ranch on left. The Williams family has been ranching along Largo Creek for several decades. White cliffs of pumiceous sandstone in Cañon del Leon at 11:00. **1.0**

147.0 Cross arroyo of lower Cañon del Leon. **1.1**

148.1 Roadcut on left in light brownish-gray andesitic sandstone and red mudstones of the volcaniclastic unit of Largo Creek. Just beyond is entrance to Apache National Forest; mostly private property along Largo Creek from here to Quemado Lake. **0.2**

148.3 Largo Creek sandstones and mudstones on left. **0.2**

148.5 Upper(?) Pleistocene gravel in channel over Largo Creek beds in cut on left. **0.2**

148.7 Volcaniclastic unit of Largo Creek in cut on left is unconformably overlain by Quemado Formation gravel and colluvium. **0.5**

149.2 Slight angular unconformity visible on mesa side at 9:00. Largo Creek sandstone beds are noticeably truncated by the gently north-sloping Fence Lake conglomerates. **0.1**

149.3 Low roadcuts expose sands and gravels of Quemado Formation **0.4**

149.7 Enter zone of gently folded and faulted Largo Creek beds, apparent in road cuts for next 1.7 mi. Gentle anticlinal warp on left is cut by numerous calicite-filled tension fractures (joints) that are nearly vertical and strike N20°E. Intersection of joints and bedding gives outcrop a pseudo-cross-bedded appearance. **0.6**

150.3 Note open synclinal fold in interbedded sandstones and mudstones of the unit of Largo Creek well exposed in roadcut on right. (Fig. 1.45). **0.2**

150.5 White crest of local landmark known as the Hub is barely visible through low saddle in foreground at 9:15 (Fig. 1.46). **0.3**

150.8 Roadcuts ahead for next 0.2 mi in essentially horizontal Largo Creek sandstones unconformably overlain by Quemado Formation gravels; note channel fill on left. **0.6**

FIGURE 1.45. Road logger points to sandstone-mudstone contact that defines open synclinal fold in the unit of Largo Creek. View is to west-northwest along axis of fold. Slope wash was scraped from part of the exposure to improve visibility of bedding.

FIGURE 1.46. View of the west shoulder of Escondido Mountain, on skyline, and the barely visible white crest of the Hub, in middle slope through saddle in left foreground. Cliff-forming outcrops on Escondido Mountain consist of Squirrel Springs Canyon Andesite disconformably overlain by Bearwallow Mountain Andesite, capping ridge crest. The Hub is topped by pumiceous sandstones of the volcaniclastic unit of Cañon del Leon. Large scar on right exposes subhorizontal Largo Creek beds unconformably overlain by cliff-forming Fence Lake Formation.

151.4 Enter large roadcut that provides a cross section view on left of two cut and fill units of the Quemado Formation that unconformably overlie anomalously north-tilted Largo Creek beds locally cut by a low-angle fault (Fig. 1.47). Close examination of the low-angle fault at the southeast end of the roadcut shows that it is a dip-slip thrust fault (Fig. 1.48). Total displacement on this southwest vergent thrust is probably not more than 30-100 ft. Moderate deformation in Largo Creek beds observed over the last 1.7 mi (folds, tension fractures and a small thrust) are all consistent with local northeast-directed shortening and northwest extension. This is a typical orientation for late Laramide compressional deformation in the southern Colorado Plateau (Davis, 1978, Chapin and Cather, 1981). These northwest-trending structures appear to be relatively local and were not observed along trend in the Cañon del Leon area. The orientation of these structures and the pre-35.3 Ma age (late Eocene) of the volcaniclastic unit of Largo Creek would allow this area to be interpreted as a minor zone of late Laramide deformation. **0.6**

152.0 Route enters narrow valley cut by Largo Creek. Bouldery colluvium caps unit of Largo Creek in roadcut on left. **0.4**

152.4 Paleocanyon fill of Quemado Formation is well exposed in roadcut on left. Pebble imbrications indicate both westward and northward paleotransport, suggesting that this exposure contains both axial deposits of ancestral Largo Creek and tributary deposits from the adjacent canyon that leads eastward to Leon Tank. This paleo- canyon was cut through the Fence Lake Formation and the volcaniclastic unit of Largo Creek to a depth of about 300 ft and then backfilled by about 100 ft of Quemado Formation in Pliocene to late Pleistocene time. **0.3**

FIGURE 1.47. Roadcut along NM-32 showing two cut and fill units of the Quemado Formation (Pliocene to upper Pleistocene), unconformably overlying north-dipping andesitic sandstones and mudstones of the volcaniclastic unit of Largo Creek; view to east. A low-angle fault juxtaposes dark-colored muddy Largo Creek beds over light-colored Largo Creek sandstone at the south end (right side) of the cut. A 6-ft-thick basaltic gravel bed of the Quemado Fm. caps the south end of the roadcut; note south-facing channel wall at north end of gravel bed. Another northeast-facing channel wall is locally defined by a stone-line, which cuts obliquely down to the left, from just above the left edge of the gravel bed. This younger(?) channel is back filled with 30-40 ft of buff sand and basaltic gravel of the Quemado Formation. The latter channel wall is exposed on both sides of this roadcut and trends N37°W. Both of these channels probably represent tributary arroyos to ancestral Largo Creek that were entrenched and back filled in Pliocene to early Pleistocene time.

152.7 Roadcut on left in light gray, well-sorted, fine-grained, moderately quartz-rich (10-20% quartz) andesitic sandstone assigned to a transition zone in the upper part of the unit of Largo Creek. This transition zone, approximately 200 ft thick, is locally characterized by the upsection appearance of moderately quartz-rich andesitic to rhyolitic sandstones, thin pumiceous mudstone and a general coarsening upward into light brown and blue-green (celadonite cemented) andesitic conglomerate and conglomeratic sandstone. This transition is exposed in roadcuts for the next 4 mi. **1.4**

154.1 Roadcut on left in subhorizontal moderately quartzose andesitic sandstones in transitional zone in the upper part of the volcaniclastic unit of Largo Creek. **0.3**

FIGURE 1.48. Close-up view of low-angle fault in the unit of Largo Creek (at mileage 153.5; about 7 ft above top of slope wash), view to east. The fault strikes N50°W and dips 32°NE. Fibrous calcite along the fault plane shows a lineation plunging 31° to N35°E, and indicates dip slip. Small secondary shears below the main fault surface displace thin mudstone beds with a hanging-wall up (reverse) sense of shear. Collectively these observations confirm that this is a dip-slip thrust fault.

154.4 Concrete retainer wall protects roadway on left for next 0.1 mi. Conglomeratic sandstone beds near middle of roadcut are offset by a small high-angle fault. Approximately 3 ft of reverse throw is indicated by displacement of a cobble conglomerate bed. Fault surface lacks striations necessary to evaluate slip characteristics. Both andesitic and minor rhyolitic tuff clasts are present in the conglomeratic beds. A thin pumiceous mudstone occurs near the base of the roadcut. This outcrop reflects the coarsening upward and local first appearance of rhyolitic pumice within the transition zone of the upper unit of Largo Creek. Bouldery Quemado Formation fills channel at top of cut and grades into colluvium. **0.6**

155.0 Roadcut on right in brownish to greenish gray (celadonitic) conglomeratic sandstones with abundant pebbles of andesite porphyry and minor rhyolitic tuff clasts, the transition zone at the top of the unit of Largo Creek. Route temporarily leaves valley of Largo Creek. **0.2**

155.2 Transition zone conglomeratic sandstones in cut on right. **0.3**

155.5 Roadcut and gully ahead on right exposes light gray andesitic sandstone and pumiceous mudstone of the transition zone of the unit of Largo Creek. At 9:00 on skyline, bluish-green sandstones of the transition zone are unconformably overlain by coarse Fence Lake Formation conglomerates. **0.2**

155.7 Agua Fria Mountain on skyline at 10:30 is capped by Squirrel Springs Canyon Andesite and Bloodgood Canyon Tuff. **0.2**

155.9 **Slow down for left turn.** Unnamed butte at 1:00 is capped by Fence Lake Formation. Castle Rock in foreground at 11:00 is capped by cliff-forming basal pumiceous and tuffaceous sandstone of the volcaniclastic unit of Cañon del Leon. **0.1**

156.0 Junction with NM-103. **Turn left and head south on NM-103 to Quemado Lake. 0.5**

156.5 The cliff-forming basal marker bed of the volcaniclastic unit of Cañon del Leon is well exposed on the east side of Castle Rock at 8:30 (Fig. 1.49). For practical mapping purposes the base of the volcaniclastic unit of Cañon del Leon is placed at the bottom of the angular, jointed, cliff-forming marker bed, just above the thin white tuffaceous sandstone bed. Although locally offset by faults, this pumiceous marker bed can be traced for 9 mi to the east along the north foot of Escondido Mountain and for 12 mi to the west along the north foot of Fox Mountain where it lies below the andesite of Dry Leggett Canyon. Phenocrysts of sanidine from pumice in muddy siltstone near the base of the marker bed at Castle Rock have yielded an $^{40}Ar/^{40}Ar$ age of 35.33 ± 0.06 Ma. This bed is interpreted as the sedimentary response to rhyolitic pyroclastic eruptions that saturated the regional fluvial system with ash and pumice; this then lead to a rapid sheet-like aggradational event. Larger pumice fragments are preferentially concentrated in slackwater mudstones and siltstones in the marker bed. Presumably most of the pumice was freshly introduced into the stream system shortly before the time of deposition (Fig. 1.50). According to the terminology of Smith (1991) this tuffaceous marker bed would be classified as a syneruptive deposit. **0.3**

156.8 **Caution, slow for dangerous curve to right ahead. 0.1**

156.9 Roadcut on right in thin bedded pumiceous sandstones and mudstones at the base of the volcaniclastic unit of Cañon del Leon. **0.3**

157.2 Pumiceous sandstone of basal Cañon del Leon and greenish sandstones of upper transitional unit of Largo Creek in roadcuts on left. **0.7**

157.9 Road to Buz and Beverly Easterling Ranch on right. Base of the volcaniclastic unit of Cañon del Leon is well exposed in side of mesa on left. Mesas at 9:00, 12:00 and 3:00 are capped by Fence Lake Formation. **0.1**

158.0 Re-enter valley of Largo Creek. **0.3**

158.3 Road crosses perennial flow of middle Largo Creek; apparently fed by natural springs and seepage under Quemado Lake dam about 2 mi southeast of here. **0.2**

158.5 Fence Lake Formation unconformably overlies unit of Cañon del Leon at 3:00. **0.3**

158.8 Cattle guard. Red fence around Buz Easterling's polo field at 2:00. **0.1**

158.9 Road rises along cuts on right in thin-bedded pumiceous sandstones and mudstones of the volcaniclastic unit of Cañon del Leon. **0.1**

159.0 Road crosses poorly exposed north wall of a locally west-trending paleocanyon filled with as much as 200 ft of Fence Lake Formation conglomerates. Reconnaissance mapping indicates that this upper Miocene paleocanyon was in part controlled by an east-trending segment of the Spur Lake fault, which is downthrown approximately 600 ft to the south in this area. **0.1**

FIGURE 1.49. Castle Rock provides easily accessible boundary stratotype for the base of the volcaniclastic unit of Cañon del Leon. Basal contact is immediately below jointed cliff-forming marker bed of tuffaceous sandstone and pumice mudstones. Underlying thin white tuffaceous sandstone and tan to bluish green (celadonitic) andesitic conglomerate and conglomeratic sandstones are within transition zone in the upper part of the volcaniclastic unit of Largo Creek.

FIGURE 1.50. Tuffaceous sandstone and pumiceous muddy siltstone from one foot above the base of the unit of Cañon del Leon at Castle Rock (Sample CR-93-1-XR, McIntosh and Chamberlin, this volume). Tuffaceous sandstone may represent hyperconcentrated-flow deposit; muddy siltstone is slack water deposit.

159.1 Crest of hill. Snuffys steakhouse, saloon, tackle shop and gas pump on left. **0.1**

159.2 Road descends through cuts in Fence Lake Formation conglomerates and sandstones of paleocanyon fill, then crosses unnamed south-flowing tributary to Largo Creek. Paleocanyon wall is locally exposed 0.15 mi up this drainage. Here the upper Miocene canyon wall shows a 30-ft-high step as it rises northward onto the canyon rim, which is cut in lower Cañon del Leon beds. North of the paleocanyon rim the Fence Lake Formation is locally about 30 ft thick. Road crosses projected (buried) trace of Spur Lake fault near here. **0.1**

159.3 Water wells on right intersected 20 ft of alluvium, 80 ft of Fence Lake conglomerate (local aquifer) and 10 ft of light gray Oligocene sandstone (Buz Easterling, oral commun., 1993). Correlation of the Oligocene sandstone is uncertain, but local field relationships suggest that it is down-faulted sandstone of Escondido Mountain. As much as 7 mi of water lines have been installed to service the Quemado Lake Estates from these wells (Fig. 1-51). Nearly all of the lower 80 ft of Fence Lake Formation is saturated with water here. One of these wells is a stand by, the other is reported to produce approximately 100 gallons per minute with a submersible pump. (Buz Easterling, oral commun., 1993). **0.1**

159.4 Fence Lake Formation above road level on left is approximately 120 ft thick. This exposed thickness plus 80 ft intercepted in the water wells below the valley floor sum to a total of about 200 ft of fill along the axis of the paleocanyon. Fence Lake conglomerate is well exposed in the south wall of the modern valley at 4:00. **0.1**

159.5 Cattle guard. Enter Apache National Forest. **0.2**

159.7 Enter Quemado Lake Recreation Area. **0.3**

160.0 Parking lot (also spillway) on north side of Quemado Lake Dam. Well-imbricated boulder conglomerate of Fence Lake paleocanyon fill is exposed in cuts on left (Fig. 1.52). Construction of this earthen dam in 1970-1971 is summarized in the following minipaper. As stated on the official sign near the dam, Quemado Lake and the surrounding recreation area are managed by the U.S. Forest Service in cooperation with the New Mexico Department of Game and Fish. The lake has a surface area of approximately 131 acres. **0.1**

FIGURE 1.51. Trench in Fence Lake Formation at Quemado Lake Estates during development of a local water system, June 1992. Field assistant C. G. Chamberlin is only slightly larger than local maximum grain size of Fence Lake Formation (~5 ft).

FIGURE 1.52. Roadcut in proximal Fence Lake Formation on NE side of Quemado Lake dam. Imbricated boulders indicate right to left (NW) paleoflow in this paleocanyon-fill deposit.

QUEMADO LAKE DAM

William C. Haneberg

New Mexico Bureau of Mines and Mineral Resources, Socorro, NM 87801

Quemado Lake Dam is an earthen embankment along Largo Creek, a tributary of the Little Colorado River, and lies within the Apache-Sitgreaves National Forest. Completed in 1971 and intended to promote recreation, the dam was constructed by the New Mexico Game and Fish Department under a special use permit from the US Forest Service. To this end, parking and camping are permitted in the spillway. Before receiving its present name, Quemado Lake Dam was referred to as Largo Creek Dam.

The dam is 924 ft long, 73 ft high, and 20 ft wide at its crest. It is zoned, with an impervious core flanked by semi-pervious transitional zones and pervious outer zones (Fig. 1.53). Up- and down-stream slopes are both 2.25:1 (horizontal to vertical) and covered with large riprap. Foundation materials are basaltic gravels and conglomerates of the middle to late Miocene Fence Lake Fm., which locally occurs as a 200-ft-thick paleo-valley fill into which are inset Quaternary alluvial deposits (R. M. Chamberlin, personal commun., 1993). Seepage problems were encountered during construction of the cutoff trench, which was quickly filled with muck. When the trench was cleaned, foundation materials were found to consist of volcanic boulder conglomerate with a poorly- to well-cemented clayey sand matrix. Maximum boulder diameter was 3 ft. Inspectors reported an absence of porous zones that might be conducive to piping, although gravelly stringers issuing groundwater were described near the top of the conglomerate. In addition, an abrupt change from a boulder conglomerate to a pebble conglomerate was observed in the cutoff trench near the right abutment. A State Engineer Office technical report dated 1958 includes logs of 10 shallow borings and cross-sections inferring a buried reverse fault along the axis of the Largo Creek valley. The fault was believed to be inactive, and the report further stated that a thick alluvial blanket beneath the dam would inhibit seepage along the fault. Cross-sections included on the plans for the current dam, however, do not include a fault and it is not known whether any seismic slope stability analyses were performed. The dam has performed well to date, with only minor problems such as rodent burrowing and plant growth along the embankment.

Outlet works at the Quemado Lake Dam consist of a 24-in. diameter concrete pipe with a capacity of 60 cfs, and the service spillway, which controls the reservoir elevation at 7628 ft, and has a capacity of 33,000 cfs. The emergency spillway (crest elevation 7634 ft) consists of a box weir drop inlet structure leading to a 5-ft-diameter concrete pipe, which in turn discharges into an open concrete channel and stilling basin along the right abutment. The emergency spillway was designed to pass in conjunction with the service spillway the 6 hr probable maximum precipitation (PMP) without overtopping the dam.

Drainage area of the Quemado Lake Dam is 62 mi^2. Quemado Lake normally stores 2000 acre-ft of water, with a maximum storage capacity of 2550 acre-ft. During design, evapotranspiration and seepage losses were anticipated to be on the order of 440 acre-ft per year. Enlargement of the reservoir was contemplated during the mid-1970s. The proposed increase in reservoir elevation would have decreased emergency storage (i.e., the amount of storage available between operating and emergency spillway elevations) from 860 acre-ft to an unacceptably low 152 acre-ft. Although a catastrophic failure would release some 152,000 cfs, the Quemado Lake Dam is classified as a low-hazard structure because of its remote location. Flood routing calculations predict that a catastrophic failure would increase water levels only a foot or so where Largo Creek crosses US-60 some 18 mi downstream from the dam.

Information for this note came primarily from reports and construction documents on file at the New Mexico State Engineer Office (SEO file 3026). The assistance provided by David Quintana and Larry Ferns of the SEO is greatly appreciated.

160.1 Road makes loop to right around small bay at north end of Quemado Lake. **0.2**

160.3 Road crosses northeastern paleocanyon margin; unconformity at base of Fence Lake Formation is poorly exposed in roadcut on left. **0.1**

160.4 High-angle, cross-bedded sandstone of Escondido Mountain is well exposed in roadcut on left (Fig. 1.54). Steepest cross bedding dips 30°E. **0.2**

160.6 Unconformity at base of Fence Lake Formation is well exposed about 30 ft above road on left. Variably cross-bedded sandstone of Escondido Mountain in lower third of cut. **0.1**

160.7 Road to new campground (under construction in summer of 1993) on right. **Continue straight ahead,** to old campground in valley floor at southeast end of lake. Sandstone of Escondido Mountain moderately well exposed on left for next 0.2 mi. **0.2**

160.9 Cross unconformity onto Fence Lake Formation. **0.1**

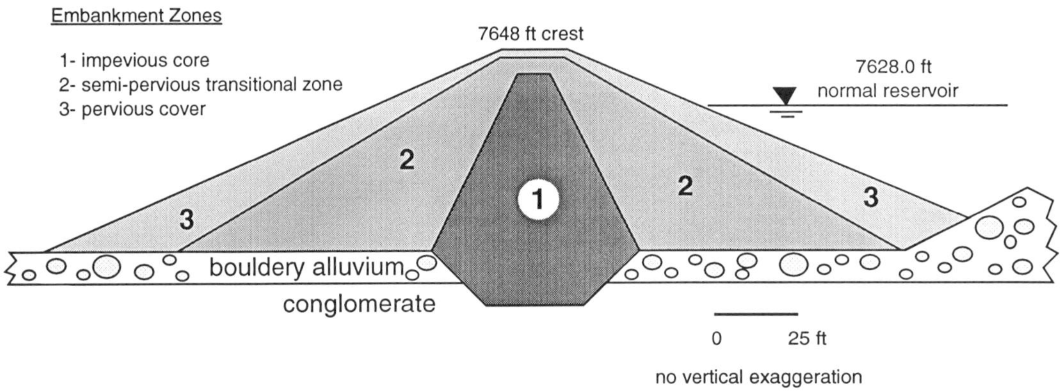

FIGURE 1.53. Generalized cross-section through Quemado Lake Dam, illustrating the geometry of the zoned earthen embankment. The material identified as bouldery alluvium in exploratory borings corresponds to Quarternary alluvium, and the conglomerate corresponds to the Fence Lake Formation.

FIGURE 1.54. High-angle cross bedding in the Oligocene sandstone of Escondido Mountain. Cross beds in this eolian sandstone dip predominantly to the east and imply prevailing westerly winds in Oligocene time.

FIGURE 1.55. El Caso Peak, which stands well above adjacent mesas capped by inset deposits of the Miocene Fence Lake Formation. Old Quemado Lake campground is located in valley floor to left of El Caso Peak.

161.0 Road crests above Quemado Lake; Agua Fria Mountain on skyline above lake at 4:00. **0.1**

161.1 Road to new campground on left. El Caso Peak at 1:00 (Fig. 1.55) is capped by Oligocene Squirrel Springs Canyon Andesite overlying sandstone of Escondido Mountain. The latter is largely covered by colluvium that is not mapped on regional reconnaissance maps. **0.2**

161.3 Road cut on left exposes bouldery conglomerate overlying pedogenic(?) carbonate zone over cobble conglomerate, all in Fence Lake Formation. **0.1**

161.4 Fence Lake Formation unconformably overlies sandstone of Escondido Mountain in cut on left. **0.2**

161.6 First entrance to the older Quemado Lake campground on left. Additional entrances to campground on right for next mile. Find a campsite you like, park, set up camp and then enjoy catered BBQ in the pines.

End of first day road log.

William C. McIntosh (left) instructs Robert M. Chamberlin (right) on the technique of collecting oriented core samples for paleomagnetic analysis. Paleomagnetic analysis and $^{40}Ar/^{39}Ar$ geochronology confirm that this light gray crystal-poor ignimbrite at Saulsberry Ranch is the Rock House Canyon Tuff. Sample NM 1096, collected July 14, 1993.

SECOND-DAY ROAD LOG FROM QUEMADO LAKE TO MANGAS MOUNTAINS, OMEGA, QUEMADO, TEJANA MESA AND RED HILL, NEW MEXICO, AND SPRINGERVILLE AND ALPINE, ARIZONA

STEVEN M. CATHER, RICHARD M. CHAMBERLIN, WILLIAM C. MCINTOSH, JAMES WITCHER, JAMES C. RATTÉ and ORIN J. ANDERSON

FRIDAY, SEPTEMBER 30, 1994

Assembly Point:	East end of Quemado Lake Campground, 2.6 mi SE of Quemado Lake Dam on Forest Road 13.
Departure Time:	8:00 a.m.
Distance:	133.3 mi
Stops:	4

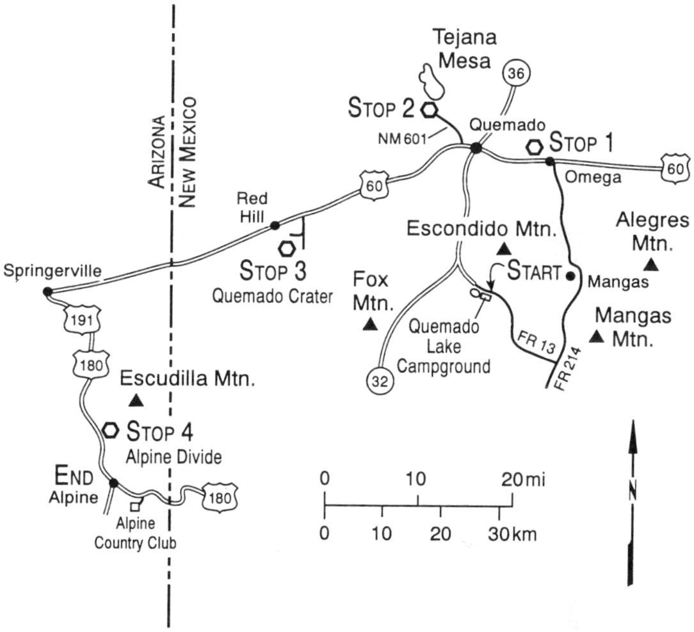

SUMMARY

Today's route traverses part of the northwestern margin of the late Eocene to late Oligocene Mogollon-Datil volcanic field in west-central New Mexico and eastern Arizona. Volcanic and sedimentary rocks of late Cenozoic age are also exposed along the field-trip route and will be discussed.

The route begins with a scenic loop through the Mangas Mountains. Stop 1 is an overlook area along US-60 west of Omega, where we will discuss petroleum exploration and regional seismic reflection profiles in the Mangas-Alegres Mountain area, and the hydrology of the Largo Creek drainage basin. At Stop 2 we will visit a fissure vent at El Porticito near the south end of Tejana Mesa to discuss the geochemistry and petrology of upper Tertiary volcanic rocks near Quemado and the planned coal mine east of Tejana Mesa. Quemado crater, a Quaternary maar/cinder cone complex just south of US-60 near Red Hill, is Stop 3 and the lunch stop. There we will discuss the volcanology of the crater and role of faults and fractures in the localization of late Cenozoic volcanoes in the region. Stop 4 is on the Alpine Divide north of Alpine, Arizona, where we will discuss the recent Alpine 1/Federal corehole and regional heat flow. After Stop 4, we will travel south to Alpine, Arizona to end the day's tour. Much of the geologic information presented here is from reconnaissance studies by Chamberlin et al. (1994b).

Mileage

0.0 East end of Quemado Lake Recreation Area, 2.6 mi southeast of Quemado Lake Dam on Forest Road (FR) 13. Head east, upstream, along El Caso Springs Canyon toward the Alpine meadows of Slaughter Mesa, approximately 5 mi from here. **0.1**

0.1 El Caso Peak at 1:30 is capped by lava flows of the Squirrel Springs Canyon Andesite, and the underlying andesite of San Antone Canyon. The andesite of San Antone Canyon is an aphanitic flow-banded mafic lava about 60 ft thick. Two coarsely porphyritic flows, each approximately 60 ft thick, comprise the Squirrel Springs Canyon Andesite. Vicks Peak Tuff (28.5 Ma), which locally occurs between these formations, is apparently absent here. **0.2**

0.3 Sign for Slaughter Mesa loop to NM-32, 26 mi. This road log takes the undesignated Mangas Mountain loop to US-60, 38 mi. **0.2**

0.5 Culvert; road crosses Canyon de Lolo which drains the south flank of Escondido Mountain. Sandstone cliff at 2:30 consists of massive light gray, crossbedded eolian sandstone assigned to the upper Oligocene sandstone of Escondido Mountain (Chamberlin and Harris, this volume). **0.1**

0.6 Junction with Baca Road on left. **Continue straight ahead on FR-13**. Baca Road parallels Canyon de Lolo to the south flank of Escondido Mountain. East-northeast- and east-trending normal faults that cut across Escondido Mountain locally *appear* to control the distribution of Squirrel Springs Canyon Andesite and the overlying Bloodgood Canyon Tuff, which suggests normal movement on these faults may have initiated as early as 26 to 28 Ma. Alternatively, these volcanic flows may have filled east-trending swales in the underlying dune deposits assigned to the sandstone of Escondido Mountain. In the latter case, local stratigraphic omission on the structurally high block would be fortuitous. **0.1**

0.7 Canyon wall exposure on left in sandstone of Escondido Mountain. Private road to El Caso Springs on right, just ahead. Low mesa at 2:00 is capped by Miocene Fence Lake Formation. High mesa at 12:00 is capped by upper Oligocene (ca. 26 Ma) basaltic andesite lava flows assigned to the Mangas Mountain Member of the Bearwallow Mountain Andesite. **0.2**

0.9 Outcrop at road level on left in upper Oligocene sandstone of Escondido Mountain, unconformably overlain by middle to upper Miocene Fence Lake Formation. Basal Fence Lake conglomerate contains abundant cobbles and boulders derived from the Bearwallow Mountain Andesite (dark gray basaltic andesite). Sparse clasts of light gray Bloodgood Canyon Tuff and coarsely porphyritic, Squirrel Springs Canyon Andesite are also present. **0.2**

1.1 Road crosses drainage of El Caso Springs Canyon at culvert. **0.4**

1.5 Road ascends onto upper surface of Fence Lake Formation, which here is only 40 ft above Quaternary valley fills along El Caso Springs Canyon. **0.5**

2.0 Road descends onto Quaternary valley fill that is inset against Fence Lake Formation. The Fence Lake is here locally inset against upper Oligocene volcanic strata that define its source area. **0.3**

2.3 Canyon wall on left is cut in flow-banded andesite of San Antone Canyon. **0.2**

2.5 Road rises onto poorly exposed terrace gravels assigned to the Fence Lake Formation. **0.2**

2.7 Jeep trail on left leads to Gooseberry Pit and upper reaches of El Caso Springs Canyon. **0.3**

3.0 Cattle guard. Road enters unnamed V-shaped canyon which heads near McKibben Tank on the north side of Slaughter Mesa. Fence Lake gravels terminate here, marking the local up-canyon limit of middle to late Miocene aggradation in this area. **0.2**

3.2 Low roadcut on right in medium gray, coarsely porphyritic and vesicular Squirrel Springs Canyon Andesite. The Squirrel Springs lavas typically contain about 20% phenocrysts of tabular plagioclase, 1–3 cm wide. **0.1**

3.3 Bloodgood Canyon Tuff, largely mantled with basaltic colluvium, is poorly exposed in low roadcut on right. Just ahead, road crosses colluvium-covered trace of east-northeast striking normal fault, downthrown approximately 60 ft to north. **0.1**

3.4 Cut on right in Squirrel Springs Canyon Andesite defines upthrown footwall of normal fault. Note light-colored bench of Bloodgood Canyon Tuff partially hidden in trees about 60 ft above road level on right. **0.4**

3.8 Vertical roadcut on right is in coarsely porphyritic Squirrel Springs Canyon Andesite (Figs. 2.1, 2.2). This locality may be used as an **optional stop** to view three regional volcanic formations locally well exposed along the western (right) side of the valley (Fig. 2.3). Whole-rock total-gas $^{40}Ar/^{39}Ar$ analyses of the Squirrel Springs Canyon Andesite and the basal Bearwallow Mountain Andesite collected from this site have yielded isochron ages of 28.8 ± 0.1 Ma and 26.1 ± 0.1 Ma, respectively (McIntosh and Chamberlin, this volume). The intervening Bloodgood Canyon Tuff, which was sampled 6 mi north of here in the same stratigraphic position, has yielded an $^{40}Ar/^{39}Ar$ age of 27.98 ± 0.09 Ma from sanidine. Field conference caravan will **not** stop here. **0.2**

4.0 Road on right leads to a small borrow pit 300 ft to west, which exposes 15 to 20 ft of trough-crossbedded granular sandstone apparently derived almost entirely from the underlying Squirrel Springs Canyon Andesite (Fig. 2.4). Local detailed mapping indicates that this granular sandstone fills an east-southeast trending (flowing?) paleovalley as much as 40 ft deep and a half mile wide. Except at the borrow pit, the granular sandstone is generally masked by gentle soil-covered slopes between resistant outcrops of the underlying Squirrel Springs Canyon Andesite and the overlying Bloodgood Canyon Tuff. The sandy matrix of the granular sandstone (Fig. 2.5) may have been recycled from contemporaneous eolian deposits, namely the sandstone of Escondido Mountain. **0.1**

4.1 On hillside to right, outcrops of the Squirrel Springs Canyon Andesite rise abruptly to meet with the base of the Bloodgood Canyon Tuff, about 30 ft higher on the slope. This relationship apparently defines the steeper

FIGURE 2.1. Roadcut in lava flow of the Squirrel Springs Canyon Andesite, looking northwest.

FIGURE 2.2. Close up of Squirrel Springs Canyon Andesite, showing large tabular plagioclase phenocrysts glistening in the sun.

FIGURE 2.4. Bulldozer cut in trough crossbedded granular sandstone near Mile 4.0.

southern margin of the granulestone-filled paleovalley. A similar outcrop pattern is found on the east side of the modern canyon. **0.3**

4.4 In trees upslope at 9:00 and 4:00, narrow ledges of Bloodgood Canyon Tuff abruptly wedge out against the underlying Squirrel Springs Canyon Andesite, a relationship that apparently defines the outer margin of the previously mentioned east-southeast-trending paleovalley filled by the granular sandstone. From here to the south for the next mile, dark gray aphanitic basaltic-andesite lavas, of the Bearwallow Mountain Andesite, unconformably overlie weathered and eroded outcrops of the Squirrel Springs Canyon Andesite. **0.5**

4.9 Culvert; road crosses tributary drainage from Antelope Tank. On the north side of this drainage, at 3:30, the basal contact of the Bearwallow Mountain Andesite (on Squirrel Springs Canyon Andesite), is approximately 40 ft above road level. On the south side of the drainage, at 10:00, basaltic Bearwallow outcrops form the modern valley floor, as much as 30 ft below road level. Similar outcrop patterns nearby indicate the road is now crossing the north wall of a late Oligocene basalt-filled paleocanyon at least 70 ft deep that was cut in the underlying Squirrel Springs Canyon Andesite. This locally east-northeast-trending paleocanyon probably merged with the previously described sandstone-filled paleovalley approximately 1 mi east of here. **0.1**

5.0 Roadcut on right in paleocanyon fill consists of dark gray olivine basalt or basaltic andesite of the Bearwallow Mountain Andesite. **0.1**

5.1 Culvert. Road crosses McKibben Tank drainage. **0.1**

5.2 Roadcut on left in Bearwallow basaltic flow, which contains a large, rounded mass of tan sandstone, possibly ripped up from the floor or walls of the underlying paleocanyon. Just beyond, road again crosses McKibben Tank drainage; here the modern valley transects the south wall of the basalt-filled paleocanyon. Approximately 40 ft of relief is visible here at the base of the Bearwallow lavas. **0.1**

5.3 Cut on right in porphyritic Squirrel Springs Canyon Andesite that defines the south margin of the paleocanyon. The Bearwallow-Squirrel Springs contact forms a break in slope 20 ft above the road at 2:00. **0.1**

5.4 Roadcut on right exposes a 10 ft zone of red clay developed on the top of the Squirrel Springs Andesite. The red clay contains angular clasts of altered andesite porphyry. This clayey zone probably represents a paleoweathering profile buried by the Bearwallow basaltic lavas at 26 Ma. Base of Bearwallow is 10 ft above road at 2:00. **0.1**

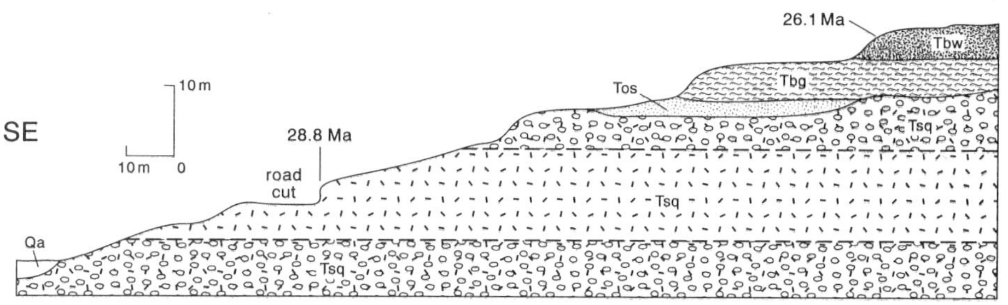

FIGURE 2.3. Sketch cross section illustrating upper Oligocene volcanic stratigraphy exposed on the west side of the western fork of El Caso Springs Canyon, Slaughter Mesa 7.5′ quadrangle. Open circle pattern in lower and upper Squirrel Springs Canyon flows represent vesicular zones. Unit abbreviations are: Tsq=Squirrel Springs Canyon Andesite; Tos=andesitic granular sandstone; Tbg=Bloodgood Canyon Tuff; Tbw=Bearwallow Mountain Andesite. Approximate locations of $^{40}Ar/^{39}Ar$ age samples are indicated.

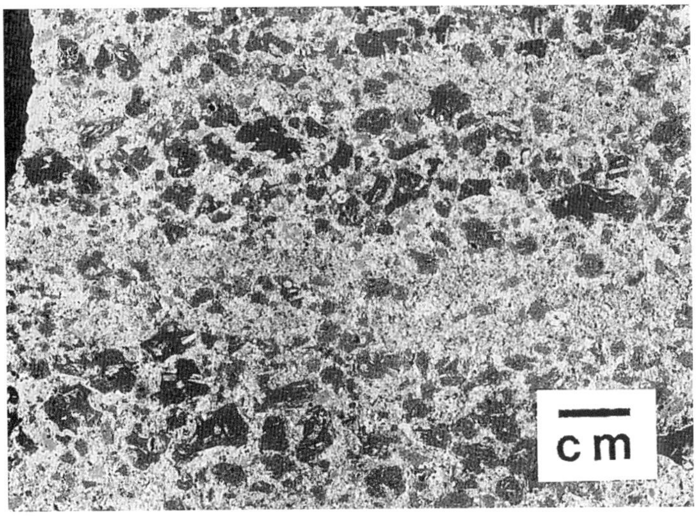

FIGURE 2.5. Close-up view of granular sandstone showing highly angular fragments of andesite porphyry in a sandy matrix.

5.5 Basal contact of Bearwallow Mountain Andesite at road level on right. **0.1**

5.6 Roadcut on right in dark gray vesicular basaltic-andesite lava containing sparse fine-grained phenocrysts of yellow-brown olivine. Bearwallow lavas are intermittently exposed for next 1.6 mi, from here to Slaughter Mesa. **1.2**

6.8 McKibben Tank is visible at valley bottom on left. **0.4**

7.2 Road crests at elevation of approximately 8690 ft. Open meadows from 12:00 to 2:00 define the shallow basin known as Slaughter Mesa. Several small borrow pits (Fig. 2.6) and local incision along upper Largo Creek Canyon indicate this open flat is underlain by 20 to 120 ft of poorly consolidated gravels derived entirely from basaltic-andesite lavas of the Bearwallow Mountain Andesite. These upper Tertiary(?) gravels are tentatively assigned to the Fence Lake Formation and appear to fill an incipient fault-block basin at the head of ancestral Largo Creek. **0.2**

7.4 From 10:30 to 1:00, a 300-ft-high Ponderosa-covered fault-line escarpment marks the trace of a Neogene normal fault that locally forms the southeast margin of the Slaughter Mesa basin. From 2:00 to 3:00 another fault-line scarp, which trends north-northwest about 200 ft above the flats, locally forms the western side of the Slaughter Mesa basin. This high-angle normal fault is well exposed 2 mi farther south, at Indio Canyon, where it displaces the Bloodgood Canyon Tuff 300 ft down to the east. Ancient colluvial breccias derived from the Bloodgood Canyon Tuff (buried by Bearwallow lavas) occur locally along the Indio Canyon fault and suggest that it may have been active as early as 28 to 26 Ma. **0.3**

7.7 Culvert, road crosses drainage of upper Largo Canyon, headwaters of the ephemeral stream that feeds Quemado Lake. **0.3**

8.0 Slow down for four-way intersection. **Continue straight ahead on FR-13**. Watch for cross traffic, most likely to be big game hunters in the early fall. The Mangas Mountains and Slaughter Mesa are a favorite area for zealous elk hunters who come great distances (Fig. 2.7) to hunt the bellowing bulls. FR-93, to right, is continuation of Slaughter Mesa loop to NM-32. Roadcuts on FR-93 approximately 8 mi from here expose 40-ft-high dune faces in the sandstone of Escondido Mountain. **0.5**

8.5 Begin ascent of Pleistocene fan surface that appears to have prograded over dissected Fence Lake gravels. **0.4**

8.9 Cattle guard. Route crosses fault-line scarp at southeast margin of Slaughter Mesa basin and enters narrow canyon of upper Largo Creek, which has cut deeply into Bearwallow Mountain Andesite on the upthrown side of the fault. **0.2**

9.1 Roadcut on left in basaltic andesite lava of Bearwallow Mountain Andesite. **0.2**

9.3 Platy cleavage resulting from flow foliation in basaltic-andesite lava is well exposed in cut on left. **0.4**

9.7 Vesicular basaltic-andesite lavas in roadcuts on left and right. **0.4**

10.1 Crest of hill. Cross "Rim Divide" of Cather et al. (this volume). **0.1**

FIGURE 2.6. View to west of borrow pit in Fence Lake Formation gravels that underlie Slaughter Mesa near intersection of FR-13 and FR-93. Tree line along west end of Slaughter Mesa delineates fault-line scarp of the Indio Canyon fault.

FIGURE 2.7. This avid hunter came a long way to "GO 4 ELK" in the Mangas Mountains.

10.2 Drill pad on left for Shell-Elf Aquitane-BP Mangas Mountains Federal Well No. 1, a wildcat well drilled in the autumn of 1987 to Precambrian basement and a total depth of 7808 ft. This is one of several hydrocarbon test wells drilled in the Mangas region during the late 1980s, as part of the "Mogollon Frontier" or "Magic Area" play (Kopacz, et al., 1989; Garmezy, 1990). General stratigraphy and some structural implications of the Mangas Mountains Federal No. 1 Well are discussed in the following minipaper. **0.3**

STRATIGRAPHIC AND STRUCTURAL IMPLICATIONS OF THE MANGAS MOUNTAINS FEDERAL NO. 1 WELL

Richard M. Chamberlin

New Mexico Bureau of Mines and Mineral Resources, Socorro, NM 87801

According to records of the New Mexico Bureau of Mines and Mineral Resources Library of Subsurface Data, the Mangus (sic, Mangas) Mountain Federal No. 1 Well intersected 4246 ft of Tertiary strata (Mogollon-Datil volcanic pile and Baca Fm.) unconformably overlying 558 ft of Triassic Chinle, unconformably overlying 2311 ft of Permian strata, unconformably overlying Precambrian basement rocks. Permian strata include 402 ft of San Andres Fm., 235 ft of Glorieta Sandstone, 1060 ft of Yeso Fm. and 614 ft of Abo Fm. In meters, the above thicknesses are 1291, 170, 703, 122, 71, 322 and 187, respectively.

The Mangas Mountains Federal No. 1 Well is collared in Bearwallow Mountain Andesite (~ 150 m thick here) at the top of the middle Tertiary volcanic pile. Thus, the total thickness of middle to early Tertiary strata, 1291 m, may be compared with the total thickness of middle to early Tertiary strata, 2137 m, estimated for the Sun Oil Co. No. 1 Well, approximately 55 km east of the Mangas Mountain well (see First-Day road log, mile 49.2). This westward decrease in thickness of about 40% probably occurs mainly within middle to late Eocene strata of the Baca Fm. and/or the lower Spears Group. Younger middle Tertiary strata (35-26 Ma) in the Sun Well (930 m) and that exposed in Mangas Mountains area (690 m, Chamberlin et al., 1994b) show a similar westward thinning trend but can only account for about 28% of the observed total thickness change (846 m). The westward thinning of middle and early Tertiary strata can be attributed to differences in relative proximity to late Eocene volcanic centers and/or relative proximity to Laramide transpressional welts (cf. Garmezy, 1990). Generally southeastward coarsening of the lower Spears group suggests the Sun Well is more proximal to an inferred late Eocene volcanic center (ca. 40-36 Ma) now buried under the southern San Agustin Plains (S. M. Cather, oral commun., 1994).

A recently published structure contour map of the base of Upper Cretaceous strata in west central New Mexico and eastern Arizona (Rauzi, 1994), including well data in the Mangas region, suggests the presence of a northwest-trending(?) Laramide syncline east of the Mangas Mountains between Alegres Mountains and Wallace Mesa. The detailed geometry of buried Laramide structures in the Mangas region is presently uncertain and may involve multiple transpressional welts along major regional structures now expressed as normal faults in the Tertiary volcanic pile (Cather, 1989; Garmezy, 1990, Chamberlin, et al., 1994b). The presence of 345 m of upper Cretaceous strata in the Sun Well (R. Foster, unpub. log, NMBMMR Subsurface Library) and the total absence of Cretaceous strata in the Mangas Mountains Federal well also implies the latter was relatively proximal to a Laramide highland. Detailed logging of cuttings (as yet not available) from the Tertiary intervals in the "Magic Play" wells could provide significant insight as to the geometry and timing of Laramide structures in the Mangas region. For a summary of Tertiary stratigraphy in the Hunt Oil Co. No. 1-16 well, approximately 23 km east of here near Wallace Mesa, see minipaper by Broadhead and Chamberlin (this road log). Broadhead (this volume) has summarized data concerning the petroleum source rock potential of pre-Tertiary rocks in the Hunt well.

10.5 Del Buey Vista. Slow down to look through breaks in tree tops at majestic view across the San Agustin depression from continental divide to continental divide (Fig. 2.8). **0.2**

10.7 Begin 400 ft descent along south-facing escarpment that locally defines the Cañon del Buey fault. This breakaway fault forms the north rim of the Sand Flat Canyon graben (see Third Day road log); it is also a scissors fault that decreases in displacement eastward to where it dies out near the north-northeast-trending Hickman fault. Topographic relief across the Cañon del Buey fault increases westerly to a maximum of 800 ft. **0.5**

11.2 At 1:00 is good view of the northeast end of the Reserve graben near Aragon. **0.1**

11.3 Sharp curve to left. Roadcuts on right and left in Bearwallow Mountain Andesite. **0.1**

11.4 For next 0.1 mi roadcuts on left provide excellent exposure of near-vent basaltic agglomerates (scoriaceous cinders and bombs) overlain by channelized flows of platy basaltic andesite lava (Fig. 2.9). Also present, below the road level on the right, are large blocks of variably oxidized (reddish orange to black) basaltic spatter that show well-developed flow structure similar to eutaxitic foliation in ash-flow tuffs (Fig. 2.10). **0.6**

12.0 Road junction on right. **Continue straight ahead on FR-13**. FR-13 crosses the continental divide here at an elevation of 8613 ft. FR-13 is now paralleling the base of the Cañon del Buey fault-line scarp; thus the continental divide crosses the road at a right angle and ascends the scarp to the ridge line at an elevation of about 9000 ft. In contrast, the continental divide ridge line south of the del Buey fault (3:00) is at an average elevation of about 8650 ft. **0.2**

FIGURE 2.8. View to southeast across the San Agustin depression from Del Buey Vista. Ridge line in foreground is on continental divide; Cerro Caballo in low middle ground is at the west end of the San Agustin depression, and Pelona Mountain on the skyline at right is also on the continental divide. Exposures of the Horse Springs dacite at Cerro Caballo will be visited at Stop 3 on the third day of this field conference.

FIGURE 2.9. Platy to massive basaltic andesite lava fills channel cut in cinders and bombs of basaltic andesite that are interpreted as near-vent agglomerate. Road logger is standing at base of channel which rises about 16 ft to the right. Elongate vesicles at channel bottom indicate flow parallel to channel walls (S15°E).

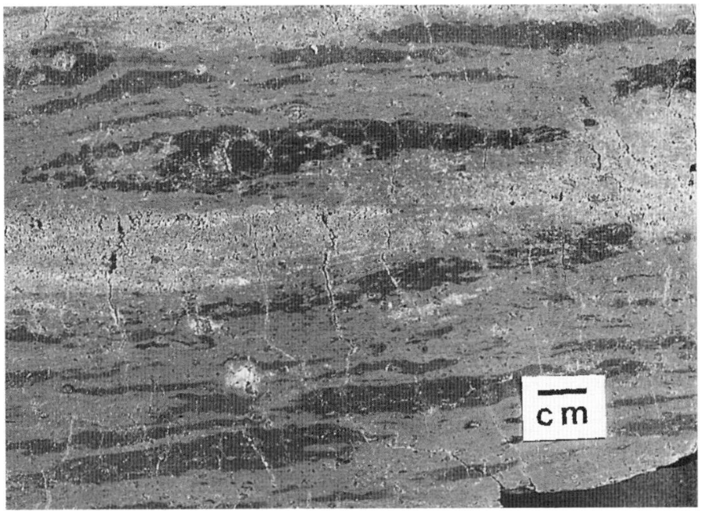

FIGURE 2.10. Flow structure developed in basaltic andesite spatter. Lighter colored matrix is dense, red and glassy. Dark streaks apparently represent unoxidized cores of lava bombs and spatter.

12.2 Steep slopes and bulldozer cuts on left in carbonate-cemented basaltic colluvium. Small blocks of white opal appear to wash out of the colluvium. Projected trace of the Cañon del Buey fault now follows the floor of Caballerzia Canyon at 3:00. **0.1**

12.3 Cut on left in massive core of basaltic-andesite lava flow. **1.4**

13.7 Dusky red vesicular top of basaltic-andesite lava flow in roadcut on left. **0.2**

13.9 Northwest flank of Mangas Mountain at 12:00 on skyline. In early autumn, look for yellow aspen groves here. **0.1**

14.0 Cross projected trace of Cañon del Buey fault where it bends north and its apparent displacement rapidly decreases to less than 100 ft. **0.6**

14.6 Descend onto older Pleistocene alluvium filling Valle Tio Vinces, a small fault-controlled depression on the downthrown western side of the north-northeast striking Hickman fault (main strand of Hickman fault zone). Surface waters exit through gap at 2:00, which is the head of Patterson Canyon. Hill slope to right of water gap (and stock tank) is underlain by Bloodgood Canyon Tuff on the upthrown side of the Hickman fault. **0.2**

14.8 Stop at intersection of FR-13 and FR-214. **Turn left on FR-214**, which leads to Mangas and US-60. Mangas Mountain, elevation 9691 ft, on skyline at 1:30. Reconnaissance mapping indicates that Mangas Mountain is underlain by as much as 1100 ft of basaltic andesite lavas, the Mangas Mountain Member of the Bearwallow Mountain Andesite. In comparison, the average thickness of Bearwallow basaltic andesite lavas in the Slaughter Mesa area is approximately 500 ft. These observations suggest the high peaks of Mangas Mountain represent the eroded apex of a basaltic-andesite shield(?) volcano of late Oligocene age (~25–26 Ma). **1.0**

15.8 Junction with FR-11 on right leads to Mangas Mountain lookout tower. **Continue straight ahead on FR-214**. Note Forest Service horse corrals on left. Water tanks at corrals are fed by a siphon from Valle Tio Vinces spring, which discharges from basaltic-andesite lava just below the tree line at 2:00. **0.5**

16.3 Cattle guard. Cross continental divide at elevation of approximately 8280 ft. Road now descends on the west side of divide following San Antone Canyon and the ephemeral drainage of Mangas Creek. Numerous cuts and canyon wall exposures in Bearwallow Mountain Andesite can be observed over next 1.9 mi. **0.4**

16.7 Sleeping Steer Tank on right. **1.5**

18.2 Basaltic andesite lava at base of Bearwallow Mountain Andesite is exposed in roadcut on left, with low roadcut just beyond in iron-stained Bloodgood Canyon Tuff. **0.2**

18.4 Canyon wall on right exposes 20 ft ledge of porphyritic Squirrel Springs Canyon Andesite overlain by 15 ft of slabby Bloodgood Canyon Tuff. **0.1**

18.5 Cut on left in Squirrel Springs lava. **0.5**

19.0 Low cut on left in dark gray flow-banded aphyric andesite of San Antone Canyon. **0.2**

19.2 Highwall (roadcut) on left in flow-banded andesite of San Antone Canyon (Fig. 2.11). A whole rock sample from this roadcut has yielded a $^{40}Ar/^{39}Ar$ isochron age of 29.2 ± 0.1 Ma (McIntosh and Chamberlin, this volume). The canyon wall on right exposes 60 ft of the andesite of San Antone Canyon, which is overlain by 10 ft of Vicks Peak Tuff and in turn by 100 ft of Squirrel Springs Canyon Andesite. This new $^{40}Ar/^{39}Ar$ age for the andesite of San Antone Canyon is in good agreement with the overlying Vicks Peak Tuff, which is well dated at 28.5 Ma. This locality may be regarded as an informal type section for this informal map unit. **0.1**

19.3 Platy base of the andesite of San Antone Canyon overlies unusually red-colored sandstone of Escondido Mountain in cut on left. **0.4**

19.7 Crossbedded sandstone of Escondido Mountain is well exposed in roadcut on left. Steepest crossbeds dip 28° to northeast. Pleistocene terrace gravels unconformably overlie the upper Oligocene sandstone here. **0.4**

20.1 Mangas Work Center on right. Reconnaissance mapping of high ridge at 1:00 to 2:30 has defined late Oligocene paleotopographic relationships similar to those observed in the west fork of El Caso Springs Canyon. These important stratigraphic relationships are well hidden by vegetation and soils (Fig. 2.12) and best illustrated in a cross section along this ridge line (Fig. 2.13). **0.1**

20.2 Large exposure of sandstone of Escondido Mountain across valley at 3:00. **0.2**

20.4 Road to Cassidera Springs on left; cattle guard beyond. **0.1**

20.5 Gallegos Ranch homestead on right. **0.2**

20.7 Cattle guard. Junction with FR-93 on left; **continue straight ahead on FR-214**. On the butte at 2:30, Bearwallow Mountain Andesite disconformably overlies the sandstone of Escondido Mountain. As illustrated in the cross section (Fig. 2.13), four regional volcanic formations are missing here. These paleotopographic relationships *may* reflect the beginning of normal fault displacement along Hickman Fault zone as early as 26 to 28 Ma. However, it is also evident that the Vicks Peak Tuff and Bloodgood Canyon Tuff fill swales between dunes in the sandstone of Escondido Mountain in the Mangas region, and that the underlying, eolian paleotopography may be the principal control on the pinch-out of these tuffs. Documenting possible growth-faults that may control the thickness of upper Oligocene volcanic units will require detailed regional mapping. **0.4**

21.1 FR-214 crosses concealed trace of a normal fault downthrown approximately 300 ft to the north; the concealed fault is a locally east-trending strand of the Hickman fault zone. **0.1**

21.2 Road bends to right and crosses Mangas Creek drainage at culvert. Road now on upper Quaternary alluvium in Mangas Creek. **0.2**

21.4 Cattle guard. Antique Gabaldon Windmill on left was operational as of autumn 1993. **0.2**

21.6 Ruins of old Gabaldon homestead in cottonwood and ash grove on left (Fig. 2.14). The building walls are slabs of fine-grained, buff colored sandstone and adobe mortar. Presumably, these slabs were quarried from a nearby outcrop of the sandstone of Escondido Mountain. **0.2**

21.8 Leaving Apache National Forest. Cattle guard. **0.2**

22.0 Pleistocene gravels underlie northeast-sloping graded surface on skyline at 9:00. **0.6**

22.6 Road-metal quarry at 3:00 was developed in low hill of Bloodgood Canyon Tuff as much as 80 ft thick. This quarry, elevation 7600 ft, lies at the structurally lowest point on the west side of the Hickman fault zone. On the northeastern corner of the Mangas Mountains, 4:00 on skyline, the Bloodgood Canyon Tuff is present on the upthrown eastern side of the Hickman zone at an elevation of 8700 ft. The apparent 1100 ft displacement is distributed on three mappable down-to-the-northwest normal faults in this 2 mi wide zone (Fig. 2.15) **0.8**

23.4 Low mesa at 8:00 consists of 300 ft of basaltic-andesite lavas of the Bearwallow Mountain Andesite overlying lenticular outcrops of Bloodgood Canyon Tuff and the sandstone of Escondido Mountain. **0.2**

23.6 Cattle guard. Stone walls of abandoned homestead at 1:00 and old schoolhouse at 12:00 are made from hand-hewn slabs of Bloodgood Canyon Tuff with an adobe mortar. On the skyline from 11:45 to 12:45, all of the low conical hills east of Mangas and the high flat-topped peaks of Alegres Mountain are capped by Bearwallow Mountain Andesite, disconformably overlying 28.5-Ma Vicks Peak Tuff and the sandstone of Escondido Mountain (Fig. 2.16). Alegres Mountain, elevation 10,229 ft, is the highest landmark on the Mogollon slope of west-central New Mexico. Alegres Mountain lies on a northeast-trending horst that defines the structurally highest block in the Mangas-Pie Town area. From Alegres Mountain to the conical hills at 11:45 the total Neogene (post 26 Ma) stratigraphic throw across the Hickman fault zone is approximately 2000 ft down to the west. As much as 500 ft of the Neogene displacement postdates the middle to upper Miocene Fence Lake Formation. **0.4**

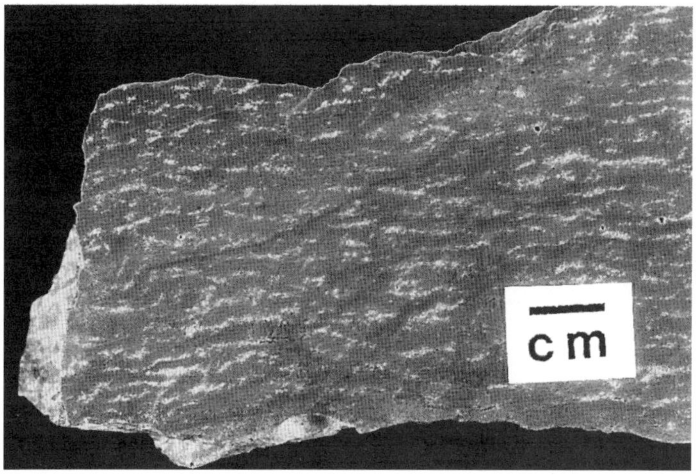

FIGURE 2.11. Close-up view showing typical discontinuous flow bands in mafic aphyric lava of the andesite of San Antone Canyon.

FIGURE 2.12. Road logger attempts binocular remote sensing to decipher Oligocene volcanic stratigraphy on ridge line east of the Mangas work center. Dark iron-strained tuffs that look like lavas from a distance and many trees make this a futile effort. Bearwallow Mountain Andesite caps butte at center and Bloodgood Canyon Tuff caps the ridge on the right.

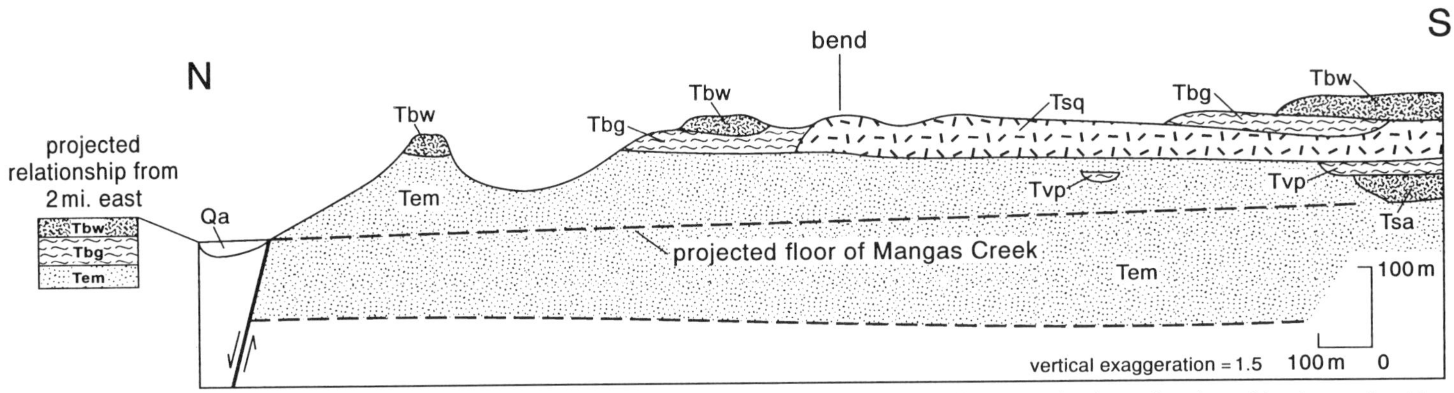

FIGURE 2.13. Cross section along ridge line east of Mangas work center. Cross section illustrates upper Oligocene volcanic stratigraphy and local unconformities that may reflect nascent (late Oligocene) fault-block topography within the Hickman fault zone. Stratigraphic relationships and present elevation of a downthrown block located 2 mi farther east (near road metal quarry) are projected to the line of section. Unit symbols: Tbw=Bearwallow Mountain Andesite, Tbg=Bloodgood Canyon Tuff, Tsq=Squirrel Springs Canyon Andesite, Tvp=Vicks Peak Tuff, Tsa=Andesite of San Antone Canyon, Tem=sandstone of Escondido Mountain.

24.0 Cross Mangas Creek to its eastern bank. For the next 14 mi, this route follows Mangas Creek valley and adjacent piedmont slopes descending from Alegres Mountain and the eastern flank of Escondido Mountain. Coarse bouldery piedmont-slope deposits on the upper rim of the broad valley are correlated with the Fence Lake Formation. Along the east rim of the valley, Fence Lake deposits are faulted and down-dropped to the west. Pliocene and early to middle(?) Pleistocene alluvial fans and bajadas (Quemado Formation of Cather and McIntosh, this volume) are inset against Fence Lake Formation on both sides of the valley. The finer grained, pebbly to cobbly sands of the Quemado Formation are graded to levels 40 to 100 ft above upper Quaternary alluvium underlying the broad floodplain of Mangas Creek. **0.1**

24.1 Road rises 30 ft onto toe of middle (?) Pleistocene fan fed by Killion Canyon, which heads on the east flank of Mangas Mountain. Excellent exposures of crossbedded sandstone of Escondido Mountain and a swale-filling outcrop of 28.5-Ma Vicks Peak Tuff are present at the head of Killion Canyon (Figs. 2.17, 2.18). **0.8**

24.9 Route ascends onto Plio-Pleistocene piedmont deposits of the Quemado Formation. **0.1**

25.0 Junction, road to Greens Gap on right. **Continue straight ahead to north on county road leading to US-60** (13 mi). The Hunt No. 16-1 State Well, located about 5 mi southwest of Greens Gap, was drilled to Precambrian basement in 1987. A reevaluation of the Tertiary stratigraphy intersected in this well is presented in the following minipaper. Also see Broadhead (this volume) for a discussion of petroleum source rock potential and source rock maturity in the Hunt Well. **0.4**

FIGURE 2.14. Ruins at Gabaldon homestead.

FIGURE 2.15. View to southeast across the Hickman fault zone. High wall in left foreground is cut in Bloodgood Canyon Tuff. The northeast corner of Mangas Mountain forms a high ridge on the skyline. At north end of ridge a dark ledge of Bearwallow Mountain Andesite directly overlies the Bloodgood Canyon Tuff on the upthrown eastern side of the Hickman fault zone. The main strand of the Hickman fault parallels the base of this ridge.

FIGURE 2.16. Panorama to northeast across the Hickman fault zone from abandoned school house south of Mangas. Flat-topped peaks of Alegres Mountain on right are capped by 600 ft of Bearwallow Mountain Andesite. Conical hills beyond the schoolhouse are also capped by Bearwallow lavas at an elevation as much as 1700 ft below equivalent lavas on Alegres Mountain.

HUNT OIL CO. NO. 1-16 STATE WELL WIRELINE LOG RESPONSES, TERTIARY AND CRETACEOUS SECTIONS

Ronald F. Broadhead and Richard M. Chamberlin

New Mexico Bureau of Mines and Mineral Resources, Socorro, NM 87801

Petrographic sections of cuttings from depths of 480, 540, 750, 900, 1200, 1350, 1650, 1860, 2250, 2700, 2850, 3150, 3300 and 3600 feet in the Hunt Oil Co. No. 1-16 State well (Sec. 16, T3S, R13W) were analyzed by the junior author to evaluate lithologic breaks in the Tertiary section in this drill hole. The senior author has characterized these lithologic units in terms of their wireline log response. The results of this brief study are as follows.

The contact between plagioclase- and hornblende-rich andesitic sandstones of the lower Spears Group (formerly lower Datil Group) and quartz-rich sandstones of the Baca Formation is at 1370 ft. This contact is delineated by a sharp decrease in resistivity in the Baca as compared to the overlying Spears. This effect is seen on both the deep induction log and the spherically focused log. There is also a slight increase in radioactivity in the Baca. Scout cards for this well on file at the New Mexico Bureau of Mines and Mineral Resources petroleum records library report a pick of 510 ft depth as the top of the Baca Formation. Thin section data do not support a lithologic break at this depth.

From 1370 to 1830 ft, the Baca is characterized by uniformly low resistivity and high radioactivity and appears to be composed dominantly of shaly or muddy sediments with thin (<10 ft) sandstone/conglomerate beds. From 1830 to 2710 ft, however, the Baca consists of thickly-bedded (10-40 ft) units interbedded with the shales. These thickly-bedded units are characterized by high resistivity and low radioactivity and are most likely sandstones and conglomerates (channels?). The total thickness of *typical* Baca Formation in this hole is 1340 ft.

The interval from 2710 ft to 3290 ft may belong to a previously unrecognized pre-Baca Tertiary unit, the lowermost Baca Formation or a Tertiary zone of oxidation and weathering developed on Upper Cretaceous rocks prior to Baca deposition (Chamberlin, 1981b). Cuttings in this interval consist of uniformly red mudstones, siltstones and fine-grained sandstones. These are significantly finer grained than the overlying sediments. Compared to the overlying part of the Baca, the interval from 2710 to 3290 ft is marked by a decrease in resistivity

FIGURE 2.17. A view to north of cross bedded eolian sandstone of Escondido Mountain, which is well exposed at the head of Killion Canyon on the northeast flank of Mangas Mountain.

FIGURE 2.18. Wedge of 28.5-Ma Vicks Peak Tuff within the sandstone of Escondido Mountain at the head of Killion Canyon. The base of the unwelded tuff is banked against a late Oligocene dune face.

on both the deep induction log and the spherically focused log. Uniformly low resistivities within this part of the Tertiary are indicative of a shale/mudstone section. Thick beds of sandstones or conglomerates are conspicuous by their absence. Mottled colors typical of the pre-Baca weathering zone in outcrop (Chamberlin, 1981b) are not observed in cuttings; which places the third alternative in disfavor. The lower Baca Formation in outcrop is typically coarse grained and conglomeratic; thus the first alternative listed above might be preferred. However, the total thickness of Tertiary red beds in this well is 1920 ft, which is in good agreement with the 1830 ft of Baca Formation measured by Cather and Johnson (1984) on the north flank of the Datil Mountains. Alternatively, this lower fine-grained Tertiary unit could simply be interpreted as an unusually fine-grained basal facies of the Baca Formation.

The contact between Tertiary red beds and gray Cretaceous rocks is at 3290 ft. This contact is delineated by a marked increase in radioactivity within the organic-rich Cretaceous shales and a corresponding decrease in resistivity on both the deep induction log and the spherically focused log.

FIGURE 2.19. Abandoned church at Mangas. Bearwallow Mountain Andesite caps east flank of Escondido Mountain at skyline on left.

25.4 Entrance to Davila Ranch headquarters on right. Slow down for sharp section-corner bend in road. **0.1**

25.5 Ninety degree bend to left. **0.3**

25.8 Weather-worn adobe church at 11:00 marked the center of Mangas during the 1930s great depression era (Fig. 2.19). Mangas is Spanish for sleeve or fringe of land; and is also used in a figurative sense for "sleeves" of rain pouring from clouds in an otherwise open sky. One of the Apache chieftains who roamed this country in the mid-19th century was called Mangas Coloradas, "Red Sleeves", but it is unlikely that the place name commemorates this Apache leader. The Mangas Creek, aka Pinoville or Piñonville (?), post office was active from 1909 to 1943 (Pearce, 1965). **0.7**

26.5 Slow for sharp curve to left past cattle guard. Cross upper Quaternary floodplain of Mangas Creek for next 0.4 mi. **0.4**

26.9 Slow for sharp curve to right. For next 8.1 mi the Mangas-Omega road crosses the toes of lower to middle Quaternary alluvial fans that grade to a level about 40-60 ft above modern Mangas Creek. East of Mangas Creek, the toes of upper Pliocene to lower Pleistocene fan deposits rise 80 to 120 ft above Mangas Creek. **1.0**

27.9 Cross tributary arroyo of Mangas Creek. Good view at 3:00 of conical hills capped with Bearwallow Mountain Andesite that lie between strands of the Hickman fault zone. On northern hill, approximately 300 ft of eolian sandstone of Escondido Mountain is present between the Vicks Peak Tuff (28.5 Ma) and Caballo Blanco Tuff (31.5 Ma) (McIntosh and Chamberlin, this volume). Another 60 ft of eolian sandstone underlie the Caballo Blanco, which in turn overlies fluvial volcaniclastic conglomerates that lie above and below the 33.7-Ma Blue Canyon Tuff. **1.2**

29.1 Headquarters of McKinley ranch on right. Ranch road on left leads to upper Cañon del Macho where a large, lenticular, cliff-forming outcrop of Bloodgood Canyon Tuff as much as 100 ft thick is present within the eolian sandstone of Escondido Mountain. **0.7**

29.8 Slow down for sharp curve to left. **0.1**

29.9 Cattle guard. **0.5**

30.4 Road crests on Quemado Formation. Good view at 8:30 of cliff-forming Bloodgood Canyon Tuff in middle slope on east flank of Escondido Mountain. **0.6**

31.0 Low hill across Mangas Creek at 2:45 is capped by coarse Fence Lake gravels. Just north of this hill another isolated Fence Lake outcrop unconformably overlies pumiceous sandstones correlative with the upper Eocene (34-35 Ma) volcaniclastic unit of Cañon del Leon (Chamberlin and Harris, this volume). This bedrock exposure demonstrates that the alluvial fill of the Mangas Valley is locally not more than 100 ft thick. Road crosses projected trace of concealed northeast-trending normal fault (down to northwest) near here. From here northward, the thin Plio-Pleistocene fill of the Mangas Valley grades into the thicker (as much as 600 ft) middle Miocene to Pleistocene fill of the structural depression referred to here as the Omega Basin syncline. **1.0**

32.0 Escondido Mountain at 9:30 (Fig. 2.20). Note decreasing northward slope (primary dip) on upper surface of Fence Lake Formation at 10:00. **0.5**

32.5 Cuestas on skyline from 11:30 to 1:30, capped with Fence Lake Formation, slope gently away from each other and appear to outline the crest of a breached anticline of Pliocene or Pleistocene age (Fig. 2.21). Grassy hills underlain by Pliocene-Pleistocene alluvium (Quemado Formation) in the middle foreground are inset against piñon-juniper-covered Fence Lake deposits on the sky line. **0.2**

32.7 Snakeweed herbicide spray test area of Quemado Soil Conservation District on right. **1.8**

34.5 Old adobe homestead at 2:00. Nearby stock well probably taps an unconfined aquifer within Plio-Pleistocene alluvium or possibly the underlying Fence Lake Formation. These permeable units most likely rest unconformably on andesitic sandstones of the lower Spears Group, which tend to act as an aquitard. Estimated thickness of alluvial fill here is 200-300 ft.

FIGURE 2.20 View southwest toward Escondido Mountain (peak on right). Snow outlines dual cliff-forming flows of Squirrel Springs Canyon Andesite overlain by Bearwallow Mountain Andesite. Bearwallow is much thicker on Escondido Mountain, but reconnaissance shows no clear indication of a vent here.

FIGURE 2.22. View across the Omega Basin eastward to the Datil-Sawtooth Mountains on the skyline. Note gentle southward dip slope defined by the crest of the Datil Mountains.

Datil Mountains on skyline at 3:00 to 4:00 form a broad south-sloping cuesta and locally define the gentle south dip of the Mogollon slope (Fig. 2.22). **1.0**

35.5 Ascend hill of Plio-Pleistocene alluvium (Quemado Formation) on northeast side of Mangas Creek valley. Roadcuts in buff-colored carbonate-cemented sandstone and gravels–typical Plio-Pleistocene alluvial deposits. **1.1**

36.6 Road continues to traverse broad plain underlain by Plio-Pleistocene alluvial sands and silts of the Quemado Formation. These deposits accumulated in the Omega Basin, the easternmost of several paleovalley/basin systems in western New Mexico that are incised into Fence Lake and older strata in the upper reaches of tributaries to the ancestral Little Colorado River. The deposits in the westernmost basin (Cow Springs Basin; see below) are Pliocene and are demonstrably correlative with the upper part of the Bidahochi Formation. However, some of the alluvium in the Red Hill basin to the east is Pleistocene (McIntosh and Cather, this volume) and thus postdates the end of Bidahochi deposition in Arizona. The Omega basin differs from basins to the west in that it is partly of structural origin due to subsidence along the Hickman fault zone that bounds the basin on the east. The western basins appear to be largely erosional in origin. **1.6**

FIGURE 2.21. View to north across Omega Basin to area southwest of Adams Diggings, where cuestas capped by Fence Lake Formation appear to define a breached anticline.

38.2 Stop Sign, Junction with US-60. **Turn left onto US-60. 1.1**

39.3 Milepost 42 at Omega (pronounced Omeega) Village limit. **0.1**

39.4 Road to Adams Diggings on right. **1.0**

40.4 Low roadcuts on right and left in sandy Quemado Formation alluvium, locally inset against the Fence Lake Formation. **0.3**

40.7 Road rises onto moderately indurated conglomeratic sandstones and gravels of the Fence Lake Formation, which are well exposed in roadcuts for next 0.2 mi. **0.5**

41.2 **STOP 1.** Slow and park along right side of highway, as directed by flag persons. Walk past borrow pit at 1:00 to top of hill for discussion of petroleum exploration wells in the Mangas-Alegres Mountain area (see Fig. 2.23 for orientation of landmarks), regional seismic reflection profiles (minipaper by Armstrong and Chamberlin, First Day road log), and the geology and hydrology of the Largo Creek drainage basin (minipaper by Newcomer, below). Borrow pit is on BLM land, but adjacent land is private property. Later, walk to roadcuts just west of here on Highway 60 and examine the base of the Fence Lake Formation (Fig. 2.24), where it rests unconformably on the lower Spears Group (Fig. 2.25). These gray sandstones and pale red mudstones are assigned to the volcanclastic unit of Largo Creek of the Spears Group (Chamberlin and Harris, this volume). **0.2**

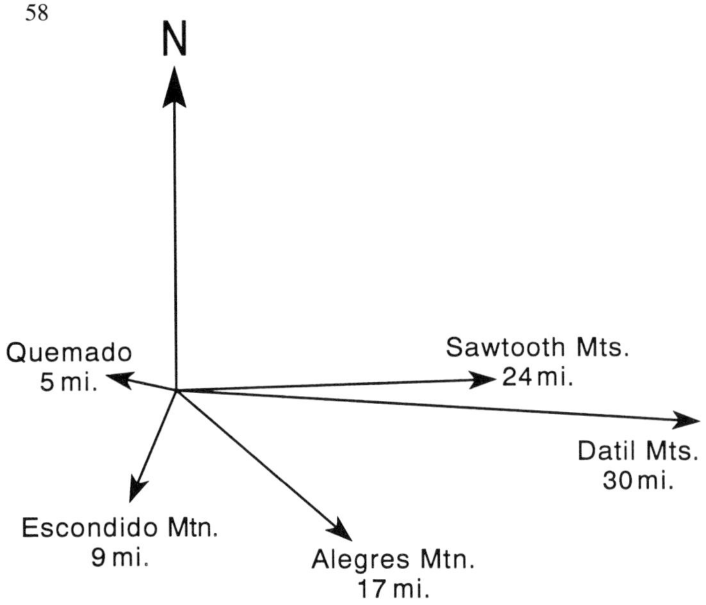

FIGURE 2.23. Panoramic index at Stop 1.

FIGURE 2.24. Basalt cobble conglomerate of basal Fence Lake Formation (upper Miocene) unconformably overlying argillaceous andesitic sandstones of the volcaniclastic unit of Largo Creek, of the lower Spears Group. Cobble imbrication suggests northwesterly paleoflow (right to left). View to east along US-60.

FIGURE 2.25. Angular unconformity consisting of horizontal Fence Lake conglomerate (upper Miocene) and associated cap of laminar calcrete overlying moderately west-tilted argillaceous andesitic sandstones and siltstones of the lower Spears Group. Small down-to-the-east normal faults cut the lower Spears Group but not the overlying Fence Lake Formation. Known regional fault zones do not project into this area; thus the significance of this pre-Fence Lake deformation is uncertain. Lower Spears Group beds are also folded and thrust faulted approximately 10 mi south of Quemado along NM-32 (see First Day road log). That deformation may be late Laramide (ca. 36-39 Ma).

HYDROGEOLOGY AND GROUND-WATER QUALITY, LARGO CREEK BASIN, CATRON COUNTY, NEW MEXICO

Robert W. Newcomer, Jr.

John Shomaker & Associates, Inc., Albuquerque, NM 87107

The Largo Creek basin is a somewhat circular drainage basin, which includes the Omega structural basin. The drainage area (Fig. 2.26) is about 660 mi^2. Its boundaries include the Continental Divide along the eastern and southern sides; the drainage divide separating Agua Fria and Largo Creeks on the west; and the drainage divide separating Frenchs Arroyo and Largo Creek on the north. Elevations range from about 10,200 to 6700 ft; the highest points are along the eastern and southern edges of the basin.

The major drainage in the basin is Largo Creek, with Rito, Escondido, San Ignacio and Mangas Creeks as tributaries. Surface water flow from the basin is to the northwest in Largo Creek. Shallow ground-water flow directions mimic surface-water flow directions, generally to the west-northwest, tributary to the Little Colorado drainage system.

Annual precipitation at Quemado and Pie Town is 15 and 9.5 in., respectively; the wettest months are during the "monsoon season" in July, August and September (Gabin and Lesperance, 1976). Precipitation is greater at higher elevations in the basin. The average annual precipitation at measuring stations above 8000 ft at several places in Catron County was greater than 21 in.

The Largo Creek basin has not been recognized by the State Engineer, who has responsibility for the supervision of the State's water resources. His jurisdiction over the State's surface water includes ground water in declared ground-water basins. A ground-water basin is generally declared when rapid development of the area might impair existing rights (Harris, 1984). The State Engineer, therefore, would have no jurisdiction concerning the use of ground water in the Largo Creek basin, except to prevent waste. The principal uses of ground and surface waters in the area are livestock watering, domestic supply and some irrigation.

The Largo Creek drainage basin includes the Omega structural basin, which consists of a down-dropped block within and west of the Hickman fault zone. The Hickman fault zone has experienced Laramide, right lateral reverse fault movement and younger Neogene normal movement (Chamberlin et al., 1994). The Eocene Baca Formation and Eocene to Oligocene volcanic and volcaniclastic rocks crop out within the drainage basin.

Little is known of the hydrogeology of the basin area. Myers (1992) described the hydrogeology of San Agustin Coal Area and the area southeast of it, which includes the northern part of the Largo Creek basin. Stone (this volume) summarized the hydrogeology of the Nations Draw area, in the San Agustin Coal Area. These areas have been studied primarily due to active exploration and development of coal resources.

The Fence Lake Formation unconformably overlies the Baca Formation in the northwestern part of the basin and overlies the Spears Group volcaniclastic rocks in the central and southern part of the Largo Creek basin. Spears Group volcaniclastic rocks were formerly assigned to the Datil Group; see Cather et al. (this volume) for revisions to Tertiary stratigraphic nomenclature in western New Mexico.

The Fence Lake Formation is a fluvial boulder to cobble conglomerate with gravelly sandstone in the upper part. The clasts are dominantly upper Oligocene basalt plus sparse Oligocene tuff and Eocene dacitic sandstones derived from the lower Spears Group near Pie Town (Chamberlin et al., 1994).

The aquifers in the Largo Creek basin are in permeable units of the Baca Formation, Spears Group and the lower part of the Fence Lake Formation. Depths to water in wells completed in these units range from within a few feet of the surface, along Rito and Largo Creeks, to greater than 200 ft at higher elevations between drainages. Quaternary alluvial fill may contain water locally, along the bottoms of major

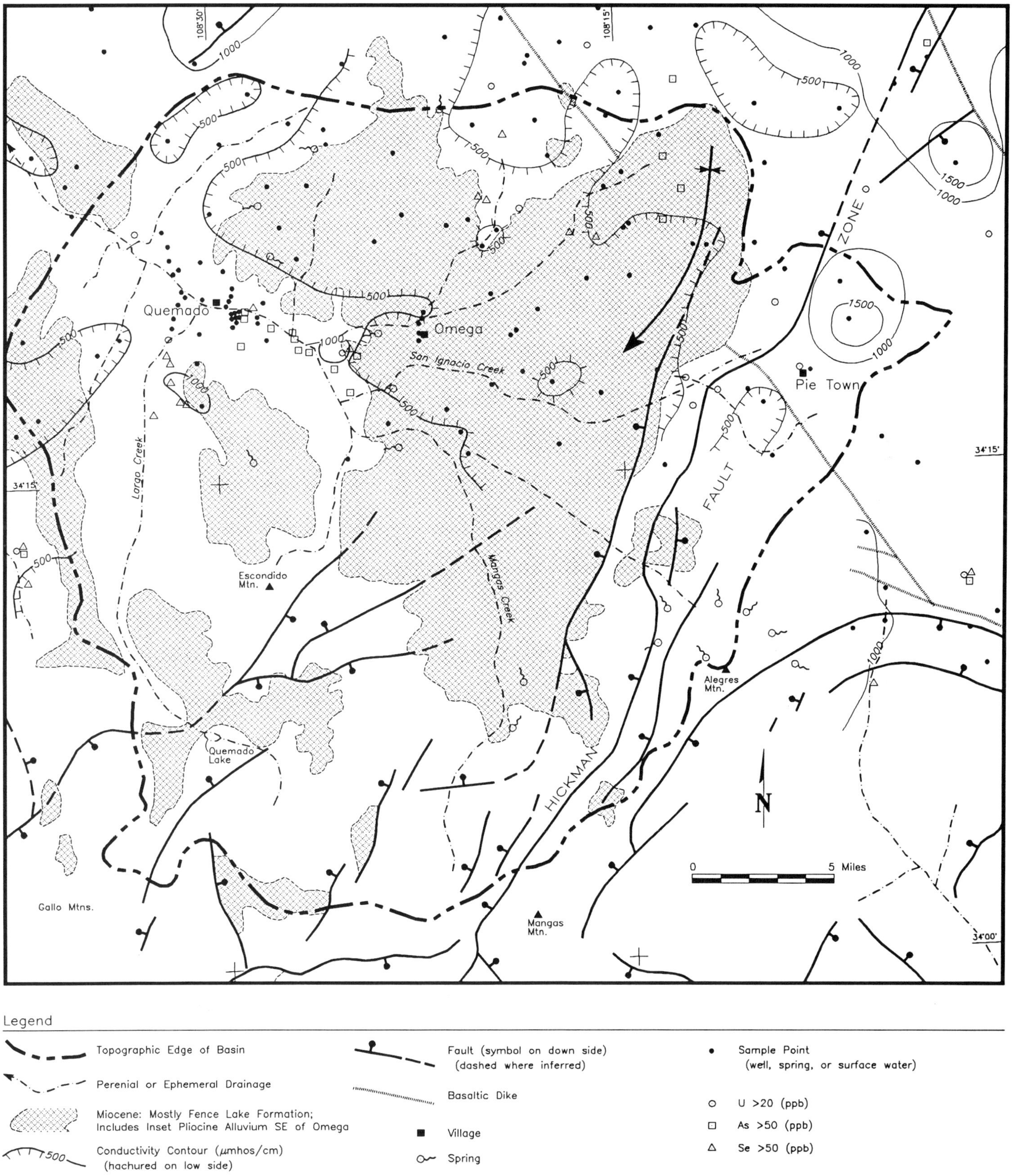

FIGURE 2.26. Map of the Largo Creek drainage basin showing geologic structure, outcrop of the Fence Lake Formation, water-sample points and water-quality data. Geology is modified after Chamberlin et al. (1994); water-quality data is from Morgan (1980).

drainages; water levels probably fluctuate seasonally in response to precipitation events and snow melt. Deeper rocks may yield ground water to wells, but the drilling depths would be excessive, and the water quality is likely to be poor.

The Fence Lake Formation and Quaternary alluvium appear to be the most important aquifers. The saturated thickness of these units vary from 0 to greater than 200 ft, and are significant in the synclinal structure west of the Hickman Fault zone, and in a down-dropped fault block near Quemado Lake (Fig. 2.26). In most other areas, the Fence Lake Formation is above the regional water table.

The shallow aquifers in the Fence Lake Formation and Spears Group appear to be recharged primarily by infiltration through the overlying soils and Quaternary alluvium. Some recharge occurs directly to these units where they crop out. As shown in Myers (1992, fig. 9), water-level contours of the potentiometric surface of the aquifer in the Datil Group (now Spears Group) indicate an area of recharge about 10 mi northeast of Quemado. The "Datil Group" aquifer of Myers probably includes unrecognized saturated zones in the Fence Lake Formation, east of Omega. Other major areas of recharge are near the eastern and southern margins of the basin, in areas that receive greater amounts of precipitation.

Ground-water flow in the aquifer(s) in the Baca Formation, Spears Group, Fence Lake Formation and Quaternary alluvium is generally to the west and northwest, parallel to the surface drainages (Myers, 1992). Discharge from the Fence Lake Formation and Quaternary alluvium is evident as springs, which are typically near or above the contact with the underlying Baca Formation and Spears Group. This suggests that these rocks are less permeable than overlying Fence Lake strata and the younger alluvium. The steepening of contours of water levels in the Baca Formation, west of Quemado, reflect a lower hydraulic conductivity of these rocks.

The nature of ground-water quality in the Largo Creek basin was inferred from a study by Morgan (1980), which involved a detailed geochemical survey of well waters in the area as part of a hydrogeochemical and stream-sediment sampling program to evaluate potential uranium resources. Several hundred water samples were collected from wells in the Quemado area and analyzed for uranium and 23 other elements, together with alkalinity.

Ground water in the Fence Lake Formation in the northern half of the basin mostly has chemical conductivities less than 500 micromhos/cm. According to analyses in Myers (1992), the ground water is a sodium-bicarbonate type. Water-quality data was lacking for the southern half of the basin. Ground water from underlying Spears Group and Baca Formation, for the most part, had higher conductivities, 500 to more than 1,000 micromhos/cm (Fig. 2.26).

Elevated levels of uranium were measured in some the ground-water samples from the Baca Formation, particularly in the vicinity of the Hickman Fault zone near Pietown (Fig 2.26). According to Morgan (1980), uranium concentrations in ground water ranged from less than 0.02 to 133 ppb.

The elevated arsenic and selenium concentrations in ground water appear to be associated with volcaniclastic rocks of the lower Spears Group. Arsenic concentrations range from less than 43 to 169 ppb, and selenium concentrations range from less than 47 to 108 ppb. The lower values of the two ranges were the reported detection limits for the analyses.

The distribution of elevated concentrations of arsenic, selenium and uranium in ground waters corresponds to the areal distribution of geologic units. Arsenic and selenium are associated with andesitic sandstones in the lower Spears Group (R. M. Chamberlin, personal commun., 1994). Uranium concentrations coincide with the Baca Formation, which is known to contain uranium mineralization in the Datil Mountains region (Chamberlin, 1981).

41.4 Milepost 40. Begin descent of steep grade into valley of Rito Creek and Escondido Creek. Roadcut ahead on right exposes Fence Lake Formation unconformably overlying upper Eocene sandstones and mudstones of the volcaniclastic unit of Largo Creek. **1.2**

42.6 Road on upper Quaternary fill of Escondido Creek valley. Roadcut on left exposes older axial stream deposits (Quemado Formation, upper Pliocene to lower Pleistocene?), unconformably deposited on upper Eocene volcaniclastic unit of Largo Creek. **0.7**

43.3 Cuts on right and left in light gray andesitic sandstones and pale red mudstones of the volcaniclastic unit of Largo Creek, the lowest stratigraphic unit of the Spears Group in the Quemado region. **0.3**

43.6 Roadcut on left in tan, well-sorted pebbly sand deposited by an axial stream of late Pliocene to early Pleistocene age. Elongate hills capped by reworked Fence Lake gravels on both flanks of Rito Creek Valley are interpreted as high-energy arroyo fills graded to a late Pliocene or Pleistocene axial stream 60 to 80 feet above the road level. **0.2**

43.8 Round knob on right exposes mudstones and andesitic sandstones of the volcaniclastic unit of Largo Creek. **1.0**

44.8 Cut on left in 20 ft of weakly indurated light brown sandstones and conglomeratic sandstones of Quemado Formation. Pebble imbrications indicate westerly paleoflow of ancestral Rito Creek in the finer beds. Cobble imbrications in basal bed indicate northerly flow, presumably in a tributary arroyo. These alluvial deposits are probably late Pliocene to early Pleistocene in age. This roadcut is a reference area for the Quemado Formation (Cather and McIntosh, this volume). **0.5**

45.3 Cliffs of lower Spears Group capped by Fence Lake Formation 9:00 to 10:00. **1.3**

46.6 Roadcuts on right in buff coarse-grained sandstones and red mudstones of the upper Baca Formation (middle to upper Eocene). **0.2**

46.8 Junction, NM-36 to Fence Lake on right. **Continue straight ahead on US-60. 0.1**

46.9 Quemado Post Office on left. Quemado (Spanish, "burned") is a ranching community established about 1880 by Jose Antonio Padilla and his family, of Belen (Pearce, 1965). **0.6**

47.5 Junction with NM-32 on left. **Continue west on US-60. 0.1**

47.6 Crossing Largo Creek. Upper Quaternary alluvium forms broad, weakly dissected floor to valley. **1.0**

48.6 Junction, US-60 and NM-601. **Turn north on NM-601. 0.1**

48.7 Mariano Mesa (1:00 to 2:00) consists of Baca Formation disconformably overlain by Fence Lake Formation. The Fence Lake is 200 ft thick on Mariano Mesa (Guilinger, 1982). Flat-top mesa to south beyond Quemado is capped by Fence Lake Formation overlying volcaniclastic sandstone of lower Spears Group (Fig. 2.27). **0.7**

49.4 Tejana Mesa (11:30 -12:30) consists of capping upper Miocene basaltic lavas underlain by Fence Lake, Baca and Moreno Hill Formations. **0.8**

50.2 Pavement ends. Red and gray strata on hill at 3:00 are Baca Formation. **0.6**

50.8 Red hills at 10:00 to 12:00 are Baca Formation. Mesas on skyline 10:30 to 11:30 are capped by upper Miocene basalts that overlie Fence Lake Formation,

FIGURE 2.27. View south toward Quemado. Mesa on skyline is capped by Fence Lake Formation (middle-upper Miocene), which overlies upper Eocene volcaniclastic unit of Largo Creek.

which, in turn, overlies Baca Formation. Watergap at El Porticito ahead at 12:00. **0.8**

51.6 North-trending fault parallels road and juxtaposes pinkish-gray Baca sandstones against green and gray shales of Upper Cretaceous Moreno Hill Formation in roadcut on left. Conglomeratic deposits on north and south ends of roadcut are inset Plio-Pleistocene(?) fluvial deposits consisting largely of recycled Baca clasts. **0.9**

52.5 Baca Formation in roadcuts. **0.1**

52.6 Crossing arroyo that drains Amen Canyon. **0.2**

52.8 El Porticito fissure vent at 12:00. Note dramatic thickening of lava flows on Tejana Mesa toward fissure vent (Fig. 2.28). A recently obtained $^{40}Ar/^{39}Ar$ age for intrusive rocks at the El Porticito vent (7.92 ± 0.20 Ma; McIntosh and Cather, this volume) is somewhat older than the age obtained for a lava exposed near the south end of Tejana Mesa (6.73 ± 0.18 Ma; Dethier et al., 1986). El Porticito is one of nearly a dozen vents developed along a northeast-trending fault zone that can be traced for at least 25 mi.

53.2 **STOP 2.** Park along right side of road and walk to fissure vent at 1:00 to discuss the late Cenozoic volcanology and petrology of the Quemado area (see minipaper by Baldridge), soil development on Tejana Mesa (see minipaper by Love et al.), and a planned coal mine nearby to the northeast (see minipaper by Anderson). **Retrace route to US-60. 4.6**

FIGURE 2.28. View north toward El Porticito fissure vent (left) and Tejana Mesa (right). Note thickening of lavas visible on escarpment of Tejana Mesa.

PETROLOGIC SUMMARY OF LATE MIOCENE VOLCANIC ROCKS OF EL PORTICITO

W. S. Baldridge

Earth and Space Sciences Division, Los Alamos National Laboratory, Los Alamos, NM 87545

The lava exposed on the southeastern face of El Porticito, as well as the lowermost flow on the southern end of Tejana Mesa, exhibits numerous light-colored sill- and dike-like structures ranging in size from several centimeters to over five meters in thickness (Guilinger, 1982; Baldridge et al., 1989). The host rock contains phenocrysts of olivine and titanaugite in a holocrystalline groundmass of clinopyroxene, sanidine, and nepheline(?). Based on its mineralogy, which includes no identifiable plagioclase, the rock appears to be phonolite. The sill- and dike-like structures, which appear to be intrusions into or segregations from the host rock, are composed of coarse-grained nepheline syenite, characterized by abundant centimeter-sized clinopyroxene grains surrounded by nepheline crystals, all characterized by bostonitic texture, in which K-feldspar forms irregular interlocking laths, as well as by a poikilitic texture. The host and sill/dike rocks appear to be of similar mineralogical composition. The host rock has been dated at 7.92 ± 0.20 Ma (McIntosh and Cather, this volume).

The unusual sill- and dike-like features exposed here may have been produced as segregations directly from the enclosing host magmas, or by intrusion into the (partially molten?) host of melts from a separate, deeper magma source. If the coarse-grained nepheline syenite represents segregation of an evolved melt fraction produced by in situ differentiation, then the minerals of these segregations would be similar to but compositionally more evolved than the minerals of the host melts. The bulk composition of the syenite would be much more enriched in incompatible elements than the host. In contrast, if the syenitic structures were emplaced into the host by intrusion from below, they could simply represent the coarser-grained crystallization products of a similar (or the same) magma. Either possibility could suggest that the sill- and dike-like structures reflect the dynamic processes within a lava lake. Since, at present, data are not available to determine the exact origin of these unusual petrological features, further detailed work is required.

SOIL DEVELOPMENT WITHIN AND AT THE TOP OF THE FENCE LAKE FORMATION AT TEJANA MESA

David W. Love[1], Bruce Harrison[2] and John W. Hawley[1]

[1]New Mexico Bureau of Mines and Mineral Resources, Socorro, NM 87801; [2]Geoscience Department, New Mexico Institute of Mining and Technology, Socorro, NM 87801

The purpose of this minipaper is to call attention to some unusual soils within the Fence Lake Formation exposed on the southwest side of Tejana Mesa, with the hope that further study will follow. The Fence Lake Formation (Miocene, Barstovian and younger, Lucas and Anderson, this volume) consists of conglomeratic and sandstone units and at least one petrocalcic horizon in its designated but yet-to-be-measured-and-described type section (McLellan et al., 1982; J. Hawley, unpubl.). In exposures of the uppermost Fence Lake Formation near Quemado and at Moreno Hill, an impressive, typical stage VI petrocalcic zone (Btkm and Bkm horizons in soil nomenclature; see Hawley, 1993, for summary) caps the local mesas. The overlying horizons (such as Bw, Bt and Av horizons) appear to overlie the well-developed petrocalcic and may be of much more recent origin than the Btkm horizon. The Btkm horizon at these localities generally is light gray to creamy white (10 YR 8/1).

The soil exposures on Tejana Mesa are just northwest of El Porticito vent. They are preserved magnificently by burial beneath base-surge deposits and 6.7 Ma (Dethier et al., 1986) basalt flows that cap the mesa. The Fence Lake Formation at this locality is about 20 m thick with at least 2 m of relief at its base where cobble-bearing gravelly channels scoured into the underlying upper Baca Formation. The soils near the top of the Fence Lake Formation exhibit peculiarities in color, texture and structure (and probably mineralogy) that indicate different circumstances affected soil development here (Figure 2.29). Beneath the base surge deposits, the top of the Fence Lake Formation is described as follows:

0-4.5 cm	Av horizon; baked red (Munsell color 5 YR 5/4) prismatic sand and silt with mudcracks filled by base surge ash; small crystals (calcite?) partially fill vesicles.
4.5-12.5	Bk1 horizon; similar red, pebbly sand with 8-cm long, 2-cm wide, vertical calcium carbonate nodules.
12.5-19	Bk2 orange (reddish yellow 7.5 YR 7/6) pebbly sand.
19-35	K1 (Stage III); similar orange sand with large (to 4 cm wide), white, vertical, calcium carbonate nodules that grade down into krotovina with concave-up carbonate menisci marking burrow slumps.
35-50	Bk3 horizon; carbonate-bearing krotovina continue downward in orange pebbly sand.
50-75	K2 (Stage IV); degraded, hard, brecciated, calcium carbonate clasts and carbonate-cemented krotovina in orange sandy matrix.
75-95	K3 (Stage V); cross-cutting unit beneath; hard, brecciated clasts of calcium carbonate in "fine earth", soft, granular carbonate matrix; vestiges of underlying pipe-margin structure penetrates upward to 80 cm; white zone of K fabric continues in massive pebbly sand across top of underlying soil pipe in typical late stage III carbonate morphology.
85-100	K4; hard, brecciated clasts of calcium carbonate cut by overlying unit and by underlying vertical pipe-margin structure.
100-120(?)	K5; massive, faintly laminar gray calcium carbonate; vertical pipe-margin laminar structure better preserved.
120-200	K5; structureless, less carbonate cemented, tan, pebbly sand; soil pipes overhung and appear to connect below 200 cm.

FIGURE 2.29. Soil development in uppermost 6 ft of Fence Lake Formation near Porticito. Base-surge deposits cover soil at top of photograph. Stage III calcium carbonate accumulation at level of tape measure degrades older Stage V Btkm horizon and plugs older soil pipe formed in lower horizon.

Below 200 to 400 cm are three stacked, somewhat corroded, gray, laminar (Stage IV-V) Bkm horizons with well churned, weakly carbonate-cemented, tan, pebbly sand beneath each Bkm horizon. Below 300 cm to the top of the basal gravelly channels are multiple (at least 15) intervals of massive (well churned) tan pebbly sand with variable amounts of carbonate impregnation. These structureless units commonly cap cross-bedded sandy gravels or coarse gravel channels. At least four of the massive intervals have indistinct or incipient soil pipes ranging to 2 m in vertical dimension.

At 5.5 m below the top of the Fence Lake Formation is an extremely well indurated ledge of medium gray carbonate- (and silica-?) cemented sandstone that displays micro-laminations in the upper part. On closer inspection, however, the laminations are small concentrations of manganese and silica(?) and are not the typical pedogenic laminations seen in the upper soils. Manganese staining also coats pebbles in the gravelly units.

The complex calcium-carbonate horizons below 20 cm indicate multiple episodes of carbonate accumulation. The K1 and Bk1 horizons contain carbonate nodules that have the morphology of krotovinas, similar to those described by McDonald and Busacca (1990) for carbonate nodules forming around cicada burrows in the Palouse loess. The stacked sequence of carbonate horizons could represent either an aggrading soil profile with the carbonate horizons developing during hiatuses in the depositional record, or climatic change with increasingly more arid conditions resulting in carbonate precipitation closer to the surface through time. The thick, indurated unit below 100 cm is a well-developed Stage V petrocalcic horizon (Btkm and Bkm), complete with laminated vertical pipes, but it is degraded at the top and grayer than similar horizons elsewhere in New Mexico. The laminar-carbonate-coated pipes are reminiscent of features illustrated by Gile et al. (1981, figs. 37 and 38). However, genetic pathways producing these features at Tejana Mesa are probably more complex and reflect an age three to six times older. The degradation (brecciation, invasion of fine-grained carbonate) appears to have taken place when a stage IV carbonate accumulated in the interval from 50 to 100 cm and plugged the previously formed soil pipe.

Several lines of evidence point to at least episodic water saturation and enhanced Fe, Mn and $CaCO_3$ mobility and precipitation in the uppermost Fence Lake Formation and overlying base-surge deposits: (1) cementation and red-purple color changes upward within the base-surge deposits, (2) presence of small calcite(?) crystals in vesicles in the Av horizon and calcium carbonate nodules within 5 cm of the former surface, and (3) manganese staining in most laminar horizons and gravel units at depth. The odd colors within the base-surge deposits and particularly in the Av horizon may indicate heat as well as water (steam?) altered the iron oxides within these units.

UPPER CRETACEOUS ROCKS OF THE TEJANA MESA AREA AND THE SALT LAKE COAL FIELD

Orin J. Anderson

New Mexico Bureau of Mines and Mineral Resources, Socorro, NM 87801

Upper Cretaceous rocks exposed in the low areas surrounding Tejana Mesa are the upper part of the Moreno Hill Formation. Named by McLellan et. al. (1983) for exposures at Moreno Hill, some 14 mi northwest of here, The Moreno Hill Formation is a non-marine, coal-bearing unit that approaches 600 ft in thickness. Outcrops of the formation define the Salt Lake coal field. Sandstone and shale (or mudstone) predominate. The sandstone is generally fine grained, crossbedded, grayish orange and quartzose, although facies have been described as feldspathic litharenite or arkosic (Johansen, 1986). The mudstones, which predominate locally, consist of gray or dusky yellow to dark gray, carbonaceous facies up to 100 ft in thickness. The mudstone or shale intervals are, however, interspersed with thin siltstone or very fine-grained sandstone. Coal constitutes a mere 2 to 3% of the section, but it is low-sulfur (< 1% S) and in beds of minable thickness (see Rodgers, this volume).

Depositional environments represented by the Moreno Hill Formation are fluvial channel sandstone, and associated extensive floodplain deposits in a lower coastal plain setting. The sequence was deposited on a sandstone platform, the Atarque Sandstone, that prograded out into the Mancos seaway during middle Turonian time. Based on stratigraphic position and geometry (Fig. 2.30), the coal beds represent accumulation in paralic peat swamps.

The regression that initiated this sequence was limited to the New Mexico area of the Western Interior seaway. It was probably not associated with eustatic change in sea level. It is pre-Ferron Sandstone (Utah) and demonstrably pre-Gallup Sandstone, on the basis of molluscan fauna and stratigraphic position. Both the Ferron and Gallup (late Turonian-early Coniacian) have been related to eustatic fall of sea level and to renewed activity in the Sevier overthrust belt in Utah.

The Atarque-Moreno Hill sequence, and the partly equivalent Tres Hermanos Formation in a more seaward (northeast) direction, probably represent a response to a major climatic change in the southwestern part of the continent. The climate perhaps became much more maritime and humid with the encroachment from the east of the Western Interior seaway; maximum extent of the seaway occurred during the Greenhorn cycle of sedimentation when the shoreline coincided with a line from southwestern Utah to southwestern New Mexico. Temporally this occurred at the Cenomanian-Turonian boundary. The climatic impact of this transgression was extreme and the fluvial response resulted in the progradation of a lower coastal plain environment out into the Mancos Seaway. The process was self-limiting in the sense that withdrawal of the sea allowed the return of a

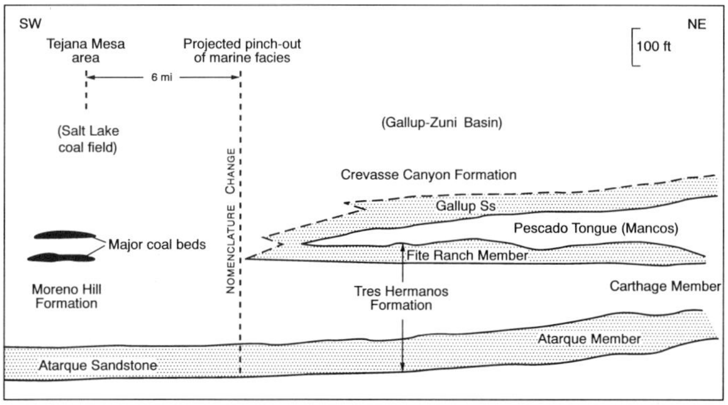

FIGURE 2.30. Stratigraphic diagram of Upper Cretaceous strata in the Salt Lake coal field.

more moderate climate in the source area, with an attendant decrease in runoff and sediment transport. This in turn was reflected in an interruption in sediment supply to the shoreline, and progradation ceased. Lithostratigraphic expression of this progradational event (shoreline regression) is recognized as far eastward as the Sierra Blanca-Capitan area in central to southeastern New Mexico. Further evidence of the interruption in clastic supply at this time is the presence of a calcareous interval in the Mancos Shale, called the Juana Lopez Member, in a stratigraphic position correlative with subsequent (late Middle Turonian) shoreline transgression.

Thus, the lithostratigraphy, timing and extent of the Moreno Hill-Atarque progradation is relatively well understood, even as to first cause. It is, however the subsequent transgression that complicates the stratigraphy and nomenclature of Turonian rocks at Tejana Mesa and areas to the immediate northeast. This transgression brought the strandline back into west-central New Mexico, into the southernmost part of the Gallup-Zuni basin to the Fence Lake area. Projection of the high-stand shoreline along depositional strike (southeast) brings marine water to within 6-8 mi northeast of Tejana Mesa. Thus, Tejana Mesa was within the coastal strip where paralic peat swamps were likely to exist. Coal resource evaluation work by the New Mexico Bureau of Mines and Mineral Resources (Campbell and Roybal, 1984; Anderson, 1987; Campbell, 1989), and the Salt River Project have determined that the thickest coals in this area occur in pods whose axes parallel this paleoshoreline. Stratigraphically these thicker coal beds are correlative with the marine strata to the northeast, which represent high-stand conditions.

The presence of the marine facies to the northeast warranted a separate nomenclature (Fig. 2.30) to properly address the lithostratigraphic expression of a major change in depositional environments (marine encroachment into a coastal plain environment). In areas where the marine influence is present and mappable the name Tres Hermanos Formation is applied to the regressive-transgressive wedge of sediments that is partly equivalent to the Moreno Hill-Atarque sequence (Hook et al., 1983). The Atarque Sandstone is the direct equivalent of the basal sandstone member of the Tres Hermanos Formation, which is called the Atarque Member. The overlying coastal-plain unit (Carthage Member) is equivalent to the lower part of the Moreno Hill Formation. The uppermost Tres Hermanos unit is a fossiliferous, littoral sand of highly variable thickness called the Fite Ranch Member. The exact point of the landward pinchout of the Fite Ranch is not known because the area where this feature would crop out has been truncated by Tertiary erosion; subsurface information has been of little help in defining the pinchout. Based on rate of thinning of the overlying Pescado Tongue of the Mancos Shale, Anderson (1987) projected a landward pinchout of the offshore mudstone facies near the Atarque area (9 mi north of Fence Lake). Depending on width of the littoral zone and shoreline trend the seaway could have been within 6 mi of Tejana Mesa during latest middle to early late Turonian time. An additional significant aspect of this landward pinchout is that it represents the point at which the Gallup regression began. The Gallup Sandstone, a more widespread regressive littoral sand, is a well-known, cliff-forming unit in the southern San Juan basin. The Gallup and its associated, overlying coastal plain depositional sequence, the Crevasse Canyon Formation, represent a response to a eustatic fall of sea level and/or renewed tectonism in the Sevier Orogenic belt in Utah.

The arbitrary line between the two sets of nomenclature, Atarque-Moreno Hill in the Tejana Mesa area and the Salt Lake coal field, versus Tres Hermanos Formation-Pescado Tongue-Gallup-Crevasse Canyon Formations to the northeast, is best drawn at the prominent northwest-trending dike system that runs from just west of Pie Town to the abandoned community of Techado. From there northwestward to the Atarque area the presence of the Pleistocene North Plains basalt flow separates the areas in which the contrasting nomenclature is used. Arbitrary lines between contrasting sets of nomenclature for correlative rocks is a necessity in areas with significant lateral facies change. These areas commonly include intertongued marine-nonmarine rocks such as we find in the Upper Cretaceous in west-central New Mexico and adjacent areas.

57.8 Junction with US-60. **Turn right (west) on US-60. 0.1**

57.9 Low-relief upper Quaternary alluvium in valley bottom is inset against older alluvium (Plio-Pleistocene Quemado Formation) along valley margin ahead. **0.2**

58.1 Road ascends older valley-fill deposits (Quemado Formation), which are exposed in roadcuts next 1.1 mi. **1.2**

59.3 Roadcut exposes Baca Formation on left. Quemado Formation is erosionally inset against the Baca Formation here. **0.1**

59.4 Quemado Formation in right roadcut. **0.4**

59.8 Baca Formation in roadcuts. **0.3**

60.1 Light-colored sandstone of lower Spears Group (upper Eocene) overlain by conglomerate of the Fence Lake Formation (middle to upper Miocene) at 1:00. Highway curves to left and enters Armstrong Canyon. **0.3**

60.4 Exposures of Spears Group at 9:00. **0.1**

60.5 Exposures of Spears Group in roadcuts next 1.5 mi. **1.7**

62.2 Highway crosses covered contact between Spears Group and Fence Lake Formation. **0.1**

62.3 Picnic area on right. **0.4**

62.7 Gravels of Fence Lake Formation capped by prominent calcareous paleosol in roadcuts (Fig. 2.31). **0.1**

62.8 Microwave tower on right. Highway traverses remnant of Fence Lake constructional surface next 0.4 mi. This surface is part of a drainage divide between two Quemado-age paleovalleys (Largo Creek Basin to east; Red Hill Basin to west; see Cather and McIntosh, this volume). **0.7**

63.5 Enter closed basin floored with Quaternary alluvium. Fox Mountain on skyline at 10:00 consists of volcaniclastic deposits of Spears Group overlain by andesite of Dry Leggett Canyon. **0.3**

63.8 Quaternary alluvium in roadcuts. **0.7**

64.5 Road ascends western margin of closed basin. **0.5**

65.0 Roadcuts expose calcareous paleosol developed on Fence Lake Formation. **0.2**

65.2 Highway traverses poorly exposed Fence Lake Formation. **0.5**

65.7 Fence Lake Formation in roadcuts. **0.1**

65.8 Approximate location of contact between Fence Lake Formation and underlying Spears Group. **0.3**

66.1 Roadcuts in Spears Group. Fence Lake Formation caps hill to right. **0.3**

66.4 Highway descends into broad, broken lowland ahead. This is a good vantage point to view part of a large, heretofore unrecognized, erosional basin (or broad paleovalley) of Plio-Pleistocene age in which large volumes of sedimentary and volcanic rocks have accumulated (Cather and McIntosh, this volume). The basin (Red Hill Basin) is geographically quite large. The highlands along the southern boundary occupies much of the skyline to the south and southwest, including the Gallo Mountains (9:00), Fox Mountain (9:30) and Black Peak (10:00). Part of the northern boundary of the basin is visible along the northwestern skyline between about 1:00 and 2:00. The basin (paleovalley) exited to the northwest near Red Hill and Cimarron Mesa. The sedimentary rocks in the basin are partially equivalent to the upper Bidahochi Formation and have been termed the Quemado Formation (Cather and McIntosh, this volume). **0.4**

66.8 Roadcut on left exposes Quemado Formation. **0.3**

67.1 Quemado Formation sandstone and siltstone in roadcuts. **0.5**

67.6 Quemado Formation sandstone in roadcuts, next mile. **1.0**

68.6 Quemado Formation containing minor basaltic gravels exposed in roadcuts. **0.4**

69.0 Most hills in the middle distance 9:00 to 11:00 consist of Plio-Pleistocene basaltic volcanic rocks that overlie or interfinger with the Quemado Formation. **0.2**

69.2 Sands and gravels of Quemado Formation in roadcuts contain weakly developed calcareous paleosoil (Fig. 2.32). These roadcuts have been designated a reference area for the Quemado Formation by Cather and McIntosh (this volume). **0.9**

70.1 Ridge 12:30 to 2:00 consists of Baca Formation overlain by Fence Lake Formation and capped by upper Miocene(?) basalt. This ridge was eroded approximately

FIGURE 2.31. Roadcut exposure of caliche-coated pebbles of Fence Lake Formation.

FIGURE 2.32. Calcareous paleosol in sandstone and siltstone of Quemado Formation in north roadcut of US-60.

to its present morphology before deposition of the Quemado Formation, which laps around the hill on three sides. The low-relief area in the foreground 12:00 to 3:00 is underlain by Quemado Formation and intercalated basalts. **0.1**

70.2 Quemado Formation in left roadcut. **0.2**

70.4 Quemado Formation and intercalated basalt in roadcuts. This basalt has yielded an ^{40}Ar/^{39}Ar age of 1.51 ± 0.01 Ma (McIntosh and Cather, this volume). **0.3**

70.7 Milepost 20. **0.1**

70.8 Quemado Formation in roadcuts contains abundant well-rounded quartzite, chert, and other pebbles recycled from the Eocene Baca Formation. Such pebbles are widespread in the Quemado Formation near the northern margin of the basin. **0.4**

71.2 Red beds exposed beneath basalt at 2:00 are Baca Formation. **0.1**

71.3 Quemado Formation in roadcuts. **0.4**

71.7 Quemado Formation in roadcuts overlain by thin mantle of Quaternary cinders. **0.4**

72.1 Upper Miocene(?) (pre-Quemado Formation) basalt overlies lower Spears Group on ridge and in slumped blocks 9:30 to 11:30. **0.9**

73.0 Quemado Formation sandstone and mudstone in roadcuts. **0.3**

73.3 Weakly dissected Quaternary alluvium forms floor of valley. Low hills 8:00 to 11:00 are poorly exposed Quemado Formation. **0.2**

73.5 Exposures of Baca Formation red beds 1:00 to 2:00. **0.4**

73.9 Quemado Formation sandstone and conglomerate in roadcuts. **0.5**

74.4 Cimarron Mesa (12:30-2:00) consists of upper Cretaceous Moreno Hill Formation and Baca/Eagar Formations overlain by Fence Lake gravels and capped by upper Miocene basalt. Preliminary petrographic study of volcanic cobbles, which compose ~ 15% of clasts in the Baca/Eagar Formations here, suggests that many are derived from the Proterozoic metavolcanic terrane of central Arizona (Fig. 2.33). The basal basalt flow along the southwest margin of Cimarron Mesa yielded a ^{40}Ar/^{39}Ar age of 6.05 ± 0.02 Ma (McIntosh and Cather, this volume). The area around Cimarron Mesa contains the poorly understood interfingering (?) of volcanic-bearing sandstone and conglomerate of the Eagar Formation and the laterally equivalent, volcanic-free Baca Formation (Cather et al., this volume). Cimarron Mesa (Fig. 2.34) forms an "island" within the Quemado Formation, which surrounds and laps against it on all sides. Road traverses low-relief surface developed on Quemado Formation next 2.0 mi. **0.3**

74.7 Milepost 16. **1.1**

75.8 Roadcuts in Quemado Formation capped by Quaternary cinders. **0.4**

76.2 Quarry for volcanic cinders at 9:00 (Fig. 2.35). Cinders were erupted from a Quaternary cinder cone with a maar about 0.6 mi southwest of here (Quemado crater of Hoffer and Corbitt, 1989). **0.2**

76.4 **Turn left** on road to Luna. **0.1**

76.5 Cinder quarry at 9:00. **0.4**

76.9 Fox Mountain (11:00) is a horst of andesite of Dry

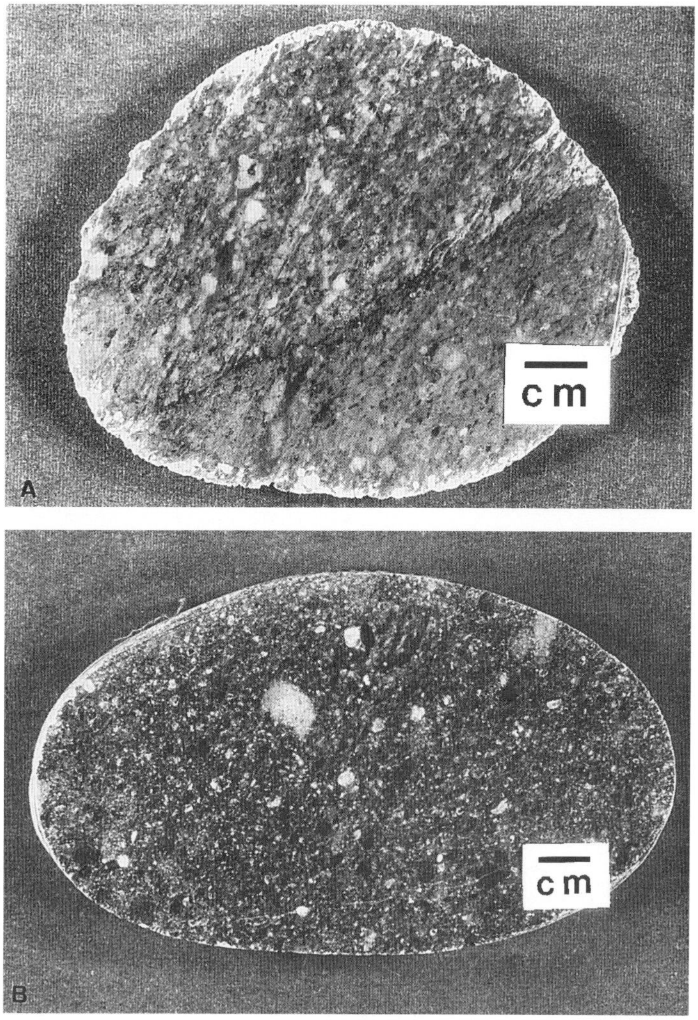

FIGURE 2.33. Slab of siliceous metavolcanic cobbles collected from the Eagar Formation on the east flank of Cimarron Mesa, (SE¼ sec 33, T1N, R19W) A, foliated and sheared metarhyolite lava that contains phenocrystic quartz and recrystallized potassium feldspar. B, rhyolitic metaignimbrite containing dark pumice fiamme (top, right of center), small quartz eyes and angular rock fragments. The Proterozoic metavolcanic terrane in central Arizona is the most likely source.

FIGURE 2.34. View north of Cimarron Mesa from north rim of Quemado crater. Cimarron Mesa is capped with late Miocene (6.05 ± 0.02 Ma) vents and basalts, which overlie Cretaceous and Eocene strata. Note cinder cone within unnamed maar in foreground.

FIGURE 2.35. Quarry developed in Quaternary cinder deposit. Cinders were derived from cone about 0.6 mi to southwest (Quemado crater).

Leggett Canyon overlying volcaniclastic strata (gray cliffs) of the Spears Group. Isolated exposures of Baca Formation occur northeast of Fox Mountain along Demetrio Creek. Cerro la Mula (Fig. 2.36) at 11:45 is a late Pliocene eruptive center that appears to lie along the same fracture system as the vent at El Porticito (Stop 2). A flow on the south flank of Cerro la Mula has been dated 1.947 ± 0.052 Ma (McIntosh and Cather, this volume). **0.4**

77.3 Turn right on two-track road. **0.4**

77.7 Cinder cones at 1:00 and 2:30. **0.2**

77.9 Quemado crater, a Quaternary maar/cinder cone complex, on left. A basalt flow exposed in the southwest wall of the crater predates the maar and has been dated 1.55 ± 0.01 Ma (McIntosh and Cather, this volume). **0.1**

78.0 **STOP 3.** Park on left to discuss Quemado crater (Fig. 2.37) and fracture controls on late Cenozoic eruptive centers. **Retrace route to US-60. 1.6**

79.6 Stop sign. **Turn left (west) on US-60. 0.5**

80.1 Quaternary cinder cones at 9:00 (see Hoffer and Corbitt, 1989). **0.2**

FIGURE 2.36. View south of Cerro la Mula, a late Pliocene eruptive center.

80.3 Quemado Formation and Quaternary cinders in roadcuts. **0.3**

80.6 Slight westerly tilts on Plio-Pleistocene(?) basalt flows 9:30 to 11:00 are related to nearby north-northeast-trending normal faults, which are part of a major fault zone of regional extent. These faults commonly control the location of late Tertiary-Quaternary vents throughout this area. **0.9**

81.5 Sandstone and pebbly sandstone of Quemado Formation in roadcuts. **0.3**

81.8 Escudilla Mountain near Alpine, Arizona on skyline at 10:30. **0.7**

82.5 Red Hill cinder cone on skyline at 3:00 is possibly the youngest volcanic feature in the Red Hill-Quemado volcanic field (Fig. 2.38). A basalt that flowed westward from the base of the cone has been dated by $^{40}Ar/^{39}Ar$ geochronometry at 0.071 ± 0.012 Ma (McIntosh and Cather, this volume). **0.1**

82.6 Sandstone and conglomerate of Quemado Formation in roadcuts. Much of the gravel in the Quemado has been reworked from the Fence Lake and Eagar Formations. **0.3**

82.9 Milepost 11. **0.1**

83.0 Quemado Formation in roadcuts. The coarseness of the Quemado here reflects the proximity of the western basin margin (on skyline ahead). **0.2**

83.2 Red Hill, New Mexico. Much of the low-lying topography south and southeast of Red Hill consists of Quemado Formation, which interfingers with and is overlain by Plio-Pleistocene basalts. **0.6**

83.8 Gravel pits at 2:00 to 3:00 expose Eagar Formation and a thick calcrete (Fig. 2.39). Low, rolling topography in foreground is Quemado Formation. **0.8**

84.6 Basalt interbedded with Fence Lake Formation in roadcuts. A sample from this roadcut yielded a $^{40}Ar/^{39}Ar$ age of 5.20 ± 0.03 Ma (McIntosh and Cather, this volume). About 6 mi to the southeast, apparently correlative basalts have been dated at 6.4 ± 0.3 Ma (Baldridge et al., 1989). These upper Miocene basalts and intercalated gravels of the Fence Lake Formation form a paleodivide between the Quemado-age Red Hill Basin to the east and the Cow Springs Basin to the west. **0.5**

85.1 Rolling topography from 8:00 to 4:00 is underlain by upper Miocene basalts and intercalated Fence Lake Formation. Exposures of the Fence Lake Formation extend nearly 12 mi to the north-northwest where they overlie Upper Cretaceous strata. These Fence Lake exposures have been incorrectly mapped as Baca Formation by some previous workers. **0.9**

86.0 Rest area on left. Roadcuts expose gravels of Fence Lake Formation. Upper Miocene (?) basalts cap ridges 1:00 to 3:00. **0.3**

86.3 Entering Cow Springs Basin. Low-lying area at 9:00 to 12:00 is Quemado Formation and interbedded Pliocene basalts. US-60 traverses northeast margin of the Cow Springs Basin next 4 mi. **1.6**

87.9 Milepost 6. Volcaniclastic sandstones in roadcuts. Correlation of these sandstones is ambiguous; they may represent either a fine-grained facies of the Fence Lake Formation or part of the lower Spears Group. **1.1**

89.0 Escudilla Mountain at 10:00. White Mountains at 11:00. Microwave tower on left. **0.7**

FIGURE 2.37. Aerial view of the Quemado crater, looking south. County road to Luna passes left (east) of the crater rim and US-60 passes from lower left to center right. A narrow graben trends south-southwest from the west rim of the crater. An early Pliestocene basalt flow is offset by the graben. A second, unnamed crater is next to US-60 in foreground. Fox Mountain, on the middle skyline, is capped by the andesite of Dry Leggett Canyon (upper Eocene; ~ 34 Ma).

89.7 Basalt-capped peak at 1:30 exposes 30 ft of volcaniclastic conglomerate of the Fence Lake Formation on south slope (Fig. 2.40). Correlation of the buff volcaniclastic sandstones beneath the Fence Lake is ambiguous; these sediments may pertain to the lower Spears Group or to an anomalously fine-grained facies within the Fence Lake Formation. The basalt that caps the Fence Lake conglomerate here was dated 6.03 ± 0.02 Ma using $^{40}Ar/^{39}Ar$ geochronometry (McIntosh and Cather, this volume). **0.3**

90.0 Quemado Formation overlain by Pliocene(?) basalt in left roadcut. **0.6**

90.6 Mesa Redonda on skyline at 11:00 to 1:00 formed a "paleoisland" within the Cow Springs Basin, which exited around both sides of the mesa to the northwest. **0.5**

91.1 Road traverses low-relief landscape underlain by Pliocene basalts and Quemado Formation. **1.5**

92.6 Roadcuts in basalt overlying variegated Quemado Formation beds (Fig. 2.41). The uppermost basalt in the south roadcut yielded a $^{40}Ar/^{39}Ar$ age of 2.46 ± 0.04 Ma (McIntosh and Cather, this volume). This is one of a series of related flows that cap the Quemado Formation type section 6 mi to the north as well as the Bidahochi Formation near Mesa Parada in Arizona about 22 mi to the north-northwest. A basalt that caps a prominent calcareous paleosol developed on the top of the Bidahochi 1.85 mi north-northwest of Mesa Parada has been dated at 2.37 ± 0.03 Ma by the $^{40}Ar/^{39}Ar$ method (McIntosh and Cather, this volume). These relations indicate that the Quemado Formation is at least in part equivalent to the Bidahochi Formation, although the lower Bidahochi is older (late Miocene) and some of the Quemado beds to the east are Pleistocene and appear to post-date the end of Bidahochi deposition. **0.3**

92.9 Roadcut on right in Eagar Formation, which comprises lower part of Mesa Redonda to north. Mesa is capped by Fence Lake Formation and upper Miocene(?) basalt. **0.3**

93.2 Thin sequence of Quemado Formation overlies Eagar Formation in right roadcuts. Quemado is, in turn, overlain by Pliocene basalt, both of which are inset against "paleoisland" mesa on skyline to north (2:00 to 4:00). Mesa-capping Pliocene basalts to west and south are locally underlain by Quemado Formation. **0.4**

93.6 Eagar Formation in right roadcuts. **0.3**

93.9 Rest area on left. Mesa Redonda on skyline at 3:00 is capped by upper Miocene(?) basalt that overlies volcaniclastic conglomerate of Fence Lake Formation

FIGURE 2.38. Northward view of Red Hill cinder cone.

which, in turn, disconformably overlies pebbly volcaniclastic sandstone of the lower Spears Group. **0.1**

94.0 Entering Arizona. With the exception of the upper Cenozoic volcanic rocks of the White Mountains and Springerville volcanic fields, the geology of the remainder of today's field trip has been little studied. As such, the geologic interpretations presented in the remainder of this road log should be regarded as preliminary. **0.7**

94.7 Roadcuts expose variegated Upper Cretaceous mudstone disconformably overlain by conglomerate and sandstone of Eagar Formation (Eocene) and Pliocene basalt. The base of the Eagar here is very coarse; slabs of Cretaceous sandstone about 3 ft long are present in the right roadcut (Fig. 2.42). The top of the Cretaceous section along US-60 in eastern Arizona is highly oxidized and commonly displays reddish hues, particularly in the sandstones. Because of this anomalous coloration, the upper part of the Cretaceous section in this area has been confused with Eocene strata by some workers (see following minipaper). Chamberlin (1981b, 1989) described similar reddened Cretaceous rocks beneath the Baca Formation in west-central New Mexico and ascribed their coloration to the effects of early Tertiary oxidation in a deeply weathered zone prior to the onset of late Laramide sedimentation. Sparse pebbles of Laramide (?) andesite porphyry are present in the basal Eagar beds here. **0.5**

FIGURE 2.39. Stage VI calcrete developed on siliceous cobble conglomerates of the Eagar Formation that are exposed in the north rim of a borrow pit approximately 1.0 mi northwest of Red Hill. This thick calcrete is probably correlative with a late Miocene calcrete commonly found at the top of the Fence Lake Formation.

LOCAL MISAPPLICATION OF THE TERM EAGAR TO UPPER CRETACEOUS STRATA IN EASTERN ARIZONA

Steven M. Cather

New Mexico Bureau of Mines and Mineral Resources, Socorro, NM 87801

Sirrine (1956) named the Eagar Formation and designated the type section in Antelope Valley about 16 km northeast of Springerville, Arizona (N½ sec. 28, T10N, R30E). The type section consists of 473 ft (144 m) of mostly red sandstone and conglomerate. At the type section, the basal part of the Eagar section consists of conglomerate and pebbly sandstone that overlies the Upper Cretaceous Mancos Shale. The top of the Eagar is not exposed at the type section, but elsewhere grades transitionally up-section into mid-Tertiary volcaniclastic sedimentary rocks (termed Spears Group by Cather et al., this volume).

During the course of his dissertation mapping, Sirrine (1956) extended the Eagar Formation east and south of the type locality. In the type area and in other parts of his study area where Upper Cretaceous rocks were not anomalously reddened, Sirrine placed the base of Eagar at the Cretaceous-Tertiary unconformity. In his geologic mapping near the Arizona-New Mexico line in the vicinity of US-60, however, Sirrine (1956) erroneously included a non-conglomeratic sequence of Upper Cretaceous red sandstones and variegated, locally lignitic, mudstones in the basal part of the Eagar Formation. Samples collected from a lignitic coal and overlying shale within this sequence yielded spores and pollen of Late Cretaceous age (R. Y. Anderson, written commun. in Sirrine, 1956). The age implications for these pollen and spores, together with the transitional contact relations with the overlying volcaniclastics, led Sirrine to assign a Late Cretaceous-Paleocene age to the Eagar Formation. Subsequent to the work of Sirrine (1956), paleontological and radiometric studies have indicated an Eocene age for Eagar-equivalent units (Baca and Mogollon Rim Formations) in eastern Arizona and west-central New Mexico (see summaries in Lucas, 1983; Cather and Johnson, 1984; Potochnik, 1989). No further age data have been obtained for the Eagar Formation.

The mapping error near the Arizona-New Mexico line and resultant incorrect age assignment for the lower Eagar by Sirrine has been

FIGURE 2.40. Basalt capping this butte overlies Fence Lake conglomeratic sandstone in exposure on left. Other sandstones below the conglomerates may represent the Fence Lake Formation or Spears Group.

FIGURE 2.41. Late Pliocene basalt flow overlies Quemado Formation at Mile 92.6.

promulgated for many years in the regional literature. As recently as 1971, Snyder attributed a possible Late Cretaceous age to basal Baca-Eagar beds in western New Mexico and eastern Arizona based on Sirrine's work. B. D. Johnson (1978, p. 5) was the first to note the erroneous inclusion of Upper Cretaceous strata in the Eagar Formation. A road log along US-60 just west of the Arizona-New Mexico state line (Anderson et al., 1987) noted the presence of unusually colored Upper Cretaceous strata in the roadcuts; these rocks were mapped as Eagar by Sirrine (1956). Anomalously red Upper Cretaceous rocks have been noted in other localities in western New Mexico beneath the basal Tertiary unconformity (Chamberlin, 1981b; Guilinger, 1982). The red coloration was interpreted by Chamberlin (1981b) to have formed in an oxidized, lateritic weathering zone prior to the onset of Baca-Eagar sedimentation in the Eocene.

In April, 1994, I visited the type section of the Eagar Formation and remapped a small part of Sirrine's study area (Fig. 2.43), being careful not to be misled by anomalous coloration. The Upper Cretaceous Moreno Hill Formation crops out in the lower elevations along Coyote Creek, and is disconformably overlain by the basal conglomerate or conglomeratic sandstone of the Eagar Formation. The unconformity beneath the Eagar shows as much as 100 m of relief within the map area, and is locally angular. Elevation differences between Upper Cretaceous strata along Coyote Creek (all of which was mapped as Eagar by Sirrine, 1956) and volcaniclastics of the Spears Group near the state line indicate an approximate maximum thickness for the Eagar of only 120 m. I also revisited Sirrine's (1956) pollen locality; it is

FIGURE 2.42. Boulder of Upper Cretaceous sandstone marks conglomeratic base of Eagar Formation at Mile 94.7.

clearly within what I regard as the Upper Cretaceous Moreno Hill Formation (Fig. 2.43). Chamberlin (1981b) and Guilinger (1982) described other Late Cretaceous pollen assemblages from within anomalously reddened Mesaverde Group outcrops beneath the basal Tertiary unconformity in areas nearby to the east.

I have found the following lithologic criteria are useful in distinguishing red Upper Cretaceous strata from the overlying Eocene rocks in west-central New Mexico and eastern Arizona. *Presence of conglomerate:* Conglomerate is common in the Eocene strata, whereas Cretaceous rocks, with the exception of intraformational mudclasts, are not coarser than very coarse sandstone. *Coloration of mudstones:* Cretaceous mudstone units are commonly variegated and preserve remnants of "typical" Cretaceous gray to maroon coloration (Chamberlin, 1981b) and lignite. The great majority of Eocene mudstones are uniformly red. *Sandstone characteristics:* Cretaceous sandstones are typically more thickly bedded and more texturally uniform than Eocene sandstones. Cretaceous sandstone beds typically contain large-scale trough cross-stratification, whereas small- to medium-scale trough and tabular cross-stratification and horizontal stratification is more common in Eocene sandstones.

95.2 Anomalously red Upper Cretaceous sandstone of Moreno Hill Formation in quarry at 9:00. **0.3**

95.5 Variegated mudstone and red sandstone of Moreno Hill Formation in roadcuts (Fig. 2.44). **0.1**

95.6 Crossing Coyote Creek. Many of the red rocks in the lower elevations along Coyote Creek 1:00 to 4:00 are Upper Cretaceous. **0.3**

95.9 Milepost 400. **0.1**

96.0 Roadcuts expose red conglomerate, sandstone and mudstone of Eagar Formation. **0.3**

96.3 Conglomerate, sandstone, and mudstone of Eagar Formation in roadcuts (Fig. 2.45). **0.6**

96.9 Milepost 399. Escudilla Mountain at 9:00. White Mountains 10:00 to 1:00. **0.6**

97.5 Roadcuts expose red mudstone and grayish pink conglomerate and conglomeratic sandstone of Eagar Formation, next 1.5 mi. **2.1**

99.6 Quaternary alluvium exposed in roadcuts. **1.2**

100.8 Red sandstone and mudstone of Eagar Formation exposed in roadcuts. **0.7**

101.5 Roadcuts expose red mudstone of Eagar Formation overlain by buff sandstone and conglomerate of the Quemado(?) Formation. These are, in turn, overlain by a basalt flow dated 3.67 ± 0.12 (whole-rock K/Ar age, Laughlin et al., 1980). A prominent calcareous paleosol intervenes between the Quemado(?) and the basalt (Fig. 2.46). The buff-colored strata beneath the basalt are tentatively correlated with the upper Bidahochi Formation on the basis of their position in the landscape, lithology, and the presence of a capping Pliocene basalt. **0.5**

102.0 Milepost 394. Coyote Hills (Fig. 2.47), a Pleistocene shield volcano (0.82 ± 0.04 Ma; Crumpler et al., this volume), is at 1:00. **0.6**

102.6 Pliocene(?) basalt in roadcuts. **0.7**

103.3 Quemado(?) Formation in right roadcut. Gravel and cinder quarry on right. View to southwest (10:00) shows progressively older basalts occupying higher positions in the landscape (Fig. 2.48). Mesa slightly below road level at 9:00 (elevation 7285 ft; middle distance on right of photo) is capped by two flows dated 2.94 ± 0.14 Ma and 0.824 ± 0.040 Ma (Laughlin et al.,

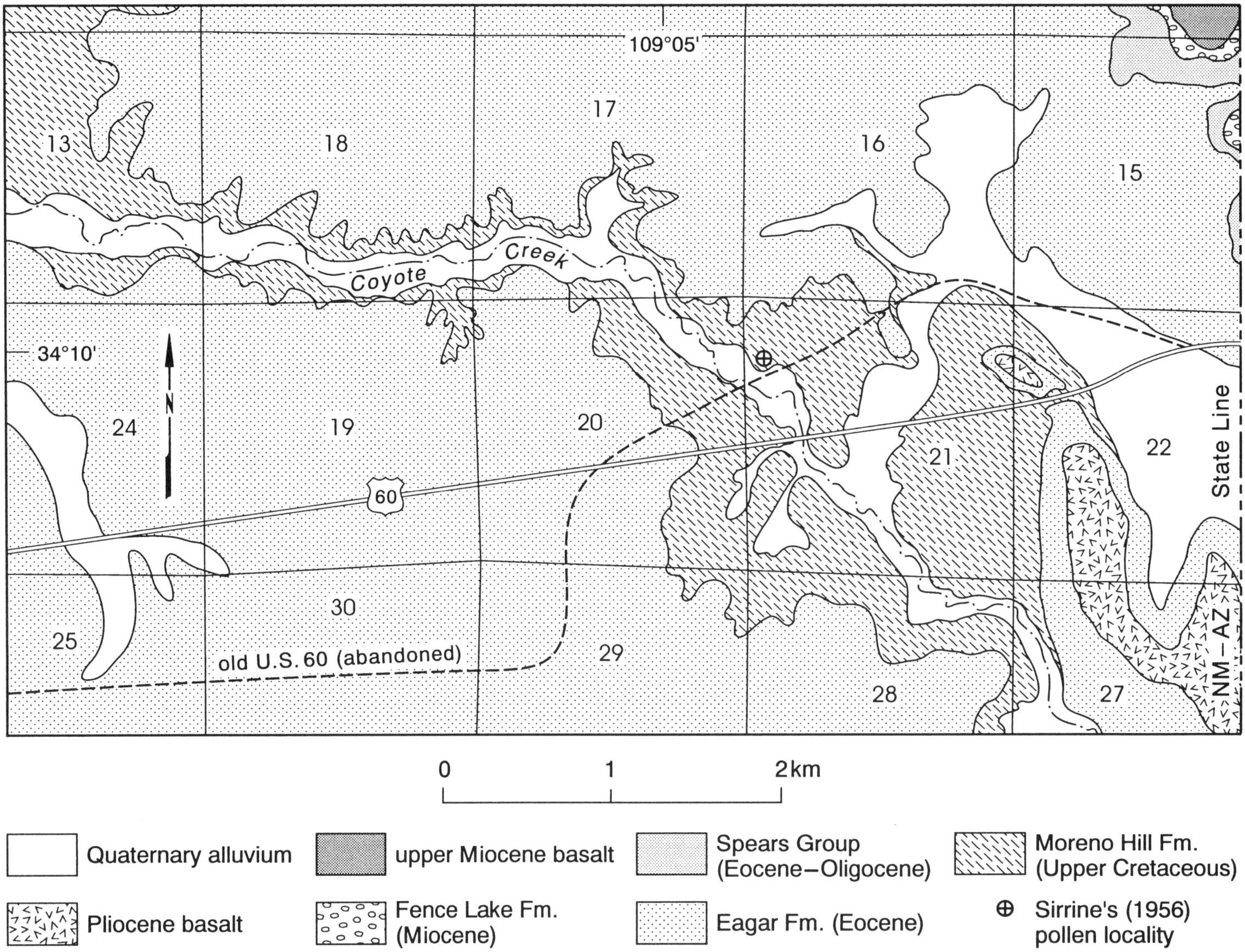

FIGURE 2.43. Reconnaissance geological map of the Coyote Creek area, east-central Arizona. Mapped by S. M. Cather during March 1994.

1979). The unnamed mesa in middle ground to left at 9:00 to 10:30 (elevation 7580 ft) is capped by 3.87 ± 0.10 Ma basalt where it is surmounted by US-180 about 3.7 mi south-southwest of here (Laughlin et al., 1980). The mesa at 10:30 (Flat Top, in center of photo, elevation 8070 ft) is capped by 6.80 ± 0.02 Ma basalt (McIntosh and Cather, this volume). The White Mountains (elevation 11,150 ft) on the distant skyline at 11:00 are composed of volcanic rocks of mostly late Miocene age. The inset nature of the progressively younger basalts visible from here illustrate the net degradation which has occurred in this area at least since the late Miocene. **0.2**

103.5 Springerville, Arizona visible through canyon at 11:00. **0.4**

103.9 Basalt in roadcuts. **0.2**

104.1 Rest area on left. Mesa on skyline at 9:30 exposes Eagar Formation red beds and thin Quemado(?) Formation overlain by Pliocene basalts. Begin downgrade into Round Valley. **0.4**

104.5 Roadcuts expose two basalt flows separated by a thin sandstone and conglomerate unit (Quemado Formation?). The lower basalt here was dated at 2.94 ± 0.14 Ma; the upper flow is 0.824 ± 0.040 Ma (Laughlin et al., 1979). **0.4**

104.9 Red beds of Eagar Formation in roadcuts. **0.3**

105.2 Eagar Formation in roadcuts. **0.3**

105.5 Quaternary alluvium, deposited along the upper reaches of the Little Colorado River and its tributaries, comprises the low-relief floor of Round Valley. **0.5**

106.0 Quaternary alluvium in roadcuts. **0.3**

106.3 Entering Springerville. Western part of Springerville volcanic field (11:00 to 1:00) consists dominantly of alkali olivine basalt erupted mainly from monogenetic cinder cones during latest Pliocene to Pleistocene time (Condit et al., 1989; Crumpler et al., this volume). The White Mountains volcanic field occupies the high ground at 10:00 to 11:00 and is distinctly older (late Miocene) than the Springerville field. **0.9**

107.2 **Turn left** on US-180/US-191. **0.8**

108.0 Flat Top Mesa (Fig. 2.49) at 1:00 consists of a thick sequence of Eagar Formation overlain by late Miocene

FIGURE 2.44. Variegated mudstone and red sandstone of Upper Cretaceous Moreno Hill Formation in north roadcut of US-60. Reddening of permeable sandstones occurred within deeply weathered zone prior to onset of Eocene deposition (Chamberlin, 1981).

FIGURE 2.46. Calcareous paleosol developed at top of Quemado(?) Formation in north roadcut of US-60, overlain by Pliocene basalt.

FIGURE 2.45. Eagar Formation conglomerate and conglomeratic sandstone fills paleochannel incised into red Eagar mudstone in roadcut at Mile 96.3.

FIGURE 2.47. View northwest of Coyote Hills, a Pleistocene shield volcano.

6.80 ± 0.02 Ma basalt (McIntosh and Cather, this guidebook). On the basis of float found along the northeast side of the mesa, it appears that a thin (<30 ft), gravelly, upper Tertiary unit intervenes between the basalt and the Eagar Formation. This gravel may correlate with the Fence Lake Formation of west-central New Mexico, based on its lithology, position in the landscape, and the age of the capping basalt. **1.3**

109.3 Stop sign at intersection. **Turn left on US-180/US-191. 1.0**

110.3 Entering Apache National Forest. **0.7**

111.0 Roadcuts expose red beds of Eagar Formation disconformably overlain by upper Tertiary volcaniclastic gravels (Quemado Formation?). Conglomerate in the Eagar Formation here contains subequal volcanic and non-volcanic clasts. A significant but unknown part of the volcaniclastic detritus in the Eagar may be derived from older volcanics (Laramide, Jurassic, or Proterozoic) to the south, as opposed to being the record of incipient mid-Tertiary volcanism. In the eastern Baca basin, the transition between the largely non-volcanic, orogenic deposits of the middle to upper Eocene Baca Formation and the overlying andesitic volcaniclastics of the upper Eocene Spears Group is commonly defined by the first up-section occurrence of megascopically visible volcanic detritus (Cather, 1980; Cather and Johnson, 1984). Such a demarcation is not feasible, however, in the western part of the Baca basin, where the synorogenic Eagar and Mogollon Rim Formations contain significant amounts of volcanic detritus throughout. This situation led Potochnik (1989) in areas to the west to draw the upper contact of the Mogollon Rim Formation where volcanic detritus becomes volumetrically dominant (>50%). A similar approach probably would be best in the Springerville area. Because the lithologic transition between the Eagar Formation and the overlying Spears Group Formation is stratigraphically quite broad and relatively fine-grained, definitive placement of the contact between these units must await detailed mapping. The placement of the contact in this road log is quite approximate and is based solely on cursory hand-lens estimation of the detrital composition of Eagar-Spears sandstones. **0.6**

FIGURE 2.48. View to southwest showing progressively older basalts occupying increasingly high positions in the landscape. See text for age of basalt flows.

FIGURE 2.50. Two basalt flows exposed in canyon of Nutrioso Creek. Upper basalt is early Pliocene in age, as it appears to correlate with dated basalt nearby to west (3.87 ± 0.10 Ma; Laughlin et al., 1980). Lower basalt is undated and appears to have flowed down paleocanyon of ancestral Nutrioso Creek.

111.6 A thin sequence of Quemado(?) gravels intervenes between the red beds of the Eagar Formation and the capping basalt of Pliocene age (3.87 ± 0.10 Ma; Laughlin et al., 1980). **0.6**

112.2 Eagar Formation and Pliocene basalt in left roadcut. **0.6**

112.8 Milepost 406. Left roadcuts expose Eagar Formation. **0.5**

113.3 Two basalt flows are visible across canyon at 11:00 to 12:00 (Fig. 2.50). The upper, mesa-capping flow is of Pliocene age. The lower (younger) flow appears to be inset along a paleocanyon of ancestral Nutrioso Creek, but has not been dated. **0.5**

113.8 Highway follows Nutrioso Creek. Roadcuts on right in Eagar Formation (Fig. 2.51) next 1.0 mi. **1.2**

115.0 Roadcut ahead on right is Eagar Formation. Young basalts near road are inset against older basalt, which caps mesa at 11:00. **0.7**

115.7 Approximate contact between Eagar Formation and overlying, dominantly volcaniclastic Spears Group in left roadcut. **0.3**

116.0 Roadcuts on left in Spears Group next 1.4 mi. Nelson Reservoir Recreation Site on right. **1.8**

117.8 Basalt-capped mesas on left and right. **0.2**

118.0 Spears Group in left roadcut. **0.2**

118.2 Escudilla Mountain at 11:00 is capped by Bearwallow Mountain Andesite (24–26 Ma). Prominent pine-covered bench about half way up the mountain consists of andesite of Dry Leggett Canyon (~34 Ma). Mesa in foreground is capped by Pliocene(?) basalt. Light-colored cliffs at 12:30 are Spears Group; this unit forms the gentle slopes on Escudilla Mountain and is largely eolian in its upper part above the andesite of Dry Leggett Canyon (Wrucke, 1961). Clasts of Precambrian red granite and Paleozoic limestone occur in the Spears Group below the andesite of Dry Leggett

FIGURE 2.49. View of Flat Top Mesa, which consists of a thick sequence of Eagar Formation capped by 6.80 Ma basalt. A thin sequence (<10 m) of gravels underlies the basalt and may correlate with the Fence Lake Formation.

FIGURE 2.51. Sandstone, siltstone and lenticular conglomerate in upper Eagar Formation exposed at Mile 113.8.

Canyon unit and may indicate part of the transition with the underlying Eagar Formation. Road is on Quaternary alluvium in Nutrioso Creek valley. **0.7**

118.9 Milepost 412. Light-gray cliffs at 11:00 and 12:00 are Spears Group. Skyline cliffs at 11:30 are andesite of Dry Leggett Canyon. **1.8**

120.7 Spears Group in left roadcut and in light gray cliffs at 11:00. **1.6**

122.3 Light gray cliffs of Spears Group at 10:00 to 11:00 and 2:00 to 3:00. **1.3**

123.6 Road to Nutrioso on left. Continue straight ahead on US-180/US-191. **0.9**

124.5 Roadcuts contain Spears Group (locally capped by Quaternary gravels) next 0.5 mi. **0.3**

124.8 Sawmill on left. Cliffs beyond sawmill are Spears Group (light gray) capped by andesite of Dry Leggett Canyon. **0.3**

125.1 Milepost 418. **0.3**

125.4 Roadcuts expose Spears Group. **0.4**

125.8 Road to Nutrioso on left. **0.7**

126.5 Roadcuts in Spears Group next 2.7 mi. **2.7**

129.2 Highway surmounts saddle at Alpine Divide (part of Rim Divide of Cather et al., this volume). Drill pad for Alpine 1 Federal corehole on left. **STOP 4. Park on shoulder** to discuss stratigraphy in corehole and regional petroleum and geothermal implications. **Obey flagmen and beware of traffic.**

The Alpine 1 Federal (Fig. 2.52), collar elevation of 8556 feet, was a slim-hole test cored to 4505 feet depth to evaluate the eastern White Mountains region as a site for near-term commercial Hot Dry Rock (HDR) geothermal utilization. Temperature gradients between 60 and 75°C/km characterize the hole in the upper 2500 feet in sediments of relatively low thermal conductivity. Temperature gradients below 2500 feet in relatively high conductivity rocks range from 48 to 32°C/km. A bottom-hole temperature of 173°F (78°C) was recorded. A preliminary heat-flow estimate is between 90 and 100 mWm^{-2} (John Sass, personal commun., 1993). This heat flow is more typical of the Basin and Range rather than the Colorado Plateau. However, the heat flow may be a part of a pattern of local high anomalies, superimposed on a region of intermediate heat flow, similar to the observations of Minier and Reiter (1991) and Minier et al., (1988) on the Mogollon slope in adjacent New Mexico.

A qualitative potential for petroleum in the area is indicated by shows of oil in the core. Carbon-rich Cretaceous sandstones and fetid and petroliferous dolomites and limestones in the Permian San Andres and the Supai/Yeso strata beneath the Glorieta indicate source-rock potential. Important formation breaks (Table 2.1) include the base of the Spears Group at 2052 ft, base of the Eagar Formation at 3139 ft, top of the Cretaceous Dakota (?) at 3246 ft, and the top of the San Andres at 3362 ft. Basalt intrusions obscure formation tops of underlying Permian units. The deepest observed Glorieta Sandstone is at 3639 ft in contact with a basaltic intrusion. The top of the Fort Apache Limestone member (Winters, 1963) is intruded by a basaltic intrusion. The base of the Permian Fort Apache Limestone occurs at 4405 ft. At total depth, the Alpine 1

FIGURE 2.52. Drill rig for geothermal test well at the Alpine Divide, east side of US-191 at an elevation of 8556 ft, August, 1993.

TABLE 2.1. Formation summary of the Alpine 1/Federal corehole (see Mile 129.2).

Eocene-Oligocene lower and middle Spears Group
- 0 to 2052 feet (0-625 m)
- thickness 2052 feet (625 m)

Eocene-Eagar Formation
- 2052 to 3,139 feet (625 to 957 m)
- thickness 3,139 feet (957 m)

Unnamed Tertiary (?)/Cretaceous(?) unit
- 3,139 to 3,226 feet (957 to 989 m)
- thickness 107 feet (32 m)

Cretaceous Dakota(?) Sandstone (upper 20 ft [6 m] is reddened)
- 3,226 to 3,362 feet (989 to 1,025 m)
- thickness 116 feet (36 m)

Permian San Andres Formation
- 3,362 to 3,436 feet (1,025 to 1,047 m)
- thickness 74 feet (22 m)

Permian Glorieta Sandstone
- 3,436 to 3,639 feet (1,047 to 1,109 m)
- thickness 203 feet (62 m)

Quaternary(?)/Tertiary(?) basaltic intrusion
- 3,639 to 3,751 feet (1,109 to 1,143 m)
- thickness 112 feet (34 m)

Permian Corduroy member Supai Formation
- 3,751 to 4,266 feet (1,143 to 1,298 m)
- thickness 515 feet (157 m)

Quaternary(?)/Tertiary(?) basaltic intrusion
- 4,260 to 4,322 feet (1,298 to 1,317 m)
- thickness 62 feet (19 m)

Permian Fort Apache Limestone member Supai Formation
- 4,322 to 4,327 feet (1,317 to 1,319 m)
- thickness 5 feet (2 m)

Quaternary(?)/Tertiary(?) basaltic intrusion
- 4,327 to 4,362 feet (1,319 to 1,330 m)
- thickness 35 feet (11 m)

Permian Fort Apache Limestone member Supai Formation
- 4,362 to 4,405 feet (1,330 to 1,343 m)
- thickness 43 feet (13 m)

Permian Big A Butte member Supai Formation
- 4,405 to 4,454 feet (1,343 to 1,358 m)
- thickness 49 feet (15 m)

Quaternary(?)/Tertiary(?) basaltic intrusion
- 4,454 to 4,505 feet (1,358 to 1,373 m)
- thickness 51 feet (15 m)

Federal corehole is in a basaltic intrusion. It is uncertain if the intrusions are the result of late Tertiary magmatic activity, associated with the White Mountains volcanic field centered 20 mi west of the Alpine Divide location, or with older mid-Tertiary activity of the Mogollon-Datil field, or with even younger Plio-Pleistocene activity of the Springerville field. The following minipapers by Witcher et al. describe the geothermal potential of the Alpine area and delineate the constraints on regional maturation of hydrocarbons. **0.2**

THERMAL REGIME OF ALPINE DIVIDE AND PETROLEUM IMPLICATIONS

James C. Witcher[1], Chandler A. Swanberg[2], and W. Richard Hahman[3]

[1]New Mexico State University, Southwest Technology Development Institute, Las Cruces, NM, 88003; [2]Consultant, 5478, E. Cholla St., Scottsdale, AZ 85254; [3]Consultant, 780 Suzanne Ave., Las Cruces, NM 88005

Numerous shallow (<0.5 km) heat-flow measurements have been reported in the region around Alpine Divide (Minier et al., 1988; Minier and Reiter, 1991; Reiter and Shearer, 1979; Sass et al., 1982; and Stone, 1980). Minier and Reiter (1991) described the shallow heat flow in the region as a pattern of local high (>90 mW/m^2) anomalies interspersed over a region of low-to-intermediate (<70 mW/m^2) heat flow. No regional trend in heat flow is apparent. However, high heat-flow measurements tend to be located near local zones or trends of folding and faulting. Minier and Reiter (1989) discussed the thermal regime in terms of fairly uniform coal maturation indices at different locations across the region. A preliminary heat flow for the Alpine 1/Federal corehole at Alpine Divide is 90 mW/m^2 (J. H. Sass, personal commun., 1994)

Heat-flow information for the area allows a provisional assessment of petroleum potential with respect to the results of the Alpine 1/Federal corehole. Oil shows in Permian carbonate rocks may indicate petroleum resource potential in the region (Rauzi, 1994a b; Witcher et al., this volume). To assess the importance of the oil shows, it is necessary to understand the burial and thermal history of potential source rocks. Time and temperature are key factors in oil generation and destruction. For each 10°C increase in temperature, maturation rates roughly double (Waples, 1980). Oil shows may indicate that the Permian source rocks are in the oil-generation window, or that oil has migrated into these rocks from elsewhere in the area or from greater depth. A third possibility is that the oil shows represent limited-volume, local generation from the thermal effects of the basaltic intrusions.

Lopatin's method provides insight into the nature and degree of maturation of potential source rocks (Waples, 1980). Two Lopatin models are presented. Model 1 (Fig. 2.53) investigates the time-temperature index (TTI) of hydrocarbon maturity for the Alpine Divide source rocks. Model 2 (Fig. 2.54) investigates oil generation in a hypothetical late Tertiary extensional basin that is superimposed on the pre-Miocene geology of Alpine Divide. The second model may provide an analog for conditions in grabens to the south and east, provided Permian and Pennsylvanian rocks have not been removed by erosion beneath a Laramide unconformity, and provided heat flow and thermal history are similar to Alpine Divide.

Several assumptions are used in both models. First, heat flow and surface temperature varied with time. The average present Colorado Plateau heat flow is about 68 mW/m^2 (Eggleston and Reiter, 1984; Morgan and Gosnold, 1988; Reiter and Clarkson, 1983). The average worldwide average heat flow is 60 mW/m^2. A 60 mW/m^2 heat flow is used from the Pennsylvanian until the end of the Oligocene in the maturation models. Actual heat flow probably varied between 50 and 70 mW/m^2 during this time frame as a result of uplift, sedimentation, and changes in tectonic setting. Today's heat flow of 90 mW/m^2 is used

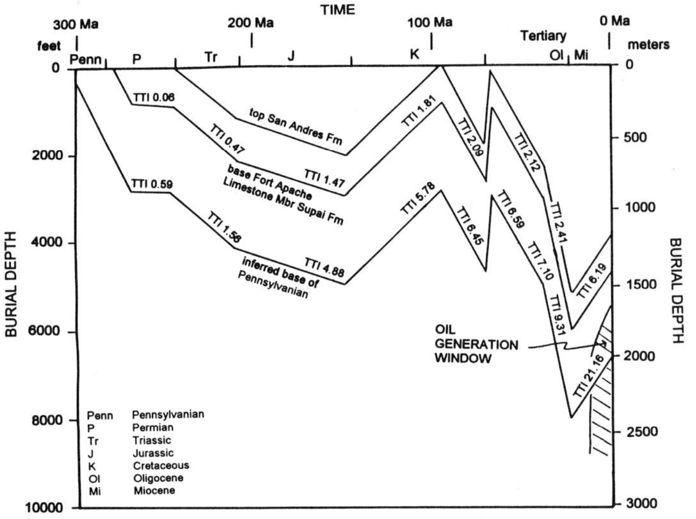

FIGURE 2.53. Lopatin maturation model for Alpine Divide. Lopatin time-temperature index (TTI) of organic matter maturity is shown at various times for the Fort Apache Limestone Member burial curve and the burial curve for the base of inferred Pennsylvanian rocks. Temperatures were calculated from the heat flow, thermal conductivity and surface temperature assumptions. Pennsylvanian to mid-Tertiary heat flow is 60 mW/m^2. Post-Oligocene heat flow is 90 mW/m^2. Thermal conductivity estimates are 2.8 W/mK for Paleozoic and Mesozoic carbonate and quartzose sandstones, 2.2 W/mK for Jurassic rocks, 1.8 W/mK for Eocene clastic rocks, and 1.4 W/mK for the Triassic Chinle and Oligocene volcaniclastic rocks. Surface temperatures are 30°C for the Paleozoic and Triassic, 25°C for the Jurassic and Cretaceous, 15°C for the Eocene and Oligocene, and 10°C for Miocene to present.

from the end of Oligocene until present. This value may be high for the early and mid Miocene, but it provides an upper constraint on late Tertiary maturation history. Surface temperatures are allowed to range from 30°C in the Paleozoic to 10°C in late Tertiary. Sabkha depositional environments during Permian indicate a hot dry climate (Witcher et al., this volume). Present mean annual temperature at Alpine Divide is less than 10 C.

While no Triassic or Jurassic rocks exist in the subsurface of Alpine Divide, we assume that 305 m (1000 ft) of Triassic Chinle Group and

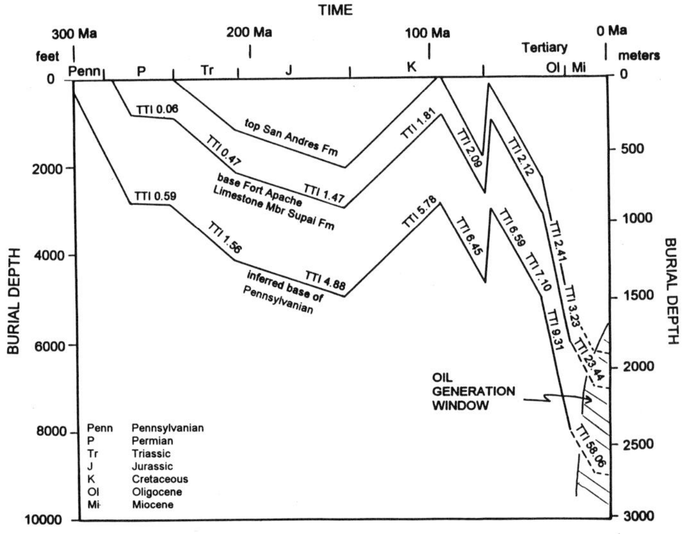

FIGURE 2.54. Lopatin maturation model for a hypothetical late-Tertiary graben with pre-Miocene Alpine Divide subsurface geology. Time-temperature index, heat flow, thermal conductivity, and surface temperature assumptions as in Fig 2.53.

305 m (1000 ft) of Jurassic sediments were deposited over the area. Uplift and erosion associated with the sub-Late Cretaceous unconformity stripped these rocks from the section along with more than two thirds of an estimated 91 m (300 ft) thickness of Permian San Andres Formation (Witcher, et al., this volume). Most of the inferred 610 m (2000 ft) thickness of Cretaceous rocks were stripped beneath a Laramide unconformity. Also, 610 m (2000 ft) of pre-Fort Apache Limestone Member of the Supai Formation rocks were included to investigate possible maturation indices in Pennsylvanian rocks, if present in the subsurface below the total depth of the Alpine 1/Divide corehole at 1373 m (4505 ft).

In Model 1, rapid burial occurred from the beginning of the Eocene until the end of the Oligocene, followed by uplift and erosion (Witcher, et al., this volume) (Fig. 2.53). Waples (1980) presented evidence that the onset of oil generation begins at a TTI of 15. Base of the Fort Apache Limestone Member of the Supai Formation reached a TTI value of 6. If the model thicknesses for the Triassic, Jurassic and Cretaceous rocks are doubled, the Fort Apache Limestone TTI remains below the TTI 15 oil-generation threshold at TTI 9.7. Model 1 temperature at the base of the Fort Apache is 74.2° C, compared to the measured temperature of 77.2° C in the Alpine 1/Federal. If a thick Pennsylvanian section is present, Model 1 predicts a TTI of 21, indicating potential for oil generation beginning after 15 to 10 Ma. No oil generation is predicted for the cored section in the Alpine 1/Federal. Oil shows in the corehole are either related to intrusive heating or represent migrated oil.

Model 2 is the same as Model 1 until the end of the Oligocene. Instead of uplift and erosion, the entire Eocene to Oligocene sequence is preserved and buried under an additional 305 m (1000 ft) of Miocene clastic sediments (Fig. 2.54). In Model 2, the TTI for the base of the Fort Apache Limestone Member is within the oil-generation window at 23.4. Potentially more significant, the Pennsylvanian TTI is 58. Waples' (1980) research indicates that peak oil generation occurs at a TTI of 75. If a heat flow of 70 mW/m^2 is used for the post-Oligocene instead of the 90 mW/m^2 value for Alpine Divide, the Fort Apache Limestone TTI is 19 and the Pennsylvanian TTI is 28. Permian and Pennsylvanian source rocks enter the oil-generation window after 15 to 10 Ma.

Three different basaltic intrusions are present in the Supai Formation (Witcher et al., this volume). Two intrusive bodies are near 30.5 m (100 ft) thickness. The third intrusion thickness is unknown. Each intrusion is assumed to have cooled to roughly ambient temperature before the next intrusion. A conductive model is used to investigate the potential role in oil generation by the basaltic intrusions (Carslaw and Jaeger, 1962, p. 58-62; Lachenbruch, 1976, p. 775, fig. 7b). A sill, 30.5 m (100 ft) thick, of semi-infinite lateral extent is instantaneously intruded with a top at 1298 m (4260 ft) depth. The sill is allowed to cool conductively with a constant pre-intrusive land surface temperature. Thermal diffusivities of the intrusion and the country rock are assumed to be the same. This assumption is adequate for a good approximation (Jaeger, 1968, p. 505). In order to account for latent heat of fusion, an additional 300°C is added to the initial intrusive temperature of 1200°C (Jaeger, 1964, 1968). No magma convection is allowed in the model. The model is not suitable for intrusions that represent conduits for volcanism on the surface. A no-flow assumption after the initial magma intrusion is reasonable because chilled margins are observed in the core. Magma convection or flow would tend to melt country rock (and chilled margins) and result in more intense thermal disturbance of the country rock (Huppert and Sparks, 1989).

The thermal histories of several sites, ranging from 15 to 213 m (50 to 700 ft), over the top of the cooling sill are modeled (Fig. 2.55). Around 30,000 yrs after the intrusion event, the country rock cools to within 10°C of pre-intrusion ambient temperature. Application of the Lopatin maturation model indicates oil-generation potential during the first 100 years in a rock volume, extending upward between 0 and 30.5 m (0 and 100 ft) above the sill. An initial country rock temperature of 70°C was added to the anomalous intrusive temperature profile in order to estimate maturation indices. If hydrothermal convection in the country rock above the sill occurred, cooling times would be faster and temperatures would be lower, possibly eliminating oil generation. A 30.5 m (100 ft) thick model for a vertical dike at 1298 m (4260 ft) would give temperatures similiar to the sill in locations situated perpendicular to the dike (parallel to land surface). However, a concordant sill has potential to mature a much greater volume of source rock compared to a cross-cutting dike.

The Lopatin oil maturation models are first-order approximations. Hunt and Hennet (1992) showed that kerogen type has a strong influence on maturation rates. For instance, high-sulfur kerogens have faster kinetics, resulting from the lower activation energies of carbon-sulfur bonds. Geochemical analysis of kerogen types in the Paleozoic source rocks will allow a more rigorous Arrhenius-equation approach to determine petroleum maturation indices. Hydrothermal convection can enhance petroleum genesis. At high temperatures (>100°C), hydrocarbons become exponentially more soluble in water, which allows rapid transport away from destructive thermal processes near intrusions (Simoneit, 1983). Also, Barker (1989) showed that Arrhenius reaction rates for hydrothermal systems and in zones along intrusive contacts are much faster than burial generation rates modeled by the Lopatin approach.

Preliminary thermal maturation models indicate the best potential for petroleum genesis is in Pennsylvanian and Permian rocks, which may be preserved in the deeper pre-mid-Tertiary basins and in the late Tertiary grabens along the Colorado Plateau (Mogollon slope) and the Transition zone (Datil-Mogollon section) of the Basin and Range Province. The models indicate that Permian and Cretaceous source rocks outside deeper basins and grabens probably have not entered the oil-generation window. Limited-volume maturation zones may include eruptive centers for late Tertiary basaltic volcanism. Conduits (volcanic

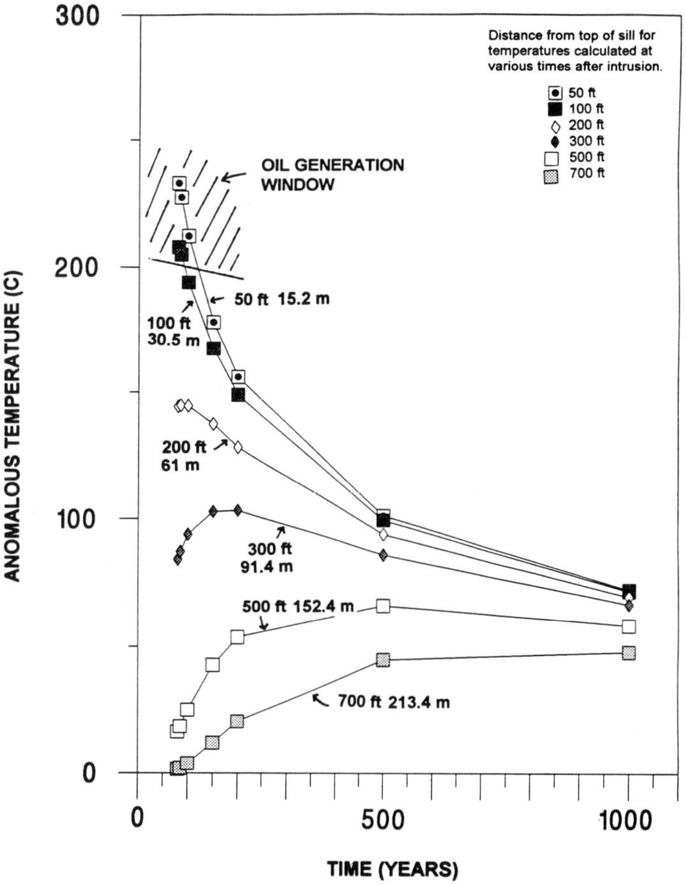

FIGURE 2.55. Temperature versus time cooling model for a sill, 30.5 m (100 ft) thick. Curves show time-temperature relations at varying distances from sill. Intrusion temperature is 1500°C (1200°C plus 300°C to correct for latent heat released during magma crystallization). Thermal diffusivity is 0.0001 m^2/sec. Model based upon Carslaw and Yeager (1959, p. 58-62) and Lachenbruch et al. (1976, p. 775, fig. 7b).

vents) for magma flow could give rise to sustained thermal disturbance for maturation of Paleozoic source rocks and possibly local hydrothermal oil generation. Preliminary analysis indicates that the oil shows in the Alpine 1/Federal borehole are probably the result of very local maturation due to basaltic intrusions. Even though piezometric levels in the corehole appear to decrease with depth, upward migration of oil from depth is not ruled out. In any case, oil generation in the region began after 15 to 10 Ma.

GEOTHERMAL POTENTIAL OF THE ALPINE DIVIDE AREA

James C. Witcher[1], Chandler A. Swanberg[2] and W. Richard Hahman[3]

[1]New Mexico State University, Southwest Technology Development Institute, Las Cruces, NM, 88003; [2]Consultant, 5478, E. Cholla St., Scottsdale, AZ 85254; [3]Consultant, 780 Suzanne Ave., Las Cruces, NM 88005

Geothermal energy as hydrothermal resources or naturally occurring hot water and steam currently provides commercially competitive electrical power and direct heat for space heating, greenhouses, industrial processing, aquaculture, food processing and many other uses that require low-grade heat. Excluding the use of heat pumps, direct-use geothermal heating in the United States at 1026 sites accounted for 18,270 trillion Btu/yr with an installed capacity of 5862 million Btu per hr (MMBTU) in 1988 from mostly low-to-intermediate temperature (<150°C) hydrothermal resources (Lienau, 1988). The installed capacity of geothermal electrical power generation in the United States was 2770 MWe in 1989 from mostly high-temperature (>150°C) hydrothermal resources (Rannels and McLarty, 1990).

However, the largest potentially accessible terrestrial energy resource is hot dry rock (HDR) geothermal. The higher grade (basement gradients > 45°C/km) HDR resource base in the United States is estimated at 650,000 Quads or the oil equivalent of about 11,000 Prudhoe Bays (118 trillion barrels of oil) (Tester et al., 1989). Hot dry rock (HDR) geothermal energy is a method of extracting natural heat from "dry" impermeable rock (conductive thermal regime). As originally conceived by scientists and engineers at Los Alamos National Laboratory, heat is extracted from relatively impermeable rock by artificially fracturing a hot volume of rock and introducing water into the man-made fracture system with an injection well. A second well produces the heated water from the fracture reservoir for use at the surface. The HDR concept and a degree of technical feasibility has been demonstrated at Fenton Hill, New Mexico. On the other hand, commercial feasibility has not been demonstrated for the general use of HDR.

Electrical power production and direct-use applications are possible end uses of HDR. Considering fossil fuel costs, a site with near-term HDR potential requires an area with currently high electrical or utility fuel costs. Because drilling costs increase non-linearly with depth, high temperatures at relatively shallow depth are required for economic energy extraction.

A temperature versus depth plot for the Alpine 1/Federal borehole shows that temperature gradients decrease with depth (Fig. 2.56). Increasing rock thermal conductivity accounts for the lower temperature gradients. Clay-rich sediments in the upper portion of the hole have relatively low thermal conductivity (<2.0 W/mK). Quartzose sandstones and carbonate rocks in the lower portion of the hole have high thermal conductivity (>2.0 W/mK). A preliminary heat flow of 90 mW/m^2 is reported for the hole (J. H. Sass, personal commun., 1994). Thermal conductivity constraints indicate that a gradient of over 35°C/km probably cannot occur in Precambrian basement. For instance, granitic rocks typically have a thermal conductivity ranging from 2.6 to 3.6 W/mK (Roy et al., 1981, p. 454, fig. 12-47).

Tester and Herzog (1990) defined three grades of HDR resources for the purpose of investigating feasibility. A high-grade resource has a gradient above 80°C/km; a mid-grade resource has a gradient of 50°C/km; and a low-grade resource has a gradient of 30°C/km. Within this framework, the Alpine area falls into the low-grade category.

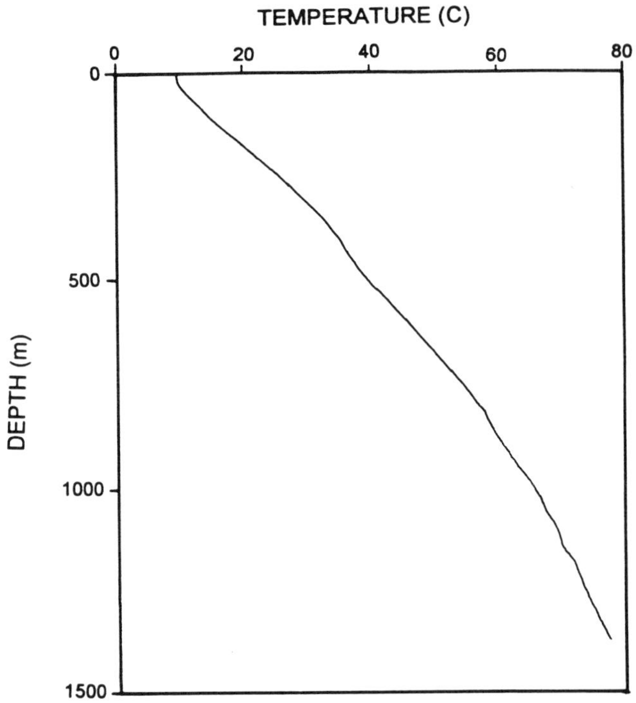

FIGURE 2.56. Temperature versus depth for the Alpine 1/Federal corehole at Alpine Divide.

Tester and Herzog (1990) modeled the cost of electricity with current technology on a per-KWe installed basis. With a 40°C/km gradient, costs for electricity are 12 to 18 cents per kilowatt-hour. These costs are unlikely to be economic for the foreseeable future in eastern Arizona.

Tester and Herzog (1990) also modeled the costs for direct-use HDR, based upon a heat supply of one million Btu per hour (MMBTU). Costs for direct use are a little more attractive. The key parameters involved with direct-use HDR in the Alpine area are heating loads and natural gas availibility. Climate in the area requires space heating for much of the year so that potentially large heating loads may exist. Inexpensive natural gas is not available. Direct-use space heating, using the 40°C/km gradient model of Tester and Herzog (1990) indicates a $4 to $7 per MMBTU cost. This cost may be marginally economic for very large space-heating requirements. However, with natural gas costs in the $3 to $5 MMBTU range ($5 to $7 MMBTU with boiler inefficiencies), it is unlikely that commercial enterprises would relocate for industrial process heat or space heat.

The best potential HDR geothermal application is a district heating system for Alpine. Costs per MMBTU for HDR will be higher than model costs because additional costs would be incurred for distribution lines. Loss of circulation during coring of the Permian San Andres, Glorieta and Supai formations indicates potential for production of 70°C water (conventional geothermal) from depths between 915 and 1220 m (3000 and 4000 ft), if these units are present beneath Alpine (Witcher et al., this volume). However, the costs to drill, complete and test a large diameter production well to 1220 m (4000 ft) could approach a million dollars. Additional capital costs would be incurred for distribution lines and possibly an injection well. A study of heating loads and engineering feasibility is required to assess the direct-use economics at Alpine. However, for HDR and conventional geothermal to compete with current propane use in Alpine, the system would probably have to operate as a utility and have initial capital costs subsidized.

129.4 Roadcuts in Spears Group next 2.0 mi. Sandstones and debris-flow deposits resemble Pueblo Creek Formation above andesite of Dry Leggett Canyon in Bull Basin quadrangle (Ratté, 1989), and likely are correlative of same. **2.2**

131.6 Road to Big Lake on right. Continue straight on US-666. **0.5**

132.1 Valley and lake on right is in headwater drainage of San Francisco River (Fig. 2.57). **0.2**

132.3 Spears Group on left. **1.0**

133.3 Junction with road to Luna, New Mexico (US-180) in Alpine, Arizona.

End second-day road log.

FIGURE 2.57. View south of town of Alpine, Arizona and headwater drainage of San Francisco River.

Morning fog rolls over the central Sawtooth Mountains, view to north from US-60. Black and white print by De Mar Co. Photographics from color slide © Kelly D. Gatlin, La Luz Photography.

THIRD-DAY ROAD LOG, FROM ALPINE, ARIZONA, TO LUNA, RESERVE, APACHE CREEK, HORSE SPRINGS AND DATIL, NEW MEXICO

J. C. RATTÉ, D. J. BOVE, S. M. CATHER, R. M. CHAMBERLIN, S. G. CREWS and W. C. MCINTOSH

SATURDAY, OCTOBER 1, 1994

Assembly Point: Alpine Country Club
Departure Time: 7:00 a.m. MST (Arizona Time)
8:00 a.m. MDT (New Mexico Time)
Distance: 130.7 mi
Stops: 4

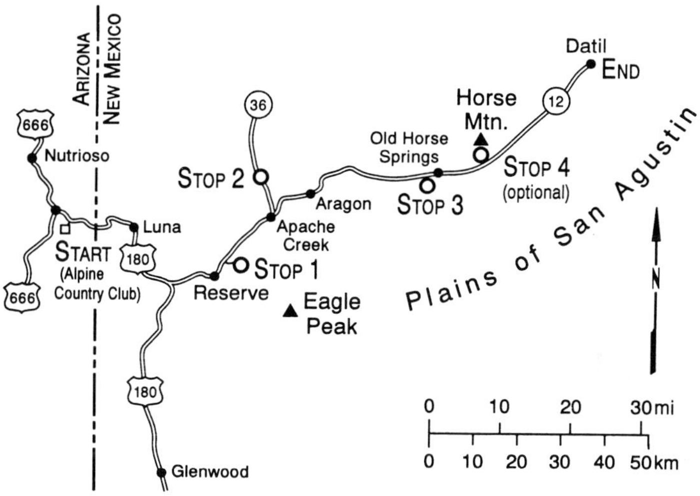

SUMMARY

On Day 3, the trip departs from the Alpine, Arizona Country Club, where we will have assembled for breakfast, starting at 6:00 a.m. Arizona Time! (7:00 a.m., NM time). Starting at Alpine, in the footwall of the Reserve graben system, the Third-Day route follows US-180 east, via Luna, New Mexico, descending into the Reserve graben across several tilted, half-graben blocks. Between Luna, and the junction of US-180 and NM-12, we pass through the recent landslide areas in Mail Hollow and Dry Leggett Canyon, which disrupted traffic for several weeks in Fall 1992. At the junction of US-180 and NM-12, we turn north and follow NM-12 through Reserve and north to the Eagle Peak road, where we turn east. Stop 1 is in the deepest part of the Reserve graben, along the Tularosa River north of Reserve, where we will discuss the evolution of the graben structure, sedimentology of the Gila Formation (Oligocene to Pliocene), and the local volcanic stratigraphy. Stop 1A, nearby, provides an opportunity to see a basalt flow interlayered in the Gila Formation and to examine a small reverse fault that is related to a small- scale buttress unconformity.

Upon returning to NM-12, we continue north to Apache Creek, where we take NM-32 northwest for about 9 mi to Stop 2, where relatively new roadcuts expose volcaniclastic sediments of the Spears Group (Eocene and Oligocene) beneath andesite lava flows of Dry Leggett Canyon (~34 Ma). Here we discuss the provenance and sedimentary structures in this sequence of well-sorted sandstones, the origin of numerous, thin, pink ash-beds interlayered with the sandstones, and the origin of the black fiamme and the pink, bedded deposits in which they occur, between andesite flows.

After we return to Apache Creek, we continue east on NM-12 to Old Horse Springs, Stop 3. At this stop, we examine pyroclastic deposits at the eruptive center of the Horse Springs dacite (informal unit), and *mineralized* jasperoid inclusions (silicified pre-Tertiary limestone, quartzite and argillaceous rocks), with skarn-type reaction rims. From this vantage point, we can also discuss the structure of the San Agustin half graben as revealed by seismic profiles, the small hogback of pre-Tertiary rocks (Laramide high?) at the foot of the Horse Mountain volcano on the northern margin of the graben, the alignment of Bearwallow Mountain volcanoes on its southern margin, and the results of the recent oil play in this region.

Stop 4 is at the small hogback of Permian sedimentary rocks at the base of Horse Mountain on the north side of NM-12, about 6 mi east of Old Horse Springs. This is an optional stop, and there may not be sufficient time on this field trip to examine these outcrops that are believed to represent the Mogollon Rim, a relict of the Laramide uplift at the southeast margin of the Colorado Plateau in this area. Tertiary stratigraphic nomenclature used in this third-day road log follows the proposals of Cather et al. (this volume).

Mileage

0.0 Alpine Country Club Dining Room and Clubhouse parking lot. Return to US-180. **0.2**
0.2 **Keep left**, past "SLOW" sign. **0.7**
0.9 **Turn left** onto road from Blue River and proceed to US-180. **0.9**
1.8 Junction with US-180; **turn right**. **0.2**
2.0 MP (Milepost) 430. Luna Lake north of highway, on

left. Entrance to Luna Lake Recreation Site and Campground about 0.25 mi ahead on left.

Intermediate to mafic composition lava flows and volcaniclastic sedimentary rocks exposed along the highway between here and the Luna quadrangle have not been mapped in detail, and little is known of the local structural geology. The closest detailed mapping is in the Alpine-Nutrioso area to the north (Wrucke, 1961), the Blue Range Primitive area to the south (Ratté et al., 1969), and the Luna quadrangle to the east (T.L. Finnell, unpubl. geologic map, 1985). Based on these studies and roadside reconnaissance, we know that the volcanic sequence in this area is essentially the same as that in adjacent parts of New Mexico. Bloodgood Canyon Tuff (Oligocene) from the Bursum caldera in the Mogollon Mountains, about 30 mi to the southeast, is present both north and south of Alpine, and provides a structural benchmark for our descent into the Reserve graben system to the east.

In terms of the new stratigraphic terminology proposed for the volcanic and volcaniclastic rocks of the Mogollon-Datil volcanic field (Cather et al., this volume), the sedimentary rocks exposed along US-180 between here and the Luna quadrangle are all within the Eocene-Oligocene Spears Group, and the intermediate to mafic lava flows are probably all andesite of Dry Leggett Canyon (~ 34 Ma) in the upper part of the Eocene-Oligocene Datil Group. **0.6**

2.6 Luna Lake Dam on left; brown sandstone above andesite of Dry Leggett Canyon(?) flows exposed in spillway. **0.4**

3.0 MP 431. Andesite flows across valley north of highway. Brown sandstone in roadcuts on south side of highway. From the highway at about MP 431, views of Turner Peak, to northeast (left), on the New Mexico-Arizona border, about 4 mi away. Light-colored rocks forming Matterhorn-like edifice on south side of Turner Peak (Figs. 3.1, 3.2) is a volcanic neck that probably fed the andesitic lava flows that cap Turner Peak. The andesitic flows, of uncertain age, overlie volcaniclastic rocks in the upper part of the Spears Group, above the andesite of Dry Leggett Canyon. In addition, the neck includes a vent-breccia margin that contains blocks, up to several meters long, of silicic ash-flow tuff (Davis Canyon(?) Tuff), and is cut by thin, irregular rhyolite dikes that contain sanidine and biotite. **2.0**

FIGURE 3.1 View of Turner Peak and andesitic eruptive center along Arizona-New Mexico state line from US-180 at about Mile Post (MP) 431. Turner Peak and eruptive center are about 4 mi away.

FIGURE 3.2. View of Turner Peak and eruptive center from point about 1.8 mi southeast of the peak.

5.0 MP 433. Arizona-New Mexico state line just ahead. **0.3**

5.3 Arizona-New Mexico state line. **0.5**

5.8 Porphyritic lava flows along both sides of the highway are correlated with andesite of Dry Leggett Canyon; they have phenocrysts of plagioclase and both clinopyroxene and orthopyroxene. **1.1**

6.9 Base of andesite flows of Dry Leggett Canyon overlying brown mudflow breccias in lower part of Spears Group. Watch for 3-7 ft of interlayered ash-flow tuff in roadcuts on left just ahead, between here and MP 2. Pink, densely welded tuff is overlain by light-gray, punky, partly welded tuff; both are probably tuff of Bishop Peak (34.7 Ma) which was mapped near here by T.L. Finnell (unpubl., 1982). **0.4**

7.3 MP 2. More outcrops, or slumped blocks, of ignimbrite just west of MP-2. Highway crosses approximately located, northwest-trending normal fault in the next 0.25 mi. On the east side of the fault, the andesite of Dry Leggett Canyon and the overlying sandstones are downdropped about 425 ft. **1.2**

8.5 San Francisco River. Andesite of Dry Leggett Canyon on both sides of river. East of bridge, road follows approximately along contact between andesite and overlying brown sandstone. A few miles north of the road are small, anomalous exposures of Pennsylvanian strata, discussed in the following minipaper. **0.8**

FAUNA FROM A SMALL PENNSYLVANIAN EXPOSURE NEAR LUNA, NEW MEXICO

Barry S. Kues

Department of Earth & Planetary Sciences, University of New Mexico, Albuquerque, NM 87131-1116

Two small exposures of Pennsylvanian strata occur within the Tertiary volcanic terrane near Luna. First reported by Foster (1957), they were mapped by Weber and Willard (1959), and the largest (Trout Creek) outcrop was described by Kottlowski (1959, 1960). This outcrop is about 7 mi northwest of Luna, along Forest Road 220, about 5.3 mi north of its intersection with F.R. 219, in the NE¼ NW¼ sec. 10, T5N, R21W, Catron County. The section, as measured by Kottlowski (1960, p. 26-27, fig. 5) consists primarily of steeply-dipping beds of gray or pink calcarenite and limestone, totaling 350-400 ft thick, which contain crinoid columnals, unbroken brachiopods, fusulinids (*Fusulina*, now *Beedeina*), corals and algae. The fusulinids suggested a probable Desmoinesian (Middle Pennsylvanian) age. This small exposure shows no evidence of baking or intrusion by the surrounding Tertiary igneous rocks, and is either a float block in the volcanic terrane or a small, upthrown fault block (Kottlowski, 1960). Similar small, isolated exposures of Virgilian or Wolfcampian limestones, assigned questionably to the Naco Formation, occur about 7 mi to the west, just inside Arizona, and were interpreted as probable inclusions in andesites of the Datil Formation (Wrucke, 1961). I visited the larger Luna exposure in 1988 to collect macrofossils that might verify the suggested Desmoinesian age, and here report briefly on the nature of the invertebrate fauna. These fossils were collected mainly from Kottlowski's unit 8, near the top of the section, a pinkish, coarsely crystalline, bioclastic limestone unit containing chert nodules locally. Preservation of the fossils is generally poor, owing to weathering and difficulty in extracting complete specimens from the hard limestone matrix.

The fauna consists chiefly of brachiopods, bryozoans, and crinoid stem segments. Among bryozoans, compact, massive to encrusting, fistuliporoid zoaria are most abundant; *Fenestella* and ?*Rhombopora* are much less common. Brachiopods are represented by at least eight species. The unornamented, biconvex athyrid *Composita subtilita* (Hall) is the most abundant brachiopod. Specimens are up to about 32 mm long, and display a wide variability in shape, as is typical of this species (Figs. 3.3A-C). Specimens range from relatively narrow to slightly wider than long, and possess a conspicuous fold and sulcus on the brachial and pedicle valves.

A second athyrid species, *Cleiothyridina pecosii* (Marcou), is smaller and has less convex valves than *C. subtilita*. The valves are slightly to moderately wider than long; a typical specimen is 13.5 mm wide and 11.5 mm long. The valves display many conspicuous growth lamellae, each with numerous, simple, small, narrow, closely-spaced spines extending posteriorly along the valve surface to or past the next lamella. On most specimens only short, somewhat flattened spine bases are preserved, and even these are obscure on worn specimens (Fig. 3.3E). However, one specimen (Fig. 3.3D) displays numerous complete spines, some more than 2 mm long. Dunbar and Condra (1932) noted of *C. orbicularis* (now known to be a synonym of *C. pecosii*; see Sutherland and Harlow, 1973), that the spines "when preserved from a mat that hides the body of the shell," which would apply equally well to unworn examples from the Luna outcrop.

Neospirifer alatus Dunbar and Condra (Figs. 3.3G-H) is common in all growth stages. Specimens attain a maximum width of more than 70 mm (twice the length), and are characterized by up to 15 ribs within the sulcus, 25 to 30 plications on the lateral slopes, of which those nearest the sulcus display early trifurcation, and by conspicuous fasciculation of the lateral plications. Smaller, relatively alate spiriferids with 10-13 bold, rounded, uniformly simple lateral plications, three to (more commonly) five plicae within the sulcus, and four (less commonly six) plicae on the fold are assigned to *Anthracospirifer* aff. *A. curvilateralis chavezae* Sutherland and Harlow (Fig. 3.3F). The largest specimen measures about 25 mm wide and 21 mm long. These specimens appear to fall within the range of variation documented for the subspecies by Sutherland and Harlow (1973), but no complete specimens are available that would allow definite identification.

A single small spiriferid pedicle valve (Fig. 3.3I), about 8 mm wide by 5.5 mm long, displays a rather high cardinal area and eight strong, broadly rounded plications, of which the central two are largest and most widely spaced. This specimen is quite similar to, although smaller than, New Mexico Morrowan specimens of *Spiriferellina campestris* (White) (see especially Sutherland and Harlow, 1973, pl. 18, fig. 1b). The surface of the valve is weathered and partially exfoliated, and does not display the tiny pustules typical of the genus. Its strong plications and absence of a well-defined sulcus immediately distinguish this specimen from juveniles of *N. alatus* and *A.* aff *A. curvilateralis chavezae*.

Productoid brachiopods are moderately common, but very incomplete and too poorly preserved to allow positive identification. *Linoproductus* sp. (Figs. 3.3J-K) attains an estimated length of about 45 mm, lacks a median sinus on the umbo, possesses relatively fine costallae (8-10/5 mm about 20 mm surface length from the beak), and virtually lacks spine bases. Specimens referred to *Antiquatonia* cf. *A. hermosana* (Girty) (Fig. 3.3L) have relatively sharp radial costae crossed by concentric wrinkles, producing elongate nodes at the intersection points on the anterior surface of the pedicle valve. The number of radial costae is about 10-12/10 mm at a distance of about 25-30 mm from the beak, within the range for Desmoinesian specimens of *A. hermosana* (Sutherland and Harlow, 1973). Shape, convexity, and hingeline features could not be accurately determined from the specimens at hand, however. A medium-sized (width to 35+ mm) species of

FIGURE 3.3. Pennsylvanian brachiopods from north of Luna, New Mexico; all figures natural size unless otherwise indicated. A-C, *Composita subtilita*, pedicle valve, UNM 11,972; and pedicle and brachial valves, UNM 11,973. D-E, *Cleiothyridina pecosii*, pedicle valve showing well preserved fine spines, UNM 11,974, and somewhat worn brachial valve, UNM 11,975. F, *Anthracospirifer* aff. *A. curvilateralis chavezae*, pedicle valve, UNM 11,976, x 1.5. G-H, *Neospirifer alatus*, pedicle valve of a large specimen, UNM 11,977; and hinge view, UNM 11,978. I, *Spiriferellina* cf. *S. campestris*, pedicle valve, UNM 11,979, x2. J-K, *Linoproductus* sp., pedicle valve, UNM 11,980; and oblique posterior view of a fragmentary pedicle valve, UNM 11,981. L, *Antiquatonia* cf. *A. hermosana*, oblique posterior view of incomplete pedicle valve, UNM 11,982.

Derbyia, possibly *D. haesitans*, is common in this fauna. The valves are significantly wider than long, and bear sharp, high, fairly widely spaced radial costellae with smaller ones in the interspaces. The brachial valve is moderately convex and the pedicle valve nearly flat. As none of the specimens is complete, specific identification was not attempted.

Two incomplete bivalve steinkerns, one a myalinid, the other unidentifiable, were also collected at this locality.

Most of the identifiable brachiopods are relatively long-ranging taxa, that together are consistent with a Desmoinesian or Missourian age. The younger age is slightly favored by the presence of *Neospirifer alatus*, which is known only from the Missourian in New Mexico (Sutherland and Harlow, 1973) and ranges from the Missourian to Virgilian in Kansas (Spencer, 1967).

I thank Frank E. Kottlowski for providing some information relating to this locality.

9.3 MP 4. Brown sandstone on north side of highway. Andesite of Dry Leggett Canyon is faulted up above highway on an approximately east-west fault that is tangent to the highway curve just ahead. **0.6**

9.9 Approximate western edge of the Luna 7-1/2' quadrangle (Finnell, unpubl., 1985). Outcrops along highway are mainly andesite flows of Dry Leggett Canyon. **0.4**

10.3 MP 5. Andesitic flows on both sides of highway; downhill, ahead, volcaniclastic sedimentary rocks have sub-horizontal dips that change to 10-15° westward approaching the Luna fault. **1.0**

11.3 MP 6. Highway descends back into valley of San Francisco River. Sharp curve to left, ahead; highway now along Luna fault, within the northeast-trending Blue-Luna-Spur Lake fault zone (Fig. 3.4). **0.9**

12.2 Cattleguard. Western margin of the Luna half graben, where the Luna fault controls the northwest edge of Luna Valley, site of the historic town of Luna, directly ahead. Volcaniclastic rocks in the lower part of the Spears Group and Gila Formation conglomerate are juxtaposed across the fault. Gila Formation crops out left of cattleguard about 300 ft ahead. **0.6**

12.8 Road junction in the town of Luna; **stay on US-180**, which curves sharply right to bridge across San Francisco River. **0.4**

HISTORICAL VIGNETTE OF THE TOWN OF LUNA, CATRON COUNTY, NEW MEXICO

J.C. Ratté

U.S. Geological Survey, P.O. Box 25046, MS 905, Denver, CO 80225

The Luna Valley was first settled by Mormon families from southern Utah in 1883 (Luna Ward-The Church of Jesus Christ of Latter-day Saints, 1983), although the brothers Luna had been grazing sheep in the valley for more than 10 years prior. The settlers voted to call their town "Grant," but Grants, New Mexico had precedence, and the Post Office Department finally accepted Luna, named for Don Solomon Luna, who was the leader of the Republican party in New Mexico

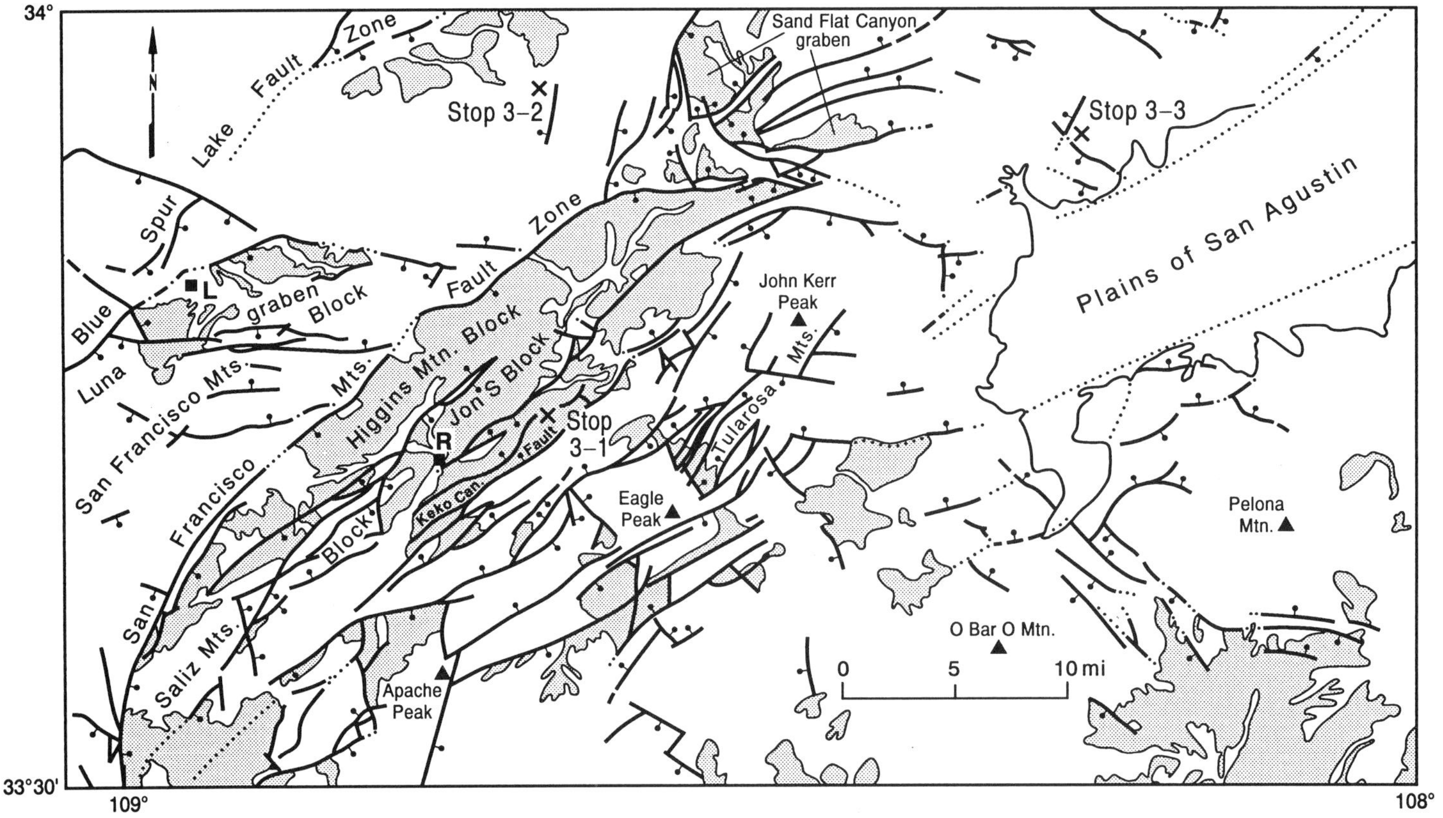

FIGURE 3.4. Generalized geologic map of the Reserve graben system between Luna and the Tularosa Mountains east of Reserve, New Mexico, showing major tilted fault blocks, and distribution of Gila Formation volcaniclastic rocks (shaded) **L**, Luna; **R**, Reserve.

Territory and in the State for more than 25 years (Pearce, 1965). There is a somewhat indirect geological connection here in that Estella and Luna Leopold, who both were once members of the U.S. Geological Survey, are descendants of the Luna and Leopold families. This is an interesting union between the New Mexico sheep ranchers who gave their name to the town of Luna, and the Leopold family of Wisconsin, which is perhaps best known through Aldo Leopold, one of the earliest conservation advocates, author of "A Sand County Almanac" (Leopold, 1966), and "father" of the first Wilderness Area, the Gila, established by the U.S. Forest Service in the Gila National Forest in southwestern New Mexico in 1924. Estella, a professor in the Department of Botany at the University of Washington in Seattle, was formerly a palynologist with the Branch of Paleontology and Stratigraphy, USGS in Denver. Estella's brother, Luna, now on the faculty of the University of California at Berkeley, was formerly Chief Hydrologist of the U.S. Geological Survey.

We thank Eloisa Brown of Albuquerque, New Mexico, cousin of Estella Leopold, for verifying the family history included here.

FIGURE 3-5. Crossbedded dune sands and torrential conglomerate lenses in Spears Group in Mail Hollow, Luna Quadrangle.

13.2 MP 8. Near south end of bridge across San Francisco River. View down Luna Valley to east, left, where San Francisco River disappears from view in box canyon on the southeast side of the Luna graben. Low ridge across valley about 2 mi east of bridge is basalt flow, ~2.5 Ma (Marvin et al., 1989, p. 49), which is representative of numerous young basalt flows and cinder cones in this area on the southeast margins of the Colorado Plateau. The Luna basalt flow follows a relict drainage alignment that crossed the Luna graben prior to the establishment of the present San Francisco River in this area.

From bridge, US-180 continues south across alluvium and low ridges of Gila Formation graben-fill, which laps onto the northwest-tilted San Francisco Mountains block of the Reserve graben. Dips in the Gila Formation here in the Luna Valley are generally about 5-6° northward. **3.0**

16.2 MP 11. Leaving Gila Formation basin-fill deposits at eastern margin of Luna half graben. Highway starts ascent of Mail Hollow to the top of the San Francisco Mountains tilted fault block. Outcrops along the road from here to divide between Mail Hollow and Dry Leggett Canyon consist mainly of distal outflow of rhyolitic ash-flow tuffs from the caldera-related eruptive centers in the Mogollon Mountains, about 25 mi to the southeast, and interlayered volcaniclastic sedimentary rocks in the Spears Group. The San Francisco Mountains block is intricately broken by minor east-west-trending normal faults (Fig. 3.4) in the Luna and Dillon Mountain quadrangles. **0.5**

16.7 Light-gray, nearly white, Davis Canyon Tuff (29 Ma) in slopes along left side of highway is overlain by Shelly Peak and Bloodgood Canyon Tuffs (28 Ma) and interlayered volcaniclastic rocks. Tuffs are 16-50 ft thick. Davis Canyon Tuff is underlain by crossbedded dune sands and lenticular torrential conglomerate beds (Fig. 3.5) that emphasize the desert setting in this area in late Oligocene time. **1.1**

17.8 Southern border of Luna quadrangle; enter Bull Basin quadrangle (Ratté, 1989). **0.4**

18.2 MP 13. Approximate center of highway slide-repair area (see Gallagher, this volume). **1.0**

19.2 MP 14. Road to Saddle Mountain Lookout Tower on right. **0.1**

19.3 Mail Hollow divide; Potato Patch road on left. Begin descent of Dry Leggett Canyon to the San Francisco Mountains fault zone at the foot of the canyon. Outcrops in roadcuts for the next 1.5 mi are part of the alluvial facies of the Pueblo Creek Formation (Ratté, 1989), now redefined as part of the lower Spears Group. These rocks grade into the proximal andesitic vent facies (now included with the upper part of the Datil Group), which dominates the formation farther southwest, where central vent andesitic volcanoes erupted prior to deposition of the Davis Canyon Tuff. **0.5**

19.8 Heifer Basin overlook on left, at new highway scenic rest stop, with views across Heifer Basin to Tularosa Mountains on the east side of the Reserve graben (Fig. 3.6). Rocks in Heifer Basin and along highway ahead are brown, mudflow breccias of the lower part of the Spears Group, overlain at the head of the basin by the ash-flow tuff sequence of the Miocene and Oligocene

FIGURE 3.6. Heifer Basin overlook; view northeast to Tularosa Mountains on the east side of the Reserve graben, on skyline at far right. Thin, discontinuous tuff lenses, **t**, are interlayered with mudflow breccia layers in upper part of Spears Group, which is overlain by ash-flow tuffs of the Mogollon Group, capping the ridge at the head of the basin.

Mogollon Group (Davis Canyon, Shelley Peak and Bloodgood Canyon tuffs). Other thin, white, discontinuous tuffs are interlayered in the upper part of the mudflow sequence and can be seen from the overlook. Hells Mesa Tuff (32 Ma), from the Socorro cauldron, may be one of the thin tuffs in this section, but efforts to date and correlate these discontinuous and sometimes reworked tuffs with regional tuffs in the upper part of the Datil Group have been largely unsuccessful. Although some of these unidentified tuffs have quartz and sanidine phenocrysts like Hells Mesa Tuff, the only dated samples, in the Luna quadrangle, have Ar/Ar ages of 33.29 ± 0.14 Ma and 33.62 ± 0.1 Ma, in the age range of the Box Canyon Tuff, derived from uncertain source(s) in the southern part of the Mogollon-Datil field (McIntosh et al., 1991; 1992, p. 864). Two outcrops of the thin tuffs are exposed on the left side of the road just ahead. **0.3**

20.1 MP 15; cattleguard and road on left. One of the thin tuffs is exposed beside the highway just beyond the cattleguard. A small fault offsets this 8-in.-thick tuff about 7 ft. **0.6**

20.7 Cattleguard on right. Contact between mudflows in Spears Group and underlying andesitic lava flows of Dry Leggett Canyon. About 250 ft up road from contact, another thin tuff (less than 3 ft thick) may be exposed, if not obscured by recent road work.

The pyroxene-plagioclase andesite flows in Dry Leggett Canyon are about 1000 ft thick beneath Prairie Point Peak, north of the highway. Flow boundaries are irregular and obscure. These andesite flows are widespread in the northwest part of the 1:100,000-scale Tularosa Mountains quadrangle and adjacent parts of the Quemado 1:100,000 quadrangle. Sources for these flows are unknown in the Bull Basin quadrangle or elsewhere, except for one probable dike in the slopes above the highway, below Prairie Point Peak, and a possible domal eruptive center in the Saliz Pass quadrangle (Ratté, 1969) to the south. The andesite is highly altered locally to yellowish-brown clay, carbonate and silica by fumarolic activity; the flows may have been erupted onto water-saturated ground or into standing water. "Luna blue" agate, much sought by rockhounds, is found largely within and near the base of these andesite flows, and probably formed as a result of the fumarolic activity. Irregular altered zones are well exposed in fresh roadcuts in the canyon, but distinct flow boundaries are less common. **1.4**

22.1 MP 17. Site of overrun of blasted material produced during widening of highway (see Gallagher, this volume). **0.3**

22.4 Around sharp curve to left, basal contact of andesitic lava flows above lower part of volcaniclastic sequence of Spears Group. Clasts in these sedimentary rocks are commonly red and well rounded in contrast to clasts in the gray to brown mudflows that characterize much of the sequence above the Dry Leggett Canyon flows. Some of the sedimentary beds beneath the Dry Leggett Canyon flows contain clasts of pre-Tertiary rocks (Precambrian granite, quartzite and fossiliferous Paleozoic limestone) as well as Tertiary volcanic rocks. Thus the lower part of the Spears Group here is probably transitional to the Eocene Baca Formation, and where the conglomerate contains less than 50% volcanic clasts it should be mapped as Baca. However, no Baca Formation was mapped in the Bull Basin quadrangle (Ratté, 1989) because of the limited area of Baca-type outcrops, and the uncertain location of the gradational contact. **0.4**

22.8 Entering paleolandslide area (Fig. 3.7). This old landslide on the north side of Dry Leggett Canyon is the location of two of the highway repair projects (Gallagher's 18.54 and 18.66) described in the article by Gallagher (this volume). Clean-up work and new highway construction here has largely eliminated the evidence of the recent mobilization of this slide area. The problem of landslides and highways in New Mexico is discussed in the following minipaper. **0.5**

LANDSLIDES AND HIGHWAY MAINTENANCE IN NEW MEXICO

Susan Gallagher

New Mexico Department of Highways and Transportation, Santa Fe, NM 87504

Every year New Mexico experiences landslides throughout mountainous areas of the State. They usually occur in the spring when the winter snows melt, or in the summer once the summer rains begin. The addition of water can increase soil moisture pore pressure to create an unstable soil condition and cause a landslide or mudflow. Evidence of this appears as a jagged scarp face at the top of the slide area and a noticeable hummock of soil above the toe of the slide. Leaning trees are an obvious indicator of an active or potential slide area.

Typically, special crews of the New Mexico State Highway Department repair smaller slide areas, while larger ones are contracted out, depending on the type and amount of equipment needed. The slide repair designs are done in-house by the Geo-technical Section of the New Mexico State Highway Department.

Designing a landslide correction is an involved process. It includes site investigation, inclinometer installation, soil sampling, and field and laboratory testing. Factors considered in design include unit weight of soil, cohesion, phi angle, depth to slip surface, and depth to water table. Various computerized slope stability programs are used to determine stability. The landslide, at the time of failure, is assumed to have a safety factor equal to or less than 0.9. A slide correction is designed for a minimum safety factor of 1.25. This means that the forces resisting a failure are at least 25% greater than the driving forces, i.e. those attempting to create another slide. Once a 1.25 safety factor is achieved, the most economical design is selected.

Over the past five years, New Mexico has spent an average of $1,250,000 per year on slide repairs. This usually takes care of one or two large remediations that are contracted out, and materials for two or three small ones that are repaired with our own special crews.

The types of landslides that we have to deal with most frequently in New Mexico include rotational, translational and mudflows. The slides generally occur in either roadway fills or cutslopes. Inclinometers are routinely installed at slide areas to monitor the movement. Where a high water content is present, piezometers are installed to monitor water levels. Remedies vary from installing horizontal drains and french drains for dewatering a slope, to building a berm at the toe to stabilize the mass, to excavating the failed material and replacing it with a select fill, which may be layered with reinforcing geogrid. Corrective action has not always produced the desired result, in which

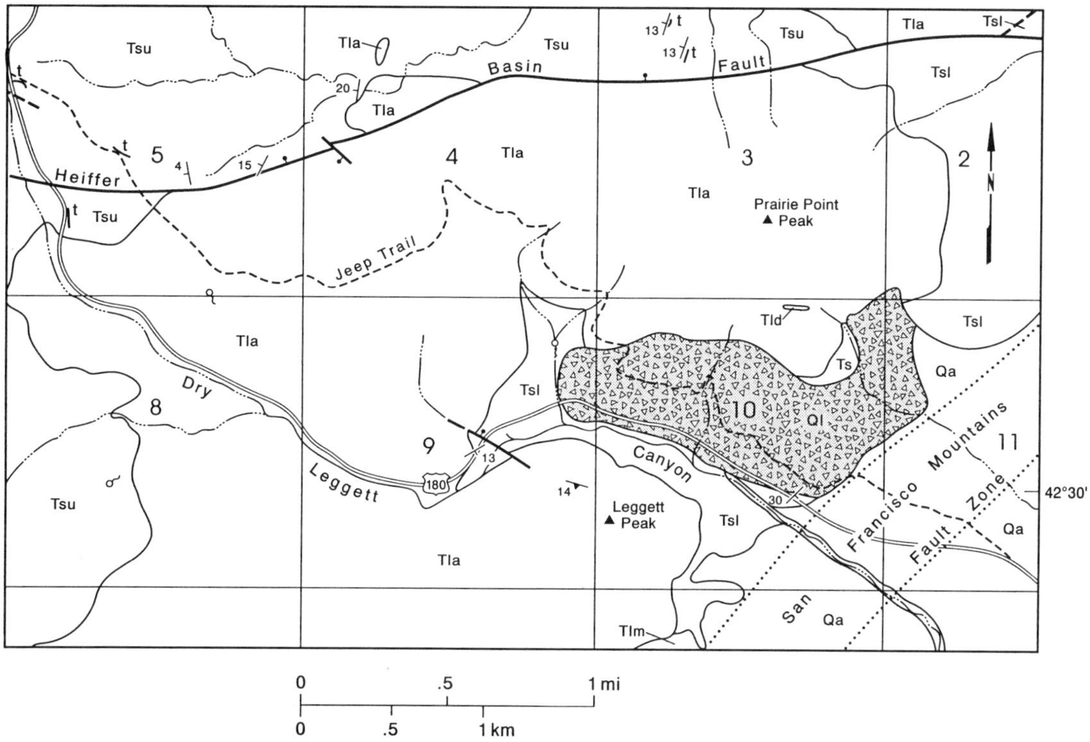

FIGURE 3.7. Geologic map showing the Prairie Peak paleolandslide on the north side of Dry Leggett Canyon and US-180. Landslide was remobilized during highway repairs in Fall 1992. **Qa**, Quaternary alluvium; **Ql**, Quaternary paleolandslide debris (patterned); **Tlm**, porphyritic quartz diorite intrusion (~ 15 Ma) south of Leggett Peak; **Tsu**, upper Spears Group and **Tsl**, lower Spears group volcaniclastic sedimentary rocks; **Tla**, andesite of Dry Leggett Canyon; **t**, discontinuous tuffs interlayered in upper Datil Group. Adapted from Ratté (1989).

case more monitoring is performed and the slide is reevaluated. In most cases the slide is stabilized. Although landslide remediation in highway construction is still more of an art than a science, improved instrumentation for data collection and computer-aided design have increased our capability for coping with this perennial problem.

23.3 Note drain pipes in slide area. **0.4**

23.7 Roadcut on north (left) side of highway in rounded boulder conglomerate containing abundant pre-Tertiary clasts, as well as Tertiary volcanic clasts (Fig. 3.8). Moderate to steep dips are probably related to deformation along the San Francisco Mountains fault zone, which bounds the footwall of the San Francisco Mountains tilted block, just ahead. The conglomerate in this roadcut represents a small exposure of the Baca Formation in the Bull Basin quadrangle. Prominent topographic nose, or buttress, about 0.6 mi south of highway at mountain front, east of Leggett Peak (Fig. 3.7) is a small intrusion of Miocene quartz diorite (~ 16 Ma), one of several silicic volcanic or sub-volcanic intrusive and eruptive centers within the Reserve graben; the intrusion here is along the San Francisco Mountains fault zone. **0.2**

23.9 Highway crosses trace of west strand of San Francisco Mountains fault zone. **0.8**

24.7 East border of Bull Basin quadrangle; enter Reserve 7½' quadrangle. MP 20 just ahead. **0.9**

25.6 Highway junction, US-180 and NM-12; **turn left on NM-12. Continue north** on alluvium-mantled surface. **0.6**

26.2 Entrance to Reserve airport on right. Good exposures of Gila Formation in roadcuts ahead. **1.3**

27.5 Cross bridge; dip slope of Brushy-Saliz Mountains tilted block to east (right). San Francisco Mountains fault zone at base of range-front scarp to west. **1.1**

28.6 MP 3. Cross S U Canyon. Gila Formation along highway fills half graben east of San Francisco Mountains fault zone and laps onto westward-dipping Brushy-Saliz Mountains tilted block east of highway. Brushy-Saliz Mountains block bifurcates into Higgins

FIGURE 3.8. East-dipping beds of conglomerate in Baca Formation at lower end of Dry Leggett Canyon on the north side of US-180, adjacent to the San Francisco Mountains fault zone near the east edge of the Bull Basin quadrangle.

Mountain and Jon S Mountain blocks of Crews (1990) north of Starkweather Canyon, just ahead. **1.5**

30.1 Cross Starkweather Canyon. Roadcuts in Gila Formation, on left, dip about 13°, somewhat less than the 15-20° dips of the underlying volcanic rocks, ahead in Starkweather Canyon. This change in attitudes is interpreted to be the result of differential tilt of the sedimentary rocks that were deposited during rotation of the Brushy-Saliz Mountains block, but an alternative explanation could be an additional fault that displaces Gila beds at the base of the dip slope east of the highway. North of Starkweather Canyon, the complex Brushy-Saliz Mountains tilted block was divided into the Higgins Mountain and Jon S-Enid sub-blocks by Crews (1990, p. 123-181), who used the Reserve graben as a field model for his doctoral dissertation at the University of Wyoming. **0.4**

30.5 Higgins Ranch road on left. Roadcuts ahead, from here to the highway bridge at the lower end of Starkweather Canyon, provide a stratigraphic descent through a 100-ft-thick sequence of typical, thin (16-33 ft) aa flows of Bearwallow Mountain Andesite (Oligocene and Miocene). Beneath the Bearwallow Mountain Andesite, Bloodgood Canyon Tuff varies from nearly white, cavernous weathering, poorly welded tuff to eutaxitic, densely welded tuff. The Bloodgood Canyon Tuff is separated from the underlying, red Shelley Peak Tuff (exposed at edge of highway, on right, at west end of bridge) by red, crossbedded, eolian sandstone. Davis Canyon Tuff is beneath the sandstone about 100 ft down-creek from the bridge. **0.1**

30.6 MP 5. Roadcut in Bearwallow Mountain Andesite. Cliffs of Bloodgood Canyon Tuff on both sides of canyon just beyond roadcut. **0.5**

31.1 **Slow**! Highway bridge over Starkweather Canyon. Beyond the bridge, in the bushes on the north (left) side of the road, coarse talus facies of the Gila Formation marks the trace of the Saliz Mountains-Higgins Mountain fault (Fig. 3.4), and we cross into Gila Formation on the hanging wall dip slope of the Jon S Mountain half graben tilted block (Jon S-Enid block of Crews, 1990). This coarse, angular talus deposit is a typical colluvial wedge of Gila Formation facies association no. 5 (Crews, this volume), adjacent to an intrabasin fault scarp. **0.4**

31.5 Reserve Ranger Station on right. MP 6 just ahead. **1.0**

32.5 Junction in downtown Reserve. **Turn left with NM-12.** The road to the right follows the San Francisco River valley south for about 7 mi, where the Bloodgood Canyon Tuff, in the deepest part of the Reserve graben, is at river level at about 5600 ft elevation at the southern border of the Reserve quadrangle (J. C. Ratté, unpubl. 1989). The position of the Bloodgood Canyon there represents a drop of about 2625 ft from exposures near Alpine, Arizona, where this trip began.

East of the San Francisco River, here in the Reserve quadrangle, northeast-trending faults dip mainly northwestward, and provide some symmetry to the otherwise asymmetrical system of half grabens. The Keko Canyon fault, in the Milligan Mountain quadrangle (Ratté and Bove, in press, 1994) seems to mark a change from rotated fault blocks to a series of narrow, non-rotated blocks, by which the Bloodgood Canyon Tuff is re-elevated to an elevation of greater than 8000 ft on the east side of the Reserve graben, at the south end of the Tularosa Mountains. Thus, the Reserve graben system, from the Luna fault on the west to the vicinity of Eagle Peak to the east, is about 25 mi wide. **0.5**

33.0 Bridges across San Francisco River; continue north through roadcuts in Gila Formation, in the hanging wall slope of the Jon S tilted block. **0.9**

33.9 Cross east boundary of Reserve quadrangle into Milligan Mountain quadrangle (Ratté and Bove, in press, 1994). Views to east (right) of the south end of the Tularosa Mountains (Fig. 3.9) on the east side of the Reserve graben. Eagle Peak is the central vent for the 11-12 Ma dacitic Eagle Peak volcano, and Milligan Mountain is a satellitic eruptive center (see following minipaper by Bove, et al).

To the west is the Higgins Mountain half-graben tilted block; meanwhile we are crossing the dip slope of the Jon S half graben, the next rotated block to the east. The Saliz Mountains-Higgins Mountain fault, along the base of Higgins Mountain, separates the two half grabens (Fig. 3.10). Fine-grained Gila Formation adjacent to the fault is among the finest-grained beds observed in the Reserve area, and is interpreted as axial-fluvial facies deposited along the fault, and it contrasts with the coarser-grained, dip-slope canyon-fill facies and distal fan deposits in the roadcuts that we have just driven through. Note the broad syncline in Gila beds along the Saliz Mountains- Higgins Mountain fault (Fig. 3.10). The syncline may indicate some strike-slip component of movement along this fault. The San Francisco River enters valley through the San Francisco Box, where it has cut through flows of the andesite of Dry Leggett Canyon at about 11 o'clock. **1.2**

35.1 Junction, road to Eagle Peak. **Turn right (east).** Between this junction and Tularosa river ahead, we are crossing the Jon S tilted block. Jon S Mountain consists of the two cuesta ridges east of NM-12 and north of the Eagle Peak road. The local volcanic stratigraphic sequence is largely repeated by a fault between the ridges. Along this part of the Eagle Peak road, we are crossing Gila Formation and alluvium. **0.5**

THE EVOLUTION OF THE MID-TERTIARY EAGLE PEAK VOLCANO, CATRON COUNTY, NEW MEXICO

Dana J. Bove[1], James C. Ratté[1], William C. McIntosh[2], Lawrence W. Snee[1], and Kiyoto Futa[1]

[1]U.S. Geological Survey, Denver, CO 80225;
[2]New Mexico Bureau of Mines and Mineral Resources, Socorro, NM 87801

Eagle Peak is a well-preserved, low-angle, calc-alkaline volcano in the mid-Tertiary Mogollon-Datil volcanic field of southwestern New Mexico. Geologic mapping (Bove et al., 1989; Bove, 1990), in conjunction with petrologic studies allows the differentiation of several distinct andesitic to dacitic eruptive phases (Fig. 3.11). These phases can be subdivided into either central vent or satellitic facies, depending on the site of eruption or emplacement. The central vent series includes andesitic to dacitic flows, agglomerate vent breccias, and a dacitic plug, whereas the

FIGURE 3.9. View east (to right) to Tularosa Mountains on the east side of the Reserve graben. Eagle Peak and Milligan Mountain are eruptive centers of the 11-Ma Eagle Peak dacite (see Bove and Ratté, this volume).

FIGURE 3.10. View west from NM-12 across dip slope of Gila Formation, Tg, to Higgins Mountain. Note broad synclinal fold in Gila beds on near side of Higgins Mountain fault; fold may be an indication of at least some strike-slip component on the fault.

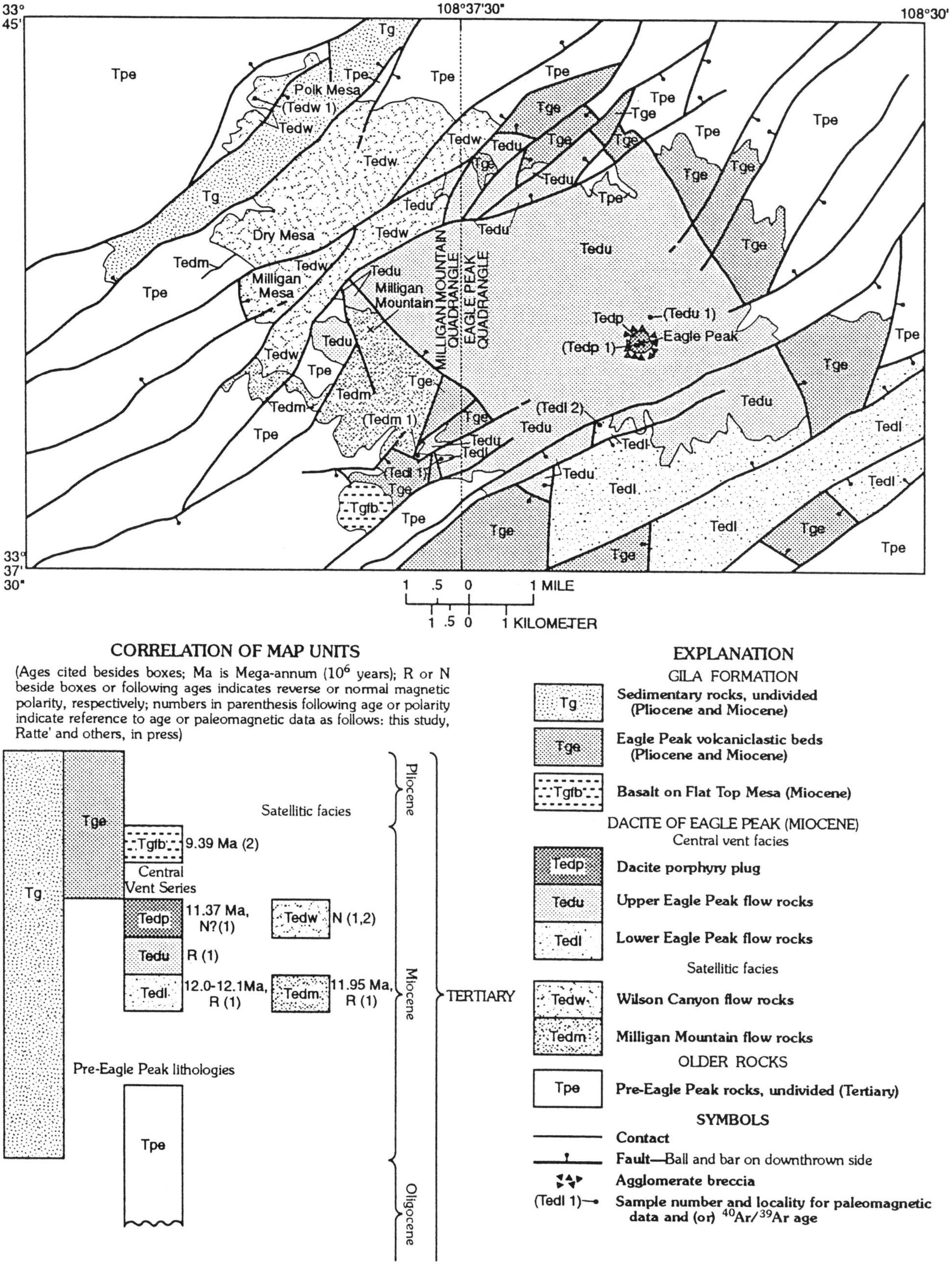

FIGURE 3.11. Geologic map of Eagle Peak Volcano area, Catron Country, New Mexico.

satellitic facies consists of thick andesitic flows and a feeder dike. Lavas erupted from the central vent area are separated into upper (Tedu) and lower (Tedl) flow units; upper flows are distinguished from lower flows by the low percentage of phenocrysts, absence of hydrous mineral phases and ilmentite, and more evolved compositions. Satellitic flows comprise two distinct bodies of lava erupted from centers on the western (Milligan Mountain unit, Tedm) and northwestern (Wilson Canyon unit, Tedw) margins of the volcano (Fig. 3.11). The Milligan Mountain and Wilson Canyon satellitic lavas, although chemically indistinguishable, are petrographically different. Milligan Mountain lavas contain a hydrous mineral-bearing assemblage, similar to that of the upper unit central vent flows, whereas Wilson Canyon lavas are essentially anhydrous but contain enclaves of finely porphyritic andesite (?) containing disequilibrium hornblende.

$^{40}Ar/^{39}Ar$ and paleomagnetic studies indicate that the Eagle Peak volcano was active between 12.1 and 11.4 Ma (Ratté et al., in press) and spanned at least one reversal in magnetic polarity. A weighted mean age of 11.73 ± 0.04 Ma was calculated for the reversely magnetized central vent and Milligan Mountain satellitic flows using the inverse of the sample error or variance (Taylor, 1982). These ages were averaged because they are analytically indistinguishable at the 95% confidence level, and geologic and paleomagnetic data indicate that flows represented by these samples were emplaced in rapid succession. The weighted mean age of 11.73 Ma, which takes into account the accuracy of the individual analysis, is analytically distinguishable at the 95% confidence level from the 11.37 ± 0.09 Ma central plug (Tedp). Geologic and petrogenetic constraints suggest that the normally magnetized Wilson Canyon lavas formed very late in the Eagle Peak volcanic sequence.

Major oxide, trace-element, and isotope geochemistry indicates that the central vent and satellitic eruptives define two separate and distinct magmatic series. The satellitic series, although similar in modal mineralogy and REE patterns to the central vent series, is slightly more alkaline and relatively enriched in the high field strength elements Nb, Ta, P and Ti. Satellitic series rocks ($^{87}Sr/^{86}Sr_i$ =0.70839, ϵNd =-4.8) also show less upper crustal contamination and are probably more primitive than rocks of the central vent series ($^{87}Sr/^{86}Sr_i$ =0.70958, ϵNd =-8.5).

Constraints based on major and trace element data support petrogenetic models that require a periodic tapping of the central vent and satellitic series magmas, which individually became more evolved by removal of crystals from the magma chamber. Crystal fractionation modelling of the central vent series suggests that magma forming the hydrous lower unit lavas was parental to the anhydrous upper unit flows. Accordingly, initial eruptions (62-64% SiO_2) from the central vent tapped a deep level magma reservoir in which hydrous mineral phases and ilmentite were stable. Subsequent anhydrous central vent eruptions (64-67% SiO_2) tapped a higher level reservoir that was established after eruption from the deep reservoir. The upper level reservoir was probably fed repeatedly from the deeper magma chamber. Transfer of magma into the shallow reservoir favored conditions of lower P_{H2O} and fO_2 resulting in the destabilization of the hydrous mineral phases and ilmentite, respectively. This scenario is supported by the rare occurrence of disequilibrium hornblende in the otherwise anhydrous upper unit and the presence of magnetite without ilmentite, which are present together in the lower unit. The near uniformity in parent-daughter orthopyroxene and plagioclase core compositions in the upper unit is consistent with a system in which the bulk magma chemistry evolved mainly by crystal removal without evolution in the major phenocrysts or in the residual magma. In contrast, least-squares modelling results paired with the presence of evolving phenocryst compositions within rocks of the satellitic series suggest that the residual magma evolved in conjunction with crystal fractionation.

Regionally, the Eagle Peak volcanic rocks, although temporally associated with <21-Ma bimodal basalt-rhyolite volcanism (Group II of Ratté and Futa, 1989), have strong geochemical affinities for Group I rocks of Ratté and Futa (1989), which mainly consist of 27-23 Ma-Bearwallow Mountain Andesite. Sr and Nd isotope chemistry demonstrates that Group II mafic rocks, which have Sr and Nd isotopic ratios ranging from depleted mantle to bulk earth values ($^{87}Sr/^{86}Sr_i$ = 0.7030-0.7057, ϵNd = 0 to +9.12), are substantially more primitive than Eagle Peak rocks ($^{87}Sr/^{86}Sr_i$ = 0.70839-0.70958, ϵNd =-8.5 to -4.8). In contrast, Eagle Peak rocks are geochemically more similar to Group I Bearwallow Mountain rocks ($^{87}Sr/^{86}Sr_i$ = 0.7070-0.7102, ϵNd =-8.15 to -5.95). Eagle Peak and Group I volcanic rocks are strongly depleted in the HFS elements relative to Group II rocks and are enriched by two to four times in the elements Rb, Zr, Ba, Th and Hf. Eagle Peak rocks and Group I andesites were affected significantly by upper crustal contamination and were probably derived from similar magmatic sources.

35.6 Crossing Largo Creek; views to left of Jon S composite fault block, with 20-25° dip slopes on the two fault-separated hogbacks (Fig. 3.12). **0.8**

36.4 Road crosses crest of low north-south ridge capped by Group II "true" basalt of Futa and Ratté (1989, p. 100) as distinguished from older, contaminated mafic rocks such as the Bearwallow Mountain Andesite, Group I of Futa and Ratté. Basalt barely crops out beyond side trail on left, and as a dark patch on right side of road, but caps ridge on north (left) side of road for 0.6-1.2 mi. An Ar/Ar age of ~ 13.7 Ma has been determined for plagioclase from this flow. These young Group II basalts are associated with basin- and- range extensional faulting around the margins of the Mogollon-Datil volcanic field.

The Group II basalts also are sometimes described as "big feldspar" basalt because they commonly contain plagioclase phenocrysts 0.4-1.2 in. or larger (Figs. 3.13, 3.14). The clarity and citrine yellow color of the plagioclase phenocrysts make them much prized by "rockhounds" as semi-precious stones, which are known popularly and sold commercially as "bytownite" (Murphy, 1971), and are commonly faceted for rings and other jewelry. However, the feldspar crystals are actually labradorite (An 67-68), as indicated by their refractive indices and x-ray pattern. Some centimeter-size phenocrysts have been found near here, but much better collecting can be found elsewhere, as in the Saliz Pass quadrangle (Ratté, 1980). **0.8**

37.2 Sharp switchback to left. Tuff of Triangle C Ranch, a few meters thick, exposed at left edge of road, about 300 ft beyond curve. This tuff, which is indistinguishable in age and composition from the Bloodgood Canyon Tuff, is distinct in having normal paleomagnetic polarity, and locally, as here, is separated from the underlying Bloodgood Canyon Tuff (28.1 Ma) by a few feet of volcaniclastic sedimentary rock. **0.2**

37.4 Outcrops on left side of road at river level were mapped as Bearwallow Mountain Andesite because they lack the big feldspar phenocrysts characteristic of the younger basalt flows, but their relatively fresh appearance, abundance of fresh olivine, and microdiabase texture suggest that they may, in fact, be the younger basalt. See further discussion in the following minipaper.

Gila Formation outcrops ahead are faulted down by the east-northeast-trending Kimball Spring fault (Fig. 3.12); Bloodgood Canyon Tuff and older rocks form the footwall behind the fault, to the ridge crest. The Gila Formation here is the prime example of Crews' (1990) facies association 4, massive, channeled sandstone (see Crews, this volume, fig. 10). **0.2**

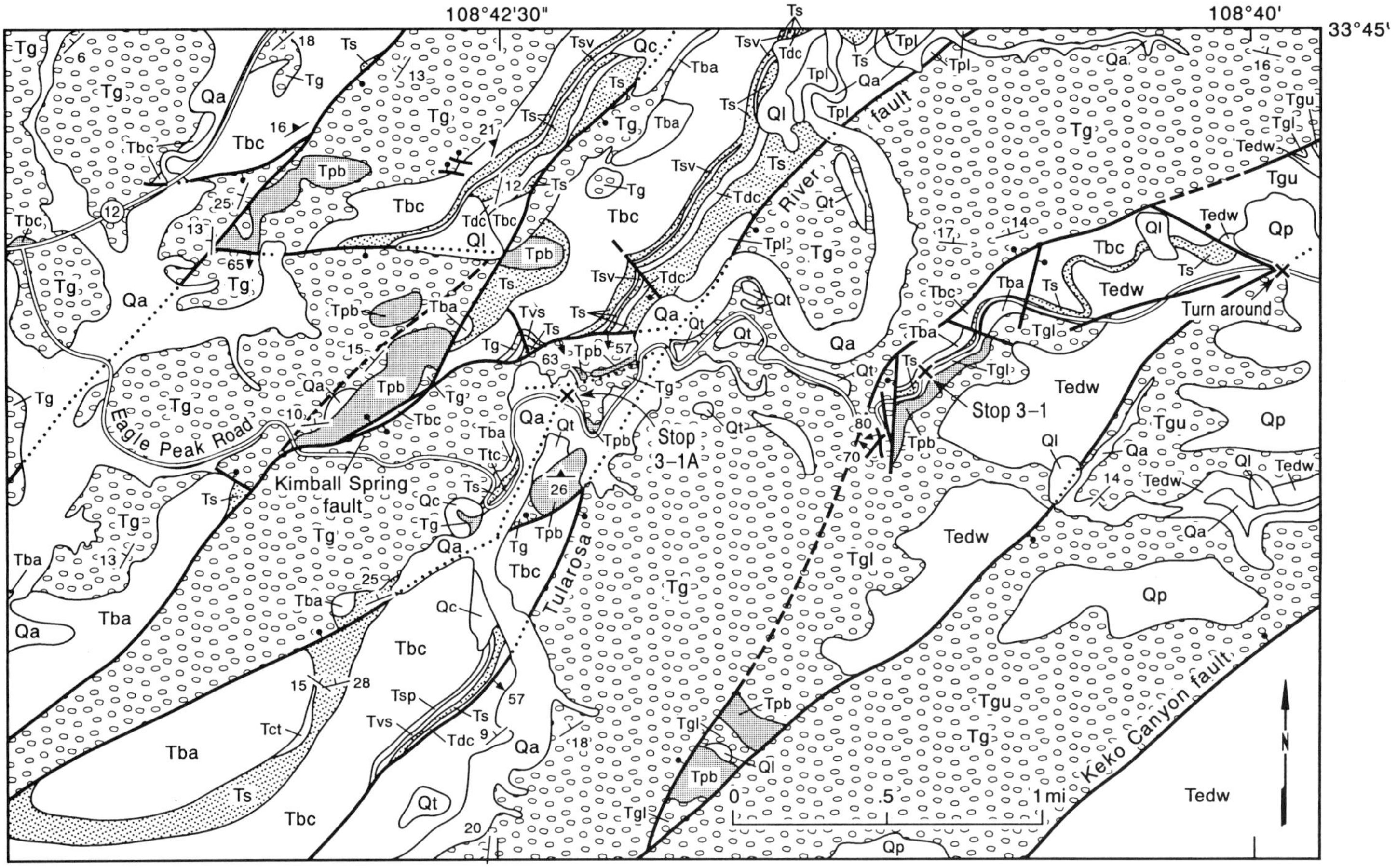

Correlation of Map Units

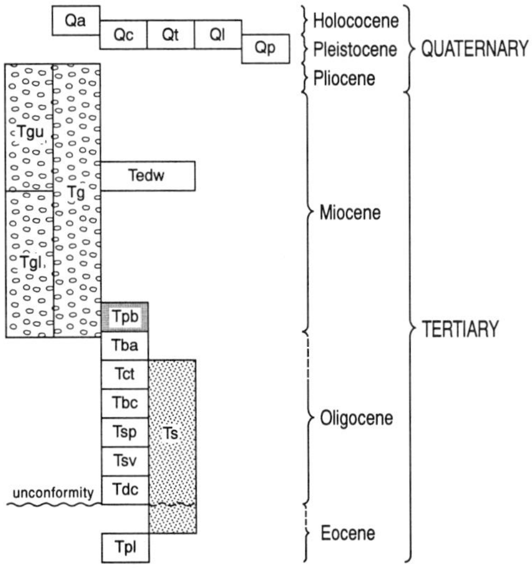

Map Units

Surficial Deposits (Quaternary)
- Qa—alluvium
- Qc—colluvium and talus
- Ql —landslide deposits
- Qt —terrace gravels
- Qp—pediment gravels

Gila Formation (Pliocene–Oligocene)
- Tg —Gila Formation undivided
- Tgu—Upper Gila Formation (Pliocene and Miocene)
- Tgl —Lower Gila Formation (Miocene and Oligocene)

Tedw— Eagle Peak dacite (Miocene)
Tpb — Basalt of Pueblo Park (Miocene)
Tba — Bearwallow Mountain Andesite (Miocene and Oligocene)
Tct — Tuff of Triangle C Ranch (Oligocene)
Tbc — Bloodgood Canyon Tuff (Oligocene)
Tsp — Shelley Peak Tuff (Oligocene)
Tsv — Squirrel Springs Canyon Andesite and/or Vicks Peak Tuff (Oligocene)
Tdc — Davis Canyon Tuff (Oligocene)
Ts — Spears Group volcaniclastic sedimentary rocks (Oligocene and Eocene)
Tpl — Dry Leggett andesite (Eocene)

FIGURE 3.12. Geologic map of the Jon S half graben in the vicinity of Stop 1. Adapted from Ratté and Bove (1994).

FIGURE 3.13. Labradorite megacryst in porphyritic Miocene basalt.

FIGURE 3.14. Labradorite crystals from the weathered surface of a Miocene basalt flow in the Saliz Pass quadrangle southwest of Reserve (Ratté, 1980).

DISTINGUISHING YOUNG 'TRUE' BASALTS FROM BEARWALLOW MOUNTAIN ANDESITE

J.C. Ratté, D.J. Bove and Kiyoto Futa

U.S. Geological Survey, P.O. Box 25046, MS 905, Denver, CO 80225

Distinguishing Basin-and-Range-age basalt (group II) and largely pre-Basin-and-Range-age Bearwallow Mountain Andesite (group I) can be a difficult mapping problem. In most outcrops, the relatively fresh appearance and the presence of large but generally sparse plagioclase and augitic pyroxene phenocrysts, or interlayering with Gila Formation sedimentary rocks serve to identify the group II basalts of Futa and Ratté (1989) from older group I basaltic andesite to dacitic flows of Bearwallow Mountain Andesite. Major oxide and trace element analyses usually provide clear distinctions between the two groups, but not always. The use of zirconium:niobium ratios or Zr:Nb plotted against SiO_2 contents, can be useful in distinguishing the mafic flows of the two groups, but it also has its limitations. Most group I basalts have Zr:Nb ratios greater than 10, and most group II rocks have Zr:Nb less than 10, but some rocks that have Zr:Nb in the 8-12 range are ambiguous. Rb-Sr and Sm-Nd isotope analyses also tend to separate these mafic rocks into two distinct groups in most cases, and provide evidence of different magmatic source regions related to the structural evolution of the Mogollon-Datil region (Futa and Ratté, 1989).

The recognition problem is well illustrated in the vicinity of Kimball Spring in the Milligan Mountain quadrangle near Stops 1 and 1A. Four outcrops of mafic lava flows, labelled 1-4 on Figure 3.12, were mapped as either group I or group II on their field characteristics, and then sampled for chemical comparisons (Table 3.1). The rocks at localities 2 and 4 were identified as young basalts during mapping, whereas the flow at locality 3 was mapped as Bearwallow Mountain Andesite; their distinguishing lithologic characteristics are consistent with stratigraphic sequence. At locality 1, however, where the geologic context favors correlation of the outcrops on opposite sides of the river, the mafic flow east of the river is easily identified as belonging in group II by its sparse, big plagioclase phenocrysts, but the flow along the road at locality 1, west of the river, gives mixed signals. It seems to lack large plagioclase phenocrysts, and it has a Zr:Nb ratio of 12, nearly identical to that of the Bearwallow Mountain Andesite sample from locality 3 (Table 3.1); these characteristics imply a group I affinity. However, the bulk chemistry of the rock at locality 1 (Table 3.1), shows a basaltic composition, with SiO_2 content intermediate between that of group II samples from localities 2 and 4; the highest MgO and CaO, by far, of any of the four samples, and the lowest K_2O of the four samples, all indicating a more likely affinity for the young basalts of group II. Additional chemical analyses or dating, or both, will be necessary to ascertain if the Bearwallow Mountain Andesite mapped at locality 1 on the published map (Ratté and Bove, in press) is correctly identified.

37.6 Cross cattleguard. Corrals on left. Return here later and park for Stop 1A. **0.3**

37.9 Sharp curve with young, "big feldspar" basalt outcrops on left. This is locality 2 on Figure 3.12 and in Table 3.1. Basalt contains 0.4-0.8 in. labradorite, and 0.2-0.4 in. augite phenocrysts, and quartz-diorite? crustal xenoliths. **0.3**

38.2 Ford across Tularosa River above Kimball Spring. Jon S tilted block to west, with Kimball Spring-Tularosa River volcanic sequence exposed in slopes above river. Road now crosses through Gila Formation between Tularosa River fault, at river level, and unnamed fault to east (Fig. 3.12). Unnamed fault defines the west side of a narrow, wedge-shaped horst within the also wedge-shaped, symmetrical, Tularosa graben, which is at the eastern footwall of the Reserve graben, marked by the Keko Canyon fault. **0.9**

39.1 Road is now on uppermost of three Tularosa River terraces, here on the east side of the river. As road curves to right, approaching fault on west side of intragraben horst, view through opening in trees to left shows Bloodgood Canyon Tuff, at about this level within the horst. **0.2**

39.3 Switchback to left. Before turn, view is up Wilson Canyon, where Eagle Peak dacite flows (11-12 Ma) can be seen *overlying* lower Gila Formation. **0.4**

39.7 Curve overlooks big bend in Tularosa River. Roadcuts on right show young basalt above Gila Formation. **This is Stop 1, but continue to designated turn-around on mesa top. 0.2**

TABLE 3.1. Comparative chemical analyses of Bearwallow Mountain Andesite (Tba) and younger basalt of the bimodal basalt-rhyolite suite (Tb) in the vicinity of Stop 1. X-ray spectrography: laboratory chief, J.S. Mee; analysts, D.F. Siems and J.S. Mee. Zr and Nb by Ratté.

Map no. (Fig. 3-35.55)	1	2	3	4	5
Formation	Tb?	Tb	Tba	Tb	Tb
SiO_2*	50.5 (50.8)	50.8 (51.7)	51.9 (53.4)	49.8 (50.9)	54.9 (55.1)
Al_2O_3	15.4	16.2	14.5	16.6	17.4
$FeTO_3$	11.5	11.3	9.13	11.8	8.34
MgO	6.91	4.84	5.13	4.28	4.60
CaO	8.51	7.48	7.30	7.70	7.73
Na_2O	3.01	3.35	3.09	3.45	4.00
K_2O	1.20	1.81	3.23	1.67	1.71
TiO_2	1.90	1.94	1.73	2.04	1.35
P_2O_5	0.35	0.37	0.92	0.39	0.44
MnO	0.15	0.17	0.13	0.15	0.13
LOI (925°C)	0.41	1.66	2.44	1.99	0.31
SUM	99.84	99.92	99.50	99.87	100.01
Zr:Nb	12	5	13	5	5

*SiO_2 values in parentheses are normalized volatile free.

39.9 Bearwallow Mountain Andesite crops out along road; compare this fine-grained basaltic andesite, having tiny, orange, iddingsite-after-olivine grains, with the similar but coarser grained "big feldspar" basalt at the previous point. The Bearwallow Mountain Andesite here dips southward beneath the Gila Formation and younger basalt down the road. **1.1**

41.0 Road tops out on Polk Mesa. **Make U-turn, utilizing space off road to right, around tree.** Retrace route to overlook at Stop 1. Park along edge of road behind lead vehicle. **1.3**

42.3 **STOP 1.** The purpose of this stop is to present an overview of the Tularosa and San Francisco river valleys, as a basis for a discussion of the Reserve graben. This stop also affords an excellent view of the volcanic sequence present in the Jon S half graben block as exposed in the footwall of the Tularosa River fault on the west side of the river (Fig. 3.15). After a discussion of what you can see from here, and the regional context, we'll go down into the valley to Stop 1A and crawl around on the rocks a bit.

At this viewpoint, we are standing on a narrow horst within the deepest part of the Reserve graben. Behind us, about 1 mi to the east, is the westward-dipping Keko Canyon fault, along which stratigraphic units are uplifted eastward within the Tularosa Mountains block on the eastern side of the Reserve graben. To the west, we are looking across the Tularosa River fault at the obsequent face of the Jon S Mountain tilted block. Here, along the road, thin Bearwallow Mountain Andesite flows overlie thin volcaniclastic sandstone, which is above Bloodgood Canyon Tuff in the slopes below us. The fault on the west side of this narrow horst drops the Gila Formation down to river level. However, deposition of the Gila beds, according to the half-graben model presented here for the Reserve graben system, accompanied rotational (listric) faulting and was confined to individual basins within the half graben, at least initially (Crews, 1990). On the slopes above us, on the east side of the road, thin Gila Formation is overlain by thin, young, basalt flows, more Gila beds, and dacite flows of Eagle Peak from the Eagle Peak volcano in the Tularosa Mountains to the east (Fig. 3.12).

The volcanic sequence exposed across the Tularosa River (Fig. 3-15) consists of the following units: at river level, barely visible from here, are porphyritic flows of red-gray andesite of Dry Leggett Canyon (~ 34 Ma), overlain by mostly red volcaniclastic sedimentary rocks about 525 ft thick, of the Oligocene Pueblo Creek Formation (Ratté, 1989), which is now included in the Eocene and Oligocene Spears Group. Above, are the approximately 100 ft-thick, bold, columnar jointed cliffs of 29-Ma Davis Canyon Tuff, which was erupted from the Mogollon Mountains cauldron complex, in turn overlain by volcaniclastic sedimentary rocks and 16-33 ft of 28.6-Ma Vicks Peak Tuff, which was erupted from the Nogal Canyon cauldron at the south end of the San Mateo Mountains about 75 mi to the east. The Vicks Peak Tuff is discontinuously overlain by 0-16 ft of Squirrel Springs Canyon Andesite. Above, is another 80-100 ft of volcaniclastic sandstone, capped by about 200 ft of 28.1-Ma Bloodgood Canyon Tuff, erupted from the Bursum caldera in the Mogollon Mountains about 12 mi to the southeast. The total thickness of the volcanic sequence as seen from here is about 1000 ft, and on top of that are patchy outcrops of Bearwallow Mountain Andesite and Gila Formation that mantle the dip slope, out of sight from here, on the back side of the ridge.

A down-to-the-south, high-angle normal fault offsets this sequence about 65-100 ft in the face of the Jon S block, and the much larger, east-northeast-trending Kimball Springs fault 1000-1300 ft to the left of the smaller fault, drops the Gila Formation down against

FIGURE 3.15. View of Kimball Spring volcanic sequence near Stop 1. **Tls**, lower Spears Group volcaniclastic rocks; **Tdc**, Davis Canyon Tuff (29 Ma) with 1-2 m of non-welded pyroclastic flow, **pf**, at base; **Tus**, upper Spears Group volcaniclastic rocks; **Tvt**, Vicks Peak Tuff (28.5 Ma) with a few meters of non-welded pyroclastic flows, **pf**, at base; **Tsq**, Squirrel Springs Canyon Andesite; **Tbc**, Bloodgood Canyon Tuff (28 Ma).

the volcanic sequence. This east-northeast-trending Kimball Springs fault links the Tularosa River fault and the next northeast-trending graben fault to the west, which offsets the two Jon S Mountain ridges.

Looking to the southwest, down the Tularosa River valley (Figs. 3.16, 3.17) we see the double, west-dipping cuestas with the saddle between, which constitute the continuation of the Jon S block (Fig. 3.15). To the right of the double cuesta, the ridge with the open areas along its base is the Saliz Mountains tilted block on the west side of the San Francisco River. The Saliz Mountains block is the southward continuation of the Jon S and Higgins Mountain sub-blocks, which lose their identity south of Starkweather Canyon, which can be seen at the break in slope at the north end of the Saliz Mountains ridge. The next ridge to the west, on the skyline, is the San Francisco Mountains on the west side of the San Francisco Mountains fault zone. On a clear day, the stretch of US-180 that is on the east side of Mail Hollow divide near the Heifer Basin overlook (Fig. 3.6) can be seen from here.

The Reserve graben has been known as a major structural feature at least since the reconnaissance geologic map of the Reserve 30-minute quadrangle (Weber and Willard, 1959). It was described as one segment of the outer graben system surrounding the Mogollon plateau, which has been interpreted as the surface expression of a large composite pluton about 75 mi in diameter (Elston et al., 1976), and a genetic relationship between the graben system and the batholith has been implied. Recent, detailed USGS geologic mapping of 7-1/2 minute quadrangles along the Morenci-Reserve fault zone, including the Reserve graben, has led to the proposal that the Reserve graben is actually part of a separate structural trend (~ N35°E), more northerly than the Morenci lineament (~ N55°E), which includes the San Agustin graben (Ratté, 1989). It has been speculated that the N35°E Morenci-Reserve trend might extend northward beyond the Reserve area, possibly intersecting the N52°E Jemez lineament (Aldrich et al., 1981; Aldrich and Laughlin, 1984) in the vicinity of Mount Taylor (Futa and Ratté, 1989), rather than bending around the north end of the Mogollon plateau, as proposed by Elston et al. (1976). Reconnaissance geologic mapping of the Quemado 1:100,000 quadrangle by Chamberlin et al. (1994b), has reinforced the idea that the Morenci-Reserve fault zone extends northward as

FIGURE 3.16. View southwest across Reserve graben from Stop 1.

represented by the Hickman and Spur Lake faults discussed in the Day 1 road log of this volume.

In summary, the Reserve graben, or graben system, is considered to be about 25 mi wide between the Luna fault on the northwest side and the Keko Canyon fault on the southeast side. Bloodgood Canyon Tuff, the 28-Ma outflow sheet of the Bursum caldera, was downdropped from an elevation of about 8000 ft or more, near Alpine about 15 mi west of Luna, to about 5500 ft elevation along the San Francisco River south of Reserve, and was uplifted back to more than 8000 ft again in the Tularosa Mountains near Eagle Peak (Bove, 1990), 5-6 mi east of the Keko Canyon fault (Fig. 3.12). Thus the aggregate stratigraphic throw across the graben system is about 2625 ft.

The current structural interpretation of the Reserve graben as a series of tilted half grabens formed by rotation on shallow normal faults that flatten with depth is an outgrowth of the work of R.M. Chamberlin, C.E. Chapin, G.R. Osburn and S.M. Cather et al. on structures along the Rio Grande rift, and the northeastern part of the Morenci lineament in the vicinity of Socorro. Crews applied modern concepts of sedimentology and listric faulting specifically to the Reserve graben (Crews, 1990; this volume).

FIGURE 3.17. Oblique aerial view across the Reserve graben showing the major tilted fault blocks: 1, Jon S Mountain block; 2, Higgins Mountain block; 3, Brushy Mountains-Saliz Mountains block; 4, San Francisco Mountains block. R, Reserve.

An additional regional aspect of the north-northeast Morenci-Reserve zone is its apparent segmentation by nearly east-west-trending oblique transfer(?) faults, such as the Kimball Springs fault. One of these east-west zones breaks the dip slope of the San Francisco Mountains block and appears to coincide with a dextral offset of the San Francisco Mountains fault zone (Fig. 3.4). Another east-west zone, the Sand Flat graben, marks a similar dextral offset of structure at the north end of the Tularosa Mountains near Aragon, but somewhat diminished north-northeast structures continue to the north into the Quemado 1:100,000 quadrangle. The full meaning of the fault pattern is currently unknown.

Return to cars and continue retracing route to ford across Tularosa River, and 0.4 miles beyond to optional stop 1A. 2.1

44.4 **OPTIONAL STOP 1A.** At this stop we will examine a small-scale buttress unconformity and a small related reverse fault within the Gila Formation. But first, notice the changing dip of the Gila beds from sub-horizontal near the Kimball Springs fault to 10-15° or more east of the fault. Do these changing dips represent decreasing rotation of younger beds, or drag on the Kimball Springs fault? If the change in dip is related to drag, then the Gila beds in this place have been faulted and are not merely deposited against a preexisting fault scarp.

From the parking area, walk about 600 ft northeast to a small gulch where basalt on the west side is in contact with Gila on the east side, along an apparent fault contact, but Gila on the ridge crosses the projected

FIGURE 3.16. View southwest across Reserve graben from Stop 1.

trace of the fault without offset. Closer examination of the Gila-basalt contact shows basalt blocks included in the Gila along the contact, and thus it appears that this is a buried fault scarp that shed basalt cobbles into Gila beds that overlap basalt on the ridge crest without offset, thus a buttress unconformity. Observe the small reverse fault in the Gila Formation (Fig. 3.18) in the "hanging wall" of this outcrop, where the Gila beds have been deformed, but barely offset, during renewed movement on the buried fault.

From the parking area for stop 1A, one can walk back along the road about 1000 ft to the east and examine outcrops of young, "big feldspar" basalt (locality 2, Fig. 3.12; analysis 2, Table 3.1), at the point of the low ridge, at road level. After returning to stop 1A, walk south along the road about 1600 ft to examine outcrops

FIGURE 3.18. Massive, channelled sandstone beds (Crews; 1990 facies association 4) of Gila Formation, Tg, in hanging wall of Kimball Springs fault. Gila beds on right are deposited against faulted young basalt in a buttress unconformity. Tbc = Bloodgood Canyon Tuff.

of Bearwallow Mountain Andesite or young basalt (locality 1, Fig. 3.12; analysis 1, Table 3.1) for comparison. What would you call this outcrop? See specimens at parking area.

If time permits, some may wish to climb around to the Kimball Springs fault to see the oblique slickensides and look at some of the rocks in the older part of the volcanic sequence.

Return to cars and back-track to NM-12. 2.3

46.7 Junction of Eagle Peak road and NM-12. **Turn right on NM-12.** Directly opposite junction is a roadcut in young basalt, which is interlayered in the Gila Formation (see analysis 5, Table 3.1). Bove obtained a whole rock, Ar/Ar age of 13.5 ± 0.12 Ma on this basalt, nearly identical to the age of 13.7 Ma for the big feldspar basalt on the ridge crest to the east at mile 36.4. After right turn onto NM-12, Gila Formation in roadcut on right dips beneath young basalt opposite Eagle Peak road. **0.5**

47.2 MP-10. Bridge across Largo Canyon; outcrops ahead on right are typical densely welded Bloodgood Canyon Tuff on about 15° dip slope in hanging wall of Jon S Mountain block. Bloodgood Canyon Tuff here was called Railroad Canyon Rhyolite Tuff by Elston and Damon (1970, p. 140), with a suggested origin in the Corduroy Canyon cauldron in the Black Range (Elston, 1984, table 1). All Railroad Canyon Rhyolite Tuff is now correlated with Bloodgood Canyon Tuff (Ratté et al.,1984; McIntosh et al., 1991), and the name was abandoned by Lawrence and Richter, 1986. The Corduroy Canyon cauldron is no longer necessary, because the distribution of Railroad Canyon Rhyolite Tuff was the only evidence of its existence. Gila conglomerate unconformably overlies Bloodgood Canyon Tuff locally on this dip slope, and provides good examples of the dip-slope canyon-fills in the coarse proximal deposits of Crews' (this volume) facies association 1. **0.6**

47.8 North edge of Milligan Mountain quadrangle; entering Cruzville quadrangle (T.L.Finnell, unpubl., 1989). **0.4**

48.2 MP 11. Ahead, in roadcut on right, young basalt (Zr:Nb =5) overlies red, oxidized Gila Formation. **1.0**

49.2 MP 12. Ahead, 0.2 mi on left, junction with J.T.S.Park-Torriette Lakes road.

50.2 MP 13. Through trees on right, young (Miocene) basalt flow and underlying Gila Formation dip conformably about 15-20° to west toward highway. Crossing Jon S Flat between here and Cruzville, at MP 16, the hogback ridges east of the highway expose both basalt flows that dip conformably with the underlying volcaniclastic rocks, and younger, flat-lying basalt that unconformably overlies tilted Gila Formation sedimentary rocks, interlayered Miocene basalt, Bearwallow Mountain Andesite and Bloodgood Canyon Tuff. The flat-lying basalt probably all correlates with the basalt flow at Apache Creek, which has been dated at about 1 Ma (Bikerman, 1972; Marvin et al., 1987). The 1-Ma olivine basalt fills topography, appears not to be faulted, and is virtually flat (Fig. 3.19). Northwest of highway (on left), Gila beds locally dip 5-10° toward the San Francisco Mountains fault zone, which parallels highway about 2.5 mi away. Volcaniclastics of Spears Group and andesite of Dry Leggett Canyon crop out in low hills west of fault zone. **3.0**

53.2 MP 16. Passing through settlement of Cruzville. Roadcut on right, at curve, is Gila Formation overlying Miocene olivine basalt, which is exposed on the back side of the cut. Continue on NM-12 with excellent views of Apache Creek water gap ahead (Fig. 3-20), where Tularosa River has cut through basalt flow dated at 0.9-1.0 Ma by conventional K-Ar whole rock method (Bikerman, 1972; Marvin et al., 1987). Downcutting by Tularosa River thus represents at least 330 ft of erosion since eruption of the basalt flow at Apache Creek. More precise Ar/Ar dating of the basalt may indicate a longer time for this amount of erosion. Older basalt flows, exposed at river level in this area, have Miocene K-Ar ages of about 13 Ma (Marvin et al., 1987), and an Ar/Ar age by Bove of 7 Ma. **1.0**

54.2 MP 17. 1-Ma basalt flow caps ridge to right. **2.0**

56.2 MP 19. Cross bridge over Apache Creek; this crossing was not always so easy (Fig. 3.21) Now in western part of Squirrel Springs Canyon quadrangle (Rhodes and Smith, 1976, fig. 2). **0.1**

56.3 Apache Creek store and junction with NM-32 on left. **Turn left onto NM-32**. Gila beds beneath flat-lying Apache Creek basalt flow strike about N35°E and dip about 10° northwest in roadcut on right (Fig. 3.22). Petroglyphs in basalt flow at top of mesa across from Apache Creek Store. See following minipaper by Wells. **2.0**

APACHE CREEK ARCHAEOLOGY

Stermon Wells
New Mexico State Parks, Santa Fe, NM 87504

Prehistoric village sites in the Apache Creek and Tularosa River drainages were cataloged and sketched by Walter Hough in 1907. In those days walls were still standing on some of the terraces and hills beside the streams. Some of these ruins may be about a thousand years old.

Along NM-12 and NM-32 today, most early village sites are hard to see and may only be noticed by traces of ground disturbance or recent digging. Where once stood multi-level apartment houses with hundreds of small cell-like rooms and adjoining underground kiva chambers now only piles of rocks with a sprinkling of small ceramic shards remains. Not a single Tularosa Phase (1100-1200 AD) ruin has been preserved or restored. No road sign indicates these sites along the way, or speaks of these old people at all. Generations of treasure hunters and archaeologists have combed these ruins, and carried all their specimens elsewhere.

FIGURE 3.19. View east from NM-12 near Cruzville showing basalt of Apache Creek (~ 1 Ma), Tb, unconformably on gently tilted beds of the Gila Formation, Tg. Tbc, Bloodgood Canyon Tuff; Tba, Bearwallow Mountain Andesite. Basalt of Apache Creek was probably deposited against older fault scarp, a buttress unconformity?

The basalt bluffs above the Apache Creek store have an interesting collection of petroglyphs pecked into them, and it is a worthwhile climb to see them. Many designs may pertain to Pueblo religion, as well as a very nice looking buffalo!

From the top of this mesa you probably would have been able to see more buildings in 1200 AD than you can see today. Farming may have been more intensive also, even though it was all done by hand. Paul Sidney Martin (1959), of the Chicago Natural History Museum and a leading authority on the local archaeology, found much evidence of agriculture in these valleys. He excavated Tularosa Cave, about 10 mi to the northeast, up the Tularosa River, and discovered that several different kinds of corn, kidney and tepary beans, two kinds of squashes, as well as bottle gourds were being grown in these fields.

Martin and his associates wrote numerous papers and books describing their findings in this area. Their work shed significant light on the antiquity of maize agriculture in the Southwest. Corn may have been introduced as early as 2500 BC, and by perhaps 200 AD the local inhabitants had become quite dependent on it as a staple crop (Martin, 1959). Evidently by 1200 AD people in these valleys had become excellent farmers, as well as being remarkable experts in craft arts such as weaving, pottery and basketry. Local residents also hunted and foraged a wide variety of wild animals and plants to complement their agricultural products.

From the hill overlooking the Apache Creek store, the geography of these stream valleys suggest that they might be excellent natural trail routes, as they were. The modern highways still follow the old trails and arrive at the most significant ancient ruins (or go through them). The distribution of larger ruins along stream drainages throughout southwestern New Mexico in many cases defines the old trail networks. Most of these large ruins were less than a day's walk apart from each other along the natural watercourses, resulting in a very different settlement and land use pattern than occurs today. It allowed people to make maximum use of the arable land and available water resources with the level of technology they had at that time. Trails to the next town stopped at all the farms along the way, thus maintaining a web of trade and social network among the people.

Cross-finds of pottery have shown that people were in trade contact from Chaco Canyon to the Mimbres country. Although the extent and frequency of these contacts is unknown, geography dictates that Apache Creek was on one of the most important trails between the two areas.

About 3 mi up Apache Creek stand the ruins of a stone pueblo containing pottery and remains datable to 1000-1200 AD and related to the Tularosa culture. This pueblo directly overlies the pithouses of an earlier period from 800-900 AD, having pottery and architecture identical with that being used at the same time in the Mimbres region, over 150 mi to the south. This similarity strongly suggests common roots in the pithouse period for these two later groups and lends credence to the idea of a "Mogollon Culture" in these mountains. There has always been much academic discussion of whether these people were more "Mogollon" or "Anasazi", since they were influenced a great deal by both groups.

During the rise of the Chaco civilization, black on white pottery styles became prevalent throughout the entire region, and Tularosa pottery is one "Anasazi" style that has been found in many parts of the San Juan basin. Chacoan "roads" may be traced as far south as Zuni Salt Lake, about 50 mi north of Apache Creek, and trails run south from there through the mountain passes.

The inhabitants of the Apache Creek area employed Chacoan ideas of masonry in their wall construction, but were less adept in their efforts. Ruins 50 to 60 mi north of Apache Creek were more obviously part of the Chaco network in terms of architecture and artifacts. This area may well represent a prehistoric border of some kind between larger cultural groups, with a consequent mixing of cultural traits.

Oddly enough, there has been hardly any recognizable evidence in the archaeological record for contact between the Tularosa region and the Rio Grande. And even though Apache Creek was named for the Apache tribe, there is little recognizable evidence that they were here either. Around the Apache Creek store the map shows canyons named First, Second and Third Apache Canyons. There is an Apache Mountain nearby. Obviously the Apache figure importantly in the long

FIGURE 3.20. Apache Creek water gap is at the confluence of Apache Creek and the Tularosa River. The 1 Ma basalt flow which caps the mesas on opposite sides of the gap probably flowed down an ancestral Apache Creek, to the left of the photograph, into the Tularosa River valley, where the Tularosa eventually cut through the flow into the underlying Gila Formation to form the water gap.

FIGURE 3.21. Pat Birmingham (horseback) and Ambrocio Carrejo crossing Apache Creek in 1929. Courtesy of Buz and Beverly Easterling, Quemado Lake Estates.

FIGURE 3.22. View northward up Apache Creek and NM-32 across the San Francisco Mountains fault zone (SFZ) where it turns nearly east-west near the northeast end of the Reserve graben. Rocks north and west of the fault zone are andesite of Dry Leggett Canyon and younger units of the Spears and Mogollon groups as in Apache Mountain and Tularosa Mountain.

history of this area. In terms of the archaeological record, it is surprising that the 400-year tenure of the Apache left so little evidence of occupation, compared to the Tularosa people who left their subtle, but definite, mark on this beautiful land.

A good collection of local artifacts may be seen at the Western New Mexico University Museum in Silver City. The library at New Mexico State University in Las Cruces contains copies of most of the archaeological reports published about the area.

58.3 MP 2. Approximate west edge of Squirrel Springs Canyon quadrangle; return into Cruzville quadrangle at ranch road just ahead on left. **1.0**

59.3 MP 3. Ranch road and cattleguard on right. Now crossing approximate trace of San Francisco Mountains fault zone, as we move from the edge of the Cruzville quadrangle into the Queens Head quadrangle, and from Gila Formation volcaniclastic rocks on the southeast side of the fault zone into Spears Group volcaniclastic rocks in the footwall on northwest side. Here is about where the San Francisco Mountains fault zone bends from a northeast trend to nearly east-west.

Although there are several hundred feet of throw on the San Francisco Mountains fault zone, the Quaternary basalt flow of Apache Creek continues to crop out in patches along the ridge crest on the east side of Apache Creek with little or no discernible offset (T.L.Finnell, unpubl., 1987). This flow must have entered the ancestral Tularosa River valley essentially at grade along an ancestral Apache Creek drainage from a source area farther northwest. **0.4**

59.7 Outcrops of Spears Group include white sandstone beds and brown conglomerate beds that contain boulders of fossiliferous Paleozoic limestone and Precambrian crystalline rocks. **0.6**

60.3 MP 4. Andesite of Dry Leggett Canyon caps ridge on left, above beds of lower part of the Spears Group. **0.7**

61.0 Approaching curve to left; cliff-forming unit in slopes ahead on right is 15-30-ft-thick tuff of Bishop Peak (34.7 Ma), interlayered with Spears Group volcaniclastics and overlain by andesite of Dry Leggett Canyon. **1.3**

62.3 MP 6. New Mexico Ranch for the Deaf on left at mouth of Lee Russel Gulch. Good collecting up this gulch for Luna blue agate within andesite of Dry Leggett Canyon. **2.0**

64.3 MP 8. Mouth of Whiskey Canyon. Old NM-32, as shown on 1965 Queens Head topographic quadrangle map, goes up Whiskey Canyon. Stay left on new alignment of NM-32. Roadcuts ahead, mainly on east side of road, consist of Spears Group volcaniclastic sedimentary rocks beneath the andesite of Dry Leggett Canyon. The volcaniclastic rocks include a lower, white, pumice-bearing sandstone sequence and an upper, green sandstone sequence. Both sequences locally contain distinctive, thin, centimeter-scale, pink, plagioclase-bearing tuff beds that apparently are juvenile and/or reworked ash layers. The ash layers probably represent premonitory eruptions associated with the source of the overlying lava flows of Dry Leggett Canyon. Plagioclase was collected for dating from several of the ash layers in the belief that it was sanidine, but only a couple of sanidine crystals were recovered from several age samples, and the plagioclase has given erratic Ar/Ar ages.

About 1 mi above the mouth of Whiskey Creek, Apache Creek branches into Hardcastle Canyon on the left and Deep Canyon on the right. NM-32 follows Deep Canyon (Fig. 3.23). Near the junction of the

FIGURE 3.23. Oblique aerial view overlooking Hardcastle Canyon and Stop 2 in Deep Canyon. Ts, Eocene-Oligocene Spears Group; Tla, Andesite of Dry Leggett Canyon.

FIGURE 3.24. Roadcut along NM-32 shows green and white sandstone in lower Spears Group, overlain by columnar jointed andesite of Dry Leggett Canyon at right side of photograph.

FIGURE 3.26. Partly welded surge beds with black fiamme beneath geology pick.

canyons, and in the spectacular roadcuts ahead, on the left, in Deep Canyon, the green sandstone intertongues with the white sandstone.

Drive through roadcut with green, bedded sandstone to left (Fig. 3.24), to where the overlying andesite lava flows of Dry Leggett Canyon come down to road level. We will try to park everyone in the wide pull-out on the west (left) side of the road; **prepare to make a U-turn into the parking area** so that we will be headed correctly to return to Apache Creek store and NM-12, when we leave Stop 2. **1.7**

66.0 **STOP 2. Make U-turn into parking area on west side of road**. At this stop, we shall focus on the pink beds between lava flows of the andesite of Dry Leggett Canyon at the parking area (Fig. 3.25), and the green, volcaniclastic sandstone sequence and interlayered pink ash beds down the road beneath the lava flows (Fig. 3.24).

Here above the parking area, what initially appears to be a thin sedimentary interval between lava flows is complicated by the presence of dark layers (Fig. 3.26) having lenses of perlitic glass as much as several centimeters long (Figs. 3.27, 3.28). These appear to be fiamme, i.e. collapsed or fused pumice fragments (Fig. 3.29), which implies a volcanic rather than a fluviatile sedimentary origin for these beds. The presence of fine-grained, graded and crossbedded sedimentary layers and dune forms mixed with the pyroclasts (Figs. 3.30, 3.31) rules out a simple pyroclastic flow origin. However, at least one ignimbrite (tuff of Bishop Peak) is interlayered in the sandstone sequence more or less immediately beneath the andesite of Dry Leggett Canyon, in the lower part of Apache Creek canyon. That the bedded interval near the base of the andesite flows at this Stop might be the tuff of Bishop Peak was considered, but the latter is characterized by sanidine and biotite, and minor clinopyroxene phenocrysts rather than by the plagioclase, clinopyroxene and orthopyroxene phenocrysts that dominate the phenocryst assemblage in the beds here.

Our collective, but tentative, conclusion is that the various volcanic and sedimentary features exhibited here represent some type of pyroclastic surge deposit

FIGURE 3.25. Pink, partly welded, pyroclastic surge deposits between andesite lava flows of Dry Leggett Canyon along NM-32 in Deep Canyon.

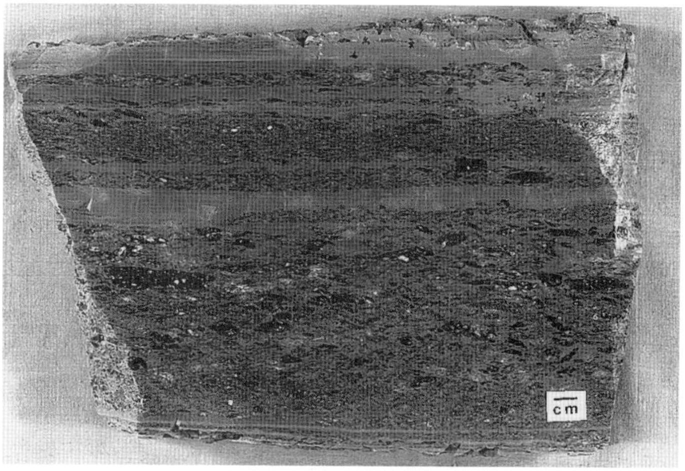

FIGURE 3.27. Polished slab of welded surge showing centimeter-scale, welded glass lenses (fiamme).

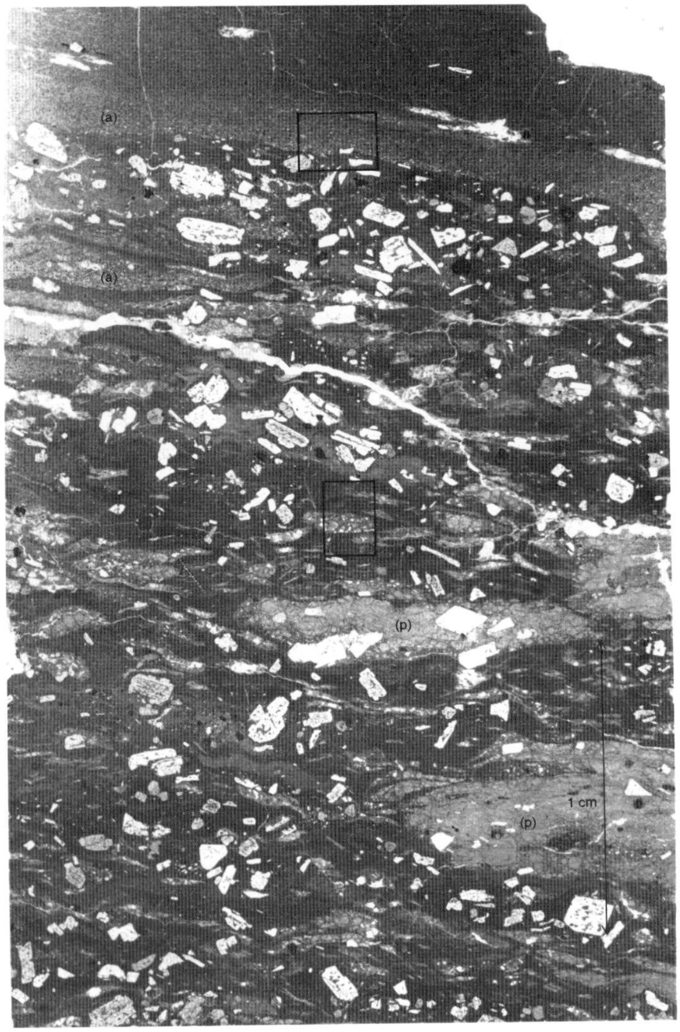

FIGURE 3.28. Photomicrograph, in plane polarized light, of an entire thin section (~ 3.5 x 2 cm.) of welded pyroclastic surge materials showing perlitic glass lenses (fiamme), p; fine, graded ash, a; and plagioclase and pyroxene phenocrysts. See Figs. 3.29 and 3.31 for closeup views of areas in lower and upper rectangles, respectively.

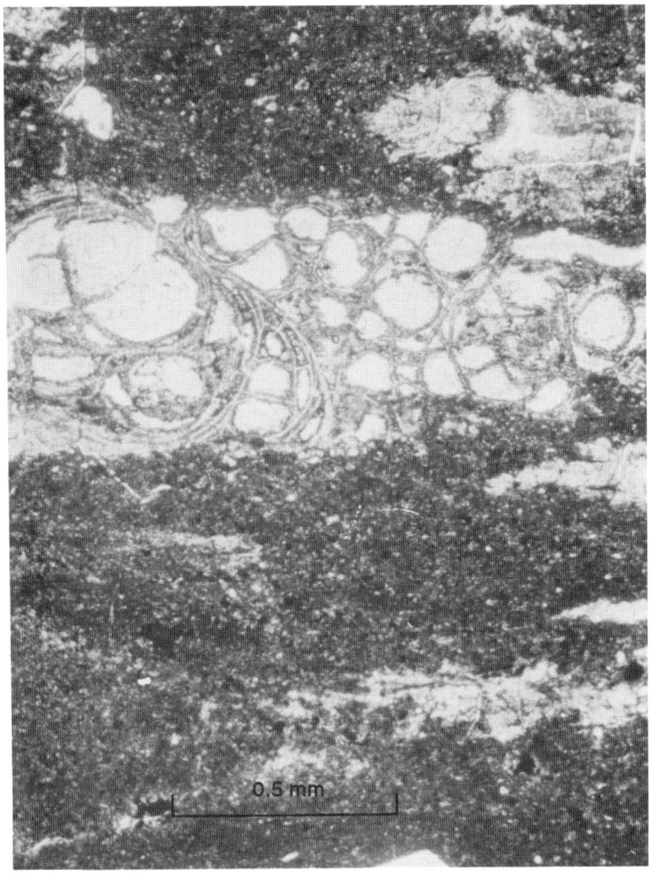

FIGURE 3.29. Welded glass lens (fiamme) with peritic hydration cracks surrounded mainly by fine ash.

that originated in conjunction with the eruption of the andesite lava flows with which the bedded deposits are intercalated.

Pyroclastic surge deposits have been divided into three types: base surge, ground surge and ash-cloud surge (Fisher and Schmincke, 1984; Cas and Wright, 1987), but it is not perfectly clear which type is present here. All three types of surge deposits are related in one way or another to the eruption columns associated with pyroclastic flows, or to processes related to the pyroclastic flows themselves. However, Cas and Wright (1987, p. 120) cited evidence of pyroclastic surge deposits, which they call ground surge, produced by eruption column collapse associated with *fallout* deposits that were *not* accompanied by pyroclastic flows. No discrete pyroclastic flow or ignimbrite is present here, and thus the bedded interval may comprise a pyroclastic surge deposit associated with fall deposits related to the eruption of andesite flows of Dry Leggett Canyon. If so, in order to maintain heat sufficient for welding to occur, the surge must have been deposited while the underlying flow was still hot and then covered almost immediately by the overlying flow.

The presence of small-scale dunes and antidunes supports the surge hypothesis, and the phenocrysts present in the glassy fiamme, mainly plagioclase and pyroxene, suggests comparable chemical compositions for the

FIGURE 3.30. Dune structure in pyroclastic surge beds at Stop 2.

FIGURE 3.31. Photomicrograph showing normally graded ash layers in pyroclastic surge beds at Stop 2.

flows and the bedded surge. However, chemical analyses of the underlying flow and a glassy lens show about 7.0% more silica in the perlitic glass than in the flow (Table 3.2). The chemistry of the glass seems to indicate different sources for the andesite flows and the fiamme, but turbulent flow processes during deposition of the pyroclastic surge could have rendered the glass analyses essentially meaningless, and thus the flows and surge deposits may still be chemically related.

TABLE 3.2. Chemical analyses of andesite of Dry Leggett Canyon and glassy clots (fiamme) from pyroclastic surge beds between Dry Leggett Canyon andesite flows at Stop 2. X-ray spectrography; laboratory chief, J.S. Mee; analysts, D.F. Siems and J.S. Mee.

	glassy clots	porphyritic andesite
SiO_2*	60.2 (64.2)	56.0 (57.2)
Al_2O_3	15.9	16.2
$FeTO_3$	4.26	8.03
MgO	3.13	4.08
CaO	4.12	6.78
Na_2O	3.14	3.29
K_2O	2.06	2.28
TiO_2	0.71	0.99
P_2O_5	0.21	0.23
MnO	0.08	0.11
LOI (925°C)	5.51	1.52
SUM	99.32	99.51

*SiO_2 values in parentheses are normalized volatile free.

Ground surge deposits are generally found within a few miles of their source, indicating that Stop 2 is near a vent for the andesite of Dry Leggett Canyon. Similar surge(?) deposits are present on the northeast flanks of Prairie Point Peak, north of Dry Leggett Canyon, which suggests another possible source for the Dry Leggett flows.

Keep these deposits in mind when we examine pyroclastic surge beds of probable ash-cloud origin at Stop 3. Walk back along the highway to the thick sandstone sequence, where you will find sedimentary structures and other features pertinent to the interpretation of the origin and provenance of these deposits. Keep your eyes and ears open for rock falls from these cliffs; the concrete barriers along the edge of the road are not without an obvious purpose. If you brought a hard hat, it might not be a bad idea to wear it.

The white and green sandstone beds in the Spears Group are at least 800 ft thick in this area; the base is not exposed. Sedimentary structures, such as massive to thin bedding sequences of well-sorted sandstone and silt, fine-grained concentrations of heavy minerals (Fig. 3.32), and shallow channels (Fig. 3.33), are interpreted as overbank deposits on a broad alluvial floodplain. The volcaniclastic origin of these fluvial deposits is well documented by the presence of sand-sized and larger clasts of white to gray pumice throughout most of the sequence, but particularly in the white sandstone. Even more conspicuous are the multiple pink beds, from less than 0.5 in. to greater than 7.5 in. thick, of very fine-grained to medium-grained, (reworked?) ash containing pumice and plagioclase and biotite crystals. These ash beds have phenocryst assemblages more like those of the overlying lava flows than like the tuff of Bishop Peak, which also occurs near the base of the Dry Leggett Canyon flows in the lower part of the canyon of Apache Creek. The tuff of Bishop Peak seems to be absent in this immediate vicinity, either having pinched out or having been eroded before the andesite flows were erupted. Argon 40/39 ages for biotite from some of the ash layers could provide information about the minimum age of the sediments beneath each of the dated ash layers.

Chamberlin and Harris (this volume) describe a correlative interval of pumiceous sandstones, the volcaniclastic unit of Cañon de Leon, exposed on the north flank of Escondido Mountain, 23 mi to the northeast. Sanidine ages of pumice clasts indicate that these beds range from 35.33 ± 0.06 Ma at the base to 34.17 ± 0.13 Ma at the top of this unit, which is about 600 ft thick. This age range includes the tuff of Bishop Peak and the andesite of Dry Leggett Canyon (McIntosh and Chamberlin, this volume).

Return to cars and retrace route to junction of NM-32 and NM-12. 9.4

75.4 Junction, NM-32 and -12. **Turn left onto NM-12** and drive northeast through the Tularosa River water gap, where the Gila Formation is capped by the 1-Ma Apache Creek basalt flow on both sides of the highway. We are now back in Squirrel Springs Canyon quadrangle. **0.7**

FIGURE 3.32. Green sandstone beds in lower Spears Group at Stop 2 showing streaks of heavy minerals. Sandstone beds are separated by pink, shaly and pumiceous layers, interpreted to be reworked? volcanic ash. Pencil is about 5.5 in. long.

FIGURE 3.33. Channel(?) in green and white sandstone in lower Spears Group at Stop 2. Left side of channel appears to be a slip surface.

76.1 Ranch road on right. To left, Gila Formation is capped by the 1-Ma basalt flow at Apache Creek, but on right at river level, beside the various colored pick-up truck beds being used as cattle feeders, and in roadcut just ahead, more young (Miocene) basalt, in lower part of Gila Formation, has been dated at 7.0 ± 0.4 Ma by whole rock Ar 40/39 by D.J. Bove. **0.2**

76.3 MP 20. Ahead, north of the highway, the San Francisco Mountains fault zone swings across our view and trends nearly east-west as the Reserve graben is squeezed down to its apparent termination near the head of the Tularosa River valley. **4.0**

80.3 MP 24. Approximate north edge of the Squirrel Springs Canyon quadrangle; entering Aragon quadrangle (T. L. Finnell, unpubl., 1989). Outcrops in roadcuts, and on both sides of the highway, are Gila Formation within the narrowing Reserve graben. As the highway and the valley swing to a more nearly east-west trend, the San Francisco Mountains fault zone is only about 1300 ft or less north of, and parallel to, the highway. **1.0**

81.3 MP 25. Road on left. San Francisco Mountains fault zone north of here includes a fault about 1300 ft north of the highway that is down to the south, and a parallel, vertical fault about 0.6 mi north of highway that is down to the north. Spears Group volcaniclastic rocks crop out in a narrow horst between faulted Gila beds on both sides. Richard Chamberlin found both sub-horizontal and dip-slip (near vertical) slickensides in gouge along this nearly vertical fault, indicating a complex history of movement for this fault zone. **1.0**

82.3 MP 26. Santo Nino Catholic Church in Aragon on left. San Francisco Mountains fault zone is about 650 ft behind the church. A 15- to 30-ft-thick ignimbrite beneath the andesite flows of Dry Leggett Canyon in the footwall of the fault zone appears to be Kneeling Nun Tuff with a 40/39 Ar age of 34.8 ± 0.54 Ma. Aragon is named for a family that arrived in New Mexico in 1693 (Pearce, 1965). **0.8**

83.1 Cross San Francisco Mountains fault zone. Here at the north end of the Tularosa Mountains, off to the east and south (right), the Reserve graben is en echelon with the San Agustin graben, the south end of which is about 10 mi to the southeast. Thus, the Reserve graben seems to be terminated on the north by a zone of east-west-trending faults, and major extension is shifted to the San Agustin graben. On the other hand, the north-northeast fault trend continues into the Quemado area to the north with the Hickman and associated faults. A small trapdoor like graben, the Sand Flat graben, (Fig. 3.4) appears to represent the eastward continuation of the Reserve and Luna grabens depression that links with the southwest end of the San Agustin depression near Horse Springs. See following road log entries. **0.2**

83.3 MP 27. Andesite of Dry Leggett Canyon crops out on both sides of the highway. Flows are within a horst that marks the San Francisco Mountains fault zone. Around curve to left, and straight ahead in lower slopes of Tularosa Mountain, on the north side of the highway, is the Tularosa Cave archeological site (Fig. 3.34; see minipaper by Wells, this roadlog). The caves are excavated in crossbedded, brown sandstone that normally underlies Squirrel Springs Canyon Andesite, but here the sandstone is overlain unconformably by coarse debris flows of unknown age within the Gila Formation. The debris flows, which contain blocks of Bloodgood Canyon Tuff, Squirrel Springs Canyon Andesite, and other clasts of local origin, as much as 10 ft long, are graded almost to the level of the present drainage, which suggests that the debris flows could be approximately as young as the 1 Ma basalt flow at Apache Creek. However, the apparent induration of the debris flows indicates that they may be much older.

The rocks in the upper part of Tularosa Mountain consist of volcaniclastic sandstone overlain by a volcanic sequence that includes Vicks Peak Tuff, Squirrel Springs Canyon Andesite, Shelley Peak Tuff, Bloodgood Canyon Tuff, and Bearwallow Mountain Andesite. **1.0**

84.3 MP 28. Up side canyon to right, about 0.5 mi, in Tularosa Canyon quadrangle (Ratté, 1992), Bearwallow Mountain Andesite contains abundant, small, upper(?) crustal xenoliths, a few centimeters

FIGURE 3.34. Tularosa Cave archeological site west of NM-12 at about MP 27 east of Aragon, New Mexico. Caves are in brown, crossbedded sandstone in upper part of Spears Group at unconformable contact with overlying debris deposits in Gila Formation, which contains 3-6-ft-wide clasts of Bloodgood Canyon Tuff and Squirrel Springs Canyon Andesite and other older rocks (see minipaper by Wells, this road log.).

across, that include gabbroic and granitic rock types, gneissic hornblende diorite?, and pyroxenite.

Around the curve ahead on left, crossbedded, brown sandstone in roadcut underlies landslide deposits coming off Tularosa Mountain. Across the Tularosa River on the right, the same brown sandstone underlies thin Squirrel Springs Canyon Andesite and Bloodgood Canyon Tuff, overlain by Bearwallow Mountain Andesite. Patchy outcrops of Gila Formation on the ridge top, above the Bearwallow Mountain Andesite, were prospected for uranium by means of bulldozer cuts in the 1960s. This accounts for the prospects shown on the Tularosa Canyon and Aragon quadrangles, according to U.S. Forest Service officials at Reserve. **1.0**

85.3 MP 29. Between the mobile home on the right near Tularosa Spring, and MP 29, we crossed a fault that is one of several approximately east-west-trending faults on the northeast side of Tularosa Mountain that drops the section down to the north. In combination with the arcuate, down-to-the-east Cañon del Buey fault (see First Day road log) these down-to-north faults form the Sand Flat graben. The reversals in fault block asymmetry from the Reserve graben (down to southeast) to down to northwest in the Sand Flat graben is apparently facilitated by a north-trending, arcuate accommodation zone on the east flank of Apache Mountain. The east end of the Sand Flat graben rises to form a ramp against the south end of the Hickman fault zone. **0.6**

85.9 Sand Flat Road on left. Note buff-colored Gila Formation cropping out in grass above road junction, at eastern edge of Aragon quadrangle. **0.2**

86.1 Highway bridge across Cañon del Buey in Tularosa Canyon quadrangle. The andesite lava flows exposed in Cañon del Buey are correlated with the Bearwallow Mountain Andesite, although they contain hornblende, biotite and bright green pyroxene phenocrysts, not usually present in the Bearwallow Mountain Andesite. However, they have chemical compositions similar to Bearwallow Mountain Andesite flows on the lower slopes of Mangas Mountain in the northern part of the Tularosa Mountain quadrangle. **0.2**

86.3 MP 30. Highway approximately follows the contact between Gila Formation and the underlying Bearwallow Mountain Andesite flows. The Gila Formation in this part of the Tularosa River valley is essentially a clastic wedge on the south-facing dip slope of the small Sand Flat half graben (Fig. 3.4). The footwall of this half graben lies 2.5 mi south of the highway, where Spears Group volcaniclastic sedimentary rocks are uplifted along more or less east-west faults between the highway and the lower slopes of Wagontongue Mountain, at the south edge of the Tularosa Canyon quadrangle. Wagontongue Mountain is the main ridge on the skyline south of the highway; the ridge crest is in the John Kerr Peak quadrangle, which has not been mapped in detail, but is the site of the John Kerr Peak dacite dome complex (Smith, 1976; Smith et al., 1976), about 12-14 Ma (Marvin et al., 1987). **2.8**

89.1 Junction on right. A combination of ranch roads and forest trails will take one close to the unique locality where the Bearwallow Mountain flows contain small centimeter-size crustal xenoliths, including granitic, gabbroic, quartz dioritic and pyroxenitic types of inclusions. The xenolith-bearing rocks are in the creek bed in a small box canyon, on the downstream side of a tight horseshoe loop, near the north edge of sec. 10, T5S, R16W, about 0.5 mi from the west edge of the Tularosa Canyon quadrangle. This locality can also be reached from the mouth of the canyon, in the Aragon quadrangle, in the northwest corner, sec. 10, as earlier described at 84.3 miles (MP 28) in this road log. **0.2**

89.3 MP 33. Bloodgood Canyon Tuff underlies Gila Formation in roadcuts from here to the Continental Divide. **1.6**

90.9 Continental Divide, elevation 7312 ft. **1.4**

92.3 MP 36. About 16 ft of Shelley Peak Tuff (28 Ma), from a source in the Mogollon Mountains, is sandwiched between Squirrel Springs Canyon Andesite, below, and Bloodgood Canyon Tuff above, in low ridges on both sides of highway between here and east edge of Tularosa Canyon quadrangle. **1.5**

93.8 East edge of Tularosa Canyon quadrangle. Entering Bell Peak quadrangle (Ratté, unpubl. 1986-1987). Hills projecting above alluvium and colluvium on both sides of highway are Squirrel Springs Canyon Andesite overlain by Shelley Peak and Bloodgood Canyon Tuffs and Bearwallow Mountain Andesite. **0.5**

94.3 MP 38. Views on right (about 12:00 to 3:00) of southwest part of Plains of San Agustin (San Agustin graben). **2.0**

96.3 MP 40. Junction; north end of Bursum Road on right just south of milepost. Highway is on older alluvial fans that are dissected by the present Patterson Canyon

FIGURE 3.35. View northeast along NM-12 to Cerro Caballo near vent outcrops of the Horse Springs dacite (Ratté et al., this volume). Horse Mountain is on skyline to left.

drainage. Mangas Mountains north of highway (on left) are mainly Bearwallow Mountain Andesite. **2.0**

98.3 MP 42. Views of Horse Mountain on skyline at about 11:00, and Cerro Caballo (Stop 3) straight down road (Fig. 3.35). Cross main Patterson Canyon, just ahead; Rincon Canyon road is on left between Patterson Canyon and Rincon Canyon. Bloodgood Canyon Tuff is overlain by Bearwallow Mountain Andesite on nose of ridge on east side of Rincon Canyon, less than 0.3 mi north of highway; low hills behind that ridge are capped by Miocene(?) "big feldspar" basalt, which overlies Squirrel Springs Canyon Andesite, which overlies rhyolite lava flows of Bat Cave Wells on both sides of the highway. The rhyolite flows were called Rhyolite of Wye Hill by Bornhorst (1976), but Wye Hill is not identified on modern maps, so flows are now named for tentatively correlative outcrops along south side of Plains of San Agustin at Bat Cave Wells. **2.0**

100.3 MP 44. Ranch road on left. A pull-off just ahead on right is a good place to park to examine outcrops of the Horse Springs dacite along the north side of the highway for about the next half mile. Along the right side of the highway, the outcrops are of Eocene tuffs (Blue Canyon, Rock House Canyon), which underlie the Horse Springs pumiceous dacite cropping out farther south of the road in the cavernous, columnar jointed cliffs. Stop 3 is on the back side of this ridge, where the exposures are somewhat more informative. However, float of the jasperoid and quartz monzonite inclusions to be seen in the dacite at Stop 3 can also be found here along the north side of the highway. **1.0**

101.3 MP 45. East edge of Bell Peak quadrangle. Enter Horse Mountain West quadrangle (Ratté et al., 1991). Blue Canyon Tuff in roadcut on north side of highway. Here this phenocryst-rich tuff shows a sharp change from densely welded to poorly welded material. There is no known source for this apparent regional ash-flow tuff. A $^{40}Ar/^{39}Ar$ analysis for a sample from this outcrop indicates an age of 33.7 Ma (McIntosh et al., 1991). **0.5**

101.8 **STOP 3. Turn right at ranch road**; there may be a steel cable across cattleguard; the cable may be locked. We are able to enter here through the courtesy of Dave and Karin Farr. Follow lead car to parking area about 2 mi from cattleguard at highway. Outcrops in low ledges along route are mainly ash-flow tuffs of the upper part of the Datil Group, mostly Rock House Canyon and Blue Canyon Tuffs, 34.4 and 33.7 Ma, respectively.

The primary purpose at this Stop is to examine the Horse Springs dacite in the outcrops on the east-facing slopes of Cerro Caballo, an informal name for the ridge west of us. To the east are the San Agustin plains, a.k.a. the San Agustin basin and the San Agustin graben, and the site of prehistoric (Quaternary) lake San Agustin and its numerous abandoned shorelines. Across the plains is a line of 25-Ma Bearwallow Mountain Andesite volcanoes, including Luera Peak, Pelona Mountain and O Bar O Mountain. On a clear day, one can see Tres Montosa peaks over by the Very Large Array (VLA), the National Radio Astronomy Laboratories site at the east end of the Plains of San Agustin. The mid-Miocene, dacitic Horse Mountain volcano is at the north edge of the plains, and at the base of Horse Mountain, a 0.6-mi-long hogback of Permian sedimentary rocks is believed to be a remnant of the Laramide Morenci uplift (Cather and Johnson, 1984; Chamberlin and Anderson, 1989), and the location of the Mogollon Rim in this area. The low hills south of NM-12, at New Horse Springs, contain the most complete sequence of upper Datil Group ignimbrites and interlayered Spears Group volcaniclastic rocks in this area. The site of the first of several wells drilled during recent petroleum exploration in this region is just west of the lava flows on the west ridge of Horse Mountain, beside the road to Dutch Oven Pass, several miles north of here.

The Horse Springs dacite is largely a proximal pyroclastic vent deposit, which here consists of a lower sequence of pyroclastic pumice flows, capped by discontinuous remnants of ash-cloud surge beds, and overlain by a thick block-and-ash deposit (Fig. 3.36). The lower pumice flows contain pumiceous dacite blocks, which show reverse size grading particularly near the top of the two upper flows, which we will traverse. The surge beds are present at the base of the cliffs. The cliffs also exhibit ancient fumarole tubes, or elutriation pipes, at the base of the block-and-ash flows, which is unusual in that such features are usually at the top of flow units.

Both the pumice flows and the block-and-ash flows contain cognate, porphyritic quartz monzonite inclusions. These are generally lapilli size or smaller in the

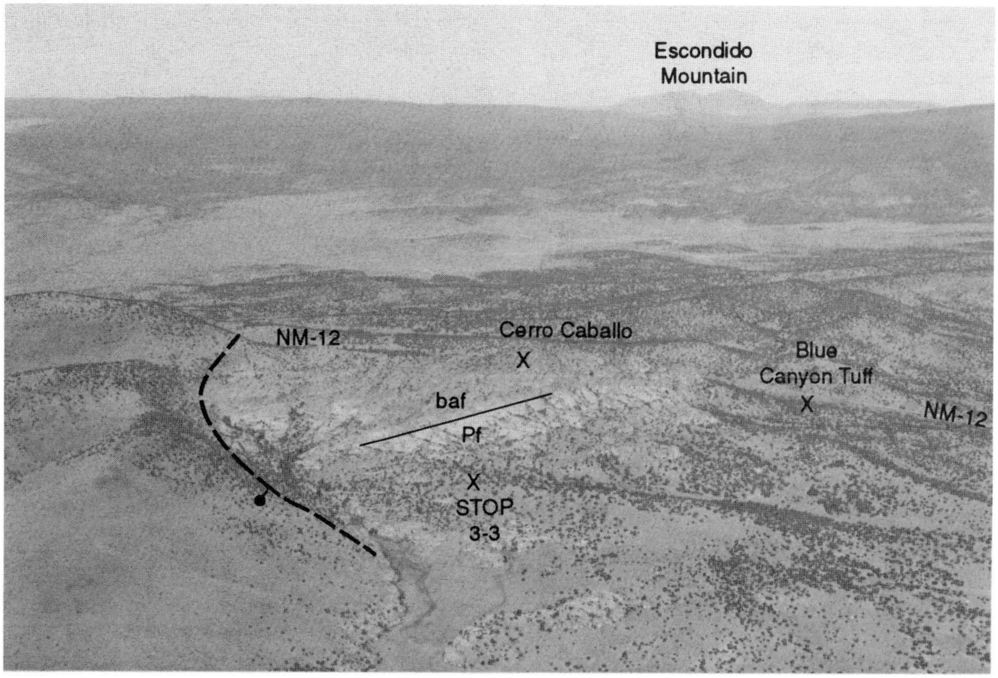

FIGURE 3.36. Aerial photograph looking northwest across Cerro Caballo showing location of Stop 3 and contact between pumice flows, pf, and overlying block-and-ash-flows, baf, of Horse Springs dacite.

pumice flows, but attain dimensions of 6-10 ft in the block-and-ash flows, as do the juvenile pumiceous dacite blocks. The block-and-ash flows also contain accidental inclusions of Paleozoic sedimentary basement rocks, which apparently represent the same sedimentary sequence that crops out in a 0.6-mi-long hogback about 6 mi away at the base of Horse Mountain. The accidental inclusions are largely jasperoidized (silicified) blocks of limestone, calcareous argillite and quartzite from the Lower Permian Yeso Formation, Glorieta Sandstone and San Andres Limestone. These inclusions also are mineralized (iron and zinc), and where in place, many are enclosed in skarn-like reaction rims; "skarn-like" because the reaction rims contain clay minerals as well as calc- silicate minerals, which are dominantly diopside and minor garnet. Our present interpretation of the dacite deposit and its inclusions is described in more detail in the article on the Horse Springs dacite (Ratté et al., this volume), but is summarized here as follows:

1. A shallow body of quartz monzonite magma initially rose into contact with, and probably stoped its way into, the Paleozoic sedimentary rocks, and jasperoidization and mineralization may have occurred at this time, while a crystalline hood of coarsely porphyritic quartz monzonite formed at the top of the magma (Ratté et al., this volume, fig. 20).

2. Confining pressure in the magma chamber was eventually relieved by fracturing in the crystallized hood, and pumice flows were deposited from an eruption column that entrained small fragments of the ruptured quartz monzonite hood. These eruptions ended with the ash-cloud surge deposits that settled atop the pumice flows.

3. The premonitory pumice flow eruptions were followed closely by a massive, paroxysmal block-and-ash eruption, which not only carried 6-10 ft blocks of pumiceous dacite, but also equally large blocks of cognate porphyritic quartz monzonite. In addition, the block-and-ash flows entrained accidental inclusions of the jasperoidized Paleozoic sedimentary rocks, which appear to be concentrated near the top of the greatest concentration of large pumiceous dacite and quartz monzonite blocks.

4. The skarn-like rims around the accidental calcareous inclusions must have been acquired after eruption and during cooling of the block-and-ash deposit, because it seems unlikely that the fragile encompassing rims of friable diopside crystals could have survived eruption if they had been formed adjacent to the magma prior to eruption. Furthermore, the clay minerals in the skarn-like rims probably were derived by alteration of the enclosing dacite rather than from the inclusions.

5. The venue for mineralization is somewhat less certain. Iron and zinc oxides, which characterize the mineralization, could have formed adjacent to the magma chamber in reduced forms such as magnetite, that were erupted as part of the jasperoid inclusions, and later oxidized to hematite, magnesioferrite and other minerals now found in the surface deposits. Alternatively, mineralization and jasperoidization, as well as the skarn-like rims could have formed in place within the cooling dacite.

The reasons for suggesting that jasperoidization and mineralization took place at depth rather than after eruption of the dacite are:

1. The apparent absence of *unaltered* (non-jasperoidal) accidental inclusions, which might be anticipated in cooler parts of the deposit, although it may be

that the inclusions were all entrained in the earliest phase of the block-and-ash eruption, as suggested by their apparent concentration in its lower part.

2. The lack of mineralization associated with the ancient fumaroles at the base of the block-and-ash flows.

3. The Paleozoic sedimentary rocks adjacent to the magma chamber might have been more susceptible to alteration and mineralization than inclusions of the same rocks in the erupted dacite because of higher temperature and more trapped volatiles at depth.

One consequence of ruling out alteration and mineralization having occurred next to the magma chamber is to lessen expectations for possible economic deposits at depth.

The Horse Springs dacite also has implications relative to the origin of some of the regional ignimbrites in the Datil Group in this area. New Ar/Ar ages for the dacite and its quartz monzonite inclusions of 33.8 ± 0.03 Ma, and 33.6 ± 0.06 Ma, respectively, indicate that it is indistinguishable in age from the immediately underlying Blue Canyon Tuff. The Horse Springs dacite and Blue Canyon Tuff are also largely indistinguisable in chemistry and phenocryst mineralogy.

A caldera buried beneath the San Agustin Plains is a possible source for both the Blue Canyon Tuff and the Horse Springs dacite. This is an alternative to the proposed subsidence of the ~ 25x30 mi Crosby Mountains depression in response to the eruption of the Horse Springs dacite (Elston, 1984, table 1; Bornhorst, 1976; Jones, 1980). See minipaper by Chamberlin in First Day road log for additional discussion of the "Crosby Mountains depression" as proposed by Elston, 1984.

Return to NM-12 and turn right; resume road log. In the trees north of the highway, opposite this ranch road, are two large springs, each about 150 ft across, for which Horse Springs was named. **0.5**

102.3 Settlement of Old Horse Springs. Horse Springs was the setting for one of Louis LaMour's western novels, called "Conagher," but use of the name is about as close as the locale comes to being recognizable in "Conagher", so maybe it was another Horse Springs that the author had in mind! **2.3**

104.6 New Horse Springs, site of now defunct Horse Springs store and gas station. Ridges south of highway here offer one of the most complete sections of the upper Datil Group ash-flow tuff outflow sequence. Ar/Ar dates and paleomagnetic data are available for most of the units in this section (McIntosh et al., 1991,1992; Ratté et al., 1991). Tuffs present include Datil Well Tuff, tuff of Farr Ranch, Kneeling Nun Tuff, tuff of Lebya Well, Rock House Canyon Tuff, and Blue Canyon Tuff, interlayered with Spears Group volcaniclastic sedimentary rocks.

Road on left goes to Dutch Oven Pass and Collinsville. The Shell Oil Co. exploration well (SWEPI State No. 1), is located on the slopes above the road on its west side, about 5 mi from NM-12. **3.8**

108.4 East edge of Horse Mountain West quadrangle. Entering Horse Mountain East quadrangle (J. C. Ratté, unpubl. 1990). Highway traverses shoreline deposits of prehistoric lake San Agustin (see Weber, and McFadden et al. minipapers in First Day road log). South of highway, shoreline deposits give way to playa and lake beds. A drill hole in the southwestern part of the playa penetrated 2000 ft of lacustrine silt, clay and alluvial gravel beneath the playa; Tertiary pollen appear at a depth of 920 ft. Paleoenvironmental analysis indicates that prehistoric lake San Agustin dried up about 5000 years ago (Markgraf et al., 1983). **0.9**

FIGURE 3.37. Outcrops of pre-Tertiary rocks at the foot of Horse Mountain north of NM-12 at about MP-53. Pre-Tertiary rocks include Permian Yeso, Glorieta and San Andres formations.

109.3 MP 53. **OPTIONAL STOP 4**. About 0.6 mi north of highway, pre-Tertiary rocks crop out in a ridge about 400 m long (Fig. 3.37). The rocks include the Lower Permian Yeso Formation, Glorieta Sandstone, and San Andres Limestone (see following minipaper by Lucas and Kues). The pre-Tertiary rocks are overlain by float of round, pebble conglomerate like that in the Baca Formation to Spears Group transition, but the nature of the contact of these rocks with the underlying pre-Tertiary rocks is uncertain. However, seismic data supports the interpretation of a Laramide high below the Permian rocks on the flanks of Horse Mountain. The pre-Tertiary rocks probably represent a relic of a Laramide structural high that is now faulted beneath the Plains of San Agustin. This interpretation is supported by seismic data (L. Garmezy, personal commun., 1994). **1.0**

PERMIAN STRATA AT HORSE MOUNTAIN

Spencer G. Lucas[1] and Barry S. Kues[2]

New Mexico Museum of Natural History and Science, 1801 Mountain Road NW, Albuquerque, NM 87104; Department of Earth & Planetary Sciences, University of New Mexico, Albuquerque, NM 87131-1116

Charles Stearns (1962; see also Foster, 1964, and Willard and Stearns, 1971) first noted and mapped Permian strata near Horse Mountain, and Kottlowski and Foster (in Stearns, 1962) described a stratigraphic section through these strata in sec. 20, T4S, R12W, assigning them to the Yeso, Glorieta and San Andres formations and to an overlying Triassic(?) sandstone. Megafossils collected from a gray limestone on the crest and back slope of a hogback south of Horse Mountain indicated, according to Rousseau Flower, the presence of the San Andres Formation there. Dane and Bachman (1965), on the geologic map of New Mexico, showed about 2 mi^2 of Permian strata to the south, overlain by Triassic rocks in the north. No further work on or mention in the literature of the Horse Mountain Paleozoic strata occurred until Lucas and Hayden (1989, p. 203, fig. 10) described and illustrated these units and concluded that no Triassic strata are present near Horse Mountain.

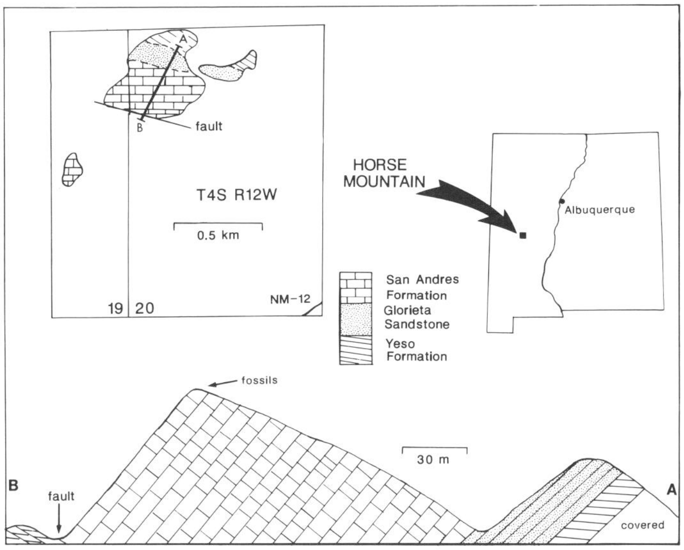

FIGURE 3.38. Geologic map and cross section of Permian strata south of Horse Mountain, Catron County, New Mexico.

Stearns' (1962, pl. 1) geologic map accurately depicts the distribution of Paleozoic strata south of Horse Mountain, which mainly crop out in the NW¼ sec. 20, T4S, R12W, with a small inlier in the SE¼ NE¼ sec. 19. In the larger outcrop area, the San Andres Formation forms a prominent, southeasterly dipping hogback; the stratigraphic section measured by Kottlowski and Foster in this area and our own observations lead to recognition of (in ascending order) the Permian Yeso, Glorieta and San Andres formations.

Kottlowski and Foster measured 61 m of faulted Yeso Formation, mostly gray and light red sandstones and thin limestone beds, with more than half of this thickness consisting of covered intervals. Our examination of the Yeso section began at the top of Kottlowski and Foster's section, where an oolitic dolomite about 10 m thick crops out. This dolomite is medium bluish gray (5B5/1) but weathers medium light gray (N6), moderate yellowish brown (10YR5/4) and dark yellowish brown (10YR4/2). Its resistant, ledgy beds dip 40° to S45°E, and are the uppermost Yeso strata exposed (Fig. 3.38).

Overlying beds of the Glorieta Sandstone have the same strike and dip and are about 30 m thick. These sandstones are fine-grained, supermature quartz arenites that display abundant trough crossbeds. Unmetamorphosed Glorieta beds are moderate orange pink (5YR8/4) to light brown (5YR6/4) and weather dark yellowish brown (10YR4/2). Beds that have been metamorphosed to metaquartzites are darker, grayish hues.

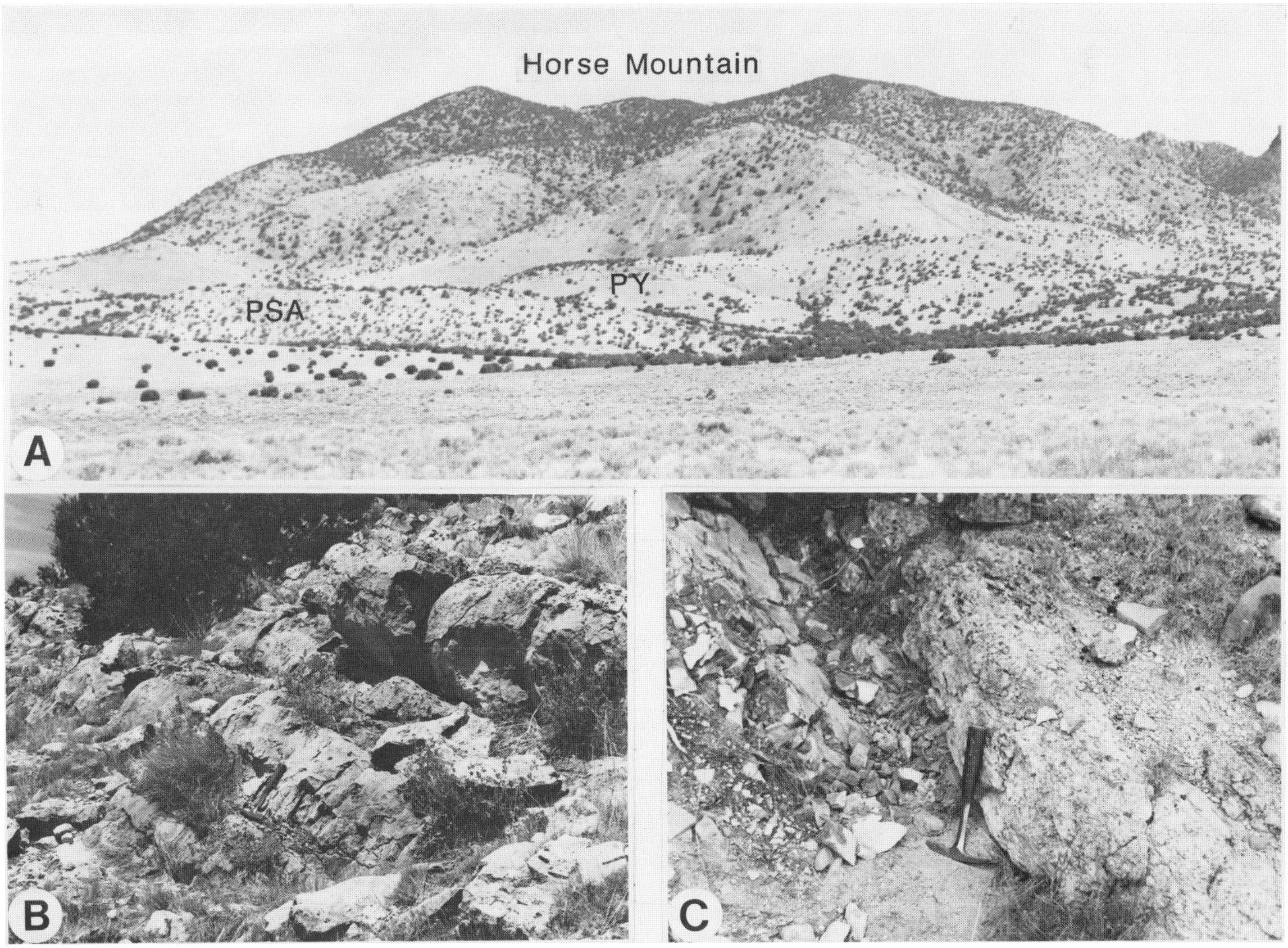

FIGURE 3.39. Permian strata south of Horse Mountain. A, Overview of Horse Mountain, looking north-northwest with San Andres Formation (PSA) hogback in foreground, underlain by Glorieta (not seen) and Yeso (PY) strata. B, San Andres limestones and dolomites on the hogback crest. C, Fault breccia (to right of hammer) in streamcut on north end of San Andres hogback dipslope.

The great majority of Permian strata south of Horse Mountain belong to the San Andres Formation, which is 115 m thick here. Most San Andres units are light brownish gray (5YR6/1) and pinkish gray (5YR8/1) micrites and dolomites (Figs. 3.38, 3.39B). Some limestone strata, especially those at the crest of the hogback, contain dark gray chert nodules. The San Andres fossils described below were collected from the crest of the hogback in a 3-m-thick bioclastic limestone interval about 83 m above the base of the formation.

At the base of the north dip slope of the hogback is a distinctive fault that tops the section. In the streamcut to the southwest, the hogback strata dip 30° to N20°W and are a mixture of grayish orange pink (10R8/2) to pale red (5R6/2) beds of fine-grained, gypsiferous, mature quartz sandstone and fault breccia consisting of limestone and sandstone clasts cemented by euhedral calcite crystals and gypsum veinlets (Hayden and Lucas, 1989). These sandstones belong to the Yeso Formation (Lucas and Hayden, 1989; O. Anderson, oral commun., 1993), and do not resemble Triassic sandstone to the north.

The Horse Mountain Permian strata show many similarities to Permian units 80 km to the north, in the Zuni Mountains of Cibola County. Indeed, the Horse Mountain outcrops can be added to the fence line for Permian strata extending from the Zuni Mountains to north of Pie Town (Colpitts, 1989, fig. 3). Yeso strata exposed at Horse Mountain are dolomites and sandstones referable to the Torres Member. The overlying Joyita Member of the Yeso present in the Zuni Mountains has pinched out to the south between about Pie Town and Horse Mountain. The Glorieta Sandstone is continuous across this area, although relatively thin at Horse Mountain. The tripartite division of the San Andres Formation—lower calcareous dolomite, middle cross-bedded sandstone, and upper dolomitic limestone—can no longer be discriminated at Horse Mountain (Fig. 3.40).

The San Andres Formation at Horse Mountain is locally fossiliferous, but preservation of specimens is poor. Most bivalves are represented by steinkerns and molds, and the shells of other groups are coarsely recrystallized or adhere closely to the hard limestone matrix, making extraction of intact, nonexfoliated specimens for study nearly impossible. These factors precluded precise identification of most taxa. The fauna is strongly dominated by molluscs, mainly bivalves and gastropods, with smaller numbers of nautiloids and the scaphopod *Prodentalium canna* (White). Moderate numbers of brachiopods (unidentified small productoids and *Composita*) and sparse echinoid and trilobite fragments are also present.

Bivalve taxa include at least nine species. Small to medium-size steinkerns of *Schizodus* are common; the shape and proportions of some specimens suggest *S. supaiensis* Winters, but others are higher and subtrigonal, closer in shape to *S. altus* Newell and Boyd. None of the specimens can be referred to *S. canalis* Branson, a common constituent of San Andres faunas to the northwest, near St. Johns, Arizona (Newell and Boyd, 1975) and Ojo Caliente, New Mexico (Kues and Lucas, 1989). *Leptodesma* (*Leptodesma*) aff. *sulcata* (Geinitz), *Promytilus retusus* Chronic, and *Sanguinolites*? sp. (of Chronic, 1952 and Winters, 1963) are each represented by one to a few specimens. Steinkerns of several small, laterally compressed, elongate species with anterior beaks are fairly abundant, but lacking preservation of hinge structures and shell ornamentation, they cannot be identified. These probably belong to *Permophorus*, *Edmondia* and possibly *Prothyris*. A small subtrigonal nuculanid, probably *Phestia perumbonata* (White) (see Yancey, 1978) displays the fine concentric step-rib ornamentation that characterizes this group.

Several species of low- and high-spired gastropods are present in this fauna, but available specimens are too severely weathered for identification. Specimens of *Bellerophon* (*Bellerophon*) sp. are moderately common, as are relatively large specimens (up to 40 mm wide) of *Straparollus* (*Euomphalus*) *kaibabensis* Chronic. This species is distinguished by large, well-spaced, rounded nodes along the dorsolateral margin of the whorls.

Insofar as can be determined from these rather sparse and poorly preserved remains, this fauna suggests a Leonardian age, consistent with that of the San Andres Formation in central New Mexico. The scaphopod *Prodentalium canna* is common in Lower Permian strata across the western U.S., and other molluscs from the Horse Mountain locality, such as *S.* (*E.*) *kaibabensis*, are conspecific with or closely related to species described from the Kaibab (Chronic, 1952) and/or upper Supai (Winters, 1963) formations of eastern Arizona, both of which are of Leonardian age.

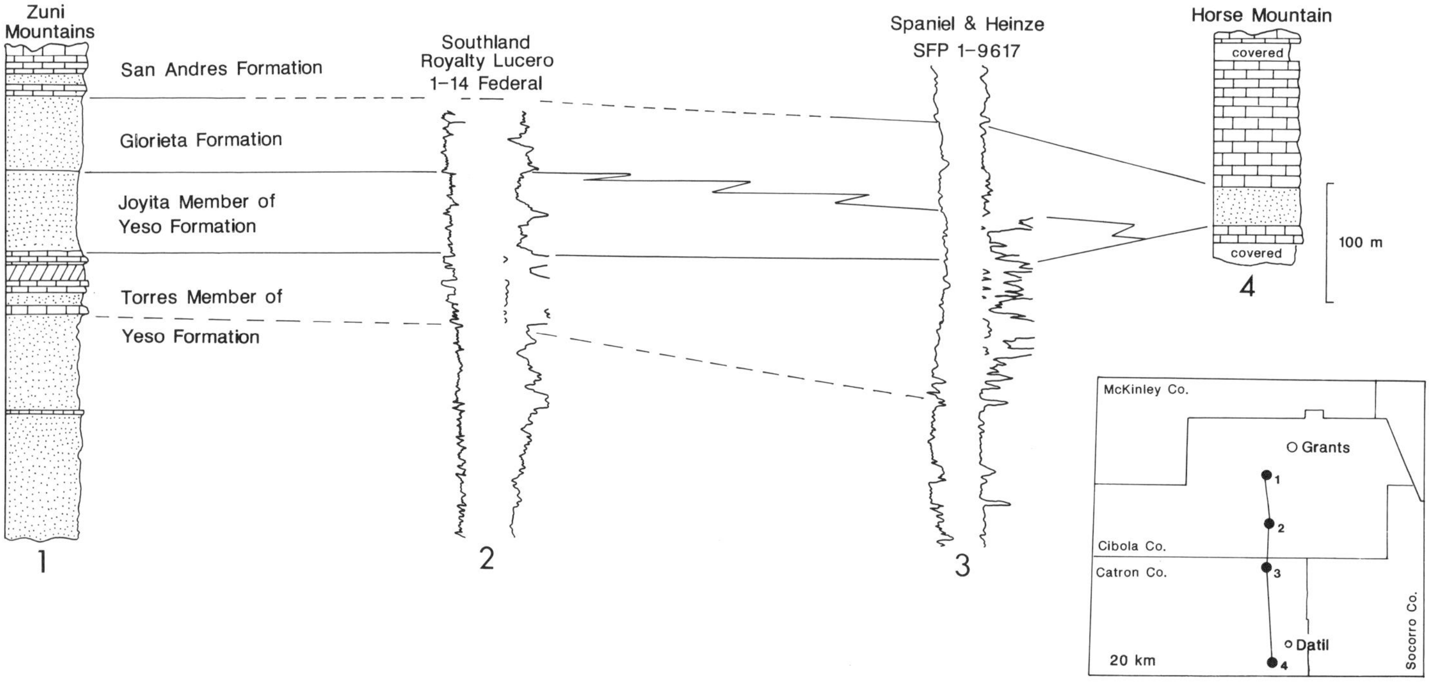

FIGURE 3.40. Correlation of Permian strata between the Zuni Mountains and Horse Mountain, west-central New Mexico (modified from Colpitts, 1989).

110.3 MP 54. Deep canyon north of highway slices into the central part of the Horse Mountain volcano, where a dacitic plug is surrounded by vent breccia and lava flows. Incomplete detailed geologic mapping in the Horse Mountain East quadrangle (Ratté, unpubl. 1991), old and new isotopic ages, and paleomagnetic analysis indicate that younger eruptive rocks may fill a crater surrounded by an older flow sequence in the crater walls. **2.0**

112.3 MP 56. Small isolated butte north of highway (Fig. 3-41) consists of Horse Mountain lava flows that are the same as those in the northeast ridge of Horse Mountain. This small hogback might be interpreted as a landslide block, but because there seems to be no obvious place from which it has slid, it is interpreted here as down-dropped on a buried northeast-trending fault along the mountain front. Conventional K-Ar and fission track ages of rocks related to the Horse Mountain volcano all indicate mid-Miocene eruption. For the central plug, we have a K-Ar plagioclase age of 12.0 ± 3.0 Ma, a zircon fission-track age of 10.4 ± 1.3 Ma, and a spurious K-Ar hornblende age of 35.3 ± 6.1 Ma. From the east ridge flow, we have K-Ar ages of 14.4 ± 0.6 Ma (plagioclase), and 13.7 ± 0.5 Ma (biotite), and a new Ar/Ar hornblende age of 12.64 ± 0.06 from the flow at the downdropped butte. These ages are all roughly consistent with the fission-track age from the central intrusion of 13.7 ± 2.1 Ma, reported earlier by Jones (1980). Another K-Ar age for biotite in tuff breccia, poorly exposed in a ravine beneath the east ridge flow, is 13.4 ± 1.1 Ma. The tuff breccia may be part of the initial cone of the volcano. A distinctive trachyte flow that has large (0.4-1.2 in.) sanidine phenocrysts crops out along the mountain front southwest of the east ridge flow (Fig. 3.41); it has a conventional K-Ar sanidine age of 11.2 ± 0.6 Ma. Beneath it is a basalt flow with a preliminary Ar/Ar age of 13.55 Ma. In addition, a mafic dike on the north side of the east ridge has a K-Ar whole rock age of 17.5 ± 2.1 Ma. Most ages cited here are from Marvin et al. (1987), and unpublished ages from Harold Mehnert (personal commun., 1991); the two new Ar/Ar ages are by Bill McIntosh. Additional Ar/Ar dating and paleomagnetic analysis arc currently in progress for support of a more complete interpretation of the evolution of this volcano, and its regional volcano-tectonic relationships. **1.0**

113.3 MP 57. Good outcrops north of highway, on the nose of the east ridge of Horse Mountain, expose upper Datil Group ash-flow tuffs beneath the east ridge flow of the Horse Mountain volcano. Tuffs include the Datil Well Tuff and the tuff of Farr Ranch(?), separated only by a cooling break, and Rock House Canyon Tuff and Blue Canyon Tuff. Northeast of Horse Mountain, and separated from it by a northwest-trending buried fault, the low hills capped by Datil Well Tuff and tuff of Farr Ranch are largely buried by fanglomerate. **1.0**

114.3 MP 58. Luera Mountain, across the plains east of highway at 2:00-3:00 (Fig. 3.42), is a Bearwallow Mountain Andesite shield volcano at the northeast end of the line of andesitic volcanos that parallels the southeast edge of the Plains of San Agustin. This andesitic volcano has a central plug(?) of rhyolitic composition on the east side of Luera Peak, that has given a whole rock conventional K-Ar age of 25 Ma (Harold Mehnert, oral commun., 1988). Pelona Mountain and O Bar O Mountain are other andesitic shield volcanoes in this alignment along the south side of the plains that are visible from here at about 4:00 and 5:00, respectively. Both volcanoes have central plugs of medium-grained andesite and are about 25-26 Ma (Marvin et al., 1987). **2.0**

116.3 MP 60 is about 0.1 mi beyond the north edge of Horse Mountain East quadrangle. Highway cuts across southeast corner of Log Canyon quadrangle and into Sugarloaf Peak quadrangle about 1.85 mi ahead in the Quemado 1:100,000 quadrangle. Sugarloaf Mountain, north of highway, is at the southern end of the Crosby Mountains (Fig. 3.43), which consist mainly of volcaniclastic sedimentary rocks of the Oligocene South Crosby Peak Formation and tuffs and lava flows of the Mogollon Group. **5.0**

121.3 MP 65. Low, rounded outcrops and quarry on right are quartz diorite intrusion of Cerrito Viejo (Fig. 3.44), for which we have a conventional K-Ar biotite age of 30.2 ± 1.1 Ma, and a zircon fission- track age of 21.4 ± 2.3 Ma (Marvin et al., 1987), which are supported by a new Ar/Ar plagioclase age of 30.6 ± 1.4 Ma (McIntosh and Chamberlin, this guidebook). However, most of the roadcuts between here and Datil, about 9 mi ahead, are in the older Hells Mesa Tuff (32 Ma), which was previously mapped as the tuffs of Rock Tank and Ary Ranch (Lopez and Bornhorst, 1979), at the top of the Datil Group.

The South Crosby Peak Formation has been suggested as the fill of the proposed Crosby Mountains cauldron (Elston, 1984; Bornhorst, 1976; Jones, 1980), and the distribution of the South Crosby Peak Formation is within the outline of the proposed cauldron on the State Highway Geologic Map (NMGS, 1982). Other structural elements cited as part of the Crosby Mountains cauldron by Elston, Bornhorst and Jones consist of a proposed peripheral graben and a resurgent dome centered just southwest of Horse Mountain.

As discussed at Stop 3, we suggest that a caldera related to eruption of the Blue Canyon Tuff and the Horse Springs dacite may be buried beneath the Plains of San Agustin, but the eruption of the Horse Springs dacite as the principal cause of collapse for such a caldera seems unlikely because of its apparent small volume. The proposal that the South Crosby Peak Formation fills such a caldera seems to be untenable because the South Crosby Peak Formation is approximately 3.9 Ma younger (McIntosh and Chamberlin, this volume) than the proposed caldera-forming rocks (Blue Canyon Tuff and Horse Springs dacite). The suggested resurgent dome (Bornhorst, 1976, p. 81) is based on Stearns' description of "a broad, gentle dome underlying the lowlands southwest of Horse Mountain" (Stearns, 1962, p. 33). The area referred to is one of largely isolated outcrops of thin, upper Datil Group ignimbrites surrounded by alluvium (Ratté et. al., 1990). Most of the ignimbrites there are older than the Blue Canyon Tuff, and thus if they are part of a

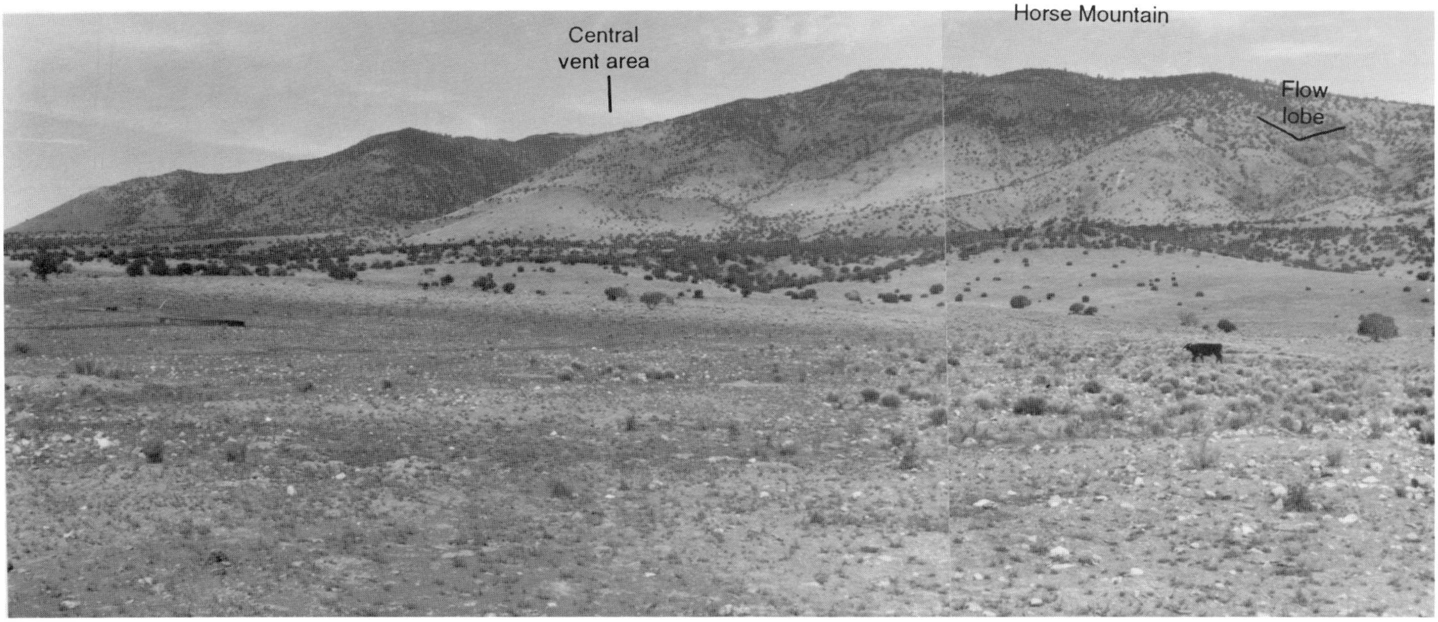

FIGURE 3.41. Panoramic view of Horse Mountain volcano from NM-12 at about MP-56.

resurgent dome, they must represent either the floor of the related caldera, or megabreccia from the caldera wall. The idea of a megabreccia from the *wall* (presumably northern wall) of a Blue Canyon Tuff caldera is appealing because some of the ignimbrite outcrops are broken into blocks that are separated by slickensided fractures. Thus a caldera, if present, and its resurgent dome, if any, would seem to have to be quite different from the geometry of the Crosby Mountains depression as proposed by Elston (1984). Chamberlin (see First Day road log) suggests that the "Crosby Mountains caldera" as defined by Elston (1989) represents a disparate collage of temporally unrelated structural elements. **3.0**

124.3 MP 68. Approaching roadcuts in Hells Mesa Tuff (32 Ma), which is the ledge-forming unit west of the highway. **1.0**

125.3 MP 69. Northern part of Crosby Mountains on left (northwest of highway); Anderson Mountain at about 10:00; East Sugarloaf Mountain at about 11 o'clock (Fig. 3.45). **2.5**

127.8 Roadcut in uppermost of three ash-flow tuffs interlayered with volcaniclastic rocks of South Crosby Peak Formation. Oldest of the three tuffs has an Ar/Ar plagioclase age of 29.7 ± 0.1 Ma. **1.0**

128.8 Road junction on left. Roadcut ahead on left is Hells Mesa Tuff. **1.9**

130.7 Road junction, NM-12 and US-60. Socorro, New Mexico is approximately 60 mi to the east on US-60. **End of road log.**

FIGURE 3.42. Luera Peak in the Luera Mountains, on the east side of the Plains of San Agustin east of NM-12 at about MP 58. Luera Peak marks the approximate eruptive center of the northeasternmost of the Bearwallow Mountain Andesite volcanoes that are along the southeast margin of the San Agustin graben.

FIGURE 3.43. Aerial view, from south, of Sugarloaf Mountain at the southern end of the Crosby Mountains; Sawtooth Mountains in background left. Sugarloaf Mountain and mesa to right are underlain mainly by volcaniclasic rocks of South Crosby Peak (out of photo to right) and capped by basaltic andesite of Crosby Mountains and thin white Vicks Peak Tuff (28.5 Ma) beneath the capping flows.

FIGURE 3.41. Panoramic view of Horse Mountain volcano from NM-12 at about MP-56.

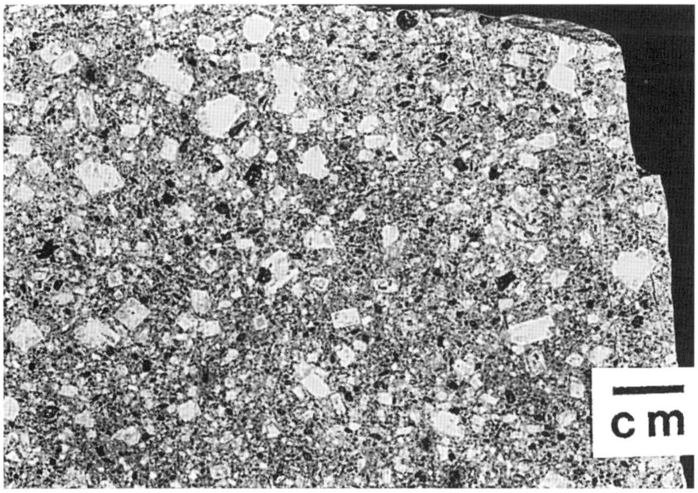

FIGURE 3.44. Slab of Cerrito Viejo intrusive rock showing seriate porphyritic texture consisting of fine- to coarse-grained phenocrysts of plagioclase, biotite and hornblende in a microcrystalline groundmass of equivalent composition. Sample NM-1114 of McIntosh and Chamberlin (this volume); collected from southeast side of road-metal quarry.

FIGURE 3.45. Aerial view of northern Crosby Mountains; Sawtooth Mountains in background center; Datil Mountains in background right. Crosby Mountains consist mainly of South Crosby Peak volcaniclastic rocks, capped by basaltic andesite of Crosby Mountains. Hells Mesa Tuff (32 Ma) crops out in low terrane on both sides of NM-12, from left to right across center of photo, and in highway cuts.

San Francisco Mountains and canyon of San Francisco River north of Reserve, New Mexico. View is N28°W. Gila Formation in foreground and middle ground is fill of Jon S half graben deposited against and faulted along the Saliz Mountains-Higgins Mountain fault of the Reserve graben system. In background, cliffs of the San Francisco River Box are formed mostly by andesite of Dry Leggett Canyon in the footwall of the San Francisco Mountains fault zone. Dillon Mountain, on skyline behind the box, is another upthrown block of mostly andesite of Dry Leggett Canyon. Camera station is alongside Highway 12 in SW ¼ sec. 36, T6S, R19W, approximately 3 km north of Reserve. Altitude is about 1825 m. Wayne Lambert photograph No. 93L66. August 14, 1993, 3:24 p.m. MDT.

SUPPLEMENTAL ROAD LOG, TEJANA MESA (EL PORTICITO) TO US-60 VIA ZUNI SALT LAKE MAAR

ORIN J. ANDERSON, SPENCER G. LUCAS and WILLIAM A. COBBAN

SUMMARY

Tejana Mesa is capped by a late Tertiary basalt flow that has yielded a K-Ar date of 6.7 ± 0.2 Ma (Dethier et al. 1986). The main vent for this flow lies 0.2 mi north of NM-601, as indicated by the dramatic increase in thickness of the flow. The vent area, called El Porticito, lies on a N25° to 35°E trending chain of vents, as much as 10 mi long. New $^{40}Ar/^{39}Ar$ data indicate an age of 7.92 ± 0.20 Ma for the intrusive rock at El Porticito (McIntosh and Cather, this volume). Thus the Tejana Mesa vent complex may represent multiple episodes of eruption in late Miocene time.

At the latitude of Tejana Mesa (35°25'N), two important changes affect topography and landform. First, here, near the southern margin of the Colorado Plateau, we are also at the southern limit of continuous exposure of Cretaceous and older Mesozoic rocks. Second, from this latitude southward, Neogene volcaniclastic sediments (Fence Lake Formation) and pre-volcanic sediments of late middle Eocene age (Baca Formation) comprise most of the surface rock. In addition, the area southward is characterized by late Oligocene and Miocene basalt and andesite flows, by ignimbrites of Oligocene age, and volcaniclastic sedimentary rocks of late Eocene age.

Mileage

0.0 Milepost 5, on NM-601. At 3:00, note exposures on the flanks of Tejana Mesa, which consist of the Baca Formation (reddish color), and the overlying coarse-grained Fence Lake Formation, approximately 100 ft thick locally, in turn overlain by the 6.7 Ma basalt flow. Proceed northwest on NM-601. **0.3**

0.3 Cross cattleguard; **0.5**

0.8 Crest of hill; at 1:00 note fluvial channel sandstones of the Upper Cretaceous Moreno Hill Formation capping small hill. The Moreno Hill Formation consists of sandstone, shale, carbonaceous shale and thin coal beds deposited in a coastal plain and deltaic environment during middle Turonian to Coniacian time. **0.2**

1.0 Cattleguard; road to left leads to Mesa Ranch. Mesa behind the ranch consists of Baca Formation capped with a Miocene basalt flow, presumably of similar age to the flow capping Tejana Mesa. **0.2**

1.2 Roadcut on left and brown ledgy sandstones to the right of road are outcrops of Moreno Hill Formation. **0.6**

1.8 Bridge over tributary of Largo Creek; at 2:00 note fluvial channel sandstone in Moreno Hill Formation. **0.3**

2.1 Roadcut at crest of hill exposes typical overbank-floodplain deposits in Moreno Hill Formation; the dusky yellow colors are characteristic of the finer-grained facies often associated with carbonaceous zones. **0.1**

2.2 At 10:30-11:00 note excellent outcrops of Moreno Hill Formation on slope of hill. **0.4**

2.6 Cross culvert; the flat-topped mesa at 12:00 to 1:00 in the distance is Santa Rita Mesa; it is capped by a coarse-grained facies of the Fence Lake Formation (Miocene) above strata of the Moreno Hill Formation. **0.4**

3.0 Milepost 8; the sandstone bench north of the road is developed in the Moreno Hill Formation. **0.6**

3.6 Junction; **follow main road to left**. The prominent notch in Tejana Mesa at 2:30 is Augustine Canyon. **0.2**

3.8 Cattleguard. **0.4**

4.2 At 2:00 note small west-facing cliff and bench formed by typical fluvial channel sandstone of Moreno Hill Formation. **1.1**

5.3 Cross culvert; road now ascends onto dip slope developed on the middle sandstone member of the Moreno Hill Formation. The very coarse-grained facies in the middle sandstone plus its stratigraphic position strongly suggest that it is correlative with the Torrivio Member of the Gallup Sandstone. **0.7**

6.0 Crest of hill; milepost 11; at 2:30 in the distance the dark-colored hill is Cerro Prieto, a basaltic andesite neck. The northwest end of Tejana Mesa is also visible on the right; just below the edge of Tejana Mesa is the site of the Salt River Project and proposed Fence Lake Coal Mine. The geology and coal resources of the area immediately to the north were described by Campbell (1989). **0.3**

6.3 Road curves to the left; at 2:30 to 3:00 the prominent bench is defended by the middle sandstone member of the Moreno Hill Formation. **0.2**

6.5 From 12:00 to 2:00 is Santa Rita Mesa in the distance. The south and east flanks of the mesa offer excellent exposures of the entire thickness of the Moreno Hill Formation, which locally exceeds 800 ft; type section of the Moreno Hill Formation is on the south end of the mesa at Moreno Hill. **1.1**

7.6 Hills and mesas from 9:00 to 12:00 are capped by Neogene and, locally, perhaps Quaternary basalts or basaltic andesites. These overlie Tertiary sediments of the Fence Lake Formation, which here contains abundant well-rounded siliceous cobbles derived from the Baca Formation that are mixed with subrounded volcanic clasts. **0.4**

8.0 Milepost 13. **0.8**

8.8 Cross cattleguard. **1.5**

10.3 Hubble Draw at 1:30 to 2:00 is the major drainage line in area. Outcrops to right beyond the windmill are Moreno Hill Formation. **0.5**

10.8 Crest of hill; at 1:00 in distance the Dakota Sandstone (Upper Cretaceous) is visible overlying the red mudstones and sandstones of the Upper Triassic Painted Desert Member of the Petrified Forest Formation (Chinle Group). **0.3**

11.1 After crossing cattleguard note the increase in eastward dip in Moreno Hill Formation strata at 3:00. Volcanic cinder derived from Zuni Salt Lake maar now begins to appear on both sides of road. **0.7**

11.8 Low mound on right and much of the immediate countryside is draped with volcanic cinder from Zuni Salt Lake Maar. **1.1**

12.9 Road forks; **bear left and leave NM-601;** at 1:00 is a good view of the Petrified Forest Formation (Upper Triassic) overlain by the Dakota Sandstone. For a detailed measured section of the Painted Desert Member of the Petrified Forest Formation at this locality, on the north side of Hubbell Draw, see Lucas and Hayden (1989). **0.1**

13.0 Junction, **bear left** and descend into Zuni Salt Lake maar. **0.2**

13.2 Stop and park. Zuni Salt Lake occupies the northern half of the crater, which is approximately 1 mi in diameter (Fig. S1.1). Three small cinder cones can be seen rising from the floor of the southern half of the crater; the largest of these cones has its own, more concentrated, brine lake.

A maar is defined as a coneless volcanic crater formed by a single eruptive event and is commonly filled with water. The view from this stop makes it obvious that Zuni Salt Lake fits the definition of a maar. This crater is the result of a phreatomagmatic eruption, meaning that significant steam was involved as well as the common magmatic gases. The source of the steam was magma contact with groundwater, specifically a good aquifer. These conditions are commonly met where a magma conduit develops in an area of porous, flat lying sedimentary rocks such as we have here in the form of Upper Cretaceous sandstones. Zuni Salt Lake maar has been dated at 22,900 ± 1400 yrs by ^{14}C (Bradbury, 1966). Recent $^{40}Ar/^{39}Ar$ data (McIntosh and Cather, this volume), however, indicate activity near 100,000 yrs, 114 ± 38 ka for a juvenile bomb, and 86 ± 31 ka for a ring intrusive. For more information on the origin of this crater see Baldridge et al. (1989, p. 218).

The crater has been determined by recent geologic mapping (Anderson, 1994) to lie on a northeast-trending normal fault. The fault is down to the southeast as much as 600 ft, although the throw varies considerably along strike. It can be traced southwestward for 18 mi to the state line, but beyond that point it is concealed by Neogene basalt flows. The fault cannot be demonstrated to cut the Fence Lake Formation (Miocene) anywhere along its trace. Thus, we assume the faulting is related to mid-Tertiary crustal extension, the onset of which has been determined to be approximately 31 Ma (Baldridge, et. al., 1989, p. 187). There is little reason to speculate an older, classic Laramide, age for the fault, as the apparent style of deformation does not fit well with a model of northeast-directed compression.

The throw on the Zuni Salt Lake fault can be calculated on the basis of the stratigraphic units juxtaposed. Contrary to what is shown on the New Mexico State Geologic map (Dane & Bachman, 1965) no Mesaverde Group (Upper Cretaceous) units are present. Instead, the Atarque Sandstone (middle Turonian) is downfaulted against the lower Rio Salado Tongue of the Mancos Shale. This suggests up to 350 ft of stratigraphic separation at this location.

Large blocks of Atarque Sandstone occur in the northeastern rim of the crater, 600 ft. west of the road descending to the north shore of the lake. Fossils present in the Atarque here (UTM 38140725N, 704665E, zone 12) include *Pleuriocardia pauperculum, Crassatella excavata, Trigonarca* sp., *Phelopteria* sp., *Inoceramus* sp., *Plicatula ferryi, Ostrea* sp., *Parmicorbula* sp., *Gyrodes* sp., *Rostellites ambigua* and locally *Collignoniceras woolgari;* all indicate a middle Turonian age.

FIGURE S1.1. Aerial view to east of Zuni Salt Lake (left center) and Tejana Mesa (middle ground). Low rim of Zuni Salt Lake maar is approximately 1 mi in diameter. Note dark cinder cones in maar to the right of the salt lake. El Porticito vent at the south end of Tejana Mesa (far right) is location of Stop 2 on second day.

Return to main road. 0.2

13.4 Junction with main road; **bear left and proceed southward** along west side of maar. **0.4**

13.8 Tracks to left lead to west rim of the maar, for a good photo opportunity. Road is very near the trace of the Zuni Salt Lake fault, with poorly exposed lower Rio Salado Tongue of the Mancos Shale on the right, and the Moreno Hill Formation (which overlies the Atarque) on the left. **0.4**

14.2 Crater in the cinder cone at south end of the lake has its own small lake with more highly concentrated brine than main lake. The plumbing system responsible for the brines is not well understood, but it likely is tapped into the evaporite and brine-bearing Supai Formation (Yeso equivalent) which may lie as shallow as 600-700 ft on the footwall block. **1.0**

15.2 Crest of hill; road on cinders from Zuni Salt Lake maar. Tree-covered hills ahead are developed on Moreno Hill Formation. **1.5**

16.7 Cattleguard. **0.3**

17.0 At 10:00, closed depression is frequently inundated and is known as Cheap John Lake (Zuni Salt Lake USGS 7.5' quadrangle). The depression would not appear to be a deflation basin, and thus some karsting or structural control is likely. **1.7**

18.7 Cattleguard. **0.1**

18.8 Pass under power line. **0.6**

19.4 At 10:00 note Cerro Pomo, a Quaternary cinder cone, breached on the north side. **0.6**

20.0 Cattleguard and Cox Ranch; outcrops and toreva blocks on right consist of Moreno Hill Formation. **0.4**

20.4 On right, note the vertical stacking of fluvial channel sandstones in Moreno Hill Formation. **1.6**

22.0 On right at 4:00, note outcrops of fine-grained overbank and floodplain facies in Moreno Hill Formation. **2.0**

24.0 Ranch on right; low sandstone ledges around ranch are in lower part of Moreno Hill Formation. Immediately to the northwest of the ranch the top of the fossiliferous Atarque Sandstone is exposed. **0.1**

24.1 Cross Agua Fria Creek. **0.7**

24.8 Note carbonaceous shales of the lower Moreno Hill Formation in roadcuts on left as we proceed southward. Prominent mesa on right (to west) is capped by Neogene basalt flow. **0.4**

25.2 Cattleguard. **1.7**

26.9 Quaternary Red Hill cinder cone visible at 2:30; mesas on both sides of road ahead are Moreno Hill Formation capped with residual gravels derived from Baca Formation (Eocene). **1.4**

28.3 Cattleguard. **0.5**

28.8 Road curves to left; note good exposures of Moreno Hill Formation sandstone beds on left. **0.4**

29.2 Crest of hill; at left the gentle slopes are developed on erosional remnants of the Baca Formation resting on Moreno Hill Formation. At 1:00 to 2:00 is Cimarron Mesa, which is underlain by both Baca and Fence Lake Formations, at least in places, and is capped by a Miocene basalt flow. Straight ahead is the shallow, late Neogene Red Hill basin, which contains a thin, poorly exposed section of recycled older Tertiary rocks. With respect to several criteria, we are here leaving the Colorado Plateau and entering the Mogollon Datil volcanic field (or subprovince). The criteria are (1) disappearance of flat-lying Mesozoic rocks, (2) volcanic features from this area southward dominate or strongly influence the landscape, (3) thick sequences of volcaniclastic sediments may be present, and (4) generally higher heat flow values southward, however, regional trends in the transition zone at the southern margin of the Colorado Plateau are not profound (Minier, 1987). **0.2**

29.4 At 11:00 in the distance is Escondido Mountain, thick Bearwallow Mountain Andesite (late Oligocene) overlying Spears Group sediments. **1.3**

30.7 Crest of hill; to left thin, poorly sorted, basin-fill deposits of Late Tertiary to Pleistocene age conceal Baca Formation southward into the shallow Red Hill basin (see McIntosh and Cather, this volume). **1.1**

31.8 Cattleguard; at 2:00 note the small notch in east face of Cimarron Mesa which exposes the reddish Baca Formation beneath the basalt cap. **1.6**

33.4 Intersection with US-60. End of supplemental log; corresponds with mile 76.4 of Day 2 road log.

ROAD-LOG REFERENCES

Aldrich, M. J., Jr. and Laughlin, A. W., 1984, A model for the tectonic development of the southeastern Colorado Plateau boundary: Journal of Geophysical Research, v. 89, p. 10207-10218.

Aldrich, M. J., Jr., Ander, M. E. and Laughlin, A. W., 1981, Geological and geophysical signatures of the Jemez lineament: a reactivated Precambrian structure: International Basement Tectonics Association Publication no. 4, p. 77-85.

Anderson, O. J., 1987, Geology and coal resources of the Atarque Lake 1:50,000 quadrangle (NW quadrant of Fence Lake 1:100,000): New Mexico Bureau of Mines and Mineral Resources, GM-61, 2 sheets, scale 1:50,000.

Anderson, O. J., 1994, Geologic map of the Zuni Salt Lake quadrangle, Catron County, New Mexico: New Mexico Bureau of Mines and Mineral Resources, Open-file Report 405, 1:24,000.

Anderson, O. J., Chamberlin, R. M. and Roybal, G. H., 1987, Road log from Fence Lake, New Mexico to Springerville, Arizona: New Mexico Bureau of Mines and Mineral Resources, Bulletin 121, p. 26–30.

Armstrong, R. M. and Chamberlin, R. M., 1994, Preliminary geologic interpretation of the ARMA Baca Basin seismic reflection profile, northeastern Catron County, New Mexico: New Mexico Geological Society, Guidebook 45.

Austin, G. S. and Barker, J. M., 1993, Production and marketing of perlite of the western United States: 29th Forum on the Geology of Industrial Minerals: Division of Mines and Geology, California Department of Conservation, Long Beach, California, April 25–30, 1993, in press.

Bachman, G. O. and Mehnert, H. H., 1978, New K-Ar dates and late Pliocene to Holocene geomorphic history of the central Rio Grande region, New Mexico: Geological Society of America Bulletin, v. 89, p. 283–292.

Baldridge, W. S., Perry, F. V., Nealey, L. D., Laughlin, A. W. and Wohletz, K. H., 1989, Field guide to excursion 8A: Oligocene to Holocene magmatism and extensional tectonics, central Rio Grande rift and southeastern Colorado Plateau, New Mexico and Arizona: New Mexico Bureau of Mines and Mineral Resources, Memoir 46, p. 202–230.

Baldridge, W. S., Perry, F. V., Vaniman, D. T., Nealy, L. D., Leavy, B. D., Laughlin, A. W., Kyle, P., Bartor, Y., Steinitz, G. and Gladney, E. S., 1989, Magmatism associated with lithospheric extension; middle to late Cenozoic magmatism of the southeastern Colorado Plateau and central Rio Grande rift, New Mexico and Arizona: New Mexico Bureau of Mines and Mineral Resources, Memoir 46, p. 187.

Barker, C. E., 1989, Temperature and time in the thermal maturation of sedimentary organic matter; *in* Naeser, N. D. and McCulloh, T. H., eds., Thermal history of sedimentary basins–methods and case histories: Springer-Verlag, New York, p. 71–98.

Bikerman, M., 1972, New K-Ar ages on volcanic rocks from Catron and Grant Counties, New Mexico: Isochron/West no. 3, p. 9-12.

Black, B. A., 1964, The geology of the northern and eastern parts of the Ladron Mountains, Socorro County, New Mexico [M.S. thesis]: Albuquerque, University of New Mexico, 117 p.

Bobrow, D. J., Kyle, P. R. and Osburn, G. R., 1983, Miocene rhyolitic volcanism in the Socorro area of New Mexico: New Mexico Geological Society, Guidebook 34, p. 211–226.

Bornhorst, T. J., 1976, Volcanic geology of the Crosby Mountains and vicinity, Catron County, New Mexico [M.S. thesis]: Albuquerque, University of New Mexico, 113 p.

Bove, D. J., 1990, Preliminary geologic map of the Eagle Peak quadrangle, Catron County, New Mexico: U.S. Geological Survey, Open-file Report 90-0546, scale 1:24,000.

Bowie, M. R. and McLemore, V. T., 1987, Clay mineralogy of selected sedimentary and volcanic rocks, Socorro County, New Mexico; *in* McLemore, V. T. and Bowie, M. R., compilers, Guidebook to the Socorro area, New Mexico: New Mexico Bureau of Mines and Mineral Resources, p. 46–54.

Bradbury, J. P. 1966, Pleistocene-Recent geologic history of Zuni Salt Lake: New Mexico Geological Society, Guidebook 17, p. 119.

Broadhead, R. F., 1994, Geology and petroleum source rocks in the Hunt No. 1-16 State well: New Mexico Geological Society, Guidebook 45.

Brouillard, L. A., 1984, Geology of the northeastern Gallinas Mountains, Socorro County, New Mexico: New Mexico Bureau of Mines and Mineral Resources, Open-file Report 219, 172 p.

Brown, L. D., Chapin, C. E., Sanford, A. R., et al., 1980, Deep structure of the Rio Grande rift from seismic reflection profiling: Journal of Geophysical Research, v. 85, p. 4773–4800.

Bryan, K., 1926, Ground water reconnaissance in Socorro County, New Mexico: New Mexico State Engineer, 7th Biennial Report, p. 81–87.

Bucknam, R. C. and Anderson, R. E., 1979, Estimation of fault scarp ages from a scarp-height—slope-angle relationship: Geology, v. 7, p. 11-14.

Bull, W. B, 1991, Geomorphological responses to climatic change: Oxford University Press, New York, 326 p.

Campbell, F., 1989, Geology and coal resources of the Fence Lake 1:50,000 quadrangle: New Mexico Bureau of Mines and Mineral Resources, Geologic Map GM-62.

Campbell, F., and Roybal, G., 1984, Geology and coal resources of the Fence Lake quadrangle, Catron and Cibola Counties, New Mexico: New Mexico Bureau of Mines and Mineral Resources, Open-file Report 207, 37 p.

Carslaw, H. S. and Jaeger, J. C., 1959, Conduction of heat in solids: Oxford University Press, New York, 510 p.

Cas, R. A. F. and Wright, J. V., 1987, Volcanic successions—modern and ancient: London, Allen and Unwin Ltd., 528 p.

Cather, S. M., 1980, Petrology, diagenesis and genetic stratigraphy of the Eocene Baca Formation, Alamo Navajo Reservation and vicinity, Socorro County, New Mexico [M.A. thesis]: Austin, The University of Texas at Austin, 243 p.

Cather, S. M., 1989, Post-Laramide volcanic and tectonic transition in west-central New Mexico: New Mexico Geological Society, Guidebook 40, p. 91–97.

Cather, S. M. and Chapin, C. E., 1989, Day 2: Field guide to upper Eocene and lower Oligocene volcaniclastic rocks of the northern Mogollon-Datil volcanic field: southern Rocky Mountain region: New Mexico Bureau of Mines and Mineral Resources, Memoir 46, p. 61–68.

Cather, S. M. and Johnson, B. D., 1984, Eocene tectonics and depositional setting of west-central New Mexico and eastern Arizona: New Mexico Bureau of Mines and Mineral Resources, Circular 192, 33 p.

Cather, S. M. and McIntosh, W. C., 1994, The Plio-Pleistocene Quemado Formation of west-central New Mexico: New Mexico Geological Society, Guidebook 45.

Cather, S. M., Chamberlin, R. M. and Ratté, J. C., 1994, Tertiary stratigraphy and nomenclature for western New Mexico and eastern Arizona: New Mexico Geological Society, Guidebook 45.

Cather, S. M., McIntosh, W. C., and Chapin, C. E., 1987, Stratigraphy, age and rates of deposition of the Datil Group (Upper Eocene-lower Oligocene), west-central New Mexico: New Mexico Geology, v. 9, no. 3, p. 50–54.

Cather, S. M., Chamberlin, R. M., Chapin, C. E. and McIntosh, W. C., in press, Stratigraphic consequences of episodic extension in the Lemitar Mountains, central Rio Grande rift: Geological Society of America Special Paper 291.

Chamberlin, R. M., 1980, Cenozoic stratigraphy and structure of the Socorro Peak volcanic center, central New Mexico: New Mexico Bureau of Mines and Mineral Resources, Open-file Report 118, 532 p.

Chamberlin, R. M., 1981a, Cenozoic stratigraphy and structure of the Socorro Peak volcanic center, central New Mexico: a summary: New Mexico Geology, v. 3, p. 22–24.

Chamberlin, R. M., 1981b, Uranium potential of the Datil Mountains-Pietown area, Catron County, New Mexico: New Mexico Bureau of Mines and Mineral Resources, Open-file Report 138, 51 p.

Chamberlin, R. M., 1989, Criteria for the recognition of a pre-Eocene lateritic weathering profile, west-central New Mexico: New Mexico Geological Society, Guidebook 40, p. 34–35.

Chamberlin, R. M., 1991, A micro-indentation-extrusion tectonic model for the Laramide Zuni uplift, west-central New Mexico, USA: Geological Society of America Abstracts with Programs, v. 23, no. 5, p. A482.

Chamberlin, R. M., 1993, Neogene sinistral transtension along the Hickman fault zone, southeastern Colorado Plateau, New Mexico: Geological Society of America Abstracts with Programs, v. 25, no. 5, p. 19.

Chamberlin, R. M. and Anderson, O. J., 1989a, The Laramide Zuni uplift, southeastern Colorado Plateau: a microcosm of Eurasian-style indentation extrusion tectonics?: New Mexico Geological Society, Guidebook 40, p. 81–90.

Chamberlin, R. M. and Anderson, O. J., 1989b, The southern margin of the Colorado Plateau: a ragged edge bounded by Neogene domains of crustal extension: New Mexico Geological Society, Guidebook 40, p. 37-38.

Chamberlin, R. M. and Harris, J. S., 1994, Upper Eocene and Oligocene volcaniclastic sedimentary stratigraphy of the Quemado-Escondido Mountain area, Catron County, New Mexico: New Mexico Geological Society, Guidebook 45.

Chamberlin, R. M. and Osburn, G. R., 1986, Tectonic framework, character and evolution of upper crustal extensional domains in the Socorro area of the Rio Grande rift, New Mexico: Arizona Geological Society Digest, v. 16, p. 464.

Chamberlin, R. M., Barker, J. M. and Jenkins, D. A., 1994a, Economic geology of perlite production in New Mexico: Society for Mining, Metallurgy, and Exploration, Inc., Preprint No. 94–225, 24 p.

Chamberlin, R. M., Cather, S. M., Anderson, O. J. and Jones, G. E., 1994b, Reconnaissance geologic map of the Quemado 30 x 60' quadrangle, Catron County, New Mexico: New Mexico Bureau of Mines and Mineral Resources Open-file Report 406, scale, 1:100,000.

Chapin, C. E., 1971, The Rio Grande rift, Part I: Modifications and additions: New Mexico Geological Society, Guidebook 22, p. 191–201.

Chapin, C. E., 1983, An overview of Laramide wrench faulting in the southern Rocky Mountains with emphasis on petroleum exploration; *in* Lowell, J. D., ed., Rocky Mountain forelands and uplifts: Rocky Mountain Association of Geologists, p. 169–180.

Chapin, C. E., 1989, Volcanism along the Socorro accommodation zone, Rio Grande rift, New Mexico: Southern Rocky Mountain Region: New Mexico Bureau of Mines and Mineral Resources, Memoir 46, p. 46–57.

Chapin, C. E. and Cather, S. M., 1981, Eocene tectonics and sedimentation in the Colorado Plateau–Rocky Mountains area: Arizona Geological Society Digest, v. 14, p. 173–198.

Chapin, C. E. and Cather, S. M., in press, Tectonic setting of the axial basins of the northern and central Rio Grande rift: Geological Society of America Special Paper 291.

Chapin, C. E. and Lindley, J. I., 1986, Potassium metasomatism of igneous and sedimentary rocks in detachment terranes and other sedimentary basins: economic implications: Arizona Geological Society Digest, v. 16, p. 118–126.

Chapin, C. E., Chamberlin, R. M., Osburn, G. R., White, D. W. and Sanford, A. R., 1978a, Exploration framework of the Socorro geothermal area: New Mexico Geological Society, Special Publication no. 7, p. 115–129.

Chapin, C. E., Jahns, R. H., Chamberlin, R. M. and Osburn, G. R., 1978b, First day road log from Socorro to Truth or Consequences via Magdalena and Winston: New Mexico Geological Society, Special Publication no. 7, p. 1–31.

Chapin, C. E., Osburn, G. R., Hook, S. C., Massingill, G. L. and Frost, S. J., 1979, Coal, uranium, oil and gas potential of the Riley-Puerticito area, Socorro County, New Mexico: New Mexico Bureau of Mines and Mineral Resources, Open-file Report 103, 33 p.

Chronic, H., 1952, Molluscan fauna from the Permian Kaibab Formation, Walnut Canyon, Arizona: Geological Society of America Bulletin, v. 63, p. 95–166.

Cleaveland, A. M., 1941, No life for a lady: Boston, Houghton Mifflin Co., 356 p.

Clisby, K. H. and Sears, P. B., 1956, San Agustin Plains-Pleistocene climatic changes: Science, v. 124, p. 537–538.

Coffin, G. C., 1981, Geology of the northwestern Gallinas Mountains, Socorro County, New Mexico: New Mexico Bureau of Mines and Mineral Resources, Open-file Report 159, 214 p.

Colpitts, R. M., Jr., 1989, Permian reference section for southeastern Zuni Mountains, Cibola County, New Mexico: New Mexico Geological Society, Guidebook 40, p. 177–180.

Condit, C. D., Crumpler, L. S. and Aubele, J. C., 1989, Fourth-day field trip: Springerville volcanic field: New Mexico Bureau of Mines and Mineral Resources, Memoir 46, p. 33–38.

Crews, S. G., 1990, Structure, syntectonic sedimentation, and stratigraphy of asymmetrical non-marine basins [PhD thesis]: Laramie, University of Wyoming, 196 p.

Crews, S. G., 1994, Tectonic control of synrift sedimentation patterns, Reserve graben, southwestern New Mexico: New Mexico Geological Society, Guidebook 45.

Crumpler, L. S., Aubele, J. C. and Condit, C. D., 1994, Volcanoes and neotectonic characteristics of the Springerville volcanic field, Arizona: New Mexico Geological Society, Guidebook 45.

Dane, C. H. and Bachman, G. O., 1965, Geologic map of New Mexico: U.S. Geological Survey, scale 1:500,000.

Davis, G. H., 1978, Monocline fold pattern of the Colorado Plateau: Geological Society of America, Memoir 151, p. 215–233.

Decker, P. L., 1990, Style and mechanics of liquefaction-related deformation, lower Absaroka Volcanic Supergroup (Eocene), Wyoming: Geological Society of America, Special Paper 240, 71 p.

Dethier, D. P. and Demsey, K. A., 1984, Erosional history and soil development on Quaternary surfaces, northwest Española Basin, New Mexico: New Mexico Geological Society, Guidebook 35, p. 227-234.

Dethier, D. P., Aldrich, M. J., Jr. and Shafiqullah, M., 1986, New K-Ar ages for Miocene volcanic rocks from the northeastern Jemez Mountains and Tejana Mesa, New Mexico: Isochron/West, no. 57, p. 12–14.

Dethier, D. P., Harrington, C. D. and Aldrich, M. J., 1988, Late Cenozoic rates of erosion in the western Española Basin, New Mexico: Evidence from geological dating of erosion surfaces: Geological Society of America Bulletin, v. 100, p. 928-937.

Dick, H. W., 1965, Bat Cave: School of American Research, Monograph No. 27, Santa Fe, 114 p.

Dunbar, C. O. and Condra, G. E., 1932, Brachiopoda of the Pennsylvanian System in Nebraska: Nebraska Geological Survey, Bulletin 5, 2nd series, 377 p.

Dunbar, N. W., Chapin, C. E., Ennis, D. J. and Campbell, A. R., 1994, Trace element and mineralogical alteration associated with moderate and advanced degrees of K-metasomatism in a rift basin at Socorro, New Mexico: New Mexico Geological Society, Guidebook 45.

Eggleston, R. E. and Reiter, M., 1984, Terrestrial heat-flow estimates from petroleum bottom-hole temperature data in the Colorado Plateau and the eastern Basin and Range Province: Geological Society of America Bulletin, v. 95, p. 1027–1033.

Eggleston, T. L., Osburn, G. R. and Chapin, C. E., 1983, Third day road log from Socorro to San Antonio, Nogal Canyon, Chupadera Mountains, Luis Lopez manganese district and the MCA mine: New Mexico Geological Society, Guidebook 34, p. 61–81.

Elston, W. E., 1978, Mid-Tertiary cauldrons and their relationship to mineral resources, southwestern New Mexico: a brief review: New Mexico Geological Society, Special Publication no. 7, p. 107–113.

Elston, W. E., 1984, Mid-Tertiary ash-flow tuff cauldrons, southwestern New Mexico: Journal of Geophysical Research, v. 89, p. 8733-8750.

Elston, W. E., 1989, Overview of the Mogollon-Datil volcanic field: Southern Rocky Mountain region: New Mexico Bureau of Mines and Mineral Resources, Memoir 46, p. 43–46.

Elston, W. E. and Damon, P. E., 1970, Significance of 4 new K-Ar dates from the Mogollon Plateau volcanic province southwestern New Mexico; *in* Annual Progress Report COO-689-130, Contract AT (11-1)-689, U.S. Atomic Energy Commission: Geochronological Laboratories, University of Arizona, Tucson, AVI-1-AVI-9.

Elston, W. E., Rhodes, R. C., Coney, J. P. and Deal, E. J., 1976, Progress report on the Mogollon Plateau volcanic province, southwestern New Mexico: New Mexico Geological Society, Special Publication no. 5, p. 3-28.

Eveleth, R. W., 1983, Gustav Billing, the Kelly mine, and the great smelter at Park City, Socorro County, New Mexico: New Mexico Geological Society, Guidebook 34, p. 89–95.

Ferguson, C. A., 1991, Stratigraphic and structural studies in the Mt. Withington caldera, Grassy Lookout quadrangle, Socorro County, New Mexico: New Mexico Geology, v. 13, p. 50–54.

Fisher, R. V. and Schmincke, H.-U., 1984, Pyroclastic rocks: New York, Springer-Verlag, 472 p.

Fisher, R. V. and Smith, G. A., 1991, Volcanism, tectonics and sedimentation; *in* Sedimentation in volcanic settings: SEPM Special Publication no. 45, p. 1–5.

Fitzsimmons, J. P., 1959, Structure and geomorphology of west-central New Mexico: New Mexico Geological Society, Guidebook 10, p. 112–116.

Foreman, F., 1956, San Agustin Plains—the sediments: Science, v. 124, p. 539.

Foreman, F., Clisby, K. H. and Sears, P. B., 1959, Plio-Pleistocene sediments and climates of the San Agustin Plains, New Mexico: New Mexico Geological Society, Guidebook 10, p. 117–120.

Foster, R. W., 1957, Stratigraphy of west-central New Mexico: Geology of Southwestern San Juan Basin, Four Corners Geological Society Guidebook, Second Field Conference, p. 62–72.

Foster, R. W., 1964, Stratigraphy and petroleum possibilities of Catron County, New Mexico: New Mexico Bureau of Mines and Mineral Resources, Bulletin 85, 55 p.

Futa, K. and Ratté, J. C., 1989, Petrogenetic implications of Rb-Sr and Sm-Nd isotopes related to post-caldera volcanism in the western Mogollon-Datil volcanic field, New Mexico: New Mexico Bureau of Mines and Mineral Resources, Bulletin 131, 100 p.

Gabin, V. L. and Lesperance, L. E., 1976, New Mexico climatological data, precipitation, temperature, evaporation, and wind, monthly and annual means, 1850–1975: W. K. Summers and Associates.

Gallagher, S., 1994, Landslide repairs on US-180 between mileposts 13 and 19, near Luna, Catron County, New Mexico: New Mexico Geological Society, Guidebook 45.

Garmezy, L., 1990, Strike-slip deformation along the southeastern margin of the Colorado Plateau: Geological Society of America Abstracts with Programs, v. 22, p. A-276.

Gile, L. H., Hawley, J. W. and Grossman, R. B., 1981, Soils and geomorphology in the Basin and Range area of southern New Mexico: New Mexico Bureau of Mines and Mineral Resources, Memoir 39, 222 p.

Gile, L. H., Peterson, F. F. and Grossman, R. B., 1966, Morphological and genetic sequences of carbonate accumulation in desert soils: Soil Science, v. 101, p. 347-360.

Godson, R. H., 1981, Digital terrain map of the United States: U.S. Geological Survey, Miscellaneous Investigations Map I-1318.

Guilinger, D. R., 1982, Geology and uranium potential of the Tejana Mesa-Hubbell Draw area, Catron County, New Mexico: New Mexico Bureau of Mines and Mineral Resources, Open-file Report 176, 137 p.

Harris, L. G., 1984, New Mexico water rights: New Mexico Water Resources Research Institute, Las Cruces, Miscellaneous Report no. 15, 53 p.

Harrison, R. W., 1980, Geology of the northeastern Datil Mountains, Socorro and Catron Counties, New Mexico: New Mexico Bureau of Mines and Mineral Resources, Open-file Report 136, 146 p.

Hawley, J. W., compiler, 1978, Guidebook to the Rio Grande rift in New Mexico and Colorado: New Mexico Bureau of Mines and Mineral Resources, Circular 163, 154 p.

Hawley, J. W., 1993, Some principles of calcisol development and soil-geomorphic relationships in southeastern New Mexico: New Mexico Geological Society, Guidebook 44, p. 4–5.

Hoffer, J. M. and Corbitt, L. L., 1989, A preliminary note on some unusual explosion-collapse craters, Quemado basalt field, Catron County, New Mexico: New Mexico Geological Society, Guidebook 40, p. 157–158.

Hook, S. C., Molenaar, C. M. and Cobban, W. A., 1983, Stratigraphy and revision of nomenclature of Upper Cenomanian to Turonian (Upper Cretaceous) rocks of west-central New Mexico: New Mexico Bureau of Mines and Mineral Resources, Circular 185, p. 7–28.

Hough, W., 1907, Antiquities of the upper Gila and Salt River valleys in Arizona and New Mexico: Smithsonian Institution, Bureau of American Ethnology, Bulletin 35, p. 70–76.

Houser, B. B., 1994, Geology of the late Cenozoic Alma Basin, New Mexico and Arizona: New Mexico Geological Society, Guidebook 45.

Hunt, J. M. and Hennet, R. J-C., 1992, Modeling petroleum generation in sedimentary basins; *in* Whelan, J. K. and Farrington, J. W., eds., Organic matter: productivity, accumulation, and preservation in recent and ancient sediments: Columbia University Press, New York, p. 20–52.

Huppert, H. E. and Sparks, R. S., 1989, Chilled margins in igneous rocks: Earth and Planetary Science Letters, v. 92, p. 397–405.

Hurt, W. R., Jr. and McKnight, D., 1949, Archaeology of the San Agustin Plains, a preliminary report: American Antiquity, v. 14, p. 172–194.

Jaeger, J. C., 1964, Thermal effects of intrusions: Reviews of Geophysics, v. 2, p. 443–466.

Jaeger, J. C., 1968, Cooling and solidification of igneous rocks; *in* Hess, H. H. and Poldervaart, A., eds., Basalts, the Poldervaart treatise on rocks of basaltic composition: John Wiley and Sons, New York, v. 2, p. 503–536.

Jenny, H., 1980, The soil resource-origin and behavior: Springer-Verlag, New York, 377 p.

Johansen, S. J., 1986, Provenance of the Mesaverde Group of west-central New Mexico [Ph.D. dissertation]: Austin, University of Texas, 309 p.

Johnson, B. D., 1978, Genetic stratigraphy and provenance of the Baca Formation, New Mexico and the Eagar Formation, Arizona [M.A. thesis]: Austin, University of Texas, 150 p.

Johnson, P. S., 1992, A paleohydrologic and paleoclimatic reconstruction for the San Agustin Basin, New Mexico, based on Oxygen-18 variation in ostracod shells {Ph.D. dissertation]: Socorro, New Mexico Institute of Technology.

Jones, D. P., 1980, Volcanic geology of the Alegres Mountain area, Catron County, New Mexico [M.S. thesis]: Albuquerque, University of New Mexico, 76 p.

Keller, G. R. and Cordell, L., 1983, Bouguer gravity map of New Mexico: National Geophysical Data Center, NOAA, Washington, D.C., scale 1:500,000.

Kelley, S. A., Chapin, C. E. and Corrigan, J., 1992, Late Mesozoic to Cenozoic cooling histories of the flanks of the northern and central Rio Grande rift, Colorado and New Mexico: New Mexico Bureau of Mines and Mineral Resources, Bulletin 145, 39 p.

Kelley, V. C., 1957, Tectonics of the San Juan Basin and surrounding area: Four Corners Geological Society, Second Field Conference Guidebook, p. 44–52.

Kopacz, M. A. Zarrow, L. and Garmezy, L., 1989, The Mogollon subvolcanic frontier play: Abstracts, Volume 2 of 3, 28th International Geological Congress, Washington, D.C., p. 2-213.

Kottlowski, F. E., 1959, Pennsylvanian rocks on the northeast edge of the Datil Plateau: New Mexico Geological Society, Guidebook 10, p. 57–62.

Kottlowski, F. E., 1960, Summary of Pennsylvanian sections in southwestern New Mexico and southeastern Arizona: New Mexico Bureau of Mines and Mineral Resources, Bulletin 66, 187 p.

Kues, B. S. and Lucas, S. G., 1989, Stratigraphy and paleontology of a San Andres Formation (Permian, Leonardian) outlier, Zuni Indian Reservation, New Mexico: New Mexico Geological Society, Guidebook 40, p. 167–176.

Lachenbruch, A. H., Sass, J. H., Munroe, R. J. and Moses, T. H., Jr., 1976, Geothermal setting and simple heat conduction models for the Long Valley caldera: Journal of Geophysical Research, v. 81, p. 769–784.

Laughlin, A. W., Aldrich, M. J. and Vaniman, D. T., 1983, Tectonic implications of mid Tertiary dikes in west-central New Mexico: Geology, v. 11, p. 45–48.

Laughlin, A. W., Damon, P. E. and Shafiqullah, M., 1980, New K-Ar dates from the Springerville volcanic field, central Jemez zone, Apache County, Arizona: Isochron/West, no. 29: 3–4.

Laughlin, A. W., Brookins, D. G., Damon, P. E. and Shafiqullah, M., 1979, Late Cenozoic volcanism of the central Jemez zone, Arizona-New Mexico: Isochron/West, no. 25, p. 5–8.

Leinau, P. J., 1988, Direct-use developments in the United States: Oregon Institute of Technology Geo-heat center, Quarterly Bulletin, v. 11, p. 1–4.

Leopold, A., 1966, A Sand county almanac: Oxford University Press, 295 p.

Lipman, P. W., 1984, The roots of ash flow calderas in western North America: windows into the tops of granitic batholiths: Journal of Geophysical Research, v. 89, p. 8801–8841.

Lipman, P. W., 1992, Magmatism in the Cordilleran United States; progress and problems; *in* Burchfiel, B. C., Lipman, P. W. and Zoback, M. L., eds., The Cordilleran Orogen: conterminous U.S.: Geological Society of America, The Geology of North America, v. G-3, p. 481–514.

Lopez, D. A., 1975, Geology of the Datil area, Catron County, New Mexico [M.S. thesis]: Albuquerque, University of New Mexico, 72 p.

Lopez, D. A. and Bornhorst, T. J., 1979, Geologic map of the Datil area, Catron County, New Mexico: U.S. Geological Survey, Miscellaneous Geologic Investigations Map I-1098, scale 1:50,000.

Love, D. W., 1989, Bidahochi Formation: an interpretive summary: New Mexico Geological Society, Guidebook 40, p. 273–280.

Lucas, S. G. and Anderson, O. J., 1994, Miocene Proboscidean from the Fence Lake Formation, Catron County, New Mexico: New Mexico Geological Society, Guidebook 45.

Lucas, S. G. and Hayden, S. N., 1989, Triassic stratigraphy west-central New Mexico: New Mexico Geological Society, Guidebook 40, p. 191-211.

Luna Ward—The Church of Jesus Christ of Latter-day Saints, 1983, Do You Remember Luna? 100 Years of Pioneer History: Albuquerque, New Mexico, Adobe Press, 301 p.

Machette, M. N., 1982, Quaternary and Pliocene faults in the La Jencia and southern part of the Albuquerque-Belen basins, New Mexico: evidence of fault history from fault scarp morphology and Quaternary geology: New Mexico Geological Society, Guidebook 33, p. 161–169.

Machette, M. N., 1986, History of Quaternary offset and paleoseismicity along the La Jencia fault, central Rio Grande rift, New Mexico: Seismological Society of America Bulletin, v. 76, p. 259-272.

Machette, M. N., 1988, Quaternary movement along the La Jencia fault, central New Mexico: U.S. Geological Survey, Professional Paper 1440, 82 p.

Machette, M. N. and McGimsey, R. G., 1982, Quaternary and Pliocene faults in the Socorro and western part of the Fort Sumner 1° x 2° quadrangles, New Mexico: U.S. Geological Survey, Miscellaneous Field Studies Map MF-1465A.

Mack, G. H., Nightengale, A. L., Seager, W. R. and Clemons, R. E., 1994, The Oligocene Goodsight-Cedar Hills half graben near Las Cruces and its implications to the evolution of the Mogollon-Datil volcanic field and to the southern Rio Grande rift: New Mexico Geological Society, Guidebook 45.

Markgraf, V., Bradbury, J. P., Forester, R. M., Singh, G. and Sternberg, R. S., 1984, San Agustin Plains, New Mexico: age and paleoenvironmental potential reassessed: Quaternary Research, v. 22, p. 336–343.

Markgraf, V., Bradbury, J. P., Forester, R. M., McCoy, W., Singh, G. and Sternberg, R., 1983, Paleoenvironmental reassessment of the 1.6-million-year-old record from San Agustin basin, New Mexico: New Mexico Geological Society, Guidebook 34, p. 291-297.

Martin, P. S., 1959, Digging into history: Chicago Natural History Museum, 157 p.

Marvin, R. F., Naeser, C. W., Bikerman, M., Mehnert, H. H. and Ratté, J. C., 1987, Isotopic ages of post-Paleocene igneous rocks within and bordering the Clifton 1° x 2° quadrangle, Arizona–New Mexico: New Mexico Bureau of Mines and Mineral Resources, Bulletin 118, 63 p.

Massingill, G. L., 1979, Geology of the Riley-Puerticito area, southeast margin of the Colorado Plateau, Socorro County, New Mexico: New Mexico Bureau of Mines and Mineral Resources, Open-file Report 107, 316 p.

Maxwell, C. H., 1976, Stratigraphy and structure of the Acoma region, New Mexico: New Mexico Geological Society, Special Publication no. 6, p. 95–101.

McDonald, E. V. and Busacca, A. J., 1990, Interaction between aggrading geomorphic surfaces and the formation of a late Pleistocene paleosol in the Palouse loess of eastern Washington state: Geomorphology, v. 3, p. 449–470.

McGrath, D. G. and Hawley, J. W., 1987, Geomorphic evolution and soil geomorphic relationships in the Socorro area, central New Mexico; *in* McLemore, V. T. and Bowie, M. R., compilers, Guidebook to the Socorro area, New Mexico: New Mexico Bureau of Mines and Mineral Resources, p. 55–67.

McIntosh, W. C. and Cather, S. M., 1994, $^{40}Ar/^{39}Ar$ geochronology of basaltic rocks and constraints on Late Cenozoic stratigraphy and landscape development in the Red Hill-Quemado area, New Mexico: New Mexico Geological Society, Guidebook 45.

McIntosh, W. C. and Chamberlin, R. M., 1994, $^{40}Ar/^{39}Ar$ geochronology of middle to late Cenozoic ignimbrites, mafic lavas, and volcaniclastic rocks in the Quemado-Datil region, New Mexico: New Mexico Geological Society, Guidebook 45.

McIntosh, W. C., Kedzie, L. L. and Sutter, J. F., 1991, Paleomagnetism and $^{40}Ar/^{39}Ar$ ages of ignimbrites, Mogollon-Datil volcanic field, southwest New Mexico: New Mexico Bureau of Mines and Mineral Resources, Bulletin 135, 99 p.

McIntosh, W. C., Chapin, C. E., Ratté, J. C. and Sutter, J. F., 1992, Time-stratigraphic framework for the Eocene-Oligocene Mogollon-Datil volcanic field, southwest New Mexico: Geological Society of America Bulletin, v. 104, p. 851-871.

McIntosh, W. C., Geissman, J. W., Chapin, C. E., Kunk, M. J. and Henry, C. D., 1992, Calibration of the latest Eocene-Oligocene geomagnetic polarity time scale using $^{40}Ar/^{39}Ar$ dated ignimbrites: Geology, v. 20, p. 459–463.

McLellan, M., Haschke, L., Robinson, L., Carter, M. D. and Medlin, A., 1983, Middle Turonian and younger rocks, northern Salt Lake coal field, Cibola and Catron Counties, New Mexico: New Mexico Bureau of Mines and Mineral Resources, Circular 185, p. 41–46.

McLellan, M., Robinson, L., Haschke, L., Carter, M. D. and Medlin, A., 1982, Fence Lake Formation (Tertiary), west-central New Mexico: New Mexico Geology, v. 3, p. 53–55.

McLemore, V. T., 1994, Volcanic-epithermal deposits in the Mogollon-Datil volcanic field, west-central New Mexico: New Mexico Geological Society, Guidebook 45.

Menges, C. M. and Perthree, P. A., 1989, Late Cenozoic tectonism in Arizona and its impact on regional landscape evolution: Arizona Geological Society Digest, v. 17, p. 649-680.

Menges, C. M., Kawaguchi, G. H., Ritter, J. B., McFadden, L. D. and Lozinsky, R. P., 1984, Rates and amounts of Quaternary faulting on the VLA fault scarp, northeastern San Agustin Plains: New Mexico Geology, v. 6, p. 85.

Minier, J., 1987, A geothermal study in west-central New Mexico [Ph.D. dissertation]: Socorro, New Mexico Institute of Mining and Technology, 112 p.

Minier, J. and Reiter, M., 1989, Coal maturation and geothermal history, west-central New Mexico: New Mexico Geological Society, Guidebook 40, p. 127–133.

Minier, J. and Reiter, M., 1991, Heat flow on the southern Colorado Plateau: Tectonophysics, v. 200, p. 51–66.

Minier, J., Reiter, M., Shafiqullah, M. and Damon, P. E., 1988, Geothermal studies in the Quemado area, New Mexico: Geothermics, v. 17, p. 735–755.

Morgan, T. L., 1980, Uranium hydrogeochemical survey of well waters from an area around Pie Town, Catron County, west-central New Mexico, including concentrations of twenty-three additional elements: Los Alamos LA-8017-MS, 23 p.

Morgan, P. and Gosnold, W. D., 1989, Heat flow and thermal regimes in the continental United States: Geological Society of America, Memoir 172, p. 493–522.

Murphy, M. O., 1971, Bytownite bonanza: Lapidary Journal, v. 25, p. 170-178.

Myers, R. G., 1992, Geohydrology and potential hydrologic effects of surface coal mining of the San Agustin coal area and adjacent area, Catron and Cibola Counties, New Mexico: U.S. Geological Survey, Water-Resources Investigations Report 92-4004, 52 p.

Neal, J. T., 1965, Giant desiccation polygons of Great Basin playas: Air Force Cambridge Research Laboratories, Environmental Research Papers no. 123, 30 p.

Neal, J. T. and Motts, W. S., 1975, Recent geomorphic changes in playas of western United States; *in* Neal, J. T., ed., Playas and dried lakes—occurrence and development: Dowden, Hutchinson, and Ross, p. 327–342.

New Mexico Geological Society, 1982, New Mexico Highway Geologic Map: scale 1:1,000,000.

Newell, N. D. and Boyd, D. W., 1975, Parallel evolution in early trigoniacean bivalves: Bulletin of the American Museum of Natural History, v. 154, p. 55–162.

Osburn, G. R. and Chapin, C. E., 1983a, Nomenclature for Cenozoic rocks of the northeast Mogollon-Datil volcanic field, New Mexico: New Mexico Bureau of Mines and Mineral Resources, Stratigraphic Chart 1.

Osburn, G. R. and Chapin, C. E., 1983b, Ash-flow tuffs and cauldrons in the northeast Mogollon-Datil volcanic field: a summary: New Mexico Geological Society, Guidebook 34, p. 197–204.

Osburn, G. R., Laroche, T. M. and Weber, R. H., 1993, Geology of Lion Mountain and northern Arrowhead Well quadrangles, Socorro County, New Mexico: New Mexico Bureau of Mines and Mineral Resources, Geologic Map 68, 11 p.

Palmer, A. R., compiler, 1983, Decade of North American geology 1983 time scale: Geology, v. 11, p. 503–504.

Pearce, T. M., ed., 1965, New Mexico place names—a geographical dictionary: The University of New Mexico Press, 187 p.

Phillips, F. M., Campbell, A. R., Kruger, C., Johnson, P. S., Roberts, R. and Keyes, E., 1992, A reconstruction of the response of the water balance in western United States lake basins to climate change: New Mexico Water Resources Research Instiute, Technical Completion Report (Volume 1) on Project Nos. 11345662 and 1423687, WRRI Report No. 269, 167 p.

Pierce, H. W., 1985, Arizona's backbone: the transition zone: Arizona Bureau of Geology and Mineral Technology, Field Notes, v. 15, p. 1–6.

Potochnik, A. R., 1989, Depositional style and tectonic implications of the Mogollon Rim Formation (Eocene), east-central Arizona: New Mexico Geological Society, Guidebook 40, p. 107–118.

Potter, L. D., 1957, Phytosociological study of San Agustin Plains, New Mexico: Ecological Monographs, v. 27, p. 113–136.

Powers, W. E., 1939, Basin and shore features of the extinct lake San Agustin, New Mexico: Journal of Geomorphology, v. 2, p. 345–356.

Powers, W. E., 1941, Volcanic rocks of the western San Agustin Plains district, New Mexico: Journal of Geology, v. 49, p. 207–217.

Rannels, J. E. and McLarty, L., 1990, Geothermal power generation in the United States: 1985 through 1989: Geothermal Resources Council, Transactions, v. 14, part 1, p. 293–305.

Ratté, J. C., 1980, Geologic map of the Saliz Pass quadrangle, Catron County, New Mexico: U.S. Geological Survey, Miscellaneous Field Investigations Map MF-1203, scale 1:24,000.

Ratté, J. C., 1989a, Geologic map of the Bull Basin quadrangle, Catron County, New Mexico: U.S. Geological Survey, Geologic Quadrangle Map GQ-1651, scale 1:24,000.

Ratté, J. C., 1989b, Days 3 and 4: selected volcanic features of the western Mogollon-Datil volcanic field: New Mexico Bureau of Mines and Mineral Resources, Memoir 46, p. 68-91.

Ratté, J. C., 1992, Preliminary geologic map of the Tularosa Canyon quadrangle, Catron County, New Mexico: U.S. Geological Survey, Open-file Report 92-187, scale 1:24,000.

Ratté, J. C. and Futa, K., 1989, Petrogenetic implications of Rb-Sr and Sm-Nd isotopes related to post-caldera volcanism in the western Mogollon-Datil volcanic field, New Mexico: New Mexico Bureau of Mines and Mineral Resources, Bulletin 131, p. 100.

Ratté, J. C., Bove, D. J. and McIntosh, W. C., 1994, Geologic map of the Milligan Mountain quadrangle, Catron County, New Mexico: U.S. Geological Survey, Geologic Quadrangle Map GQ-1736, scale 1:24,000.

Ratté, J. C., McIntosh, W. C. and Houser, H.B., 1991, Geologic map of the Horse Mountain West quadrangle, Catron County, New Mexico: U.S. Geological Survey, Geologic Quadrangle Map GQ-1685, scale 1:24,000.

Ratté, J. C., Landis, E. R., Gaskill, D. L. and Raabe, R. G., 1969, Mineral resources of the Blue Range Primitive Area, Greenlee County, Arizona, and Catron County, New Mexico: U.S. Geological Survey, Bulletin 1261-E, p. E1-E91, with geologic map, scale 1:62,500.

Ratté, J. C., Modreski, P. J., McIntosh, W. C. and Chamberlin, R. M., 1994, The Horse Springs dacite: an enigmatic pyroclastic deposit at the edge of the Plains of San Agustin: New Mexico Geological Society, Guidebook 45.

Rauzi, S. L., 1994a, Geothermal test hints at oil potential in eastern Arizona volcanic field: Oil and Gas Journal, Jan. 3, 1994, p. 52–54.

Rauzi, S. L., 1994b, Implications of live oil shows in eastern Arizona geothermal test: Arizona Geological Survey, Open-file Report 94-1, 16 p.

Reiter, M. and Clarkson, G., 1983, A note of terrestrial heat flow in the Colorado Plateau: Geophysical Research Letters, v. 10, p. 929–932.

Reiter, M. and Shearer, C., 1979, Terrestrial heat flow in eastern Arizona: a first report: Journal of Geophysical Research, v. 84, p. 6115–6120.

Rhodes, R. C. and Smith, E. I., 1976, Stratigraphy and structure of the northwestern part of the Mogollon Plateau volcanic province, Catron County, New Mexico: New Mexico Geological Society, Special Publication no. 5, p. 57-62, geologic map scale ~1:100,000.

Ritter, J. B., Kawaguchi, G. H., Menges, C. M., McFadden, L. D. and Lozinsky, R. P., 1984, A preliminary soils chronosequence for the San Agustin Plains, west-central New Mexico: New Mexico Geology, v. 6, p. 86.

Roberts, D. G., Adams, D. C. and Keller, G. R., 1994, Crustal structure of west-central New Mexico: a preliminary seismic interpretation: New Mexico Geological Society, Guidebook 45.

Robinson, B. R., 1981, Geology of the D-Cross Mountain quadrangle, Socorro and Catron Counties, New Mexico: New Mexico Bureau of Mines and Mineral Resources, Open-file Report 147, 225 p.

Rodgers, K. W., 1994, Salt River Project's proposed Fence Lake mine, Catron County, New Mexico: New Mexico Geological Society, Guidebook 45.

Roy, R. F., Beck, A. E. and Touloukian, Y. S., 1981, Thermophysical properties of rocks; *in* Touloukian, Y. S., Judd, W. R. and Roy, R. F., eds., Physical properties of rocks and minerals: McGraw-Hill CINDAS Data Series on Material Properties, v. II-2, chapter 12, p. 409–502.

Sanford, A. R., 1978, Characteristics of Rio Grande rift in vicinity of Socorro, New Mexico, from geophysical studies: New Mexico Bureau of Mines and Mineral Resources, Circular 163, p. 116–119.

Sanford, A. R., Mott, R. P., Jr., Shuleski, P. J., Rinehart, E. J., Carvella, F. J., Ward, R. M. and Wallace, T. C., 1977, Geophysical evidence for a magma body in the crust in the vicinity of Socorro, New Mexico: American Geophysical Union, Geophysical Monograph 20, p. 385–403.

Sass, J. H., Stone, C. and Bills, D. J., 1982, Shallow subsurface temperatures and some estimates of heat flow from the Colorado Plateau of northeastern Arizona: U.S. Geological Survey, Open-file Report 82-994, 112 p.

Simoneit, B. R. T., 1983, Organic matter maturation and petroleum genesis: geothermal versus hydrothermal: Geothermal Resources Council, Special Report 13, p. 215–241.

Sirrine, G. K., 1956, Geology of the Springerville-St. Johns area, Apache County, Arizona [Ph.D. thesis]: Austin, The University of Texas at Austin, 247 p.

Smith, E. I., 1976, Structure and petrology of the John Kerr Peak dome complex southwestern New Mexico: New Mexico Geological Society, Special Publication no. 5, p. 71-78.

Smith, E. I., Aldrich, M. J., Jr., Deal, E. G. and Rhodes, R. C., 1976, Fission-track ages of Tertiary volcanic and plutonic rocks, Mogollon plateau, southwestern New Mexico: New Mexico Geological Society, Special Publication no. 5, p. 117-118.

Smith, G. A., 1991, Facies sequences and geometries in continental volcaniclastic sediments; Sedimentation in volcanic settings: SEPM, Special Publication no. 45, p. 109–121.

Smith, T., 1983, Pie Town: A trip back: Impact, Albuquerque Journal Magazine, p. 4–8 (18 January).

Snyder, D. O., 1971, Stratigraphic analysis of the Baca Formation, west-central New Mexico [Ph.D. thesis]: Albuquerque, University of New Mexico, 158 p.

Spence, W. and Gross, R. S., 1990, A tomographic glimpse of the upper mantle source of magmas of the Jemez lineament, New Mexico: Journal of Geophysical Research, v. 95, p. 10,829–10,850.

Spencer, R. S., 1967, Pennsylvanian Spiriferacea and Sprifernacea of Kansas: University of Kansas Paleontological Contributions, Paper 14, 35 p.

Stearns, C. E., 1962, Geology of the north half of the Pelona quadrangle, Catron County, New Mexico: New Mexico Bureau of Mines and Mineral Resources, Bulletin 78, 46 p., with map, scale 1:63,360.

Stewart, J. H., 1980, Regional tilt patterns of later Cenozoic basin-range fault blocks, western United States: Geological Society of America Bulletin, v. 91, p. 460–464.

Stone, C., 1980, Springerville geothermal project—geology, geochemistry, geophysics—final report: Arizona Bureau of Geology and Mineral Technology, Open-file Report 8-4, 23 p.

Sutherland, P. K. and Harlow, F. H., 1973, Pennsylvanian brachiopods and biostratigraphy in southern Sangre de Cristo Mountains, New Mexico: New Mexico Bureau of Mines and Mineral Resources, Memoir 27, 173 p.

Taylor, J. R., 1982, An introduction to error analysis: Mill Valley, California, University Science Books, 270 p.

Tester, J. W. and Herzog, H. J., 1990, Economic predictions for heat mining: a review and analysis of hot dry rock (HDR) geothermal energy technology: Massachusetts Institute of Technology Energy Laboratory, Report MIT-EL 90-001, 180 p.

Tester, J. W., Brown, D. W. and Potter, R. M., 1989, Hot dry rock geothermal energy—a new energy agenda for the 21st century: Los Alamos National Laboratory, Report LA-11-11514-MS, 30 p.

Wallace, R. E., 1977, Profiles and ages of young fault scarps, north-central Nevada: Geological Society of America Bulletin, v. 88, p. 857-868.

Waples, D. W., 1980, Time and temperature in petroleum formation: application of Lopatin's method to petroleum exploration: American Association of Petroleum Geologists Bulletin, v. 64, p. 916–926.

Weber, R. H., 1963, Geologic features of the Socorro perlite deposit: New Mexico Geological Society, Guidebook 14, p. 144–145.

Weber, R. H., 1980, Geology of the Ake Site; *in* Beckett, P. H., ed., The Ake Site—collection and excavation of LA 13423, Catron County, New Mexico: New Mexico State University, Department of Sociology and Anthropology, Cultural Resources Management Division, Report 357, p. 221–238.

Weber, R. H., 1982, Quaternary geology of the Plains of San Agustin: New Mexico Journal of Science, v. 22, p. 77.

Weber, R. H., 1994, Pluvial lakes of the Plains of San Agustin: New Mexico Geological Society, Guidebook 45.

Weber, R. H. and Willard, M. E., 1959, Reconnaissance geologic map of the Reserve 30 minute quadrangle: New Mexico Bureau of Mines and Mineral Resources, Geologic Map 12, scale 1:126,720.

Wengerd, S. A., 1959, Regional geology as related to the petroleum potential of the Lucero region, west-central New Mexico: New Mexico Geological Society, Guidebook 10, p. 121–134.

Wernicke, Brian, 1992, Cenozoic extensional tectonics of the U.S. Cordillera; *in* Burchfiel, B. C., Lipman, P. W. and Zoback, M. L., eds., The Cordilleran orogen: conterminous U.S.: Geological Society of America, The Geology of North America, v. G-3, p. 553–581.

Wernicke, B. and Axen, G. J., 1988, On the role of isostacy in the evolution of normal fault systems: Geology, v. 16, p. 848–851.

Wilkinson, W. H., Jr., 1976, Geology of the Tres Montosas-Cat Mountain area, Socorro County, New Mexico: New Mexico Bureau of Mines and Mineral Resources, Open-file Report 39, 169 p.

Willard, M. E., 1959, Tertiary stratigraphy of northern Catron County, New Mexico: New Mexico Geological Society, Guidebook 10, p. 92–99.

Willard, M. E. and Stearns, C. E., 1971, Reconnaissance geologic map of the Pelona thirty-minute quadrangle: New Mexico Bureau of Mines and Mineral Resources, Geologic Map 23, scale 1:126,720.

Wills, W. H., 1988, Early prehistoric agriculture in the American Southwest: School of American Research Press, Santa Fe.

Wills, W. H., Ford, R. I. and Speth, J. D., 1984, Preliminary report of the University of Michigan investigations at Bat Cave (LA 4935): Manuscript on file at the Bureau of Land Management, Socorro, New Mexico.

Winchester, D. E., 1920, Geology of the Alamosa Creek Valley, Socorro County, New Mexico: U.S. Geological Survey, Bulletin 716A, 15 p.

Winters, S. S., 1963, Supai Formation (Permian) of eastern Arizona: Geological Society of America, Memoir 89, 99 p.

Witcher, J. C., Hahman, W. R. and Swanberg, C. A., 1994, Alpine 1/Federal—Subsurface stratigraphy of the eastern White Mountains, Apache County, Arizona: New Mexico Geological Society, Guidebook 45.

Wrucke, C. T., 1961, Paleozoic and Cenozoic rocks in the Alpine-Nutrioso area, Apache County, Arizona: U.S. Geological Survey, Bulletin 1121-H, p. H1-H26.

Yancey, T. E., 1978, Brachiopods and molluscs of the Lower Permian Arcturus Group, Nevada and Utah, Part I: brachiopods, scaphopods, rostroconchs and bivalves: Bulletin of American Paleontology, v. 74, no. 303, p. 257–363.

GEOLOGY OF THE LATE CENOZOIC ALMA BASIN, NEW MEXICO AND ARIZONA

BRENDA B. HOUSER

U.S. Geological Survey, Tucson Field Office, Gould-Simpson Building #77, University of Arizona, Tucson, AZ 85721

Abstract—The Alma Basin is an irregularly shaped late Cenozoic basin located just south of the Colorado Plateau at the intersection of the Morenci-Reserve fault zone and the Mangas trench. The sedimentary rocks of the Alma Basin record four stages in the history of the basin: (1) a pre-Basin-and-Range-faulting stage (23 to 18.7 Ma), when the basin had a different shape than the modern basin, (2) an early Basin-and-Range stage (18.7 to about 6.0 Ma), marked by only minor deposition, during which the basin began to take on its present shape, (3) a third stage (6.0 to about 1.5 Ma), marked by significant subsidence and deposition of a relatively thick section of basin-fill fluvial sedimentary rocks, and (4) waning subsidence during the Quaternary and erosion of basin-fill deposits by the San Francisco River and its tributaries. Structurally, the Alma Basin is an asymmetric graben composed of about six non-rotated fault blocks that step progressively downward toward the Mogollon Mountains on the east. Nearly all the faults are high-angle normal faults. A depositional hiatus between about 16 and 6 Ma may have been caused by shallow emplacement of an intrusive body beneath the west side of the Mogollon Mountains. The presence of such an intrusion at this time is inferred from the age of hydrothermal alteration in the Mogollon mining district.

INTRODUCTION

The Alma Basin lies immediately west of the Mogollon mining district (a silver district with important past production) within a remote mountainous region of great scenic beauty (Fig. 1). Although geologic studies of the region have focused chiefly on the mountains, they have also supplied significant structural data for the Alma Basin. Geologic mapping of the Alma Basin was undertaken primarily to provide a better understanding of the Morenci-Reserve fault zone.

Preliminary analysis of mapping data from the Alma Basin has given insight into the late Cenozoic history of extension of the southern Basin and Range province just southeast of the Colorado Plateau. Because of its location at the intersection of two important structural trends and because of exceptionally good exposures of basin-fill rocks, study of the Alma Basin can provide information not available in the surrounding ranges.

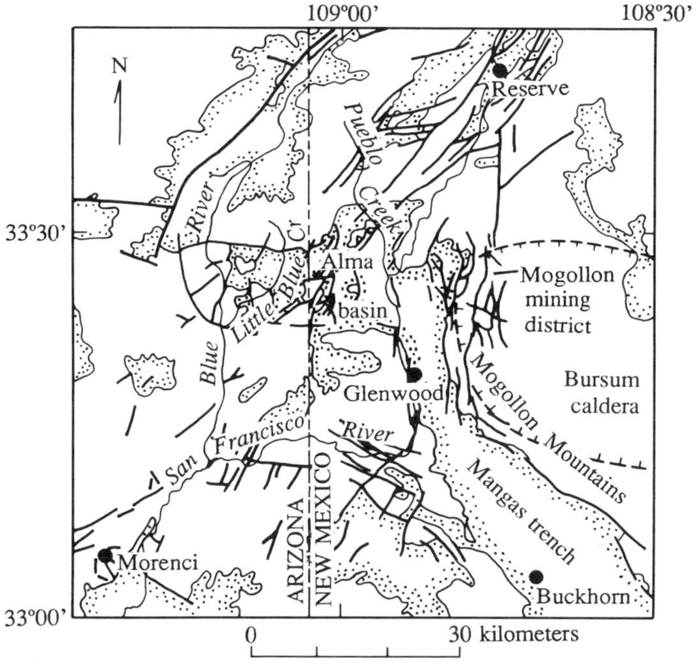

FIGURE 1. Index map showing the Alma Basin in relation to major streams and other geographic and geologic features of the region. Stippled pattern indicates late Cenozoic basins. Faults are shown by heavy lines. Modified from Dane and Bachman (1965), and Wilson et al. (1969).

GEOLOGIC SETTING

The Alma Basin is west of the Mogollon Mountains, at the intersection of the northeast-trending Morenci-Reserve fault zone and the northwest-trending Mangas trench (Fig. 1). Ratté (1989) described the Morenci-Reserve fault zone as a complex graben with intragraben horsts that trends approximately N30°E and is 40 to 50 km wide. He interpreted the zone to be distinct from, and in part younger than, the N60°E San Agustin trend (Morenci lineament of Mayo, 1958; Chapin et al., 1978). Based on mapping in southern Arizona, Houser (1994) inferred that the Morenci-Reserve fault zone can be traced into Sonora, Mexico.

The Morenci-Reserve zone is apparently a long-lived feature. Estimated ages of mid-Tertiary sedimentary deposits associated with the zone in Arizona suggest that it may be as old as 30 Ma (Houser, 1994), while fault scarps in the Alma Basin indicate that the zone was still active during the late Pleistocene.

The Mangas trench is one of about seven northwest-trending basins that are truncated along the southeastern margin of the Colorado Plateau. These basins formed during the Basin-and-Range extensional event that began shortly after 19 Ma in this part of the southern Basin and Range province, as indicated by ages of the oldest tholeiitic or alkalic basalt flows (Richter et al., 1983; Marvin et al., 1987; Houser, unpubl.)

Within this tectonic setting, late Cenozoic uplift and subsidence has been controlled by both of these structural trends; thus, the Alma Basin has an irregular shape. In spite of the complexities of its tectonic setting, the basin is a fairly simple asymmetric graben composed of about six unrotated fault blocks, stepped down on the east side adjacent to the Mogollon Mountains.

The rocks of the surrounding ranges are intermediate to silicic Oligocene and Miocene volcanic and hypabyssal intrusive rocks of the Mogollon-Datil volcanic field (Ratté et al., 1969, 1979; Ratté, 1980, 1981, 1989). The 28-Ma Bursum caldera, immediately east of the Alma Basin, and other calderas of the Mogollon Mountains were the source of distinctive and voluminous Oligocene ash-flow tuffs (Ratté et al., 1984). These tuffs (for example, the Davis Canyon, Shelley Peak and Bloodgood Canyon Tuffs) are important regional stratigraphic markers in the volcanic rocks. Additionally, the first occurrence of these tuffs as clasts in basin-fill deposits documents and provides time constraints for uplift and unroofing of the ranges during the Miocene.

Bedrock highs within the Alma Basin are composed of 27-to-23 Ma basaltic andesite to dacite lava flows of the Bearwallow Mountain Andesite (Marvin et al., 1987; Houser, 1987). The sedimentary rocks present in the basin (Fig. 2) are the lower to middle Miocene Dog Gulch Formation (Ratté, 1981), conglomerate of Little Blue Creek, and

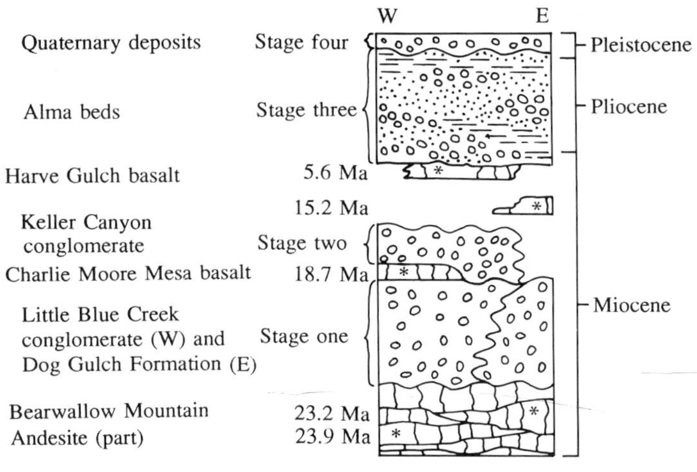

FIGURE 2. Correlation of stratigraphic units for the Alma Basin, New Mexico and Arizona.

conglomerate of Keller Canyon; the chiefly Pliocene Alma beds; and Quaternary alluvium and other surficial deposits (Houser, 1987). Miocene tholeiitic or alkalic basalt flows separate the basin-fill sedimentary units and serve as time constraints on their deposition.

PREVIOUS WORK

The geology of the central part of the Alma Basin was mapped by Houser (1987). Quadrangle mapping of most of the rest of the basin and adjacent ranges has been completed by Houser and Ratté (unpubl.). Studies of the Mogollon mining district include Ferguson (1927), Ratté (1980), Kamilli (1994), and Senterfit et al., (1994). Reconnaissance geologic maps were prepared for mineral-resource assessments of the Gila Wilderness east of the Alma Basin (Ratté et al., 1979) and the Blue Range Primitive Area west of the basin (Ratté et al., 1969). Detailed quadrangle mapping north of the basin was done by Ratté (1980, 1989).

STRATIGRAPHY

The correlation of the stratigraphic units in the Alma Basin is shown in Figure 2. The sedimentary deposits of the Alma Basin (mostly Gila Group) are interpreted to record four stages in the history of the basin: (1) a pre-Basin-and-Range-faulting stage (23 to 18.7 Ma), when the basin had a different shape than the modern basin, (2) an early Basin-and- Range stage (18.7 to about 6.0 Ma), during which the basin began to take on its present shape, (3) a third stage (6.0 to about 1.5 Ma), which was marked by significant subsidence and deposition of a relatively thick section of basin-fill fluvial sedimentary rocks, and (4) waning basin subsidence during the Quaternary, accompanied by erosion of basin-fill deposits by the San Francisco River and its tributaries.

Stage one

The earliest stage of deposition in the Alma Basin is represented by the conglomerate of Little Blue Creek on the west side of the basin and by the Dog Gulch Formation (Ratté, 1981) on the east side. The ages of volcanic rocks below and above these units indicate that they were deposited between about 23 and 19 Ma. Both units are chiefly conglomeratic and both overlie an erosional surface cut on the 27-to-23-Ma Bearwallow Mountain Andesite, a compositionally variable volcanic unit 100 to more than 300 m thick (Ratté et al., 1979; Ratté, 1981; Marvin et al., 1987). The conglomerate of Little Blue Creek overlies a rhyolite ash-flow tuff that, in turn, overlies a silicic dome complex dated at 23.9 Ma (Ratté et al., 1969; Marvin et al., 1987). The Dog Gulch Formation overlies the 23.2-Ma Last Chance Andesite (Ratté, 1981). The minimum age of the Little Blue Creek conglomerate is 18.7 + 2.0 Ma based on a whole-rock K-Ar date on the overlying basalt of Charlie Moore Mesa (R. F. Marvin, written commun., 1986). The minimum age of the Dog Gulch Formation is 15.2 Ma based on the age of an overlying basalt flow near Bearwallow Mountain (Ratté, 1981).

The thicknesses of the Little Blue Creek conglomerate and Dog Gulch Formation are more than 300 m and as much as 200 m, respectively; both are fairly well indurated and consist chiefly of poorly-sorted fluvial and debris-flow conglomerate facies, although minor mudstone and sandstone facies occur in both. The facies changes in the Little Blue Creek and Dog Gulch units (particularly the distribution of basin center facies) bear no relationship to the modern topography, suggesting that the shape of this early basin was probably different than that of the modern basin. Also, although the clasts in the conglomerates consist entirely of volcanic rocks having a wide range of compositions, there are no clasts of the Davis Canyon, Shelly Peak and Bloodgood Canyon Tuffs, indicating that these regional ash-flow tuff sheets were not yet exposed to erosion in the vicinity of the basin during stage-one deposition.

Stage two

During the second stage of the Alma Basin (about 18.7 to 6.0 Ma), the conglomerate of Keller Canyon (Houser, 1987) was deposited. This conglomerate unit is as much as 85 m thick and becomes coarser grained adjacent to the modern bedrock boundaries of the basin. This indicates that the basin configuration at the time of deposition was similar to the modern basin.

As with the Little Blue Creek conglomerate and Dog Gulch Formation, there are no clasts of the regional ash-flow tuff sheets in the Keller Canyon conglomerate. These tuffs are present along the northern edge of the basin and, if they had been exposed at that time, presumably would have been carried into the basin and deposited along with the rest of the varied volcanic clasts of the Keller Canyon conglomerate. Their absence in the Keller Canyon conglomerate is evidence that the ash-flow tuffs were still covered by the younger volcanic pile of the 100-to-300-m thick Bearwallow Mountain Andesite.

The conglomerate of Keller Canyon is similar in appearance to the conglomerate of Little Blue Creek, but is slightly less indurated. Poorly-sorted, pebbly to bouldery, volcanic conglomerate is the dominant facies, much of it matrix supported, indicating deposition by debris flows. Other facies are very poorly sorted mudstone, and poorly sorted to well-sorted sandstone and conglomerate.

The Keller Canyon conglomerate overlies the 18.7-Ma basalt of Charlie Moore Mesa and the conglomerate of Little Blue Creek; it is capped by the 5.6-Ma basalt of Harve Gulch (Ratté and Finnell, 1978) and the Alma beds of stage three (Houser, 1987). The ages of the lower and upper basalts indicate that deposition of the Keller Canyon conglomerate could have spanned a time interval of about 13 Ma. This is unlikely, however, because it is a relatively thin unit (85 m) and its conglomeratic character implies a high rate of deposition. It was probably deposited in only a few million years. The similarity of its lithologic character to that of the underlying Little Blue Creek conglomerate suggests that it is probably closer in age to that unit than to the Alma beds.

The Keller Canyon is exposed in the central and western part of the Alma Basin and around all the margins except for the eastern side next to the Mogollon Mountains, where it is probably present at depth, but has been covered by the mostly Pliocene Alma beds.

Stage three

The Alma beds (Houser, 1987) were deposited during the third stage in the development of the Alma Basin (about 6.0 to 1.5(?) Ma) and are the youngest basin-fill unit. They are more than 300 m thick in exposures throughout the basin. They overlie the basalt of Harve Gulch and the conglomerate of Keller Canyon, and underlie gravel deposits of the San Francisco River and other Quaternary surficial deposits. The minimum age of the Alma beds is not well constrained. The age of 1.5 Ma (early Pleistocene) was chosen because skeletal remains of *Paenemarmota*, a Pliocene giant ground squirrel, have been found in the Alma beds at several locations in the basin (Houser, 1987 and unpubl.). Pleistocene faunal remains, however, have been found at only one locality in the southern part of the basin east of Pleasanton (Ratté and Finnell, 1978). The paucity of Pleistocene faunal localities may indicate that deposition in the Alma Basin did not continue very long into the Pleistocene. The stratigraphic context of a Pleistocene mammoth occurrence near San Francisco hot springs at the southern end of the basin (Ratté, personal commun., 1983) has not been investigated.

In the Alma quadrangle (Houser, 1987), the Alma beds were separated into three units overlain by a fourth unit, the Whitewater Mesa beds. For the present purposes, these four units are combined as the Alma beds. They document a transition from deposition dominated by conglomeratic mudflows, through distal alluvial fan and braid plain deposits, to proximal and medial fan deposits of streams having seasonal(?) sustained flow. Stream flow was to the south and southeast through the Mangas trench toward the lacustrine depocenter of this trench in the vicinity of the village of Buckhorn.

The Alma beds are well indurated at the base and semi-indurated at the top of the section. Clasts of the regional ash-flow tuffs are prominant in all the units indicating that a significant interval of erosion had occurred between deposition of the Keller Canyon conglomerate and the Alma beds. The beginning of integration of the headwaters of the San Francisco River is shown by the rare occurrence in the upper Alma beds of red granite clasts reworked from Tertiary conglomerate to the west and north (Ratté et al., 1969).

Stage four

The fourth stage in the history of the Alma Basin probably was confined to the Quaternary. Subsidence waned, basin-fill deposition ceased, and erosion of the basin fill began during this stage. Although fault scarps as high as 15 m provide clear evidence for continued fault movement into Quaternary time, the rate of subsidence apparently has not been rapid enough to cause ponding of the San Francisco River and result in deposition of basin fill.

Integration of the headwaters of the San Francisco River continued, as indicated by the common occurrence in river gravel deposits of red granite and other exotic clasts derived from Tertiary conglomerate sources to the north and west. Two deposits of San Francisco River gravel in the northeastern part of the basin show that the ancestral San Francisco flowed 3-5 km east of its present course when it was about 150 m higher than it is now. The general elevations of pediment surfaces in this area are consistent with the slope of the surfaces being at grade with the course of the ancestral San Francisco River between the two gravel deposits. The pediments were probably abandoned soon after the course of the river was diverted westward, presumably by stream capture, although the evolution of the course of the San Francisco River has not been studied.

STRUCTURE

Three main sets of faults exist in the Alma Basin (Fig. 1). The northwestern part of the basin is dominated by N30°E-trending faults of the Morenci-Reserve fault zone. Faults in the central and eastern part of the basin trend generally north-south and are probably part of the fault system of the Mangas trench. Discontinuous, generally east-west-trending faults are present throughout the basin and along the northern and southern basin margins.

Nearly all the faults are high-angle normal faults (70° to 85°) that dip to the southeast or east. The lowest dip measured is 43° and only a few are in the range of 45° to 60°. Striations indicating oblique-slip and essentially strike-slip movement on faults were noted at five locations along the margins of the Alma Basin in the course of geologic mapping. Three of the locations are in late Oligocene and early Miocene volcanic rocks at the west, north and southeast sides of the basin. The remaining two locations are in sedimentary rocks at the east side of the basin near Mineral Creek. At one of these, a north-trending vertical fault showing striae at a rake angle of 13°S displaces conglomerate of the Miocene Dog Gulch Formation, and at the other, the Pliocene Alma beds are displaced by a north-trending normal fault dipping at 60°E with striae at a rake angle of 30°S.

Except for a complex horst block that cuts diagonally across the basin and apparently bounds the southeast side of the Morenci-Reserve fault zone in this area, the sense of displacement on the majority of both the northeast-trending faults of the Morenci-Reserve fault zone and the north-trending faults of the Mangas trench is down to the southeast—toward the deepest part of the Alma Basin next to the Mogollon Mountains. The total structural displacement of the basin (which may be only moderate) is distributed across about six nonrotated fault blocks.

Models of extensional deformation predict domino-style rotation of fault blocks back toward the master fault. This is the deformation style observed by Crews (1990) in his study of the Reserve graben about 20 km northeast of the Alma Basin. In the Alma Basin, however, most fault blocks are either unrotated or are rotated less than 10° (away from the faults), opposite from the direction predicted in domino-style rotation. In two areas mapped by Ratté (1980, 1989) between the Alma Basin and Crews' study area, Ratté showed strata within fault blocks tilted toward the normal faults, away from the normal faults, or unrotated. The lack of correspondence of the tilt direction of the Alma Basin fault blocks to those of extensional models (and the partial correspondence in the areas mapped by Ratté) may be a result of the tectonic interaction of the Morenci-Reserve trend with the Basin-and-Range trend, or it may indicate that there has not been very much extension in the area.

DISCUSSION

Sedimentation and the Mogollon mining district

The characteristics of the four stages of basin-fill sedimentary rocks of the Alma Basin indicate that there was a significant hiatus between deposition of the conglomerate of Keller Canyon and the Alma beds. The Keller Canyon conglomerate is inferred to be closer in age to the underlying 18.7-Ma Charlie Moore Mesa basalt and Little Blue Creek conglomerate than to the overlying 5.6-Ma Harve Gulch basalt and Alma beds. The reasons for this inference are, first, that the contacts of the Keller Canyon conglomerate with the Charlie Moore Mesa basalt and the Little Blue Creek conglomerate are for the most part conformable or only slightly unconformable. Second, in general appearance and sedimentologic character, the Keller Canyon is much more similar to the Little Blue Creek conglomerate than to the overlying Alma beds. Finally, the Keller Canyon is only about 85 m thick, implying that it was deposited in a relatively short time period, perhaps only 1 or 2 Ma.

Thus, there may have been a hiatus spanning about 10 Ma (16 Ma to 6 Ma) between deposition of the Keller Canyon conglomerate and the Alma beds. There must have been significant erosion during this time interval, because the cover of the 100- to 300-m thick Bearwallow Mountain volcanic rocks over the ranges surrounding the Alma Basin was breached and the regional Oligocene ash-flow tuffs were exposed to erosion. Presumably, this detritus was carried by an axial stream southward out of the Alma Basin to the Mangas trench where it was deposited.

It is tempting to speculate that the apparent lack of deposition in the Alma Basin during the latter half of the Miocene may have been associated with the hydrothermal event that resulted in mineralization of the nearby Mogollon mining district (Ratté, 1981). The timing of the two events is appropriate. An $^{40}Ar/^{39}Ar$ date on vein adularia from the district indicates that hydrothermal alteration occurred in the Mogollon Mountains at about 16.7 Ma (W.C. McIntosh, personal comm., 1993). A possible source of the hydrothermal fluids may have been a shallow intrusion beneath the western side of the Mogollon Mountains. Emplacement of the intrusion is likely to have caused uplift of the surrounding area or at least to have minimized subsidence.

Renewed deposition and subsidence at about 6 Ma in the Alma Basin may have occurred after the intrusive body cooled, or may have been associated with tectonic events related to the eruption of the the basalt of Harve Gulch.

REFERENCES

Chapin, C.E., Chamberlin, R.M., Osburn, G.R., White, D.W. and Sanford, A.R., 1978, Exploration framework of the Socorro geothermal area, New Mexico: New Mexico Geological Society, Special Publication 7, p. 115-129.

Crews, S.G., 1990, Structure, syntectonic sedimentation, and stratigraphy of asymmetrical non-marine basins [Ph.D. dissertation]: Laramie, University of Wyoming, 196 p.

Dane, C.H. and Bachman, G.O., 1965, Geologic map of New Mexico: U.S. Geological Survey, scale 1:500,000.

Ferguson, H.G., 1927, Geology and ore deposits of the Mogollon mining district, New Mexico: U.S. Geological Survey, Bulletin 787, 100 p.

Houser, B.B., 1987, Geologic map of the Alma quadrangle, Catron County, New Mexico: U.S. Geological Survey, Geological Quadrangle Map GQ-1610, scale 1:24,000.

Houser, B.B., 1994, Santa Ana-Reserve extension zone — a proposed zone of interactive northeast-and northwest-directed extension during the Cenozoic, New Mexico, Arizona, and Sonora: U.S. Geological Survey, Circular 1103-A, p. 49-52.

Kamilli, R.K., 1994, Geologic studies of the Mogollon mining district, New Mexico; does a porphyry system lie below?: Arizona Geological Society, Digest 20, in press.

Marvin, R.F., Naeser, C.W., Bikerman, Michael, Mehnert, H.H. and Ratté, J.C., 1987, Isotopic ages of post-Paleocene igneous rocks within and bordering the Clifton 1°x2° quadrangle, Arizona-New Mexico: New Mexico Bureau of Mines and Mineral Resources, Bulletin 118, 63 p.

Mayo, E.B., 1958, Lineament tectonics and some ore districts of the southwest: American Institute of Mining, Metallurgical, and Petroleum Engineers Transactions, v. 211, p. 1169-1175.

Ratté, J.C., 1980, Geologic map of the Saliz Pass quadrangle, Catron County, New Mexico: U.S. Geological Survey, Miscellaneous Field Studies Map MF-1203, scale 1:24,000.

Ratté, J.C., 1981, Geologic map of the Mogollon quadrangle, Catron County, New Mexico: U.S. Geological Survey, Geological Quadrangle Map GQ-1557, scale 1:24,000.

Ratté, J.C., 1989 Geologic map of the Bull Basin quadrangle, Catron County, New Mexico: U.S. Geological Survey, Geological Quadrangle Map GQ-1651, scale 1:24,000.

Ratté, J.C. and Finnell, T.L., 1978, Road log from Silver City to Reserve via Glenwood and the Mogollon mining district: New Mexico Geological Society, Special Publication 7, p. 49-63.

Ratté, J.C., Gaskill, D.L., Eaton, G.P., Peterson, D.L., Stotelmeyer, R.B. and Meeves, H.C., 1979, Mineral resources of the Gila Primitive Area and Gila Wilderness, New Mexico: U.S. Geological Survey, Bulletin 1451, 229 p., scale 1:200,000.

Ratté, J. C., Landis, E. R., Gaskill, D. L. and Raabe, R.G., 1969, Mineral resources of the Blue Range Primitive Area, Greenlee County, Arizona, and Catron County, New Mexico, with a section on aeromagnetic interpretation by Eaton, G.P.: U.S. Geological Survey , Bulletin 1261-E, 91 p.

Ratté, J.C., Marvin, R.F., Naeser, C.W. and Bikerman, M. 1984, Calderas and ash-flow tuffs of the Mogollon Mountains, southwestern New Mexico: Journal of Geophysical Research, v. 89, No. B10, p. 8713-8732.

Richter, D.H., Houser, B.B. and Shafiqullah, M., 1983, Geologic map of the Guthrie quadrangle, Graham and Greenlee Counties, Arizona: U.S. Geological Survey, Miscellaneous Investigations Series Map I-1455, scale 1:48,000.

Senterfit, R.M., Kamilli, R.J., Abrams, G.A., Klein, D.P. and Ratté. J.C., 1994, Audio-magnetotelluric and gravity study of the Mogollon mining district, southwest New Mexico: U.S. Geological Survey, Circular 1103-A, p. 94-95.

Wilson, E.D., Moore, R.T. and Cooper, J.R., 1969, Geologic map of Arizona: U.S. Geological Survey, Special Geologic Map, scale 1:500,000.

TECTONIC CONTROL OF SYNRIFT SEDIMENTATION PATTERNS, RESERVE GRABEN, SOUTHWESTERN NEW MEXICO

STEVEN G. CREWS

ARCO Exploration & Production Technology, 2300 West Plano Parkway, Plano, TX 75075

Abstract—The Reserve graben is part of a Miocene and younger extensional corridor that lies between the Mogollon Plateau volcanic center and the Colorado Plateau; it comprises a series of distinct half-graben sub-basins, each associated with one or more bounding normal faults, and each filled with alluvial deposits of the Gila Group. Topography and uplift/subsidence patterns created by extensional faulting directly influenced nearly every aspect of synrift sedimentation. Thickness, texture, sediment-transport direction and depositional environment of the synrift sediments were all directly affected by episodic movement on basin-bounding normal faults. The primary depositional environments represented in the deposits in these half grabens are hanging wall dip-slope alluvial fans, axial fluvial systems and footwall-derived alluvial fans. The footwall-derived fans are coarser grained, but areally smaller and less common than the dip-slope fans, which dominate the basin. Mass-flow deposits filling arroyos (which probably fed the hanging wall fans), and small colluvial wedges originating from fault scarps, have also been recognized. No lacustrine deposits have been identified. Two tectonic influences stand out. First, paleoflow was mostly either down structural dip or along strike—i.e., dip-slope fans and axial fluvial systems dominate the basin fill. This suggests that fault kinematics directly affected drainage patterns. Second, the finest-grained deposits in each half graben often occur adjacent to the fault scarps, in spite of the high topographic relief found there. Many individual processes probably contributed to this situation, but most can be traced back ultimately to the influence of fault kinematics on topography and on the distribution of uplift and subsidence within the basin. In general, footwall uplifts in the Reserve graben do not act as prolific line sources of sediment. Large influxes of footwall sediment occur only at a few special points, such as topographic gaps associated with transfer zones in the border fault system.

INTRODUCTION

Tectonics and sedimentation in continental rifts

The distinctly asymmetrical "half graben" morphology that characterizes most extensional basins can exert a strong influence on drainage, distribution of sedimentary environments and, ultimately, on stratigraphic architecture (Davis, 1925; Blair, 1987; Leeder and Gawthorpe, 1987; Frostick and Reid, 1987; Mack and Seager, 1990; Paton, 1992; Leeder and Jackson, 1993). Sedimentation patterns are also influenced by other related structural/tectonic features characteristic of extensional basins, including (1) the formation of low- or high-relief "accommodation zones" at transfer zones between alternating-polarity sub-basins; (2) transfer zones or other discontinuities in border fault systems; (3) the asymmetry of rift-shoulder or footwall uplifts; (4) the episodic nature of faulting; and (5) the geometry and kinematics of normal faults (Rosendahl, 1987; Hamblin and Rust, 1989; Morley, 1989; Crews and McGrew, 1990; Mack and Seager, 1990; Cohen, 1991; Lambiase, 1991; Schlische, 1991; Scholz and Rosendahl, 1991; Gawthorpe and Hurst, 1993).

A typical half graben is characterized by an abruptly faulted, high-relief margin on one side, contrasting with a hinged or more gradually faulted opposing margin with proportionately lower topographic relief. If the structural asymmetry of a half graben is mirrored in the surface topography during tectonically active periods, the basin floor will slope in the same direction as the structural dip (i.e., towards the fault-bounded margin of the basin). Leeder and Gawthorpe (1987) termed this concordance between land surface and structure "tectonic slope".

Depositional systems having deposits contributing to the fill of a continental half graben can be divided into four basic groups: (1) transverse alluvial systems with a source on the hanging wall dip slope and flowing down the tectonic slope; (2) transverse systems flowing into the basin from the footwall uplift (opposing tectonic slope, if they cross the basin axis); (3) longitudinal fluvial systems, flowing parallel to the axis of the basin (perpendicular to tectonic slope); and (4) low energy or slope-independent/autochthonous depositional systems such as lakes, swamps, playas and ergs that may occupy a topographic low coincident with the structural axis of the basin. Various depositional systems of these four classes will in essence compete to fill the basin, and to dominate the stratigraphic architecture of the preserved basin fill. The relative proportions of deposits from each group, and the characteristics of each group's component facies, depend on a variety of factors including rate, magnitude and episodicity of extension, fault spacing and kinematics, climate, watershed size and pre-existing geology.

Intuitively, one might expect the high relief typically found on the fault-bounded margins of a half graben to result in the construction of large alluvial fans or braidplains that prograde across the basin, dominating the stratigraphic architecture. However, this is clearly not the case in most rift basins. Davis (1925) noted "flat detrital plains, sloping gently *toward* the ranges" throughout the Great Basin. The competition between low-relief but large-area hanging wall catchments and high-relief small-area footwall catchments is usually won by the dip-slope systems—drainage area is apparently more critical than drainage basin relief in determining sediment yield. Death Valley provides an excellent modern example of this phenomenon (Fig. 1).

The contribution of footwall-uplift-source fans to the basin fill is usually limited for several reasons. The total sediment flux may be low because footwall uplifts are typically narrow, severely limiting the amount of sediment they can provide, and the asymmetry of footwall uplifts tends to create a dip slope that promotes transport *away* from the associated half graben, rather than into it (of course, in many extensional provinces, the footwall uplift of one half graben may form the dip slope of another half graben). From this it may be seen that the footwall uplift is not so much a contributor of sediment as a barrier to sediment from beyond the rift shoulder getting into the basin. Moreover, since rapid subsidence and uplift are juxtaposed at the faulted margin of a half graben, sediment that does enter the basin from the footwall-uplift will tend to be absorbed by the rapidly subsiding basin axis. Significant progradation cannot occur unless sediment flux is sufficient to overfill the axis and reverse the tectonic slope.

An important exception to the foregoing occurs when streams enter the basin from the footwall but their source is not in the footwall. Topographic gaps in footwall uplift (associated with transfer zones or other discontinuities in bounding fault system) may allow streams that access large drainage basins located beyond the rift shoulder into the half graben. Explanations of why footwall-uplift clastic wedges typically do not dominate half-graben basin fill are more fully developed by Blair (1987), Blair and Bilodeau (1988), Crews, (1993), Leeder and Gawthorpe (1987), and Leeder and Jackson (1993).

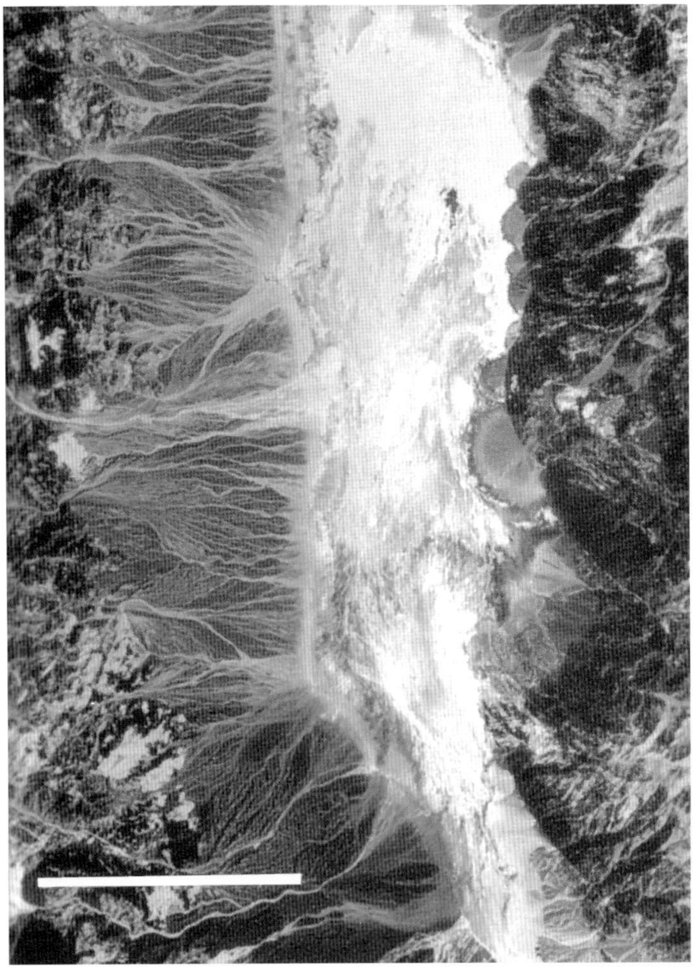

FIGURE 1. SPOT image of southern Death Valley, USA. The structure of Death Valley is that of a half graben, in which the main bounding fault zone is adjacent to the Black Mountains (right side of image), while the Panamint Range (left) forms the hinged margin. The dip-slope alluvial fans issuing from the Panamint Range cover a much larger area and coalesce more extensively than the footwall fans issuing from the Black Mountains. The relatively wide flat area adjacent to the footwall uplift is occupied by a playa lake during wet climatic periods; currently, it is a salt pan. If an axial river system existed, it would presumably flow down this corridor. Scale bar is approximately 10 km long (from cover of GSA Bulletin, v. 101, November, 1989).

Geologic setting

The Reserve graben lies between the southeastern edge of the Colorado Plateau and the northwestern rim of the Mogollon Plateau, in Catron County, New Mexico (Fig. 2). The graben forms one link in a chain of narrow mid-Miocene and younger extensional basins that encircles and helps defines the Mogollon Plateau, a topographic high that was a major center of caldera-derived mid-Tertiary ignimbrites within the larger Mogollon-Datil volcanic field (Elston et al., 1973; Ratté et al., 1984; Bornhorst, 1988).

The basins adjoining the Reserve graben are the San Agustin Basin, directly to the east-northeast, and the Mangas trench, or Glenwood graben, directly to the south-southeast (Figs. 2 and 3). The San Agustin Basin is notable because it is still a largely undissected, hydrologically closed basin, whereas most of the other basins in the region have been incised and partially exhumed by integrated Quaternary stream systems. The Reserve graben is itself drained and dissected by the San Francisco River—a principal tributary of the Gila River—and its own tributaries. The sedimentary rocks filling the Reserve graben comprise the mid-Miocene and younger Gila Group.

The main axis of the Rio Grande rift lies immediately across the Mogollon Plateau, a distance of about 100 km; it is linked to the Reserve graben by the San Agustin Basin (Fig. 2). As a result of its unique location at the intersection of the Basin and Range province, Rio Grande rift, Colorado Plateau and Mogollon-Datil volcanic field, the tectonic evolution of the Reserve graben is intertwined with that of each of these features.

Intermediate volcanism in the Mogollon-Datil volcanic field began about 40 Ma and continued until about 21 Ma, punctuated by episodes of silicic ignimbrite eruption (Elston et al., 1973; Elston, 1984; Ratté et al., 1984; Bornhorst, 1988; McIntosh et al., 1992). The Mogollon Plateau, now a topographically high region rimmed by Miocene extensional faults and defined geophysically by a distinct gravity anomaly (Krohn, 1976), was a particularly notable center of ignimbrite eruption, especially during a brief but intense burst of activity 27.4 to 29 Ma (McIntosh et al., 1992). Elston et al. (1976) suggested that the pattern of Basin and Range faulting surrounding the Mogollon Plateau may reflect an underlying batholith: "Basin and Range faults, younger than the Mogollon Plateau, tend to wrap around it ... [this] pattern may reflect adjustment to the buoyant and rigid mass of the inferred Mogollon Plateau composite pluton, modified by regional trends ..." The regional trends referred to include the Morenci-Reserve trend and San Agustin trend, or Morenci lineament, of Ratté (1989).

Mid-Tertiary andesitic volcanism ended in the region, and throughout much of the Basin and Range, by 21 Ma. Subsequent volcanism has largely consisted of a bimodal association dominated by true basalts and high-silica rhyolites, with less intermediate-composition material (Elston et al., 1973; Ratté et al., 1984; Hamilton, 1987). A correlation between this change in magma composition and the beginning of Basin-and-Range-style block faulting and sedimentation has been noted by many workers, especially—but not only—in southwestern New Mexico (Elston et al., 1976; Elston and Bornhorst, 1979; Ratté et al., 1984; Engebretson et al., 1984; Coney, 1987). However, Chapin (1988), emphasized slightly earlier basin sedimentation in the Rio Grande rift proper: "sedimentary basins were accumulating bolson-type sediments along the entire length of the Rio Grande rift by 25-27 Ma".

In the Reserve graben, Gila Group sediments that were clearly deposited in the context of the same fault blocks that define current topography overlie pre-faulting 23-27 Ma Bearwallow Mountain Andesite, and are interbedded with basalt flows having ages ranging from 1 to 21 Ma (Marvin et al., 1987; Ratté, 1989; Ratté and Bové, 1990). These relationships are consistent with a change in tectonic style during the 40-Ma history of the extensional orogen, and imply that this change occurred about 21 Ma in southwestern New Mexico.

The basic structure of the Reserve Basin is illustrated in Figures 3 and 4. The San Francisco Mountains, on the northwest, and the Tularosa Mountains, on the southeast, define the primary margins of the basin, while some smaller intrabasin ranges, notably Jon S Mountain, Higgins Mountain and the Saliz Mountains, split the basin into two or more separate half grabens. Principal faults strike north-northeast to northeast, and nearly all dip southeasterly (some down-to-the-northwest faults occur near the southeast margin of the basin). Transverse faults that strike roughly east-west are associated with mountainous zones at each end of the basin, separating the Reserve graben from its neighbors.

The San Francisco Mountains fault zone, which defines the northwest margin of the Reserve graben, appears to be the main bounding fault zone of the basin. The steep dips noted near the footwall cutoffs of the fault-bounded intra-basin uplifts suggests a domino-style, fault-rotation model for the structure of the basin (Fig. 4). However, apparent tilting of footwall rocks may not always be sufficiently diagnostic to allow discrimination between dominoes and closely spaced non-rotating listric faults (Stewart, 1971; McGrew and Crews, 1990).

The sedimentary unit that fills the basin, the Gila Group, was named by G.K. Gilbert (1875) for exposures "along the gorges of the Upper Gila and its tributaries"; it is quite possible, but not recorded, that Gilbert visited the Reserve graben itself. Gilbert (1875) defined the Gila as:

> "A system of valley beds, of which a conglomerate is the characteristic member... The bowlders of the conglomerate are of local origin, and their derivation from particular mountain flanks is often

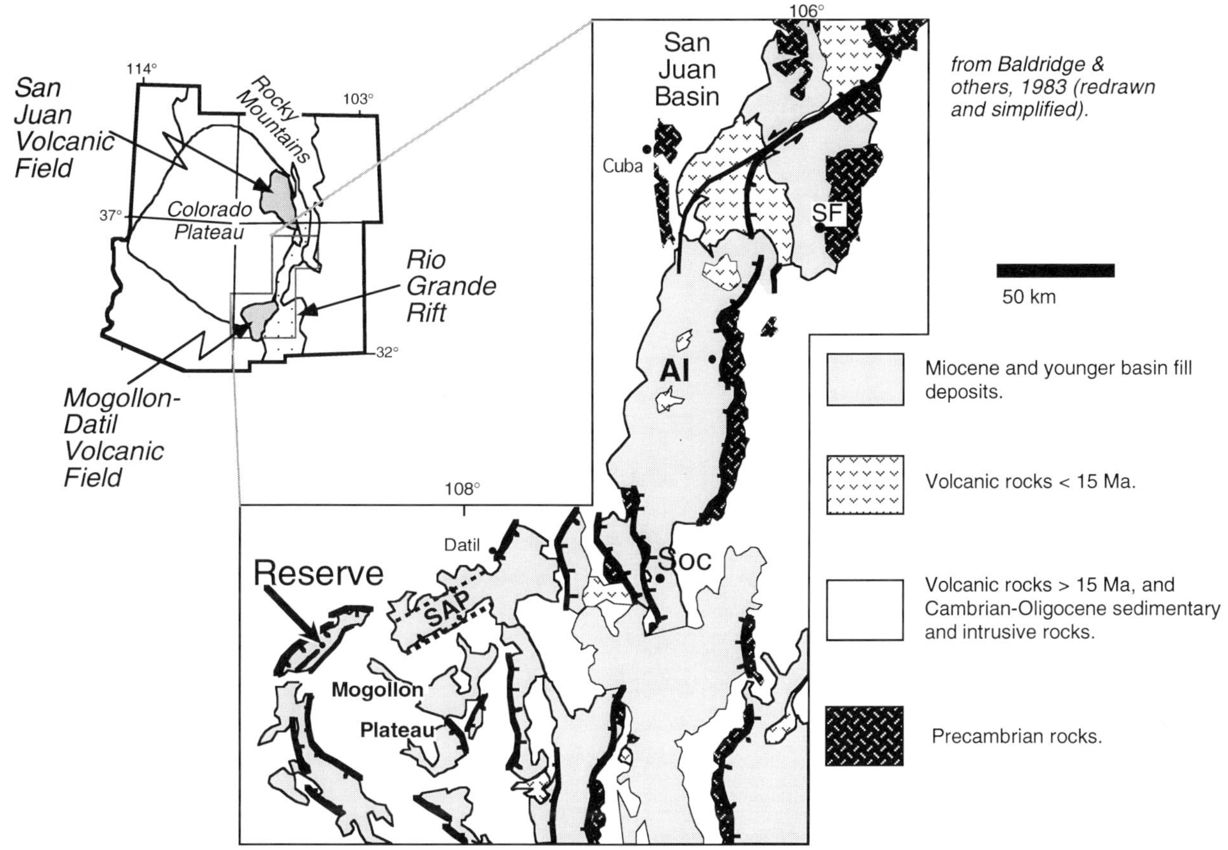

FIGURE 2. Regional location map, placing the Reserve graben in the context of the Basin and Range province, Colorado Plateau and Rio Grande rift. Abbreviations: SF = Santa Fe; Al = Albuquerque; Soc = Socorro; SAP = Plains of San Agustin.

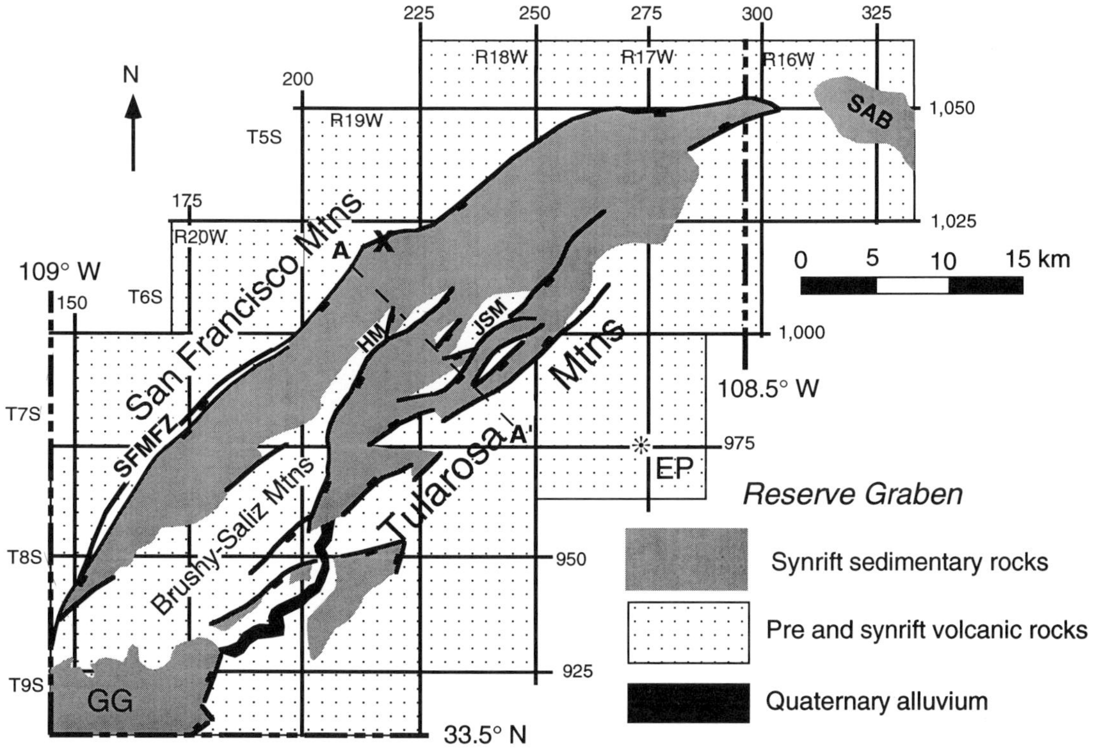

FIGURE 3. Simplified tectonic map of the Reserve graben. Abbreviations: SAB = San Agustin Basin; JSM = Jon S Mountain; EP = Eagle Peak; HM = Higgins Mountain; SFMZ = San Francisco Mountains fault zone; GG = Glenwood graben. "X" marks "the Box"—a box canyon located on a bend in the SFMZ where the modern San Francisco river enters the basin, and where a large Miocene alluvial fan complex was located (see text). Line A-A' marks the location of the cross section shown in Fig. 4. JSM, HM and Brushy-Saliz Mountains are intra-basin tilted fault blocks. Most bounding faults dip southeast; most strata dip northwest.

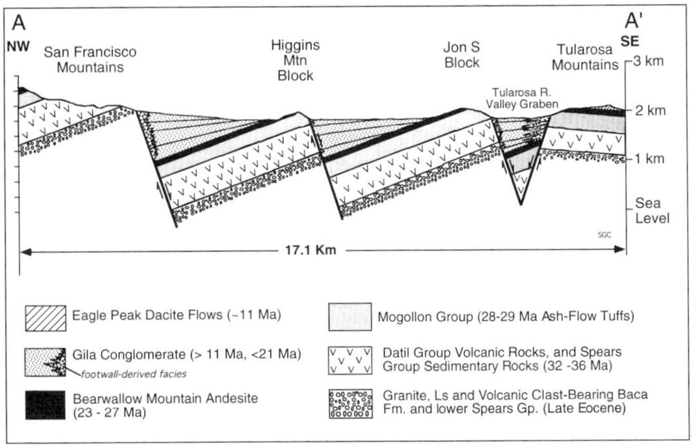

Figure 4. Diagrammatic cross section across the Reserve graben (line of section indicated on Fig. 3). Note the "domino"- like effect of a section across an area comprising several intra-basin fault blocks. Deep fault geometry is not known, nor is pre-Baca stratigraphy.

> indicated by the slopes of the beds ... Its cement is calcareous. Interbedded with it are layers of slightly coherent sand, and of trass, and sheets of basalt ... One thousand feet of the beds are frequently exposed, and the maximum exposure on the Prieto [River] is probably 1,500 feet."

Constraints on the age of the Gila Group in the Reserve graben were mentioned above—the base of the Gila overlies Bearwallow Mountain Andesite (23-27 Ma) or, where the Bearwallow Mountain Andesite has been erosionally removed, rests unconformably on tuffs derived from Mogollon plateau calderas (24-32 Ma). The Gila is interbedded with 21-1 Ma basalts (post-caldera "Group II" rocks of Futa and Ratté, 1989). Deposition of the Gila Group probably occurred episodically throughout this entire period, ending only when Quaternary stream incision began to exhume the basin. Most beds examined in this study are well-lithified conglomerates and sandstones believed to pre-date the 11 Ma Eagle Peak dacite flow (Ratté and Bové, 1989; Bové, 1990), which separates well-lithified lower Gila from poorly consolidated or caliche-cemented upper Gila.

The Reserve graben, especially its sedimentary facies, has not been extensively studied. Weber and Willard (1959) published a reconnaissance map of the thirty-minute quadrangle (on a non-topographic base) that contains the basin; more recently, detailed mapping and study of the volcanic rocks in and around the basin have been undertaken by J.C. Ratté and his colleagues at the US. Geological Survey (Ratté, 1980; Ratté et al., 1984; Ratté, 1989; Futa and Ratté, 1989; Ratté and Bové, 1990; Bové, 1990; Bové and Ratté, this volume).

FACIES ASSOCIATIONS AND DEPOSITIONAL ENVIRONMENTS

The syntectonic basin fill of the Reserve graben was divided into five distinctive facies associations in this study. Each of these sedimentary facies associations is described below, and interpreted in terms of sedimentary process, depositional environment, and relationship to tectonic features of the basin.

Facies association 1: coarse proximal deposits

This facies consists of massive beds of cobble and boulder conglomerate with little or no discernible sorting or fabric (Fig. 5). Maximum clast size is about 1.5 m; average clast size ranges from 10-50 cm, typically decreasing appreciably in the downstream direction over distances of a few 100s of m. A sandy matrix is ubiquitous and typically supports the clasts; matrix content varies from a matrix-rich end member in which clasts appear to "float" in matrix (Fig. 5B) to a matrix-poor end member in which clast-clast contact is more common than not (Fig. 5A). Deposits are usually ungraded but inverse and normal grading each occur in places. A-axis imbrication is sometimes present. Bedding is tabular, with sharp, slightly erosional bases. The upper surfaces of beds commonly show more relief than their bases—a result of clasts that project upward into overlying beds. Bed thickness ranges from <1 m to several m; bed thickness generally appears to correlate with maximum particle size, although quantitative methods such as the MPS/BTh plot described by Nemec and Steel (1984) have not yet been attempted.

Facies association 1 occurs in two modes: (1) as fill in well-defined erosional scours, 10s to 100s of m wide located high on the dip slopes of intra-basin imbricate fault blocks (Fig. 6), and (2) as larger and thicker accumulations located adjacent to the largest bounding faults, especially at a few locations along the San Francisco Mountain fault zone. The erosional scours filled by facies association 1 deposits typically cut through the Bearwallow Mountain Andesite (youngest of the pre-Gila, pre-block faulting units in the Reserve Basin) and into underlying tuff—creating erosional relief of up to 100 m (Fig. 6). Exposures of Facies 1 adjacent to major faults become interbedded with sheetflood and ephemeral stream deposits (see facies association 2 below) 10s to a few 100s of m away. In other words, grain size decreases laterally in all directions, suggesting a point source of coarse sediment along the fault scarp.

The primary sedimentary process reflected in the rocks of this facies is viscous or plastic mass flow, probably where flow was at least partially confined, either by bedrock canyons (on the dip slopes of imbricate fault blocks) or by fan-head channels (adjacent to fault-bounded basin margins). The variability of the facies suggests a range of mass-flow processes, from "fluidal flows" (Lowe, 1979) or hyperconcentrated streamflood (Smith, 1986), to true debris flows (Nemec and Steel, 1984). The environments in the Reserve Basin where this high energy mass flow took place include both small canyons or arroyos incised into the dip slopes of intra-basin tilt blocks, and incised channels and channel-mouth outflow lobes located near the heads of large, discrete, alluvial fans occurring in a few places along the fault-bounded margins

FIGURE 5. Facies association 1, coarse proximal deposits. A, Lower-matrix-content facies typical of proximal alluvial fan occurrence of facies association 1. B, Matrix-rich end member from dip-slope canyon fill occurrence. Both end members comprise mostly mass-flow deposits.

FIGURE 6. Example of dip-slope paleocanyon fill. Arrow is parallel to erosional base of canyon fill and points to a straw hat hanging just below the contact between the canyon fill and underlying Triangle C Ranch tuff.

of the basin. The paleocanyon occurrence of this facies tends to be more matrix rich than the fan-head channel occurrence.

The presence of significant erosional features on the hanging walls of an active half graben was predicted in the conceptual model of Leeder and Gawthorpe (1987), and modern analogs were described by Davis (1925) from the Oquirrh and Panamint ranges, both of which have fault-bounded western sides and dip to the east:

> "The V-gorges and the sharp clefts recently cut in the rock faces of the revealed fault surfaces on the west side of these two ranges in consequence of their upheaval may therefore be correlated with a moderate, down-grade furrowing of the detrital deposits on their eastern slopes in consequence of an increase in declivity there.... the furrows are several hundred yards wide and 100 or 200 feet deep ..."

Although the gradient of the hanging wall dip slope is gentle compared to the fault-bounded margin of a half graben, it is likely to be quite steep in stream-gradient terms. A relatively modest hanging wall dip of 1-5° will result in high-energy stream flow and is therefore likely to provoke canyon entrenchment over much of the hanging wall dip slope.

Dip-slope paleocanyon fills can be seen in many places around the Reserve graben, on both the Higgins Mountain and the Jon S Mountain intrabasin fault blocks. The best exposure is located along the San Francisco River in SE1/4 sec. 26, T6S, R19W (Fig. 6). A more accessible example of a canyon-fill deposit can be seen in a roadcut on Highway 12, near the center of sec. 29, T6S, R18W.

In many cases, as on the dip slope of the Higgins Mountain block, the paleocanyon fills occur immediately updip from mid-distal fan deposits, suggesting that the canyons fed the hanging wall fans. The deep erosion and coarse grain size associated with these dip-slope canyon fills certainly imply that the imbricate blocks existed as topographic features during the formation and fill of the basin—an indication that tectonism outstripped erosion and deposition enough to control topography and therefore drainage. Sediment produced by canyon incision and transported down the dip slope would be deposited on fans/braidplains at the canyon mouths. With time, sediment supply must have increased relative to tectonic movement, since the canyons eventually filled with sediment. Increased sediment flux relative to subsidence or tilting rate would cause the fan/sandflat environment to onlap the hanging wall as the dip-slope canyons backfilled.

Proximal alluvial-fan deposits are relatively rare in the Reserve graben, although there are some striking examples. The best and largest example is adjacent to the San Francisco Mountains fault zone (SFMFZ) in sec. 11, T6S, R19W, which is the same place where the modern San Francisco River enters the Reserve graben through a spectacular canyon known as the Box. The location of this fan also coincides with a 30° bend in the strike of the SFMFZ (Fig. 3). It seems quite likely that this bend results from segmentation of the fault zone—a displacement-transfer zone. As discussed above, this structural situation can result in a topographic gap in the footwall uplift through which an external drainage system may enter the basin, potentially delivering large volumes of sediment and water to the basin (Leeder and Gawthorpe, 1987; Jackson and White, 1989; Gawthorpe and Hurst, 1993).

Proximal fan deposits are also exposed adjacent to the SFMFZ in the bed of Pueblo Creek, in the southwestern corner of the study area (SE1/4 sec. 14, T8S, R21W; Fig. 3). It is important to note that Gila outcrops at many locations along the SFMFZ have been examined and found *not* to include proximal fan deposits. The San Francisco Mountains did not therefore constitute a major line source of sediment for the Reserve graben; rather, sediment entered the basin through discrete passages in the mountain front.

A third locus of proximal fan deposits is found on the southeast margin of the Reserve graben, adjacent to one of the down-to-the-northwest faults that border the Tularosa Mountains. These deposits can be seen in roadcuts along the Reserve-Beaverhead Road in sec. 3, T8S, R19W.

Facies association 2: mid-distal alluvial-fan deposits

This facies association is characterized by tabular-bedded poorly stratified sandstone and conglomerate, occurring in thin (10-100 cm), flat-based beds that often extend laterally for 100s of m in both strike and dip directions (Fig. 7). Grain size ranges from medium sand to large cobble but is dominated by very coarse sand through pebble-sized clasts (1-32 mm). Internal stratification, where present, consists either of horizontal or nearly horizontal sets (plane-bed stratification) or of low-angle trough cross-bed sets (Fig. 7B). Horizontal stratification is commonly defined by alternation of coarser (pebbly) and finer (sandy) laminae. Massive beds are also common. Facies association 2 is commonly interbedded with facies association 1; in such cases the ratio of facies association 2 to facies association 1 increases in the paleo-downstream direction.

FIGURE 7. Facies association 2, mid-distal alluvial fan deposits. A, Outcrop view; note the extremely continuous tabular beds, most with flat bases. B, Close-up, showing probable sheetflood, shallow channel, and distal debris flow deposits typical of this facies association.

Several processes are reflected by this facies association, including non-channelized "sheetflood" deposition, deposition in small, ephemeral, relatively wide and shallow sandy streams (streams of the sort described by Bull (1972) from arid-region alluvial fans), and relatively fine-grained (compared to facies association 1) mass flows. This collection of processes probably reflects a mid-distal alluvial fan or sandflat depositional environment. This interpretation is strengthened by the common occurrence of facies association 2 as a downstream equivalent of the proximal fan and canyon fill deposits of facies association 1.

Facies association 2 is extremely common in the Reserve Basin, constituting roughly half the outcrop exposure of the basin fill. In several places within the basin, sections 50-100 m thick that are dominated by facies association 2 are overlain by sections of similar thickness that consist largely of through-going fluvial channel deposits (see facies association 3, below). This vertical transition may reflect a widespread change in depositional style, from fan-dominated to axial-stream dominated deposition and suggests a decrease in relative subsidence rate that allowed sedimentation to "catch up" with tectonism.

Paleocurrents are difficult to obtain from the poorly stratified deposits of this environment but sparse imbricated pebbles generally indicate southwesterly transport, suggesting that the deposits reflect drainage down structural dip—down the hanging wall dip slopes of tilted intra-basin fault blocks. As mentioned above, facies association 2 is the most common type of deposit in the basin. It is likely that during the most active period of tectonic subsidence, much of the basin was covered with broad, sheetlike fans from the canyons incised into the dip slopes of tilted intrabasin blocks. Rapid subsidence would tend to encourage rapid aggradation with little sorting or reworking of sediment. Logically enough, the deposits inferred to reflect rapid subsidence are thicker than those associated with slower subsidence. This scenario is consistent with the model described above (also see Leeder and Gawthorpe, 1987; Blair, 1987; Cavazza, 1989) and with observations in Death Valley, where sheetlike hanging wall fans grade into sandflats bordering playas (Hunt, 1975; Fig. 1).

Depending on the hydrology of the basin, transverse systems originating on the hanging wall dip slope may empty into either an axial fluvial system or into a lake occupying the basin axis—in fact, an axial system may originate as a transverse system that curves into the axial trough and continues along strike. If the climate is arid, and if there is a wide low gradient zone in the axis of the basin, these systems may simply die out downstream due to loss of water and energy, forming "terminal fans" (Friend, 1978; Parkash et al., 1983; Kelly and Olsen, 1993). In Death Valley, transverse drainage systems terminate in a salt pan, or playa, which is usually dry (Hunt, 1975). In the Reserve graben, however, no mudstones or evaporites indicative of playa lake deposition have been found, but axial fluvial deposits are common.

Facies association 3: axial fluvial channel deposits

The third facies association (Fig. 8) comprises cross-bedded, channeled conglomerate and sandstone; grain size ranges from medium sand to large cobble (250 mm). In contrast to both the proximal and distal fan deposits just described, facies association 3 is marked by good sorting, abundant sedimentary structures (especially trough cross-bed sets), strong b-axis imbrication fabric, and high-relief bounding surfaces that reflect erosional scour (Fig. 8). Commonly the entire exposure of this facies at a given outcrop occurs within a channel-form erosional bounding surface; in other cases such channel fills are stacked and multi-lateral. Sedimentation units on the order of 1 m thick generally fine upward, from an erosional, conglomeratic base to a sandy top that is often reddened and devoid of primary sedimentary structures (suggesting paleosol development).

This distinctive facies association consists of fluvial channel and interchannel deposits. Detailed paleohydraulic work was not attempted, but the dominance of trough cross-bed sets, lack of lateral accretion surfaces, and apparent coarse-grained, non-cohesive nature of the bed and bank material suggest a low sinuosity bedload-dominated stream. Flow depth, estimated from the thickness of channel fill packages, was 0.5-5 m. Paleocurrents are typically parallel to the basin axis, suggesting that these were longitudinal fluvial systems, possibly flowing southwestward the length of the basin and passing into the next basin downstream.

FIGURE 8. Facies association 3, axial fluvial channel deposits. A, Trough cross-bedded channel fill cut into fine-sand interchannel deposits. Object at base of photo is a knapsack. B, Larger-scale cross bedding from Wilson Canyon axial fluvial deposits, near the Tularosa River. Note hammer (circled) for scale.

Evidence of axial fluvial systems was found in all three of the sub-basins shown on the cross section in Figure 4. In the half graben between Higgins Mountain and Jon S Mountain, the axial system appears to be confined to the "deep" side, i.e., close to the Higgins Mountain fault (the eastern part of the half graben is dominated by dip-slope paleocanyon fill and distal fan deposits). Significantly, fine-grained floodplain deposits separating channel deposits are thickest and texturally finest immediately adjacent to the mountain front fault (Fig. 9). This suggests that rapid subsidence adjacent to the fault promoted rapid aggradation—i.e., vertical accretion was rapid compared to lateral migration of channels, so that preservation of interchannel deposits was enhanced (Bridge and Leeder, 1979; Leeder and Gawthorpe, 1987).

In the Tularosa River Valley graben (on the eastern edge of the Reserve Basin, Fig. 4), which is only a few kilometers wide, the axial system appears to have occupied most of the width of the sub-basin. This sub-basin is unusual in that it is bounded by faults on both sides, making it less asymmetrical than the other sub-basins—more like a full graben. The fluvial deposits in this sub-basin are unusually coarse, with little preservation of out-of-channel fines (Fig. 8B).

The axial deposits in the westernmost half graben in the Reserve Basin, between the San Francisco Mountains and Higgins Mountain, appear to be laterally (downstream) equivalent to the proximal fan deposits associated with the large fan that issued from the San Francisco Mountains near the Box. The transition occurs over a distance

FIGURE. 9. Contrasting modes of axial fluvial channel deposits. A, Stacked channel deposits with little preservation of interchannel fines; arrow points to 3-m-high tree at base of outcrop. B, Isolated channel bodies separated by interchannel fines found adjacent to the Higgins Mountain fault scarp, where rapid subsidence apparently allowed less reworking of the floodplain, leading to greater preservation of fine-grained interchannel deposits. Staff is 1.5 m long.

of approximately 5 km over which exposures are nearly continuous in the San Francisco river valley. This unique, large fan may have been elongated longitudinally, and may have brought sufficient volumes of sediment and water into the basin to create an axial system that passed from an alluvial fan environment distributary upstream to a fluvial channel environment downstream.

Where vertical control permits, the axial fluvial deposits in this most western sub-basin are often seen to overlie transverse fan deposits. In one case where exposure is especially good, permitting horizontal control of about 1 km and vertical control of about 100 m, the fluvial channel deposits appear to onlap hanging wall dip-slope transverse fan deposits. These observations suggest that an early phase of basin subsidence and fill was dominated by sedimentation on transverse fans flowing down the dip slope of the Higgins Mountain imbricate fault block, and that with time, the ratio of sediment supply to subsidence increased, enabling the axial fluvial channel system to occupy a progressively wider area of the half graben. Important questions are left unanswered by the work done so far; some may be resolved by further work, while others cannot be addressed because of the limitations of outcrop exposures. For example, it is not known whether the suspected increase in the ratio of sediment flux to subsidence resulted from a *decrease* in subsidence as tectonic activity decreased, or from an *increase* in the rate at which sediment was brought into the basin by the large longitudinal fan system. It is also possible that no axial fluvial system existed during the initial stages of basin evolution.

Facies association 4: massive, channeled sandstones

In a few places within the Reserve Basin, beds dominated by massive coarse sandstone occur. Bed thickness is variable and bed geometry includes clear cross-cutting channel forms (Fig. 10A). However, the beds lack internal stratification. Occasional pebble- or cobble-size clasts are seen "floating" within the sandstone (Fig. 10B).

The origin of this rare but prominent facies association is problematic. It is found only in proximal settings (mainly adjacent to fault scarps) and can be interbedded with the coarse mass- flow deposits of facies association 1. This association, together with the apparent lack of sedimentary structures, suggests a high energy mass-flow origin that contrasts strangely with the fine-grained texture of the facies. On the other hand, the facies usually occurs in *rapidly subsiding* proximal settings, where rapid deposition/aggradation is expected. High sediment concentrations associated with rapid deposition in sand-dominated environments may suppress bedform development, resulting in hyperconcentrated flood flow deposits that resemble this facies association (Smith, 1986).

FIGURE 10. Facies association 4, massive sandstones. A, Outcrop view; arrows highlight erosional bounding surface with massive sandstone fill. Note dip changes; the significant decrease in dip from base to top of the exposure reflects syntectonic deposition. Steeply dipping outcrops in lower right portion of photograph are axial fluvial deposits, not massive sandstones. B, Closer view of massive sand facies association. Note cobble-sized clasts "floating" in sandy matrix.

Another possibility is that the massive sandstone facies reflects a peculiar source-terrane lithology whose breakdown products did not include gravel, or, if gravel was produced, the clasts were too soft to survive transport. A likely source lithology is poorly welded ash flow tuff, which is extremely common in the mid-Tertiary "basement" underlying the Gila (e.g., Bloodgood Canyon Tuff, see Ratté et al., 1989).

The primary locality for observing this facies association is adjacent to a prominent fault scarp several tens of meters high associated with a collapsed segment of the Jon S Mountain tilted block, in SE1/4 sec. 33, T6S, R18W.

Facies association 5: colluvial wedge deposits

Large (boulder-size, some larger than 1 m) angular blocks of homogeneous lithology (usually basaltic andesite) are occasionally found cemented together in wedge-shaped deposits adjacent to intra-basin fault scarps (Fig. 11). The most prominent and accessible example of this facies is near a water gap in the Higgins Mountain fault block called Starkweather Canyon (SW1/4 sec. 2 and SE1/4 sec. 3, T7S, R19W). There the facies can be traced for approximately 500 m along strike but only about 10 m in the down dip direction. These deposits are devoid of sorting or fabric. Distal fan deposits of facies association 2 are found within 10 m of the deposit, with no discernible fault between the contrasting deposits.

This facies is probably of colluvial origin, specifically comprising colluvial wedges resulting from rock falls or slides on active fault scarps. They are rare, typically quite localized, and in the case of the Starkweather Canyon deposit constitute the only significant accumulation of coarse-grained material found adjacent to the fault. The significance of the example along the Higgins Mountain scarp is twofold. First, it is a further indication that the Higgins Mountain intrabasin fault block existed as a topographic feature during Gila deposition, and second, the fact that this single colluvial wedge apparently constitutes the only notable accumulation of coarse-grained material along the Higgins Mountain escarpment is an indication that relief does not guarantee a large sediment flux. In other words, it is significant that sediment accumulation adjacent to an active fault scarp is dominated by the deposits of transverse fans originating on the hanging wall and flowing toward the scarp, and by deposits of an axial fluvial channel system flowing parallel to the mountain front.

DISCUSSION AND CONCLUSIONS

Figure 12 depicts the syntectonic paleogeography of the Reserve graben, as inferred from paleocurrent measurements and the spatial distribution of the depositional environments discussed above. This diagram synthesizes many of the observations and interpretations of this study, and suggests a clear pattern of structural control over topography, drainage patterns and depositional systems similar to that predicted by the conceptual model outlined in the Introduction. Specific conclusions regarding the syntectonic deposits of the Reserve graben are: (1) surface topography conforms to the "tectonic slope" (i.e., the hanging wall slopes down to the structural axis of the half graben); (2) drainage of this hanging wall dip slope results in erosion of incised canyons on the upper part of the dip slope and deposition on large, relatively fine-grained alluvial fan complexes on the lower part of the dip slope; (3) longitudinal fluvial systems flow south-southwest along the coincident structural-topographic axis of the basin. These axial streams often flowed so close to bounding-fault escarpments that rockfall deposits are found among the alluvium, and rapid subsidence rates are reflected in increased floodplain:channel deposit ratios; (4) alluvial fans along the faulted, high relief, margins of half grabens are rare, isolated features, albeit coarser-grained than the dip-slope fans. Footwall uplifts in the Reserve graben did not act as major line sources of sediment because their catchment areas were small, and because progradation was inhibited by rapid subsidence of the basin axis; and, (5) footwall fans or axial fluvial systems fed by "gaps" in the footwall uplift were more important contributors of sediment than systems fed directly by erosion of the footwall uplift, as a result of their access to larger catchment areas.

FIGURE 11. Colluvial wedge facies association. Exposure is about 8 m high. Deposit consists of chaotic accumulation of angular basalt clasts with no discernible bedding or fabric.

The presence of incised-canyon fills on the hanging wall dip slope, together with the widespread distal fan/sandflat deposits that reflect transverse drainage down the hanging wall dip slope, lead directly to the conclusion that the asymmetrical structure of the basin was mirrored in the surface topography. Similarly, the rare fans and colluvial deposits along basin-margin faults imply high relief across these faults. Finally, the concentration of longitudinal stream deposits near the structural axis of each sub-basin suggests that the structural axis of each half graben was overlain by a topographic low that naturally attracted longitudinal systems (Alexander and Leeder, 1987).

The scarcity of fans along the fault-bounded margins of each half graben, and their limited areal extent (i.e., the lack of transverse systems that reflect flow *up* the hanging wall dip slope), is still another indication that topographic and tectonic slopes dipped the same direction. It also supports two major concepts advanced above: high relief of marginal uplifts does not compensate for small drainage areas, in terms of ensuring high sediment supply. In other words, if a footwall uplift is narrow, it may not shed enough sediment to construct significant clastic wedges, even if it is a high uplift, undergoing rapid denudation. The fact that such uplifts are typically asymmetrical themselves effectively decreases their width even further, because much of the water and sediment drains down the dip slope, not into the basin. In addition, the juxtaposition of uplifts and rapidly subsiding basin axes on the active margins of asymmetrical basins further inhibits progradation, by increasing the volume of sediment required to construct a basinward slope.

The probable existence of a large longitudinal fan that issued from a topographic gap between two segments of the SFMFZ near the Box (Fig. 12) supports the assertion that drainages that originate beyond the rift-shoulder uplift, and cut across the uplift to reach the basin, constitute a more important source of basin-filling sediment than any drainages from within the footwall uplift itself. This of course is because these antecedent or superimposed systems have much larger potential watersheds than any system having a drainage basin limited to the footwall uplift. It is significant that these features can be seen, not only in Holocene settings such as the Great Basin, but in the stratigraphic record, since this justifies the extrapolation from two-dimensional models of topography and drainage pattern to three dimensional models of the basin fill.

The Gila Group was deposited in the context of the same Basin and Range fault-block topography that prevails today. To this conclusion two caveats must be added. First, the topography of today is largely erosional—fault blocks and escarpments have been exhumed as the San Francisco River and its tributaries dissected the basin, differentially eroding sedimentary and igneous rocks as they incised. As a result, although the topography of today may be broadly similar to the topography that influenced Gila deposition during the period 11-20 Ma, there may have been a subsequent period of more subdued topography that existed between the last episode of rapid uplift and subsidence and the downcutting of regional stream systems. Second, the

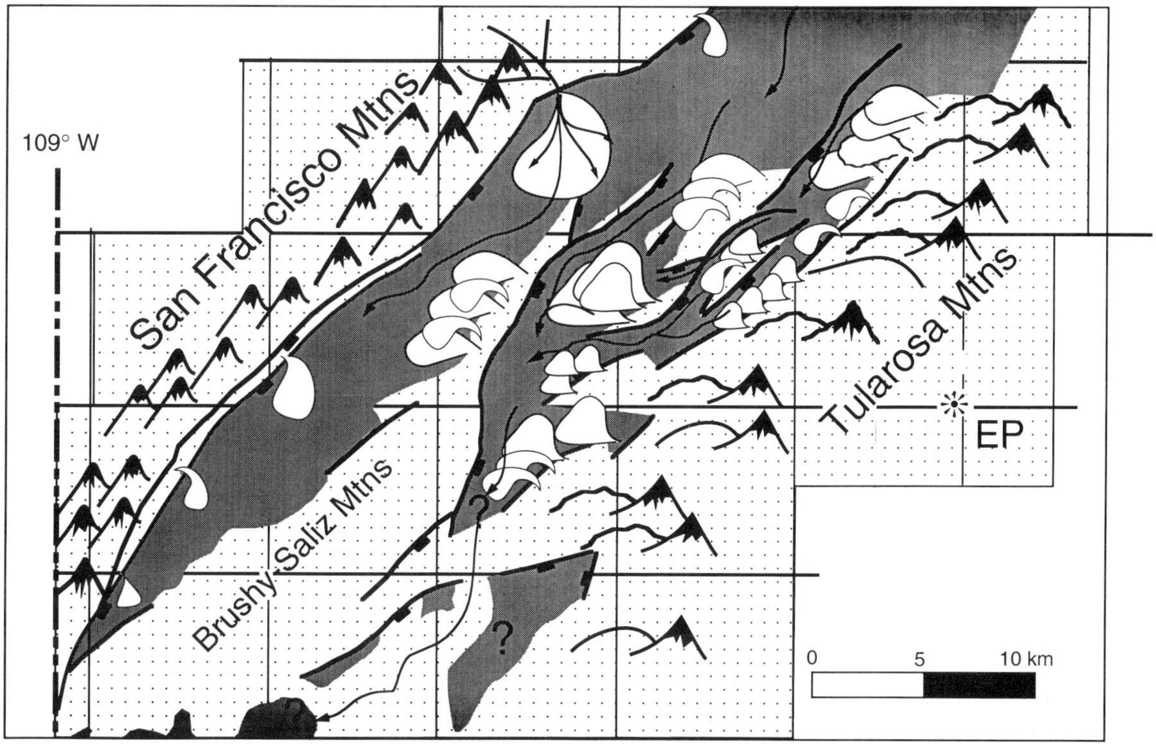

FIGURE 12. Miocene paleogeography of the Reserve graben, emphasizing drainage patterns and distribution of alluvial fans. The principal features are: (1) numerous dip-slope fans draining the uplifted intra-basin tilt blocks, and the Tularosa Mountains; (2) axial flow to the southwest (denoted by arrows), and (3) sparse fans issuing from the San Francisco Mountains, notably a large fan entering the basin at the bend/transfer zone in the main bounding fault. Note that this fan is depicted contributing to an axial-flow system.

Reserve graben is a highly faulted region, and not all faults within it share the same history—some are older than others, and many have undoubtedly experienced recurrent movement. To say that the structural framework reflected in today's topography is the same as that which affected the deposition of the Gila Group is not to say that every fault in the region was active before or during Gila deposition, nor is it meant to suggest that faults active during Gila deposition ceased moving once the section of the Gila studied here was deposited. On the contrary, the Gila is faulted in many places, and there is clear evidence of young faulting in the Milligan Mountain quadrangle, on the southeastern margin of the basin (Weber and Willard, 1959; Ratté, 1980; Ratté and Bové, 1990).

Finally, it should be noted that this investigation of sedimentation in the Reserve graben is not represented as a comprehensive study of the stratigraphy and sedimentology of the Gila Group—it is more a preliminary reconnaissance, specifically aimed at discerning the tectonic imprint on the basin fill. The principal accomplishments of this study are that it sheds some light on the manifestations of structural asymmetry in the stratigraphic record of continental extension, corroborates the essential aspects of the conceptual model sketched at the beginning of this paper, and provides a basis for future studies of the Gila Group in the Reserve graben.

ACKNOWLEDGMENTS

I thank Steve Cather, Jeff Corrigan, Jim Ratté and Lee Russell for constructive reviews that improved this paper. Funding from Arco, the National Science Foundation and Norsk Hydro is gratefully acknowledged. Finally, thanks to Steve Cather, Jim Steidtmann, Ron Steel, and especially Jim Ratté for spending time in the field with me.

REFERENCES

Alexander, J. and Leeder, M.R., 1987, Active tectonic control on alluvial architecture; *in* Ethridge, F.G., Flores, R.M. and Harvey, M.D., eds., Recent developments in fluvial sedimentology: contributions from the Third International Fluvial Sedimentology Conference: Society of Economic Paleontologists and Mineralogists, Special Publication 39, p. 243-252.

Blair, T.C., 1987, Tectonic and hydrologic controls on cyclic alluvial fan, fluvial, and lacustrine rift basin sedimentation, Jurassic-lowermost Cretaceous Todos Santos Formation, Chiapas, Mexico: Journal of Sedimentary Petrology, v. 57, p. 845-862.

Blair, T.C. and Bilodeau, W.L., 1988, Development of tectonic cyclothems in rift, pull-apart, and foreland basins: sedimentary response to episodic tectonism: Geology, v. 16, p. 517-520.

Bornhorst, T. J., 1988, Character of mid-Cenozoic andesites, basaltic andesites, and related volcanic rocks, Mogollon-Datil volcanic field, southwestern New Mexico: Colorado School of Mines Quarterly, p. 1-13.

Bové, D.J., 1990, Preliminary geologic map of the Eagle Peak Quadrangle, Catron County, New Mexico: U.S. Geological Survey, Open-file Report 90-0546.

Bridge, J. S. and Leeder, M. R., 1979, A simulation model of alluvial stratigraphy: Sedimentology, v. 26, p. 617-644.

Bull, W. B., 1972, Recognition of alluvial fan deposits in the stratigraphic record; *in* Rigby, J.K. and Hamblin, W.K. , eds., Recognition of ancient sedimentary environments: Society of Economic Paleontologists and Mineralogists, Special Publication 16, p. 63-83.

Cavazza, W., 1989, Sedimentation pattern of a rift-filling unit, Tesuque formation (Miocene), Espanola basin, Rio Grande rift, New Mexico: Journal of Sedimentary Petrology, v. 59, p. 287-296.

Chapin, C.E., 1988, Axial basins of the northern and central Rio Grande rifts; *in* Sloss, L.L., ed., Sedimentary cover—North American craton: U.S.: Geological Society of America, Decade of North American Geology, The Geology of North America, v. D-2, p. 165-170.

Cohen, A. S., 1991, Tectono-stratigraphic model for sedimentation in Lake Tanganyika, Africa; *in* Katz, B., Lacustrine exploration: case studies and modern analogues: American Association of Petroleum Geologists, Memoir 50, p. 137-150.

Coney, P. J., 1987, The regional tectonic setting and possible causes of Cenozoic extension in the North American Cordillera; *in* Coward, M. P., Dewey, J. F. and Hancock, P. L., eds., Continental extensional tectonics: Geological Society, Special Publication No. 28, p. 177-186.

Crews, S.G., 1990, Structure, syntectonic sedimentation, and stratigraphy in asymmetrical non-marine basins [Ph.D. thesis]: Laramie, University of Wyoming, 196 p.

Crews, S.G., 1993, Laramide tectonics and humid alluvial fan sedimentation, NE Uinta uplift, Utah and Wyoming: Journal of Sedimentary Petrology, v. 63, p. 420-436.

Crews, S.G. and McGrew, A.J., 1990, Influence of structural style on rift basin morphology and non-marine sequence geometry: Geological Society of America, Abstracts with Programs, v. 22, no. 7, p. A239.

Davis, W. M., 1925, The Basin and Range problem: Proceedings of the National Academy of Sciences, v. 11, p. 387-392.

Elston, W.E., 1984, Mid-Tertiary ash flow tuff cauldrons, southwestern New Mexico: Journal of Geophysical Research, v. 89, p. 8733-8750.

Elston, W.E. and T.J. Bornhorst, 1979, The Rio Grande rift in context of regional post-40 m.y. volcanic and tectonic events; *in* Riecker, R.E., ed., Rio Grande rift–tectonics and magmatism: American Geophysical Union, Washington, D.C., p. 416-438.

Elston, W.E., Damon, P.E., Coney, P.J., Rhodes, R.C., Smith, E.I. and Bikerman, M., 1973, Tertiary volcanic rocks, Mogollon-Datil province, New Mexico, and surrounding region: K-Ar dates, patterns of eruption, and periods of mineralization: Geological Society of America Bulletin, v. 84, p. 2259-2274.

Elston, W.E., Rhodes, R.C., Coney, P.J. and Deal, E.G., 1976, Progress report on the Mogollon Plateau volcanic field, southwestern New Mexico, No. 3—surface expression of a pluton: New Mexico Geological Society, Special Publication 5. p. 3-28.

Engebretson, D.C., Cox, A. and Thompson, G.A., 1984, Correlation of plate motions with continental tectonics: Laramide to Basin and Range: Tectonics, v. 3, p. 115-119.

Flemings, P.B. and Jordan, T.E., 1990, Stratigraphic modeling of foreland basins: interpreting thrust deformation and lithosphere rheology: Geology, v. 18, p. 430-434.

Friend, P.F., 1978, Distinctive features of some ancient river systems; *in* Miall, A.D., ed., Fluvial sedimentology: Canadian Society of Petroleum Geologists, Memoir 5, p. 531-542.

Frostick, L.E. and I. Reid, 1987, Tectonic control of desert sediments in rift basins ancient and modern; *in* Frostick, L. E., and Reid, I., Desert sediments: ancient and modern: Geological Society of London, Special Publication 35, p. 53-68.

Futa, K. and Ratté, J.C., 1989, Petrogenetic implications of Rb-Sr and Sm-Nd isotopes related to post-caldera volcanism in the western Mogollon-Datil volcanic field, New Mexico: New Mexico Bureau of Mines and Mineral Resources, Bulletin 131, p. 100.

Gawthorpe, R.L. and Hurst, J.M., 1993, Transfer zones in extensional basins: their structural style and influence on drainage development and stratigraphy: Journal of the Geological Society of London, v. 150, p. 1137-1152.

Gilbert, G.K., 1875, Geology of portions of New Mexico and Arizona, explored and surveyed in 1873; in Wheeler, G.M., Report upon geographical and geological explorations and surveys west of the 100th meridian, v. 3, part 5: Washington, Government Printing Office, p. 507-567.

Hamblin, A.P. and Rust, B.R., 1989, Tectono-sedimentary analysis of alternate-polarity half-graben basin-fill successions: Late Devonian–Early Carboniferous Horton Group, Cape Breton Island, Nova Scotia: Basin Research, v. 2, p. 239-255.

Hamilton, W., 1987, Crustal extension in the Basin and Range Province, southwestern United States; *in* Coward, M.P., Dewey, J.F. and Hancock, P.L., eds., Continental extensional tectonics: Geological Society of London, Special Publication No. 28, p. 155-176.

Hunt, C.B., 1975, Death Valley geology, ecology, archeology: Berkeley, University of California Press, 234 p.

Jackson, J.A. and N.J. White, 1989, Normal faulting in the upper continental crust: observations from regions of active extension: Journal of Structural Geology, v. 11, p. 15-36.

Kelly, S.B. and Olsen, H., 1993, Terminal fans; a review with reference to Devonian examples: Sedimentary Geology, v. 85, p. 339-374.

Lambiase, J.J., 1991, A model for tectonic control of lacustrine stratigraphic sequences in continental rift basins; *in* Katz, B., Lacustrine exploration: case studies and modern analogues: American Association of Petroleum Geologists, Memoir 50, p. 265-276.

Leeder, M.R. and R.L. Gawthorpe, 1987, Sedimentary models for extensional tilt-block/half-graben basins; *in* Coward, M.P., Dewey, J.F. and Hancock, P.L., eds., Continental extensional tectonics: Geological Society of London, Special Publication No. 28, p. 139-152.

Leeder, M.R. and Jackson, J.A., 1993, The interaction between normal faulting and drainage in active extensional basins, with examples from the western United States and central Greece: Basin Research, v. 5, p. 79-102.

Lowe, D. R., 1979, Sediment gravity flows: their classification and some problems of application to natural flows and deposits; *in* Doyle, L. J. and Pilkey, O. H., eds., Geology of continental slopes: Society of Economic Paleontologists and Mineralogists, Special Publication 27, p. 75-82.

Mack, G.H. and Seager, W.R., 1990, Tectonic control on facies distribution of the Camp Rice and Palomas Formations (Pliocene-Pleistocene) in the southern Rio Grande rift: Geological Society of America Bulletin, v. 102, p. 45-53.

Marvin, R.F., Naeser, C.W., Bikerman, M., Mehnert, H.H. and Ratté, J.C., 1987, Isotopic ages of post-Paleocene igneous rocks within and bordering the Clifton 1° x 2° quadrangle, Arizona—New Mexico: New Mexico Bureau of Mines and Mineral Resources, Bulletin 118, 63 p.

McGrew, A. J. and Crews, S. G., 1990, A quantitative model for the geometrical evolution of multiple-fault extensional systems: Geological Society of America Bulletin, Abstracts with Programs, v. 22, p. A273.

McIntosh, W.C., Chapin, C.E., Ratté, J.C. and Sutter, J.F.,1992, Time-stratigraphic framework for the Eocene-Oligocene Mogollon-Datil volcanic field, southwest New Mexico: Geological Society of America Bulletin, v. 104, p. 851-871.

Morley, C.K., 1989, Extension, detachments, and sedimentation in continental rifts (with particular reference to East Africa): Tectonics, v. 8, p. 1175-1192.

Nemec, W. and Steel, R. J., 1984, Alluvial and coastal conglomerates: their significant features and some comments on gravelly mass-flow deposits; *in* Koster, E.V. and Steel, R.J., eds., Sedimentology of gravels and conglomerates: Canadian Society of Petroleum Geologists, Memoir 10, p. 1-31.

Parkash, B., Awasthi, A.K. and Gohain, K., 1983, Lithofacies of the Markanda terminal fan, Kurukshetra District, Haryana, India; *in* Collinson, J.D. and Lewin, J., eds., Modern and ancient fluvial systems: International Association of Sedimentologists, Special Publication 6, p. 337-344.

Paton, S., 1992, Active normal faulting, drainage patterns and sedimentation in southwestern Turkey: Journal of the Geological Society, London, v. 149, p. 1031-1044.

Ratté, J. C. , 1980, Geologic map of the Saliz Pass quadrangle, Catron County, New Mexico: U.S. Geological Survey, Miscellaneous Field Studies Map MF-1203, scale 1:24,000.

Ratté, J. C. , 1989, Geologic map of the Bull Basin quadrangle, Catron County, New Mexico: U.S. Geological Survey, Quadrangle Map GQ-1651, scale 1:24,000.

Ratté, J. C., Marvin, R. F. and Naeser, C. W., 1984, Calderas and ash flow tuffs of the Mogollon Mountains, southwestern New Mexico: Journal of Geophysical Research, v. 89, no. B10, p. 8713-8732.

Ratté, J. C. and Bové, D. J., 1990, Preliminary geologic map of the Milligan Mountain quadrangle, Catron County, New Mexico: U.S. Geological Survey, Open-file Report 90-0268.

Rosendahl, B. R., 1987, Architecture of continental rifts with special reference to East Africa: Annual Reviews of Earth and Planetary Science, v. 15, p. 445-503.

Schlische, R.W., 1991, Half-graben basin filling models: new constraints on continental extensional basin development: Basin Research, v. 3, p. 123-141.

Scholz, C.A. and Rosendahl, B.R., 1991, Coarse-clastic facies and stratigraphic sequence models from Lakes Malawi and Tanganyika, East Africa; *in* Katz, B., Lacustrine exploration: case studies and modern analogues: American Association of Petroleum Geologists, Memoir 50, p. 151-168.

Smith, G. A., 1986, Coarse-grained non-marine volcaniclastic sediment: terminology and depositional processes: Geological Society of America Bulletin, v. 97, p. 1-10.

Stewart, J.H.,1971, Basin and Range structure: a system of horsts and grabens produced by deep-seated extension: Geological Society of America Bulletin, v. 82, p. 1019-1044.

Weber, R.H. and Willard, M.E., 1959, Reconnaissance geologic map of Reserve thirty-minute quadrangle: New Mexico Bureau of Mines and Mineral Resources, Geologic Map 12.

THE OLIGOCENE GOODSIGHT-CEDAR HILLS HALF GRABEN NEAR LAS CRUCES AND ITS IMPLICATIONS TO THE EVOLUTION OF THE MOGOLLON-DATIL VOLCANIC FIELD AND TO THE SOUTHERN RIO GRANDE RIFT

GREG H. MACK, ALICE L. NIGHTENGALE, WILLIAM R. SEAGER, and RUSSELL E. CLEMONS
Department of Geological Sciences, New Mexico State University, Las Cruces, NM 88003

Abstract—The Goodsight-Cedar Hills (GCH) depression near Las Cruces, New Mexico, was a north-northeast-trending, nonmarine, elliptical basin of Oligocene age that developed along the southeastern margin of the Mogollon-Datil volcanic field. Within the basin were deposited six ash-flow tuff outflow sheets, as well as sedimentary rocks composed of interbedded syneruption and post-eruption lithofacies. The syneruption lithofacies consists of light-colored, medium-grained tuffaceous sandstones that represent reworked fallout tephra. Epiclastic conglomerates and coarse sandstones constitute the post-eruption lithofacies, with clast composition reflecting derivation from adjacent bedrock uplifts. An asymmetrical distribution of post-eruption lithofacies, based on provenance and sediment dispersal data, suggests that the GCH basin was an eastward-tilted half graben, with the Cedar Hills vent zone as the footwall and the Ancestral Goodsight uplift as the hanging wall. The GCH half graben provides important information about the Mogollon-Datil volcanic field, not only because it was a depository for several regional outflow sheets, but also because syneruption sandstones may record explosive eruptions that occurred during or between major ignimbrite-generating events. Furthermore, initial development of the half graben at 35 Ma precedes by 2 to 7 million years previous interpretations of the onset of regional crustal extension in the southern Rio Grande rift.

INTRODUCTION

The Mogollon-Datil volcanic field of west-central New Mexico is characterized by as many as 11 latest Eocene and Oligocene silicic calderas and their associated outflow sheets (McIntosh et al., 1986, 1990, 1991, 1992). The southeastern edge of the volcanic field extends to the vicinity of Las Cruces, where three calderas, Emory, Doña Ana and Organ, were active between 36.2 and 34.9 Ma (Fig. 1; Elston et al., 1975; Seager et al., 1976; Seager, 1981; McIntosh et al., 1990, 1991, 1992). Coeval to and positioned among the active calderas was a subsiding tectonic feature referred to as the Goodsight-Cedar Hills (GCH) volcano-tectonic depression, most of which was subsequently uplifted along late Tertiary normal faults (Fig. 2; Seager, 1973, 1975). The GCH depression contains up to six mappable ash-flow tuffs and interbedded sedimentary rocks of the Bell Top Formation, but was not itself a cauldron (Seager, 1975). It is not clear from earlier studies, which concentrated primarily on the volcanic rocks of the Bell Top Formation, whether subsidence of the GCH depression was driven by volcanic processes or by movement of a basin-bounding fault (Seager, 1973, 1975; Clemons, 1976a). A cauldron origin is unlikely because the basin contains only outflow sheets, most of which are correlated with distant calderas, and there is no evidence for extrusive centers corresponding to the age of the ash-flow tuffs. This study presents sedimentologic and petrographic data from the sedimentary members of the Bell Top Formation that suggest the GCH depression was a half graben, in which were deposited epiclastic conglomerates and sandstones derived from adjacent bedrock uplifts, reworked fallout tephra, and ash-flow tuff outflow sheets. This interpretation provides information about the possible existence of Plinian eruptions between major ash-flow sheets, as well as suggests that regional crustal extension in the southern Rio Grande rift may have begun by 35 Ma.

STRATIGRAPHY OF THE BELL TOP FORMATION

The type section of the Bell Top Formation is in the Sierra de las Uvas, where it consists of six mappable ash-flow tuffs, numbered 2 through 7, and two mappable sedimentary members (Middle and Upper Sedimentary Members), which are the focus of this study (Figs. 2, 3; Kottlowski, 1953; Seager, 1973; Clemons, 1975, 1976a). The upper five tuffs have recently been dated by the $^{40}Ar/^{39}Ar$ method and range in age from 35.7 to 28.5 Ma (McIntosh et al., 1990, 1991, 1992). Ash-flow tuff 7 is thin and discontinuous in the Sierra de las Uvas and probably is the distal part of the Vicks Peak Tuff of the San Mateo Mountains area (McIntosh et al., 1990, 1991, 1992). Along the southeastern margin of the basin in the Cedar Hills and Rough and Ready Hills is a local map unit, the Coyote Canyon Conglomerate, which is correlative with the Upper Sedimentary Member of the Bell Top Formation (Fig. 2; Clemons, 1976b). The GCH depression is bounded on the east by the Cedar Hills vent zone, a northerly trending series of flow-banded rhyolite domes and related volcanic rocks exposed in the Cedar Hills and Rough and Ready Hills (Figs. 1, 2). The domes appear to be onlapped by ash-flow tuffs 4 and 5 and are roughly the same age or slightly younger than ash-flow tuff 3 (Fig. 3; Seager, 1975; Clemons, 1975).

Ash-flow tuffs 2, 3 and 4 are absent north of the Sierra de las Uvas, although ash-flow tuffs 5, 6 and 7 and the Middle and Upper Sedimentary Members are present in the Rincon Hills and at Point of Rocks (Figs. 2, 3; Seager and Hawley, 1973; Seager et al., 1982).

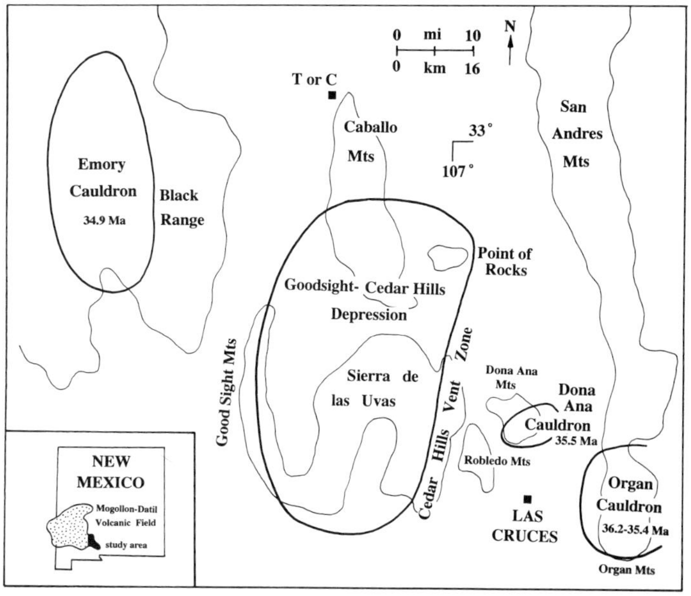

FIGURE 1. Oligocene tectonic elements in south-central New Mexico. Numbers refer to radiometric ages in millions of years (Ma) of tuffs located within calderas, based on McIntosh et al. (1990, 1991, 1992).

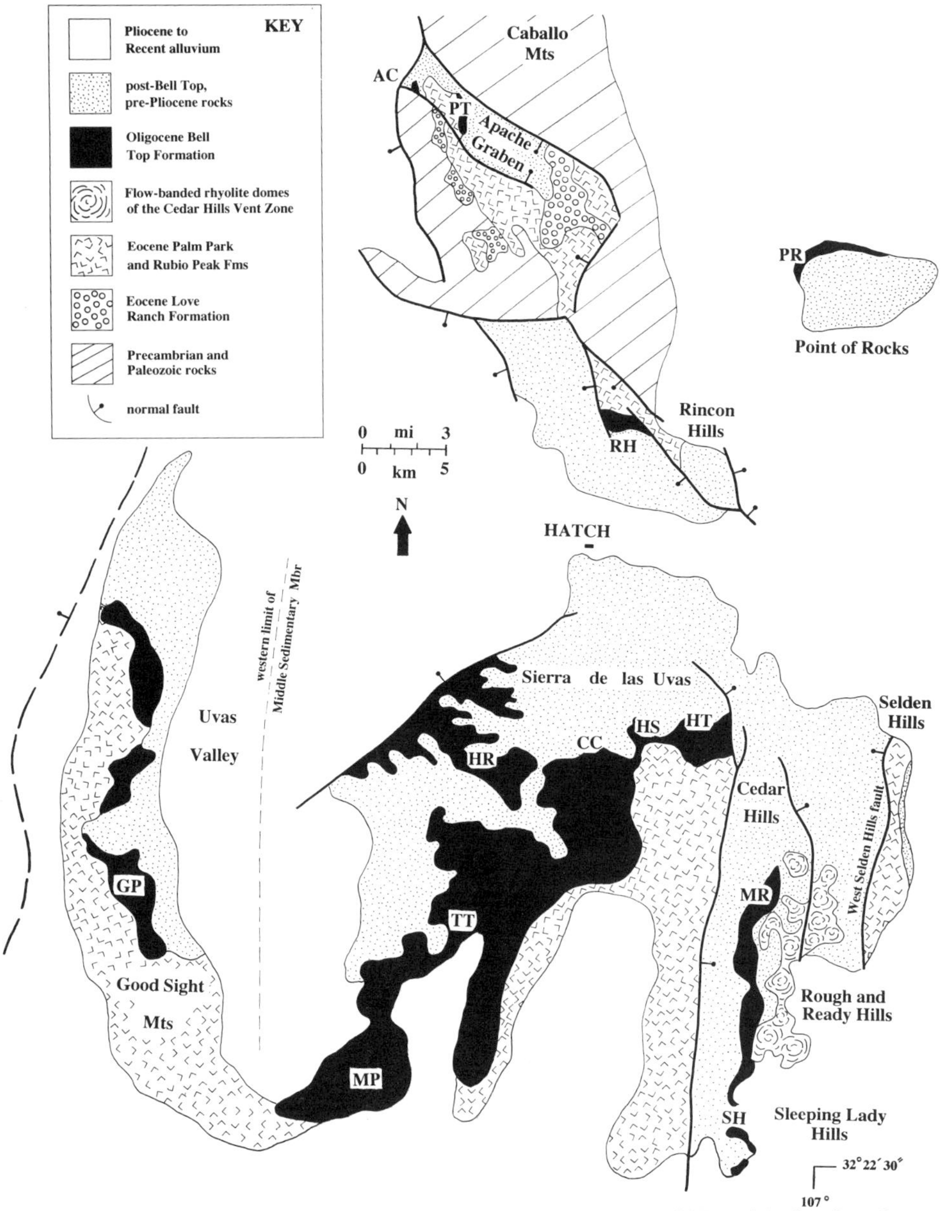

FIGURE 2. Generalized geologic map of south-central New Mexico, adapted from Seager et al., (1982, 1987), and the location of measured sections of the Bell Top Formation. Western limit of the Middle Sedimentary Member is from Clemons (1979). GP=Good Sight Peak, MP= Massacre Peak, TT=Tajanio Pinto Tank, HR= Horse Canyon Ranch, CC= Choases Canyon, HS= Hermantano Springs, HT= Hersey Tank, MR= McCall Reservoir, SH= Sleeping Lady Hills, RH= Rincon Hills, PR= Point of Rocks, AC= Apache Canyon, PT= Pass Tank.

Farther north, in Apache Graben of the Caballo Mountains, ash-flow tuff 6 is absent and Bell Top sedimentary rocks overlie the Kneeling Nun Tuff, which was derived from the Emory cauldron (Figs. 1-3; Elston et al., 1975; Seager and Mack, in press). The Kneeling Nun Tuff is very close in age to ash-flow tuff 5 and the two tuffs may be the same (McIntosh et al., 1990, 1991, 1992).

The Bell Top Formation unconformably overlies andesitic volcanic and volcaniclastic rocks of the Rubio Peak Formation in the Goodsight Mountains and the Palm Park Formation elsewhere (Fig. 3; Seager et al., 1982, 1987). Conformably underlying the Palm Park Formation is the Love Ranch Formation, which is a synorogenic Laramide conglomerate (Fig. 3; Seager, 1983; Seager and Mack, 1986; Seager et al., 1986). The Love Ranch Formation is thickest (up to 600 m) in the southern Caballo Mountains, where it and the lowermost part of the Palm Park Formation are composed of clasts of Paleozoic sedimentary rocks, Precambrian granitic and metamorphic rocks, and Cretaceous(?) andesitic rocks (Seager et al., 1986).

Conformably overlying the Bell Top Formation south of Apache Graben is the Uvas Basaltic Andesite, which has been radiometrically dated (K-Ar) at 27 to 28 Ma (Seager et al., 1984). At the Good Sight Peak, Point of Rocks, and McCall Reservoir measured sections, Uvas-type basalt flows are interbedded within the Upper Sedimentary Member of the Bell Top Formation. In Apache Graben, ash-flow tuff 7 of the Bell Top Formation is conformably(?) overlain by tuffaceous sandstones of the Thurman Formation, the normally intervening Uvas Basaltic Andesite having pinched out to the south (Figs. 2, 3; Seager and Mack, in press).

METHODS

Thirteen stratigraphic sections of the Bell Top Formation between ash- flow tuff 5 or Kneeling Nun Tuff and ash-flow tuff 7 or the Uvas Basaltic Andesite were measured in the study area. Eight of these sections had been measured previously, but were re-examined as part of this study (Fig. 2, SH, MR, RH, HR, HS, CC, TT, MP; Seager and Hawley, 1973; Clemons and Seager, 1973; Seager and Clemons, 1975; Seager et al., 1975; Clemons, 1976b, 1977, 1979). Sedimentologic data collected include thickness of syneruption and post-eruption lithofacies, measurements of the length of the A-axis of the 10 largest clasts within 39 post-eruption conglomerates, and imbrication measurements of 35 post-eruption conglomerates, involving measuring the strike and dip of the AB-plane of 10 to 30 clasts per bed. Compositional data were also collected from post-eruption conglomerates and syneruption sandstones. Three hundred boulder-, cobble- and pebble-sized clasts were counted at each of 39 beds of post-eruption conglomerate, with compositional categories including andesite, basalt, ash-flow tuff, flow-banded rhyolite, granite, foliated metamorphic rocks and sedimentary rocks. In addition, 14 medium-grained, syneruption sandstones were point counted, utilizing 300 points per thin section. Principal grain categories include monocrystalline quartz, polycrystalline quartz, potassium feldspar, plagioclase feldspar, volcanic rock fragments, glass shards and pumice.

LITHOFACIES

The Middle and Upper Sedimentary Members of the Bell Top Formation and the correlative Bell Top Formation in Apache Graben are divided into syneruption and post-eruption lithofacies (Figs. 4-6). Syneruption lithofacies, based on Smith (1991), refers to largely sand-sized fallout tephra that is reworked by sedimentary processes. As a result of Plinian eruptions, large volumes of relatively fine-grained fallout tephra mantles the countryside tens to hundreds of kilometers from the eruptive center. This material is quickly redistributed into local depocenters by gravity flows, running water and/or wind processes. Walton (1979, 1986) called this type of sediment 'active-eruption" or 'active-apron" facies, based on analysis of Oligocene rocks in west

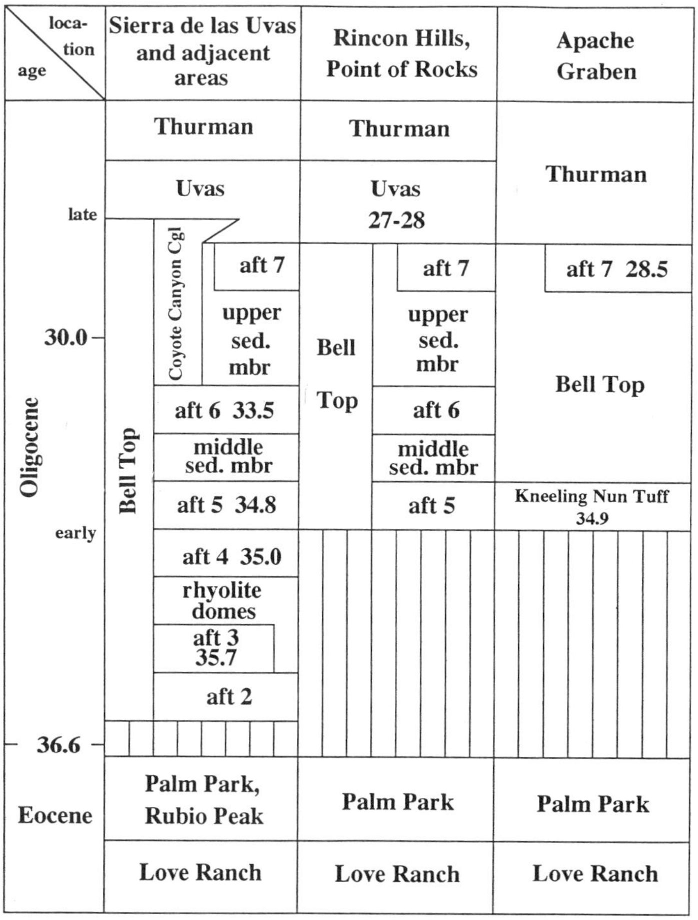

FIGURE 3. Correlation chart of Eocene and Oligocene stratigraphic units in south-central New Mexico. Numbers associated with ash-flow tuffs (aft), Kneeling Nun Tuff, and Uvas Basaltic Andesite are radiometric ages in million of years from Seager and others (1984) and McIntosh and others (1990, 1991, 1992). Age of Eocene-Oligocene boundary is from DNAG time scale (Palmer, 1983).

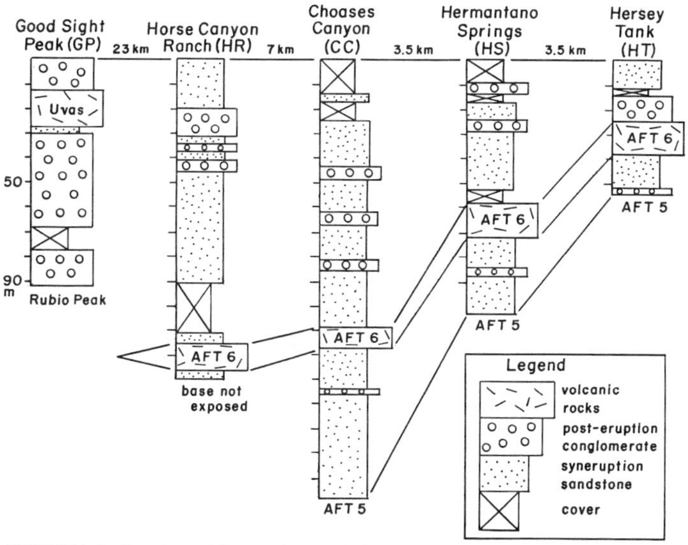

FIGURE 4. Stratigraphic sections of the Middle and Upper Sedimentary Members of the Bell Top Formation in the Good Sight Mountains and northern Sierra de las Uvas. Note differences in scale of Figures 4, 5 and 6.

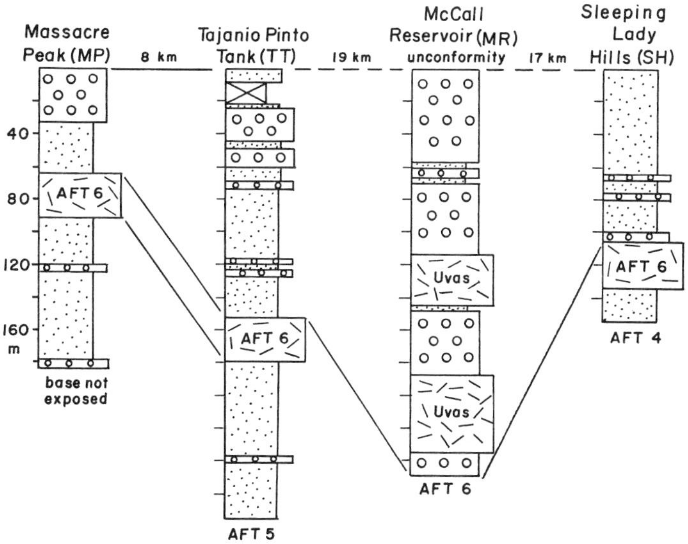

FIGURE 5. Stratigraphic sections of the Middle and Upper Sedimentary Members of the Bell Top Formation in the southern Sierra de las Uvas, Cedar Hills and Sleeping Lady Hills; see Fig. 4 for legend.

Texas. The studies of Walton (1979, 1986) are especially applicable to the Bell Top Formation, because both the volcaniclastic sediment in west Texas and in the present study area were derived from silicic calderas.

Once the fallout tephra is removed from upland regions, bedrock is exposed and supplies epiclastic sediment to adjacent basins. The epiclastic detritus is, in most cases, distinctly different than the syneruption sediment in terms of grain size and composition. Smith (1991) called these deposits "inter-eruption facies", whereas Walton (1979, 1986) used the terms "inactive-apron" or "post-eruption" phase; the term 'post-eruption lithofacies" is used in this study.

Syneruption lithofacies

Syneruption lithofacies of the Bell Top Formation consist of whitish to light gray or tan, fine- and medium-grained sandstone containing scattered granule- and pebble-sized clasts of white pumice. Sandstones generally have in excess of 50% glass shards and pumice, some of which are altered to clay and/or have devitrified to quartz or chert.

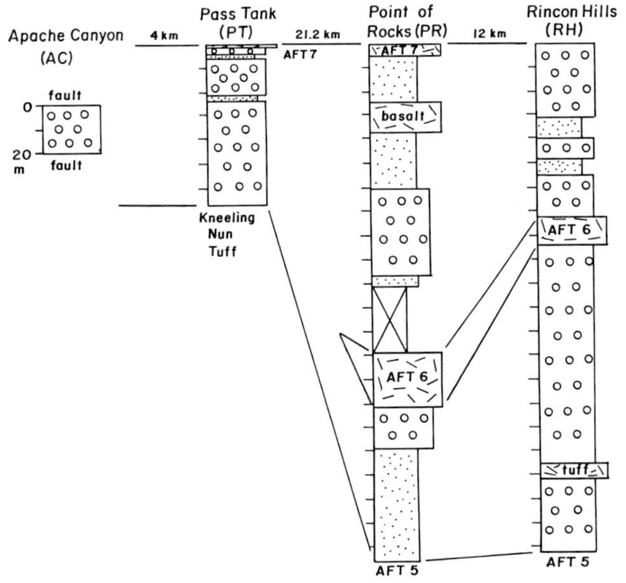

FIGURE 6. Stratigraphic sections of the Middle and Upper Sedimentary Members of the Bell Top Formation at Point of Rocks and in the Rincon Hills and of the Bell Top Formation at Apache Canyon and Pass Tank in the Caballo Mountains; see Fig. 4 for legend.

Shards are typically medium- or fine-sand sized, thin, and curved or Y-shaped, which are common products of rhyolitic or dacitic eruptions (Heiken, 1972). The remainder of the sand grains are primarily sanidine and plagioclase; volcanic rock fragments and quartz are rare (Fig. 7A).

The majority of syneruption sandstones are thin bedded and display horizontal laminae and/or broad (meters), shallow (centimeters) channels. Crossbeds are rare. Locally, pumice granules and pebbles are concentrated along the base of channels or constitute the base of small-scale (<20 cm thick) normally graded beds. The thin-bedded variety of syneruption lithofacies is interpreted to represent deposition by sheetfloods or by low sinuosity streams.

Less commonly, the syneruption lithofacies consists of thick-bedded sandstones that display no internal layering or sedimentary structures. In many cases, gravel-sized pumice clasts are randomly distributed in the sandstone matrix. Thick-bedded, structureless syneruption sandstones are interpreted to be debris-flow deposits.

The syneruption lithofacies is most abundant in the southern and central part of the basin, where it constitutes up to 80% of the sedimentary rocks (Figs. 4, 5). In the northern part of the basin, at Pass Tank and Rincon Hills, the syneruption lithofacies makes up less than 10% of the sedimentary strata (Fig. 6). Geographic variation in abundance of syneruption lithofacies may reflect a southerly source of the fallout tephra or more effective erosion of tephra from the northern part of the basin. The syneruption lithofacies is also less common along the margins of the basin (eg. Apache Canyon, Good Sight Peak, McCall Reservoir) than within the central part of the basin. This trend is most likely related to more effective removal of tephra along the basin margins due to steeper gradients.

Post-eruption lithofacies

The post-eruption lithofacies of the Bell Top Formation consists primarily of conglomerate with minor interbeds of sandstone and mudstone. Gravel-sized clasts are rounded to subrounded and vary in size from granules to boulders up to 1.5 m long dimension.

Conglomerates can be separated into five petrofacies based on composition of the clasts (Fig. 7B). The Coyote Canyon Conglomerate, exposed along the southeastern margin of the basin, constitutes the flow-banded rhyolite-clast petrofacies. The dominant clast type is dark brown to maroon flow-banded rhyolite and rhyolite, but also present are small amounts (<5 %) of basalt. The clasts of flow-banded rhyolite are probably derived from flow-banded rhyolite domes exposed in the Cedar Hills vent zone.

The most widespread petrofacies is the andesite-clast petrofacies, which consists primarily of clasts of purple or green andesite and andesite porphyry and secondarily of ash-flow tuff clasts (15%). The sources of the andesite clasts are the Palm Park and Rubio Peak Formations and the tuff clasts are probably reworked from ash-flow tuffs 5 and 6, based on hand-sample and petrographic comparison. Two stratigraphic sections, Hermantano Springs and Hersey Tank, contain conglomerates composed of subequal amounts of flow-banded rhyolite and andesite clasts and these define the mixed andesite-flow-banded rhyolite-clast petrofacies (Fig. 7B).

In the northern part of the study area, post-eruption conglomerates belong to the mixed igneous-sedimentary-metamorphic-clast petrofacies and consist of a mixture of andesite, andesite-porphyry, ash-flow tuff, granite, gneiss, schist, sandstone and carbonate (Fig. 7B). Andesite and andesite porphyry clasts are most likely derived from the Palm Park Formation and/or from Cretaceous andesitic rocks, whereas the tuff clasts resemble the Kneeling Nun Tuff. The remainder of the clasts are probably second-cycle clasts of Precambrian (granite, gneiss, schist) and Paleozoic (sandstone, carbonate) age derived from the Love Ranch Formation or from the conglomeratic base of the Palm Park Formation. A polycyclic origin for the Precambrian and Paleozoic clasts is supported by excellent rounding of the clasts.

The least common petrofacies, the tuff-clast petrofacies, is exposed in the Sleeping Lady Hills and consists primarily of ash-flow tuff clasts (90%), together with minor amounts of flow-banded rhyolite (5%) and basalt (5%).

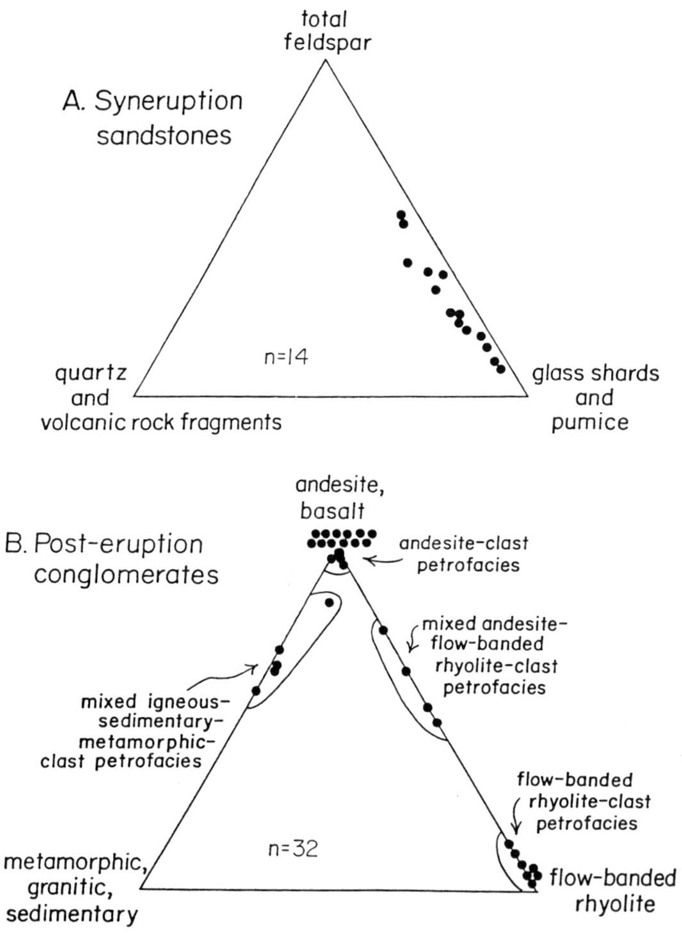

FIGURE 7. Composition of Bell Top sedimentary rocks. A, Triangular plot of the composition of syneruption sandstones, based on 300 points per thin section. B, Triangular plot of principal boulder, cobble and pebble clast types of post-eruption conglomerates, based on 300 counts per bed.

The majority of post-eruption conglomerates are grain supported, moderately well sorted, and locally exhibit imbrication. Meter-scale fining-upward sequences are present, as are thin lenses of coarse sandstone that display horizontal laminae and trough crossbeds. This type of conglomerate is waterlaid and probably represents deposition in high-gradient, low-sinuosity streams. Rare thin (<1 m) beds of red mudstone are interbedded with conglomerates, although a 10-m-thick mudstone, which was probably deposited on an alluvial flat or in a lake, is exposed at the Rincon Hills section.

Less commonly, the post-eruption lithofacies consists of thick-bedded, very poorly sorted conglomerate with no evidence of internal stratification or imbrication. This type of conglomerate, which is interpreted to represent debris flows, is most common along the margins of the basin at Goodsight Peak, McCall Reservoir, and Apache Canyon (Fig. 2).

RECONSTRUCTION OF THE GOODSIGHT-CEDAR HILLS HALF GRABEN

Asymmetrical distribution of lithofacies

The areal distribution of lithofacies in a half graben is markedly asymmetrical, in response to asymmetrical subsidence (Leeder and Gawthorpe, 1987; Blair and Bilodeau, 1988). Footwall-derived alluvial fans are small in area, because they are derived from small, steep drainage basins occupying the footwall scarp. In contrast, hanging wall drainage basins are much larger in area than footwall drainage basins, resulting in large hanging wall-derived alluvial fans that extend far down the hanging wall dip slope. The basin-axis lithofacies, which in a nonmarine depositional setting is commonly a lake or an axial drainage system, occupies a narrow belt above the zone of maximum subsidence near the footwall uplift. The asymmetrical distribution of lithofacies in half grabens has been observed in Neogene sedimentary basins and is a powerful tool in reconstructing the asymmetry of ancient extensional basins (Leeder and Alexander, 1987; Alexander and Leeder, 1987; Frostick and Reid, 1987; Mack and Seager, 1990; Leeder et al., 1991).

Provenance and dispersal directions of post-eruption conglomerates of the Bell Top Formation suggest that the GCH basin was an eastward-tilted half graben, with the Cedar Hills vent zone as the footwall and the Ancestral Goodsight uplift as the hanging wall (Figs. 8, 9). Outcrops of footwall-derived fanglomerate are represented by the Coyote Canyon Conglomerate, exposed in the east-central part of the study area. Extremely coarse grain size, westerly paleocurrents, and the predominance of flow-banded rhyolite clasts (flow-banded rhyolite-clast petrofacies) suggest that the Coyote Canyon Conglomerate is a proximal fan lithofacies derived from flow-banded rhyolite domes in the Cedar Hills vent zone. The flow-banded rhyolite-clast petrofacies cannot be traced westward into the Sierra de las Uvas, supporting the interpretation that this petrofacies was deposited on areally restricted, footwall-derived alluvial fans.

Most of the GCH basin is occupied by the andesite-clast petrofacies, which displays evidence of eastward paleocurrents and an eastward decrease in maximum clast size (Fig. 8). The andesite-clast petrofacies is interpreted to represent hanging wall-derived sediment, from a source of andesitic rocks of the Rubio Peak Formation in the Ancestral Goodsight uplift (Fig. 9). The hanging wall-derived fans extended at least 30 km down the hanging wall dip slope. The mixed igneous-sedimentary-metamorphic-clast petrofacies in the northern part of the basin represents a minor variation in provenance of the hanging wall-derived detritus. Precambrian granitic and metamorphic clasts and Paleozoic sedimentary clasts are of second-cycle origin and were derived from the base of the Palm Park Formation and/or from the Love Ranch Formation, which were not exposed farther south in the Ancestral Goodsight uplift. Thus, the change in provenance of the hanging wall-derived detritus reflects a change in the Eocene source rocks exposed on the hanging wall.

Outcrops at Hersey Tank and Hermantano Springs in the northern Sierra de las Uvas represent a basin-axis lithofacies, because of relatively fine grain size, paleocurrents that indicate northward paleoflow, a direction that roughly parallels the basin axis, and a composition that is a subequal mixture of both flow-banded rhyolite clasts derived from the footwall and andesite clasts derived from the hanging wall (Figs. 8, 9). The fact that the basin-axis lithofacies is not present in the Rincon Hills or at Point of Rocks suggests either that the drainage system changed to a northeasterly course or that the drainage system emptied into a lake located east of the Rincon Hills. No outcrops are available in this area to test these hypotheses.

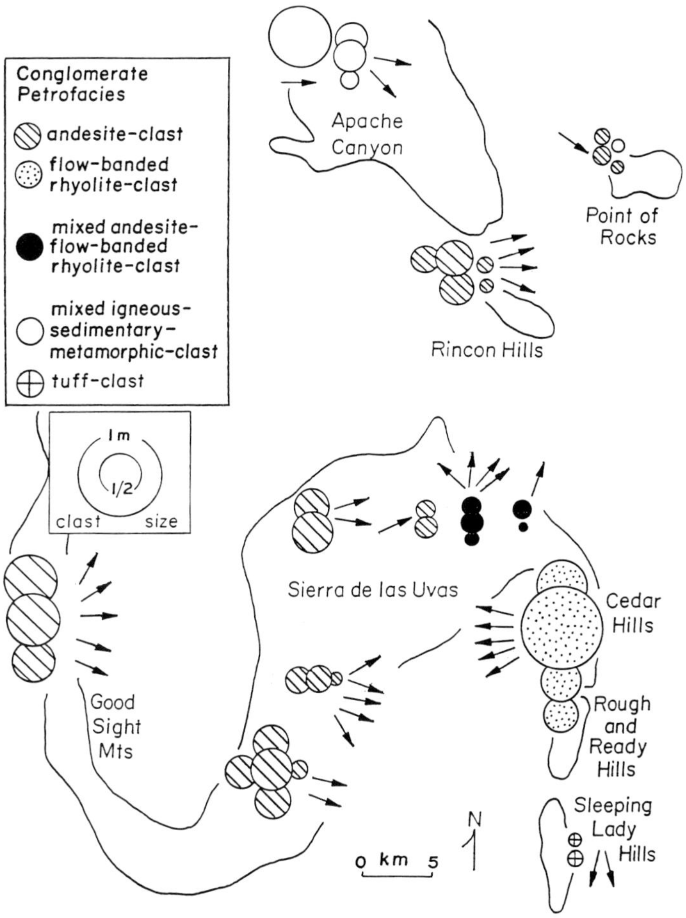

FIGURE 8. Sedimentologic and compositional data of post-eruption conglomerates of the Bell Top Formation. Size of circles represents average maximum clast size and pattern of circles corresponds to conglomerate petrofacies. Arrows represent vector mean paleoflow directions based on imbrication measurements.

The western extent of the Middle and Upper Sedimentary Members also provides evidence in support of a half graben. According to the model of Leeder and Gawthorpe (1987), progressive extension of a half graben causes sedimentary onlap of the hanging wall. This relationship is demonstrated in the Goodsight-Cedar Hills half graben by the fact that the Middle Sedimentary Member exhibits a depositional pinchout in the Uvas Valley, whereas the Upper Sedimentary Member extends at least 8 km farther up the hanging wall into the Good Sight Mountains (Fig. 2; Clemons, 1979).

Post-eruption conglomerates of the Bell Top Formation in the Sleeping Lady Hills do not easily fit the half graben model (Figs. 2, 9). These conglomerates are unique because of relatively fine grain size, paleocurrents indicating southward paleoflow, and a composition primarily of tuff clasts and secondarily of flow-banded rhyolite and basalt clasts. Perhaps the conglomerates in the Sleeping Lady Hills were deposited by axial drainage systems controlled by faults synthetic to the basin-bounding fault, as described in the East African rift by Frostick and Reid (1987).

Nature of faults

The boundary fault along the eastern margin of the GCH half graben is not presently exposed and is thus inferred in Figure 9.

However, the existence of a westward-dipping normal fault or fault zone at this location was suggested previously by Seager (1973, 1975) and Clemons (1976b), in order to account for the coarse grain size and provenance of the Coyote Canyon Conglomerate, for the absence of most ash-flow tuffs east of the inferred fault, and for the northerly alignment of the flow-banded rhyolite domes in the Cedar Hills vent zone.

The fault east of the boundary fault, known as the West Selden Hills fault, is a mappable westward-dipping normal fault (Figs. 2, 9; Seager et al., 1971; Seager and Clemons, 1975). The West Selden Hills fault experienced post-late Miocene movement, because it cuts the Rincon Valley Formation, which contains the 9.6 Ma Selden basalt. The fault is also inferred to have experienced pre-early Miocene movement, because the Eocene Palm Park Formation on the footwall is onlapped by the early Miocene Hayner Ranch Formation. Thus, the Palm Park Formation was uplifted and younger Oligocene volcanic and volcaniclastic rocks either were not deposited above or were eroded off of the Palm Park Formation prior to early Miocene onlap. The age, trend and dip of the West Selden Hills fault suggest that it may be synthetic to the boundary fault of the GCH half graben.

Also shown on Figure 9 is an inferred fault that marks the western boundary of the Ancestral Goodsight uplift. The existence of this fault is based on a linear zone of steep Bouguer gravity gradient displaying a range of up to 200 milligals (Decker and Smithson, 1975; Daggett and Keller, 1982). In its southern part, the gravity gradient corresponds to the margin of the late-rift (post-late Miocene) Good Sight Mountains, but its northern segment cuts across the late-rift Palomas basin. This latter relationship, together with the large magnitude of the gravity anomaly, suggests that the gravity gradient defines the western margin of an early-rift uplift, the Ancestral Goodsight uplift. However, the exact age of the northern part of the gravity gradient cannot be determined at this time and the possibility exists that it postdates development of the GCH half graben.

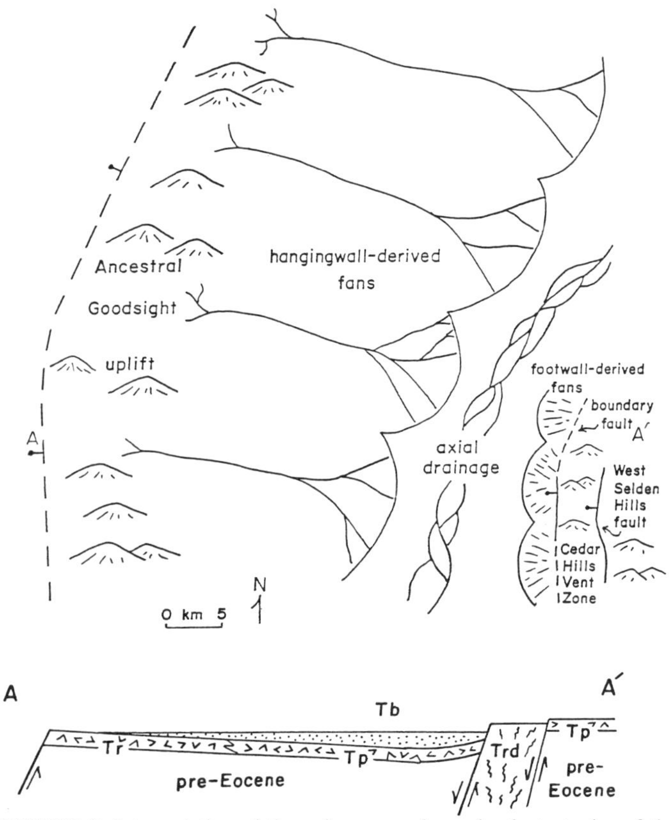

FIGURE 9. Interpretation of the paleogeography and paleotectonics of the Goodsight-Cedar Hills half graben from 35 to 28 Ma. Tp=Palm Park Formation; Trd=flow-banded rhyolite domes; Tr=Rubio Peak Formation; Tb=Bell Top Formation.

Alternative model

We evaluate here an alternative model that does not involve a half graben to explain the distribution of epiclastic conglomerates and sandstones of the Bell Top Formation. In this model, the east-southeastward dispersal system is the distal part of a volcanic apron derived from Palm Park/Rubio Peak andesitic volcanic centers, located approximately 30 km west of the western edge of the GCH basin in the Black Range and Cookes Range. Furthermore, exposure of the rhyolite domes along the eastern margin of the basin represents original relief associated with intrusion, rather than relief created by faulting.

There are several problems with this model that make it less viable than the half graben model. Specific problems related to derivation of Bell Top conglomerates from Palm Park/Rubio Peak volcanic centers are: (1) Bell Top epiclastic conglomerates along the western edge of the GCH basin are too coarse grained to have been derived from the volcanic centers in question; (2) A volcanic source cannot easily account for the Precambrian and Paleozoic clasts reworked from the Love Ranch Formation; (3) There appears to have been a tectonic/topographic reorganization between Palm Park/Rubio Peak time and Bell Top time, as indicated by an unconformity between the units, which in the Good Sight Mountains is angular (Clemons, 1979); and (4) Palm Park/Rubio Peak volcanic aprons contain abundant lahars, which are absent in Bell Top andesite-clast conglomerates.

Unfaulted rhyolite domes along the eastern flank of the GCH basin also present problems. Unfaulted domes should be bordered by a broad, gently-sloping apron of detritus that extends many tens of kilometers into the basin, when in fact the flow-banded rhyolite petrofacies is restricted to a narrow belt adjacent to the domes. Also it would appear to defy normal denudation rates for rhyolite domes to remain topographically high enough to supply coarse gravel to the GCH basin for up to 7 Ma after their intrusion.

IMPLICATIONS TO THE HISTORY OF THE MOGOLLON-DATIL VOLCANIC FIELD

Volcanic and volcaniclastic rocks of the Bell Top Formation in the GCH half graben provide important information about the history of the Mogollon-Datil volcanic field. Because of their position along the southeastern margin of the volcanic field, ash-flow tuffs of the Bell Top Formation in the GCH half graben delimit the maximum lateral extent of regional outflow sheets. Based on $^{40}Ar/^{39}Ar$ ages and paleomagnetism, McIntosh et al., (1990, 1991, 1992) correlated Bell Top tuffs with other ignimbrites and their source cauldrons in the Mogollon-Datil volcanic field. These correlations expand the geographic range of inferred individual outflow sheets by up to 25 percent, particularly in the cases of ash-flow tuffs 5, 6 and 7. Ash-flow tuff 5 is equated to the Kneeling Nun Tuff, which was derived from the Emory cauldron. Ash-flow tuff 6 is correlated with the Box Canyon Tuff, the source of which was probably the Schoolhouse Mountain cauldron, approximately 150 km west-northwest of the GCH basin. Ash-flow tuff 7, which is the distal part of the Vicks Peak Tuff, erupted from the Nogal Canyon cauldron, located about 100 km north of the GCH basin. However, uncertainties remain regarding these correlations, especially with respect to ash-flow tuffs 5 and 6. Mineralogic and geochemical data are needed to test existing correlations.

Syneruption, tuffaceous sandstones of the Bell Top Formation in the GCH half graben also provide potentially important information about the history of the Mogollon-Datil volcanic field. Tuffaceous sandstones, which represent reworked fallout tephra, record Plinian-type eruptions that occurred contemporaneous with or between ignimbrite-generating events. These sandstones may have been related to major ignimbrite events or to eruptions that did not generate major ignimbrite sheets, but rather generated fallout tephra, which was reworked as tuffaceous sandstones in the GCH basin. Systematic radiometric dating and geochemical analyses of the tuffaceous sandstones will probably be required to fully understand their significance to the evolution of the Mogollon-Datil volcanic field.

IMPLICATION TO EVOLUTION OF THE SOUTHERN RIO GRANDE RIFT

The interpretation of the GCH basin as a half graben has important implications to the evolution of the southern Rio Grande rift. The GCH half graben may be an indicator of regional crustal extension, because it does not appear to be spatially or temporally related to caldera development, although the half graben periodically received ash-flow tuffs and fallout tephra from extrabasinal sources. The asymmetrical distribution of lithofacies of the Middle and Upper Sedimentary Members of the Bell Top Formation and their correlatives suggests that the half graben had already developed at the time of or immediately after eruptions of ash-flow tuff 5 / Kneeling Nun Tuff at about 35 Ma. The interpretation of Seager (1973, 1975) that the flow-banded rhyolite domes of the Cedar Hills vent zone, which are roughly coeval with ash-flow tuff 3 (Fig. 3), were emplaced along a fault or fault zone may indicate that extension began about a million years before eruption of ash- flow tuff 5 / Kneeling Nun Tuff. This idea cannot easily be tested by sedimentologic data, however, because Bell Top sedimentary rocks older than ash-flow tuff 5 are poorly exposed and restricted to the southern part of the study area.

Interpretation of the initiation of regional extension at approximately 35 Ma is 2 to 7 million years older than previous interpretations of the onset of extension in the southern and central Rio Grande rift (Chapin and Seager, 1975; Seager et al., 1984; Newcomer and Giordano, 1986). However, the interpretation presented here corresponds closely to the recent model of Cather (1986, 1990) and Cather and Chapin (1989), who suggested that the change from regional Laramide compression to extension in the central Rio Grande rift occurred at about 36 Ma, based on displacement reversals across relict Laramide fault zones and the change from intermediate to bimodal volcanism.

Interpretation of paleostress orientation during development of the GCH half graben is not possible because the boundary fault is not exposed. It does seem safe to say, however, that the GCH half graben does not represent reactivation of Laramide structures. The axis of GCH half graben is almost perpendicular to the west-northwest-trending late Laramide Rio Grande uplift and complementary Love Ranch basin in the same region (Seager, 1983; Seager and Mack, 1986; Seager et al., 1986). It also appears clear, based on the lack of exposure of pre-Eocene rocks in the hanging wall and footwall uplifts and relatively slow sediment accumulation rates (maximum, uncompacted rate of 40 m/Ma), that extension rates were probably not rapid during development of the half graben.

CONCLUSIONS

The Oligocene Goodsight-Cedar Hills basin in south-central New Mexico is interpreted to have been an eastward-tilted half graben, in which several hundred meters of ash-flow tuffs, syneruption tuffaceous sandstones and post-eruption conglomerates of the Bell Top Formation were deposited. Exposed in the footwall of the half graben were flow-banded rhyolite domes of the Cedar Hills vent zone, whereas exposures in the hanging wall, referred to as the Ancestral Goodsight uplift, consisted primarily of Eocene andesitic rocks and locally in its northern part of Eocene synorogenic conglomerates.

The GCH half graben is an important component of the Mogollon-Datil volcanic field, not only because regional ignimbrite outflow sheets accumulated within it, but because syneruption, tuffaceous sandstones, which represent reworked fallout tephra, may provide a record of explosive volcanism that occurred during or between other major ignimbrite-generating events. Regional crustal extension responsible for development of the half graben began about 35 Ma, 2 to 7 Ma earlier than previous interpretations of the onset of crustal extension in the southern Rio Grande rift.

ACKNOWLEDGMENTS

W.C. James, E. Timothy Wallin, Steven M. Cather and Richard M. Chamberlin read an earlier version of this manuscript and made helpful suggestions for its improvement.

REFERENCES

Alexander, J.A. and Leeder, M.R., 1987, Active tectonic control of alluvial architecture: Society of Economic Paleontologists and Mineralogists, Special Publication 39, p. 243-252.

Blair, T.C. and Bilodeau, W.L.,1988, Development of tectonic cyclothems in rift, pull-apart, and foreland basins: sedimentary response to episodic tectonism: Geology, v. 16, p. 517-520.

Cather, S.M., 1986, Volcano-sedimentary evolution and tectonic implications of the Datil Group (latest Eocene-early Oligocene), west-central New Mexico [Ph.D. dissertation]: Austin, University of Texas, 484 p.

Cather, S.M., 1990, Stress and volcanism in the northern Mogollon-Datil volcanic field, New Mexico: effects of post-Laramide tectonic transition: Geological Society of America Bulletin, v. 102, p. 1447-1458.

Cather, S.M. and Chapin, C.E., 1989, Field guide to upper Eocene and lower Oligocene volcaniclastic rocks of the northern Mogollon-Datil volcanic field: New Mexico Bureau of Mines and Mineral Resources, Memoir 46, p. 623-652.

Chapin, C.E. and Seager, W.R., 1975, Evolution of the Rio Grande rift in the Socorro and Las Cruces areas: New Mexico Geological Society, Guidebook 26, p. 297-321.

Clemons, R.E., 1975, Petrology of the Bell Top Formation: New Mexico Geological Society, Guidebook 26, p. 123-130.

Clemons, R.E., 1976a, Sierra de las Uvas ash-flow field, south-central New Mexico: New Mexico Geological Society, Special Publication No. 6, p. 115-121.

Clemons, R.E., 1976b, Geology of East Half Corralitos Ranch quadrangle, New Mexico: New Mexico Bureau of Mines and Mineral Resources, Geologic Map 36.

Clemons, R.E., 1977, Geology of West Half Corralitos Ranch quadrangle, New Mexico: New Mexico Bureau of Mines and Mineral Resources, Geologic Map 44.

Clemons, R.E., 1979, Geology of Good Sight Mountains and Uvas Valley, southwest New Mexico: New Mexico Bureau of Mines and Mineral Resources, Circular 169, 32 p.

Clemons, R.E. and Seager, W.R., 1973, Geology of Souse Springs quadrangle, New Mexico: New Mexico Bureau of Mines and Mineral Resources, Bulletin 100, 31 p.

Daggett, P.H. and Keller, G.R., 1982, Complete Bouguer anomaly map of northwest part of Las Cruces 1° X 2° sheet; *in* Seager, W.R., Clemons, R.E., Hawley, J.W. and Kelley, R.E., Geology of northwest part of Las Cruces 1° X 2° sheet: New Mexico Bureau of Mines and Mineral Resources, Geologic Map 53.

Decker, E.R. and Smithson, S.B., 1975, Heat flow and gravity interpretation across the Rio Grande rift in southern New Mexico and west Texas: Journal of Geophysical Research, v. 80, p. 2542-2552.

Elston, W.E., Seager, W.R. and Clemons, R.E., 1975, Emory cauldron, Black Range, New Mexico, source of the Kneeling Nun Tuff: New Mexico Geological Society, Guidebook 26, p. 283-292.

Frostick, L.E. and Reid, I., 1987, Tectonic control of desert sediments in rift basins ancient and modern: Geological Society, Special Publication No. 35, p. 53-68.

Heiken, G., 1972, Morphology and petrography of volcanic ashes: Geological Society of America Bulletin, v. 83, p. 1961-1988.

Kottlowski, F.E., 1953, Tertiary-Quaternary sediments of the Rio Grande valley in southern New Mexico: New Mexico Geological Society, Guidebook 4, p. 144-148.

Leeder, M.R. and Alexander, J., 1987, The origin and tectonic significance of asymmetrical meanderbelts: Sedimentology, v. 34, p. 217-226.

Leeder, M.R. and Gawthorpe, R.L., 1987, Sedimentary models for extensional tilt-block/half-graben basins: Geological Society, Special Publication No. 28, p. 139-152.

Leeder, M.R., Seger, M.J. and Stark, C.P., 1991, Sedimentation and tectonic geomorphology adjacent to major active and inactive normal faults, southern Greece: Journal of the Geological Society, London, v. 148, p. 331-343.

Mack, G.H. and Seager, W.R., 1990, Tectonic control on facies distribution of the Camp Rice and Palomas Formations (Pliocene-Pleistocene) in the southern Rio Grande rift: Geological Society of America Bulletin, v. 102, p. 45-53.

McIntosh, W.C., Sutter, J.F., Chapin, C.E., Osburn, G.R. and Ratté, J.C., 1986, A stratigraphic framework for the Mogollon-Datil volcanic field based on paleomagnetism and high-precision ^{40}Ar/^{39}Ar dating of ignimbrites—a progress report: New Mexico Geological Society, Guidebook 37, p. 183-195.

McIntosh, W.C., Sutter, J.F., Chapin, C.E. and Kedzie, L.L., 1990, High-precision ^{40}Ar/^{39}Ar sanidine geochronology of ignimbrites in the Mogollon-Datil volcanic field, southwestern New Mexico: Bulletin of Volcanology, v. 52, p. 584-601.

McIntosh, W.C., Kedzie, L.L. and Sutter, J.F., 1991, Paleomagnetism and $^{40}Ar/^{39}Ar$ ages of ignimbrites, Mogollon-Datil volcanic field, southwestern New Mexico: New Mexico Bureau of Mines and Mineral Resources, Bulletin 135, 79 p.

McIntosh, W.C., Chapin, C.E., Ratté, J.C. and Sutter, J.F., 1992, Time-stratigraphic framework for the Eocene-Oligocene Mogollon-Datil volcanic field, southwest New Mexico: Geological Society of America Bulletin, v. 104, p. 851-871.

Newcomer, R.W. and Giordano, T.H., 1986, Porphyry-type mineralization and alteration in the Organ mining district, south-central New Mexico: New Mexico Geology, v. 8, p. 83-86.

Palmer, A.R., 1983, (compiler), Decade of North American Geology, 1983 Time Scale: Geology, v. 11, p. 503-504.

Seager, W.R., 1973, Resurgent volcano-tectonic depression of Oligocene age, south-central New Mexico: Geological Society of America Bulletin, v. 84, p. 3611-3626.

Seager, W.R., 1975, Cenozoic tectonic evolution of the Las Cruces area, New Mexico: New Mexico Geological Society, Guidebook 26, p. 241-250.

Seager, W.R., 1981, Geology of the Organ Mountains and southern San Andres Mountains, New Mexico: New Mexico Bureau of Mines and Mineral Resources, Memoir 36, 97 p.

Seager, W.R., 1983, Laramide wrench faults, basement-cored uplifts, and complementary basins in southern New Mexico: New Mexico Geology, v. 5, p. 69-76.

Seager, W.R. and Clemons, R.E., 1975, Middle to late Tertiary geology of Cedar Hills-Selden Hills area, New Mexico: New Mexico Bureau of Mines and Mineral Resources, Circular 133, 24 p.

Seager, W.R. and Hawley, J.W., 1973, Geology of the Rincon quadrangle, Doña Ana County, New Mexico: New Mexico Bureau of Mines and Mineral Resources, Bulletin 101, 42 p.

Seager, W.R. and Mack, G.H., 1986, Laramide paleotectonics of southern New Mexico: American Association of Petroleum Geologists, Memoir 41, p. 669-685.

Seager, W.R. and Mack, G.H.(in press), Geology of McLeod Tank quadrangle, Sierra and Doña Ana Counties, New Mexico: New Mexico Bureau of Mines and Mineral Resources.

Seager, W.R., Clemons, R.E. and Hawley, J.W., 1975, Geology of Sierra Alta quadrangle, New Mexico: New Mexico Bureau of Mines and Mineral Resources, Bulletin 102, 56 p.

Seager, W.R., Hawley, J.W. and Clemons, R.E., 1971, Geology of San Diego Mountain area, Doña Ana County, New Mexico: New Mexico Bureau of Mines and Mineral Resources, Bulletin 97, 38 p.

Seager, W.R., Kottlowski, F.E. and Hawley, J.W., 1976, Geology of Doña Ana Mountains, New Mexico: New Mexico Bureau of Mines, Circular 147, 36 p.

Seager, W.R., Clemons, R.E., Hawley, J.W. and Kelley, R.E., 1982, Geology of northwest part of Las Cruces 1° X 2° sheet, New Mexico: New Mexico Bureau of Mines and Mineral Resources, Geologic Map 53.

Seager, W.R., Shafiqullah, M., Hawley, J.W. and Marvin, R.F., 1984, New K-Ar dates from basalts and the evolution of the southern Rio Grande rift: Geological Society of America Bulletin, v. 95, p. 87-99.

Seager, W.R., Mack, G.H., Raimonde, M.S. and Ryan, R.G., 1986, Laramide basement- cored uplift and basins in south-central New Mexico: New Mexico Geological Society, Guidebook 37, p. 123-130.

Seager, W.R., Hawley, J.W., Kottlowski, F.E. and Kelley, S.A., 1987, Geology of the east half of Las Cruces and northeast El Paso 1° X 2° sheets, New Mexico: New Mexico Bureau of Mines and Mineral Resources, Geologic Map 57.

Smith, G.A., 1991, Facies sequences and geometries in continental volcaniclastic sediments: Society of Economic Paleontologists and Mineralogists, Special Publication 45, p. 109-121.

Walton, A.W., 1979, Volcanic sediment apron in the Tascotal Formation (Oligocene?), Trans-Pecos Texas: Journal of Sedimentary Petrology, v. 49, p. 303-314.

Walton, A.W., 1986, Effect of Oligocene volcanism on sedimentation in the Trans-Pecos volcanic field of Texas: Geological Society of America Bulletin, v. 97, p. 1192-1207.

CRUSTAL STRUCTURE OF WEST-CENTRAL NEW MEXICO: A PRELIMINARY SEISMIC INTERPRETATION

DONALD G. ROBERTS, DONALD C. ADAMS and G. RANDY KELLER
Department of Geological Sciences,The University of Texas at El Paso, El Paso, TX 79968

Abstract—A new, unreversed seismic profile was recorded in the region from Socorro to Datil, New Mexico in June 1993. These data, when interpreted in the context of previously existing data, provide a new picture of the transition zone between the Colorado Plateau and Rio Grande rift in western New Mexico. Crustal structure characteristic of the rift extends at least as far west as Magdalena, New Mexico before the crust appears to gradually thicken into the Colorado Plateau. This broad transition zone is generally similar to that found in recent studies of the boundary between the Basin and Range province and Colorado Plateau in western Arizona.

INTRODUCTION

As part of the University of Texas at El Paso's ongoing investigations of crustal structure in the Rio Grande rift region, our group has been recording seismic refraction data along a profile extending between the White Sands Missile Range (WSMR), New Mexico, and the Nevada Test Site (NTS). As part of this effort, we deployed a seismic array in June 1993 to record the Minor Uncle conventional surface explosion at WSMR. Sixty-eight stations were deployed at 1 mi intervals along US Highway 60 beginning 4 mi west of Socorro, New Mexico, and ending 10 mi west of Datil, New Mexico (Fig. 1). Each station consisted of a 1 Hz, vertical component seismometer and a USGS Seismic Cassette Recorder (SCR). The Minor Uncle experiment went off as scheduled and 61 SCRs recorded a strong signal. The resultant record section shows excellent signal-to-noise ratios and exhibits several prominent arrivals (Fig. 2). Taken alone, this unreversed profile would be of limited utility. However, when combined with other seismic data in the region it provides a useful constraint on crustal structure in the area. In this paper, we present a preliminary interpretation of these new data and update our analysis of regional crustal structure. We hope to gather additional data along this profile over the next few years.

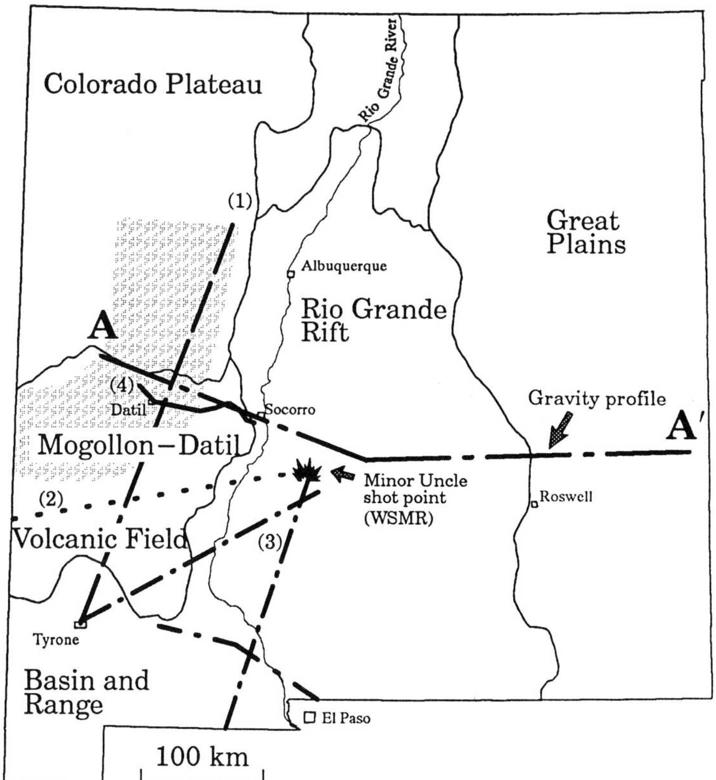

FIGURE 1. Index map of New Mexico showing locations of seismic profiles and geophysical models. 1 - Seismic profile of Schneider and Keller (in press); 2 - seismic profile of Jaksha (1982); 3 - seismic profiles of Sinno et al. (1986); 4 - seismic profile of this study. Shaded area is the zone of transition in crustal structure of Schneider and Keller (in press) as modified in this study. Profile A-A' represents the gravity data modeled in Figure 5.

PREVIOUS WORK

Previous surveys in the region have helped to constrain our seismic model. Sinno et al. (1986) conducted a series of seismic surveys in the southern Rio Grande rift utilizing explosions at WSMR and at the copper mines near Tyrone, New Mexico (Fig. 1). Their reversed NE-SW trending profile indicated a relatively flat Moho at a depth of 32 km between Tyrone and the northern portion of WSMR, with a slight increase in the depth to the Moho to the southwest.

Jaksha (1982) monitored another WSMR explosion with a widely spaced seismic array extending westward from WSMR to Morenci, Arizona. He reversed his profile by recording blasts from the copper mine at Morenci, and interpreted his results in terms of flat-lying horizontal layers. These data were reinterpreted by Schneider (1990) and Keller and Schneider (in press) who combined the Jaksha (1982) data with their seismic and gravity data gathered along a profile extending from Tyrone to Grants, New Mexico. These two profiles cross in the

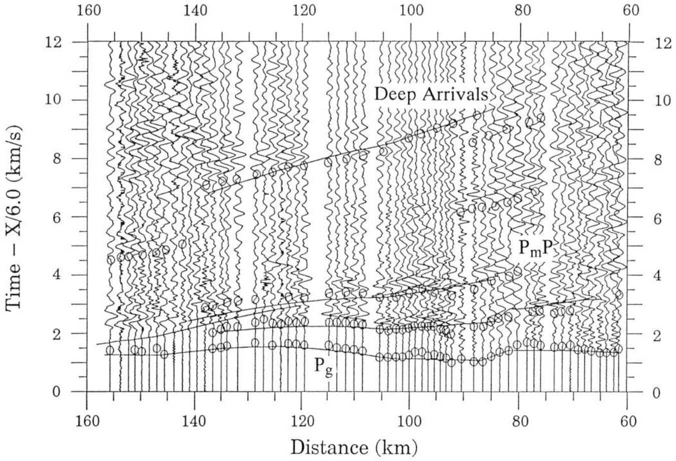

FIGURE 2. Seismic record section of the Minor Uncle data. The picked arrival times of the main phases modeled are identified by circles and the modeled arrival times are shown as the line. These arrivals are also shown in Figure 4 and are explained in the text.

central portion of the Mogollon-Datil volcanic field. They used two-dimensional modeling techniques to interpret the new seismic and gravity data and to reinterpret Jaksha's data. Their work indicated the presence of an upper crustal intrusive complex beneath the Mogollon-Datil volcanic field (MDVF) and under at least part of the San Agustin plains, as well as a mostly gradual south to north thickening of the crust (Fig. 3). Their analysis suggests a gradual crustal transition between the Colorado Plateau and the Rio Grande rift which extends under the MDVF (Fig. 1).

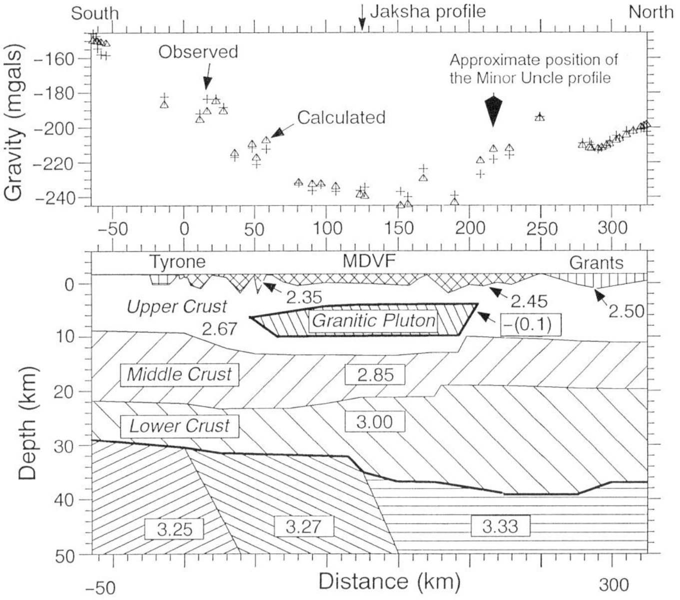

FIGURE 3. Earth model derived by Schneider and Keller (in press) for the Tyrone to Grants profile (Fig. 1). Numbers indicate densities in g/cc. Modified from Schneider and Keller (in press).

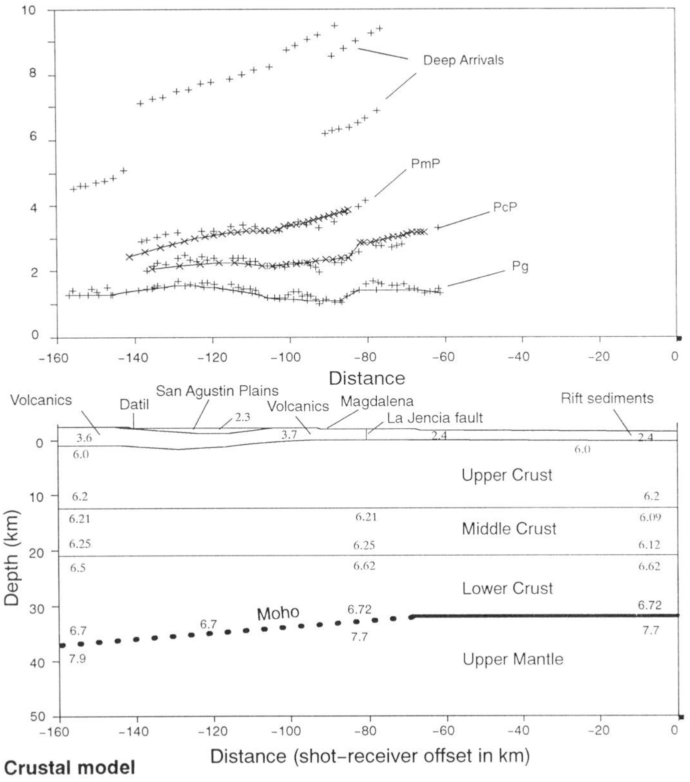

FIGURE 4. Picked arrival times (from Fig. 2) and preliminary earth model derived from ray tracing the Minor Uncle seismic data. Numbers indicate P-wave velocities in km/s.

Adams and Keller (in press) investigated crustal structure in central New Mexico by integrating seismic, gravity and geologic data. They delineated the general basin structure of the region and found that the transition in crustal structure from the Rio Grande rift to Colorado Plateau across the Datil region was gradual.

DISCUSSION

The data presented here complement the previous studies in the area. The seismic record section (Fig. 2) shows clear phase arrivals for P_g (refraction from the upper crust), P_cP (reflection from the middle crust), and P_mP (reflection from the Moho). A P_n (refraction from the upper mantle) arrival is also present on the westernmost traces. Each of these phases were modeled by ray tracing employing the previous work of Sinno et al. (1986) and Schneider and Keller (in press) as independent constraints. A preliminary earth model is shown in Figure 4.

With a maximum source-receiver offset of 155 km, these data do not provide any constraint on the upper mantle velocity structure and contain only a limited amount of information on the structure of the crust/mantle transition. However, these data do suggest that the relatively thin crust (~ 32 km) interpreted by Sinno et al. (1986) within the central part of the rift extends westward to an offset of at least 70 km in this survey. This is approximately the position of the La Jencia fault (Fig. 4). The seismic profile of Schneider and Keller (in press) shows that the crustal thickness is about 38 km at Datil, and thus requires the westward increase in Moho depth shown in Figure 4.

The shallow crustal features in Figure 4 are noticeably simplistic due to the lack of fine resolution contained in the data. Each of the crustal phases, especially the P_g arrival, are affected by several shallow velocity contrasts. The largest contrasts are associated with the transition from thick, central-rift, basin-fill sediments in the east to Tertiary volcanic sequences and thin basin-fills in the west; and the subsidence and sedimentation that form the San Agustin plains. The first transition occurs at a source-receiver offset of approximately 80 km and shows up as a sharp decrease in phase arrival time with increasing offset. This transition is located just east of the Magdalena Mountains at the basin-bounding La Jencia fault. The San Agustin plains show up as a broad travel time delay centered at an offset of 130 km. This anomaly was modeled with approximately 1 km of basin sediments and a slight thickening of the volcanic sequences to the west. No compelling evidence for an upper crustal batholith was found, nor was expected based on the interpreted extent of the Mogollon-Datil batholith presented by Schneider and Keller (in press). Two other regions of the section indicate the presence of shallow crustal features that have not been incorporated into the model at this time. Two subtle phase arrival delays are present between 70 and 80 km (Figs. 2, 4). These are interpreted to be associated with the sediments of the La Jencia basin in the west and the Socorro basin in the east, which are separated from each other by the Socorro-Lemitar Mountains. West of the La Jencia fault (at approximately 100 km offset), a third phase delay is interpreted as the Abbey Springs basin sediments (R. Chamberlin, personal commun., 1994).

These data also indicate one and possibly two deep phases arriving significantly later than the crustal phases (Fig. 2). The timing of these arrivals is inconsistent with a P_mP crustal multiple but can be fitted with a west-dipping reflector in the upper mantle. The data suggest that there may be at least two such upper mantle reflectors.

In order to place these results in a regional context, a crustal model that crosses the state of New Mexico was constructed by gravity modeling (Fig. 5). This model is consistent with the seismic results presented here, and corroborates the conclusions of Schneider and Keller (in press) that the transition in crustal structure between the Rio Grande rift and the Colorado Plateau occurs over a broad region in western New Mexico. This model also illustrates our interpretation that the eastern (Roberts et al., 1991) and western margins of the rift are quite different, with the change in structure across the western margin being more gradual.

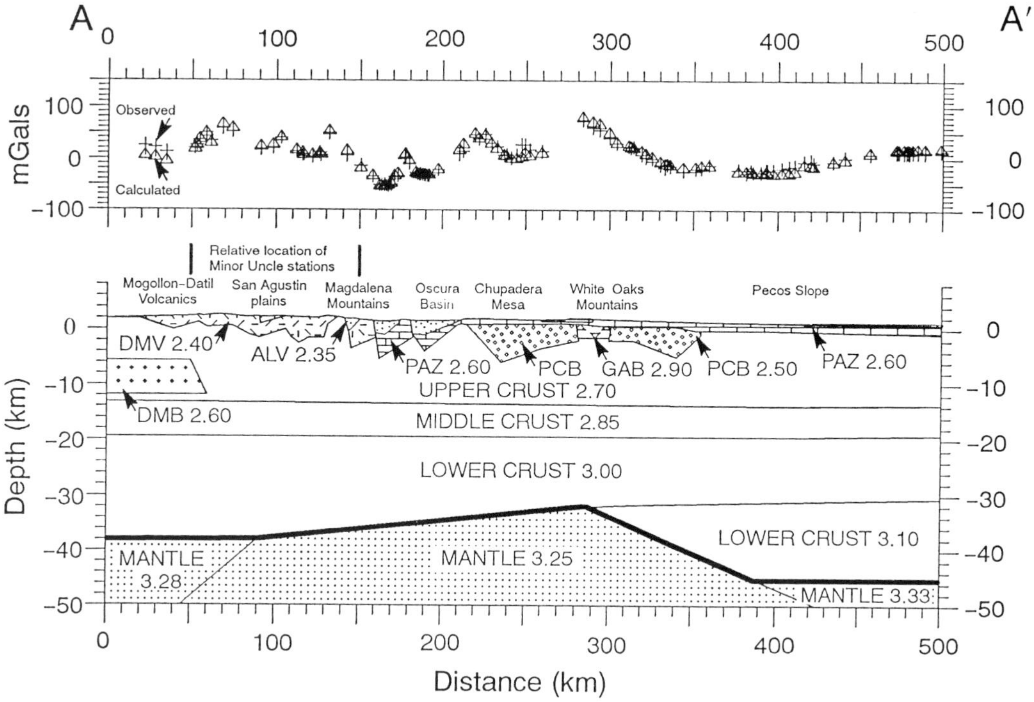

FIGURE 5. Free air gravity model across New Mexico (A-A', Fig. 1). This model is constrained by deep petroleum exploration wells and the results of seismic refraction modeling of Roberts et al. (1991), Sinno et al. (1986), and Olsen et al. (1979). MDV - Mogollon-Datil volcanics, MDB - Mogollon-Datil batholith, ALV - Alluvium, PAZ - Paleozoic, PCB - Possible Precambrian or Paleozoic basin, GAB - Gabbroic intrusion.

CONCLUSIONS

The transition zone from the Rio Grande rift to the Colorado Plateau in western New Mexico is gradual in terms of crustal structure. However, thin crust associated with the rift extends westward to about the location of Magdalena, New Mexico. The preliminary crustal model shown in Figure 4 is similar to recent results across the transition zone between the Basin and Range province and Colorado Plateau in western Arizona (Jill McCarthy, personal commun., 1994), but needs further constraints to the west. Our future investigations will focus on a more detailed investigation of the seismic data shown here and on the underlying causes for variations in crustal structure along the margins of the Colorado Plateau and Rio Grande rift.

ACKNOWLEDGMENTS

This research was supported by contract number 49620-92-J-0438 from the Air Force Office of Scientific Research. We express our thanks to landowners in the region for allowing access to their property and to the White Sands operations staff for their cooperation. Helpful reviews were provided by Mark Ouimette, Michael Whitelaw, Alan Sanford, and in particular Richard Chamberlin.

REFERENCES

Adams, D. C. and Keller, G. R. (in press), Crustal structure and basin geometry in south-central New Mexico; in Keller, G. R. and Cather, S. M., eds., Basins of the Rio Grande rift: structure, stratigraphy, and tectonic setting: Geological Society of America, Special Paper 291.

Jaksha, L. H., 1982, Reconnaissance seismic refraction-reflection surveys in southwestern New Mexico: Geological Society of America Bulletin, v. 93, p. 1030–1037.

Olsen, K. H., Keller, G. R. and Stewart, J. N., 1979, Crustal structure along the Rio Grande rift from seismic refraction profiles; *in* Riecker, R. E., ed., Rio Grande rift: tectonics and magmatism: American Geophysical Union, Special Publication, p. 127–143.

Roberts, D. G., Adams, D. C. and Keller, G. R., 1991, A geophysical analysis of crustal structure in the Ruidoso area: New Mexico Geological Society, Guidebook 42, p. 191–197.

Schneider, R. B., 1990, An integrated geophysical and geological investigation of the transition zone between the Colorado Plateau, Rio Grande rift, and Basin and Range provinces: Arizona and New Mexico [PhD Dissertation]: El Paso, University of Texas at El Paso, 204 p.

Schneider, R. V. and Keller, G. R. (in press), Crustal structure of the western margin of the Rio Grande rift and Mogollon-Datil volcanic field, southwestern New Mexico and southeastern Arizona; *in* Keller, G. R. and Cather, S. M., eds., Basins of the Rio Grande rift: structure, stratigraphy, and tectonic setting: Geological Society of America, Special Paper 291.

Sinno, Y. A., Dagget, P. H., Keller, G. R., Morgan, P. and Harder, S. H., 1986, Crustal structure of the southern Rio Grande rift determined from seismic refraction profiling: Journal of Geophysical Research, v. 91, p. 6142–6156.

Day 2 — Ship-On-The-Desert Research Center

VOLCANOES AND NEOTECTONIC CHARACTERISTICS OF THE SPRINGERVILLE VOLCANIC FIELD, ARIZONA

L. S. CRUMPLER[1], JAYNE C. AUBELE[1] and C. D. CONDIT[2]
[1]Department of Geological Sciences, Brown University, Providence, RI 02912; [2]Department of Geology/Geography, University of Massachusetts, Amherst, MA 01003

Abstract—The Plio-Pleistocene Springerville volcanic field (SVF) is one of the largest basaltic volcanic fields in the western U.S.. It is petrologically and volcanologically diverse and includes tholeiitic, alkalic and evolved alkalic rock types characterized by a general evolutionary trend in petrology in which tholeiitic rocks account for 24% of the field volume and were erupted early, and alkalic and evolved alkalic rocks (hawaiite, mugearite, and benmoreite) were erupted later. The average volume rate of effusion over the history of the field (1.5×10^{-4} km^3/a) is similar to rates observed for other cinder-cone type volcanic fields throughout the world, but lower by about an order of magnitude than typical rates associated with large volcanic edifices and calderas. The magma supply in the Springerville volcanic field was long-lived, but too low and sporadic to sustain a single large volcano. Large-scale physiographic and morphologic features in the Springerville volcanic field reflect the kinematics of on-going, weak late Cenozoic, distributed tectonic deformations. The style, distribution and geometry of individual eruptions record the influence of contemporaneous tectonic stresses on the ascent and emplacement of magmas as well as environmental conditions present at the time of the eruptions. The general absence of strong patterns of fissure vents appears to be a consequence of the absence of regional tensile strain. Neotectonic structures are interpreted to indicate that left lateral shear stresses in the brittle crust occurred throughout much of the history of the Springerville volcanic field, rather than simple tensile stresses that would result in long fissures. Deeply penetrating faults in the northeast edge of the field interrupted the deep regional ground water flow and contributed to formation of spring deposits (travertine) in the northern margin of the field. The neotectonic deformation appears to record a transtensional style of lithospheric extension across this part of the southern margin of the Colorado Plateau in response to continued late Cenozoic clockwise rotation of the Colorado Plateau.

INTRODUCTION

The Springerville volcanic field is an areally extensive late Pliocene to Pleistocene basaltic volcanic field in east-central Arizona and western New Mexico, lying on the inboard edge of the Colorado Plateau and along the easternmost Arizona segment of the Mogollon Rim. It is the southernmost of the predominantly basaltic, late Cenozoic volcanic fields located around the margin of the Colorado Plateau (Figs. 1,2) and one of the largest basaltic cinder cone-type volcanic fields in the western United States. It is also one of the best characterized in the detail of geologic mapping, chemical sampling, and absolute and relative age determinations (Condit et al., 1994). The combination of large size and relative youth of the volcanism means that a great number of individual vents and diverse styles of basaltic volcanism and vent morphologies are present, representing a natural museum of physical volcanological characteristics exposed in one setting. For these reasons, the Springerville field represents an excellent opportunity to understand the characteristics and origin of cinder cone fields.

In this study, following a brief review of general characteristics of the field, we discuss some salient volcanological features not generally included in previously published accounts of the petrology, lithology and chronology of the Springerville field. Our purpose is to characterize features of general interest in understanding the regional physiography, their relationship to the origin of the individual volcanoes, and the insight they provide into the characteristics and origin of the cinder cone field style of volcanism.

OVERVIEW OF THE FIELD

The Springerville volcanic field was one of the last large basaltic volcanic fields in the U.S. to be geologically mapped in detail. Ironically, it was also the first such field to be visited by Europeans in what is now the United States, as it is likely that Coronado passed through the center of the field in 1540 on his journey northward to Zuni Pueblo and the Rio Grande (Bolton, 1974). Aside from a passing mention by Darton (1915), who first noted its volcanic origin, (and the mention of its cones and lava flows in the opening setting of one of Edgar Rice Burroughs's (1935) John Carter of Mars books!) almost nothing was known about the volcanology of the field until it was mapped by us a decade ago. Earlier geologic mapping studies in the region (e.g., Sirrine, 1958) were concerned primarily with the surrounding sedimentary rocks and with the ground water geology. Previous work on volcanic rocks in the region (Merrill and Péwé, 1977) focused primarily on White Mountain Baldy to the south, a series of mainly porphyritic trachytic rocks 5 to 6 Ma older than most of the Springerville volcanic field rocks and now exposed as a deeply dissected shield-shaped volcano (Nealey, 1989). Because of differences in age, style of volcanism and overall preservation of the two areas, we refer to the younger region of basaltic cinder cones and flows (mostly north of highway 260) as the Springerville volcanic field in order to distinguish it from the older White Mountain Baldy complex. Most local residents and vacationers from Phoenix, however, still refer to the elevated and forested region in eastern Arizona as a whole as the White Mountains, including the youthful part of the field between Show Low, Concho, Springerville, Eagar, Greer and Pine Top that is the focus of this study.

The Springerville volcanic field (SVF) is a prominent topographic feature in the southern Colorado Plateau. From Interstate Highway 40, near the entrance to Petrified Forest National Park or south of Holbrook, the SVF appears on the horizon to the south as a broad, bumpy domical rise. From the town of Springerville the view of the eastern one-third of the field presents a profile of numerous clustered cinder cones along a broad slope (Fig. 3). Some of the domical relief of the field may be attributed to volcanic construction, as the total thickness of volcanic rocks based on water well logs is estimated to average in excess of 90 m in the central northern parts of the field (near Vernon) and up to 300 m in regions of highest vent density. But the presence of pre-volcanic sedimentary outcrops at high elevations within the central field also suggests that the elevation of pre-existing surfaces contributed to the total relief. The substrate beneath much of the SVF consists of gently dipping Mesozoic sediments, including the Triassic Petrified Forest Formation of the Chinle Group on the north margin, and coarse alluvial gravels of Tertiary age. Similar alluvial gravels also occur throughout the Mogollon Rim.

The SVF covers an area of 3000 km^2 and is notable for its cinder cone field style, large volume of lava flows, and petrologic diversity (Aubele and Crumpler, 1983; Condit et al., 1989 a,b; Crumpler et al., 1990; Cooper et al., 1990; Connor et al., 1992). The field consists

FIGURE 1. Satellite view of the Southwestern United States (GOES satellite image). The Mogollon Rim appears as a dark elongated area in central Arizona. Several prominent late Cenozoic volcanic fields are clearly visible around the margins of the Colorado Plateau. Rectangle outlines the area shown in Fig.2.

primarily of basaltic lava flows, related basaltic pyroclastic cones up to several hundred meters in relief, and isolated more differentiated extrusive rocks. Although cinder cones and their flows occur throughout the western U.S., fields with the large dimension and morphological diversity of the SVF are relatively uncommon. In overall areal extent, and in volcanic and petrologic style, the Springerville volcanic field is comparable to the San Francisco volcanic field (Ulrich et al., 1989) near Flagstaff and the Mount Taylor field in western New Mexico (Crumpler, 1982,1990).

Unlike the San Francisco or Mount Taylor fields, however, the SVF does not include a topographically dominating large volcano or volcanic complex of equivalent age. Nonetheless, elevations in the field exceed 3000 m at the base of Greens Peak (Fig. 4), a moderately young cinder cone that lies at the summit of the field and attains a summit elevation of 3140 m, from which there is a view of most of the surrounding region of closely-spaced cinder cones. The upper elevations within the field are characterized by aspen, spruce and open grassy parks supported by frequent summer rainfall. In the extreme northern end of the field, near the town of St. Johns, elevations are less than 1830 m and the landscape is characterized by rolling grasslands, junipers, and related semi-arid and transition zone vegetation.

Run off and spring discharge from the higher elevations feed the two major drainages in the field, the Little Colorado River, which originates near Greer, and the North Fork of the White River, in the central part of the field. The Little Colorado skirts the eastern edge of the field, along which there are numerous paleo-Indian sites, such as the Casa Malpais site near the town of Springerville, and continues northward into the interior of the Colorado Plateau. During the course of mapping, many sites within the field were identified that preserve pre-historic as well as historic artifacts.

Over 400 individual volcanic vents were identified during the course of our mapping at a scale of 1:24,000 and compiled at a scale of 1:100,000 (Condit et al., 1987; Aubele et al., 1987; Condit et al., 1989; Condit et al., 1994). Vents include cinder cones, maars, viscous domes, a laccolith or cryptodome, related lava flows and various pyroclastic deposits. Neotectonic faults and folds (Fig. 5) were also identified by detailed geologic mapping. Identification and characterization of the basic lithologic units and tectonic deformation were constrained through chemical analyses of 257 lava flows, trace element analyses, and petrogenetic modeling, 30 K-Ar age dates, and extensive paleomagnetic determinations. Radiometric ages of volcanic rocks range mostly from ~0.3 to 3 Ma (Condit et al., 1994), only a few of which are older than 2.5 Ma. Ages of individual units not dated were constrained on the basis of combining the existing K-Ar dates (Laughlin et al.,1979; Laughlin et al.,1980; Aubele and Crumpler, 1983; Condit and Shafiqullah, 1985; Aubele et al., 1986; Cooper et al., 1990) and magnetic polarity reversal records from 90 units with detailed stratigraphic relationships determined through mapping of every flow unit.

The chemical data demonstrate that the field is petrologically diverse and includes tholeiitic, alkalic, transitional alkalic-tholeiitic

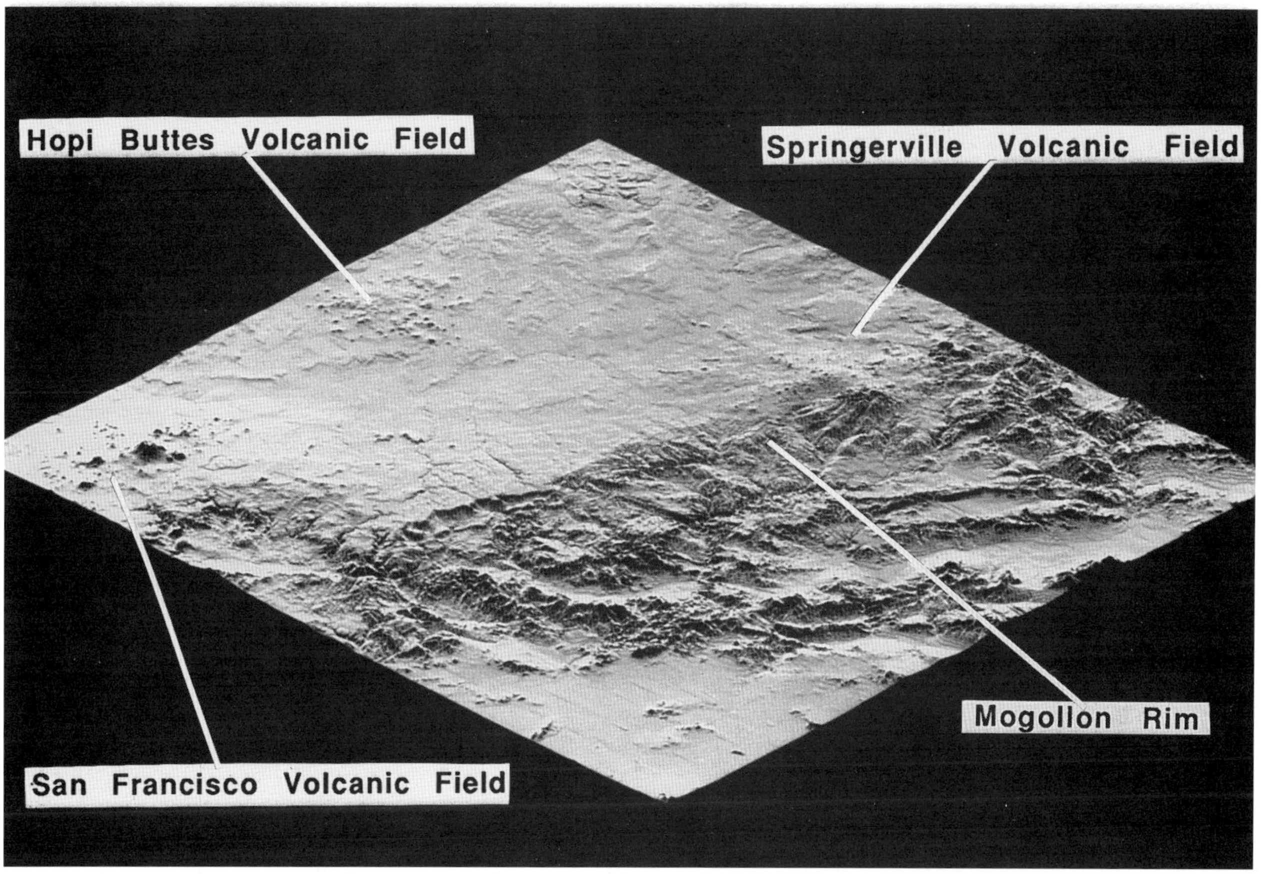

FIGURE 2. Oblique digital elevation model centered on the south and southwest margin of the Colorado Plateau. The large-scale relief across the Mogollon Rim and Transition Zone is clearly visible, as is the small scale relief associated with the San Francisco, Springerville and Hopi Buttes volcanic fields. North is toward the upper left corner.

and evolved alkalic rock types (hawaiite, mugearite and benmoreite). Tholeiitic rocks account for 24% of the field volume and were erupted early, whereas alkalic and evolved alkalic rocks were erupted later (Fig. 6). Most of the older tholeiitic rocks occur as extensive sheets exposed mostly in the northern periphery of the field and now form extensive low grasslands. The source vents for these flows are largely unexposed and probably lie buried to the south under the stack of younger alkalic lavas. Consideration of the total number of vents, the volume of volcanic rocks, and the duration of volcanism indicates that the average volume rate of effusion over the history of the field was 1.5 X 10^{-4} km^3/a (Crumpler et al., 1990). Because the volumes and ages are well-constrained by detailed mapping, we are able to estimate effusion rates from several intervals in the history of the field. These suggest rapid onset of nominal effusion rates and steady production of magmas for at least 2 Ma (Fig.7). The average recurrence interval between individual eruptions was on the order of 3000 years.

The geometric center of active vents migrated toward the east at a rate of approximately 2.9 cm/yr, such that the older vents tend to lie on the west side of the field and the youngest occur on the east side, in an area to the north and west of the town of Springerville (Condit et al.,1989a). A similar pattern of generally eastward migration of successive eruptions was noted in the San Francisco volcanic field near Flagstaff (Tanaka et al.,1986). Because the vector representing this drift in the center of activity corresponds roughly with the motion of the North American continent, it is possible that this eastward drift in volcanism reflects relative motion between the lithosphere and a fixed sublithospheric source region for most of the basaltic melts. The nature of the stationary mantle source is undetermined, but need not represent a sublithospheric plume. Other origins might include viscous heating of

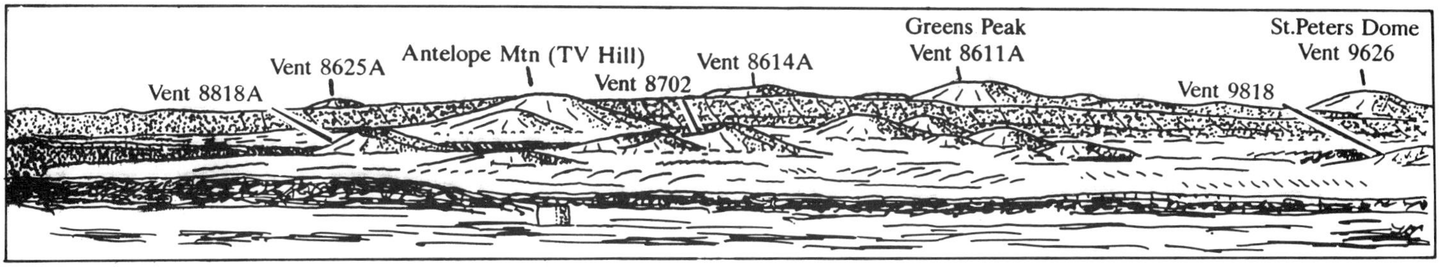

FIGURE 3. Sketch of the western skyline from the town of Springerville, showing the southern and eastern one-half of the Springerville volcanic field. The cinder cones along the top of the skyline all lie at greater than 2700 m above sea level. Vent numbers and names of some prominent cinder cones are indicated. Saint Peters Dome is dated at 0.67± 0.02 Ma and Greens Peak is dated at 0.76±0.02 Ma. Most lava flows in the nearer slopes of the field are about 1.5 Ma. The cultivated alluvial plain of the Little Colorado River and Round Valley (Springerville-Eagar) lies in the foreground. Sketch by LSC.

FIGURE 4. Greens Peak (viewed from SE) a typical cinder cone of intermediate age, lies at the topographic summit of the field. One of the longer lava flows in the field erupted from this cone and flowed to the northwest.

a lithospheric bump or keel (Condit et al., 1989a), viscous heating of a sublithospheric upwarp intruding the base of the lithosphere along the Jemez lineament, progressive eastward propagation of sublithospheric fractures opened during clockwise rotation of the Colorado Plateau (Spence and Gross, 1990), or melting of sublithospheric mantle inhomogeneities (Jemez lineament) in the presence of regional increased heat flow.

NEOTECTONIC STRUCTURES

Detailed mapping has allowed us to identify structural features, many of which appear to offset the basaltic lavas dated between 0.9 and 1.3 Ma and thus appear to be relatively young (Fig. 5) (Aubele et al.,1986; Crumpler et al.,1989). Neotectonic features are described here first because they relate to discussions of many of the individual volcanic features. These structural characteristics are also of interest because they are evidence for the kinematics of ongoing weak, late Cenozoic, distributed tectonic deformation along this part of the Mogollon Rim. Only the more prominent structural elements are described here. The overall sense of deformation is one of limited, very nearly incipient strike-slip faulting along several regional fault zones striking northwest through the field, in which strike-slip appears to be accommodated by normal faulting near the ends of the fault zones.

Tectonic deformation within the SVF is expressed at several scales, including regional topography, local faulting and folding of basalt flows, and vent alignments and clustering. The tectonic tilting of basalt flows as young as those in the SVF is relatively unusual in our experience, but is well constrained. Many of these characteristics are subtle and some are detected only as a result of apparent back-tilting of the substrate on which individual basalts have flowed. Tectonic tilting of flows is readily distinguished from primary variations in surface slope of flows, based on a knowledge of the source direction, consideration of the path taken by the flow over original topography, and thickness of the flow in relation to apparent tectonic relief on the substrate. In short,

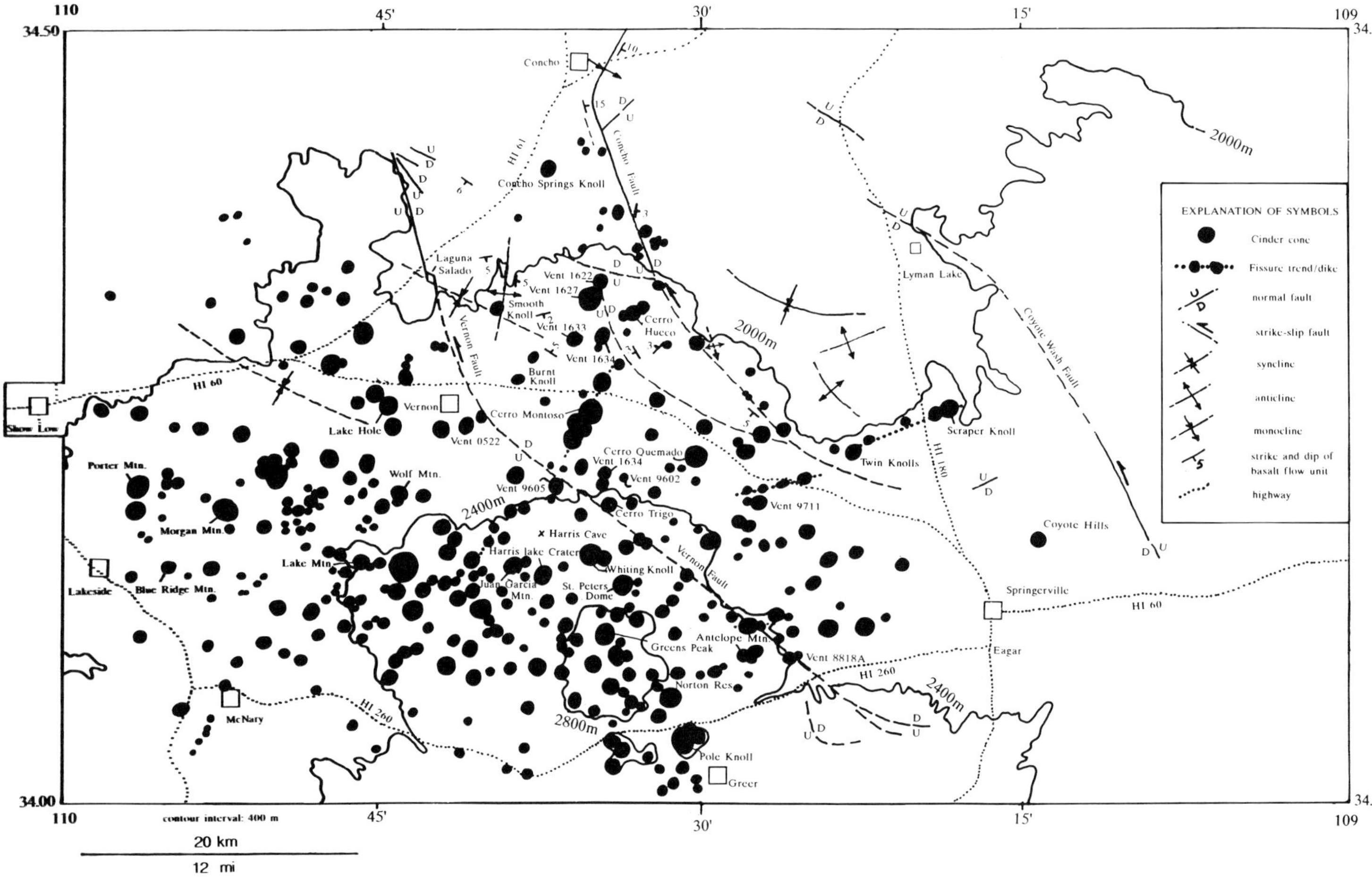

FIGURE 5. Map of the Springerville volcanic field showing the locations of all cinder cones and related vents, neotectonic features, and regional topography of the field. For the corresponding detailed lithologic, geochemical, chronological and paleomagnetic maps of the Springerville volcanic field refer to Condit et al., 1994. [Editor's note: Normal faults with an indicated component of strike slip are more appropriately labeled as oblique-slip faults.]

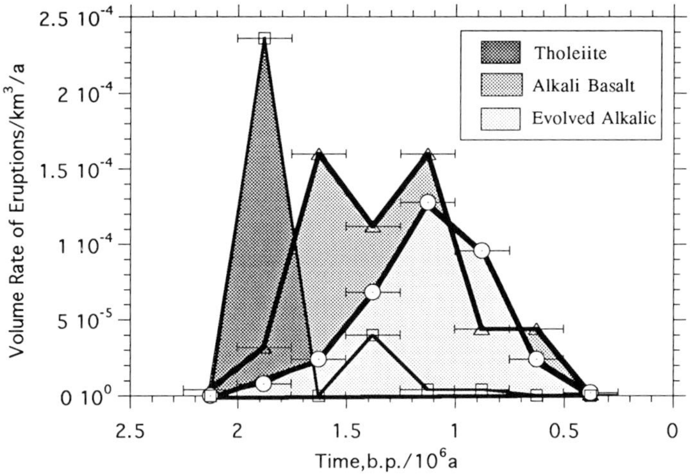

FIGURE 6. Volume rates of eruption of tholeiites, alkali olivine basalts, and evolved alkalic rocks with time. Although the three types erupted contemporaneously, there is a trend toward more evolved alkalic compositions and slightly decreased volume rates of eruption with time.

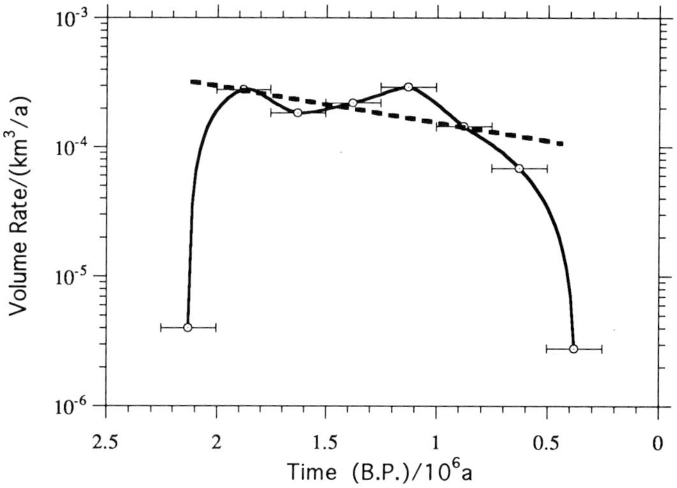

FIGURE 7. Total volume rates of eruption for all rock types in the Springerville field as a function of time. Oblique line is least squares fit to the peak activity period during which volume rates of eruption were on the order of 10^{-4} km^3/a. This is typical of the rates for cinder cone fields in general. Based on data in Condit et al.(1989a).

many basalt flows appear to have "flowed up hill" or over swells that exceed the thickness of the individual flows and have clearly been deformed subsequent to their emplacement.

Three principal deformation zones

Three principal deformation zones are defined by major topographic steps across the field, with a variety of smaller scale structural features (Fig. 5). The topographic steps include the Vernon fault zone striking northwest across the center of the field; the Concho fault zone, which is approximately parallel to the Vernon fault and extends from the eastern margin to the northern margin of the field; and the Coyote Wash fault zone, which lies mostly on the eastern margin of the field and also strikes northwest. Between these major fault zones are a number of smaller faults that splay out from the main deformation zones. Other isolated features include synclines, anticlines and a few short fissure vent trends.

Topographically the Vernon and Concho fault zones correspond to prominent down-to-the-northeast steps or scarps and the Coyote Wash fault zone is characterized by a down-to-the-southwest scarp (Fig. 8). Each of these appear at some point along their length to be related to faulting or folding of either pre-existing surfaces or the later volcanic units. Throughout much of the length of these steps, many of the of the basalt flows appear to have flowed over the steps, and the steps might otherwise be interpreted at first sight as simple pre-existing erosional scarps. Although it is likely that erosion has been important in enhancing segments of the scarps, local deformation features at various points along their length (including graben, normal faults, apparent strike-slip deformation, and pull-apart basins), imply that the scarps are fundamentally structural in origin. This varied and complex style of deformation is not consistent with an origin of the scarps through simple normal faulting and a significant component of strike slip is inferred as discussed below.

Vernon fault

The most prominent step occurs as a scarp trending from the vicinity of Greer on the southeast to Vernon, and is here referred to as the Vernon fault (Fig. 5). The southeastern end of the Vernon fault zone splays into a series of east-west trending normal faults forming high scarps in the much older (~ 6 Ma) basalts from the White Mountain Baldy complex in the vicinity of Greer. Immediately northward, within the younger Springerville basalts, the fault zone is expressed only as a slight rise in the surface between vent 8818A and Antelope Mountain (Fig. 3). Farther northwestward along the Vernon fault zone there is little direct evidence for structural control because of the dense forest cover and abundance of cinder cones. Instead, deformation along the center of this scarp is characterized by anomalous outcrops of pre-volcanic sediments and ridge-like remnants of earlier volcanic units oriented along the apparent structural trend and around which later basalts have flowed. However, at Vernon the scarp-like character disappears and the trend changes to a more northerly direction. Along this northerly trend are several structural features with observable offsets. The presence of an apparent pull-apart basin, Laguna Salado (located north of the town of Vernon and 1 km east of Highway 81) (Fig. 9), is consistent with the accommodation of small amounts of left lateral simple shear strike-slip motion. At its terminus, several kilometers northwestward, the Vernon fault appears to be accommodated by a series of straight normal faults and graben that displace older basalt units (Fig. 9; see also Fig. 5). These faults are oblique to the strike of the Vernon fault zone and are consistent with an origin as extensional Riedel fractures associated with the left-lateral shear.

Concho fault

The Concho fault zone occurs along the next prominent topographic step to the north and is characterized by an arcuate trend of discontinuous structural features from an area beginning near the junction of Highway 60 and Highway 180-666, approximately 5 km west of Springerville and traversing northwest across the field to the town of Concho. In addition to its association with a topographic step, the Concho fault zone also marks the northeastward limit of most vents in the field and therefore appears to exert an indirect control on the location of vents. Throughout the length of the Concho fault zone are numerous fault scarps, graben and local anticlinal folds that displace basalt units. On the southeast end the Concho fault is little more than a low scarp-like rise. But farther northwest, near the center of the northern field, are a series of fault scarps with normal fault displacements, basalt benches with surfaces tilted several degrees, and a gentle anticline approximately 1 km long and oriented obliquely to the lineament. At a point northwest of Cerro Hueco along the Concho fault a complex anastomosing series of arcuate fractures cuts the surface of a long ridge of older basalt for about a kilometer. Farther northwest and a few kilometers southeast of Concho the structural trend widens into a deep graben along which basalt units and a cinder cone appear to have been down dropped 30 m. The south flank of the graben consists of a series of step-like fault scarps and steeply dipping basalt flow units. Farther northwest, along the south wall of the graben, basalt units are cut by a deep wash and dip steeply into the graben (Fig. 10; see also Fig. 4). All

FIGURE 8. Oblique view to the north-northwest of the central field, centered on Highway 60 west of Cerro Quemado. Several cinder cones are visible. The large scale relief of the field is characterized by prominent topographic steps related to major structural trends. Two of the major topographic steps cross the upper half of the image from lower right to upper left and are annotated by dotted lines. The arrow points in the direction of the regional gradient and indicates the apparent sense of deformation and displacement of the surface. Aerial photo by C. Condit.

of the volcanic units deformed at this graben are at least 0.7 Ma old, but the fact that the slopes of the graben are affected very little by erosion is evidence that the faulting is considerably younger. Northward of the deep wash and along the same fault trend, the Concho fault abruptly terminates and merges with a northeastward-striking monoclinal fold. Old basalts capping a mesa east of Concho dip gently to the southeast along the west side of the mesa. The sense of deformation in many of the structures along the length of the Concho fault zone, their occurrence along a narrow belt of deformation, and the overall kinematics of the large scale structures associated with it are interpreted to be the results of oblique slip involving a slight amount of left lateral strike slip and down-to-the-northeast normal faulting. In general, normal faulting is more prominent along the southern extremities of the fault and apparent strike-slip movement appears to be dominant along the central portion of the fault zone. The mergence of the Concho fault at a right angle with a monocline suggests that in this case the proposed left-lateral strike-slip was accommodated by shortening at its northern terminus. Although no marker units or piercing points occur that would confirm strike slip or its absolute displacement, the disparate types of deformation and the relatively confined belt along which they occur is similar to deformation associated with known strike-slip faults (Sylvester, 1988).

Coyote Wash fault

The Coyote Wash fault, on the east and northeast side of the field (Fig. 5), is similar in many structural characteristics to the Concho and Vernon faults, but is curved in an opposite sense and with an opposing (down to the southwest) topographic step and component of normal faulting. The detailed structural character of this fault zone is poorly constrained compared with other fault zones in the field due to the extensive development of drainages throughout much of its southern segment, but normal fault displacements of at least 10 m are visible in sandstone outcrops east of the Coyote Hills. Relatively minor monoclinal features (not shown in Fig. 5) appear to splay to the east from this fault before the fault becomes obscured to the south. Together, the Coyote Wash fault and the Concho fault thus enclose a structural depression in the northeast quadrant of the field best described as a larger scale version of the Laguna Salado pull-apart basin. Within the structural depression, a series of anticlinal and synclinal folds in the basalts are randomly oriented, suggesting contractional strain throughout the depression. In addition, several examples of other recent anticlinal and synclinal warping of basalt surfaces are generally consistent with secondary strain patterns associated with these major deformation trends.

East of the SVF, young basalts and their vents in the Red Hill area of New Mexico (Crumpler and Aubele, 1990) appear to be cut by numerous parallel, northeast-striking normal faults. Thus, the Springerville field appears to occur at a distinct junction in tectonic trends bound on the east by northeast-striking faults and characterized throughout most of the field by northwest-striking major fault patterns.

Origin of regional deformation

The southeastern termini of the Concho and Vernon faults appear to be the sites of normal faulting or fissure vents and imply that the apparent tectonic transport to the northwest along both major structural trends is accommodated in part by extensional strain at their southeastern ends. In this sense the regional deformation is altogether similar to thin-skinned or listric-style shallow faulting, typical of some areas of

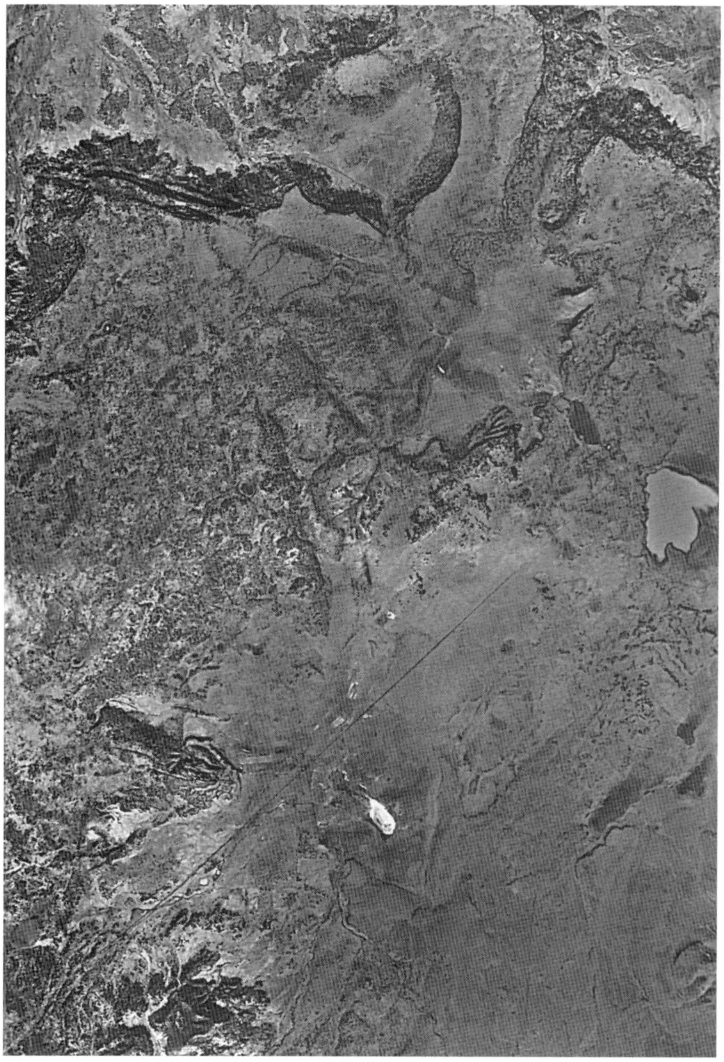

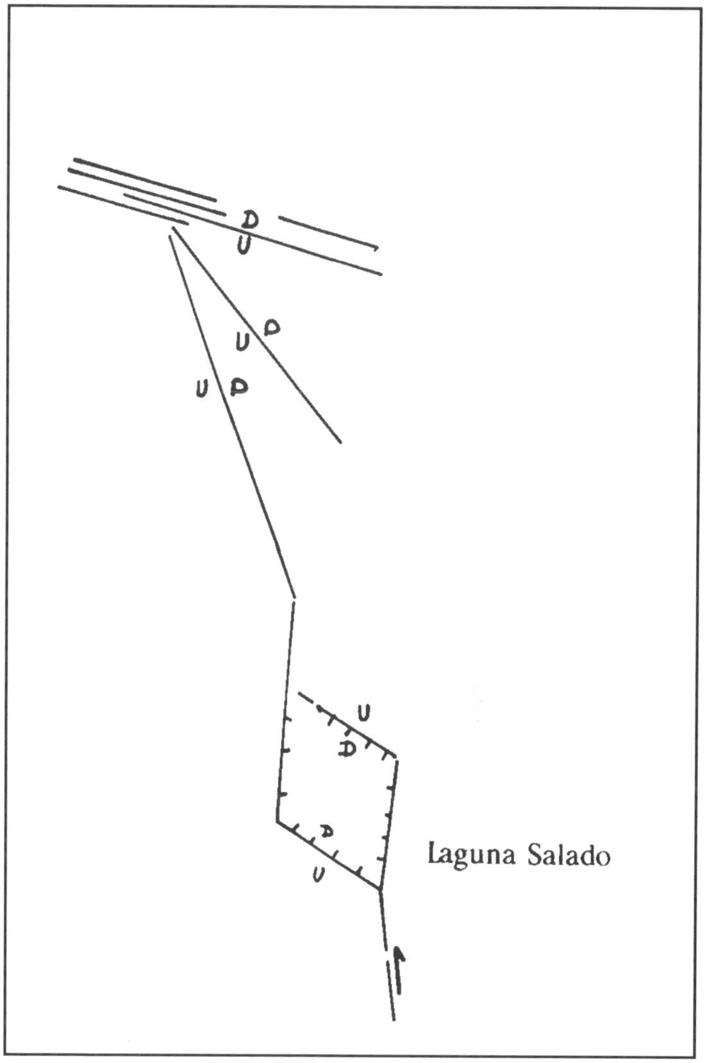

FIGURE 9. Air photo and sketch map showing a series of normal faults and graben accommodating left lateral simple shear along the Vernon fault zone trending northwest through the center of the image. A large pull-apart basin (Laguna Salado) is visible along the fault zone in the lower half of the image. Width of image (E-W) is 10 km.

the Basin and Range, and is headed at simple breakaway zones on the south. The regional tectonic stresses associated with the margin of the Colorado Plateau are likely to be complex because of the combined effects of Basin and Range deformation, the influence of the opening of the Rio Grande rift, and structural fabrics remaining from the Laramide history of the Mogollon Rim. However, relatively simple tectonic transport can account for most of the neotectonic deformation in the SVF. Northeast-southwest extension along the Mogollon Rim could account for the stepped faulting in the SVF, but slight amounts of left-lateral strike-slip better account for the diversity and style of observed structures. Although the northeast-southwest extension might reflect continued extension and encroachment of the Basin and Range on the Colorado Plateau (Brumbaugh, 1987), strike-slip faulting is not widely observed in the Colorado Plateau and the possible origin of proposed left-lateral shear strain is unclear. Strike-slip could arise in this region if large scale clockwise rotation of the Colorado Plateau has occurred as suggested by numerous authors (Steiner, 1986; Hamilton, 1987; Spence and Gross, 1990; Bryan and Gordon, 1990). Some structural elements of the Jemez lineament, the Colorado Plateau and the Rio Grande rift have been attributed to this sense of motion (Spence and Gross, 1990). Although the regional left lateral shear of the magnitude

FIGURE 10. Tectonically tilted basalt unit (approximately 1 Ma) near Concho, dipping steeply toward the alluvial-filled floor of a graben that forms part of the Concho fault zone.

and orientation predicted by clockwise rotation of the Colorado Plateau about the pole established by Bryan and Gordon (1990) in northern New Mexico is consistent with the observed orientation and sense of tectonic transport observed in the SVF (Fig. 11), the presence of this sense of tectonic transport must be considered controversial, as this sense of motion does not account for many elements of the structure in the Rio Grande rift on the eastern margin of the Colorado Plateau (e.g., Kelley, 1982). Deformation in relatively young basalts of the SVF is nonetheless evidence that at least small amounts of rotation of the Colorado Plateau might have contributed to the observed deformation over the past 2 Ma, irrespective of the additional tectonic influences that may have governed principal deformations outside the plateau and in the Rio Grande rift.

Regardless of the origin of the regional structure, there is evidence that can be interpreted as left-lateral strike-slip type deformation and a regional pull-apart basin style of regional deformation. It may be relevant to the origin of the field as a whole that the center of vent concentrations is offset from the center of the overall pull-apart style depression that characterizes the regional structure.

DESCRIPTION OF INDIVIDUAL VOLCANIC FEATURES

In the following, many of the more interesting and unusual physical volcanological characteristics of the field are discussed. Detailed discussions of petrology, chemistry and age relations may be found in Condit et al. (1989; 1994). The emphasis here is on physical features of general interest, but illustrates that the record of basaltic volcanism is of more than geochemical interest and may provide useful insights into the physical processes of the emplacement of cinder cone fields, the ascent and eruption of basaltic magmas, the influence of regional tectonics on eruption characteristics, and the regional geology.

Because many vents and their associated volcanic features are unnamed, we have assigned "vent numbers" to each volcanic center and refer to vent number frequently. Vent numbers are shorthand for location referenced to the township, range and section in which the vent occurs; the first digits of the township and range are omitted because they are repetitive throughout the field. Multiple vents within a single section are arbitrarily assigned sequential letters of the alphabet. Thus, vent 0621 is located in sec 21, T10N, R26E.

Lava flows

Most of the field volume consists of lava flows. The pyroclastic materials associated with cinder cones account for only about 10% of the total volume of the volcanic field, as is the case for the few volcanic fields for which estimates of the relative volumes of lava and pyroclastic materials have been made (Wood, 1980; Hasenaka and Carmichael, 1985). Cinder cones are prominent topographically and record the conditions of eruption, but most of the magma erupted was emplaced as lava flows. Dimension, thickness, volume, shape, response to slopes, state of preservation and source vents for individual flows are constraints for conditions of magma ascent and eruption. Characteristics of lava flows that are of general interest include exposures of primary lava flow morphology, lava tubes and flows of unusual dimensions, morphology and petrology.

Youngest lava flows

The youngest flows in the SVF are transitional basalt with moderately abundant olivine phenocrysts, erupted from the Twin Knolls (vents 0833 and 0828), two overlapping, dark, morphologically youthful cinder cones 8 km NW of the town of Springerville, along the abandoned old Highway 180-666 to St. Johns (Fig. 12). The northern cone is also somewhat more eroded and has grass-covered slopes. The primary morphology of the lava flows is of the *aa* type. Two flows of slightly different age were erupted. The older of the two flowed northward down local drainages developed in earlier basalts toward the Little Colorado River and is exposed approximately 10 km north of Springerville along the west side of Highway 180; the highway crosses the flow near its terminus about 11 km north of Springerville. The surface of this lava flow is locally overlain by alluvium and outcrops of the actual aa surface are weathered slightly reddish-brown, but pressure ridges and related larger-scale structures associated with young lava flows are preserved. Samples from a road cut in this flow yielded an

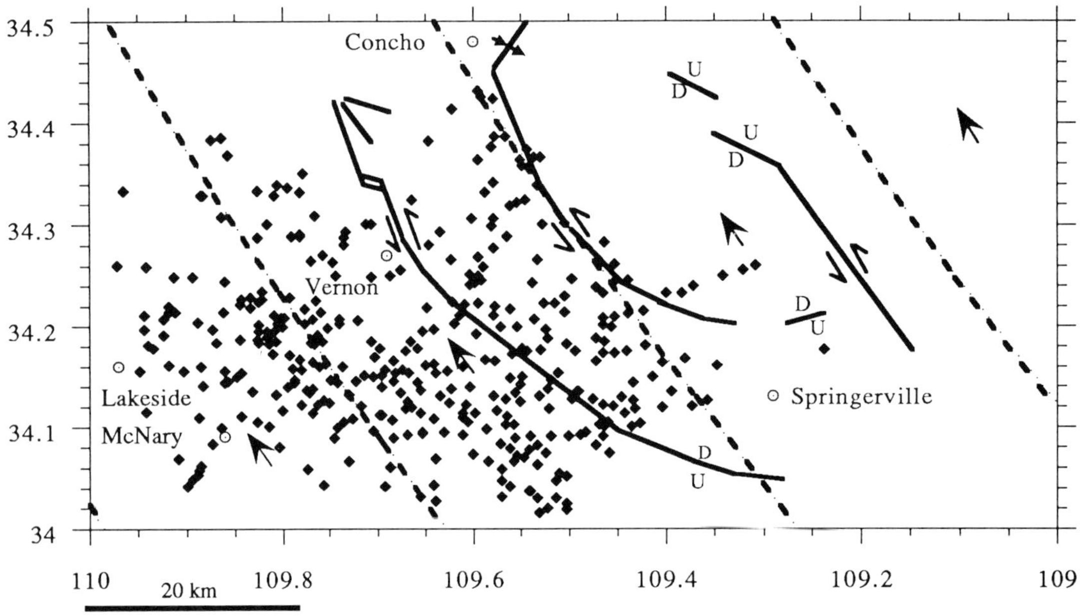

FIGURE 11. Map illustrating the location and orientation of the main faults in the Springerville volcanic field (dashed lines) in relation to small circles (lines of latitude shown by dashed lines) about the suggested (Steiner, 1986) pole of rotation (37°N latitude and 103°W longitude) for the Colorado Plateau. Arrows indicate the local sense of plateau motion with respect to the continental U.S.. Most faulting in the SVF is consistent with small amounts of left-lateral simple shear oriented parallel to the suggested rotation of the Colorado Plateau. The implied sense of rotation is consistent with many characteristics of structure within the Colorado Plateau (Spence and Gross, 1990), although some characteristics, such as the late structure of the Rio Grande rift (Kelley, 1982), are not clearly related to this sense of motion. Note that the center of the structural basin in the northeast is offset from the center of concentration of vents. Offset of most of the faulting from the center of the field suggests that the Springerville volcanic field may have resisted the northwestward translation of the rest of the Colorado Plateau and resulted in the opening of a large pull-apart basin in the Lyman Lake region between Springerville and St. Johns.

age of 0.75±0.13 Ma (Laughlin et al., 1980). A younger, petrographically similar, and morphologically better preserved flow was erupted from the southern of the two overlapping cones and spread out at the base of the cone. This flow is dated at 0.308±0.070 Ma. This appears to be a case where two eruptions, greatly removed in time, occurred from very nearly the same vent area. Although cinder cones are often considered to be "monogenetic", several studies of other cinder cones in the southwest (Renault et al., 1988; Wells et al., 1990) have shown that repeated eruptions over geologically extended periods of time from a single cone may be more common than previously thought.

These young vents must lie at the trailing end of the observed record of activity in the field. However, considering that these eruptions occurred several hundred thousand years ago and that the mean interval between eruptions for the field as a whole is approximately 3000 years, they may not represent the final eruptions. Activity of the SVF may have shifted further east, as a morphologically youthful eruption dated at 0.071± 0.012 Ma (McIntosh and Cather, this guidebook) occurred further east in the Red Hill area (Crumpler and Aubele, 1990) of western New Mexico.

Other young flows, slightly more eroded than the older of the two Twin Knolls flows, occur in the vicinity of Cerro Hueco (vent 1626A; see discussion of maars below), in the north central part of the field (Fig.11). Although no age is available for these flows, they are more degraded than the Twin Knolls flows, yet display more primary morphologic features than flows dated at about 1 Ma. This intermediate age, based on morphology, agrees with the estimated age determined strictly from stratigraphic and paleomagnetic data (0.30-0.79 Ma) (Condit et al., 1994).

Lava ponds

Two thick flow units of a slightly older flow (0.70-0.97 Ma) in the Cerro Hueco area are mugearitic in composition and were erupted from a relatively youthful-appearing cone (vent 1622) about 2 km west of Cerro Hueco. The upper of the two flow units was erupted from a deep breach in the cone and is unusual in two respects. First, scattered about the surface of this flow are small mounds of horizontally to slightly-dipping cinder and spatter up to 10 m in relief (Fig.13). The fact that the internal bedding in each of these mounds appear to be truncated around the margins, together with the unusually large breach in the source cone, suggests that the mounds are large fragments of the cone rafted away from the flank during the final phases of the eruption. Rafted cone fragments are common on lava flows erupted from cinder cones (Holm, 1987), because of the difference in density of cinders versus lava flows and the relatively low shear strength of pyroclastic cones.

FIGURE 12. Twin Knolls viewed from the SW. Only the youngest of the two cinder cones northwest of the town of Springerville is visible. This is the site of the youngest dated eruptions within the field (0.75±0.13 Ma and 0.308±0.070 Ma). The southern of the two cones was emplaced on the south exterior flank of the northern cone and appears to be several hundred thousand years younger.

The second unusual characteristic of this flow is its overall shape and relief. The flow is unusually thick, short and equant in plan view; in fact, it is similar in morphology to more viscous lavas. An even more unusual characteristic is the overall concave surface relief, forming a broad closed depression in the center of the flow that extends out to the flow margins, broken only by the isolated rafted cone fragments. Petrographically, the margins are characterized by unusually large olivine phenocrysts, whereas the interior basin is aphyric. It is chemically similar to many of the transitional-to-evolved basalts throughout the field. This morphology might represent a perched lava pond or an inflated lava flow. Ponding in lava flows is common where a sudden decrease in slope causes a flow to spread laterally, increases the area relative to the volume of lava, results in rapid cooling relative to channelized lava flows, and as a result increases the yield strength relative to that of basaltic lavas on steeper slopes. Observation of actively forming lava ponds in Hawaii (Wilson and Parfitt, 1993) showed that the effusion rate can be estimated based on the observed dimensions of the pond. Employing their method, and using the observed dimension (1.8 km pond diameter and a slope of 1.5° determined for the surface of the underlying flow unit) we estimate a rate of effusion of 3.4 m^3/sec. This may be categorized as being between average and low for most historic basaltic eruptions, so there appears to have been nothing unusual about the rate of emplacement. The relatively low, but steady rates of eruption might also have favored lava inflation, subsequently followed by draining and sagging of the lava surface.

At least three other examples of near-vent lava ponds of similar dimension and petrographic character (porphyritic margins and aphyric interiors) occur in the field, associated with vents 1633 (dated at 1.19±0.04 Ma, Aubele et al., 1986), 1634 and 9711. Vent 1622 and these three examples all occur within a narrow corridor oriented west-northwest along one of the prominent topographic steps that trend through the field. A potentially related fact is that all four flows are unusually porphyritic, although of different petrographies and whole rock chemical content. And in each case the ponded flow appears to have erupted onto the surface of an earlier flow unit, thus implying that topographic ponding in a pre-existing valley was unlikely. The unusual number and proximity of such flows, and their occurrence along one of the principal tectonic deformation zones in the field, might imply that unusual conditions of magma ascent influenced by regional tectonism were responsible for their occurrence over a restricted area. Two of the cinder cones from which these lava flows were erupted are among only a few vents in the field that are characterized by bipolar fusiform bombs and spindle bombs, an indication of relatively mild strombolian activity compared to the more fragmental characteristics of the pyroclastic deposits associated with most of the vents in the SVF. This implies that the volatile overpressures characteristic of vigorous vesiculation associated with most eruptions were relatively lower in the case of these flows and that the non-hydrostatic, possibly tectonic, component of the driving pressure may have been more significant. Alternatively, the abundant phenocrysts characterizing the margins of the flows might reflect a purging of a crystal-rich crustal magma reservoir followed by more aphyric lavas later. The high phenocryst content of initial lavas would tend to increase the bulk yield strength of the margins, whereas later aphyric and low yield strength lavas would accumulate and pond within the interior. Why the scenario should be repeated several times in this region is unclear, but it is possible that unusual conditions of tectonic over pressure induced in the mid-to-shallow crustal reservoirs during contemporaneous (oblique-slip?) faulting may have led to unusual driving pressure characteristics.

Longest lava flow/lava tube

One of the longest of the younger lava flows is an aphyric basalt of transitional petrologic type erupted from the topographic summit of the field (Greens Peak, vent 8611A). This flow traveled 18 km north-northwest, where the terminus is within sight of Highway 60 and 1.6 km west of Burnt Knoll (vent 0512).The margin of the Greens Peak flow is unusual in that it is relatively digitate and complex in outline compared with many flows in the field and includes many short subsidiary

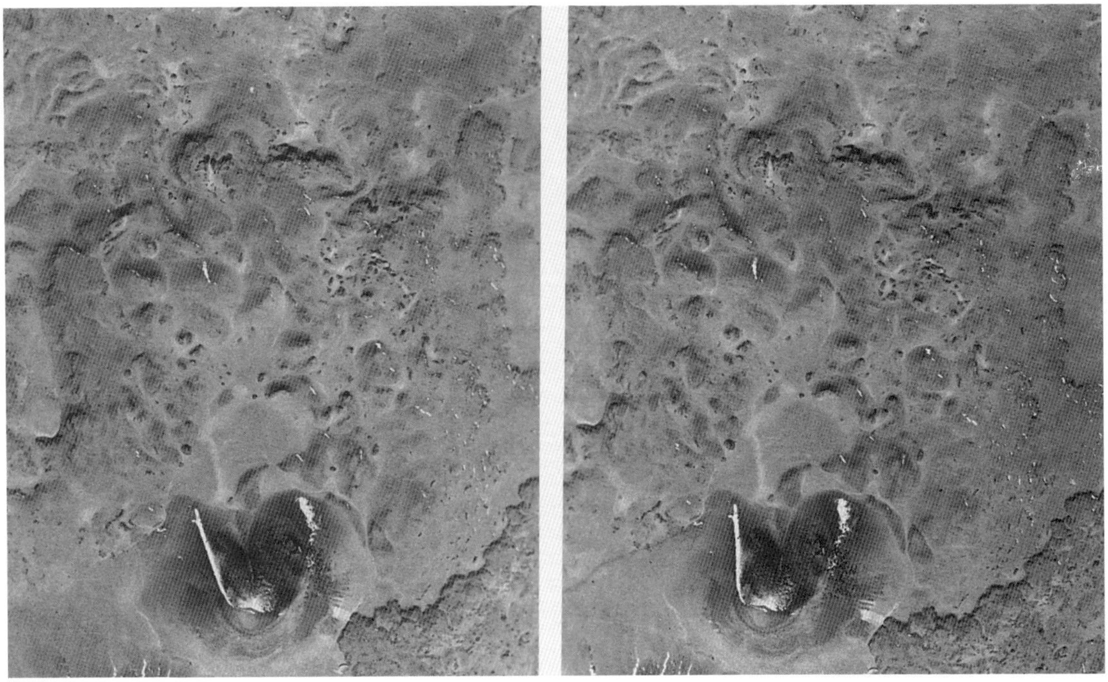

FIGURE 13. Stereo air photo pair of the ponded flow of vent 1622 west of Cerro Hueco. Hummocks distributed throughout the flow are interpreted to be rafted cone fragments torn from the main cone during formation of the large northward-facing breach in the crater. Several similar ponded flows occur throughout this part of the field. Image width is 2.5 km.

tongues and "break outs". One of the few lava tubes in the field occurs in the Greens Peak flow and the emplacement of the flow through lava tubes likely accounts for its relatively great length rather than an unusual composition or eruption dynamics. In addition, the unusual shape is consistent with tube-fed emplacement of the flow, as numerous "breakouts" and short flow pods are generally enhanced in flows emplaced by lava tubes (Taylor, 1992). The only identified opening to the lava tubes of the Greens Peak flow (Harris Cave) occurs south of the fire road connecting Vernon and the Whiting homestead (at the base of Cerro Trigo) and is 200 m west and 300 m south of where the road crosses Mineral Creek (SE 1/4 NE 1/4sec. 18, T9N, R6E). This flow is dated at 0.76±0.02 Ma (Cooper et al., 1990) (See Fig. 5 for location.) Prehistoric Indian artifacts removed from this tube by two local residents when they were children are on display in the regional museum in St. Johns.

Ultramafic nodules and megacrysts

Basalt flows and vents bearing ultramafic nodules are rare in the SVF compared with other fields around the Colorado Plateau. At the west end of Lake Boynton and just east of Negro Knoll (vent 9605), a variety of small nodules occur in an exposure near the terminus of an alkali olivine basalt flow (estimated 0.73-0.90 Ma) erupted from Whiting Knoll (vent 9621 and 9622). Smaller xenoliths also occur in the hawaiitic basalts from the Harris Lake crater vent area (vent 9630 or 9525B). Kaersutite megacrysts up to 1 and 2 cm in size also occur in the margins of a flow from vent 9603 exposed west of Cerro Trigo (in SW1/4 NW1/4 sec.4, T9N, R6E). Throughout the field, a variety of crystal-rich and megacrystic lava flows occur and are identified in unit description comment on the lithologic map (Condit et al., 1994).

Cinder cones and related vents

Cinder cones are visually arresting but constitute a small volume of the volcanic materials in the field. The general characteristics of cinder cones are remarkably similar over a wide range of settings (Kear, 1957; Colton, 1967; Porter, 1972; Wood, 1979; Hasenaka and Carmichael, 1985). In detail there are variations in morphology and these differences are important because they are frequently records of the different conditions under which each magma batch was erupted (McGetchin et al., 1974; Head and Wilson, 1989). In addition, the pattern of vent distributions and associated dikes may also be indicators of regional stress orientations and the structural controls on volcanism (Wadge and Cross, 1988).

Large scale patterns associated with fissures or dikes are relatively rare in the Springerville field (Fig. 5), but vents in the SVF are not randomly distributed. Distinct clusters, local complexes and non-fissure alignments are detected on the basis of quantitative cluster and lineament analysis (Connor et al., 1992). The highest concentration of vents (8-12 vents/100 km^2) occurs in the south-central part of the field in a corridor extending approximately from Lake Mountain on the west to Greens Peak on the east. A lower density complex extends from a cone cluster at Cerro Montoso to the vicinity of Cerro Hueco in the north-central part of the field. Connor et al. (1992) also showed that several quantitatively significant linear arrangements occur, the most prominent striking northwest, north-northeast and east-northeast. A lesser trend strikes north and east-west. Only two of these correspond to the few known fissure orientations (northeast and east-northeast). These trends define a radially-oriented pattern with respect to the center of the field, and we conclude that although the alignments are not true fissures, they might represent zones along which selective ascent and mid-crustal storage of magmas occurred along pre-existing regional structural grains. The dominant northwesterly cone alignment direction, for example, is similar to identified orientations (Fig. 5) of the more prominent anticlines and synclines. It is possible that the regional left-lateral strain interpreted from observations resulted in a relatively weak, vertically-oriented, less principal stress orientation and no single orientation was favored for true fissure eruptions. True fissures occur near the sites of pure normal faulting at the southern end of the Concho fault, or associated with fractures, interpreted as Riedel structures, in the vicinity of the major deformation trends.

Fissure vents

Because conditions of magma emplacement in the brittle upper crust imply ascent of magmas in dikes of large length with respect to their width, the initial interaction of dikes with the surface initiate most eruptions as fissures. Evidence for fissure vents is rarely seen in the SVF and implies that the length of typical fissure vents must be less than 1 km (the average diameter of the cinder cones), such that most of the initial fissures, if present, are buried by the later construction of vents. Only three prominent fissure trend lines are known to occur, and all are in regions where there is neotectonic evidence for extensional strain or where shearing strain is predicted from kinematic relations to be at a minimum relative to much of the field. Several of the fissures appear to be truncated or step left across the main fault zones in the field and are further evidence for the influence of the proposed fault zones on stress in the brittle crust throughout the region. As the driving pressures necessary to open dikes of differing length and widths in different strain environments may be calculated (Reiches and Fink, 1988), these observations ultimately may prove useful as a method of inferring the magnitude and orientation of regional tectonic stresses occurring at the time of eruption.

Character of pyroclastic materials

Many of the cinder cones consist of a variety of pyroclastic deposits, fluidal agglomerate, bombs, spatter and clinkery cinder. The more spectacular fluid types of pyroclastic deposits, such as ribbon and spindle bombs, appear to be unusually rare in the SVF compared to other basaltic fields around the Colorado Plateau. Instead, most of the Springerville cones consist of loosely-welded cinder and local clinkery agglomerate. This type of cone might develop if the fire fountains associated with eruptions forming the Springerville cones were generally more dispersed, a probable consequence of increased volatile content, vesiculation and disruption of the magmas. Gas-rich eruptions are capable of generating smaller particle sizes due to the increased fragmentation of the magma during terminal phases of ascent in the conduit. The resulting particles may be rapidly cooled below the solidus before falling to the surface of the cones such that little actual welding or formation of rootless flows occurs. Interaction of the magmas with regional water tables, especially during the past 2 million years when climates were frequently wetter in this region and aquifers were more productive may be one reason for increased incorporation of volatiles in eruptions. There is no evidence that the compositions of the magmas were themselves unusual or likely to be associated with unusual amounts of volatiles.

Young cinder cones

Varied states of preservation of primary volcanic morphologies occur in the SVF cinder cones. Unbreached craters typical of many youthful cones, however, are relatively rare. Prominent unbreached cones include Lake Mountain (vent 9423), Juan Garcia Mountain (vent 9525) and Cerro Montoso (vent 0621). The summit crater of Cerro Montoso, a prominent cinder cone in the center of the field on Highway 60 (32 km west of Springerville), is 50 m deep from rim to floor.

The southern of the two vents forming the Twin Knolls (vent 0833), the youngest center in the field (Fig. 12; see discussion of lava flows above), is characterized by dark red to black cinders, flank slopes that are at or near the angle of repose for loose cinder, and only the initial stages of erosional incision. The dark color of the cinders contrasts with most of the cones in the field, which are weathered, partially soil covered, and light reddish brown in color. Most of the cones in the SVF are slightly eroded, such that their flank slopes are less than the angle of repose typical of active cinder cones (30 ± 5°), and cut by radial drainages, as typified by Cerro Trigo (vent 9610). The cinder cones of the SVF are grass covered at elevations below 2440 m, or covered above this elevation by dense stands of aspen, ponderosa pines, or spruce, particularly on their northern flanks. Because of the latter condition, exposures of cinder cone structure are best observed at lower elevations. Scraper Knoll (vent 0919B), 2.4 km east of Highway 180 and 8 km north of Springerville (Figure 14A), is a typical example of the general state of preservation of the cones in the SVF. Based on this example, the slopes of the typical Springerville cinder cone may be divided into several elements (Fig. 14B), which correlate with different states of erosion on the upper flanks and deposition of the eroded materials on the lower flanks. Erosion has resulted in gradual reduction of the slopes relative to the angle of repose of cinders, creating an angular truncation of the bedding planes in the upper one-half of the cone flank in which the cinder and ash layers dip more steeply down slope than the surface topographic slope (Fig. 14C). As a result, the bedding planes are exposed and may be discerned as faint concentric bands around the upper flanks. The lower flanks are correspondingly buried by alluvial cinders and soils that contribute to an apron of debris that gradually diminishes in slope and blends with the more or less horizontal surface of the surrounding landscape. The slope angle is slightly less in the upper flank where the bedding planes are close to horizontal. This suggests that material is removed by erosion and transported down the cone flank from the area of exposed nearly horizontal bedding planes more rapidly than within the steeply dipping sections, and that once material is deposited in the apron it is more resistant to further

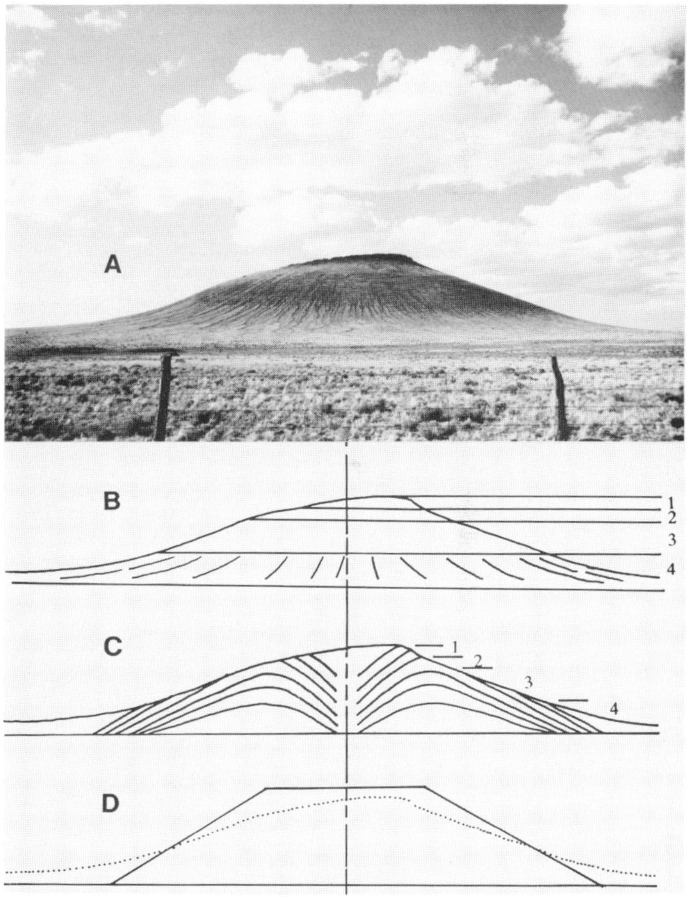

FIGURE 14. A, View of Scraper Knoll (vent 0919B) from the south. This cinder cone is typical of the state of preservation of the cinder cones of intermediate age in the Springerville volcanic field. B, Sketch map showing divisions of the flanks of Scraper Knoll defined on the basis of apparent local dip of cinder bedding planes and overall texture and color of the slopes: (1) inward-dipping, welded, near-vent cinder, spatter, and agglomerate; (2) cinder and spatter layers characterized by nearly horizontal bedding orientation; (3) outward dipping fine cinders; and (4) erosional debris apron. C, Interpretation of the relationship between erosion, secondary surface characteristics of the cone, and primary internal structure illustrating the influence of internal structure on patterns of erosion. D, Reconstruction of the original profile of Scraper Knoll compared with current profile. Reconstruction is based on assumption of an initial outer slope of ~29° (based on the average angle of repose of ash and cinder throughout the field) passing through the current contact between the encircling alluvial fan and current dissected flanks of the cone.

transport. As a consequence, there is a slight reduction of slope in the upper flank relative to lower flanks. This characteristic control of internal structure by erosion might also account for the rounded or domical profile of older cinder cones instead of a butte-like summit surrounded by a lower-sloped debris apron.

The summit rim of Scraper Knoll is not gently rounded and illustrates why many cinder cones in the SVF, and in fields of similar age throughout the Southwest, often appear morphologically more youthful to the untrained eye than they actually are. On most cinder cones at the stage of erosion in Scraper Knoll, and in slightly eroded recent cones elsewhere throughout the world, ash and spatter beds near the upper flanks dip toward the original vent crater and, together with the more strongly welded near-vent characteristic of pyroclastic materials near the summit, result in more resistant erosional behavior of the near-summit slopes. As a result, prominent outcrops of inwardly-dipping agglomerate, spatter and cinder typically occur around the uppermost flanks and summit. These "spatter rims" or "ramparts" define the rim of a type of structural crater in many cases, which mimics the original crater-shaped summit topography of more youthful cinder cones. Exposures of these fine-grained, almost glassy, and magnetite-rich materials of spatter rims in the upper flanks of many cones are frequently preferred sites for paleomagnetic sampling. Unfortunately, the outcrop faces of spatter rims are also frequently streaked by fulgarite (silicates fused into greenish glass by lightning strikes). The strong electric fields associated with lightning strikes are known to have adverse effects on remnant magnetism even without melting. Paleomagnetic sampling of these outcrops is accomplished only with careful analysis.

Although the prominent agglomerate and spatter rims represent cones that are somewhat eroded, they appear to be relatively short-lived features that occur mostly on younger cones. At this elevation, spatter rims appear to be characteristic of cinder cones between 0.3 Ma and 0.8 Ma, as exemplified by Scraper Knoll. This morphology/age sequence is climate-dependent and only locally applicable, however, as spatter rims characterize much older cones elsewhere (for example, the Mount Taylor field; Crumpler, 1980).

Vent 1627, a symmetric and elegantly shaped basalt cone that erupted flows of porphyritic hawaiite, is notable because of the presence of specular hematite in the vent area. This occurs as coatings on cinders and bombs and in the loosely welded cinder and spatter materials exposed in an erosional breach crater in the southeast summit crater rim. Post-eruptive specular hematite mineralization is observed commonly in volcanic vents and occurs in cinder cones over a wide range of ages throughout the world (Aubele, 1979; Naboko and Glavatskikh, 1984).

In a few cinder cones, erosion has exposed the connection between surrounding flows and former lava ponds. In vent 0718 (1.1 km north of Highway 60 and 30 km west of Springerville), a hawaiite lava flow forms a ramp-like ridge that extends from a bulbous outcrop in the center of the cone to the surrounding petrologically similar basalt flows. This sequence therefore records the emergence of magma from the cone and its connection with the surrounding lava flows. This is unusual because the stratigraphic relationship between most cinder cones and their erupted lava flows is rarely visible and there is rare evidence for compositionally and petrographically similar basaltic lavas within the mass of the cones. Most basalt samples recovered from within cinder cones are typically fine-grained spatter composed dominantly of microlitic, opaque, oxides-rich, rootless flow materials.

Shield volcanoes

At least two volcanic vents in the SVF are broad, shield volcano-shaped accumulations of lava flows capped with small summit pyroclastic deposits. These include Blue Ridge Mountain (vent 9328B, estimated 1.32-1.87 Ma) on the west side of the field near Pinetop, and Coyote Hills on the east side of the field about 7 km northeast of Springerville (Fig. 15). Flows from the Coyote Hills are relatively young (0.82±0.04 Ma), as dated from the upper flow unit in an outcrop in the roadcut east of Springerville (Laughlin et al., 1980). The outlines of many of the individual flows from which the shield volcano flanks are built are still clearly visible in air photos. Several similar shield-shaped vent areas occur further east, in the area south of the Red Hill region of New Mexico.

These shield-building eruptions are of interest because they illustrate a different style of eruption from that occurring throughout much of the history of the SVF. Most vents in the field consist of cinder cones from which one to several flow units erupted, usually totaling considerably less than 0.1 km^3. In contrast, the Coyote Hills consists of numerous tongue-like flows radiating from the summit area, an indication that many small effusive eruptions were repeated from this vent over an extended interval. The overall volume of lava is not much greater than that associated with single effusions of typical cinder cones. This means that the eruption proceeded in a more "staccato" style, as compared with that associated with cinder cones. The origin of these two different styles is not clear, as there is nothing anomalous about the petrology of the shield volcanoes or the morphological style of the flows with respect to the "monogenetic" cone-type eruptions. Models of the ascent and eruption of basaltic magmas (Wilson and Head, 1983) suggest that rather than resulting from gross differences in the magmas themselves, these variations in eruption style might instead simply reflect differences in the physical characteristics of the reservoirs and ascent pathways. In cinder cones and their lava flows, typically larger volumes of magma are supplied in a few batches to the surface, accompanied by more vigorous gas exsolution. In the shield volcanoes, for multiple, small volume flow units to erupt, there must have been a repetitive behavior in the supply of small volumes of magma from the underlying reservoir, as well as relatively limited magma vesiculation and pyroclastic activity at the vent.

One non-petrologic characteristic that exerts a strong influence on magma ascent, feeder dike emplacement, and eruption style is the state of stress in the crust (e.g., Delaney et al., 1986; Reiches and Fink, 1988). Evidence of considerable variation in this factor exists in the Springerville region. Long, relatively recent, normal fault scarps and faults are common in the region east of Springerville, many of which cut relatively young volcanic rocks (e.g., Crumpler and Aubele, 1990) and reflect a different stress environment than that interpreted to exist throughout most of the SVF. The structural features west of Springerville appear to be characterized more by regional strike slip than regional normal faulting. The general influence of these different styles on eruptions are such that wide variations in feeder dike characteristics and magma driving pressures may be expected and could account for the different vent morphology at the Coyote Hills. However, this may not account for Blue Ridge Mountain and it is likely that additional factors may be significant.

Clearly, more sophisticated study of the physical characteristics of volcanic deposits, on a level with our current knowledge of their petrologic characteristics, is required to better assess the origin of such differences in eruption style. The influence of these conditions is a subject of a continuing analysis using the SVF.

Maars, viscous domes, and unusual vents

Although cinder cones dominate the population of vents in the SVF, a small number of vents reflect emplacement under unusual conditions of environment or magma composition. The following unusual volcanic centers are worthy of note because they offer insight into different styles of volcanism and evidence for the varied conditions under which magmas may be emplaced.

Maars

Craters resulting from probable phreatomagmatic eruptions, or maars, occur throughout the central part of the field and attest to wetter conditions throughout much of the Southwest during the lifetime of the SVF. The youngest maar, Cerro Hueco (vent 1626A), which appropriately is Spanish for "Hole in the Hill ", is located in the north central area (Fig. 16) and is one of the younger eruptions in the field. This is a morphologically impressive east-west elongated crater situated in the saddle between two cinder cones. Total relief from the summit of the

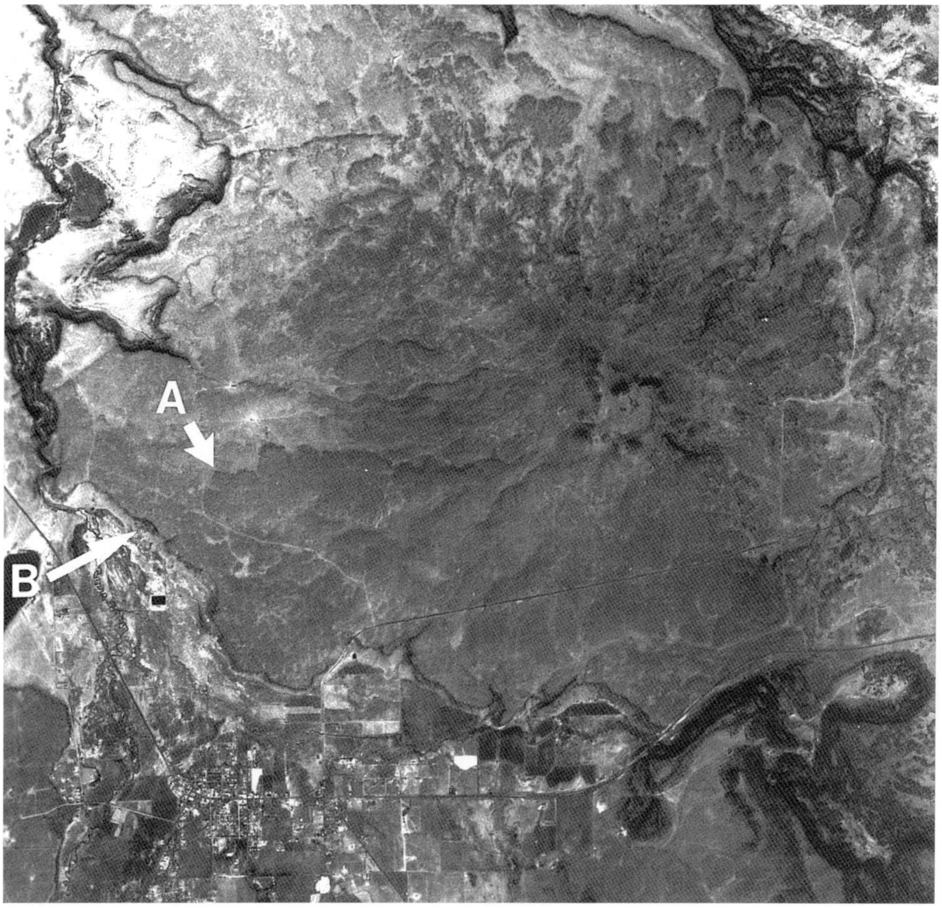

FIGURE 15. Air photo showing the Coyote Hills shield volcano and the town of Springerville (lower left) situated along Highway 60. One of the typical lava flows from which the Coyote shield volcano was built is clearly visible on the west flank (Arrow A). Arrow B indicates the Casa Malpais archeological site. Coyote Wash is in the extreme upper right. Width of image (E-W) is 11 km.

cone on the eastern rim of the maar to the bottom of the crater is 330 m; total relief from the average maar ejecta rim crest is more than 120 m. The crater was excavated through two pre-existing flow units separated by a 2-15-m-thick ash layer exposed in the the north and south crater walls (Fig. 17). The lower unit of the flow units is from a cone on the east rim, and the upper unit is from the cone on the west rim estimated to be about 0.5 Ma. Both this cone and the maar show evidence for similar minor amounts of mass wasting on their steeper slopes and may be approximately similar in age. The ejecta unit at the rim is 70 m thick and thins outward to substantially blanket surrounding lava flow surfaces for a radius of at least 1 km from the maar. The ejecta consists mostly of fine ash and cinder, but also includes accidental materials consisting commonly of blocks of basalt from the underlying flow units, less commonly of well-rounded alluvial cobbles, and rarely of angular yellow to buff-colored sandstone blocks. The cobbles are probably derived from the immediate substrate, as Tertiary conglomerates consisting of gravels and cobbles of similar lithology are exposed in the slopes of the local valley in which the maar is situated. Thus the maar-forming eruption probably originated when the Cerro Hueco magmas neared the surface and encountered the conglomerate that regional evidence suggests was likely to have been an excellent aquifer at the time of the eruption.

Several other maars are hybrid vents, the rims of which consist of interbedded palagonized tuffs and agglomerates. These probably formed in pre-existing cinder cones that subsequently erupted more phreatomagmatically. They include Harris Lake (vent 9630, estimated 0.73-0.90 Ma), Lake Hole (vent 0424, estimated 0.8-1.6 Ma; 4.8 km west of Vernon), and Smooth Knoll (vent 1526, estimated 1.15-1.45 Ma; 6.4 km N20E of Vernon). Smooth Knoll is more accurately categorized as a tuff ring, but the origin is similar in that accidental ejecta, cinders and ash, together with palagonitic materials, form a prominent crater rim around the site of former phreatomagmatic eruptions. However, in this case a subsequent cinder cone and lava pond filled in the crater. Some details of the complex intercrater stratigraphic relationships are exposed by the erosion of formerly enclosing cinders.

Accidental blocks and cinders occur on the outer flanks of Lake Hole, whereas the inner slopes consist entirely of bright reddish brown cinder and agglomerate outcrops. In this case a cinder cone was formed initially by Strombolian eruptions, but was subsequently disrupted by later, more violent phreatomagmatic activity. It is not clear whether the phreatomagmatic eruption was the ending stage of the eruption that had initially built a cone, or was an entirely later event in which magmas accidentally encountered the cone during ascent. The northeast rim of Lake Hole is considerably lower than the rest of the rim, and it appears that this side of the original cinder cone was blown away. This implies that the phreatomagmatic eruption responsible was slightly offset to the northeast from the original center of the cone. Ash and anomalously red cinder cover the pre-existing lava flows and drainages to the northeast and along Highway 60. Considering the coarse size, unusual red color typical of the interior of older cones, and unusually widespread characteristic of this ash, most of the materials likely represent cinders excavated from a pre-existing cone and blown out of the crater during the final phreatomagmatic eruption. Most ejecta may have been directed northeastward due to the location of the vent on the northeast flank of the Lake Hole cone.

Harris Lake and Norton reservoir maars are located in relatively high altitude alpine settings. The sequence of events at Harris Lake (vent 9630) is similar to that at Lake Hole; that is, a pre-existing cinder

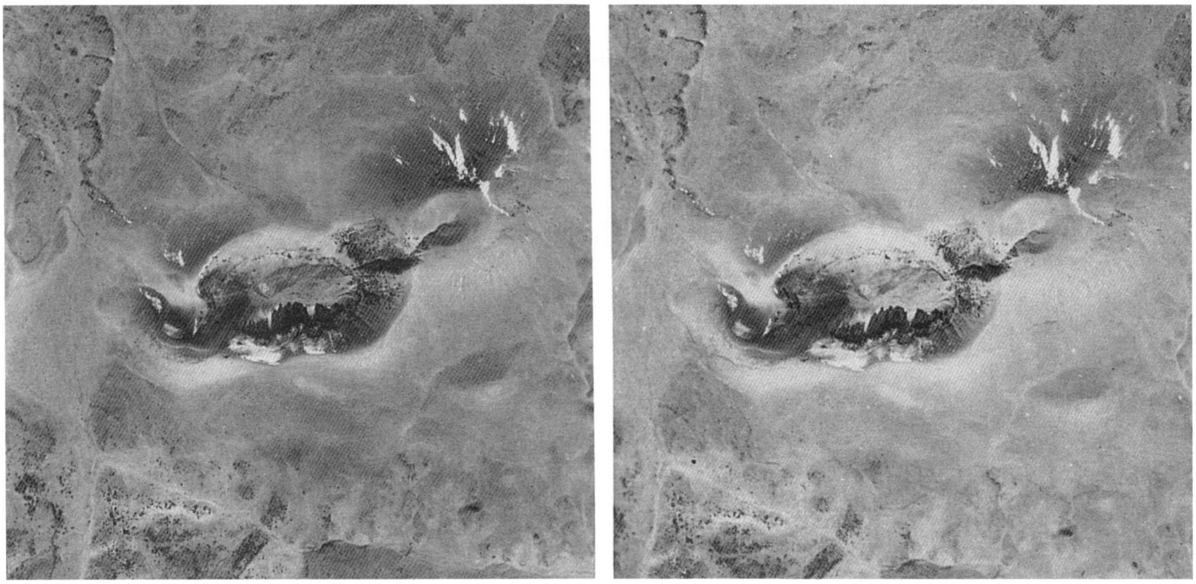

FIGURE 16. Stereo air photo pair of Cerro Hueco maar in the saddle between two cinder cones. Total relief from the summit of the eastern cinder cone to the floor of the maar is over 300 m. The cone on the west (left) end slightly predated the maar eruption and is the source of the youthful lava flow extending off the image to the northwest.

cone was partially destroyed during a final more violent phreatomagmatic eruption. Phreatomagmatic tuffs are exposed in the south rim of Harris Lake and along the outer slopes of the residual cinder cone flank forming the north rim. This sequence is likely to be governed by a balance between dike volume injection rates and ground water recharge rates. Declining volume rates and longer intervals between eruptions in later stages of the active vent history may favor increased incidents of infiltration of water into the brecciated plug of the vent area, correspondingly more violent interactions with the water table, and enhanced steam explosions. Norton Reservoir (vent 8729), located 4.9 km east of Greens Peak, is similar in many respects to the other maars in the field except that it is has a more classical maar-type morphology consisting primarily of a low circular rim of yellowish-brown pheatomagmatic tuffs surrounding a central depression.

Other maars in the SVF are generally somewhat older and more degraded. A ranch dwelling in the outskirts of Vernon, 2.4 km south and east of Highway 60 along the Vernon-McNary Road, is blithely situated in the center of what remains of a maar (vent 0522, estimated 0.91-1.53 Ma), the southern rim of which has been removed by a through-going wash and later filled with younger basalt flows. Well data indicate basalt fills the floor to a depth of at least 90 m. The prevalence of maars in the Vernon area may be attributed to the presence of well-developed drainages fed by streams from higher altitudes in this part of the field.

Viscous domes

Vents formed from eruptions of more evolved and viscous magmas are rare and in this respect the SVF is unusual, especially considering the long history and relatively great volume of volcanic rocks erupted. Eruptions of more evolved magmatic composition are common in many of the volcanic fields around the Colorado Plateau. However, at least two examples of the extrusion of more viscous flows are identified in the SVF. One striking example is Wolf Mountain (vent 9507), dated at 1.56±0.05 Ma (Aubele et al., 1986), approximately 8 km S45°W of Vernon. Here, a dome, 1 km in diameter and consisting of viscous and massive flows of porphyritic benmoreite, was emplaced on the south side of a pre-existing cinder cone that had previously erupted relatively aphyric mugearite flows. Judging from the structure and location of the dome, it erupted from a vent slightly offset from the center of the pre-existing cone. This dome is relatively degraded, with excellent exposures of massive, inward-dipping benmoreite.

A similar, although not as well exposed, occurrence of viscous benmoreite is exposed in the north flank of Pole Knoll (vent 8732, dated at 1.30±0.04 Ma, Cooper et al., 1990), a large composite cinder cone 9 km S45°E of Greens Peak (1.4 km south and 18 km west of Eagar on Highway 260). The small exposure of gray benmoreite in the north flank contrasts strikingly with the darker aphyric and olivine basalts from the main mass of Pole Knoll. In this case the benmoreite may have pre-dated the emplacement of most of the cone.

Neither of these centers are among the younger eruptions. This illustrates the point that although evolved rocks occur, in general, with greater frequency among the younger age eruptions, the most evolved rocks are not necessarily the youngest. It likewise reflects the fact that within typical volcanic fields there are a variety of different magma sources that probably evolve separately, and the degree of evolution need not reflect the timing of such eruptions with respect to the field as a whole or the evolution of a single magma source.

Other examples of viscous lavas were erupted from more typical cinder cones or did not involve the construction of domes. Most such examples are compositionally mugearites, often with unusually large phenocrysts of hornblende or plagioclase. A large flow of mugearitic composition with no apparent vent structure occurs south of Concho near Concho Springs Knoll (vent 2529, estimated 0.90-1.30 Ma).The spatter of Concho Springs Knoll immediately to the north of this flow is compositionally similar to the mugearitic flow, and the vent and flow must be considered to be related, but there is no clear connection between the cone and the flow. One possibility is that the cone grew separately but from the same reservoir, and the main mugearite flow was erupted from a related vent to the south. Other examples of similar relatively thick and apparently viscously erupted flows include those from vents 1634, 9605 (mentioned above for the unusually large size of hornblende phenocrysts in its lavas), and 0634. Burnt Knoll (vent 0512, estimated 0.55-1.10 Ma), 5 km east of Vernon on Highway 60, is a cinder cone with an associated porphyritic mugearite flow (up to 30 m thick) in which plagioclase appears as the dominant phenocryst.

Unusual vents

A particularly unusual vent (9602, estimated 0.30-1.10 Ma) occurs in the central field, 3.2 km west of the Greens Peak-to-Highway 60 forest road and 4 km south of Highway 60. In this case, a pre-existing basalt flow surface was domed up 100 m into a steep cone-shaped rise, or cryptodome, by an apparent near-surface laccolithic intrusion. Magma

FIGURE 17. Cerro Hueco maar seen from the north rim. The view is to the south. Cerro Montoso is the peak in the near distance and Greens Peak area vents at the summit of the field are in the far distance.

subsequently broke through the eastern end of a small east-west graben along the crest of the dome and erupted a thick, short flow of mugearite accompanied by a small amount of vent spatter and agglomerate. The lavas were clearly emplaced after the dome was formed, as indicated by the fact that the lava flow is diverted around the base of the dome. Thus a considerable amount of magma was injected extremely near the surface before eruption of the flow commenced. The relatively minor amounts of pyroclastic material in the vent area suggest that the eruption was relatively volatile-poor compared to many eruptions in the field, perhaps as a consequence of degassing of the magma during formation of dome.

An apparently rootless flow, 3 km long, occurs south of Cerro Hueco (sec.2, T10N, R6E). This is a typical olivine basalt, but it tapers to a narrow ridge at its apparent origin near the base of a long slope from the Cerro Montoso complex of cones and it has no obvious vent materials or uphill source. In the absence of any other obvious mechanism for its emplacement, it is interpreted to represent a relatively quiet outwelling of lava from a small fissure striking N80°E that formed in the side of the regional slope leading up to the Cerro Montoso vent complex. The surface of this flow was also tilted 2-3°NW throughout much of its length during the later tectonic deformation. Similar distortion, warping and back-tilting characterizes many flows in the northern part of the field and illustrates one of the means whereby small amplitude tectonic deformations were detected through detailed mapping.

TRAVERTINE MOUNDS

Travertine mounds occur throughout the valley of the Little Colorado, on the northeast edge of the SVF. Their origin is relevant to understanding the regional structure and the many excellent exposures illustrate an unusually symmetric form of travertine mound development.

Almost all of the travertine occurrences lie in the structural depression bounded on the southwest by the Concho lineament and on the northeast by the Coyote Wash fault trend (Fig. 18).This indicates that the travertine originated when ground water flowed down the hydrologic gradient from the central volcanic field, and was forced to emerge as springs along deep faults on the periphery of the structural depression. Carbonates acquired by the water along deep circulation pathways from the central field outward to the margins precipitated at the surface as the water emerged as springs. Few springs are active (Sirrine, 1958); springs in most of the travertine mounds are currently inactive, reflecting the general drying of the climate and decreased recharge of the deep water table throughout the region in the recent past.

Some travertine deposits in the SVF are relatively featureless, but most are simple cone-shaped features. Although travertine occurs throughout the world in various settings (Chafetz and Folk, 1984; Kendall and Warren, 1987), the unique and elegantly symmetrical structure of the travertine deposits here and at a few other sites in the southwestern U.S. mimics in many respects the morphology and formation of small shield volcanoes (Fig. 19) and appears to have gone largely unstudied. Completing the analogy with small volcanoes, each travertine mound is characterized by a circular summit depression, or spring orifice, which is frequently steep-sided and deep, or shallow and bowl-shaped. Many of the earlier travertine mounds stand as isolated travertine-topped sedimentary buttes due to the downward incision of the Little Colorado and its flood plain, but several travertine mounds,

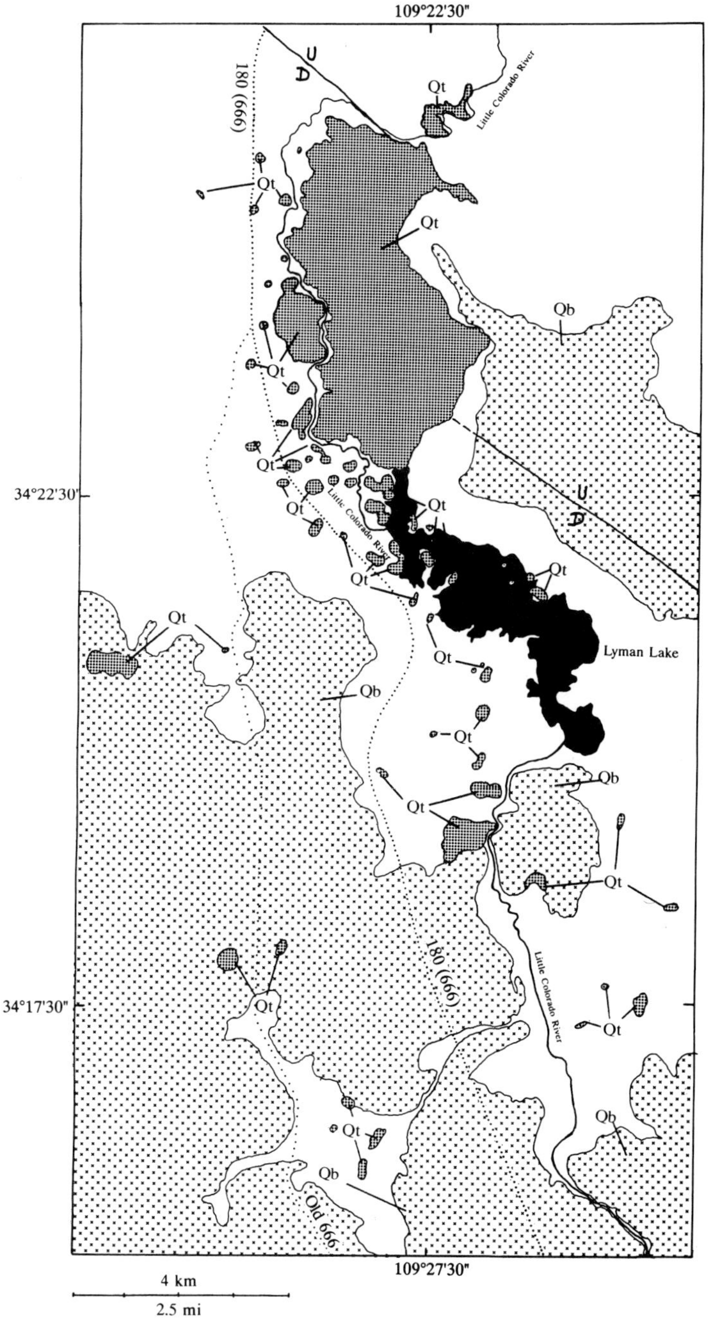

FIGURE 18. Occurrence of spring deposits (Qt) in the SVF. Most of the spring deposits occur in the valley of the Little Colorado River and lie up the hydrologic gradient from a major regional fault (Coyote Wash fault, see Fig. 5). The spring deposits probably formed here when deep carbonate-charged ground water flowing down the hydrologic gradient from the White Mountains and Springerville volcanic field emerged along the Coyote Wash fault zone. Qt - travertine deposits; Qb - basalts of the Springerville volcanic field, undivided; white - sedimentary units undivided.

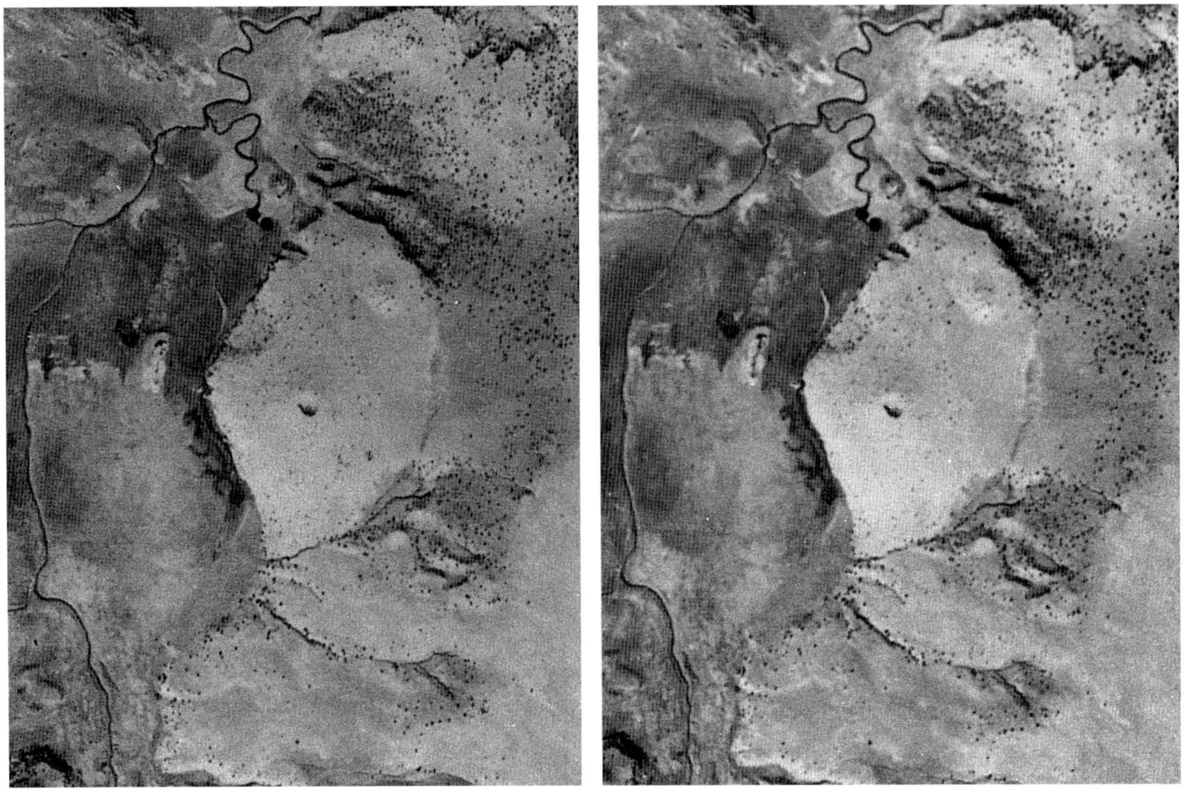

FIGURE 19. A. Stereo air photo pair of a typical small, symmetrical mound-type spring deposit in the northeast margin of the Springerville volcanic field, in the vicinity of the Little Colorado River and Lyman Lake. The travertine mounds are generally characterized by cone-shaped profiles and small summit pits resulting in an appearance similar to small volcanic cones or shield volcanoes. The similarity to volcanic cones is a consequence of the slow accumulation of precipitates from many small spring "discharge events" about a central orifice, analogous to lava flows accumulated about a volcanic vent. Image width is 1 km.

particularly those high on the slopes of the valley are relatively uneroded. One travertine mound is naturally half-sectioned on the east shore of Lyman Lake (Fig. 20). Based on this example, most of these mounds consist of gently dipping alternating dark and light layers, sweeping to a summit where the layering is disturbed by chaotic collapse structures in the immediate vicinity of the spring orifice. Others are simple straight-sloped cones. The origin of the cone or shield-like morphology is probably a consequence of repeated deposition of precipitates from azimuthally random discharges about a stable central spring. Variable rates and distances of discharge owing to seasonal and secular variation in rates of ground water flow probably control the magnitude and lateral extent of each deposition during each cycle and thus control the ultimate slope and form of individual mounds.

CONCLUSIONS

Several broad conclusions about the Springerville volcanic field can be stated on the basis of mapping, analysis of petrologic samples and regional structure, and physical volcanological characteristics. First, the field is petrologically diverse and includes tholeiitic, alkalic and evolved alkalic rock types. These are characterized by a general evolutionary trend in petrology in which tholeiitic rocks account for 24% of the field volume and were erupted early, and alkalic and evolved alkalic rocks (mugearite and benmoreite) were erupted later (Condit et al., 1989a). This might imply that the relative degree of partial melting and the total volume of melts available for eruption diminished steadily with time.

Second, the average volume rate of effusion over the history of the field was 1.5 X 10^{-4} km^3/a (Crumpler et al., 1990). This is similar to rates observed for other cinder cone volcanic fields throughout the world (Fedotov, 1984; Shaw, 1985), and is lower by about an order of magnitude than typical rates associated with large volcanic edifices and calderas characterized by single large magma reservoirs and repeated eruptions from isolated and evolved magmatic centers (Fig. 21). Estimates of the rates of magma supply necessary to maintain large magma reservoirs (Hardee, 1982; Shaw, 1985) imply that volume rates equivalent to those seen in cinder cone fields are slightly less than that necessary for establishing and maintaining a magma body in the crust over periods of several thousand to one million years. Although magma supply in the SVF was long-lived, it has been too low and sporadic to sustain a single large volcano, thus resulting in a concentration of isolated vents.

FIGURE 20. Natural section through a travertine cone (400 m in diameter) exposed on the east shore of Lyman Lake.

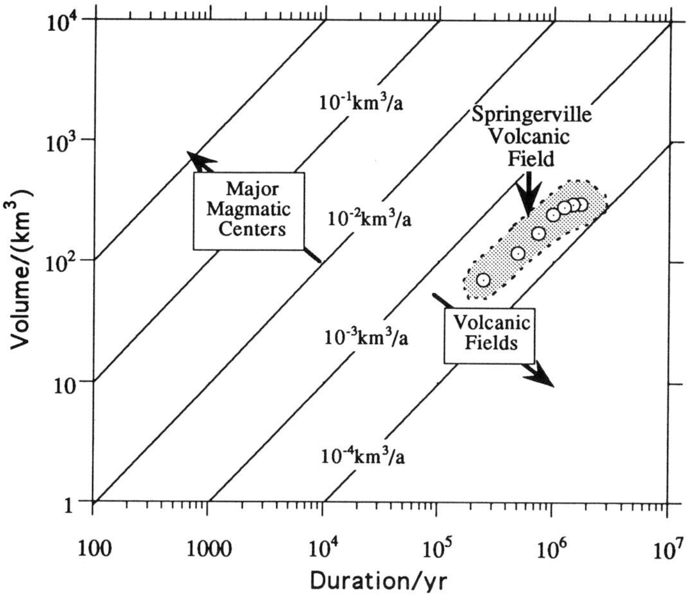

FIGURE 21. Volume-duration plot for the Springerville field, showing the volume rates of eruption for several intervals during the life of the field (open circles). Observed rates of volcanic output throughout the lifetime of the SVF are typical of volcanic fields in general. Based on consideration of the rates of magma replenishment necessary to develop central magma chambers, magma replenishment rates in the SVF, as with many cinder cone type volcanic fields, appear to be lower than that necessary to sustain a central edifice.

Third, the great variety in the style of vents reflects both environmental conditions present at the time of the eruptions and the influence of neotectonic stresses on the orientation and the dynamic characteristics of individual dikes and fissures initiating each eruption. The general absence of strong patterns of fissure vents implies that simple tensile stresses that would favor long fissures have not been present, a consequence perhaps of the interpreted left lateral shear stresses in the brittle crust throughout much of the field. Pyroclastic materials in many of the vents may be more fragmental (less spatter) than elsewhere. This may reflect greater fragmentation resulting from generally wetter near-surface conditions at the time of most of the eruptions, favoring increased participation of non-juvenile volatiles and stronger vesiculation. In addition, the higher magma driving pressures necessary to open and sustain dikes in this tectonic stress environment may have been influential.

Fourth, detailed mapping has enabled us to recognize neotectonic features in this part of the Colorado Plateau, based on folds and faults post-dating uneroded late Cenozoic basaltic lava flows (Crumpler et al., 1989). Quantitative analysis has shown distinct clusters and alignments of vents in the SVF (Connor et al., 1992) and that the boundary of many clusters of vents, as well as the termini of fissure trends and vent alignments, correspond with the location of many large-scale neotectonic faults and folds. Large-scale crustal structure appears to have played a part in localizing vents and concentrating them in clusters. Similarly, fissures appear to be sensitive to variations and discontinuities in patterns of strain in the brittle upper crust. The distance between local clusters of cones (15 to 20 km) may be significant, as this dimension is less than the crustal thickness in the region (Keller et al., 1979), but approaches the thickness of the brittle crust in the Colorado Plateau (Wong and Humphrey, 1989). Although more speculative, the location of vents could be influenced by crustal blocks that are bounded by strong through-going structures. This same regional structure probably influenced the regional ground water flow and contributed to the formation of spring deposits (travertine). Many of the spring deposits are unusually symmetric in shape and reflect steady discharge of carbonate-rich water from central spring orifices.

We interpret the large-scale physiographic and morphologic features in the SVF to be a consequence of limited, very nearly incipient strike-slip faulting along several regional fault zones striking northwest through the field in which strike-slip is accommodated by normal faulting near the ends of the fault zones. The neotectonic characteristics and the volcanism may have originated in part during lithospheric extension across the margins of the Colorado Plateau associated with late Cenozoic clockwise rotation of the Colorado Plateau. Regardless of the origin of the regional deformation responsible for these structures, it is likely that at least some characteristics of distribution and geometry and eruption dynamics of individual vents reflect the influence of contemporaneous tectonic stresses on the ascent and emplacement of magmas.

ACKNOWLEDGMENTS

Mapping of the Springerville volcanic field was supported by the U.S. Geological Survey, Los Alamos National Laboratories, and the Arizona Bureau of Geology. We thank Wolf Elston (University of New Mexico), George Ulrich (U.S.Geological Survey) and Ed Wolfe (U.S.Geological Survey) for their support in making this project a reality. Sue Gaffey (Rensselaer Polytechnic Institute) provided insight into the world of travertines and spring deposits. Steve Cather and Bill McIntosh (New Mexico Bureau Mines and Mineral Rseources) kindly reviewed the manuscript and offered many helpful revisions and insights. We also thank the people of the communities and ranches located in and on the margins of the SVF, especially those residents of Springerville who became our friends during many summers of geologic mapping.

REFERENCES

Aubele, J. C., 1979, The Cerros del Rio volcanic field, New Mexico: New Mexico Geological Society, Guidebook 30, p. 243-252.

Aubele, J. C. and Crumpler, L.S., 1983, The geology of the central and eastern parts of the Springerville-Show Low volcanic field, Arizona: Geological Society of America, Abstracts with Programs, v. 15, p. 303.

Aubele, J. C., Crumpler, L.S. and Shafiqullah, M., 1986, K-Ar ages of late Cenozoic rocks of the central and eastern parts of the Springerville volcanic field, east-central Arizona: Isochron West, v. 46, p. 3-5.

Aubele, J. C., Crumpler, L.S. and Condit, C.D., 1987, Tectonic deformation of the late Cenozoic Springerville volcanic field, southern margin of the Colorado Plateau, Arizona: Geological Society of America Abstracts with Programs, v. 19, p. 576.

Bolton, H. E., 1974, Coronado, Knight of Pueblo and Plains: University of New Mexico Press, Albuquerque, 491 p.

Brumbaugh, D. S., 1987, A tectonic boundary for the southern Colorado Plateau: Tectonophysics, v. 136, p. 125-136.

Bryan, P. and Gordon, R.G., 1990, Rotation of the Colorado Plateau: an updated analysis of paleomagnetic poles: Geophysical Research Letters, v. 17, p. 1501-1504.

Burroughs, E. R., 1935, Swords of Mars : Ballantine Books, New York, re-printed 1963, 192 p.

Chafetz, H. S. and R. L. Folk, 1984, Travertines: depositional morphology and the bacterially constructed constituents: Sedimentary Petrology, v. 54, p. 289-316.

Colton, H. S., 1967, The basaltic cinder cones and lava flows of the San Francisco Mountain volcanic field: Museum of Northern Arizona, Bulletin, v. 10, 58 p.

Condit, C. D., Crumpler, L.S. and Aubele, J.C., 1987, Volcanic record of the Springerville volcanic field, East-central Arizona: Geological Society of America, Abstracts with Programs, v. 19, p. 625.

Condit, C.D., Crumpler, L.S. and Aubele, J.C., 1989a, Patterns of volcanism along the southern margin of the Colorado Plateau: the Springerville field: Journal of Geophysical Research, v. 94, p. 7975-7986.

Condit, C.D., Crumpler, L.S. and Aubele, J.C., 1989b, Springerville volcanic field: New Mexico Bureau of Mines and Mineral Resources, Memoir 46, p.33-41.

Condit, C. D., Crumpler, L.S. and Aubele, J.C., 1994, Thematic geologic maps of the Springerville volcanic field, east-central Arizona: U. S. Geological Survey, Miscellaneous Investigation Series Map I -2431, scale 1:100,000, in press; also available in digital form.

Condit, C. D. and Shafiqullah, M., 1985, K-Ar ages of late Cenozoic volcanic rocks of the western part of the Springerville volcanic field, east-central Arizona: Isochron West, v. 44, p. 3-5.

Connor, C.B., 1990, Cinder cone clustering in the TransMexican volcanic belt: implications for structure and petrologic models: Journal of Geophysical Research, v. 95, p. 19395-19405.

Connor, C.B., Condit, C.D., Crumpler, L.S. and Aubele, J.C., 1992, Evidence for regional structural controls on vent distribution: Springerville volcanic field, Arizona: Journal of Geophysical Research, v. 97, p. 12349-12359.

Cooper, J. L. and Hart, W.K., 1990, Mantle sources in the Arizona transition zone and global mantle heterogeneity: Geology, v. 18, p. 1146-1149.

Cooper, J. L., Aronson, J. L., Condit, C. D. and Hart,W. K., 1990, New K-Ar ages of lavas from the Colorado Plateau-Basin and Range transition zone, east-central Arizona: Isochron West, v. 55, p. 28-31.

Crumpler, L. S., 1980, An alkali basalt through trachyte suite, Mesa Chivato, Mount Taylor volcanic field, New Mexico: Geological Society of America Bulletin, v. 91, part 1, p. 253-255; part 2, p. 1293-1331.

Crumpler, L. S., 1982, Volcanism in the Mount Taylor region: New Mexico Geological Society, Guidebook 33, p. 291-298.

Crumpler, L.S., 1990, Mount Taylor volcanic field, New Mexico; *in* Wood, C.A., and Kienle, J., eds., Volcanoes of North America: Cambridge University Press, p. 300-302.

Crumpler, L. S. and Aubele, J. C., 1990, Red Hill Volcanic field, New Mexico; *in* Wood, C.A., and Kienle, J., eds., Volcanoes of North America: Cambridge University Press, p. 306-308.

Crumpler, L. S., Aubele, J.C. and Condit, C.D., 1989, Influence of Quaternary tectonic deformation on volcanism in the Springerville volcanic field, Colorado Plateau, USA: New Mexico Bureau of Mines and Mineral Resources, Bulletin 31, p. 64.

Crumpler, L.S., Aubele, J.C. and Condit, C.D., 1990, Springerville volcanic field, Arizona; *in* Wood, C.A., and Kienle, J., eds., Volcanoes of North America: Cambridge University Press, p. 284-286.

Darton,, N.H., 1915, The Santa Fe route: U.S. Geological Survey, Bulletin 613, 200 p.

Delaney, P. T., Pollard, D. D., Ziony, J. I. and McKee, E. H., 1986, Field relations between dikes and joints: emplacement processes and paleostress analysis: Journal of Geophysical Research, v. 91, p. 4920-4938.

Fedotov, S.A., 1981, Magma rates in feeding conduits of different volcanic centres: Journal of Volcanology and Geothermal Research, v. 9, p. 379-394.

Hamilton, W., 1987, Plate tectonic evolution of the western U.S.A.: Episodes, v. 10, p. 271-276.

Hardee, H.C., 1982, Incipient magma chamber formation as a result of repetitive intrusions: Bulletin of Volcanology, v. 45, p. 1-49.

Hasenaka, T. and Carmichael, I.S.E., 1985, The cinder cones of the Michoacán-Guanajuato, central Mexico, their age, volume, distribution, and magma discharge rate: Journal of Volcanology and Geothermal Research, v. 25, p. 105-204.

Head, J. W. and Wilson, L., 1989, Basaltic pyroclastic eruptions: influence of gas-release patterns and volume fluxes on formation of cinder cones: Journal of Volcanology and Geothermal Research, v. 37, p. 261-271.

Holm, R. F., 1987, Significance of agglutinate mounds on lava flows associated with monogenetic cones: an example at Sunset Crater, northern Arizona: Geological Society of America Bulletin, v. 99, p. 319-324.

Kear, D., 1957, Erosional stages of volcanic cones as indicators of age: New Zealand Journal of Science and Technology, v. B38, p. 671-682.

Keller, G. R., Braile, L.W. and Morgan, P., 1979, Crustal structure, geophysical models and contemporary tectonism of the Colorado Plateau: Tectonophysics, v. 61, p. 131-147.

Kelley, V. C., 1982, The right-relayed Rio Grande rift: New Mexico Geological Society, Guidebook 33, p. 147-152.

Kendall, C. G. and Warren, J., 1987, A review of tepees and their associated fabrics: Sedimentology, v. 34, p. 1007-1027.

Laughlin, A. W., Brookins, D. G., Damon, P. E. and Shafiqullah, M., 1979, Late Cenozoic volcanism of the central Jemez zone, Arizona-New Mexico: Isochron West, v. 25, p. 5-8.

Laughlin, A. W., Damon, P. E. and Shafiqullah, M., 1980, New K-Ar dates from the Springerville volcanic field, central Jemez zone, Apache County, Arizona: Isochron West, v. 29, p. 3-4.

McGetchin, T. R., Settle, M. and Chouet, B.A., 1974, Cinder cone growth modeled after Northeast Crater, Mount Etna, Sicily: Journal of Geophysical Research, v. 79, p. 3257-3272.

Merrill, R. K. and Péwé, T. L., 1976, Late Cenozoic geology of the White Mountains, Arizona: Arizona Bureau of Geology and Mineral Technology, Special Paper no.1, 65 p.

Naboko, S. I. and Glavatskikh, S. F., 1984, Post eruptive hematite mineralization: Volcanology and Seismology, v. 5, p. 83-96.

Nealey, L. D., 1989, Geology and petrology of the Mount Baldy trachyteic volcanic complex, White Mountains volcanic field, Apache and Navajo counties, Arizona [Ph.D. dissertation]: Albuquerque, University of New Mexico, 320 p.

Porter, S.C., 1972, Distribution, morphology, and size-frequency of cinder cones on Mauna Kea volcano, Hawaii: Geological Society of America Bulletin, v. 92, p. 448-456.

Reiches, Z. and Fink, J., 1988, The mechanism of intrusion of the Inyo dike, Long Valley caldera, California: Journal of Geophysical Research, v. 93, p. 4321-4334.

Renault, C. E., Wells, S.G. and McFadden, L. D., 1988, Geomorphic and pedologic evidence for polycyclic volcanism in late Quaternary cinder cones: examples from Cima volcanic field, California and Crater Flat/Lathrop Wells volcanic field, Nevada: Geological Society of America, Abstracts with Programs, v. 20, p. A115.

Shaw, H.R., 1985, Links between magma-tectonic rate balances, plutonism, and volcanism: Journal of Geophysical Research, v. 90, p. 11275-11288.

Sirrine, G. K., 1958, Geology of the Springerville-St. Johns area, Apache County, Arizona [Ph.D. dissertation]: Austin, University of Texas, 248 p.

Spence, W. and R. S. Gross, 1990, A tomographic glimpse of the upper mantle source of magmas of the Jemez lineament, New Mexico: Journal of Geophysical Research, v. 95, p. 10829-10849.

Sylvester, A. G., 1988, Strike-slip faults: Geological Society of America Bulletin, v. 100, p. 1666-1703.

Tanaka, K. L., Shoemaker, E. M., Ulrich, G. E. and Wolfe, E. W., 1986, Migration of volcanism in the San Francisco volcanic field, Arizona: Geological Society of America Bulletin, v. 97, p. 129-141.

Taylor, G. J., 1992, Fractal properties of lkava tube systems. EOS (American Geophysical Union Transactions), v. 73, p. 648.

Ulrich, G. E., Holm, R.F. and Wolfe, E. W., 1989, San Francisco volcanic field: New Mexico Bureau of Mines and Mineral Resources, Memoir 46, p. 10-21.

Wadge, G. and Cross, A., 1988, Quantitative methods for detecting aligned points: an application to the volcanic vents of the Michoacán-Guanajuato volcanic field, Mexico: Geology, v. 16, p. 815-818.

Wells, S. G., McFadden, L. D., Renault, C. E. and Crowe, B. M., 1990, Geomorphic assessment of late Quaternary volcanism in the Yucca Mountain area, southern Nevada: Implications for the proposed high-level radioactive waste repository. Geology, v. 18, p. 549-553.

Wilson, L. and Head, J.W., 1983, A comparison of volcanic eruption processes on Earth, Moon, Mars, Io, and Venus. Nature, v. 302, p. 663-669.

Wilson, L. and Parfitt, E. A., 1993, The formation of perched lava ponds on basaltic volcanoes: the influence of flow geometry on cooling-limited lava flow lengths: Journal of Volcanology and Geothermal Research, v. 56, p. 113-123.

Wong, I. G. and J. R. Humphrey, 1989, Contemporary seismicity, faulting, and state of stress in the Colorado Plateau: Geological Society of America Bulletin, v. 101, p. 1127-1146.

Wood, C.A., 1979, Monogenetic volcanoes of the terrestrial planets: Proceedings of the 10th Lunar and Planetary Science Conference, p. 2815-2840.

Wood, C. A., 1980, Morphometric evolution of cinder cones: Journal of Volcanology and Geothermal Research, v. 8, p. 387-413.

^{40}AR/^{39}AR GEOCHRONOLOGY OF MIDDLE TO LATE CENOZOIC IGNIMBRITES, MAFIC LAVAS, AND VOLCANICLASTIC ROCKS IN THE QUEMADO REGION, NEW MEXICO

WILLIAM C. MCINTOSH and RICHARD M. CHAMBERLIN,
New Mexico Bureau of Mines and Mineral Resources, Socorro, N.M. 87801

Abstract—In order to better address stratigraphic problems identified during geologic mapping in the Quemado, New Mexico, area, ^{40}Ar/^{39}Ar dating methods were applied to 33 samples of volcanic and volcaniclastic rocks. Single-crystal laser-fusion techniques were used to determine eruptive ages of rhyolitic ignimbrites and of volcanic pumice and ash in sandstones. Mafic lavas were analyzed by step-heating of bulk samples. Sanidine phenocrysts from Quemado area ignimbrites range in age from 34.98 ± 0.19 to 27.23 ± 0.09 Ma. Age and K/Ca results allow these distal-facies outflow sheets to be confidently correlated with 11 well-known regional ignimbrites erupted from calderas in the Mogollon-Datil volcanic field. Ages for the Datil Well Tuff (34.98 ± 0.19 Ma) and Rock House Canyon Tuff (34.17 ± 0.08 Ma) are significantly younger than published values, probably reflecting xenocrystic contamination in previously analyzed bulk samples. Lithic-rich ignimbrites within the South Crosby Peak Formation contain phenocrystic plagioclase that indicates an eruptive age near 29.67 Ma. These units lack phenocrystic sanidine but contain sanidine xenocrysts derived from older rhyolitic ignimbrites. Sparse sanidine and abundant plagioclase from syneruptive volcaniclastic facies suggest depositional ages of 35 to 33 Ma for the volcaniclastic unit of Cañon del Leon and approximately 37 Ma for ash-rich beds in the Dog Springs Formation. Phenocrystic plagioclase from the Cerrito Viejo dacitic intrusion yielded low-resolution results indicating emplacement near 30.6 ± 1.5 Ma. Step heating of groundmass concentrates from Oligocene and Pliocene mafic to intermediate lavas gave disturbed spectra and approximate isochron ages for the andesite of San Antone Canyon (29.2 Ma, new unit), Squirrel Springs Andesite (28.8 Ma), Bearwallow Mountain Andesite (26.1 Ma), and basalts of Blue Mesa (5.7-4.5 Ma). In combination with recent mapping and stratigraphic data, ^{40}Ar/^{39}Ar dating results provide a valuable chronostratigraphic framework for interpretation of stratigraphy, sedimentation and tectonic history of the Quemado area.

INTRODUCTION

This study presents a chronostratigraphic framework for volcanic and volcaniclastic rocks exposed near Quemado, New Mexico, an area recently mapped by Chamberlin et al. (1994a). ^{40}Ar/^{39}Ar dating methods were applied to four rock types: ignimbrites, pumice-and-ash-rich volcaniclastic rocks, mafic to intermediate lavas, and a dacitic intrusion. Dating of ignimbrites was the primary focus of the study. Large-volume ignimbrites (ash-flow-tuffs) are emplaced nearly instantaneously over wide areas, and hence provide regional time-stratigraphic marker horizons. The Quemado region exposes numerous thin ignimbrites that represent the northern distal fringes of regional ignimbrites erupted from calderas of the Mogollon-Datil volcanic field. The stratigraphy and age of Mogollon-Datil volcanic field ignimbrites has been well established by mapping, paleomagnetic correlation, and high-precision ^{40}Ar/^{39}Ar dating of sanidine phenocrysts (McIntosh et al., 1992). However, it is not possible to make reliable lithologic correlations of Quemado distal-facies ignimbrites with established Mogollon-Datil ignimbrites, because of lithologic similarities between some units and lithologic variations within others. In this study, ^{40}Ar/^{39}Ar ages and K/Ca ratios of sanidine phenocrysts were used help establish these correlations.

Pumice-and-ash-rich volcaniclastic rocks were also investigated using ^{40}Ar/^{39}Ar laser-fusion analyses of individual feldspar crystals. Where sanidine is present in such rocks, high-precision single-crystal dates help to precisely constrain provenance and maximum depositional age. Where only plagioclase is present, the ^{40}Ar/^{39}Ar results are less precise and provide more approximate constraints. In addition, a few mafic-to-intermediate lavas and one dacitic intrusion were analyzed by step-heating and laser-fusion ^{40}Ar/^{39}Ar methods. The results of these analyses have lower precision, because high-K, highly retentive phenocrysts like sanidine were not present.

GEOLOGIC SETTING

The Quemado region lies on the northern periphery of the Mogollon-Datil volcanic field, near the northern end of a 3000-km-long middle Tertiary volcanic arc mostly defined by the Sierra Madre Occidental in old Mexico (Elston, 1978). This subduction-related magmatic arc terminates abruptly against the southeastern margin of the Colorado Plateau (Mogollon slope) and then reappears approximately 350 km farther north in the San Juan volcanic field of southwestern Colorado and the Latir field of northern New Mexico. This significant middle Tertiary magma gap is associated with a major ENE-trending crustal boundary, presently expressed by the San Agustin arm of the Rio Grande rift (Chapin, 1971) and the Morenci lineament (Ratté, 1989). Based on regional seismic reflection profile data, Garmezy (1990) described this major crustal boundary as a ENE-trending, right-lateral, wrench-fault zone of Laramide ancestry. Transpressional welts formed along the Laramide Morenci lineament were part of the source terrane for siliciclastic red beds of the middle to upper Eocene Baca Formation (Chapin and Cather, 1981; Cather and Johnson, 1984).

Andesitic volcanism began in the northern Mogollon-Datil field about 40 Ma and apparently buried these Laramide welts. Middle Tertiary volcanic apron deposits in the Quemado-Datil region, as much as 1.5 km thick, loaded and downwarped the southeastern Colorado Plateau margin to form the gently south dipping homocline of the Mogollon slope (Chamberlin and Cather, this volume).

During the late Eocene to Oligocene (35-26 Ma), the northern Mogollon-Datil volcanic field erupted numerous large-volume rhyolitic ignimbrite sheets from calderas located 50-100 km south and east of the Quemado area (Osburn and Chapin, 1983; McIntosh et al., 1992). Caldera-related catastrophic eruptions were preceded, accompanied and followed by smaller-volume silicic, intermediate and mafic volcanism. The sequence that accumulated in the Quemado region during this period includes interbedded distal ignimbrites, lavas, and volcaniclastic sediments derived from lavas and pyroclastic material rapidly eroded from the volcanic highlands to the south. This volcanogenic sequence intertongues with volcaniclastic eolian deposits representing distal alluvial fan material that was blown back over the fluviatile volcaniclastic apron by prevailing westerly winds off the Colorado Plateau in Oligocene time.

The generalized geology of the Quemado area is summarized in Figure 1. Regional normal fault zones control the distribution of Miocene fluviatile conglomerates derived from erosional remnants of the Oligocene volcanic highlands. These Neogene normal fault zones probably reflect extensional reactivation of basement zones of weakness previously activated by Laramide transpression. The predominance of NNE-trending normal faults probably reflects regional westerly extension, although the overall pattern of faulting records multiple periods of basement deformation.

Details of the Eocene-Oligocene stratigraphy in the Quemado area are discussed below after presentation of dating results.

METHODS

Hand samples, typically 1-2 kg of unweathered, unaltered rock, were collected from 33 localities representing selected ignimbrites, volcaniclastic rock, lavas, and one intrusive quartz diorite in the Quemado area (Fig. 2). Ignimbrites were collected from the densest, most welded outcrops available and lava samples were collected from holocrystalline flow interiors where exposed.

Feldspar separates were prepared from ignimbrite samples by crushing and sieving samples to 125-500 μm, then applying magnetic and density-liquid techniques followed by hand picking. Most of these separates consisted entirely of sanidine grains, although a few contained plagioclase, and one was entirely plagioclase (sample 9, an ignimbrite with sanidine-free pumice, Appendix 1). Similar methods were used to prepare feldspar separates from volcaniclastic samples, except for three samples (samples 22, 24, 25, Appendix 1), from which pumice clasts were hand-picked prior to crushing. Feldspar separates from the volcaniclastic samples are predominantly plagioclase, but include a few sanidine grains. For the lava samples, matrix concentrates of samples

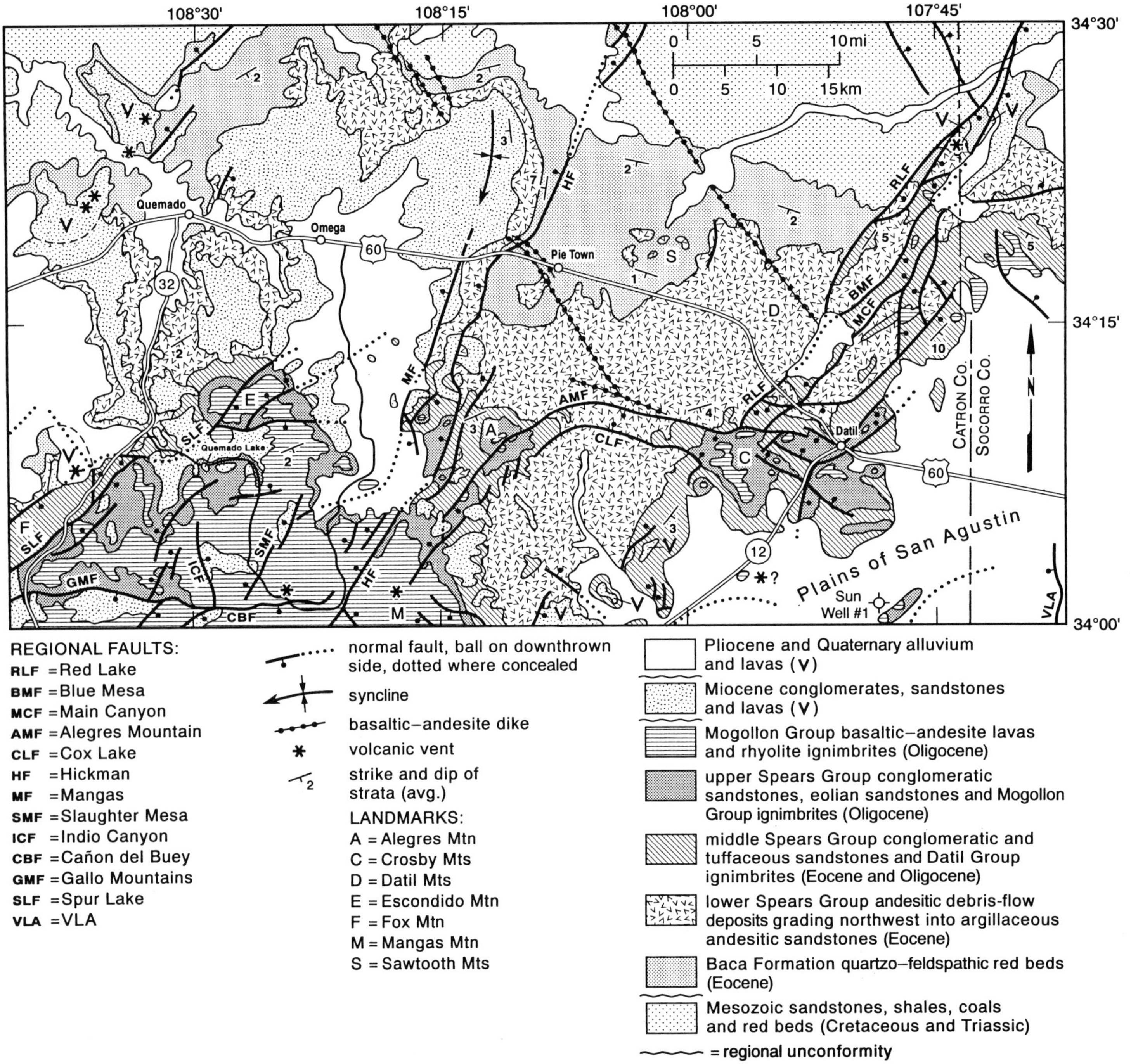

FIGURE 1. Generalized geology of the Quemado region, after Chamberlin et al. (1994a).

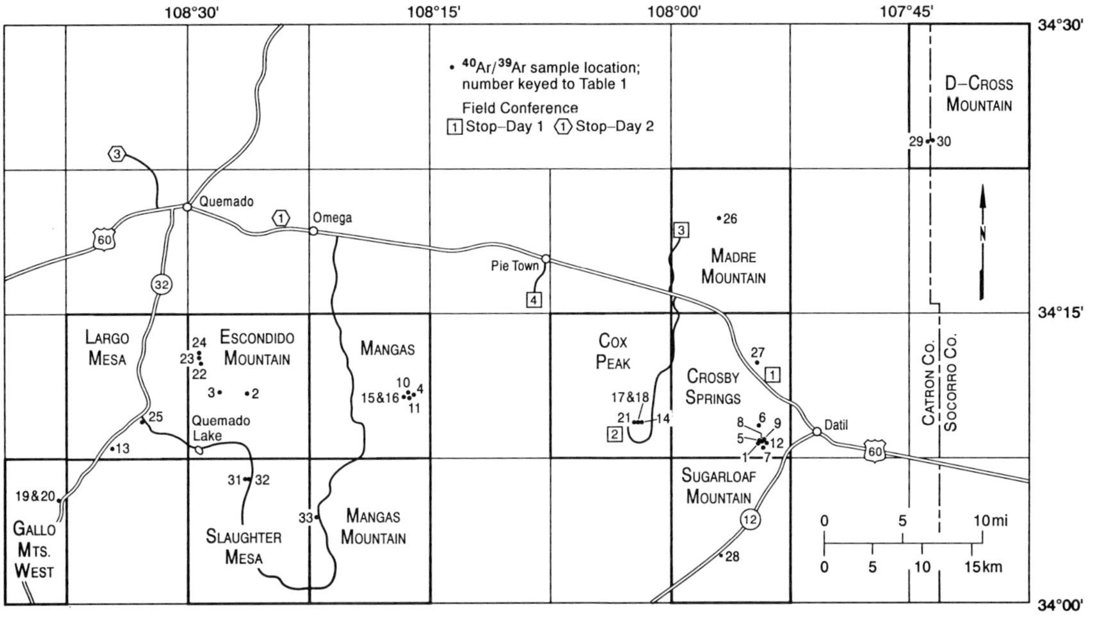

FIGURE 2. Location of $^{40}Ar/^{39}Ar$ dating samples, with numbers keyed to Table 1 and appendices.

were prepared by crushing and sieving to 250-500 μm, followed by ultrasonic cleaning in dilute (5%) nitric acid, then by removal of phenocrysts by Frantz magnetic separator and hand picking. Aliquots of feldspar separates (10-20 mg) and of groundmass concentrates (40-60 mg) were packaged with alternating flux monitors of Fish Canyon Tuff sanidine (27.84 Ma) and were irradiated for one to five hours in the L67 position of the Ford reactor at the University of Michigan.

$^{40}Ar/^{39}Ar$ analyses were performed at the New Mexico Geochronology Research Laboratory at New Mexico Tech, Socorro. This facility includes an MAP 215-50 mass spectrometer attached to a computer-automated, all-metal, argon extraction line. Grains of feldspar separates and monitor sanidines were fused by CO_2 laser for 15 seconds, then reactive gasses were removed using a SAES GP-50 prior to expansion into the mass spectrometer. Extraction line blanks during these laser analyses ranged from 5×10^{-17} to 2×10^{-16} moles ^{40}Ar and 5×10^{-19} to 2×10^{-18} moles ^{36}Ar. The pooled results of four subsamples (3-4 crystals, approximately 1 mg) of each monitor allowed the neutron flux (J-values) along the irradiation tubes to be determined to a precision of ±0.25%. A double-vacuum resistance furnace was used to heat the groundmass concentrate samples. Samples were preheated at 500°C to degas the tin foil package, and then were incrementally heated in eight to ten steps from 650°C to 1650°C. Heating times of 8-10 minutes were followed by 5-10 minutes of cleanup with SAES AP-10 and GP-50 getters. Data from each step were corrected for extraction line blanks, which for furnace analyses ranged from 3×10^{-16} to 3×10^{-15} moles ^{40}Ar and from 2×10^{-18} to 9×10^{-18} moles ^{36}Ar.

RESULTS

Results of single-crystal laser-fusion analyses of feldspar separates are detailed in Appendix 1 and displayed on ideogram and isochron diagrams in Appendix 3. Results of step-heating analyses are detailed in Appendix 2 and displayed as age spectra and isochron diagrams in Appendix 4. Age spectra are accompanied by plots of apparent K/Ca (a function of measured $^{37}Ar/^{39}Ar$) and radiogenic yield (the percent of measured ^{40}Ar that cannot be attributed to trapped argon of atmospheric composition, i.e. $^{40}Ar/^{36}Ar = 295.5$). Age data for all samples is summarized in Table 1.

Ignimbrites

Age determinations of individual sanidine grains from ignimbrites generally have high radiogenic yields (95-100%) and 1σ analytical precision of ±0.1-0.5%. Analyses of plagioclase grains (identified by K/Ca < 2.0; Appendix 1) are generally less precise (1σ = ±0.8-2%) and many have lower radiogenic yields (40-80%).

Except for the sanidine-poor tuff of South Crosby Peak (samples 7, 8, 9; Appendix 1, 3), the ages of most sanidine grains are tightly grouped within each ignimbrite sample. A few samples include one or more grains that are significantly older than the mean. In some cases (e.g., sample 2; Table 1), the aberrant grains have low radiogenic yields and are only slightly older than the majority of grains; this effect may be due to large melt inclusions, observed optically in some grains. In other cases (e.g., samples 6 and 15; Appendix 1, 3), the aberrant grains have high radiogenic yields and are as much as 3 Ma older than the majority of grains. These grains are interpreted as older xenocrysts that were incorporated into the ignimbrite during eruption or emplacement but not heated sufficiently by the ignimbrite to release all preexisting ^{40}Ar.

Mean ages were calculated for each sanidine-rich ignimbrite sample, after excluding results from plagioclase and aberrant sanidine grains discussed above. Mean ages have within-sample precision (1 σ) ranging from ±0.3 to ±0.7% (Appendix 1, 3; Table 1). These mean ages, calculated assuming a trapped component of atmospheric composition ($^{40}Ar/^{36}Ar = 295.5$), agree well with ages calculated by the isochron method (Appendix 1, 3). Isochrons for samples with relatively low radiogenic yields confirm the assumption of an atmospheric trapped component (e.g., sample 11, Appendix 3), but isochrons for many samples show variable $^{40}Ar/^{36}Ar$ intercept values with large uncertainties; this variation is merely an artifact of the high radiogenic yields (>99%) of grains from these samples. The mean ages of samples of sanidine-rich ignimbrites range from 34.98 ±0.19 Ma to 27.23 ±0.09 Ma and are interpreted as the eruption ages of these units; correlation with previously studied regional ignimbrites of the Mogollon-Datil volcanic field is considered in the discussion below. Results from sanidine-poor ignimbrite members of the tuff of South Crosby Peak are less representative of the eruption ages. The few sanidine grains analyzed from the upper and middle ignimbrite members (samples 7, 8; Appendix 3) are interpreted as xenocrysts of preexisting regional ignimbrites. Results from plagioclase grains from the lower ignimbrite member (sample 9; Appendix 3) have relatively high radiogenic yields (92-100%), but compared to typical sanidines have poorly grouped ages (1 σ = ±3%) with a mean of 29.67 ±0.86 Ma.

Volcaniclastic rocks

Age determinations of individual plagioclase and sanidine crystals separated from pumice and ash-rich layers within the volcaniclastic

TABLE 1. Summary of $^{40}Ar/^{39}Ar$ results from Quemado-Datil region samples.

Sample	Unit	Field #	7.5' Quad	Lat	Long	mineral	n	Age	±Err	comment
Ignimbrites										
1)	South Canyon Tuff	ANDSNMTN-5	Crosby Springs	34.137	107.908	sanidine	10	27.23	0.09	
2)	Bloodgood Canyon Tuff	nm1100	Escondido Mt	34.182	108.436	sanidine	6	27.98	0.09	
3)	Vicks Peak Tuff	nm1102	Escondido Mt	34.182	108.465	sanidine	7	28.56	0.09	
4)	Vicks Peak Tuff	nm1080	Mangas	34.183	108.265	sanidine	11	28.44	0.08	
5)	Vicks Peak Tuff	ANDSNMTN-6	Crosby Springs	34.138	107.907	sanidine	10	28.40	0.12	
6)	La Jencia Tuff	ESGLFMT-1	Crosby Springs	34.185	107.907	sanidine	13	28.64	0.18	
7)	tuff of S. Crosby Peak,	ANDSNMTN-4	Crosby Springs	34.134	107.902	sanidine	3	31.39	0.12	xenocrysts of Caballo Blanco
	upper ignimbrite					sanidine	6	31.97	0.17	xenocrysts of Hells Mesa
8)	tuff of South Crosby Peak,	ANDSNMTN-3	Crosby Springs	34.138	107.903	sanidine	6	31.97	0.18	xenocrysts of Hells Mesa
	middle ignimbrite					sanidine	1	35.12	0.12	xenocrysts of Datil Well
9)	tuff of South Crosby Peak, lower ignimbrite	ANDSNMTN-2	Crosby Springs	34.137	107.901	plagioclase	8	29.67	0.86	
10)	Caballo Blanco Tuff	nm1078	Mangas	34.182	108.269	sanidine	8	31.54	0.15	
11)	Caballo Blanco Tuff	nm1079	Mangas	34.184	108.268	sanidine	12	31.46	0.09	
12)	Hells Mesa Tuff	ANDSNMT-1	Crosby Springs	34.138	107.899	sanidine	10	31.93	0.07	
13)	tuff of Luna	LGMSA-1	Largo Mesa	34.133	108.576	sanidine	10	33.39	0.12	
14)	Blue Canyon Tuff	nm1098	Cox Peak	34.155	108.034	sanidine	8	33.59	0.06	
15)	Blue Canyon Tuff	nm1077	Mangas	34.183	108.271	sanidine	9	33.75	0.10	
16)	Blue Canyon Tuff	nm1076	Mangas	34.183	108.271	sanidine	9	33.71	0.10	
17)	Rockhouse Canyon Tuff	nm1097	Cox Peak	34.156	108.036	sanidine	6	34.18	0.10	
18)	Rockhouse Canyon Tuff	nm1096	Cox Peak	34.156	108.036	sanidine	5	34.16	0.07	
19)	Bishop Peak Tuff	nm1112	Gallo Mts West	34.088	107.626	sanidine	6	34.84	0.08	
20)	Bishop Peak Tuff	GMW-1-93	Gallo Mts West	34.088	107.626	sanidine	8	34.97	0.28	
21)	Datil Well Tuff	nm1095	Cox Peak	34.156	107.037	sanidine	6	34.98	0.19	
Volcaniclastic units										
22)	volcaniclastic unit of Canon	CDL-10/2-4	Escondido Mt	34.205	108.480	plagioclase	3	33.21	0.50	
	de Leon, upper					sanidine	1	34.17	0.13	Rock House Can Tuff san
23)	volcaniclastic unit of Canon de Leon, lower	CDL-10/2-9B	Escondido Mt	34.210	108.486	plagioclase	8	35.19	0.36	
24)	volcaniclastic unit of Canon de Leon, basal	CDL-10/2-11B	Escondido Mt	34.212	108.487	plagioclase	5	35.48	0.42	
25)	volcaniclastic unit of Canon	CR-93-1-XR	Largo Mesa	34.158	108.548	plagioclase	8	35.59	0.87	
	de Leon, basal					sanidine	2	35.33	0.06	
26)	Dog Springs Fm,	nm1115	Madre Mt	34.340	107.945	anorthoclase?	7	34.92	0.15	authigenic K-spar?
	tuff layer					plagioclase	8	38.70	1.18	
27)	Dog Springs Fm,	nm1121	Crosby Springs	34.207	107.912	plagioclase	6	37.57	1.29	
	tuff breccia					sanidine	1	36.94	0.07	
Intrusion										
28)	Cerrito Viejo	nm1114	Sugarloaf Mt	34.042	107.943	plagioclase	8	30.56	1.42	
Mafic Lavas										
29)	Basalt of Blue Mesa, upper lava	BLUMSA-2	D-Cross Mt	34.403	107.723	groundmass		4.48	0.02	poor isochron age
30)	Basalt of Blue Mesa, lower lava	BLUMSA-1	D-Cross Mt	34.400	107.723	groundmass		5.71	0.06	poor isochron age
31)	Bearwallow Mt Andesite, basal lava	SLTRMS-2	Slaughter Mesa	34.106	108.433	groundmass		26.1	0.1	poor isochron age
32)	Squirrel Springs Canyon Andesite	SLTRMS-1	Slaughter Mesa	34.105	108.431	groundmass		28.8	0.1	poor isochron age
33)	andesite of San Antone Canyon	MNGSMT-2	Mangas Mt	34.079	108.360	groundmass		29.2	0.1	poor isochron age

unit of Cañon de Leon and the Dog Springs Formation (samples 22-27, Appendix 1) have radiogenic yields and analytical precision similar to single crystals from ignimbrites. Of the 26 single crystals analyzed from the unit of Cañon de Leon (samples 22-25, Appendix 1) only three are sanidine; these ages (35.38-35.17 Ma) provide the best available maximum depositional age estimates for this unit. Considering the syneruptive character of this tuffaceous sandstone (Chamberlin and Harris, this volume), these ages probably closely approximate the depositional age. Ages of plagioclase crystals have between-grain variations (1σ = 1-3%) that are much larger than analytical precision for individual grains (1σ = 0.2-1.0%). Mean plagioclase ages from the unit of Cañon de Leon range from 35.48 ±0.42 Ma to 33.95 ±0.25 Ma.

Ages of plagioclase crystals from the Dog Springs Formation (samples 26, 27, Appendix 1) vary widely among grains (1σ = ±4%), probably due to low K content, small grain size, and, possibly, alteration. Mean ages of plagioclase grains from these sample (38.70 ±1.18 Ma and 37.57 ±1.29 Ma) are probably generally representative of the eruptive age of volcaniclastic material within the Dog Springs Formation and also approximate the depositional age of this syneruptive unit. This interpretation is supported by the 36.94 ±0.07 Ma age of a single sanidine grain from one of the samples of the Dog Springs Formation (sample 27, Appendix 1). The other sample (sample 26, Appendix 1) contains unusual, small, low K/Ca (<6.8) K-feldspar grains with a mean age of 34.92 ±0.15 Ma. These grains are interpreted as post-depositional authigenic feldspar, and their age determinations are rejected as maximum ages for deposition of the Dog Springs Formation.

Intrusion

Analyses of plagioclase grains from the quartz diorite Cerrito Viejo intrusion have low radiogenic yields (16-17%), low analytical precision (1σ = 1-8%) and a mean age of 30.56 ±1.42%. However, because the isochron age is nearly identical and the $^{40}Ar/^{36}Ar$ intercept value is atmospheric (Appendix 1, 3), this imprecise mean age is considered to be a reasonable estimate of the time of emplacement of the intrusion.

Mafic lavas

Age spectra from all five samples of mafic lava analyzed by incremental heating of bulk samples failed to form well-defined age "plateaus" (Appendix 2, 4). The form of the spectra ranged from nearly flat (samples 29, 31; Appendix 4), to steadily falling (samples 32, 33; Appendix 4) to highly irregular (sample 30; Appendix 4). Isochron ages are considered the best estimate of eruptive age for these mafic lavas (Appendix 2, 4); they range from 4.68 ±0.02 to 5.7 ±0.3 Ma for the two late Cenozoic lavas and from 26.1 ±0.1 to 29.2 ±0.1 for the three mid-Cenozoic lavas.

DISCUSSION

Correlation of Quemado ignimbrites with regional units

One of the main goals of this study was the correlation of distal-outflow-facies ignimbrites in the Quemado area with regional ignimbrites from the Mogollon-Datil field. The two correlation criteria generated by this study are $^{40}Ar/^{39}Ar$ age and sanidine K/Ca ratio (determined from the measured $^{37}Ar/^{39}Ar$ ratio). These criteria, added to lithologic and stratigraphic position data, allow nearly all of the sampled Quemado area ignimbrites to be confidently correlated with regional units. A paleomagnetic study to provide an additional correlation criterion is presently underway; some initial results are mentioned below.

The following sections discuss the correlation of Quemado area ignimbrites with regional outflow sheets. Figures 3 and 4 summarize comparisons of age and K/Ca data from Quemado samples with published data from regional units. In making these comparisons, it is important to consider that the samples in this study were analyzed by the single-crystal laser-fusion $^{40}Ar/^{39}Ar$ method, whereas the published data from regional units (Fig. 3; McIntosh et al., 1992) was obtained by step-heating of bulk samples containing several hundred sanidine grains, a method that tends to average variations between individual grains. Following the unit-by-unit presentation below is a discussion of some apparent systematic differences between results from the two dating methods.

Datil Well Tuff

The Datil Well Tuff, previously dated at 35.47 ±0.07 Ma by the $^{40}Ar/^{39}Ar$ bulk-sample method (Fig. 3), is the oldest rhyolitic ignimbrite in the northern Mogollon-Datil volcanic field (McIntosh et al., 1992). It is widespread throughout the Datil area as a thin (<30 m) moderately- to densely-welded ignimbrite, but no source caldera has been recognized. Sample 21, the basal ignimbrite in the Cox Peak quadrangle (Fig. 2), has lithology and sanidine K/Ca values indistinguishable from published values from the Datil Well Tuff (Figs. 3, 4). Correlation is further supported by the similarity of its paleomagnetic direction (McIntosh, unpubl. data) with the distinctive north to north-northeast normal remanence direction reported from the Datil Well Tuff elsewhere (McIntosh et. al, 1991, 1992). However, the single-crystal laser fusion age of sample 21 is about 0.5 Ma younger than the published age of Datil Well Tuff (Figs. 3, 4). This difference, significant at the 2σ level, probably reflects small amounts of incompletely degassed xenolithic feldspar in the bulk sanidine separates analyzed by McIntosh et al. (1992). The single-crystal laser fusion age from this study, 34.98 ±0.19 Ma, is now considered the best available estimate for the eruption age of the Datil Well Tuff.

Tuff of Bishop Peak

The tuff of Bishop Peak is a relatively local ignimbrite exposed in the vicinity of Luna, southwest of the present study area. It has been dated at 34.70 ±0.10 Ma (Fig. 3) and tentatively correlated with the tuff of Leyba Well, exposed near Old Horse Springs, south of the present study area. The tuff of Bishop Peak has reverse magnetic polarity that distinguishes it from the normal polarity Datil Well Tuff. Samples 19 and 20 (Fig. 2), from a thin, nonwelded ignimbrite that underlies the Leggett Canyon andesite in the southwestern corner of the study yielded age (Fig. 3) and sanidine K/Ca values (Fig. 4) that agree closely with published data from the tuff of Bishop Peak, which has mapped outcrops only 32 km to the southwest (McIntosh et al., 1991, 1992). These data would also permit correlation with the more distant Datil Well Tuff, which has indistinguishable age and sanidine K/Ca values (Fig. 4). This correlation is presently being addressed by paleomagnetic analysis.

Rock House Canyon Tuff

The Rock House Canyon Tuff is a thin regional ignimbrite widely exposed in the northern Mogollon-Datil volcanic field and best known from the Datil area. It has a distinctive crystal-poor lithology (<1% phenocrysts) and has been dated by the bulk-sample $^{40}Ar/^{39}Ar$ method at 34.42 ±0.12 Ma (Fig. 3), although intersample variations in age suggested some effects of sample contamination by xenocrystic feldspar (McIntosh et al., 1992). The source cauldron for this ignimbrite is unknown. Results from samples 17 and 18 (Fig. 2) support identification of the thin, crystal-poor ignimbrite in the Cox Peak quadrangle as Rock House Canyon Tuff. Sanidine K/Ca values from these samples overlap with published Rock House Canyon Tuff values (Fig. 4), although the published mean age is about 0.25 Ma older (Fig. 3). As is the case for the Datil Well Tuff, this discrepancy is attributed to undetected xenolithic feldspar in the bulk sanidine separates analyzed for the published age (Fig. 3). The mean (34.17 ±0.08 Ma) of the two single-crystal laser fusion ages from samples 17 and 18, is now considered to be the best available estimate for the eruption age of the Rock House Canyon Tuff.

Blue Canyon Tuff

The Blue Canyon Tuff is yet another thin, regional ignimbrite widely exposed in the northern Mogollon-Datil volcanic field and was dated by the bulk-sample $^{40}Ar/^{39}Ar$ method at 33.66 ±0.07 Ma. Results from samples 14-16, from a thin ignimbrite in the Cox Peak and Mangas quadrangles in the central study area, confirm correlation with the regional Blue Canyon Tuff. Sanidine ages (33.59-33.75 Ma) and K/Ca values are indistinguishable from published values (Figs. 3, 4).

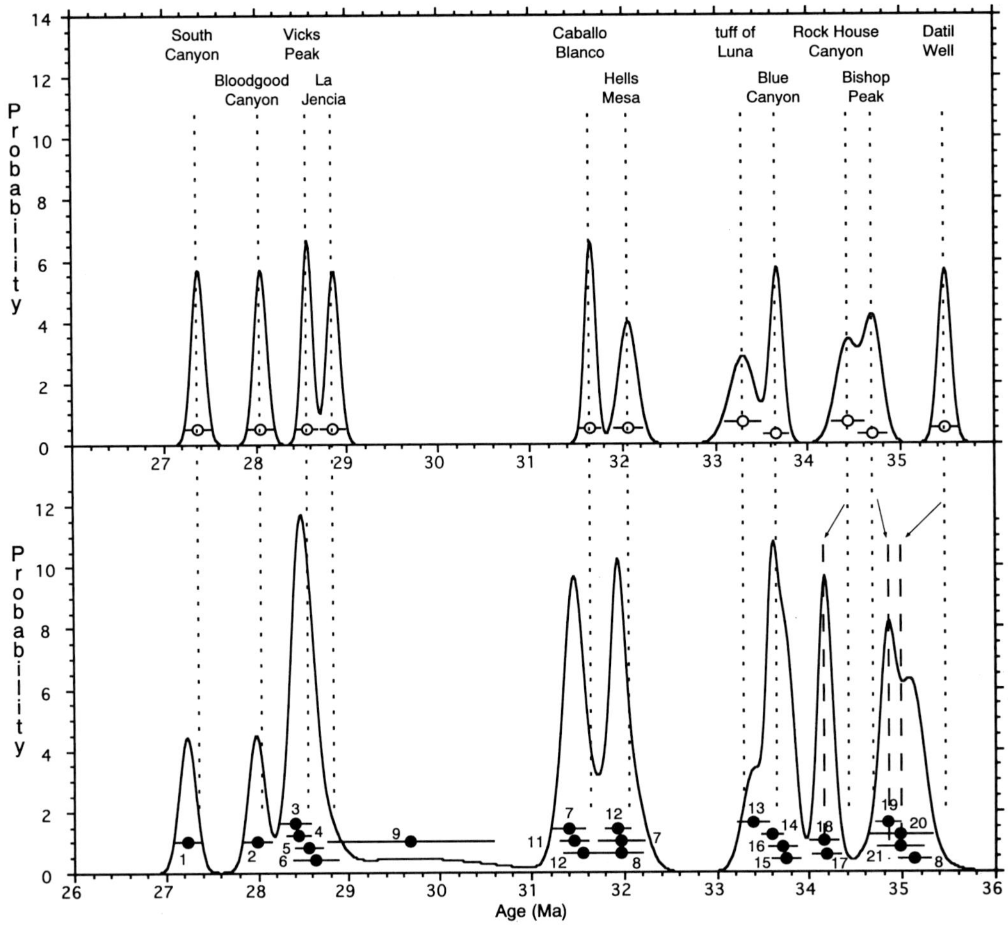

FIGURE 3. Ideogram comparing published $^{40}Ar/^{39}Ar$ ages of Mogollon-Datil regional ignimbrites (upper panel, after McIntosh et al., 1992) with ages of samples dated in this study (lower panel). Open circles and bars show published age and error (1 σ) for regional ignimbrites. Closed circles and bars show age and error of samples dated in this study, with sample numbers keyed to Appendices 1 and 2. Arrows at upper right of lower panel show where mean ignimbrite ages from this study differ significantly from published values.

Tuff of Luna

The tuff of Luna is a thin, relatively local ignimbrite exposed in the vicinity of Luna. It was dated at 33.29 ±0.14 Ma and tentatively correlated with Box Canyon Tuff in the southern Mogollon-Datil field (McIntosh et al., 1992). A thin ignimbrite in the southeastern part of the study area, represented by sample 13 (Fig. 2), yielded an age (33.39 ±0.12 Ma) that agrees at the 1 σ level with published mean ages of both the Blue Canyon Tuff and tuff of Luna. However, the high sanidine K/Ca value of sample 13 clearly supports correlation with the tuff of Luna (Fig. 4).

Hells Mesa Tuff

Hells Mesa Tuff is a large-volume (1200 km^3) regional ignimbrite exposed widely throughout the northern Mogollon-Datil volcanic field. It erupted from a cauldron near Socorro at 32.06 ±0.10 Ma (McIntosh et al., 1991, 1992). Results from sample 12 confirm that the crystal-rich ignimbrite exposed in the Crosby Mountains, mapped as tuff of Rock Tank by Lopez and Bornhorst (1979), is correlative with the Hells Mesa Tuff (Figs. 3, 4; McIntosh et al., 1991, 1992).

Caballo Blanco Tuff

Caballo Blanco is a 31.65 ±0.06-Ma, crystal-rich ignimbrite, lithologically similar to Hells Mesa Tuff, that is widely exposed in the southern Mogollon-Datil volcanic field (McIntosh et al., 1991, 1992). Although the source cauldron in not well known, distal outflow facies are known to extend to the area immediately south of the present study area (McIntosh et al., 1991). A poorly-welded crystal-rich ignimbrite exposed in the Mangas quadrangle and represented by samples 10 and 11 was originally thought to be Hells Mesa Tuff on the basis of lithology and stratigraphic position. However, the $^{40}Ar/^{39}Ar$ age determinations (31.46 ±0.09 and 31.54 ±0.15 Ma) are significantly younger than published data from the Hells Mesa Tuff, and instead support correlation with the Caballo Blanco Tuff (Fig. 3). K/Ca data cannot be used to distinguish between Hells Mesa Tuff and Caballo Blanco Tuff, because their measured sanidine K/Ca values are indistinguishable.

Tuffs of South Crosby Peak (lower, middle, and upper)

The tuffs of South Crosby Peak are exposed only in the Crosby Mountains in the east-central study area and include at least three thin, sanidine-poor, lithic-rich ignimbrites intercalated with epiclastic conglomerates of the South Crosby Peak Formation. The tuffs of South Crosby Peak are stratigraphically bracketed between the 32.0-Ma Hells Mesa Tuff and the 28.8-Ma La Jencia Tuff, but have never been precisely dated, largely because they lack sanidine phenocrysts. Although the lithology of this unit precludes correlation with any of the well-established, sanidine-bearing, rhyolitic regional ignimbrites of the Mogollon-Datil volcanic field, similar nonwelded, lithic-rich ignimbrites and other tuffs are present elsewhere between the Hells Mesa and La Jencia Tuffs. These include thin deposits east of Socorro and the thick, heterogeneous Luis Lopez Formation (Osburn and Chapin, 1983).

Three sanidine-poor, nonwelded South Crosby Peak ignimbrites (samples 7-9, Fig. 2) were analyzed in this study to better constrain their eruptive age. Sanidine crystals separated from the upper and middle ignimbrites (samples 7 and 8) yielded $^{40}Ar/^{39}Ar$ ages near 32.0 and

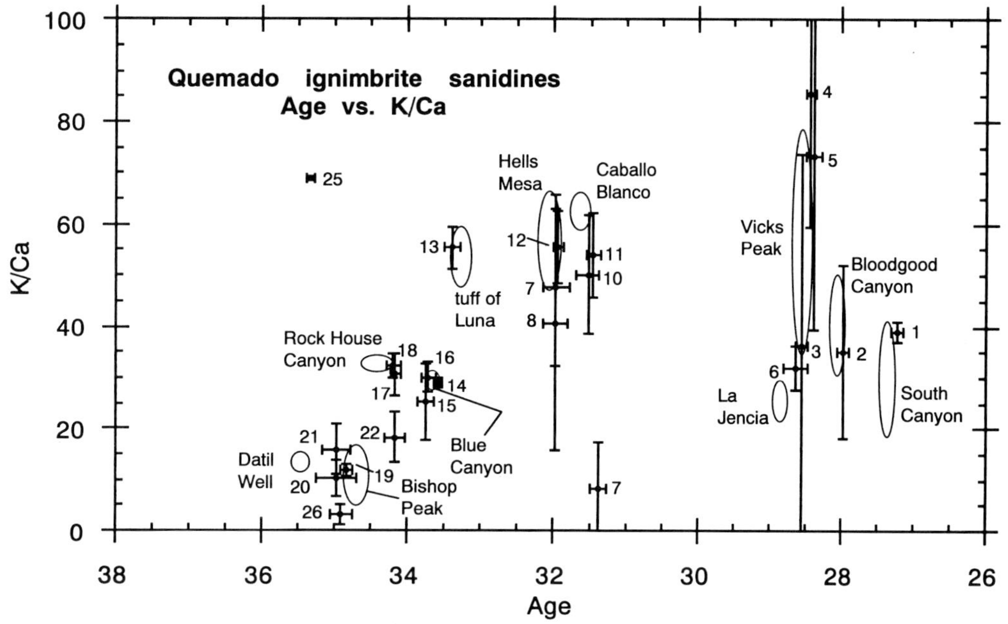

FIGURE 4. Plot comparing results from this study (points with 1 σ error bars, numbers keyed to Appendix 1 and Table 1) with published K/Ca and $^{40}Ar/^{39}Ar$ ages of regional Mogollon-Datil ignimbrites (ovals approximate ±1 σ error envelopes, after McIntosh et al., 1991, 1992).

31.4 Ma. On the basis of age and K/Ca values (Fig. 4), the 32.0-Ma sanidines are interpreted as xenocrysts derived from Hells Mesa Tuff. The 31.4-Ma sanidines are also probably xenocrysts, but their origin is less certain. They overlap in age with the Caballo Blanco Tuff, but their K/Ca values are considerably lower (Fig. 4).

Plagioclase phenocrysts from pumice in the lower ignimbrite member of the tuff of South Crosby Peak (sample 9) gave a mean age of 29.67 ±0.86 Ma. This relatively low-precision age determination is currently the best available estimate of eruption age for the lower tuff of South Crosby Peak.

La Jencia Tuff

The La Jencia Tuff is a well-known regional ignimbrite erupted at 28.85 ±0.07 Ma from the Sawmill-Magdalena cauldron west of Socorro. The crystal-poor outflow sheet is widely exposed throughout much of the northern Mogollon-Datil volcanic field (McIntosh et al., 1991, 1992). A thin ignimbrite interbedded with the South Crosby Peak Formation in the Crosby Mountains (sample 6; Fig. 2) apparently represents the northwestern distal fringe of the La Jencia Tuff outflow sheet. The 28.64 ±0.18-Ma age of sample 6 overlaps at 1 σ with the published age (Fig. 3) of the La Jencia Tuff, and the sanidine K/Ca value precludes correlation with the Vicks Peak Tuff, which has sanidine with higher and more variable K/Ca values (Fig. 4).

Vicks Peak Tuff

The Vicks Peak Tuff is one of the most voluminous (1200 km^3) and widespread ignimbrites in the northern Mogollon-Datil volcanic field. Erupted at 28.56 ±0.06 Ma from the Nogal cauldron in the southern San Mateo Mountains, its crystal-poor outflow sheet extends west nearly to the New Mexico-Arizona border (McIntosh et al., 1991, 1992). Its sanidine phenocrysts have distinctive K/Ca values that are generally high and quite variable among samples (Fig. 4). Age and K/Ca results from a thin, crystal-poor ignimbrite sheet with outcrops distributed across much of the southern study area (samples 4-6; Fig. 2) confirm correlation with the Vicks Peak Tuff (Figs. 3, 4). K/Ca results (Fig. 4) indicate that grain-to-grain variations in sanidine K/Ca values are even greater than previously established sample-to-sample variations.

Bloodgood Canyon Tuff

The Bloodgood Canyon Tuff is the most voluminous (1000 km^3) and widespread (15,000 km^2) ignimbrite in the southern Mogollon-Datil volcanic field (Ratté et al, 1984). It erupted at 28.05 ±0.06 Ma from the Bursum cauldron in the Mogollon Mountains and its outflow sheet is known to extend to the area immediately south of the present study area (McIntosh et al., 1991, 1992). Age and K/Ca results from sample 2 support identification of a thin, moderately-welded ignimbrite in the Escondido Mountain quadrangle as Bloodgood Canyon Tuff (Fig. 4).

South Canyon Tuff

The South Canyon Tuff is the youngest regional ignimbrite that is widely exposed in the northern Mogollon-Datil volcanic field. It erupted at 27.37 ±0.07 Ma from the Mount Withington cauldron in the northern San Mateo Mountains and its outflow sheet extends into the western part of the study area (McIntosh et al., 1991, 1992). Age and K/Ca data from a thin, poorly welded ignimbrite mapped as A-L Peak Rhyolite by Lopez and Bornhorst (1979) in the Crosby Mountains support its identification as the distal fringe of the South Canyon outflow sheet (Figs. 3, 4). Because of its crystal-poor lithology, this distal outflow sheet could easily be confused with La Jencia or Vicks Peak Tuff.

Systematic differences between single-crystal and bulk-sample $^{40}Ar/^{39}Ar$ ages

One striking feature of this data set is that the $^{40}Ar/^{39}Ar$ ages determined from Quemado area ignimbrites tend to be younger than published data from regional units (Fig. 3, Table 1). In two cases, Datil Well Tuff and Rock House Canyon Tuff, the published ages are significantly (0.25-0.5 Ma) older than this study's age determinations. As discussed above, these discrepancies probably reflect small amounts of incompletely degassed xenolithic feldspar in the bulk samples analyzed for the published ages. In these two cases the single-crystal laser fusion ages are probably the best estimate of eruption age. For most of the ignimbrites, however, this study's single-crystal ages are 0.05 to 0.2 Ma younger than the published bulk-sample ages, and the ages determined by the two methods overlap at the 1 σ level. This small but systematic difference in age may reflect minor xenocrystic contamination in the bulk samples, but is probably actually due to difference in technique (single crystal laser fusion vs. resistance furnace step heating) or to inexact interlaboratory calibration between the New Mexico Tech laboratory, where the single-crystal work was performed, and the U.S.G.S. laboratory in Reston, Va., where the published bulk-sample analyses were accomplished. Further single-crystal analyses of Mogollon-Datil ignimbrites will help resolve the source of these small but systematic discrepancies between dating techniques.

Age and provenance of volcaniclastic units

Volcaniclastic units were analyzed in this study in order to help constrain depositional age and provenance. Two units were studied: the volcaniclastic unit of Cañon del Leon and Dog Springs Formation.

Volcaniclastic unit of Cañon del Leon

Samples of phenocrysts hand picked from pumice clasts in pumiceous muddy siltstone samples of the volcaniclastic unit of Cañon del Leon yielded three sanidine and numerous plagioclase grains for analysis. One sanidine grain, from the uppermost pumice-rich bed (sample 22, Appendix 1), gave an age (34.17 ±0.13 Ma) and K/Ca value (Fig. 4) that suggest that the grain was derived from Rock House Canyon Tuff, and provides a relatively precise maximum depositional age for the upper part of the unit of Cañon del Leon. More abundant plagioclase grains from the same sample yielded a mean age of 33.21 ±0.50 Ma, providing a younger but less precise maximum depositional age for the bed. Samples of middle and basal beds of the unit of Cañon del Leon (sample 23-25; Appendix 1), yielded plagioclase with mean ages ranging from 35.19 ±0.42 Ma to 35.59 ±0.87 Ma and in general agreement with stratigraphic order (Table 1). Sample 23 represents a thin (5 cm), rhyolitic, ash-fall approximately 12 m above the base of the unit of Cañon del Leon. Because the ash bed is not reworked, its relatively imprecise age (35.19 ±0.42 Ma) is representative of the depositional age of the unit. Two sanidine grains from the basal pumiceous bed at Castle Rock (sample 25, Appendix 1) have a mean age of 35.33 ±0.06 Ma and a K/Ca value distinctly higher than any of the regional ignimbrites erupted near 35 Ma in the northern Mogollon-Datil volcanic field (Fig. 4). These were probably derived from a rhyolitic dome or flow, or possibly from reworked pyroclastic fall material from a distant rhyolitic ignimbrite. In any case, the age of these sanidines represents a reasonably precise maximum depositional age for the basal part of this unit. Age data from the unit of Cañon del Leon, together with the sedimentological interpretation of this unit as the product of distal reworked volcanic ejecta derived from rhyolitic to intermediate volcanism, are consistent with deposition of the unit ranging approximately from 35 to 33 Ma.

Dog Springs Formation

The two samples from the Dog Springs Formation (samples 26 and 27, Appendix 1) yielded plagioclase grains with relatively imprecise mean ages of 37.57 ±1.29 Ma and 38.70 ±1.18 Ma, and a single sanidine grain at 36.94 ± 0.07 Ma. Also present were unusual low-potassium K-feldspars with a stratigraphically unreasonable age of 34.92 ±0.15 Ma; as discussed above, these grains are believed to be authigenic and therefore only provide a minimum depositional age. Given the interpretation of the Dog Springs Formation as debris flows shed from the flanks of stratovolcanoes, the best estimate for the depositional age of sampled beds in the Dog Springs Formation is approximately 37 Ma, based on the age of the one sanidine grain analyzed.

Age of Cerrito Viejo intrusion

Plagioclase grains from the single sample of the Cerrito Viejo intrusion, a small dacitic stock near the southeastern corner of the study area (sample 28, Appendix 1), yielded a mean age of 30.56 ±1.42 Ma. This age represents a reasonable but imprecise estimate of the timing of intrusion and cooling of the stock. These data suggest that Cerrito Viejo was not related to the stratovolcano sources for the lithologically similar Dog Springs Formation; the location of these source volcanoes remains unknown. It is possible, however, that this intrusion is related to volcanism that supplied volcanic and volcaniclastic material preserved in the tuffs of South Crosby Peak.

Age of mafic lavas

The five samples of mafic lava analyzed in this study address two separate goals. The two samples of Pliocene lava from Blue Mesa (samples 29 and 30, Appendix 2) were collected to provide age control on faulting and deposition of gravels along the Red Lake fault zone. Three samples of Oligocene mafic lavas were collected to better constrain the timing of basaltic andesite volcanism. None of the samples of mafic lavas yield ages as precise or reliable as data from sanidine-bearing units. As discussed above, age spectra from these samples are not flat, indicating complexities in $^{40}Ar/^{39}Ar$ systematics that are not uncommon when analyzing multimineralic materials such as the groundmass concentrates obtained from these samples (e.g., McIntosh and Cather, this volume). The isochron ages are considered the best estimate for eruptive ages for these samples, although calculated analytical errors (Tables 1; Appendix 2) probably under-represent geological errors for these age determinations.

The two samples of alkali basalt from Blue Mesa (samples 29 and 30; Appendix 2) yielded isochron ages of 4.48 ±0.02 and 5.71 ±0.25 Ma; these data are consistent with stratigraphic order. Of these age determinations, the younger (sample 29) is considered more reliable, because the age spectrum is less complex and the radiogenic yields are higher (Appendix 2,4).

Stratigraphy and sedimentation in the Quemado area

The $^{40}Ar/^{39}Ar$ results from this study, together with recent mapping (Chamberlin et al., 1994a), provide significant new age and correlation constraints for understanding stratigraphy in the Quemado area, as follows:

1. The stratigraphically lowest andesitic volcaniclastic interval, which includes the Dog Springs Formation and volcaniclastic unit of Largo Creek (new name; Chamberlin and Harris, this volume), was deposited near 37 Ma.
2. Upper Eocene pumiceous sandstones and ash beds of the volcaniclastic unit of Cañon del Leon (new name; Chamberlin and Harris, this volume) were deposited in the Quemado area from approximately 35 to 33 Ma and are in part laterally equivalent to the Pueblo Creek Formation in the Reserve area, and to the Chavez Canyon and Rincon Windmill Formations in the Datil area.
3. The andesite of Dry Leggett Canyon in the Fox Mountain area is bracketed by the underlying 34.9 Ma tuff of Bishop Peak and the overlying 33.4 Ma tuff of Luna.
4. The eolian sandstone of Escondido Mountain (new name; Chamberlin and Harris, this volume) contains three regional ignimbrites as lenticular paleovalley fills; these include the 31.6-Ma Caballo Blanco Tuff, the 28.6-Ma Vicks Peak Tuff and the 28.0-Ma Bloodgood Canyon Tuff.
5. Fluviatile conglomeratic sandstones of the South Crosby Peak Formation are bracketed by the 32.0-Ma Hells Mesa Tuff and the 27.4-Ma South Canyon Tuff, and also contain the tuff of South Crosby Peak (ca. 29.7 Ma), the 28.9-Ma La Jencia Tuff and the 28.6-Ma Vicks Peak Tuff.
6. The Squirrel Springs Canyon Andesite is bracketed by the 28.6-Ma Vicks Peak Tuff and the 28.0-Ma Bloodgood Canyon Tuff.
7. Basaltic andesite lavas assigned to the Bearwallow Mountain Andesite began to be erupted from the Mangas Mountain center about 26.1 Ma.

Using constraints provided by $^{40}Ar/^{39}Ar$ results, the upper Eocene to Oligocene (ca. 40-26 Ma) volcanogenic strata of the Quemado-Datil area can be divided in four regional composite stratigraphic units shown on the generalized geologic map (Fig. 1). From oldest to youngest these composite units and their approximate depositional ages are (1) andesitic debris-flow deposits and andesitic sandstones of the lower Spears Group, ca. 39.6 to 36.9 Ma; (2) epiclastic basaltic andesite to rhyolitic conglomeratic and tuffaceous sandstones of the middle Spears Group intercalated with 35.0 to 31.5-Ma ignimbrites of the Datil Group; (3) epiclastic basaltic andesite to rhyolitic conglomeratic sandstones and eolian volcaniclastic sandstones of the upper Spears Group intercalated with 28.9 to 27.4-Ma ignimbrites of the

Mogollon Group; and (4) Mogollon Group basaltic andesite lavas of the Squirrel Springs Canyon Andesite and Bearwallow Mountain Andesite (ca. 28.5 and 26.1, Ma, respectively) interbedded with and overlying 28.6 to 28.0-Ma ignimbrites. Additional lithologic variations in the above regional units include the presence of eolian sandstones below the Hells Mesa Tuff (Datil Group) in the Datil area (Osburn and Chapin, 1983) and intertonguing of middle Spears Group tuffaceous sandstones with thick flows of the andesite of Dry Leggett Canyon (approximately 34 Ma) at Fox Mountain.

Additional details of Quemado area stratigraphy, as well as implications for sedimentation and tectonics, are presented in companion articles (Chamberlin et al.; Chamberlin and Harris; Armstrong and Chamberlin, this volume).

Displacement history of the Red Lake fault zone

Field relationships and $^{40}Ar/^{39}Ar$ results from basalts (samples 29 and 30, Table 1) in the Blue Mesa area of the Red Lake fault zone, north of Datil (Fig. 2), indicate most of the down-to-the-east displacement is of late Oligocene to Miocene age. Only 100 m of the total 600 m of apparent normal displacement postdates the Pliocene (4.5 Ma) basalt of Blue Mesa.

REFERENCES

Armstrong, R. M. and Chamberlin, R. M., 1994, Preliminary geologic interpretation of the ARMA Baca Basin seismic reflection profile, northeastern Catron County: New Mexico Geological Society, Guidebook 45.

Cather, S. M. and Johnson, B. D., 1984, Eocene tectonics and depositional setting of west-central New Mexico and eastern Arizona: New Mexico Bureau of Mines and Mineral Resources, Circular 192, p. 1-33.

Chamberlin, R. M. and Cather, S. M., 1994, Definition of the Mogollon slope, west-central New Mexico: New Mexico Geological Society, Guidebook 45.

Chamberlin, R. M. and Harris, J. S., 1994, Upper Eocene and Oligocene volcaniclastic sedimentary stratigraphy of the Quemado-Escondido Mountain Area, Catron County, New Mexico: New Mexico Geological Society, Guidebook 45.

Chamberlin, R. M., Cather, S. M., Anderson, O. J. and Jones, G. E., 1994a, Reconnaissance geologic map of the Quemado 30'x60' quadrangle, Catron County, New Mexico: New Mexico Bureau of Mines and Mineral Resources, Open-file Report 406, scale 1:100,000.

Chamberlin, R. M., Cather, S. M., McIntosh, W. C., Anderson, O. J. and Ratté, J. C., 1994b, First-day road log from Socorro to Magdalena, Datil, Pie Town, Quemado, and Quemado Lake: New Mexico Geological Society, Guidebook 45.

Chapin, C. E., 1971, The Rio Grande rift, part I: modifications and additions: New Mexico Geological Society, Guidebook 22, p. 191-201.

Chapin, C. E. and Cather, S. M., 1981, Eocene tectonics and sedimentation in the Colorado Plateau-Rocky Mountains area: Arizona Geological Society Digest, v. 14, p. 173-198.

Elston, W. E., 1978, Mid-Tertiary cauldrons and their relationship to mineral resources, southwestern New Mexico: a brief review: New Mexico Geological Society, Special Publication no. 7, p. 107-113.

Garmezy, L., 1990, Strike-slip deformation along the southeastern margin of the Colorado Plateau: Geological Society of America, Abstracts with Programs, v. 22, p. A-276.

Lopez, D. A. and Bornhorst, T. J., 1979, Geologic map of the Datil Area, Catron County, New Mexico: U.S. Geological Survey, Miscellaneous Investigations Map I-1098, scale 1:50,000.

McIntosh, W. C., Kedzie, L. L. and Sutter, J. F., 1991, Paleomagnetism and $^{40}Ar/^{39}Ar$ ages of ignimbrites, Mogollon-Datil volcanic field, southwestern New Mexico: New Mexico Bureau of Mines and Mineral Resources, Bulletin 135, 79 p.

McIntosh, W. C., Chapin, C. E., Ratté, J. C. and Sutter, J. F., 1992, Time-stratigraphic framework for the Eocene-Oligocene Mogollon-Datil volcanic field, southwest New Mexico: Geological Society of America Bulletin, v. 104, p. 851-871.

Osburn, G. R. and Chapin, C. E., 1983, Nomenclature for Cenozoic rocks of the northeast Mogollon-Datil volcanic field, New Mexico: New Mexico Bureau of Mines and Mineral Resources, Stratigraphic Chart 1.

Ratté, J. C., 1989, Geologic map of the Bull Basin quadrangle, Catron County, New Mexico: U.S. Geological Survey, Miscellaneous Field Investigations Map, MF-1203, scale 1:24,000.

APPENDIX 1

Sample Lab #	40/39	37/39	36/39	39K moles	K/Ca	%40*	Age	±1 σ
IGNIMBRITES								
1) South Canyon Tuff, ANDSNMTN-5, Crosby Springs quad, sanidine, J=0.0006644 ±0.000002								
1220-02	23.03	1.41E-02	7.45E-04	1.4E-13	36.2	99.0	27.11	0.07
1220-06	23.39	1.31E-02	1.87E-03	1.5E-13	39.0	97.5	27.15	0.06
1220-10	23.14	1.30E-02	9.48E-04	1.4E-13	39.4	98.7	27.17	0.04
1220-08	23.43	1.27E-02	1.89E-03	7.8E-14	40.0	97.5	27.18	0.06
1220-03	22.95	1.31E-02	2.38E-04	1.5E-13	38.8	99.6	27.19	0.06
1220-04	23.52	1.24E-02	2.03E-03	1.5E-13	41.1	97.4	27.24	0.05
1220-05	24.65	1.30E-02	5.69E-03	1.5E-13	39.2	93.1	27.30	0.07
1220-01	23.38	1.36E-02	1.37E-03	4.7E-13	37.4	98.2	27.30	0.04
1220-07	23.12	1.21E-02	4.15E-04	1.3E-13	42.1	99.4	27.34	0.04
1220-09	23.81	1.43E-02	2.68E-03	1.8E-13	35.8	96.6	27.35	0.05
mean age					n=10		**27.23**	**0.09**
isochron age	40Ar/36Ar=309.7 ±19.1		MSWD=3.2		n=10		27.22	0.09
2) Bloodgood Canyon Tuff, nm1110, Escondido Mt quad, sanidine, J=0.000909 ±0.000001								
753-08	17.27	1.17E-02	3.25E-04	4.1E-15	43.6	99.3	27.91	0.06
753-07	17.37	1.20E-02	6.42E-04	3.0E-15	42.7	98.8	27.92	0.07
753-05	17.38	3.65E-02	6.39E-04	4.5E-15	14.0	98.8	27.95	0.07
753-01	17.24	1.07E-02	1.19E-04	7.6E-15	47.6	99.7	27.96	0.06
753-06	17.66	4.06E-02	1.51E-03	5.0E-15	12.6	97.4	27.98	0.07
753-02	17.37	1.04E-02	1.83E-04	3.3E-15	49.1	99.6	28.15	0.08
753-03	20.39	2.89E-02	1.03E-02	3.3E-15	17.7	84.9	*28.18*	0.09
753-04	20.96	3.26E-02	1.20E-02	4.4E-15	15.7	83.0	*28.32*	0.10
mean age					n=6		**27.98**	**0.09**
isochron age	40Ar/36Ar=311.8 ±4.4		MSWD=1.2		n=8		27.95	0.07
3) Vicks Peak Tuff, nm1102, Escondido Mt quad, sanidine, J=0.0009079 ±0.000001								
752-03	18.04	5.01E-02	1.66E-03	1.2E-15	10.2	97.2	28.49	0.10
752-08	17.65	4.18E-02	3.22E-04	8.0E-16	12.2	99.4	28.49	0.13
752-05	17.57	5.52E-03	5.01E-05	1.6E-15	92.4	99.8	28.50	0.09
752-01	17.70	5.75E-03	3.67E-04	1.9E-15	88.7	99.3	28.55	0.09
752-02	17.71	4.01E-02	3.82E-04	1.7E-15	12.7	99.3	28.56	0.09
752-04	17.69	3.67E-02	2.65E-04	1.3E-15	13.9	99.4	28.59	0.09
752-06	17.69	2.12E-02	-6.36E-05	2.2E-15	24.1	100.0	28.75	0.07
752-07	18.00	7.75E-01	1.70E-04	3.2E-16	0.7	99.9	*29.23*	0.25
mean age					n=7		**28.56**	**0.09**
isochron age	40Ar/36Ar=241.1 ±41.4		MSWD=1.2		n=7		28.61	0.08
4) Vicks Peak Tuff, nm1080, Mangas quad, sanidine, J=0.000761 ±0.000002								
989-08	21.28	5.20E-03	1.62E-03	1.2E-13	98.0	97.6	28.30	0.07
989-01	21.38	2.16E-02	1.82E-03	2.9E-13	23.6	97.4	28.36	0.07
989-11	21.59	5.48E-03	2.52E-03	2.2E-13	93.2	96.5	28.36	0.06
989-09	21.49	6.97E-03	2.09E-03	2.8E-13	73.2	97.0	28.40	0.05
989-03	21.19	5.25E-03	1.06E-03	2.8E-13	97.2	98.4	28.41	0.06
989-06	21.57	5.40E-03	2.26E-03	1.4E-13	94.4	96.8	28.45	0.09
989-07	21.14	4.55E-03	7.30E-04	1.9E-13	112.1	98.9	28.47	0.06
989-10	21.86	5.56E-03	3.14E-03	1.5E-13	91.8	95.7	28.49	0.08
989-02	22.30	5.58E-03	4.57E-03	1.5E-13	91.4	93.8	28.50	0.08
989-05	22.02	9.40E-03	3.56E-03	2.8E-13	54.3	95.1	28.53	0.05
989-04	21.26	4.64E-03	9.71E-04	3.3E-13	110.1	98.5	28.53	0.05
mean age					n=11		**28.44**	**0.08**
isochron age	40Ar/36Ar=307.9 ±16.8		MSWD=2.0		n=11		28.40	0.09
5) Vicks Peak Tuff, ANDSNMTN-6, Crosby Springs quad, sanidine , J=0.0006614 ±0.000002								
1219-05	25.50	1.27E-02	5.46E-03	2.2E-13	40.3	93.6	28.26	0.07
1219-02	26.35	4.83E-03	8.29E-03	2.5E-13	105.6	90.6	28.27	0.07
1219-07	26.53	4.95E-03	8.82E-03	1.1E-13	103.0	90.1	28.30	0.06
1219-09	31.56	4.25E-03	2.56E-02	1.1E-13	120.1	75.9	28.37	0.10
1219-01	27.52	5.83E-03	1.19E-02	1.6E-13	87.5	87.1	28.37	0.07
1219-08	24.26	1.83E-02	8.39E-04	2.0E-13	27.9	98.9	28.41	0.04
1219-03	25.87	9.82E-03	6.24E-03	1.8E-13	51.9	92.8	28.42	0.07
1219-10	24.56	5.19E-03	1.71E-03	2.3E-13	98.4	97.8	28.46	0.05
1219-04	24.63	1.47E-02	1.71E-03	1.3E-13	34.7	97.9	28.54	0.05
1219-06	25.30	8.28E-03	3.74E-03	2.8E-13	61.6	95.5	28.61	0.05
mean age					n=10		**28.40**	**0.12**
isochron age	40Ar/36Ar=288.4 ±5.6		MSWD=4.4		n=10		28.46	0.10
6) La Jencia Tuff, ESGLFMT-1, Crosby Springs quad, sanidine, J=0.0007585 ±0.000002								
991-01	22.79	2.27E-02	6.66E-03	4.3E-13	22.5	91.3	28.25	0.10
991-08	21.00	1.65E-02	1.07E-04	1.8E-13	30.9	99.8	28.45	0.07
991-02	21.99	1.63E-02	3.35E-03	2.6E-13	31.4	95.4	28.47	0.09
991-03	27.22	1.62E-02	2.08E-02	3.0E-13	31.5	77.3	28.57	0.12
991-06	21.59	1.42E-02	1.70E-03	2.8E-13	36.0	97.6	28.60	0.05
991-09	21.21	1.46E-02	3.87E-04	2.4E-13	34.9	99.4	28.61	0.06
991-07	21.21	1.35E-02	2.84E-04	3.0E-13	37.9	99.5	28.65	0.08
991-05	21.77	1.58E-02	2.01E-03	4.5E-13	32.3	97.2	28.72	0.09
991-13	21.25	1.96E-02	1.39E-04	1.1E-13	26.0	99.7	28.76	0.07
991-04	21.81	1.40E-02	2.01E-03	2.9E-13	36.5	97.2	28.78	0.10
991-14	21.22	1.75E-02	1.43E-05	1.1E-13	29.1	99.9	28.78	0.07
991-11	21.28	1.47E-02	8.98E-05	1.7E-13	34.8	99.8	28.82	0.08
991-12	21.32	1.71E-02	1.13E-04	2.0E-13	29.8	99.7	28.87	0.06
991-10	22.40	1.28E-02	9.39E-05	1.5E-13	39.8	99.8	*30.33*	0.07
mean age					n=13		**28.64**	**0.18**
isochron age	40Ar/36Ar=285.3 ±8.6		MSWD=5.2		n=13		28.68	0.09
7) tuff of South Crosby Peak, upper ignimbrite member, ANDSNMTN-4, Crosby Springs quad, sanidine, J=0.0006609 ±0.000002								
1221-06	26.72	2.74E-02	8.78E-04	1.5E-13	18.6	99.0	31.25	0.06
1221-10	27.50	1.26E-01	3.00E-03	1.5E-13	4.1	96.7	31.45	0.06
1221-08	27.81	2.58E-01	4.02E-03	7.3E-14	2.0	95.7	31.46	0.09
1221-04	27.07	8.21E-03	4.73E-04	1.8E-13	62.2	99.4	31.80	0.05
1221-05	27.25	4.93E-02	1.08E-03	2.5E-13	10.4	98.8	31.81	0.05
1221-03	27.25	8.01E-03	8.71E-04	1.4E-13	63.7	99.0	31.88	0.08
1221-07	27.46	1.10E-02	1.12E-03	1.2E-13	46.3	98.7	32.03	0.06
1221-02	28.08	5.95E-02	3.14E-03	1.4E-13	8.6	96.6	32.06	0.07
1221-01	27.78	9.56E-03	1.69E-03	1.5E-13	53.3	98.1	32.22	0.06
1221-09	29.82	2.20E-01	7.67E-03	1.8E-13	2.3	92.4	*32.56*	0.06
mean age					n=10		31.85	0.39
mean grains 06 to 08 - Caballo Blanco Tuff xenocrysts					n=3		**31.39**	**0.12**
mean grains 04 to 01 - Hells Mesa Tuff xenocrysts					n=6		**31.97**	**0.17**
isochron age 04-01	40Ar/36Ar=428.6 ±69.7		MSWD=6.4		n=6		31.75	0.15

APPENDIX 1 continued

Sample Lab #	40/39	37/39	36/39	39K moles	K/Ca	%40*	Age	±1 σ
8) tuff of South Crosby Peak, middle ignimbrite member, ANDSNMTN-3, Crosby Springs quad, sanidine and plagioclase, J=0.0006589 ±0.000002								
1218-02	60.87	1.95E+00	1.22E-01	1.1E-14	0.3	41.0	*29.49*	0.61
1218-04	36.66	1.59E+00	3.66E-02	1.5E-14	0.3	70.8	*30.62*	0.30
1218-09	27.27	1.07E-02	9.94E-04	2.6E-13	47.5	98.8	31.76	0.05
1218-08	28.72	2.65E-02	5.90E-03	1.4E-13	19.2	93.9	31.76	0.08
1218-01	27.26	9.10E-03	4.37E-04	1.8E-13	56.0	99.4	31.94	0.05
1218-05	27.60	1.13E-02	1.23E-03	1.0E-13	45.3	98.6	32.06	0.07
1218-03	27.47	9.05E-03	5.67E-04	1.9E-13	56.4	99.3	32.14	0.07
1218-06	27.35	8.26E-03	1.58E-04	2.2E-13	61.8	99.8	32.14	0.06
1218-07	34.92	2.74E+00	2.28E-02	7.6E-15	0.2	81.3	*33.48*	0.50
1218-10	33.44	3.74E-02	1.21E-02	6.7E-14	13.6	89.2	*35.12*	0.12
mean age - Hells Mesa Tuff xenocrysts					n=6		**31.97**	**0.18**
isochron age	40Ar/36Ar=256.7 ±37.4		MSWD=11.2		n=6		31.99	0.13
9) tuff of South Crosby Peak, lower ignimbrite member, ANDSNMTN-2, Crosby Springs quad, plagioclase, J=0.0009107 ±0.000001								
743-06	18.33	1.29E+00	3.09E-03	2.8E-16	0.4	95.4	28.54	0.26
743-03	18.50	1.39E+00	3.51E-03	1.0E-16	0.4	94.8	28.62	0.60
743-07	18.31	1.37E+00	1.69E-03	7.2E-17	0.4	97.7	29.19	0.86
743-04	18.53	1.31E+00	1.34E-03	4.0E-16	0.4	98.3	29.70	0.23
743-08	18.30	1.51E+00	3.97E-04	8.8E-17	0.3	99.9	29.81	0.79
743-01	18.37	1.55E+00	2.74E-04	5.7E-16	0.3	100.1	29.99	0.16
743-05	20.39	1.77E+00	5.84E-03	3.1E-17	0.3	92.1	30.63	2.13
743-02	18.41	1.39E+00	-1.56E-03	1.7E-16	0.4	103.0	30.90	0.43
mean age					n=8		**29.67**	**0.86**
isochron age	40Ar/36Ar=89.2 ±39.8		MSWD=4.0		n=8		29.84	0.17
10) Caballo Blanco Tuff, nm1078, Mangas quad, sanidine, J=0.000763 ±0.000002								
995-05	23.92	9.66E-03	2.97E-03	4.1E-13	52.8	96.2	31.41	0.05
995-06	23.64	8.52E-03	1.99E-03	5.5E-13	59.9	97.4	31.42	0.11
995-02	25.31	1.11E-02	7.58E-03	7.0E-13	45.8	91.1	31.45	0.11
995-01	23.56	9.89E-03	1.63E-03	5.6E-13	51.6	97.9	31.46	0.07
995-08	23.36	9.30E-03	8.05E-04	4.9E-13	54.9	98.9	31.53	0.08
995-03	23.28	7.88E-03	4.52E-04	6.7E-13	64.7	99.3	31.55	0.10
995-07	24.71	1.14E-02	5.02E-03	3.7E-13	44.9	93.9	31.66	0.07
995-04	23.65	1.89E-02	9.16E-04	5.6E-13	27.0	98.8	31.87	0.06
mean age					n=8		**31.54**	**0.15**
isochron age	40Ar/36Ar=279.9 ±24.9		MSWD=7.1		n=8		31.61	0.14
11) Caballo Blanco Tuff, nm1079, Mangas Quad, sanidine, J=0.0007625 ±0.000002								
996-03	25.85	9.02E-03	9.74E-03	5.7E-13	56.5	88.8	31.30	0.17
996-01	24.79	7.74E-03	5.96E-03	3.3E-13	65.9	92.8	31.37	0.10
996-05	23.67	7.80E-03	2.07E-03	5.6E-13	65.4	97.3	31.42	0.13
996-12	25.44	1.06E-02	8.01E-03	4.5E-13	48.0	90.6	31.44	0.11
996-04	23.52	9.50E-03	1.51E-03	5.1E-13	53.7	98.0	31.44	0.05
996-10	24.78	9.24E-03	5.72E-03	4.1E-13	55.2	93.1	31.46	0.06
996-09	23.57	1.25E-02	1.61E-03	6.1E-13	40.9	97.9	31.47	0.08
996-06	23.59	1.09E-02	1.61E-03	6.3E-13	46.9	97.9	31.48	0.07
996-07	27.23	8.26E-03	1.40E-02	6.2E-13	61.8	84.8	31.48	0.12
996-11	23.65	9.31E-03	1.81E-03	5.0E-13	54.8	97.6	31.49	0.07
996-08	24.85	8.86E-03	5.87E-03	4.2E-13	57.6	92.9	31.49	0.07
996-02	23.87	1.22E-02	2.04E-03	5.5E-13	41.9	97.4	31.70	0.08
mean age					n=12		**31.46**	**0.09**
isochron age	40Ar/36Ar=291.7 ±5.5		MSWD=1.1		n=12		31.50	0.09
12) Hells Mesa Tuff, ANDSNMT-1, Crosby Springs quad, sanidine, J=0.000759 ±0.000002								
1041-04	23.60	7.88E-03	5.35E-04	4.3E-13	64.8	99.2	31.79	0.05
1041-01	23.81	8.78E-03	1.08E-03	3.9E-13	58.1	98.6	31.85	0.05
1041-10	23.61	8.44E-03	2.14E-04	4.6E-13	60.4	99.6	31.93	0.05
1041-08	23.79	8.59E-03	8.06E-04	3.5E-13	59.4	98.9	31.94	0.06
1041-02	23.69	1.11E-02	4.18E-04	4.6E-13	45.8	99.4	31.96	0.04
1041-03	24.49	1.10E-02	3.10E-03	3.3E-13	46.3	96.2	31.97	0.05
1041-09	25.99	1.01E-02	8.16E-03	3.7E-13	50.8	90.6	31.97	0.05
1041-05	23.73	1.04E-02	5.13E-04	2.6E-13	49.0	99.3	31.98	0.06
1041-06	23.63	8.05E-03	1.66E-04	3.9E-13	63.4	99.7	31.98	0.05
1041-07	24.13	8.84E-03	1.85E-03	3.6E-13	57.7	97.7	31.99	0.05
mean age					n=10		**31.93**	**0.07**
isochron age	40Ar/36Ar=300.7 ±6.6		MSWD=2.1		n=10		31.92	0.09
13) tuff of Luna, LGMSA-1,Largo Mesa quad, sanidine, J=0.0007555 ±0.000002								
1040-07	25.36	9.51E-03	2.55E-03	2.5E-13	53.6	96.9	33.21	0.06
1040-02	24.77	1.02E-02	4.17E-04	3.4E-13	50.2	99.4	33.26	0.05
1040-10	25.68	7.75E-03	3.50E-03	1.5E-13	65.8	95.9	33.26	0.06
1040-04	24.86	9.00E-03	4.29E-04	3.2E-13	56.7	99.4	33.37	0.05
1040-06	24.86	9.17E-03	4.15E-04	2.8E-13	55.6	99.4	33.38	0.06
1040-01	25.76	9.02E-03	3.39E-03	3.7E-13	56.5	96.0	33.41	0.06
1040-09	27.18	9.67E-03	8.05E-03	3.1E-13	52.8	91.2	33.46	0.06
1040-05	24.98	9.33E-03	6.09E-04	2.4E-13	54.7	99.2	33.46	0.07
1040-03	24.95	9.68E-03	2.84E-04	1.7E-13	52.7	99.6	33.55	0.06
1040-08	25.11	9.26E-03	8.06E-04	3.4E-13	55.1	99.0	33.56	0.06
mean age					n=10		**33.39**	**0.12**
isochron age	40Ar/36Ar=295.5 ±12.7		MSWD=5.6		n=10		33.38	0.10
14) Blue Canyon Tuff, nm1098, Cox Peak quad, sanidine, J=0.000909 ±0.000001								
738-04	20.65	1.67E-02	1.24E-04	8.3E-15	30.5	99.7	33.46	0.06
738-08	20.81	1.73E-02	5.14E-04	6.7E-15	29.5	99.2	33.53	0.07
738-02	20.78	1.75E-02	3.23E-04	3.7E-15	29.1	99.4	33.58	0.08
738-05	20.73	1.72E-02	8.98E-05	5.1E-15	29.7	99.8	33.61	0.07
738-03	20.99	1.86E-02	9.26E-04	5.9E-15	27.4	98.6	33.62	0.08
738-01	21.16	1.88E-02	1.51E-03	4.7E-16	27.2	97.8	33.62	0.19
738-07	20.80	1.75E-02	2.71E-04	5.3E-15	29.1	99.5	33.63	0.07
738-06	20.81	1.75E-02	2.34E-04	1.1E-14	29.2	99.6	33.66	0.07
mean age					n=8		**33.59**	**0.06**
isochron age	40Ar/36Ar=329.3 ±53.1		MSWD=1.2		n=8		33.56	0.08
15) Blue Canyon Tuff, nm1077, Mangas quad, sanidine, J=0.00076 ±0.000002								
1037-01	24.82	1.92E-02	2.38E-04	4.9E-13	26.5	99.6	33.59	0.07
1037-07	24.95	1.62E-02	5.42E-04	5.1E-13	31.5	99.3	33.65	0.08
1037-05	25.45	5.05E-02	2.11E-03	1.9E-13	10.1	97.5	33.70	0.07
1037-06	24.98	1.77E-02	4.19E-04	4.0E-13	28.8	99.4	33.74	0.08
1037-10	24.89	1.73E-02	5.97E-05	4.7E-13	29.5	99.8	33.75	0.06
1037-08	24.90	1.78E-02	3.40E-05	3.9E-13	28.7	99.9	33.78	0.07

APPENDIX 1 continued

Sample Lab #	40/39	37/39	36/39	39K moles	K/Ca	%40*	Age	±1 σ
1037-02	24.92	1.74E-02	6.63E-05	3.8E-13	29.3	99.8	33.79	0.07
1037-04	24.98	1.79E-02	1.84E-04	4.6E-13	28.4	99.7	33.83	0.06
1037-09	25.70	3.56E-02	2.36E-03	1.4E-13	14.3	97.2	33.93	0.06
1037-03	27.17	1.73E-02	3.63E-04	5.0E-13	29.4	99.5	*36.70*	0.08
mean age					n=9		**33.75**	**0.10**
isochron age	40Ar/36Ar=330.2 ±27.3		MSWD=2.6		n=9		33.73	0.10

16) Blue Canyon Tuff, nm-1076, Mangas quad, sanidine, J=0.000755 ±0.000002

Sample Lab #	40/39	37/39	36/39	39K moles	K/Ca	%40*	Age	±1 σ
1039-10	25.05	1.34E-02	6.11E-04	3.0E-13	38.1	99.2	33.54	0.05
1039-04	25.05	1.78E-02	5.75E-04	1.7E-13	28.7	99.2	33.55	0.06
1039-02	25.02	1.80E-02	1.37E-04	6.1E-13	28.4	99.8	33.69	0.05
1039-08	25.06	1.71E-02	1.80E-04	4.7E-13	29.8	99.7	33.72	0.06
1039-09	25.05	1.75E-02	7.60E-05	5.2E-13	29.1	99.8	33.74	0.06
1039-06	25.08	1.73E-02	1.23E-04	4.2E-13	29.6	99.8	33.77	0.06
1039-07	25.07	1.73E-02	4.86E-05	3.6E-13	29.5	99.9	33.78	0.06
1039-05	25.09	1.80E-02	7.62E-05	5.4E-13	28.4	99.8	33.80	0.06
1039-03	25.27	1.73E-02	6.06E-04	6.4E-13	29.4	99.2	33.83	0.06
1039-01	26.03	8.89E-03	1.21E-03	3.9E-13	57.4	98.5	*34.61*	0.05
mean age					n=9		**33.71**	**0.10**
isochron age	40Ar/36Ar=124.8 ±42.7		MSWD=3.3		n=9		33.77	0.10

17) Rockhouse Canyon Tuff, nm1097, Cox Peak quad, sanidine and plagioclase, J=0.0009085 ±0.000001

Sample Lab #	40/39	37/39	36/39	39K moles	K/Ca	%40*	Age	±1 σ
736-07	21.03	1.48E-02	7.97E-05	1.3E-14	34.4	99.8	34.07	0.08
736-08	21.22	5.64E-01	8.43E-04	2.9E-15	0.9	98.9	34.09	0.15
736-02	21.11	1.73E-02	2.53E-04	1.1E-14	29.6	99.5	34.12	0.07
736-04	21.09	1.59E-02	1.34E-04	1.2E-14	32.1	99.7	34.14	0.08
736-03	21.11	1.49E-02	1.83E-04	9.2E-15	34.2	99.6	34.15	0.11
736-06	21.18	1.63E-01	2.58E-04	2.8E-15	3.1	99.6	34.25	0.10
736-05	21.15	1.58E-02	8.15E-05	1.1E-14	32.3	99.8	34.27	0.08
736-01	21.23	1.79E-02	1.77E-04	1.2E-14	28.6	99.7	34.35	0.06
mean age					n=8		34.18	0.10
mean age - sanidines					n=6		**34.18**	**0.10**
isochron age	40Ar/36Ar=227.0 ±85.8		MSWD=2.3		n=8		34.22	0.11

18) Rockhouse Canyon Tuff, nm1096, Cox Peak quad, sanidine and plagioclase, J=0.0009107 ±0.000001

Sample Lab #	40/39	37/39	36/39	39K moles	K/Ca	%40*	Age	±1 σ
737-02	21.44	2.25E+00	3.95E-03	8.0E-16	0.2	95.3	33.30	0.48
737-03	21.02	2.07E-02	2.16E-04	6.2E-15	24.6	99.6	34.08	0.12
737-01	20.99	1.65E-02	2.39E-05	5.8E-15	31.0	99.9	34.11	0.07
737-04	21.05	1.69E-02	8.83E-05	9.6E-15	30.1	99.8	34.18	0.10
737-05	21.26	1.64E-02	7.72E-04	1.1E-14	31.2	98.8	34.19	0.09
737-06	21.10	8.05E-01	4.78E-04	5.9E-16	0.6	99.5	34.20	0.17
737-08	21.06	1.42E-02	-4.15E-05	5.6E-15	36.0	100.0	34.25	0.07
737-07	21.45	6.74E-01	1.73E-04	1.3E-15	0.8	99.9	34.88	0.11
mean age					n=8		34.15	0.43
mean age - sanidines					n=5		**34.16**	**0.07**
isochron age	40Ar/36Ar=294.3 ±71.6		MSWD=0.7		n=6		34.17	0.08

Bishop Peak Tuff

19) tuff of Bishop Peak,nm1112, Gallo Mts West quad, sanidine and plagioclase, J=0.0007575 ±0.000002

Sample Lab #	40/39	37/39	36/39	39K moles	K/Ca	%40*	Age	±1 σ
992-12	26.25	7.63E-02	2.24E-03	7.1E-14	6.7	97.4	34.62	0.09
992-01	25.76	3.87E-02	2.08E-04	1.8E-13	13.2	99.7	34.75	0.08
992-05	26.23	3.85E-02	1.80E-03	1.4E-13	13.3	97.9	34.76	0.08
992-04	26.59	3.00E-01	3.06E-03	9.8E-14	1.7	96.6	34.77	0.08
992-13	26.02	4.05E-02	9.66E-04	1.4E-13	12.6	98.8	34.81	0.08
992-09	26.07	4.60E-02	9.95E-04	1.5E-13	11.1	98.8	34.87	0.08
992-11	26.11	4.58E-02	1.07E-03	1.1E-13	11.2	98.7	34.89	0.08
992-08	26.19	4.85E-02	1.18E-03	9.6E-14	10.5	98.6	34.95	0.08
992-10	27.17	2.07E-01	4.54E-03	7.0E-14	2.5	95.0	34.96	0.13
992-07	26.23	2.67E-01	7.82E-04	4.9E-14	1.9	99.1	35.19	0.12
992-03	46.79	1.14E+00	7.06E-02	2.4E-14	0.4	*55.6*	*35.22*	0.39
992-02	27.43	6.07E-01	4.83E-03	3.6E-14	0.8	94.9	*35.24*	0.16
992-06	49.74	1.25E+00	7.77E-02	3.5E-14	0.4	*54.0*	*36.36*	0.32
mean age					n=11		34.89	0.19
mean age - plagioclase					n=3		35.61	0.66
mean age - sanidines					n=6		**34.84**	**0.08**
isochron age	40Ar/36Ar=282.0 ±32.0		MSWD=3.0		n=10		34.86	0.12

20) tuff of Bishop Peak, GMW-1-93, Gallo Mts West quad, sanidine, J=0.004019 ±0.00001

Sample Lab #	40/39	37/39	36/39	39K moles	K/Ca	%40*	Age	±1 σ
291-05A	4.85	4.06E-02	7.97E-05	7.6E-17	12.6	99.1	34.56	0.10
291-04A	4.87	4.10E-02	1.02E-04	8.2E-17	12.4	99.0	34.61	0.10
291-06A	4.88	4.67E-02	5.09E-05	1.1E-16	10.9	99.3	34.83	0.08
291-03A	4.92	4.58E-02	6.78E-05	1.0E-16	11.1	99.2	35.07	0.08
291-01A	4.93	4.42E-02	8.28E-05	4.9E-17	11.5	99.1	35.10	0.11
291-08A	4.95	2.54E-01	2.10E-04	7.9E-17	2.0	98.7	35.12	0.09
291-02A	4.93	5.40E-02	6.08E-05	8.3E-17	9.5	99.3	35.17	0.08
291-07A	4.95	4.44E-02	3.52E-05	2.1E-16	11.5	99.4	35.31	0.08
mean age					n=8		**34.97**	**0.28**
isochron age	40Ar/36Ar=601.6 ±280.7		MSWD=12.1		n=8		34.68	0.22

21) Datil Well Tuff, nm1095, Cox Peak quad, sanidine, J=0.0009068 ±0.000001

Sample Lab #	40/39	37/39	36/39	39K moles	K/Ca	%40*	Age	±1 σ
1072-02	25.95	4.23E+00	2.28E-02	7.7E-15	0.1	75.2	*31.73*	0.35
1072-04	21.66	2.49E-03	3.21E-03	4.3E-15	205.1	95.5	*33.54*	0.64
741-02	21.77	3.84E-02	1.22E-03	3.5E-15	13.3	98.3	34.66	0.09
741-03	21.56	3.30E-02	-8.67E-05	4.8E-16	15.5	100.0	34.95	0.19
1072-05	22.01	3.72E-02	1.42E-03	1.5E-14	13.7	98.0	34.96	0.21
1072-03	21.79	3.29E-02	6.70E-04	1.3E-14	15.5	99.0	34.96	0.21
1072-07	21.81	4.13E-02	3.86E-04	5.7E-14	12.4	99.4	35.13	0.08
1072-06	22.03	2.00E-02	9.31E-04	1.6E-14	25.5	98.7	35.20	0.20
1072-01	23.56	4.18E-02	4.38E-03	7.3E-14	12.2	94.4	*36.03*	0.07
mean age					n=6		**34.98**	**0.19**
isochron age	40Ar/36Ar=195.9 ±88.7		MSWD=3.4		n=6		35.09	0.17

VOLCANICLASTIC UNITS

22) volcaniclastic unit of Canon del Leon, CDL 10/2-4, Escondido Mt quad, upper, pumice clasts from siltstone, sanidine and plagioclase, J=0.0009106 ±0.000001

APPENDIX 1 continued

Sample Lab #	40/39	37/39	36/39	39K moles	K/Ca	%40*	Age	±1 σ
744-05	21.69	1.18E+00	5.82E-03	1.2E-15	0.4	92.4	32.65	0.13
744-02	20.97	1.48E+00	1.97E-03	7.0E-16	0.3	97.7	33.37	0.14
744-01	22.12	1.60E+00	5.40E-03	1.4E-16	0.3	93.2	33.61	0.45
744-04	20.93	2.79E-02	-3.09E-04	8.3E-16	18.3	100.3	34.17	0.13
744-03	34.29	2.15E+00	1.28E-02	2.3E-15	0.2	89.4	49.74	0.21
mean age					n=5		36.71	7.31
mean age - plagioclase					n=3		33.21	0.50
mean age - sanidine					n=1		**34.17**	**0.13**
isochron age	40Ar/36Ar=160.1 ±41.4		MSWD=7.2		n=4		33.95	0.25

23) volcaniclastic unit of Canon del Leon, CDL 10/2-9B, Escondido Mt quad, lower, ash bed, plagioclase, L#754: J=0.0009091 ±0.000001; L#756: J=0.0009076 ±0.000001

Sample Lab #	40/39	37/39	36/39	39K moles	K/Ca	%40*	Age	±1 σ
754-01	22.71	1.64E+00	5.27E-03	1.9E-15	0.3	93.6	34.57	0.23
754-02	27.62	1.50E+00	2.14E-02	9.1E-16	0.3	77.4	34.76	0.23
754-04	24.12	1.75E+00	8.89E-03	8.5E-16	0.3	89.6	35.14	0.17
754-03	22.37	1.75E+00	2.75E-03	1.4E-15	0.3	96.9	35.23	0.11
754-05	24.00	1.95E+00	8.17E-03	1.5E-15	0.3	90.5	35.31	0.13
754-06	23.85	1.80E+00	7.18E-03	5.0E-16	0.3	91.6	35.51	0.25
756-02	22.78	1.95E+00	3.85E-03	1.1E-15	0.3	95.6	35.35	0.14
756-01	23.52	1.85E+00	5.73E-03	1.0E-15	0.3	93.3	35.62	0.16
mean age					n=8		**35.19**	**0.36**
isochron age	40Ar/36Ar=280.8 ±12.6		MSWD=3.4		n=8		35.40	0.17

24) volcaniclastic unit of Canon del Leon, CDL 10/2-11B, Escondido Mt quad, basal, pumice clasts from siltstone, sanidine and plagioclase, J=0.0009058 ±0.000001

Sample Lab #	40/39	37/39	36/39	39K moles	K/Ca	%40*	Age	±1 σ
742-02	23.39	1.59E+00	6.46E-03	1.5E-15	0.3	92.3	35.07	0.13
742-01	22.84	1.82E+00	4.44E-03	3.4E-16	0.3	94.8	35.20	0.23
742-03	22.02	1.75E+00	1.35E-03	2.1E-15	0.3	98.7	35.33	0.11
742-07	23.26	2.63E+00	5.14E-03	2.3E-16	0.2	94.2	35.64	0.36
742-06	23.29	2.77E+00	4.28E-03	7.9E-16	0.2	95.4	36.13	0.16
742-04	36.74	3.97E+00	9.98E-03	1.5E-16	0.1	92.7	*55.17*	0.53
mean age					n=5		**35.48**	**0.42**
isochron age	40Ar/36Ar=290.2 ±64.8		MSWD=13.2		n=5		35.47	0.42

25) volcaniclastic unit of Canon del Leon, CR-93-1-XR, Largo Mesa quad, basal, pumice clasts from siltstone, plagioclase, J=0.0009088 ±0.000001

Sample Lab #	40/39	37/39	36/39	39K moles	K/Ca	%40*	Age	±1 σ
746-08	23.96	1.23E+00	8.80E-03	2.6E-15	0.4	89.4	34.72	0.24
746-03	22.75	1.77E+00	4.09E-03	2.6E-15	0.3	95.2	35.08	0.09
746-07	24.38	1.83E+00	9.44E-03	1.5E-15	0.3	89.0	35.17	0.15
746-02	22.17	1.60E+00	1.84E-03	1.8E-15	0.3	98.0	35.20	0.11
746-05	22.09	7.38E-03	9.00E-04	1.5E-14	69.1	98.7	35.29	0.07
746-04	21.90	7.44E-03	4.81E-05	1.5E-14	68.5	99.8	35.38	0.06
746-06	23.95	2.96E+00	4.97E-03	8.0E-16	0.2	94.7	36.78	0.17
746-01	31.49	1.81E+00	2.94E-02	2.4E-15	0.3	72.8	37.14	0.39
mean age					n=8		35.59	0.87
mean age - sanidines					n=2		**35.33**	**0.06**
isochron age	40Ar/36Ar=316.6 ±26.9		MSWD=21.1		n=8		35.28	0.19

26) Dog Springs Fm, nm1115, tuff layer, Madre Mt quad, plagioclase and anorthoclase, L#751: J=0.0009068 ±0.000001; L#1254: J=0.0006689 ±0.000002

Sample Lab #	40/39	37/39	36/39	39K moles	K/Ca	%40*	Age	±1 σ
751-06	22.23	1.74E-01	2.89E-03	4.5E-15	2.9	96.1	34.63	0.10
751-01	23.93	7.53E-02	8.22E-03	3.3E-15	6.8	89.8	34.82	0.18
751-03	21.79	2.89E-01	8.06E-04	2.4E-15	1.8	98.9	34.93	0.09
751-04	21.69	1.44E-01	4.07E-04	5.9E-15	3.5	99.4	34.94	0.08
751-05	22.78	1.29E-01	4.00E-03	3.0E-15	4.0	94.8	34.98	0.18
751-08	22.04	2.43E-01	1.39E-03	4.3E-15	2.1	98.1	35.04	0.13
751-02	22.32	5.36E-01	2.36E-03	2.6E-15	1.0	97.0	35.08	0.09
751-07	24.45	9.52E-02	3.93E-03	4.5E-15	5.4	95.2	*37.67*	0.13
1254-03	69.62	3.07E+00	1.30E-01	3.3E-14	0.2	45.1	37.57	0.34
1254-08	78.22	1.62E+00	1.58E-01	4.2E-14	0.3	40.6	37.98	0.36
1254-06	57.05	1.84E+00	8.53E-02	2.3E-14	0.3	56.0	38.23	0.39
1254-07	41.57	1.87E+00	3.27E-02	3.4E-14	0.3	77.1	38.30	0.16
1254-04	77.68	3.25E+00	1.55E-01	2.5E-14	0.2	41.3	38.37	0.44
1254-02	58.63	1.65E+00	8.91E-02	5.5E-14	0.3	55.3	38.74	0.23
1254-01	60.55	1.86E+00	9.50E-02	4.0E-14	0.3	53.9	38.98	0.30
1254-05	112.78	3.20E+00	2.65E-01	1.4E-14	0.2	30.7	41.42	1.07
mean age - authigenic K-spar ?					n=7		34.92	0.15
mean age - plagioclase					n=8		**38.70**	**1.18**
isochron age - plag	40Ar/36Ar=294.4 ±3.2		MSWD=2.9		n=7		38.48	0.35

27) Dog Springs Fm, nm1121, tuff breccia, Crosby Springs quad, plagioclase and sanidine, J=0.0009061 ±0.000001

Sample Lab #	40/39	37/39	36/39	39K moles	K/Ca	%40*	Age	±1 σ
740-06	26.04	3.21E+00	1.36E-02	2.8E-16	0.2	85.4	36.06	0.32
740-07	27.75	2.71E+00	1.80E-02	1.7E-16	0.2	81.5	36.64	0.58
740-08	22.95	4.82E-03	3.29E-04	8.5E-15	105.9	99.5	36.94	0.07
740-01	25.61	2.99E+00	8.61E-03	2.8E-16	0.2	90.9	37.73	0.35
740-03	27.97	2.91E+00	1.49E-02	2.6E-16	0.2	85.0	38.55	0.37
740-02	25.12	3.07E+00	3.20E-03	1.1E-16	0.2	97.1	39.51	0.79
mean age					n=6		37.57	1.29
mean age - sanidine					n=1		**36.94**	**0.07**
isochron age	40Ar/36Ar=322.5 ±36.9		MSWD=14.0		n=6		36.94	0.28

Intrusion

28) Cerrito Viejo, nm1114, quartz diorite intrusion, Sugarloaf Mt quad, plagioclase, J=0.0006652 ±0.000002

Sample Lab #	40/39	37/39	36/39	39K moles	K/Ca	%40*	Age	±1 σ
1253-07	399.83	3.33E+00	1.27E+00	2.6E-14	0.2	5.9	27.96	2.15
1253-01	40.83	4.65E+00	5.65E-02	1.4E-14	0.1	59.9	29.21	0.34
1253-03	169.62	3.60E+00	4.90E-01	3.7E-14	0.1	14.7	29.80	0.82
1253-04	159.68	3.75E+00	4.54E-01	2.5E-14	0.1	16.2	30.91	0.86
1253-08	163.48	2.97E+00	4.65E-01	1.9E-14	0.2	16.2	31.49	1.33
1253-06	98.81	3.04E+00	2.46E-01	2.6E-14	0.2	26.7	31.50	0.52
1253-02	37.69	4.25E+00	3.84E-02	1.0E-14	0.1	70.7	31.79	0.42
1253-05	59.32	4.43E+00	1.12E-01	1.8E-14	0.1	44.9	31.79	0.43
mean age					n=8		**30.56**	**1.42**
isochron age	40Ar/36Ar=295.5 ±2.7		MSWD=7.3		n=8		30.75	0.65

APPENDIX 2 - $^{40}Ar/^{39}Ar$ step-heating results from Quemado area samples are given below.

Sample Lab #	Temp	40/39	37/39	36/39	39K moles	K/Ca	Cl/K	%40*	%39Ar	Age	± 1σ
29) basalt of Blue Mesa, BLUMSA-2, upper flow, D-Cross Mt quad, groundmass concentrate, J=0.0007612 ±0.000002											
1007-01A	550	1.47E+02	8.00E-01	4.82E-01	2.2E-14	6.4E-01	1.9E-01	3.0	1.9	6.02	1.08
1007-01B	625	2.71E+01	6.96E-01	8.00E-02	8.1E-14	7.3E-01	1.5E-01	12.9	8.8	4.79	0.20
1007-01C	725	6.08E+00	5.75E-01	9.24E-03	2.2E-13	8.9E-01	8.6E-02	55.5	28.1	4.63	0.03
1007-01D	775	4.17E+00	6.04E-01	2.88E-03	1.3E-13	8.4E-01	5.3E-02	80.2	39.4	4.59	0.03
1007-01E	800	4.08E+00	7.46E-01	2.42E-03	6.1E-14	6.8E-01	4.1E-02	83.4	44.6	4.67	0.05
1007-01F	850	4.10E+00	8.59E-01	2.62E-03	5.1E-14	5.9E-01	5.0E-02	82.2	49.0	4.62	0.06
1007-01G	900	4.26E+00	8.85E-01	2.51E-03	4.6E-14	5.8E-01	7.7E-02	83.6	53.0	4.89	0.07
1007-01H	1000	4.26E+00	8.83E-01	2.93E-03	9.7E-14	5.8E-01	6.5E-02	80.8	61.3	4.73	0.04
1007-01I	1200	4.45E+00	2.78E+00	4.34E-03	1.3E-13	1.8E-01	2.3E-02	75.4	72.6	4.61	0.04
1007-01J	1400	4.43E+00	5.44E-01	5.23E-03	3.1E-13	9.4E-01	3.3E-02	65.5	99.6	3.98	0.10
1007-01K	1625	7.72E+01	3.06E+01	2.52E-01	4.5E-15	1.7E-02	-2.3E-02	6.4	100.0	6.93	1.76
total gas age					1.2E-12				n=11	4.52	0.09
mean age of near-plateau steps									B to I	4.65	0.03
isochron age						40Ar/36Ar=297.7 ±1.1		MSWD=71.8		**4.48**	**0.02**
29) basalt of Blue Mesa, BLUMSA-1, lower flow, D-Cross Mt quad, groundmass concentrate, J=0.0007643 ±0.000002											
1008-01A	550	3.58E+02	1.80E+00	1.20E+00	6.9E-15	2.8E-01	8.3E-02	1.0	0.9	4.91	4.71
1008-01B	625	1.10E+02	1.57E+00	3.59E-01	2.2E-14	3.3E-01	6.3E-02	3.3	3.9	4.95	1.11
1008-01C	725	4.29E+01	1.10E+00	1.34E-01	6.7E-14	4.6E-01	5.8E-02	7.9	13.0	4.66	0.32
1008-01D	775	3.30E+01	8.91E-01	9.97E-02	5.6E-14	5.7E-01	4.6E-02	10.9	20.7	4.95	0.24
1008-01E	800	3.85E+01	9.73E-01	1.16E-01	4.0E-14	5.2E-01	3.2E-02	11.0	26.1	5.84	0.31
1008-01F	850	5.03E+01	1.17E+00	1.56E-01	4.6E-14	4.4E-01	4.7E-02	8.7	32.3	6.04	0.39
1008-01G	900	3.58E+01	1.34E+00	1.07E-01	5.5E-14	3.8E-01	2.5E-02	11.9	39.8	5.87	0.28
1008-01H	1000	2.11E+01	1.44E+00	5.81E-02	1.1E-13	3.5E-01	3.3E-03	19.0	55.1	5.52	0.15
1008-01I	1200	1.16E+01	1.36E+00	2.72E-02	1.4E-13	3.8E-01	1.7E-03	31.2	73.8	4.97	0.07
1008-01J	1400	1.14E+01	1.66E+01	2.80E-02	1.8E-13	3.1E-02	7.1E-03	38.2	97.7	6.05	0.08
1008-01K	1636	3.69E+01	1.11E+01	1.14E-01	1.7E-14	4.6E-02	6.7E-03	10.7	100.0	5.48	0.43
total gas age					7.3E-13				n=11	5.48	0.25
disturbed spectrum									no plateau		
isochron age						40Ar/36Ar=293.8 ±0.9		MSWD=19.4		**5.71**	**0.06**
31) Bearwallow Mt Andesite, STLRMS-2, basal lava, Slaughter Mesa quad, groundmass concentrate, J=0.0007656 ±0.000002											
1011-01A	550	5.69E+01	3.21E-01	1.35E-01	7.5E-14	1.6E+00	5.1E-02	29.8	4.6	23.3	0.3
1011-01B	625	3.45E+01	5.12E-01	5.45E-02	1.5E-13	1.0E+00	3.9E-02	53.4	13.9	25.3	0.1
1011-01C	725	2.39E+01	7.77E-01	1.58E-02	2.5E-13	6.6E-01	1.8E-02	80.6	29.1	26.4	0.1
1011-01D	775	2.11E+01	8.51E-01	5.99E-03	1.8E-13	6.0E-01	6.2E-03	91.8	40.3	26.6	0.1
1011-01E	800	2.02E+01	7.68E-01	3.36E-03	1.4E-13	6.6E-01	2.5E-03	95.3	49.0	26.4	0.1
1011-01F	850	1.95E+01	7.05E-01	2.03E-03	1.8E-13	7.2E-01	1.3E-03	97.1	59.9	26.0	0.1
1011-01G	900	1.93E+01	6.77E-01	1.42E-03	1.7E-13	7.5E-01	9.1E-04	98.0	70.2	26.0	0.1
1011-01H	1000	1.93E+01	7.25E-01	1.50E-03	2.0E-13	7.0E-01	1.2E-03	97.9	82.5	26.0	0.1
1011-01I	1200	1.99E+01	7.91E-01	4.19E-03	1.9E-13	6.4E-01	3.1E-03	94.0	94.2	25.7	0.1
1011-01J	1400	2.19E+01	7.35E+00	1.21E-02	7.1E-14	6.9E-02	6.3E-03	86.1	98.6	26.0	0.1
1011-01K	1626	3.81E+01	6.37E+00	6.64E-02	2.2E-14	8.0E-02	7.2E-03	49.8	100.0	26.1	0.3
total gas age					1.6E-12				n=11	25.9	0.1
mean age of near-plateau steps									C to J	26.1	0.1
isochron age						40Ar/36Ar=286.1±1.1		MSWD=24.6		**26.1**	**0.1**
32) Squirrel Springs Canyon Andesite, STLRMS-1, Slaughter Mesa quad, groundmass concentrate, J=0.0007618 ±0.000002											
1012-01A	550	2.39E+02	8.65E-01	7.34E-01	1.7E-14	5.9E-01	5.0E-03	9.3	1.0	30.2	2.1
1012-01B	625	5.72E+01	9.21E-01	1.18E-01	5.3E-14	5.5E-01	1.4E-03	39.2	4.3	30.6	0.3
1012-01C	726	2.60E+01	7.62E-01	1.25E-02	1.9E-13	6.7E-01	5.8E-04	85.9	16.3	30.4	0.1
1012-01D	775	2.20E+01	6.28E-01	1.56E-03	1.9E-13	8.1E-01	3.4E-04	98.0	28.3	29.4	0.1
1012-01E	800	2.15E+01	5.31E-01	1.27E-03	1.6E-13	9.6E-01	4.4E-04	98.3	38.2	28.9	0.1
1012-01F	850	2.13E+01	4.71E-01	1.28E-03	2.2E-13	1.1E+00	4.5E-04	98.3	52.1	28.5	0.1
1012-01G	900	2.12E+01	4.31E-01	1.44E-03	2.2E-13	1.2E+00	6.4E-04	98.0	65.6	28.3	0.1
1012-01H	1000	2.16E+01	4.00E-01	2.10E-03	2.4E-13	1.3E+00	1.8E-03	97.2	80.4	28.6	0.1
1012-01I	1200	2.29E+01	6.01E-01	7.03E-03	2.0E-13	8.5E-01	4.1E-03	91.0	93.1	28.4	0.1
1012-01J	1400	3.77E+01	1.29E+01	5.89E-02	3.4E-14	4.0E-02	4.5E-03	56.4	95.2	29.2	0.2
1012-01K	1631	2.83E+01	7.56E+00	2.70E-02	7.7E-14	6.7E-02	2.2E-03	73.9	100.0	28.7	0.1
total gas age					1.6E-12				n=11	29.0	0.1
mean age of near-plateau steps									F-I	28.5	0.1
isochron age						40Ar/36Ar=305.1 ±1.2		MSWD=54.43		**28.8**	**0.1**
33) andesite of San Antone Canyon, MNGSMT-1, Mangas Mt quad, groundmass concentrate, J=0.0007652 ±0.000002											
1013-01A	550	1.93E+02	5.23E-01	5.86E-01	3.1E-14	9.8E-01	6.3E-03	10.4	2.0	27.6	1.6
1013-01B	625	5.53E+01	5.24E-01	1.11E-01	7.0E-14	9.7E-01	2.4E-03	40.8	6.5	30.8	0.3
1013-01C	725	2.74E+01	4.71E-01	1.76E-02	2.2E-13	1.1E+00	7.8E-04	81.1	20.4	30.4	0.1
1013-01D	775	2.27E+01	3.93E-01	2.53E-03	2.2E-13	1.3E+00	3.8E-04	96.7	34.7	30.1	0.1
1013-01E	800	2.20E+01	3.47E-01	1.74E-03	1.7E-13	1.5E+00	4.5E-04	97.7	45.4	29.4	0.1
1013-01F	850	2.18E+01	3.56E-01	1.34E-03	2.1E-13	1.4E+00	7.0E-04	98.2	58.9	29.3	0.1
1013-01G	900	2.15E+01	3.85E-01	1.30E-03	2.0E-13	1.3E+00	1.0E-03	98.3	72.0	28.9	0.1
1013-01H	1000	2.15E+01	5.60E-01	1.78E-03	2.5E-13	9.1E-01	3.1E-03	97.6	87.7	28.8	0.1
1013-01I	1200	2.21E+01	1.02E+00	5.67E-03	1.2E-13	5.0E-01	8.3E-03	92.7	95.3	28.1	0.1
1013-01J	1400	2.57E+01	7.00E+00	2.00E-02	3.2E-14	7.3E-02	2.1E-02	79.0	97.3	27.9	0.2
1013-01K	1628	3.19E+01	6.25E+00	4.04E-02	4.2E-14	8.2E-02	2.6E-02	63.9	100.0	28.0	0.2
total gas age					1.6E-12				n=11	29.3	0.1
disturbed spectrum									no plateau		
isochron age						40Ar/36Ar=300.3 ±1.2		MSWD=81.84		**29.2**	**0.1**

APPENDIX 3

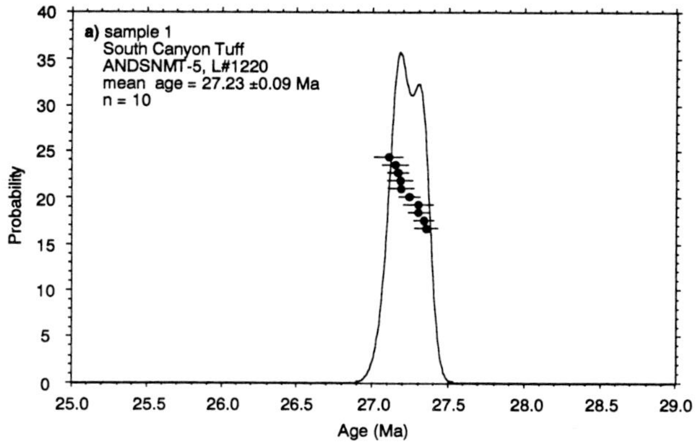
a) sample 1
South Canyon Tuff
ANDSNMT-5, L#1220
mean age = 27.23 ±0.09 Ma
n = 10
Probability
Age (Ma)

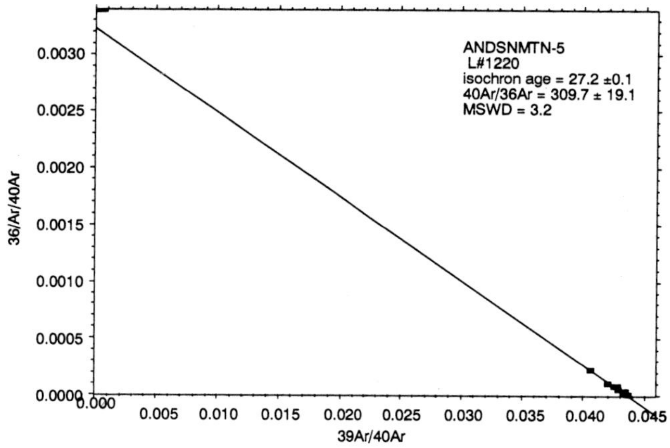
ANDSNMTN-5
L#1220
isochron age = 27.2 ±0.1
40Ar/36Ar = 309.7 ± 19.1
MSWD = 3.2
36/Ar/40Ar
39Ar/40Ar

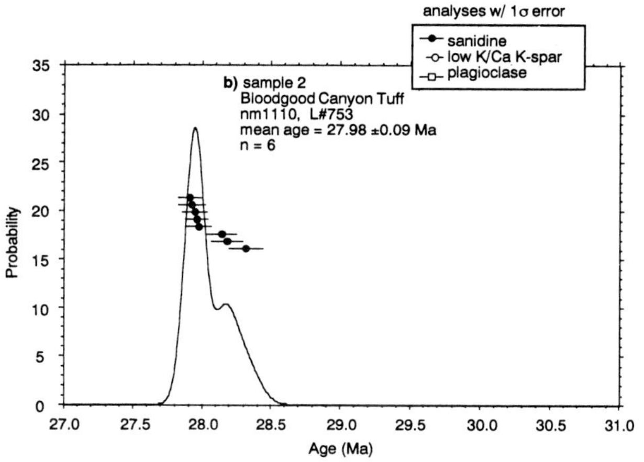
analyses w/ 1σ error
sanidine
low K/Ca K-spar
plagioclase
b) sample 2
Bloodgood Canyon Tuff
nm1110, L#753
mean age = 27.98 ±0.09 Ma
n = 6
Probability
Age (Ma)

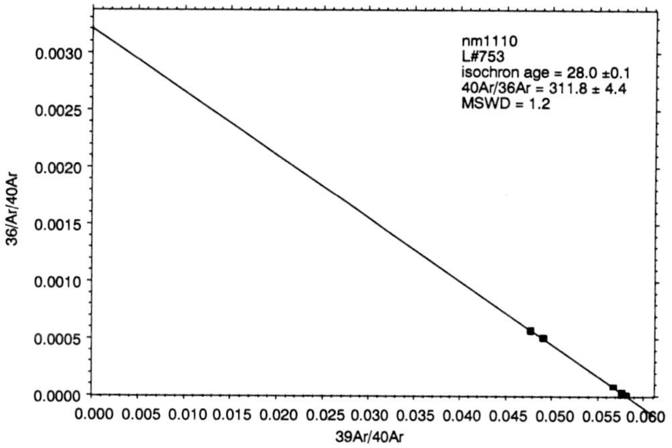
nm1110
L#753
isochron age = 28.0 ±0.1
40Ar/36Ar = 311.8 ± 4.4
MSWD = 1.2
36/Ar/40Ar
39Ar/40Ar

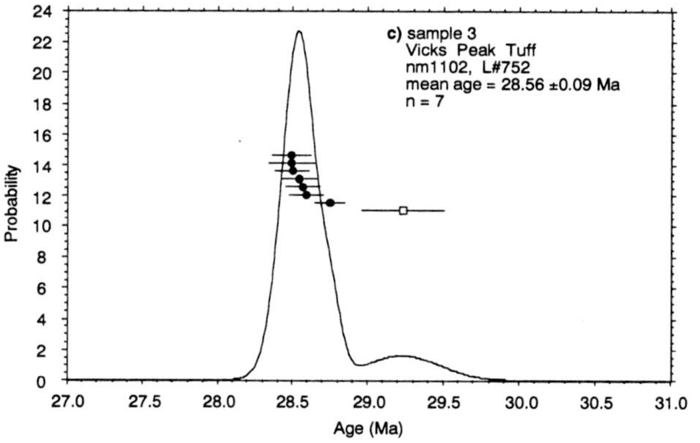
c) sample 3
Vicks Peak Tuff
nm1102, L#752
mean age = 28.56 ±0.09 Ma
n = 7
Probability
Age (Ma)

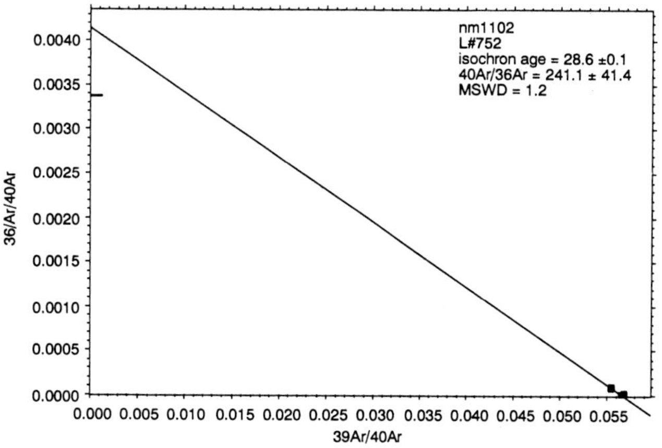
nm1102
L#752
isochron age = 28.6 ±0.1
40Ar/36Ar = 241.1 ± 41.4
MSWD = 1.2
36/Ar/40Ar
39Ar/40Ar

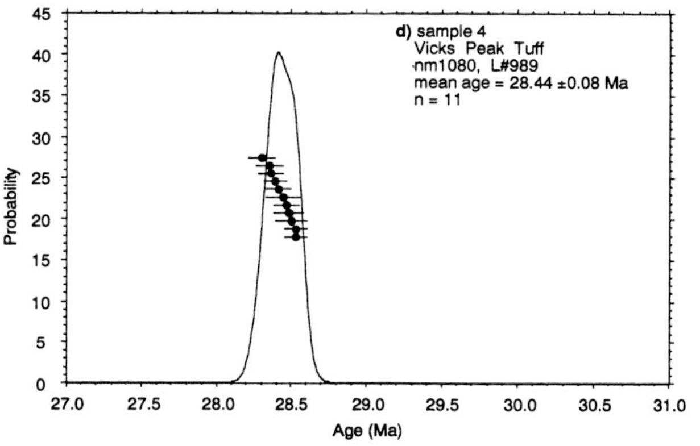
d) sample 4
Vicks Peak Tuff
nm1080, L#989
mean age = 28.44 ±0.08 Ma
n = 11
Probability
Age (Ma)

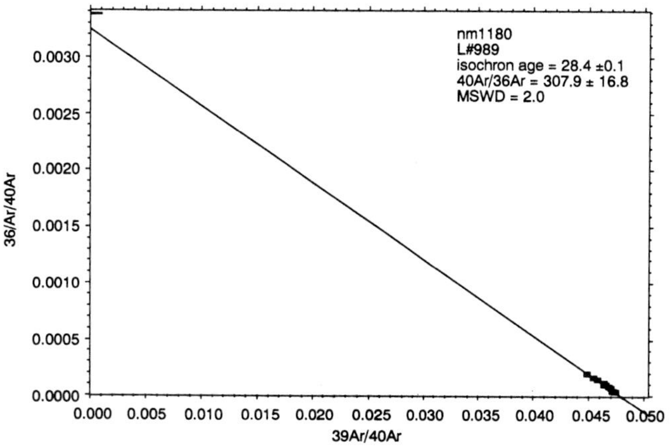
nm1180
L#989
isochron age = 28.4 ±0.1
40Ar/36Ar = 307.9 ± 16.8
MSWD = 2.0
36/Ar/40Ar
39Ar/40Ar

APPENDIX 3 continued

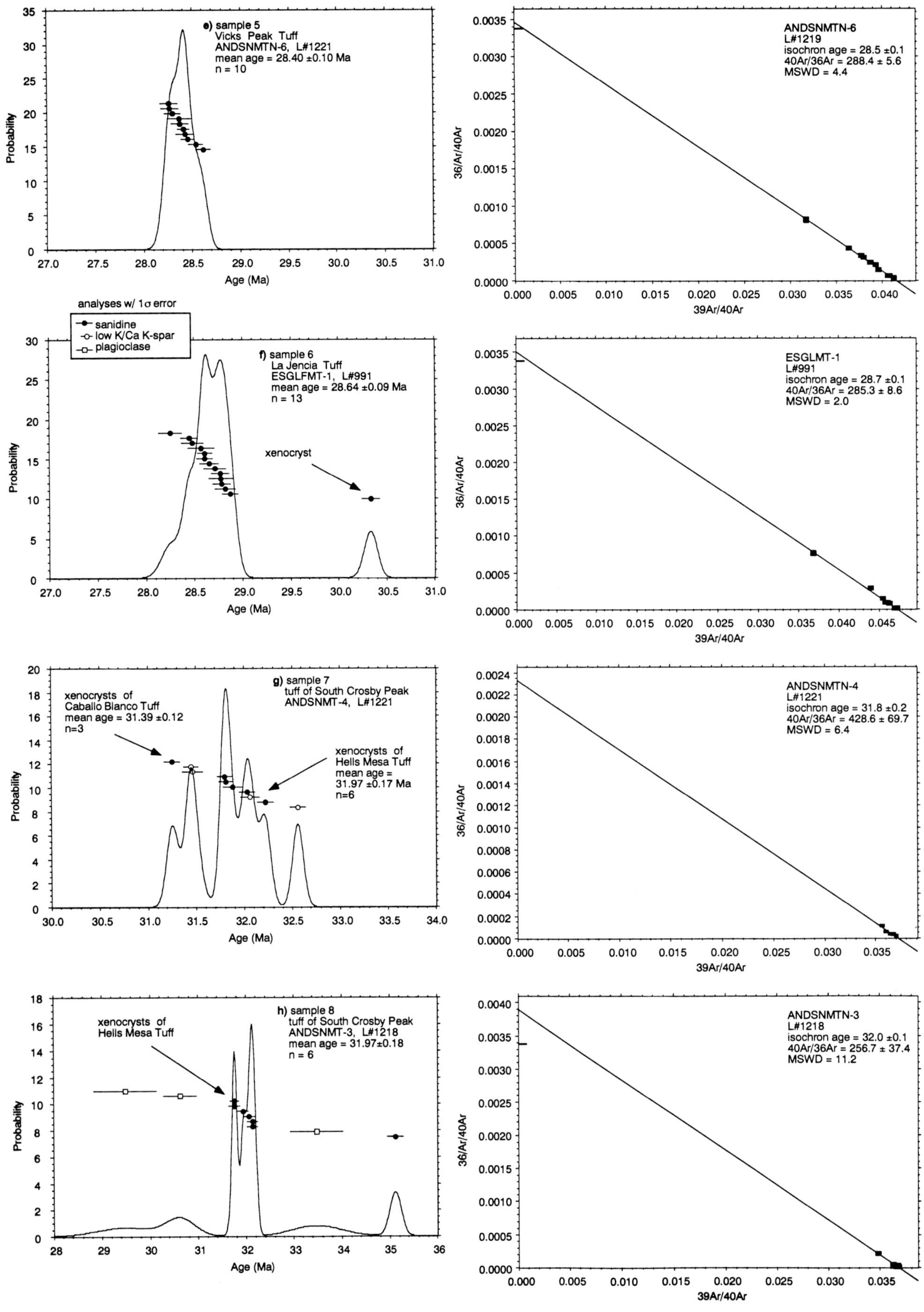

APPENDIX 3 continued

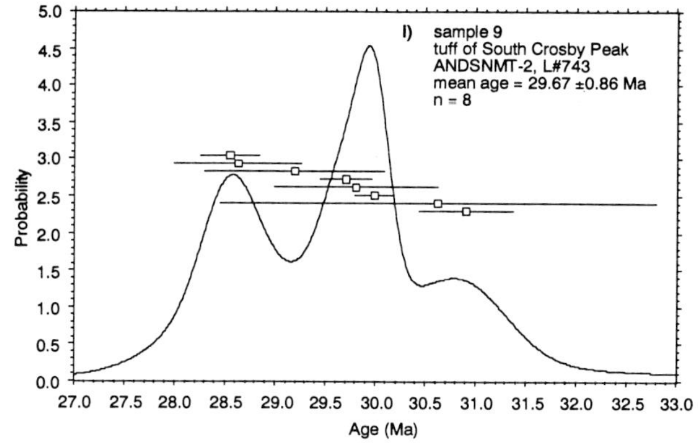

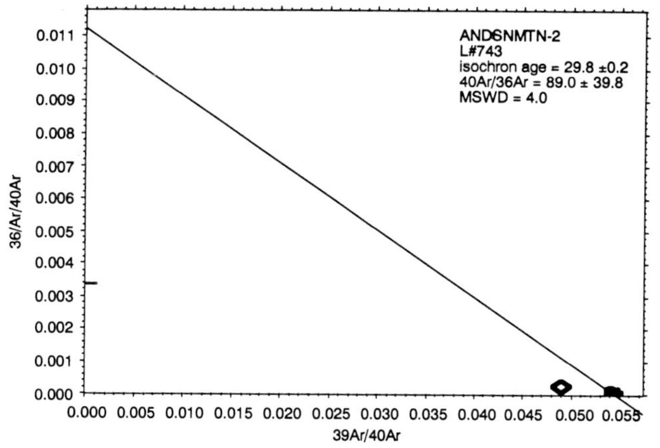

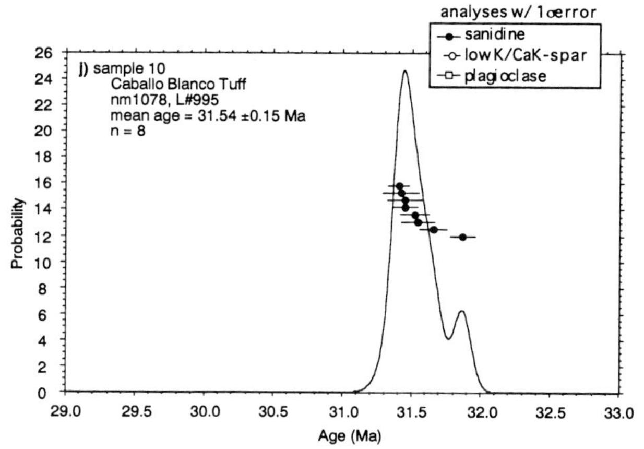

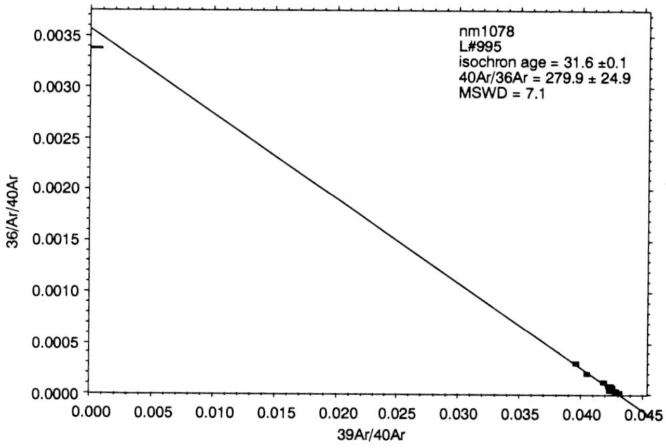

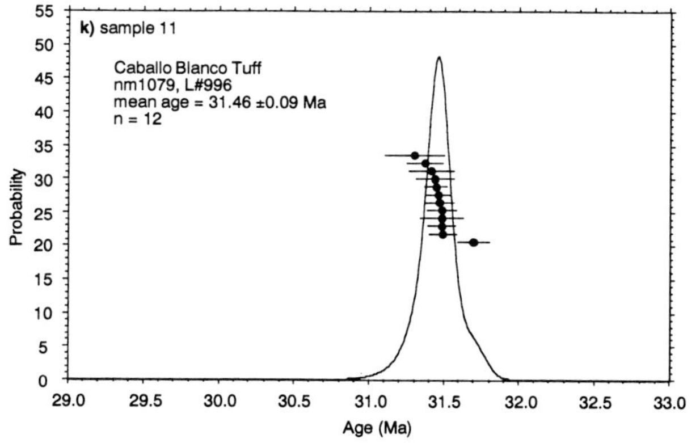

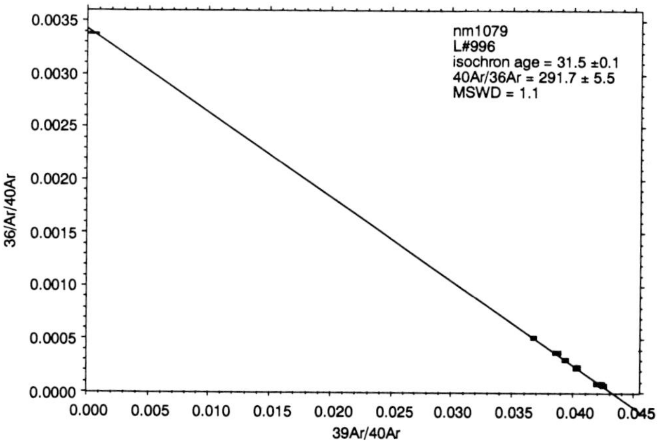

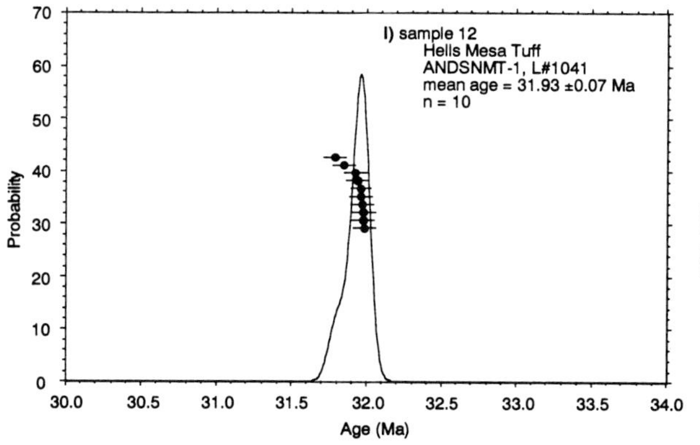

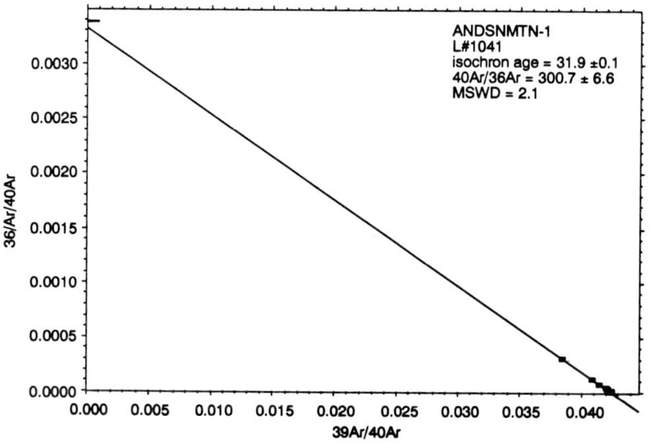

APPENDIX 3 continued

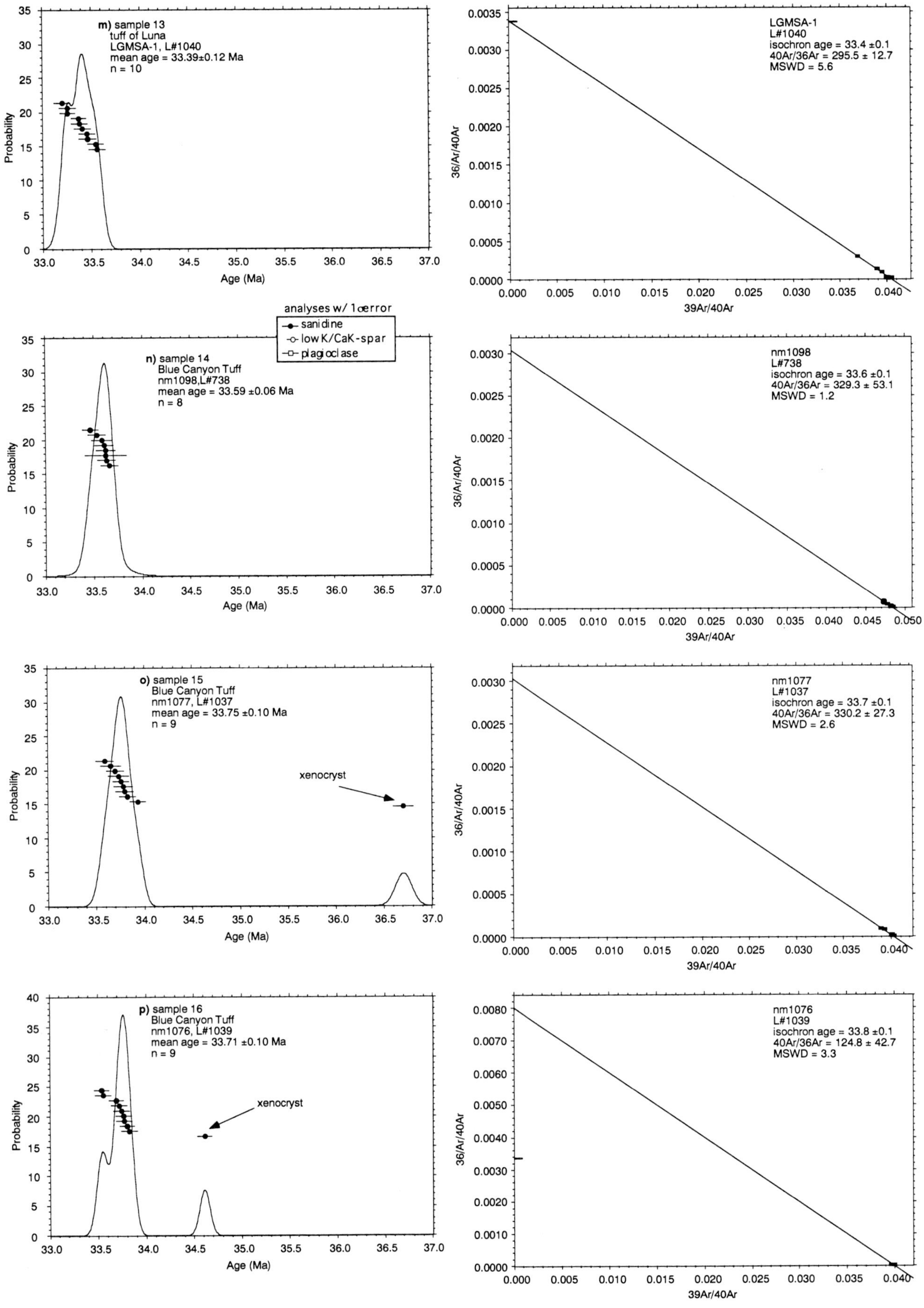

APPENDIX 3 continued

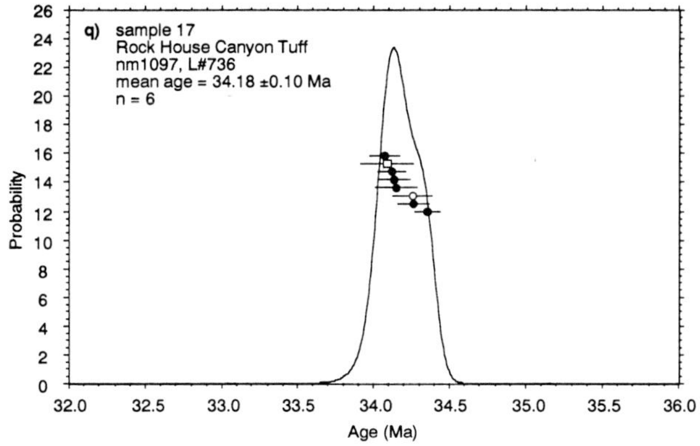

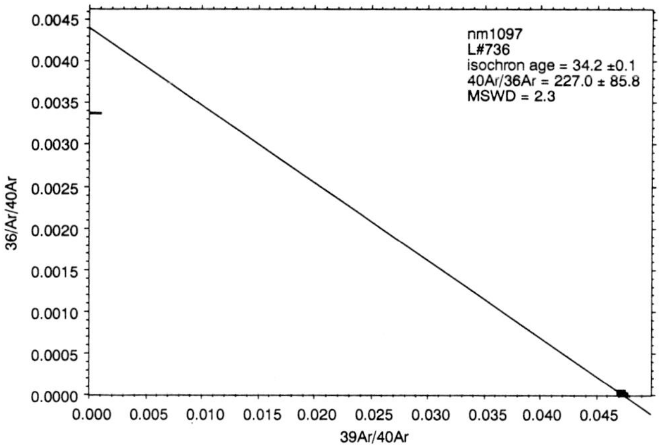

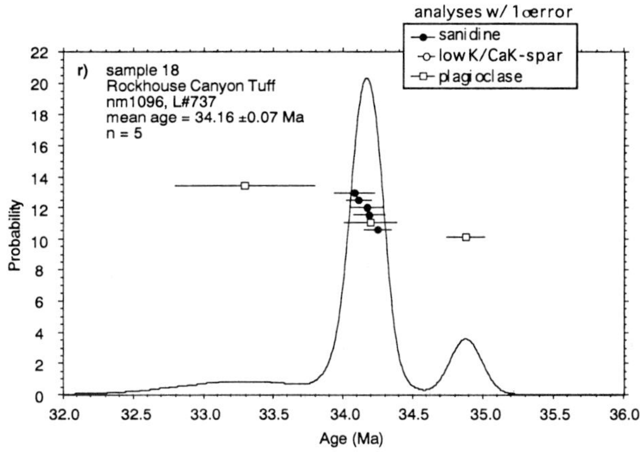

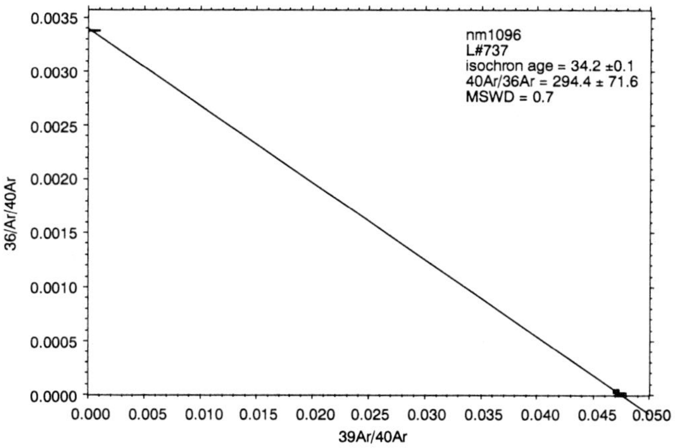

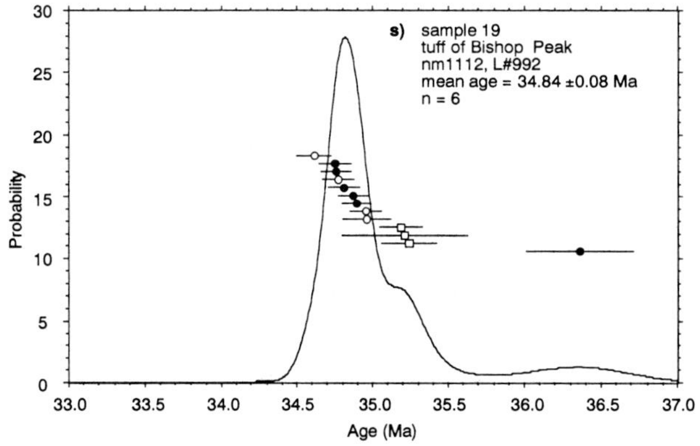

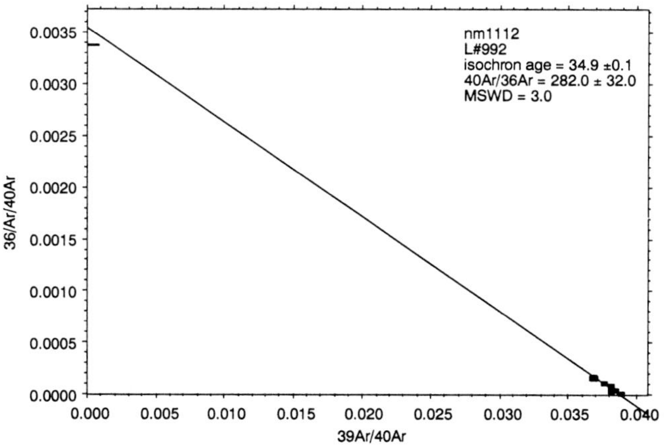

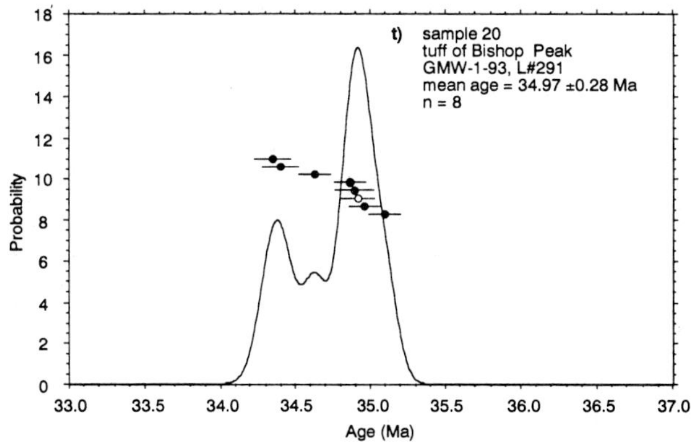

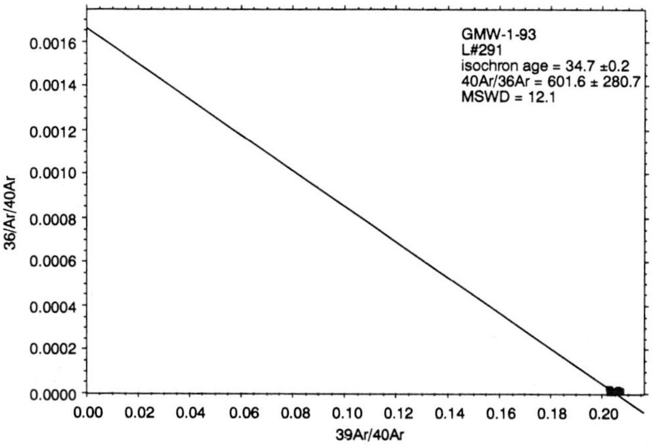

APPENDIX 3 continued

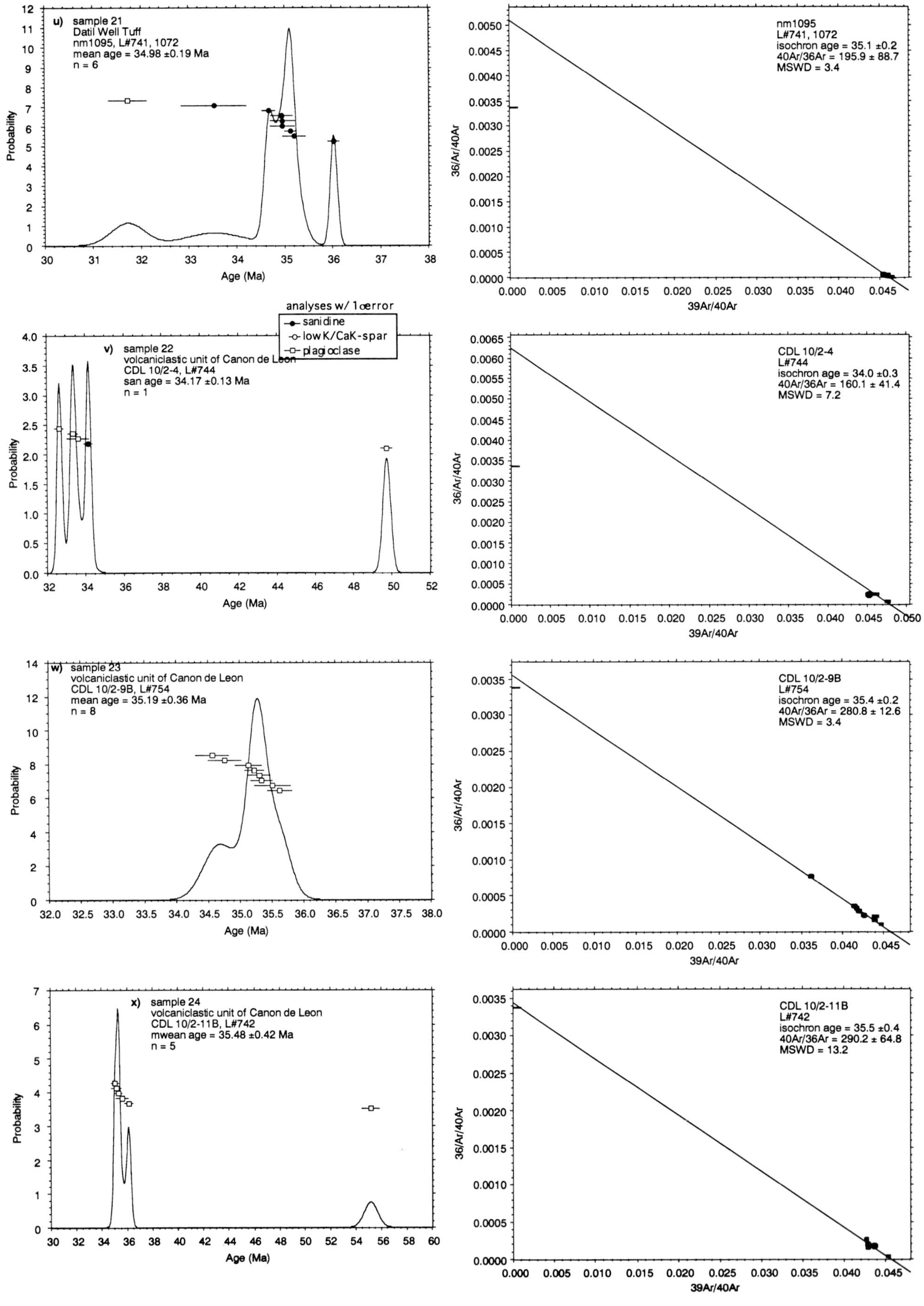

APPENDIX 3 continued

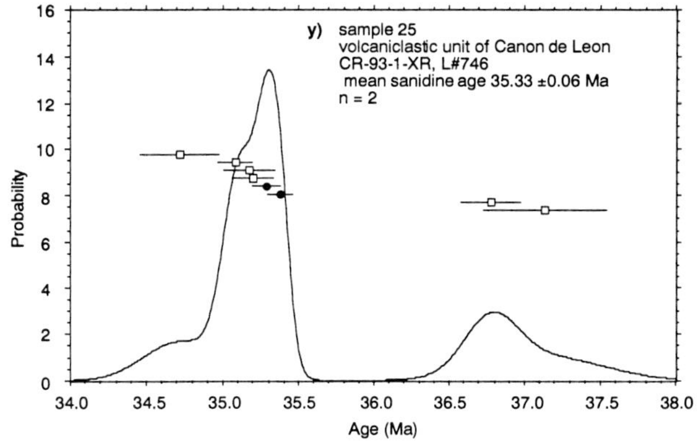

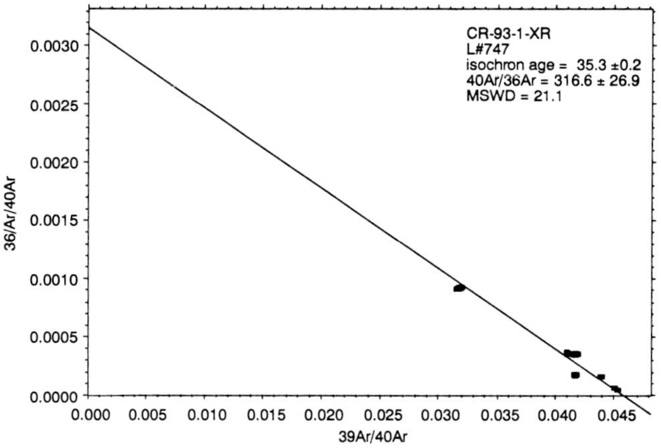

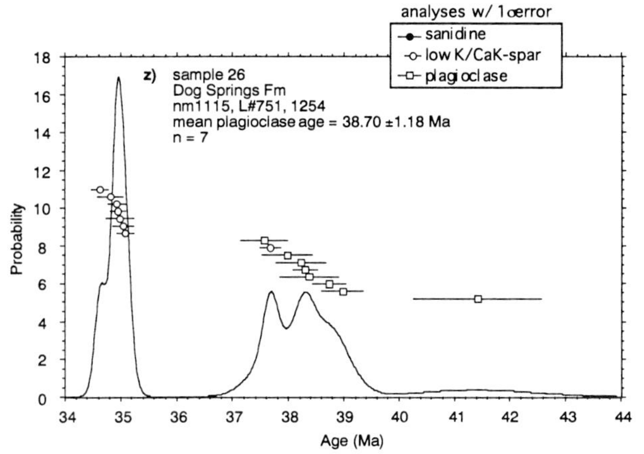

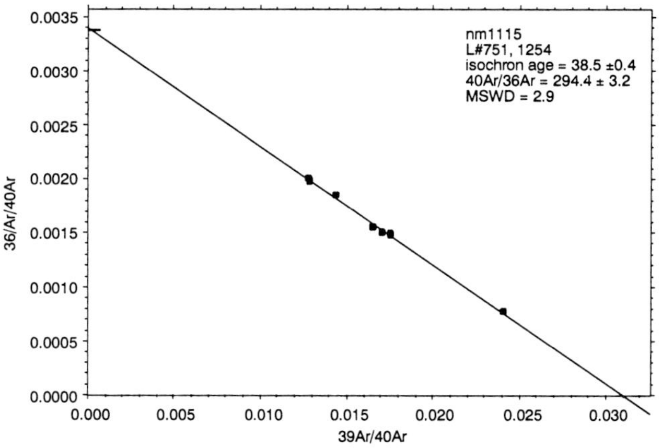

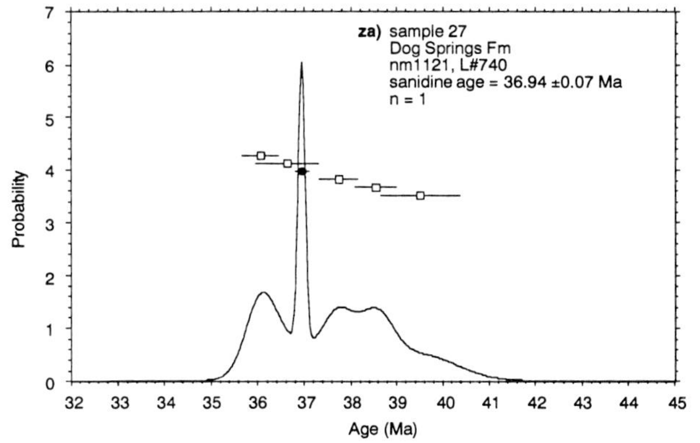

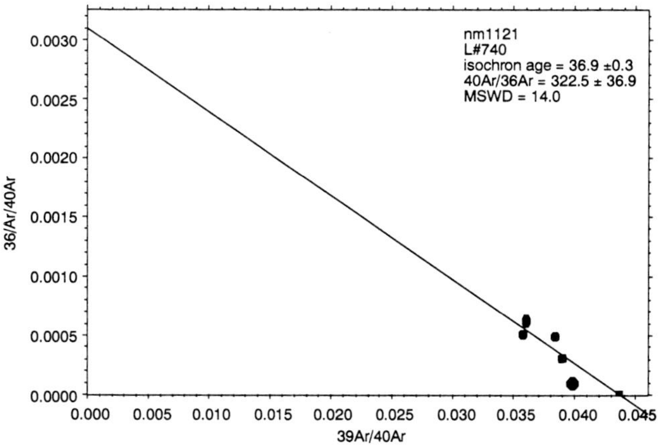

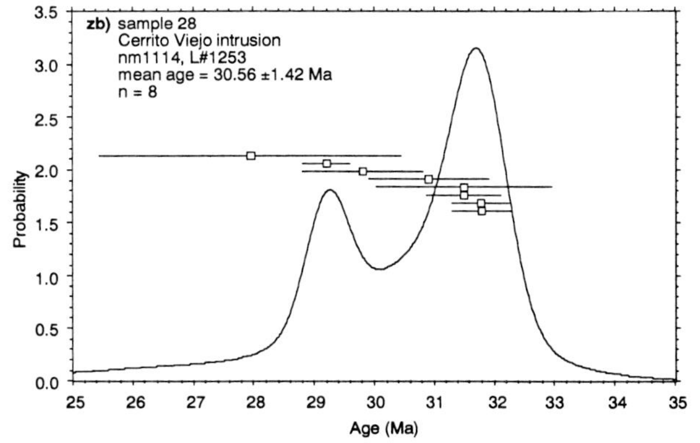

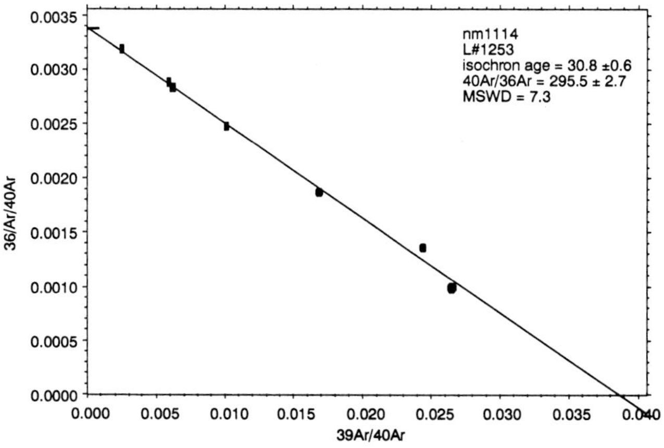

APPENDIX 4

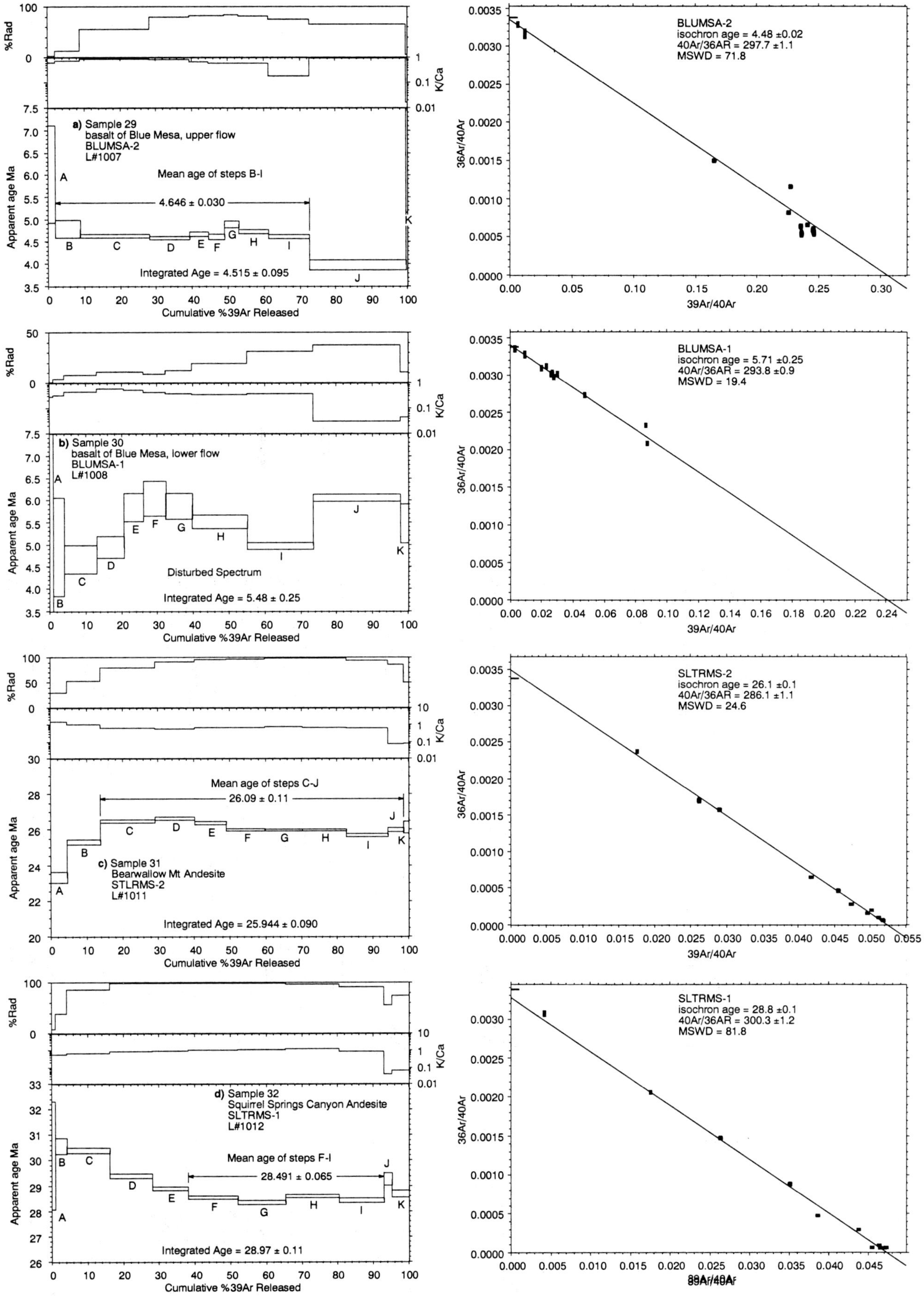

a) Sample 29
basalt of Blue Mesa, upper flow
BLUMSA-2
L#1007
Mean age of steps B-I
4.646 ± 0.030
Integrated Age = 4.515 ± 0.095
BLUMSA-2
isochron age = 4.48 ±0.02
40Ar/36AR = 297.7 ±1.1
MSWD = 71.8
b) Sample 30
basalt of Blue Mesa, lower flow
BLUMSA-1
L#1008
Disturbed Spectrum
Integrated Age = 5.48 ± 0.25
BLUMSA-1
isochron age = 5.71 ±0.25
40Ar/36AR = 293.8 ±0.9
MSWD = 19.4
c) Sample 31
Bearwallow Mt Andesite
STLRMS-2
L#1011
Mean age of steps C-J
26.09 ± 0.11
Integrated Age = 25.944 ± 0.090
SLTRMS-2
isochron age = 26.1 ±0.1
40Ar/36AR = 286.1 ±1.1
MSWD = 24.6
d) Sample 32
Squirrel Springs Canyon Andesite
SLTRMS-1
L#1012
Mean age of steps F-I
28.491 ± 0.065
Integrated Age = 28.97 ± 0.11
SLTRMS-1
isochron age = 28.8 ±0.1
40Ar/36AR = 300.3 ±1.2
MSWD = 81.8
%Rad
K/Ca
Apparent age Ma
Cumulative %39Ar Released
36Ar/40Ar
39Ar/40Ar

APPENDIX 4 continued

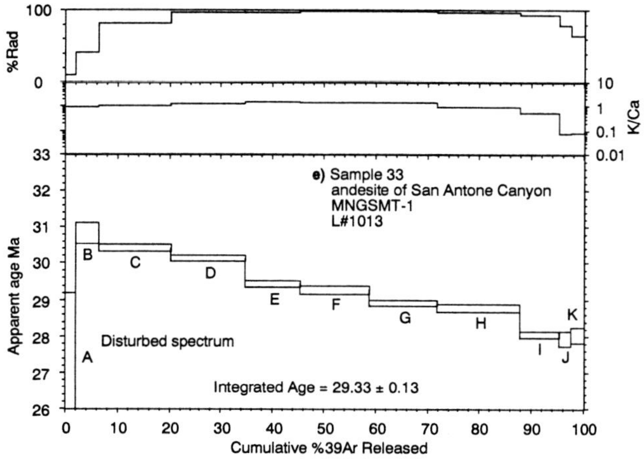

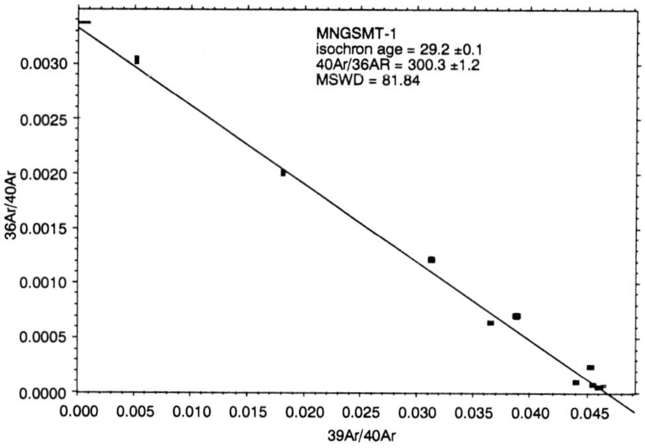

Monoliths in central Sawtooth Mountains approximately 27 km northwest of Datil, New Mexico. View is S25°W. Forest Road 6 (Cibola National Forest) is in foreground. Cliff-forming unit is the Dog Springs Formation (upper Eocene) that consists of dacitic debris-flow deposits and sandstones. Lower slopes and flats covered by piñon-juniper woodland are underlain by the Baca Formation (middle Eocene). Camera station is in NE ¼ sec. 4, T1N, R11W, Wayne Lambert photograph No. 93L52. August 11, 1993, 7:12 p.m. MDT.

ARE THE WESTERN MOGOLLON-DATIL MID-CENOZOIC ASH FLOWS COGENETIC? PEARCE ELEMENT RATIOS AND ISOTOPIC ASPECTS OF THE QUESTION

MICHAEL BIKERMAN
Department of Geology and Planetary Science, University of Pittsburgh, Pittsburgh, PA 15260

Abstract—The three major ash flows in the western Mogollon-Datil volcanic field, the Davis Canyon Tuff, Shelley Peak Tuff, and Bloodgood Canyon Tuff, and the flow-banded rhyolites with which they are associated, were produced in a period of about 1 Ma from two interlocking calderas — the Gila Cliff Dwellings (source of Davis Canyon and Shelley Peak Tuffs), and the Bursum (source of Bloodgood Canyon Tuff). Magmas parent to the tuffs could have been produced by assimilation of existing crust by a mantle-derived basaltic magma, and fractional crystallization of the resultant mix. Fractionation of pyroxene and plagioclase in varying proportions from more mafic magmas produced the present ash flows. The proportions calculated from the use of Pearce element ratios were 80:20 plagioclase:pyroxene for ash flows from the Gila Cliff Dwellings caldera and 55:45 for the Bursum caldera- derived Bloodgood Canyon Tuff.

INTRODUCTION

The western Mogollon-Datil volcanic field is dominated by outcrops of three widespread ash-flow tuffs. From the oldest to youngest these are the Davis Canyon Tuff, Shelley Peak Tuff and Bloodgood Canyon Tuff. As large scale extrusions of ash flows are commonly associated with calderas, these flows are tied to the Bursum caldera, roughly 45 km in diameter, and the slightly smaller Gila Cliff Dwellings caldera, located just to its east (Ratté et al., 1984; Elston, 1984; McIntosh et al., 1992). The ash flows have been dated by conventional K-Ar and fission tracks (Marvin et al., 1987), and more precisely by $^{40}Ar/^{39}Ar$ (McIntosh et al., 1991, 1992) as follows: Davis Canyon Tuff - 29.0 Ma; Shelley Peak Tuff - 28.1 Ma, and Bloodgood Canyon Tuff - 28.1 Ma. The ash flows are associated with their defluidized residues, the flow-banded rhyolites. Sr and Nd isotopic evidence (Bikerman, 1989; Bikerman et al, 1992) is consistent with an origin of the major ash flows by mixing of mantle-derived magma with Proterozoic upper crust in an assimilation and fractional crystallization (AFC) model (DePaolo, 1981).

Results of chemical analyses of Mogollon-Datil volcanic rocks were given by Elston and Northrup (1976), Ratté and Grotbo (1979), Bornhorst (1980, 1985) and Bikerman et al. (1992), whose values are used in the following discussion. In this paper the mineralogic changes in the evolving magma chamber purported to underlie the Bursum and Gila Cliff Dwellings calderas are examined using Pearce element ratios.

Pearce element ratios (PER) (Pearce, 1968) are petrogenetic modeling tools that use ratios of abundances of element(s) known to fractionate relative to some conservative parameter – the conserved parameter usually being a major element with an essentially unchanging composition in the evolving system. Most often these results are plotted against another similar ratio that uses the same denominator. The slope of the line through the data points reflects the stoichiometry of the fractionation (or addition) of material, with the original system controlling the intercept (Russell et al., 1990). For example, these authors used the ratios of 0.5(Mg+Fe)/Ti as the Y axis, and Si/Ti as the X axis in a test of an olivine fractionation hypothesis for variance within an Hawaiian lava. These PER axes are based on the concept that olivine fractionation will remove iron plus magnesium and silica from the melt in a 2:1 ratio. The 0.5 factor is used to get a straight line with a slope of one. Ti is used as the common denominator because it does not fractionate from a magma at low silica values. If the hypothesis is correct, then all the data points plot within experimental error of the olivine fractionation line. Similar plots can be used for fractionation of pyroxenes, feldspars, and other minerals, and have been used mostly in studies of mafic rocks. Valid trends have the additional requirements that the intercept not be zero (Pearce, 1987), and that the variance of ratios of conserved elements should not greatly exceed the experimental errors (Nicholls, 1988).

An interesting modification of PER use was by Bradshaw (1992), who adapted PERs for rhyolites and related high-silica rocks. The main change was the use of a trace element, Zr, as the common denominator. This was based on the observation that its fractionation as zircon is inhibited both by high temperature and presence of volatiles - both common factors in the magma chambers parent to ash flows.

In this study PERs using Zr (Bradshaw, 1992) as well as Cu, Ti, La and Ce as incompatible elements were calculated. All of these elements have essentially flat Harker plots of element (or oxide for TiO_2) abundances against SiO_2 (Figs. 1, 2) in the felsic range of silica contents. Flat Harker plots are used as a primary check on incompatibility. Mobile incompatible elements such as Ba and Rb were checked for their suitability, but as expected (Bradshaw, 1992), PER plots using these elements had considerable scatter, reflecting their mobility. Ni, Cr, V and Zn were checked, but yielded results similar to Cu. The technique used for elemental analyses was unlikely to dissolve zircons completely (Fabec, 1987, written commun.), so the Zr values given in Table 1 are for soluble Zr.

DATA

The chemical compositions, as oxide weight percents, used in the study are presented in Tables 1 and 2. PERs use element proportion

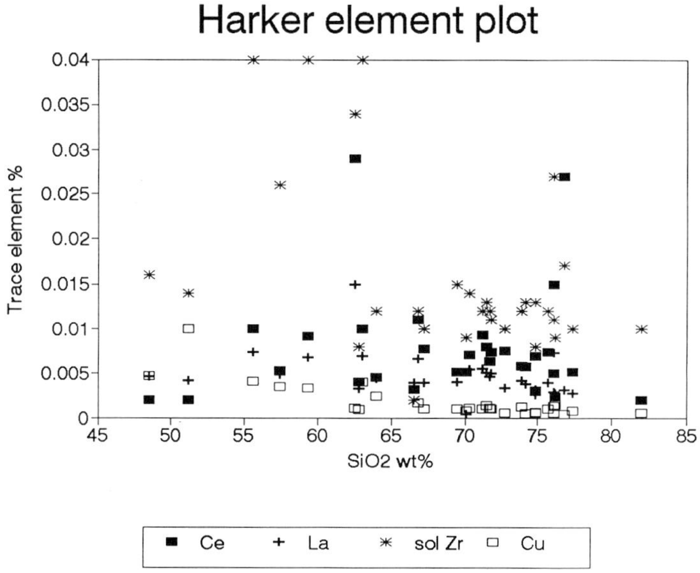

FIGURE 1. Harker element plots of Ce, La, sol Zr and Cu vs. SiO_2; symbols as shown on figure. Sol Zr is soluble zirconium under the analytical chemical technique used.

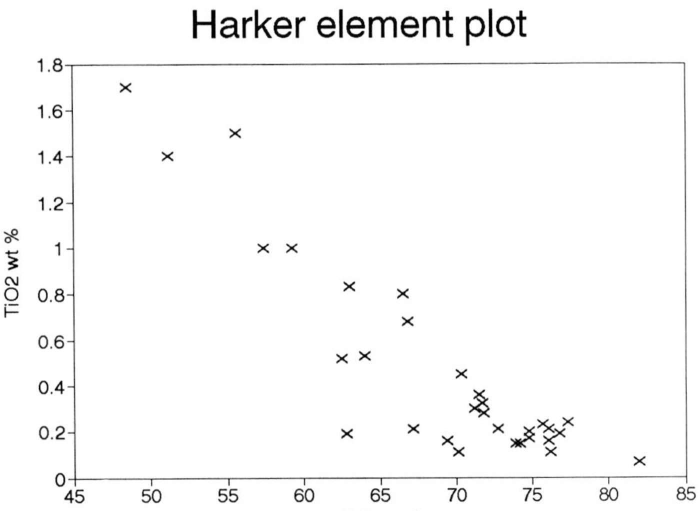

FIGURE 2. Harker element plot of TiO_2 vs. SiO_2

ratios in their calculations (Russell et al., 1990). These are obtained by multiplying the weight percent oxide by the number of cations in the oxide formula, then dividing by the molecular weight of the oxide.

The initial $^{87}Sr/^{86}Sr$ and $^{143}Nd/^{144}Nd$ ratios for the tuffs are 0.7089±0.0008 and 0.51227±0.00004 for the Davis Canyon Tuff; 0.7086±0.0004 and 0.51226±0.00004 for the Shelley Peak Tuff, and 0.7095±0.0006 and 0.51221±0.00004 for the Bloodgood Canyon Tuff, respectively; the initial $^{87}Sr/^{86}Sr$ for the flow-banded rhyolites ranged from 0.71003 for one with 32.8 ppm Sr, to high values of 0.71507 and 0.72872 for rocks with very low Sr contents of 6.2 and 4.7 ppm (Bikerman et al., 1992).

DISCUSSION

The three major ash flows erupted within a narrow interval in early Chattian (late Oligocene) time (McIntosh et al., 1992) from sources in the Gila Cliff Dwellings and Bursum calderas (Ratté et al., 1984). The short time span of the ash flow eruptions, similarity of isotopic ratios, and geographically limited possible source region, as delineated by the calderas, are consistent with eruption from a single evolving magma chamber. The Davis Canyon and Shelley Peak ash flows, which are probably from the Gila Cliff Dwellings caldera (Ratté et al., 1984), have identical $^{87}Sr/^{86}Sr$ isotopic ratios. These values are slightly lower than those of the Bloodgood Canyon Tuff, derived from the Bursum caldera. The difference could be caused by the additional million years or so of residence time in the magma chamber, which would cause an increase of *ca* 0.00001 in the $^{87}Sr/^{86}Sr$. Another possibility is that the portion of the magma chamber underlying the Bursum caldera might have encountered some more radiogenic country rock, giving rise to the slightly higher initial $^{87}Sr/^{86}Sr$ in the younger ash flow.

TABLE 1. Major element contents of rocks from the Mogollon-Datil volcanic field in weight percent.

Sample	SiO_2	TiO_2	Al_2O_3	Fe_2O_3	MnO	MgO	CaO	Na_2O	K_2O	P_2O_5
G-48	77.4	0.24	12.1	1.5	0.078	0.35	0.36	3.6	5.0	0.018
G-51	76.1	0.21	12.0	1.4	0.070	0.14	0.28	3.8	5.0	0.021
G-24	75.7	0.23	11.8	1.6	0.061	0.15	0.36	3.9	5.1	0.022
G-23	74.8	0.20	11.9	1.4	0.071	0.17	0.18	3.3	5.3	0.032
G-7	74.8	0.17	11.6	1.0	0.051	0.09	0.26	3.6	4.9	0.200
M-675	72.7	0.21	11.8	1.2	0.074	0.41	0.75	2.8	5.2	0.043
G-15B	71.8	0.28	14.5	1.7	0.054	0.31	1.00	3.9	5.1	0.077
G-15A	71.7	0.32	13.4	1.9	0.061	0.33	1.10	3.3	5.5	0.140
G-26	70.3	0.45	14.5	2.6	0.076	0.61	2.00	4.2	4.8	0.140
G-6	76.8	0.19	11.7	1.0	0.057	0.16	0.16	3.7	5.0	0.020
G-46	76.1	0.16	11.3	2.3	0.037	0.09	0.11	3.6	5.1	0.080
ARS	73.9	0.15	12.0	1.3	0.072	0.18	0.87	3.5	5.2	0.076
G-8	67.2	0.21	13.9	1.7	0.035	0.38	0.31	2.0	7.1	0.051
G-32	76.2	0.11	10.6	0.9	0.055	1.40	0.86	3.0	3.8	0.017
TTC	74.2	0.15	12.0	1.0	0.068	0.27	1.40	3.5	5.1	0.023
H9-77	71.2	0.30	14.5	2.0	0.062	0.50	1.40	4.2	4.9	0.120
G-11	82.0	0.07	9.8	0.4	0.024	0.07	0.06	0.5	6.0	0.013
TDW	70.1	0.11	11.9	0.9	0.053	0.68	1.70	0.6	5.7	0.016
G-1	69.4	0.16	12.7	1.2	0.039	0.68	1.70	2.3	3.8	0.024
G-81	66.8	0.68	15.5	3.5	0.072	0.63	1.60	4.2	4.8	0.190
G-47	66.5	0.80	16.2	5.1	0.081	1.70	4.00	3.7	2.9	0.190
G-43	64.0	0.53	15.9	4.0	0.068	0.82	3.10	3.7	3.5	0.140
G-55	62.8	0.19	14.7	1.5	0.046	3.30	4.00	0.4	1.8	0.072
G-84	62.5	0.52	15.4	2.4	0.089	0.38	1.10	4.8	5.2	0.073
G-31	71.5	0.36	13.3	2.4	0.063	0.50	1.20	2.9	5.7	0.082
A81-8	63.0	0.83	15.8	5.6	0.090	2.40	4.30	3.6	3.4	0.300
A81-12	48.5	1.70	16.4	11.8	0.170	6.10	8.50	3.2	0.8	0.290
A81-23	59.3	1.00	16.0	6.7	0.100	3.00	4.90	3.6	3.4	0.390
G-4	51.2	1.40	15.5	11.8	0.160	7.10	8.40	3.0	1.3	0.240
G-5	55.6	1.50	15.9	8.6	0.130	4.00	6.10	3.5	2.8	0.830
G-40	57.4	1.00	15.8	7.1	0.120	2.50	5.00	3.6	3.1	0.240

Sample identification:
G-7, G-23, G-24, G-48, G-51 & M-675 are Tbc;
G-15A, G-15B & G-26 are Tsp;
H9-77, TTC & G-32 are Tdc;
ARS, G-6 & G-46 are fbr;
G-8 is Cooney Tuff;
G-11 is the Sacaton quartz-latite squeeze-up;
G-1 & TDW are Fanney Rhyolite;
G-55 a white ash from Bull Basin;
G-31, G-43, G-47, G-81 & G-84 are miscellaneous felsic units;
A81-8, A81-12, A81-23, G-4, G-5 & G-40 are more mafic units.

Note: all iron is reported as Fe_2O_3

TABLE 2. Major element contents of rocks from the Mogollon-Datil volcanic field in weight percent.

Sample	Ba	Rb	Cu	Sol Zr	Sr	Zn	Cr	V	Ni	Ce	La
G-48	110	230	8	100	30	50	60	19	10	52	27
G-51	100	230	6	110	20	60	40	10	10	50	29
G-24	100	230	10	120	20	50	70	24	10	74	39
G-23	130	270	7	130	20	60	30	46	12	70	32
G-7	120	220	6	80	80	40	20	10	10	30	26
M-675	110	220	6	100	40	70	30	10	12	75	34
G-15B	650	180	10	110	150	30	20	27	13	74	50
G-15A	640	160	11	120	170	30	40	14	13	63	46
G-26	760	150	11	140	220	50	20	27	17	71	54
G-32	100	190	15	90	50	60	20	10	11	24	27
TTC	100	200	5	130	20	50	20	10	13	57	38
H9-77	700	180	10	120	170	60	30	26	17	93	55
G-6	100	280	4	170	20	50	20	10	10	270	31
G-46	100	220	13	270	20	70	60	10	16	150	73
ARS	100	240	12	120	30	30	50	10	10	58	42
G-8	450	300	10	100	90	50	20	170	9	78	39
G-11	100	400	6	100	30	20	120	10	10	20	20
TDW	100	220	8	90	90	20	20	10	13	52	4.1
G-1	180	230	10	150	370	40	20	17	10	52	40
G-81	1500	150	17	120	230	70	40	45	11	110	66
G-47	890	64	32	20	360	70	40	90	20	32	39
G-43	720	77	25	120	300	50	50	48	23	45	44
G-55	850	34	9	80	670	40	20	34	10	40	33
G-84	130	270	11	340	60	50	30	19	13	290	150
G-31	1200	180	14	130	190	40	70	34	15	80	51
A81-8	1100	100	40	400	410	80	170	94	50	100	70
A81-12	300	15	47	160	320	110	230	190	110	20	46
A81-23	910	89	34	400	450	90	120	110	50	92	68
G-4	210	13	100	140	260	120	210	180	150	20	42
G-5	1200	63	41	400	600	110	120	150	67	100	74
G-40	790	100	35	260	400	80	70	180	39	53	48

Examination of the chemical variances against silica shows that Al_2O_3 is high and has a slight negative slope for silica contents below 65% and a more pronounced negative slope for higher silica rocks. Fe, Mg, Ti, Mn and Ca oxides have a negative slope for the mafic rocks, and relatively flat and scattered plots above 65% silica. Trace elements Zr, Zn, Cr, Cu, V and Ni, have negative slopes in the sub-65% silica range, but scatter in a fairly flat-lying pattern above that. Ba, Ce and La have scatter plots with no obvious trend, with the rare earth elements plotting roughly horizontally, and Ba showing much higher scatter in the mid-range in silica. For the purpose of PER denominators, the comparatively horizontal trend of trace-element plots in the plus-65% silica range were considered adequate. Much of this scatter is experimental error in the determination of small absolute contents of the trace elements.

Using the Cox et al. (1979) alkali-silica, or the LeBas et al. (1986) total alkali-silica classifications, the Davis Canyon, Shelley Peak and Bloodgood Canyon tuffs are all rhyolites, and plot in the sub-alkaline field of Irvine and Barager (1971). Ratté et al. (1984) used the De la Roche et al. (1980) volcanic classification to place the Davis Canyon and Bloodgood Canyon tuffs in the alkali rhyolite category, and the Shelley Peak Tuff in the rhyolite class. As there is not a large difference in the chemistry between the three ash flows, changes in their parent magma would be expected to be relatively small. Such small changes should nevertheless show up as linear trends in the PER diagrams if they reflect removal of one or more minerals from the melt.

PEARCE ELEMENT RATIOS

Russell and Nicholls (1988) summarized the Pearce element ratios most suitable to various petrologic arguments. The placement of the required fraction can be done on either the ordinate or abscissa. For one or two phase segregation, one axis (usually the Y axis) used is Si/n, where n is the conserved element (Zr, Cu, Ti, La or Ce in this study) used as the denominator for the abscissa and ordinate. The other axis (usually the X axis) has the following factors, divided by n: for olivine (OL), 0.5(Mg+Fe); for plagioclase (PL), 2Na+Al; and for clinopyroxene (CPX), 2Ca+Na-Al. For two minerals concurrently involved in fractionating, the following, divided by n are used as the Y axis: OL+PL 0.5(Mg+Fe)+2Ca+3Na; PL+CPX 2Ca+3Na; and OL+CPX 0.5(Fe+Mg)+1.5Ca.

For three minerals (Russell et al., 1990) Y is (OL+CPX+PL)/n or [1.5Ca+0.25Al+0.5FM+2.75Na]/n. In all of these plots, fractionation of the indicated species will produce a line with a slope of 1.0. Another way of testing for these minerals is to use the axes Al/n and {2Ca+Na}/n (Russell and Nicholls, 1988). Here a slope of 1.0 reflects plagioclase involvement; if Al/n is the abscissa a vertical line indicates CPX, the actual slope is equal to the {CPX+PL}/PL ratio, and OL has no effect on the slope. If Al/n is the ordinate, then CPX fractionation alone will be horizontal, or a slope of ∞.

Plots and regression calculations were made using Ti and the trace elements Cu, soluble Zr, Ce and La for all the analyzed samples of Bikerman et al. (1992) and for various subsets such as the three major ash flows, or of all the high silica rocks using mainly the program Igpet 3.2 (Carr, 1992). The parameter of {4(Ca+Na)+0.5(Fe+Mg)}/n as the Y axis, selected by Bradshaw (1992) as being especially suited for silica-rich rocks, was made by summing two available Igpet 3.2 PER parameters (2Ca+3Na) and (2Ca+Na) with the also available 0.5(Mg+Fe).

Slight variations in the PER plots and their regressions occur as the element of the denominator is changed. Intercepts of the regression lines reflect the absolute abundances of the trace elements and so are not considered here. Intercepts of zero violate one of the criteria for PERs, in which case the plot is not used. Table 3 shows a series of sample calculations for the various denominator elements. There are slight differences in the slopes, but the petrologic interpretations are not significantly affected by changing the element used. As the Mogollon-Datil region is peripheral to copper mining districts, Cu was chosen as the element most commonly used in PER denominators in this study. The region's copper deposits are Laramide or older, and hence they, or their progenitors, are likely to be represented in the crust involved in the generation of the magma parent to the ash flows.

Examined first is the idea that a single phase fractionated from the parent magma. Using the axes for olivine (Fig. 3) and for clinopyroxene (Fig. 4) as listed above, plots of 15 points (6 for Bloodgood Canyon Tuff, and 3 each for Shelley Peak, Davis Canyon and flow-banded rhyolite

TABLE 3. Comparison of different elements as PER denominators. All were used in the PER test of (2Ca+Na)/n vs. A1/n.

n=	slope	R=	R^2=	t=
sol Zr	1.23	0.779	0.61	3.73
Cu	1.44	0.932	0.87	7.73
Ce	1.60	0.977	0.95	13.65
La	1.66	0.841	0.71	4.67
Ti	1.43	0.931	0.87	8.06

Note: R, R^2 and t are statistical functions for goodness of fit, square of the goodness of fit, and t- test, respectively.

analyses) yield calculated slopes of 44.8 and 11.1 respectively. These values are far from the postulated value of 1.0, indicating that neither of these minerals fractionating by itself could produce the observed trends. For plagioclase (Fig. 5) the plot of Si/Cu vs (2Na+Al)/Cu gives a calculated slope of 2.67±0.34 for 15 points, and similar values for subsets for each of the ash flows. Plagioclase fractionation is within reason.

Next, examining the pairs of minerals using the appropriate axes of Russell and Nicholls (1988) shows that either olivine + plagioclase (Fig. 6) or clinopyroxene + plagioclase (Fig. 7) give slopes of about 3.1±0.33, or roughly the same as for plagioclase alone. As olivine is unlikely to be an important factor in rhyolite magmatism, ferro-magnesian partition modeling using clinopyroxene is preferable. To determine the molar proportion of plagioclase to pyroxene the three-phase plot of Russell and Nicholls (1988) is used, where the slope of the regression line gives the PL/(PL+CPX) ratio. A slope = 1 is simple plagioclase removal, and with the axes used, a zero slope is pure CPX removal. In the 15-point set (Fig. 8) this value is 0.67, or 67% plagioclase removal to 33% clinopyroxene. The same calculation for each caldera gives 80% PL out for the Gila Cliff Dwellings Caldera tuffs and 55% PL out for the Bursum Caldera. The presence of green clinopyroxene in the younger Shelley Peak tuff is consistent with removal of plagioclase from its parent magma.

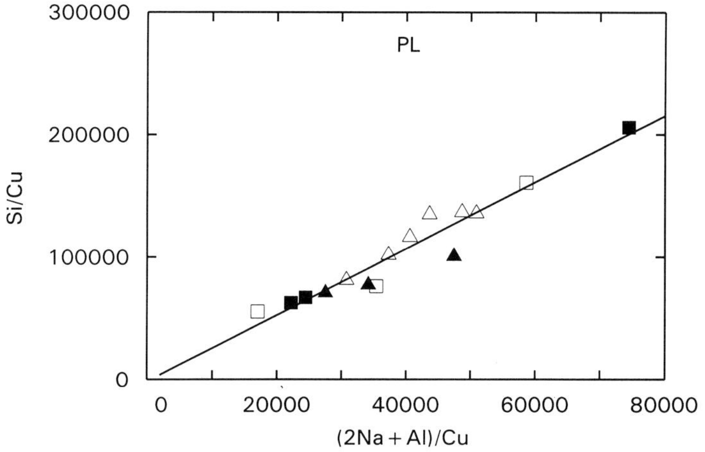

FIGURE 5. Plagioclase fractionation; symbols and line as in Fig. 4.

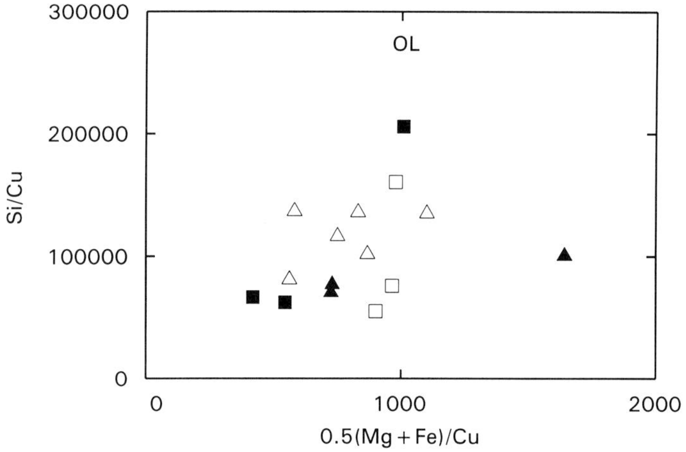

FIGURE 3. Olivine fractionation using the axes suggested by Russell and Nicholls (1988). Symbols are Tbc - open triangles; Tsp - filled triangles; Tdc - open squares; fbr - filled squares

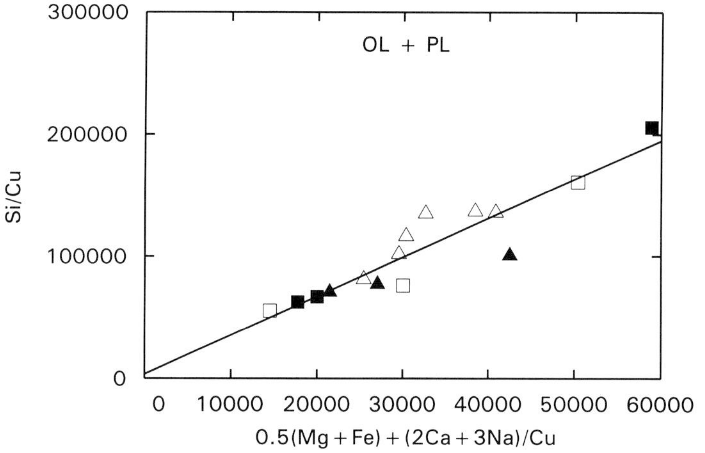

FIGURE 6. Olivine + plagioclase fractionation; symbols and line as in Fig. 4.

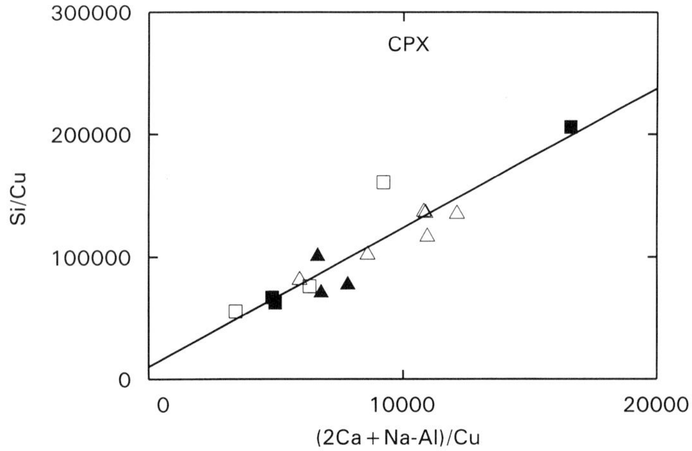

FIGURE 4. Clinopyroxene fractionation, symbols as in Fig. 3. Line is regression line calculated by Igpet 3.2 (Carr, 1992)

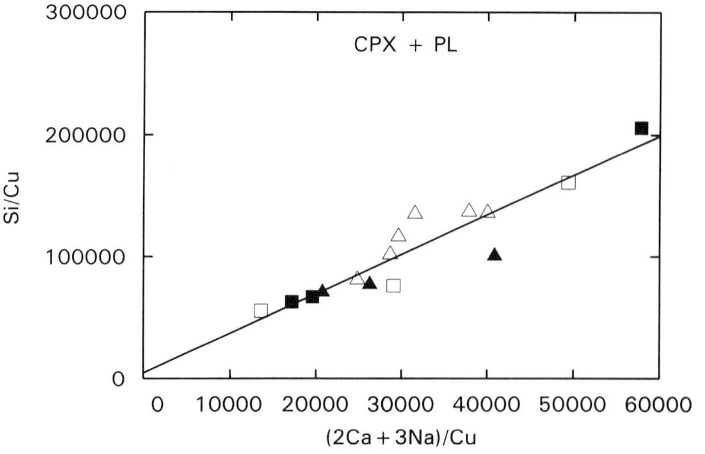

FIGURE 7. Clinopyroxene + plagioclase; symbols and line as in Fig. 4.

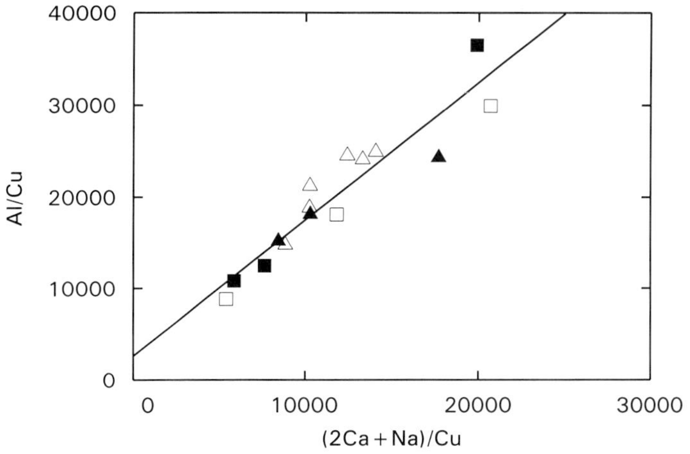

FIGURE 8. Three phase diagram with plagioclase (slope of 1) and clinopyroxene (slope ∞, horizontal); symbols and line as in Fig. 4.

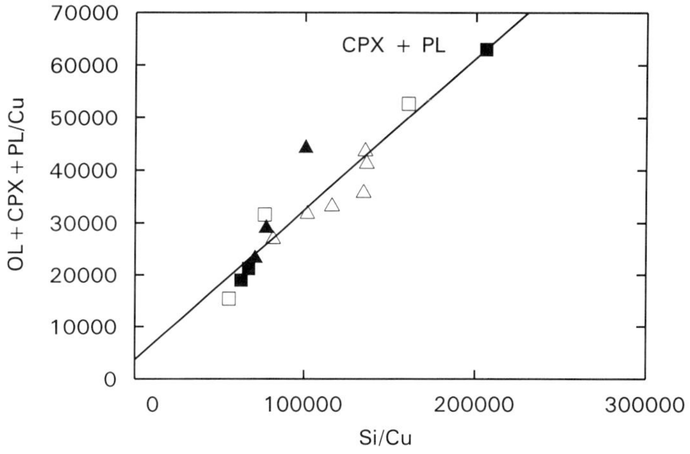

FIGURE 9. OL + CPX + PL diagram; symbols and line as in Fig. 4.

For the [OL+CPX+PL]/n diagram (Fig. 9) the slopes are less than 1.0, indicating that other phases were also involved. As sanidine is a common mineral in these ash flows, adding K-spar to the mix of minerals was attempted. In the plot suggested by Bradshaw (1992) (Fig. 10) fractionation of K-spar appears as a horizontal displacement because only Al and Si change, and enstatite has a slope of 0.5, plagioclase a slope of 1, and clinopyroxene a slope of 2.25. The 15 point regression line has a slope of 0.31 using Cu or 0.30 using Ce as the denominator. Ce was selected for use because of a better statistical fit of the points to the regression line. The slopes for the Gila Cliff Dwellings ash flows are 0.26, whereas the Bursum caldera ash flow has a slope of 0.33. These figures hint that the Bursum caldera chamber had a slightly greater pyroxene and plagioclase extraction.

The PER Assemblage Test Diagram program (Stanley and Russell, 1989) suggests using Si/n as the X axis, and [0.5Al+Fe+Mg+Ca+2.5Na+2.5K]/n as the Y axis, for a line that gives a slope of 1 for fractionation of any of the following: K- spar, AN, AB DP, HD or EN. Such plots were made, and lines with a slope of 0.24 for n=Cu, and of 0.26 for n=Ce resulted. Errors on each of the slopes are about 5%, and the R squared values are 0.909 and 0.969, respectively. Again the model was not definitive for these minerals, implying that some other phase(s) were involved.

An interesting correlation was found between the initial $^{87}Sr/^{86}Sr$ ratio and either Si or Al or both. Plots of initial $^{87}Sr/^{86}Sr$ against either SiO2 or Al2O3 produced essentially horizontal linear scatter plots, which resemble the plots of the Sr ratio against Si/Cu, Al/Cu and [Si + Al]/Cu. However, if the initial Sr ratio is treated as a PER parameter, using the initial $^{87}Sr/^{86}Sr$/Cu ratio produced tight sloping lines, especially against [Si+Al]/Cu (Fig. 11), an observation which holds for other conserved element denominators as well. The Sr ratios did not work as PER denominators themselves, emphasizing their non- conservation in the melt. The only caveat to any petrologic interpretation is that the intercepts are near zero, hence possibly spurious. If the lines have meaning, it may be that the isotopic ratios in the magma chamber parallel the mineral fractionation, possibly indicating a Rb-zoned magma chamber. The AFC-derived model for the origin of the ash flows based on the $^{87}Sr/^{86}Sr$ and $^{143}Nd/^{144}Nd$ data (Bikerman et al., 1992), could well produce a stratified magma chamber.

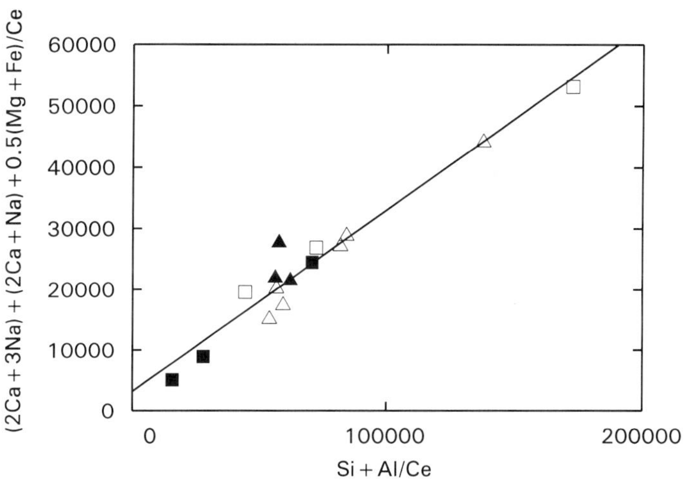

FIGURE 10. PER diagram after Bradshaw (1992); symbols and line as in Fig. 4. Ce is used rather than Cu because the regression value for Ce is 0.964 compared to 0.855 for Cu.

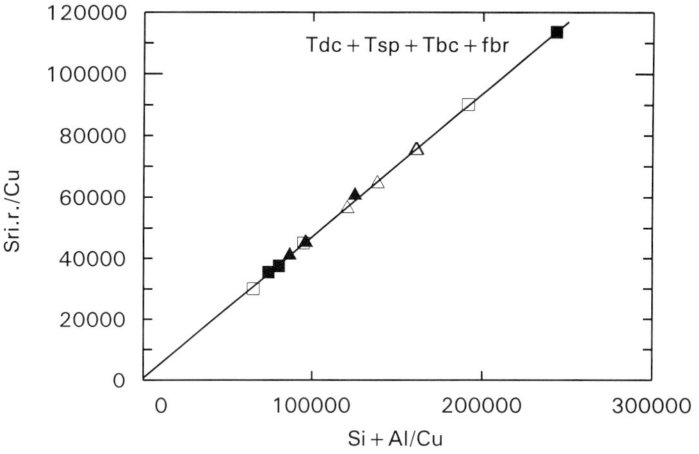

FIGURE 11. Initial $^{87}Sr/^{86}Sr$ ratio [Sr i.r.] treated as a PER factor and plotted against [Si+Al]/n. Symbols and line as in Fig. 4.

CONCLUSIONS

The chemical and $^{87}Sr/^{86}Sr$ data used in the PER models are consistent with a reasonably well mixed rhyolite magma fractionating pyroxene and plagioclase feldspar. Fractionation of 80:20 PL:CPX is found for the Gila Cliff Dwellings rhyolites, supported by the petrographic evidence of clinopyroxenes in the Shelley Peak Tuff, and of 55:45 in the Bloodgood Canyon Tuff from the Bursum caldera. Orthopyroxene fractionation could have been a major factor in the Bursum caldera also. The deviations from the simple fractionation used could be caused by biotite removal, which was not indicated by the PER models used.

The PER and isotopic data for the Davis Canyon, Shelley Peak and Bloodgood Canyon tuffs, together with the associated flow-banded rhyolites, can be explained by an evolving magma chamber derived from a mix of primary mantle-derived magma and some crustal contaminant, such as the AFC model which best fit the $^{143}Nd/^{144}Nd$ and $^{87}Sr/^{86}Sr$ determinations (Bikerman et al., 1992).

$^{87}Sr/^{86}Sr$ initial/n in the rocks plots linearly against [Al+Si]/n. If the near-zero intercept is significant, the line can be construed as a genetic relationship, rather than a stoichiometric one between the numerator parameters. The geological import of this is that more alkali-feldspar-rich rocks would have higher Al+Si, and more Rb in their source. A clearer understanding of the processes and sources leading to the rhyolites could be gained if better data on the pre-existing crust were available. Searching for crustal and mantle xenoliths has not been successful so far, but more field and laboratory time is needed.

ACKNOWLEDGMENTS

Thoughtful and helpful reviews by E. G. Lidiak and C.H. Shultz, and field support from the New Mexico Bureau of Mines and Mineral Resources, are gratefully acknowledged.

REFERENCES

Bikerman, M., 1989, Rb, Sr, Rb-Sr, and isotopic Sr values for volcanic rocks from the southwestern Mogollon-Datil volcanic field, New Mexico: New Mexico Geology, v. 11, p. 76-83.

Bikerman, M., Bell, K. and Card, J.W., 1992, Strontium and neodymium isotopic study of the western Mogollon-Datil volcanic region, New Mexico, USA: Contributions to Mineralogy and Petrology, v. 109, p. 459-470.

Bornhorst, T.J., 1980, Major- and trace-element geochemistry and mineralogy of Eocene to Quaternary volcanic rocks of the Mogollon-Datil volcanic field, southwestern New Mexico [PhD dissertation]: Albuquerque, University of New Mexico, 500 p.

Bornhorst, T.J., 1985, Abundance and variability of selected lithophile elements in high-K rhyolitic rocks, Mogollon-Datil volcanic field, southwestern New Mexico: *in* Taylor, R.P. and Strong, D.F., eds, Granite-related mineral deposits: Canadian Institute of Mining Conference, Halifax, p. 25-31.

Bradshaw, T.K., 1992, The adaptation of Pearce element ratio diagrams to complex high silica systems: Contributions to Mineralogy and Petrology, v. 109, p. 450-458.

Carr, M.J., 1992, Igpet 3.2. Terra Softa Inc., Somerset NJ, 46 p.

Cox, K.G., Bell, J.D. and Pankhurst, R.J., 1979, The Interpretation of Igneous Rocks: Allen and Unwin, London, 450 p.

De la Roche, H., Leterrier, J., Grandclaude, P. and Marchal, M., 1980, A classification of volcanic and plutonic rocks using R1R2- diagram and major element analyses - its relationship with current nomenclature: Chemical Geology, v. 29, p. 183-210.

DePaolo, D.J., 1981, Trace element and isotopic effects of combined wall rock assimilation and fractional crystallization: Earth and Planetary Science Letters, v. 53, p. 189-202.

Elston, W.E., 1984, Mid-Tertiary ash-flow cauldrons, southwestern New Mexico: Journal of Geophysical Research, v. 89, p. 8733-8750.

Elston, W.E. and Northrop, S.A., eds, 1976, Cenozoic volcanism in southwestern New Mexico: New Mexico Geological Society, Special Publication 5, 151 p.

Irvine, T.N. and Barager, W.R.A., 1971, A guide to the chemical classification of the common volcanic rocks. Canadian Journal of Earth Sciences, v. 8, p. 523-548.

LeBas, M.J., LeMaitre, R.W., Streckeisen, A. and Zanettin, B., 1986, A chemical classification of volcanic rocks based on the total alkali silica diagram: Journal of Petrology, v. 27, p. 745- 750.

Marvin, R. F., Naeser, C. W., Bikerman, M., Mehnert, H. H. and Ratté, J.C., 1987, Isotopic ages of post-Paleocene igneous rocks within and bordering the Clifton 1°X2° quadrangle, Arizona-New Mexico: New Mexico Bureau of Mines and Mineral Resources, Bulletin 118, 63 p.

McIntosh, W.C., Kedzie, L.L. and Sutter, J.F., 1991, Paleomagnetism and $^{40}Ar/^{39}Ar$ ages of ignimbrites, Mogollon-Datil volcanic field, southwestern New Mexico, New Mexico Bureau of Mines and Mineral Resources, Bulletin 135, 79 p.

McIntosh, W.C., Chapin. C.E., Ratté, J.C. and Sutter, J.F., 1992, Time-stratigraphic framework for the Eocene-Oligocene Mogollon-Datil volcanic field, southwest New Mexico: Geological Society of America Bulletin, v. 104, p. 851-871.

Nicholls, J., 1988, The statistics of Pearce element ratios and the Chayes closure problem. Contributions to Mineralogy and Petrology, v. 99, p. 11-24.

Pearce, T.H., 1968, A contribution to the theory of variation diagrams: Contributions to Mineralogy and Petrology, v. 19, p. 142-157.

Pearce, T. H., 1987, The identification and assessment of spurious trends in Pearce-type variation diagrams: a discussion of some statistical arguments: Contributions to Mineralogy and Petrology, v. 97, p. 529-534.

Ratté, J.C. and Grotbo, T., 1979, Chemical analyses and norms of 81 volcanic rocks from part of the Mogollon-Datil volcanic field, southwestern new Mexico: U.S. Geological Survey, Open-file Report 79-1435, 33 p.

Ratté, J.C., Marvin, R.F., Naeser, C.W. and Bikerman, M., 1984, Calderas and ash flow tuffs of the Mogollon Mountains, southwestern New Mexico: Journal of Geophysical Research, v. 89, p.8713-8732.

Russell, J.K. and Nicholls, J., 1988, Analysis of petrologic hypotheses with Pearce element ratios: Contributions to Mineralogy and Petrology, v. 99, p. 25-35.

Russell, J.K., Nicholls, J., Stanley, C.R. and Pearce, T.H., 1990, Pearce element ratios – a paradigm for testing hypotheses: EOS, v. 71, p. 234-6, 246-7.

Stanley, C.R. and Russell, J.K., 1989, Petrological hypothesis testing with Pearce element ratio diagrams: derivation of diagram axes: Contributions to Mineralogy and Petrology, v. 103, p. 78-89.

THE HORSE SPRINGS DACITE—AN ENIGMATIC PYROCLASTIC DEPOSIT AT THE EDGE OF THE PLAINS OF SAN AGUSTIN

JAMES C. RATTÉ[1], PETER J. MODRESKI[1],
WILLIAM C. MCINTOSH[2], and RICHARD CHAMBERLIN[2]
[1]U.S. Geological Survey, Denver, CO 80225; [2]New Mexico Bureau of Mines and Mineral Resources, Socorro, NM 87801

Abstract—The Horse Springs dacite (new, but informal, name) is a local pyroclastic deposit near the southwest end of the Plains of San Agustin and the settlement of Old Horse Springs. The regional Blue Canyon Tuff (Eocene) underlies the Horse Springs dacite, is indistinguishable from it in age (~33.7 Ma), and is petrographically and chemically similar to the dacite; the enigma is the relationship, if any, of these two units. These similarities and the distribution of the Blue Canyon Tuff, which has no known source, suggest the possibility of a genetic relationship and a possible caldera source for both formations beneath the Plains of San Agustin. The Horse Springs dacite consists of pumice flows and block-and-ash flows at least 250 m thick. Outcrops of the dacite near the edge of the plains, closest to the suspected source, contain juvenile, pumiceous dacite blocks, 2-3 m long, and cognate quartz monzonite blocks of the same size. The dacite also contains accidental inclusions of mineralized jasperoid as large as 0.5 m across, derived from pre-Tertiary sedimentary rocks, including limestone, sandstone, and calcareous argillaceous rocks. The jasperoid inclusions, where found in place, have calc-silicate, skarn-type reaction rims consisting of diopside, rare garnet, and seemingly incompatible clay minerals. Many inclusions also have iron- and zinc-enriched cores that contain as much as 1% zinc, mainly in magnesioferrite. Both the cognate quartz monzonite inclusions and the accidental mineralized jasperoid inclusions are interpreted as having been stoped from the roof of the dacite magma chamber prior to eruption. The skarn reaction rims, however, are believed to have formed after eruption, during the cooling of the enclosing block-and-ash flows, which is indicated by the delicate, granular and friable nature of the rims, the presence of clay minerals, and the absence of unaltered limestone inclusions within the dacite.

INTRODUCTION

The Horse Springs dacite is a near-vent, pyroclastic deposit that is represented by a narrow, north-south outcrop belt about 5 km wide and 20 km long, on the north side of the San Agustin plains near Old Horse Springs (Figs. 1, 2). The enigma is related to its petrographic and chemical resemblance to the Blue Canyon Tuff (Eocene), and its possible significance to a caldera source for the Blue Canyon Tuff and (or) other ignimbrites in the upper part of the Eocene to early Oligocene Datil Group, and the origin and significance of the abundant inclusions of quartz monzonite and mineralized pre-Tertiary rocks which it contains.

Here we recognize the Horse Springs dacite as an informal volcanic stratigraphic unit and update a previous discussion of its age, origin and unique inclusions (Ratté and Modreski, 1989). The Horse Springs dacite was first described as the middle tuff-breccia sequence, at least 150 m thick, of the lower rhyolite pyroclastic unit of the Datil Formation (Stearns, 1962, p. 14-16), which he considered to have been "erupted from a center at no great distance." Later, Bornhorst (1976) and Jones (1980) called these rocks the tuff breccia of Horse Springs Canyon and gave the first detailed description of their petrography, composition, distribution and approximate age. Bornhorst and Elston (1978) and Elston (1984) correlated the eruption of these tuff breccias to the ring-fracture of the proposed Crosby Mountains cauldron. On the geologic map of the Horse Springs West quadrangle (Ratté et al., 1991), these rocks were called the pumice breccia of Old Horse Springs to bring the name into conformance with the most recent geo-

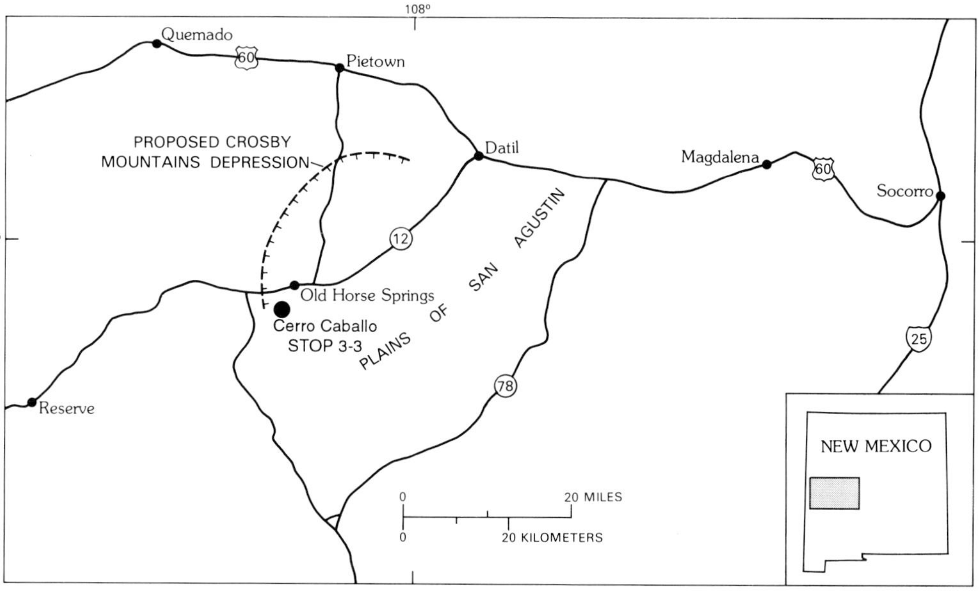

FIGURE 1. Index map showing location of Horse Springs dacite vent facies at Cerro Caballo, southwest of Old Horse Springs. Outline of Crosby Mountains depression is shown here as proposed by Bornhorst (1976) and Elston (1984).

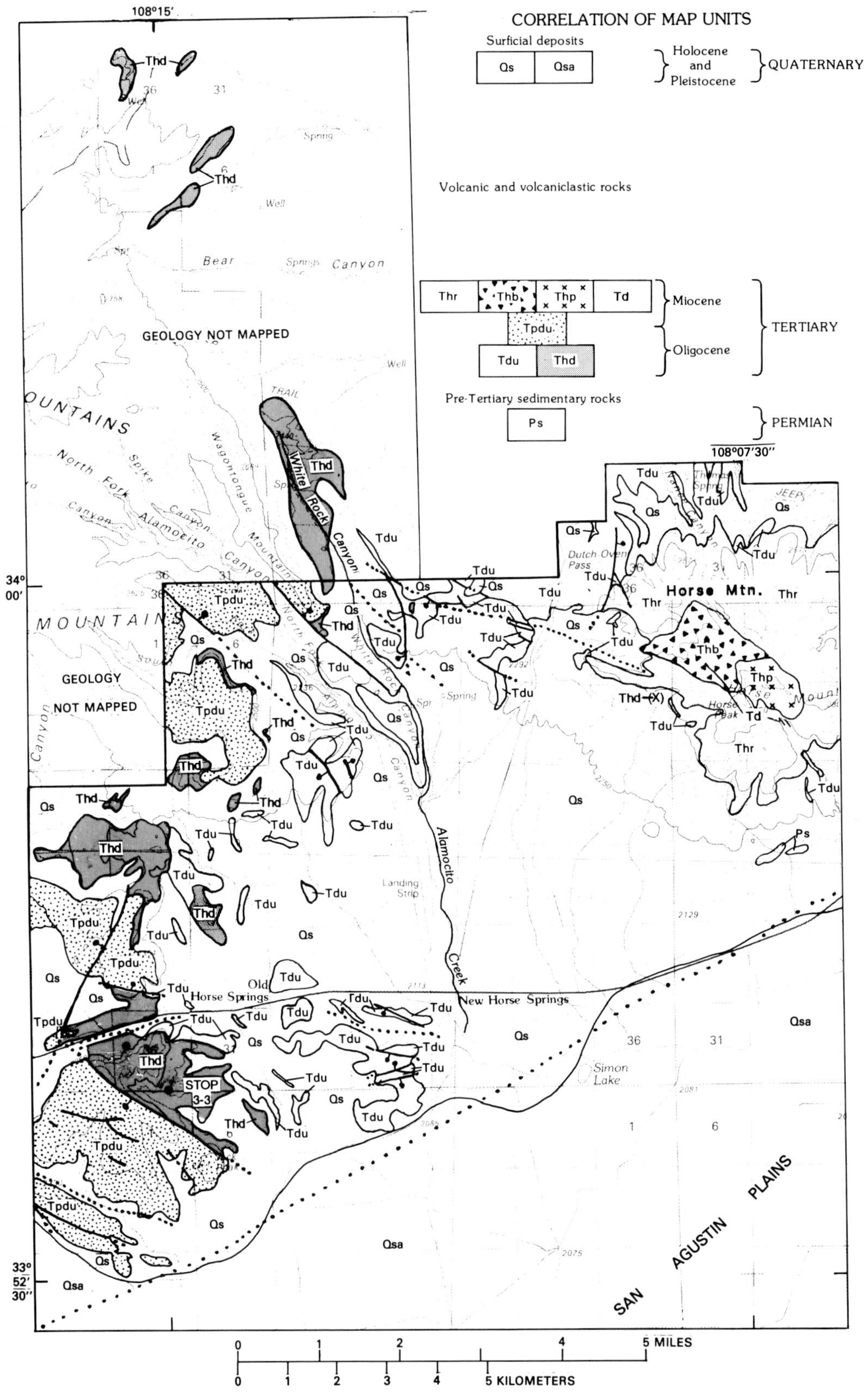

FIGURE 2. Geologic map showing distribution of Horse Springs dacite, Thd (solid black). Modified from Ratté et al., (1988, pl. 1) and Jones (1980, fig. 1). Qs = Surficial deposits, undivided (Holocene and Pleistocene); Qsa = fluviolacustrine deposits of Lake San Agustin (Holocene and Pleistocene); Rocks of the Horse Mountain volcano (Miocene, about 13 Ma): Thr = dacite and rhyolite lava flows; Thb = dacite and rhyolite domal vent breccias; Thp = rhyolite plug; Td= rhyolite dikes; Tpdu, = post-Datil Group volcanic and volcaniclastic rocks, undivided (Miocene? and Oligocene); Tdu = volcanic and volcaniclastic rocks of Datil Group, undivided (Oligocene); Ps = Permian sedimentary rocks, undivided.

graphic names in the area. However, here we propose that these rocks be called Horse Springs dacite to emphasize the composition of the rocks and in anticipation that the springs at Old Horse Springs may well outlive the historic settlements that were named for them.

STRATIGRAPHY AND LITHOLOGY

The known distribution of the Horse Springs dacite is confined to the Horse Mountain West, Bell Peak, Wallace Mesa and Mangas Mountains 7 1/2-minute quadrangles. The outcrops are mostly west of White Rock and Alamocito canyons, but one tiny outcrop on the west flank of Horse Mountain (shown by an x on Fig. 2) indicates that the dacite probably once covered a much larger area, but it is now buried by younger rocks or has been downfaulted beneath the Plains of San Agustin. These pyroclastic rocks have a maximum thickness of about 250 m southwest of Old Horse Springs on the ridge that on earlier maps (USGS 30-minute Pelona quadrangle) is called Cerro Caballo (Fig. 1), and is the site for Stop 3, Day 3 of the Field Trip described in the front section of this Guidebook. The pyroclastic deposit thins northward and is less than 30 m thick where last observed in the Wallace Mesa and Mangas Mountains quadrangles (Jones, 1980).

The Horse Springs dacite underlies Hells Mesa Tuff (32 Ma) in the northwestern part of the Horse Springs West quadrangle (Ratté et al., 1991) and thus is in the upper part of the Datil Group of Osburn and Chapin (1983), Cather (1986) and Cather et al. (1987). The dacite unconformably overlies ash-flow tuffs of the Datil Group, including Blue Canyon Tuff (33.7 Ma) and Rock House Canyon Tuff (34.4 Ma) and intertongued volcaniclastic sedimentary rocks of the Spears Group (Cather et al., this volume). Two new $^{40}Ar/^{39}Ar$ ages (sanidine laser-fusion ages, from cognate quartz monzonite blocks and pumice blocks included in the dacite) have an average age of 33.73 Ma (Table 1), virtually identical to the age of the underlying Blue Canyon Tuff. Previously reported ages for the dacite and its included quartz monzonite blocks, which ranged from about 35.5 Ma to about 31.2 Ma, are summarized by Marvin et al. (1987).

The Horse Springs dacite comprises three main lithologic facies, a basal breccia, lower pumice breccia flows with a mantle of ash-cloud surge beds, and upper block-and-ash flows.

The basal breccia, which ranges from mesobreccia to megabreccia, has been observed only locally, but it is present at the south end of the outcrop belt and may extend to the outcrops at the south end of Wagontongue Mountain (Fig. 2). The mesobreccia contains blocks of Rock House Canyon Tuff and Blue Canyon Tuff, as much as 5 m long, and smaller blocks of fine-grained red andesite and round pebble conglomerate like that in the Spears Group, all in a matrix of Horse Springs dacite. Megabreccia outcrops may actually be layers of mixed Rock House Canyon and Blue Canyon mesobreccia blocks, but some outcrops near the southern end of the Horse Springs dacite outcrop belt seem to be homogeneous slabs of tuff tens of meters long. Approximately the lower half of the Horse Springs dacite pyroclastic deposit, on the Cerro Caballo ridge southwest of Old Horse Springs, consists mainly of coarse, pumice-breccia flows. These are best exposed on the slickrock slope on the eastern side of the ridge where flows 1 and 2 (Fig. 3) are separated by a parting that probably represents a change in the size of pumice clasts produced by a pulsating eruption column. Both flow units show reverse size-grading of pumice blocks (Figs. 4, 5). The pumice flows were probably once mantled widely by surge deposits, which now are preserved erratically along the base of the cliffs at the top of the pumice flows (Fig. 6). The crossbedded deposits are probably best described as ash-cloud surge deposits (Fisher and Schminke, 1984, p. 201, 219).

Approximately the upper half of the Horse Springs dacite on the ridge southwest of Old Horse Springs is probably best described as a block-and-ash flow deposit, which may have been produced by "boiling over without a vertical eruption column" (Fisher and Schminke, 1984, p. 189). This facies of the deposit begins above the surge deposit at the base of the sharp cliff, above the lower pumice flows, and is remarkable in several respects. The 10-20-m-high fluted cliffs at the base of the block-and-ash flows display an ancient fumarole zone (Figs. 3, 7) that consists of numerous degassing pipes in which the fine-grained constituents of the flows are depleted by elutriation. The pipes here lack any evidence of mineralization accompanying the degassing and are unusual in being at the bottom of the deposit rather than at the top (Smith, 1960; Fisher and Schminke, 1984, particularly fig. 8-21).

The block-and-ash flows contain large blocks of pumiceous dacite (Fig. 8) and cognate inclusions of coarsely porphyritic quartz monzonite, commonly 2-3 m long (Fig. 9), in a somewhat finer matrix of dacite breccia. Quartz monzonite inclusions are present in all three facies of the Horse Springs dacite, but large blocks are most concentrated near the base of the block-and-ash flows, particularly on the ridge southwest of Old Horse Springs. Blocks of pumiceous dacite, having dimensions of approximately 1 m, are present in the block-and-ash flows all the way to White Rock Canyon. However, quartz monzonite inclusions in outcrops other than those on the ridge southwest of Old Horse Springs are generally ash- or lapilli-sized fragments.

TABLE 1. $^{40}Ar/^{39}Ar$ laser-fusion data for Horse Springs dacite. Each analysis is one laser fusion of a collection of 5 to 20 crystals, analyzed at the New Mexico Geochronology Research Laboratory. Ages are relative to Fish Canyon Tuff sanidine (27.84 Ma). Italicized data not used in calculation of mean age. Measured discrimination=1.0081. Correction factors: $^{39}Ar/^{37}ArCa$=0.0007, $^{36}Ar/^{37}ArCa$=0.00027, $^{40}Ar/^{39}ArK$=0.0225. Decay constants after Steiger and Jaeger (1977).

L#	40/39	37/39	36/39	39K moles	K/Ca	%40*	Age	±Err
Sample OSHBX-1, pumice block, sanidine, J + 0.004019 ± 0.00001								
292-01A	4.73	1.27E-02	1.37E-05	2.2E-16	40.2	99.5	33.763	0.072
292-04A	4.73	1.39E-02	2.79E-05	2.9E-16	36.6	99.4	33.789	0.066
292-05A	4.74	1.47E-02	2.88E-05	2.8E-16	34.8	99.4	33.806	0.070
292-07A	4.74	1.24E-02	2.43E-05	1.1E-16	41.2	99.4	33.826	0.071
292-06A	4.74	1.22E-02	8.30E-06	3.4E-16	41.9	99.5	33.857	0.063
292-08A	4.74	1.38E-02	2.74E-05	3.2E-16	37.0	99.4	33.861	0.072
292-03A	4.74	1.21E-02	2.47E-05	2.3E-16	42.2	99.4	33.865	0.066
292-02A	4.76	1.27E-02	2.89E-05	2.0E-16	40.1	99.4	33.939	0.072
					mean	**n=8**	33.838	0.055
Sample OSHBX-2, quartz monzonite, sanidine and plagioclase, J = 0.004019 ±0.00001								
293-03A	4.72	3.86E-02	9.24E-05	1.4E-16	13.2	99.0	33.572	0.071
293-06A	4.77	4.63E-02	2.47E-04	1.6E-16	11.0	98.1	33.576	0.073
293-04A	4.75	6.10E-02	1.81E-04	9.2E-17	8.4	98.5	33.627	0.076
293-01A	4.75	3.47E-02	1.55E-04	2.3E-16	14.7	98.6	33.636	0.072
293-05A	4.76	3.74E-02	1.09E-04	1.3E-16	13.6	98.9	33.768	0.071
293-02A	5.02	3.29E-02	1.99E-04	1.6E-16	15.5	98.4	*35.474*	0.077
					mean	**n=5**	33.636	0.080

FIGURE 3. Horse Springs dacite on east-facing slopes of Cerro Caballo ridge southwest of Old Horse Springs. Numbers 1-6 indicate the following: 1 and 2 = non-welded pumice flows, each on the order of 50 m thick; 3 = position of discontinuous, crossbedded, ash-cloud, surge deposits at top of pumice flow no. 2; 4 = ancient fumarole zone in cliffs at base of block-and-ash deposit; 5 = approximate position of parting ledge within zone where 2-3 m blocks of pumiceous dacite and cognate quartz monzonite are concentrated; 6 = approximate top of zone in which mineralized jasperoid inclusions are most concentrated.

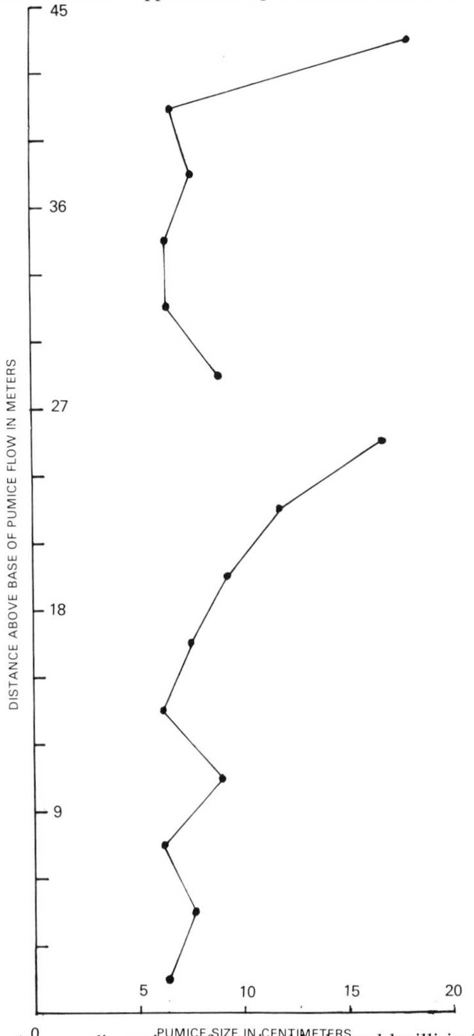

FIGURE 4. Average dimensions of pumice blocks and lapilli in 3-m intervals plotted against position above base of pumice flow no. 2 in Fig. 3. Pumice measured in plane of foliation perpendicular to a tape stretched across the outcrop.

FIGURE 5. Pumice flow no. 2, in Fig. 3, beneath fluted cliffs with ancient fumaroles. Tape stretched across outcrops was used to measure size of pumice pyroclasts, which show reverse grading.

FIGURE 6. Ash-cloud surge beds preserved locally at the top of the pumice flows in the lower part of the Horse Springs dacite, on the east side of the Cerro Caballo ridge southwest of Old Horse Springs.

FIGURE 7. Ancient fumarole degassing pipes at the base of the block-and-ash flows in the upper part of the Horse Springs dacite, on the east side of Cerro Caballo ridge southwest of Old Horse Springs (Fig. 3). Width of outcrop is about 3 m.

FIGURE 8. Meter size block of juvenile pumiceous dacite in Horse Springs dacite on west side of Cerro Caballo ridge southwest of Old Horse Springs.

Also distributed throughout the block-and-ash flows, but found in outcrop mainly on the ridge southwest of Old Horse Springs, are unique inclusions of mineralized jasperoid derived from the alteration of pre-Tertiary basement rocks. These xenoliths are sparsely concentrated above a parting ledge (no. 5 on Fig. 3), about 30 m above the base of the block-and-ash flows, and are apparently most abundant between points 5 and 6 on the photo. The zone containing jasperoidized xenoliths also appears to coincide with the upper part of the zone of greatest concentration and maximum size of cognate quartz monzonite blocks; this observation may be significant in interpreting the origin of the jasperoid. Protoliths for the jasperoidized xenoliths can probably be found in the Lower Permian sequence of the Yeso Formation (mainly calcareous to argillaceous sandstone), Glorieta Sandstone and San Andres Limestone, all of which are exposed nearby in outcrops at the foot of Horse Mountain, 12-15 km east of the Cerro Caballo ridge. The largest pre-Tertiary xenolith observed by us was a brick-red block of jasperoidized limestone (Fig. 10) about 50x15x15cm. That the majority of the jasperoid xenoliths have a limestone protolith is indicated by the presence of relict chert nodules (Fig. 11). Jasperoidized limestone inclusions generally have reaction borders composed of calc-silicate skarn a few centimeters thick (Fig. 12), which consist mainly of diopside and clay minerals, with or without garnet and phlogopite. Similar reaction rims border xenoliths that were formerly calcareous and argillaceous sandstones, but are absent on inclusions that are believed to have been sandstone or quartzite.

FIGURE 9. One to two meter diameter blocks of cognate quartz monzonite and juvenile pumiceous dacite on the east side of Cerro Caballo ridge, at the level of the main concentration of jasperoid blocks, point 5 in Fig. 3.

PETROGRAPHY AND CHEMISTRY OF THE HORSE SPRINGS DACITE

The Horse Springs dacite is a pinkish-brown, pumiceous, fine-grained, granular rock that contains between 25 and 35% 1-4 mm long phenocrysts. The phenocrysts are mainly sodic plagioclase (oligoclase-andesine) about 20-25%, subequal amounts of biotite and clinopyroxene totaling 5-10%, opaque oxides averaging about 7%, and traces of quartz, zircon and apatite. The matrix is a subvitreous, fine-grained spherulitic to granular intergrowth of feldspar and quartz that contains abundant equant to rounded crystals of quartz, plagioclase, and alkali feldspar having dimensions of about 0.1-0.3 mm that give the matrix a sieve-like texture (Fig. 13A, B). The potassic feldspar can be recognized from carlsbad twinned crystals, low refractive index, and by staining.

The quartz monzonite inclusions have a much coarser porphyritic texture than the enclosing dacite, containing abundant plagioclase phenocrysts 1-2 cm long, and having a total phenocryst content of about 40%. Otherwise, the quartz monzonite is quite similar mineralogically and chemically to the dacite (Table 2), differing mainly in having trace amounts of sanidine phenocrysts to about 1 mm long, trace amounts of brown hornblende microphenocrysts, and aphanitic groundmass that ranges from slightly devitrified brown glass to a granophyric matrix of quartz and alkali feldspar with an average grain size of about 0.1 mm. Where the matrix is not granophyric, it has the same sieve texture of rounded to equant small quartz and feldspar crystals scattered through the subvitreous groundmass as the enclosing dacite (Fig. 14). The textural and compositional similarities between the dacite and the quartz monzonite inclusions suggest that the inclusions were derived from an earlier stage of crystallization of

FIGURE 10. Accidental inclusion of jasperoidized limestone from block-and-ash flows in upper part of Horse Springs dacite on east side of Cerro Caballo ridge southwest of Old Horse Springs. Relict bedding is visible.

FIGURE 11. Accidental inclusion of jasperoidized limestone showing relict chert nodules.

the same magma that erupted to form the dacite pumice flows and block-and-ash flows.

The size and abundance of 2-3 m blocks of pumiceous dacite and cognate quartz monzonite inclusions in the Horse Springs dacite at Cerro Caballo, southwest of Old Horse Springs, indicate close proximity to an eruptive vent for the Horse Springs dacite. Further support for the location of a vent in this area is given by the concentration of degassing pipes, ash-cloud surge deposits, the presence of mesobreccia and possibly megabreccia of older rocks, and the thinning and decreased particle size of the deposits northward, away from this area, which is the only direction in which substantial outcrops of the dacite are preserved at the surface. However, there are no vent structures, such as major crossbedded or crosscutting deposits, to indicate that the dacite at Cerro Caballo is within or above the actual conduit, which is most likely buried nearby beneath the alluvium on the Plains of San Agustin.

BLUE CANYON TUFF

The petrography and composition of the Blue Canyon Tuff and the Horse Springs dacite (Table 2) are similar but not identical. The Blue Canyon Tuff could be a slightly more evolved, early eruptive facies of the same magma that produced the Horse Springs dacite. The Blue Canyon Tuff is a low-silica rhyolite (Table 2); it contains between 15-25% phenocrysts, and it differs from Horse Springs dacite mainly in containing more abundant sanidine phenocrysts relative to plagioclase (about 1:2) and a lower abundance of clinopyroxene. Eutaxitic pumice lapilli are obvious in most outcrops, and the matrix of the Blue Canyon Tuff has a vitroclastic texture of devitrified glass shards (Fig. 15) which clearly identifies its ignimbrite origin.

No source is known for the Blue Canyon Tuff, but its regional distribution from the Horse Springs area to the Socorro area indicates that it probably is a caldera-derived ash-flow tuff. At present, the only clues to its source are its distribution and thickness. Blue Canyon Tuff directly underlies the Horse Springs dacite on the east side of the Cerro Caballo ridge southwest of Old Horse Springs, where its 10-15 m thickness is similar to that elsewhere in the Horse Mountain quadrangle (Ratté et al., 1991) and the Horse Springs area. Its maximum thickness in the Datil Mountains (north of Datil) is only about 30 m (Osburn and Chapin, 1983); from the Datil Mountains, the Blue Canyon Tuff thins eastward and is present only locally in the Magdalena-Socorro area. Thus, nowhere is a thickness of Blue Canyon Tuff sufficient to suggest an intracaldera sequence in its known area of distribution, unless it is buried beneath a down-faulted basin such as the Plains of San Agustin. Because of their petrographic similarity and identical isotopic ages, it is plausible that the sources of the Blue Canyon Tuff and the Horse Springs dacite are related, and that both are hidden somewhere beneath the plains in the vicinity of Old Horse Springs (Fig. 1).

PETROGRAPHY AND CHEMISTRY OF MINERALIZED XENOLITHS

The jasperoid inclusions vary in mineral composition from nearly pure quartz, to quartz plus abundant calcite and iron oxides, to dominantly iron oxides. Many inclusions have a concentrically banded core of layers rich in quartz and iron oxides surrounded by brick-red, massive, spherulitic jasperoid (Fig. 16A). Most of the analyzed nodules were collected from float and probably represent the cores of larger jasperoid blocks (Fig. 16B-D); others, which show relict bedding, may have been originally sandstone or quartzite rather than limestone. Most of the quartz in the xenoliths exhibits fibrous or radiating (spherulitic) crystallization, indicating an origin by replacement or recrystallization (Fig. 17), but some show the relict clastic texture of the original sandstone or siltstone.

Iron in the xenoliths forms an unusual suite of ferric-iron minerals, including coarsely crystalline, black, lustrous plates of hematite and red, very fine-grained crystals of magnesioferrite (MgFe3+2O4), which are disseminated in the silica and concentrated along quartz grain boundaries. The jasperoid nodules typically have brick-red, fine-

FIGURE 12. Accidental jasperoidized and mineralized inclusion in situ in Horse Springs dacite showing narrow, calcium silicate (skarn type), diopside-rich, reaction rim, R.

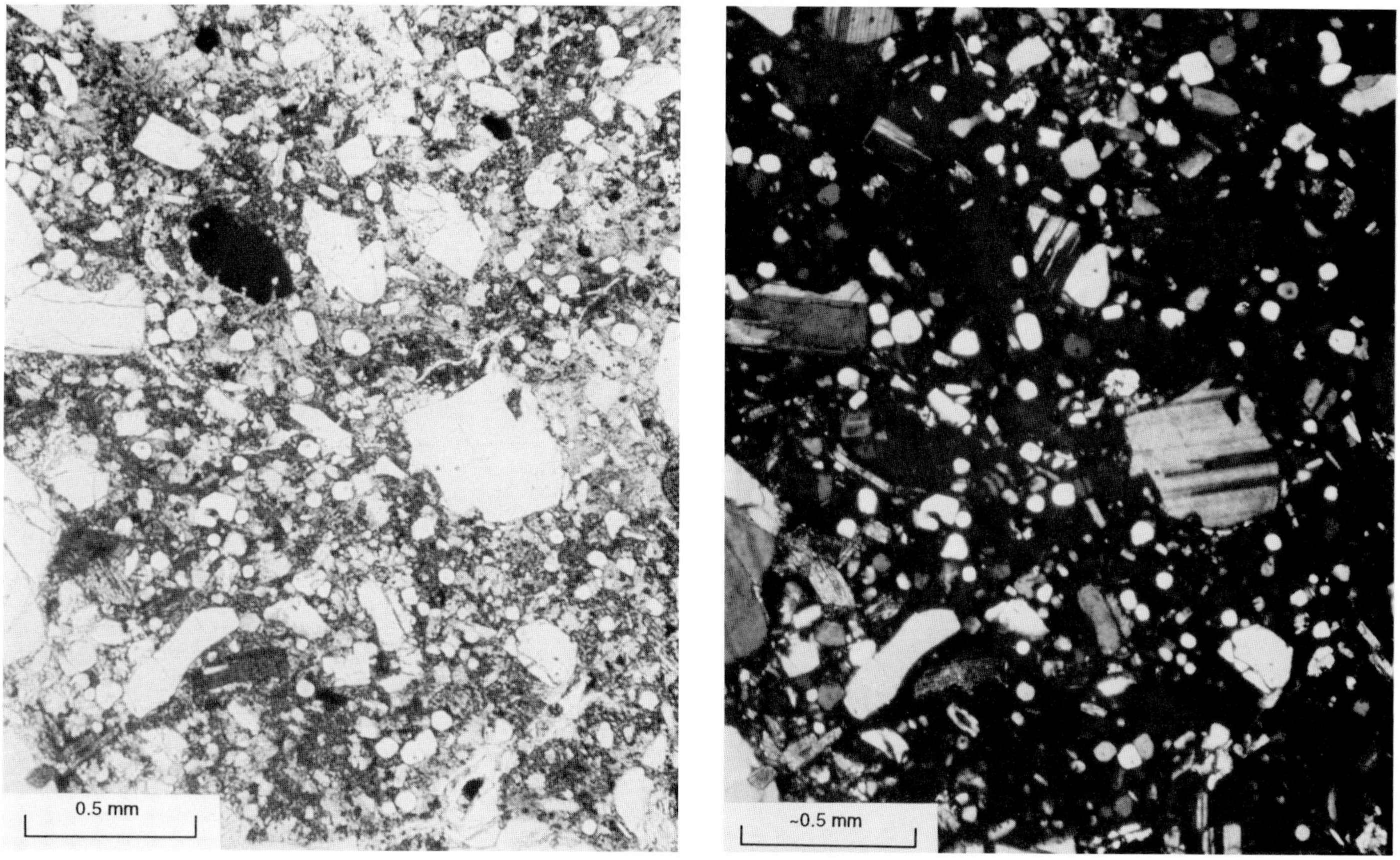

FIGURE 13. Photomicrographs showing the matrix texture of Horse Springs dacite. A, plane polarized light. B, crossed plane polarized light.

TABLE 2. Chemical and modal analyses of Horse Springs dacite, cognate quartz monzonite inclusions, and Blue Canyon Tuff. Analysts: Sample numbers 1,2,3,5 and 7 major oxides by J. Baker, A. Bartel, D. Siems, J. Taggart, J. Wahlberg; trace elements by D. Fey, R. Yeoman and D. Bove. Sample number 4 from D.P. Jones, 1980. Sample number 6 by Bornhorst, 1980. (-) indicates not analyzed or not found.

Sample No.	1	2	3	4	5	6	7
Field No.	P-5-78	BP-1-88	BP-2-88	J-218	BP-6A-85	T-377	HW-1B-82
				MAJOR OXIDES (weight percent, recalculated water-free)			
SiO_2	66.6	67.5	66.4	65.6	67.6	67.8	70.27
Al_2O_3	15.6	15.7	15.7	16.3	15.6	15.7	15.15
Fe_2O_3	3.86*	4.56*	4.64*	3.66	2.86*	3.02	0.90
MgO	1.32	1.03	1.07	1.25	1.34	1.42	0.52
CaO	3.53	2.31	3.12	3.98	3.54	3.17	1.43
Na_2O	3.51	3.16	3.60	3.89	4.19	3.34	2.90
K_2O	4.55	4.70	4.37	4.14	3.94	4.25	7.05
TiO_2	0.62	0.68	0.67	0.59	0.58	0.53	0.44
P_2O_5	0.33	0.23	0.30	0.23	0.21	0.18	0.08
MnO	0.11	0.07	0.11	0.07	0.04	0.04	0.05
FeO	—	—	—	—	—	0.19	1.21
Total	100.03	99.94	99.98	99.71	99.90	99.64	100.00
*Total iron as Fe_2O_3							
				TRACE ELEMENTS (parts per million; L, indicates detected at limit shown)			
Ba	—	898	1002	—	860	969	—
Cr	—	—	—	—	9	25	—
Cu	—	78L	78L	—	11	18	—
Ga	—	16L	20	—	17	<20	—
Ni	—	—	—	—	8	19	—
Rb	—	145	119	—	—	114	—
Sr	—	384	514	—	520	481	—
Th	—	18L	18L	—	17	19.4	—
V	—	—	—	—	45	41	—
Y	—	32	28	—	22	30	—
Zn	—	79	71	—	35	134	—
Zr	—	247	215	—	—	245	—
				MODES (volume percent)			
Groundmass	72	72	71	65.2	61	61.9	83.3
Sanidine	Trace	Trace	Trace	2.0	Trace	—	4.7
Quartz	—	Trace	Trace	0.2	—	0.3	—
Sodic plagioclase	21	24.3	22.6	25.8	30	28.9	10.7
Biotite	3.5	1.4	2.0	2.7	3.5	5.1	0.5
Clinopyroxene	3	1.1	2.7	2.2	3	2.2	Trace
Hornblende	—	—	—	—	1	—	—
Opaque oxide	0.5	0.5	1.5	1.9	1.5	1.5	0.7
Zircon	—	—	Trace	Trace	Trace	—	—
Apatite	Trace	—	Trace	—	Trace	—	—
Total	100	99.3	99.8	100	100	99.9	99.9

Sample descriptions

1. Pumice block from roadcut on north side of NM-12 west of Old Horse Springs in Bell Peak Quadrangle.
2. Pumice block from outcrops on west side of Cerro Caballo ridge south of NM-12 (elevation about 7260 ft) in Bell Peak Quadrangle.
3. Pumice block from outcrops on west side of Cerro Caballo ridge south of NM-12 (elevation about 7420 ft) in Bell Peak Quadrangle.
4. Horse Springs dacite matrix from outcrops on east side of White Rock Canyon at White Rock Spring in Wallace Mesa Quadrangle. From D.P. Jones, 1980.
5. Cognate quartz monzonite inclusion from outcrops on west side of Cerro Caballo ridge (elevation about 7500 ft) in Bell Peak Quadrangle.
6. Cognate quartz monzonite inclusion from outcrops on west side of Cerro Caballo ridge south of NM-12 in Bell Peak Quadrangle. Collected and analyzed by T.J. Bornhorst (1980, p. 593).
7. Blue Canyon Tuff from roadcut on north side of NM-12 near west edge of Horse Mountain West Quadrangle.

FIGURE 14. Photomicrograph showing quartz and alkali feldspar (twinned) in granophyric groundmass of cognate quartz monzonite inclusions in Horse Springs dacite. Phenocrysts in view are augitic pyroxene and plagioclase. Crossed plane polarized light.

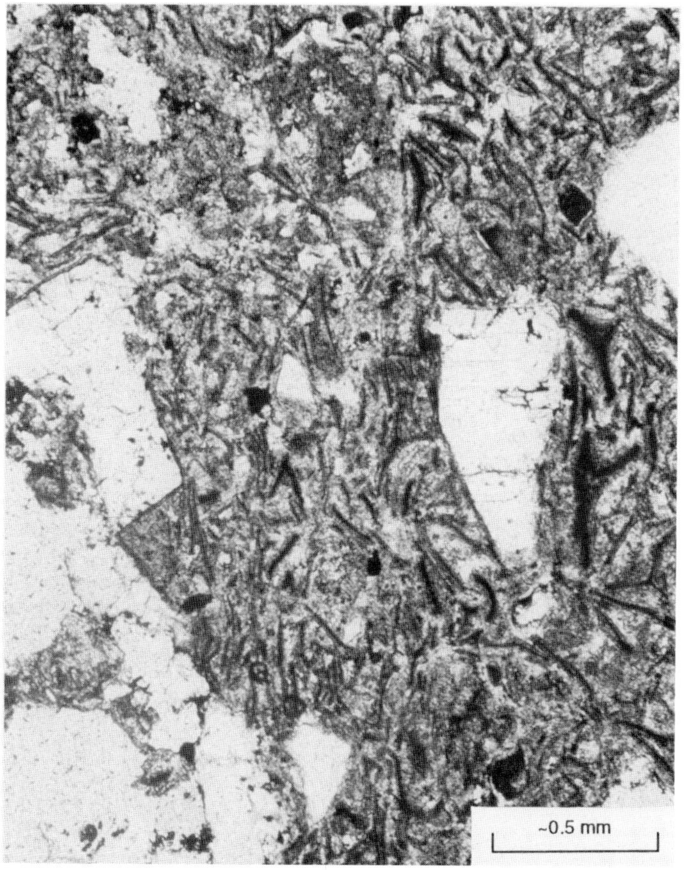

FIGURE 15. Photomicrograph of Blue Canyon Tuff showing vitroclastic matrix of devitrified glass shards between plagioclase phenocrysts. Crossed plane polarized light.

grained outer zones (colored by a fine magnesioferrite dust) and cores of coarsely crystalline quartz, hematite and calcite. Forsteritic olivine occurs as inclusions in hematite crystals; it contains approximately 1.8% total iron (Table 3). The bright-red cathodoluminescence of the forsterite suggests that all of the iron is in the ferric state. This assemblage of ferric-iron minerals indicates a high oxygen fugacity during formation, which may have involved recrystallization of preexisting iron oxides. Although no magnetite has been observed in the nodules, which contain only ferric-iron and Mn3+ phases (Table 4), textural evidence, mainly fine-grained aggregates of iron oxide that have pseudomorphic outlines, indicates probable replacement of pre-existing iron-oxide crystals.

Zinc in the jasperoid inclusions seems to be dominantly in the magnesioferrite (0.33-2.93 wgt % ZnO; Table 5). Zinc is also present in aluminum spinels (1-27 wgt %; Table 5), but the spinels are a very minor phase volumetrically. The zinc-bearing phases may have been produced by recrystallization and oxidation during thermal metamorphism of earlier-formed oxides, such as magnetite and franklinite, or by desulfidization of sphalerite. Zinc and other trace metals in the mineralized jasperoid inclusions are shown in Table 6, where zinc is seen to be as much as 1% (10,000 ppm) of some inclusions.

Reaction zones between the jasperoid inclusions and the enclosing dacite pumice breccia have a largely different mineral assemblage than the inclusions themselves (Fig. 18). Calc-silicate skarn consists of greenish-yellow, diopside-rich border zones, a few centimeters thick, around many of the in-situ jasperoid nodules. Euhedral diopside crystals are commonly zoned, and FeO ranges between 4 and 15% (Table 7). Diopside in the skarn reaction zones may be accompanied by green to brown, euhedral andradite garnet that ranges in composition from about andradite$_{90}$—grossular$_{8}$ to andradite$_{34}$—grossular$_{34}$ (Table 7). Other minerals found in the reaction zone include light-pink montmorillonite clay and phlogopite mica (Table 4). Within the tuff adjacent to the reaction zone, plagioclase phenocrysts are altered to a fine-grained mixture of clays, micas, and other unidentified K and Mg silicates.

Diopside also occurs as free-growing, transparent, yellowish-green prisms, a few tenths of a millimeter long, in vugs and fractures in the jasperoid inclusions and in miarolitic cavities in the cognate quartz monzonite blocks, where it commonly is accompanied by hematite and prismatic to bladed pseudobrookite ($Fe^{+3}_2TiO_5$; Table 5). Titanite coats some pseudobrookite crystals.

ORIGIN AND SIGNIFICANCE OF MINERALIZED INCLUSIONS

Two main problems concerning the formation of the mineralized inclusions are (1) place of origin and physical-chemical environment of jasperoid development and associated mineralization; and (2) place of origin and physical-chemical environment of calc-silicate reaction zones on the margins of jasperoid inclusions. There are two alternative explanations: (1) both mineralized jasperoid and calc-silicate reaction zones formed in the same place as parts of a single alteration process; or (2) the mineralized jasperoid and the reaction zones formed in separate places as a result of different processes.

Some of the considerations in evaluating these alternatives are:

1. The delicate, granular, calc-silicate reaction zones around jasperoid inclusions are essentially unbroken where inclusions are found in place, but are almost never found attached to the inclusions in float, attesting to their vulnerability during transport.

2. The mineral assemblage in the calc-silicate zones consists of variable proportions of diopside, garnet, phlogopite and clay minerals. Except for the clays, this assemblage is consistent with the

FIGURE 16. Accidental inclusions of mineralized jasperoid in Horse Springs dacite showing different form and composition. A, Jasperoid inclusion showing core of quartz, hematite and calcite surrounded by jasperoid containing fine-grained magnesioferrite. B, Jasperoid inclusion showing concentric layers of quartz and calcite and iron oxides. C, Jasperoid inclusion showing core of calcite, quartz, hematite and magnesioferrite surrounded by a red outer layer of quartz, hematite and magnesioferrite. D, Iron oxide-rich jasperoid inclusion containing disseminated hematite, magnesioferrite and quartz.

FIGURE 17. Photomicrograph showing spherulitic quartz and coarse, platy hematite crystals in jasperoid inclusion. Crossed plane polarized light.

FIGURE 18. Cross section of calc-silicate reaction skarn surrounding jasperoid inclusion from inside of reaction zone (top) into enclosing dacite breccia (bottom), showing: (1) 1 mm red layer of hematite and phlogopite; (2) 4-9 mm pale-green exoskarn layer of granular diopside and montmorillonite, and minor hematite and feldspar; (3) 6-11 mm dark-green endoskarn layer of diopside, minor hematite, and variably altered feldspar; (4) dacite breccia host rock showing scattered development of diopside and clay, and plagioclase phenocrysts with altered borders.

TABLE 3. Microprobe analyses of mineral phases in jasperoid, calc-silicate skarn reaction rims, and altered Horse Springs dacite. Analyses on ARL-SEMQ microprobe, 20 sec. counting times, MAGIC-IV matrix correction program. Analysis of olivine and calcite, 10 kV accelerating voltage, 5 nA sample current; standards: K,Al,Si, orthoclase; Mg, dolomite; Ca, calcite; Mn, rhodochrosite; Fe, siderite. Analysis of mica and feldspar, 15 kV accelerating voltage, 10 nA sample current; standards: Na, albite; K, orthoclase; Mg, diopside; Ca,Al,Si, grossular; Mn,Fe, spessartine-almandine; Ti, titanite; Cr, chromite; Zn, willemite. (-) indicates not analyzed.

	Phl-1	For-1	Cc-1	Pla-1	Alt-1	Alt-2
			(weight percent)			
Na_2O	0.57	—	—	7.54	—	0.29
K_2O	9.80	—	—	1.04	8.14	0.15
MgO	28.15	59.08	0.03	0.02	0.21	9.48
CaO	0.14	0.02	53.82	5.25	6.32	1.24
SrO	—	—	0.02	—	0.00	—
MnO	0.00	0.08	0.00	0.02	0.00	0.00
FeO	0.59	1.82	0.00	0.24	0.21	0.81
ZnO	0.07	0.03	—	0.01	—	0.02
Al_2O_3	10.03	0.01	0.08	22.07	25.12	16.87
Cr_2O_3	0.00	—	—	0.00	—	—
SiO_2	44.26	39.11	0.00	61.80	59.37	56.65
TiO_2	0.14	0.04	—	0.00	—	0.01
CO_2	—	—	42.28*	—	—	—
Total	93.75	100.19	96.23	98.01	99.37	85.52
			(cation proportions)			
Na	0.078	—	—	0.663	—	0.039
K	0.882	—	—	0.060	0.650	0.013
Mg	2.961	2.087	0.001	0.001	0.020	0.974
Ca	0.011	0.001	0.997	0.255	0.424	0.092
Sr	—	—	0.000	—	0.000	—
Mn^{+2}	0.000	0.002	0.000	0.001	0.000	—
Fe^{+2}	0.035	0.036	0.000	0.009	0.011	0.047
Zn	0.004	0.001	—	—	—	—
Al	0.834	—	0.002	1.179	1.855	1.370
Si	3.122	0.927	0.000	2.802	3.719	3.904
Ti	0.007	0.001	—	0.000	—	—
Σox.	11.000	4.000	1.001	8.000	11.000	11.000

Sample descriptions

Phl-1 = phlogopite mica at contact of jasperoid nodule BP-46-86 with skarn rim.
For-1 = olivine inclusion in hematite, within jasperoid nodule BP-6B-85.
Cc-1 = calcite intergrown with quartz in jasperoid nodule BP-6G-85.
Pla-1 = unaltered core (sodic andesine) of plagioclase phenocryst in tuff, within outer part of calc-silicate reaction rind around nodule BP-46-86.
Alt-1 = altered rim of plagioclase phenocryst (mineral composition unknown).
Alt-2 = Mg-Al-silicate (mineral identity unknown) surrounding an altered plagioclase phenocryst.
*calculated amount of CO_2 in calcite.

TABLE 4. List of minerals in the jasperoid inclusions, calc-silicate skarn reaction rims, and altered Horse Springs dacite.

Hematite
Magnesioferrite
Spinel (zincian)
Franklinite/gahnite
Pseudobrookite
Sphene
Forsterite
Diopside
Andradite
Phlogopite
Montmorillonite
Quartz
Calcite

pyroxene-hornfels facies of contact metamorphism, indicating a formation temperature between about 550° and 700°C. The intensity of alteration around the jasperoid decreases outward into the pumice breccia.

3. The mineral assemblage in the jasperoid, mainly quartz, calcite and hematite, is typical of low-temperature metasomatism. However, magnesioferrite and tiny forsterite inclusions in the hematite indicate a higher temperature of formation.

4. The pumice breccia deposit is not welded.

5. Unaltered limestone xenoliths have not been observed.

6. The ancient fumaroles lack evidence of iron or other sublimates deposited during degassing of the pumice breccia.

7. The anomalous abundance of several metals, notably iron, zinc and manganese,and in lesser amounts copper, vanadium, nickel and lead, is evidence of some type of active mineralizing process.

8. Quartz replacement textures observed are typically low-temperature jasperoidization.

9. Iron-oxide replacement textures in the jasperoid nodules suggest replacement by magnesioferrite of pre-existing, probably lower temperature, iron oxides.

TABLE 5. Microprobe analyses of oxide minerals in jasperoid inclusions. Analyses on ARL-SEMQ microprobe, 15 kV accelerating voltage, 15 nA sample current, 20 sec. counting times, MAGIC-IV matrix correction program. Standards for analysis of oxide minerals: Mg,Al, spinel; Ca,Ti,Si, titanite; Fe, hematite; Mn, spessartine; Zn, willemite.

	Hem-1	Hem-2	Hem-3	Mgf-1	Mgf-2	Mgf-3	Spn-1	Spn-2	Spn-3
				(weight percent)					
MgO	0.22	0.35	0.16	21.71	21.62	20.72	27.11	10.39	5.59
CaO	0.01	0.00	0.00	0.01	0.92	0.18	0.06	0.03	1.25
MnO	0.04	0.74	0.17	1.22	3.32	3.85	0.25	10.61	10.91
ZnO	0.01	0.00	0.00	0.33	0.47	2.93	1.05	11.10	27.46
Fe_2O_3	98.29	99.10	100.39	79.22	68.14	72.52	16.61	30.28	34.49
Al_2O_3	0.23	0.08	1.03	0.29	5.37	0.98	53.97	11.87	8.95
SiO_2	0.01	0.01	0.07	0.02	0.01	0.22	0.22	0.13	0.16
TiO_2	0.14	0.04	0.02	0.00	0.00	0.04	0.00	0.04	0.06
Total	98.95	100.32	101.84	102.80	99.85	101.44	99.27	96.79	88.87
				(cation proportions)					
Mg	0.009	0.014	0.006	1.037	1.029	1.002	1.029	0.537	0.329
Ca	—	—	—	—	0.032	0.001	0.002	0.001	0.053
Mn^{+3}	0.001	0.015	0.003	0.033	0.090	0.106	0.005	0.312	0.365
Zn	—	—	—	0.008	0.011	0.070	0.020	0.284	0.802
Fe^{+3}	1.981	1.973	1.958	1.910	1.637	1.770	0.318	1.374	1.026
Al	0.007	0.003	0.032	0.011	0.202	0.038	1.620	0.485	0.417
Si	—	—	0.002	0.001	—	0.007	0.006	0.004	0.006
Ti	0.003	—	—	—	—	0.001	—	0.001	0.002
Oxygen atoms	2.999	3.000	3.000	3.978	3.966	3.960	3.978	4.088	3.912

Sample descriptions

Hem-1 = coarse hematite in jasperoid nodule BP-6G-85.
Hem-2 = coarse hematite in jasperoid nodule BP-6C-85.
Hem-3 = coarse hematite, host for olivine inclusions in jasperoid BP-6B-85.
Mgf-1 = magnesioferrite, intergrown with hematite in jasperoid BP-6A-85.
Mgf-2 = magnesioferrite crystal (about 20 microns) in jasperoid BP-6G-85.
Mgf-3 = fine-grained magnesioferrite in jasperoid BP-6A-85.
Spn-1 = yellowish-brown spinel crystal in jasperoid BP-6B-85.
Spn-2 = fine-grained spinel (complex solid solution) in jasperoid BP-6E-85.
Spn-3 = very fine-grained spinel (franklinite/gahnite) in jasperoid BP-6G-85.

TABLE 6. Atomic absorption (AA) and semiquantitative spectrographic analyses (6-step D-C arc method) of mineralized jasperoid inclusions. G, greater than value shown; N( ), not detected at level of value shown; L( ), detected, but less than value shown. Analysts: D.E. Detra, R.T. Fairfield, L.S. Landon and C.D. Taylor.

	AA		Percent				Parts per million										
Field No.	Au	Ag	Fe	Mg	Ca	Ti	Mn	B	Ba	Be	Co	Cu	Ni	Pb	Sn	V	Zn
BP-2-85	N(0.05)	N(0.05)	G20	1	L(0.05)	0.15	300	30	200	N(1)	70	100	100	N(10)	N(10)	300	2000
BP-3-85	N	N	2	3	5	0.15	200	30	N(20)	2	10	15	15	N	N	15	N(200)
BP-6A-85	N	N	3	1	1.5	0.3	500	10	700	1.5	15	10	7	15	N	70	N
BP-6B-85	N	N	20	1	0.2	0.03	2000	15	20	L(1)	70	100	100	10	20	150	7000
BP-6D-85	N	N	15	2	2	0.02	3000	20	50	N	50	300	30	70	15	70	10000
BP-6E-85	N	N	G20	1	0.7	0.03	2000	10	20	N	70	50	100	N	15	300	1000
BP-6F-85	N	N	20	1	2	0.015	1500	10	20	N	50	100	70	50	N	100	5000

TABLE 7. Microprobe analyses of diopside and garnet in calc-silicate skarn reaction rims. Analyses on ARL-SEMQ microprobe, 20 sec. counting times, MAGIC-IV matrix correction program. Analyses of pyroxenes Di-1 to Di-4, 10 kV accelerating voltage, 5 nA sample current; standards: K,Al,Si, orthoclase; Mg, dolomite; Ca, calcite; Mn, rhodochrosite; Fe, siderite. Analyses Di-5 and of garnet, 15 kV accelerating voltage, 10 nA sample current; standards: Na, albite; K, orthoclase; Mg, diopside; Ca,Al,Si, grossular; Mn,Fe, spessartine-almandine; Ti, titanite; Cr, chromite; Zn, willemite. Dash (-) indicates not analyzed.

	Di-1	Di-2	Di-3	Di-4	Di-5	And-1	And-2	And-3	And-4
				(weight percent)					
Na_2O	—	—	—	—	0.35	0.00	0.01	0.01	0.02
K_2O	0.02	0.03	0.02	0.00	0.00	0.00	0.00	0.01	0.01
MgO	17.52	13.29	15.03	13.71	9.40	0.19	0.20	0.17	0.18
CaO	25.19	24.96	23.18	24.37	22.78	32.52	32.84	33.24	33.11
SrO	0.00	0.06	0.02	0.01	—	—	—	—	—
MnO	0.05	0.14	0.76	0.04	0.13	0.09	0.11	0.10	0.12
FeO	4.06	6.34	9.00	10.58	14.75	28.44*	27.09*	24.78*	20.38*
ZnO	—	—	—	—	0.03	0.00	0.00	0.00	0.03
Al_2O_3	2.95	5.43	0.64	3.58	2.17	1.73	2.58	4.41	7.20
Cr_2O_3	—	—	—	—	0.00	0.01	0.00	0.00	0.02
SiO_2	53.57	51.16	53.85	50.48	48.90	35.78	35.58	36.27	37.14
TiO_2	—	—	—	—	0.12	0.24	0.42	0.07	0.13
Total	103.36	101.41	102.50	102.77	98.63	99.00	98.83	99.06	98.34
				(cation proportions)					
Na	—	—	—	—	0.027	0.000	0.001	0.002	0.003
K	0.001	0.001	0.001	0.000	0.000	0.000	0.000	0.001	0.001
Mg	0.928	0.725	0.818	0.754	0.549	0.024	0.025	0.021	0.022
Ca	0.959	0.979	0.907	0.963	0.957	2.938	2.975	2.960	2.913
Sr	0.000	0.001	0.000	0.000	—	—	—	—	—
Mn	0.002	0.004	0.024	0.001	0.004	0.007	0.008	0.007	0.008
Fe	0.121	0.194	0.275	0.326	0.484	1.805	1.724	1.550	1.259
Zn	—	—	—	—	0.001	0.000	0.000	0.000	0.002
Al	0.123	0.234	0.028	0.156	0.100	0.172	0.257	0.432	0.697
Cr	—	—	—	—	0.000	0.001	0.000	0.000	0.002
Si	1.903	1.872	1.967	1.861	1.917	3.017	2.983	3.015	3.050
Ti	—	—	—	—	0.004	0.015	0.027	0.004	0.008
Σox.	6.001	5.999	6.000	6.000	6.000	12	12	12	12

Sample descriptions

Di-1 = unzoned diopside crystal in calc-silicate rind on nodule BP-46-86.
Di-2 = diopside crystal in calc-silicate rind on nodule BP-6F-85.
Di-3 = colorless core of diopside crystal in rind on BP-46-86.
Di-4 = yellowish-green rim of diopside crystal in rind on BP-46-86.
Di-5 = brown zone in a diopside crystal in rind on BP-46-86.
And-1 = core of andradite garnet crystal in calc-silicate rind, BP-6-86.
And-2 = rim of same garnet crystal.
And-3 = garnet crystal in calc-silicate rind, BP-6-86.
And-4 = garnet crystal in calc-silicate rind, BP-6-86.
*iron in garnet analysis is expressed as Fe_2O_3.

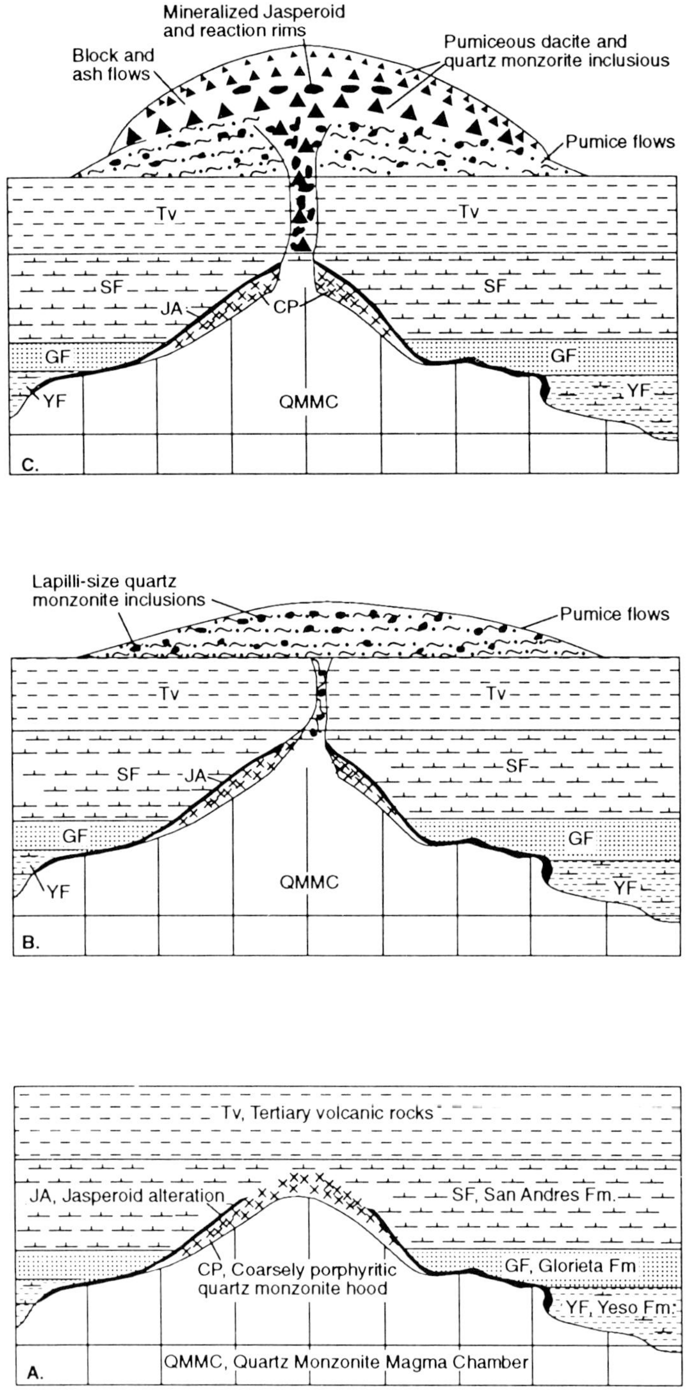

FIGURE 19. Diagrams showing the evolution of the Horse Springs dacite and the origin of the mineralized jasperoid accidental inclusions. A, Pre-eruption phase; B, Early eruptive phase, pumice flows; C, Late eruptive phase.

10. The apparent concentration of mineralized jasperoid xenoliths in the upper part of the zone in which the quartz monzonite blocks are largest and most concentrated suggests an eruptive stratigraphy progressing from the eruption of cognate quartz monzonite to jasperoidized wall rocks.

CONCLUSIONS

The observed field relationships, mineral phases, and textures suggest the following sequence of events in the evolution and eruption of the Horse Springs dacite and its inclusions:

1. Jasperoidization and iron and zinc mineralization of pre-Tertiary wall rocks and formation of iron oxides by hydrothermal solutions at the top of a dacitic magma chamber preceded eruption of the dacite pumice breccia (Fig. 19). Crystallization of the coarse-grained quartz monzonite may have occurred in the roof zone of the magma chamber at this time.

2. Volatile pressure in the magma chamber eventually exceeded the confining pressure and initiated eruption of pyroclastic pumice flows, which contain small accidental inclusions of quartz monzonite but no pre-Tertiary inclusions.

3. After the premonitory eruptions, continued volatiles exsolution increased the pressure in the magma chamber to the point that a massive failure of the chamber roof led to eruption of the block-and-ash flows and the entrainment of abundant shattered quartz monzonite roof rocks and lesser amounts of jasperoidized and mineralized pre-Tertiary wall-rock. Similarities in composition between pumiceous dacite blocks and more massive, but vesicular to miarolitic, quartz monzonite inclusions suggest gradation in the roof of the magma chamber between pumiceous dacite and porphyritic quartz monzonite.

4. During cooling of the dacite pumice breccia, reaction between the mineralized jasperoid xenoliths and the hot pumice breccia produced the diopside-rich skarn zones. Original magnetite and (or) other iron oxides in the jasperoid xenoliths may have been recrystallized to the ferric phases hematite, magnesioferrite and ferric-iron-rich forsterite at this time. Diopside and pseudobrookite, found in miarolitic cavities in the quartz monzonite, may also have formed then (according to Lindsley, 1965, pseudobrookite is unstable below 585°C). Montmorillonite clay, phlogopite and other minerals were formed by alteration of the pumice breccia at lower temperatures, as cooling of the pumice breccia proceeded.

A low to moderate potential for undiscovered resources of base and ferrous metals, particularly zinc and iron, and possibly associated silver and gold, is indicated by the anomalous metals (Table 7) in the mineralized jasperoid inclusions (Ratté et al., 1988). The most likely type of deposit might be a contact-metasomatic deposit accompanied by replacement and vein deposits related to intrusion of a silicic pluton into pre-Tertiary sedimentary rocks at shallow depths. Pre-Tertiary rocks may be within a few hundred meters of the surface in the vicinity of the probable source vent of the pumice breccia of Old Horse Springs, as indicated by outcrops of pre-Tertiary sedimentary rocks about 10 km northeast of the proposed vent. The zinc-rich middle Tertiary ore deposits in pre-Tertiary limestones in the Magdalena district, about 100 km east of the Horse Springs area, provide a possible model for the type of deposit that could be present beneath the surface here.

REFERENCES

Bornhorst, T.J., 1976, Volcanic geology of the Crosby Mountains and vicinity, Catron County, New Mexico [M.S. thesis]: Albuquerque, University of New Mexico, 113 p.

Bornhorst, T.J., 1980, Major and trace element geochemistry and mineralogy of upper Eocene to Quaternary volcanic rocks of the Mogollon-Datil volcanic field, southwestern New Mexico [Ph.D. dissertation]: Albuquerque, University of New Mexico, 1108 p.

Bornhorst, T.J. and Elston, W.E., 1978, Supplemental roadlog no. 8, Reserve to Magdalena via Horse Springs and Datil: New Mexico Geological Society, Special Publication 7, p. 99-106.

Cather, S.M., 1986, Volcano-sedimentary evolution and tectonic implications of the Datil Group (latest Eocene-early Oligocene), west-central New Mexico [Ph.D. dissertation]: Austin, University of Texas, 484 p.

Cather, S.M., McIntosh, W.C. and Chapin, C.E., 1987, Stratigraphy, age, and rates of deposition to the Datil Group (upper Eocene-lower Oligocene), west-central New Mexico: New Mexico Geology, v. 9, p. 50-54.

Elston, W.E., 1984, Mid-Tertiary ash flow tuff cauldrons, S.W. New Mexico: Journal of Geophysical Research, v. 89, no. B-10, p. 8733-8750.

Fisher, R.V. and Schminke, H.-U., 1984, Pyroclastic rocks: Springer-Verlag, 472 p.

Jones, D.P., 1980, Volcanic geology of the Alegres Mountain area, Catron County, New Mexico [M.S. thesis]: Albuquerque, University of New Mexico, 76 p.

Lindsley, D.H., 1965, Iron-titanium oxides: Carnegie Institute of Washington, 1964 Yearbook, p. 144-148.

Marvin, R.F., Naeser, C.W., Bikerman, M., Mehnert, H.H. and Ratté, J.C., 1987, Isotopic ages of post-Paleocene igneous rocks within and bordering the Clifton 1x2 degree quadrangle, Arizona-New Mexico: New Mexico Bureau of Mines and Mineral Resources, Bulletin 118, 50 p.

Osburn, G.R. and Chapin, C.E., 1983, Nomenclature for Cenozoic rocks of northeast Mogollon-Datil volcanic field, New Mexico: New Mexico Bureau of Mines and Mineral Resources, Stratigraphic Chart 1.

Ratté, J.C. and Modreski, P.J., 1989, Selected volcanic features of the western Mogollon-Datil volcanic field; New Mexico Bureau of Mines and Mineral Resources, Memoir 46, p. 68-80.

Ratté, J.C., McIntosh, W.C. and Houser, B.B., 1991, Geologic map of the Horse Mountain West quadrangle, Catron County, New Mexico: U.S. Geological Survey, Geologic Quadrangle Map GQ-1685, scale 1:24,000.

Ratté, J.C., Saltus, R. and Turner, R.L., 1988, Mineral resources of the Horse Mountain and Continental Divide Wilderness Study Areas, Catron County, New Mexico: U.S. Geological Survey, Bulletin 1734, 16 p.

Smith, R.L., 1960, Zones and zonal variations in welded ash flows: U.S. Geological Survey, Professional Paper 354-F, p. 149-159.

Stearns, C.E., 1962, Geology of the north half of the Pelona quadrangle, Catron County, New Mexico: New Mexico Bureau of Mines and Mineral Resources, Bulletin 78, p. 46.

Plains of San Agustin from southeast base of Horse Mountain, New Mexico. View is S22°E. Shaw Mountain on horizon, to right of center, consists of Bearwallow Mountain Andesite overlying Oligocene ignimbrites of the Mogollon Group. Unnamed hills left of center show wave-cut cliffs just above the plains. These cliffs are cut in thin ignimbrites of the Datil and Mogollon groups. Camera station is alongside Highway 12 in SE ¼ sec. 16, T4S, R12W. Altitude is about 2140 m. Wayne Lambert photograph No. 93L75. August 15, 1993, 12:14 p.m. MDT.

$^{40}Ar/^{39}Ar$ GEOCHRONOLOGY OF BASALTIC ROCKS AND CONSTRAINTS ON LATE CENOZOIC STRATIGRAPHY AND LANDSCAPE DEVELOPMENT IN THE RED HILL-QUEMADO AREA, NEW MEXICO

WILLIAM C. MCINTOSH and STEVEN M. CATHER
New Mexico Bureau of Mines and Mineral Resources, Socorro, New Mexico 87801

Abstract—Basaltic samples from the Red Hill-Quemado volcanic field of New Mexico and the White Mountains field of Arizona were dated by the $^{40}Ar/^{39}Ar$ incremental heating method to aid mapping and interpretation of the upper Tertiary sedimentary and volcanic rocks. The samples included 15 basaltic lavas, one volcanic bomb, and two shallow intrusive units. Step heating of groundmass concentrates from 17 of the samples yielded relatively flat age spectra from which accurate plateau or isochron ages could be determined. These results indicate two intervals of eruptive activity: 7.9-5.2 Ma and 2.5 Ma-71 ka. Vents for the late Miocene lavas define a narrow ENE-trending zone subparallel to the Jemez lineament. Vents for the late Pliocene-Pleistocene lavas occupy a broad N-S trending zone, within which $^{40}Ar/^{39}Ar$ results indicate south-to-north migration of volcanic activity. Late Pleistocene centers, including Red Hill (71 ka) and Zuni Salt Lake (86-114 ka), are confined to the northern end of the zone of Plio-Pleistocene activity; it is likely that future eruptions will occur in this area. Based in part on the ages reported here, several episodes of middle to late Cenozoic erosion and sedimentation can be discerned. The Red Hill-Quemado area aggraded until the end of Mogollon-Datil volcanism, about 24-26 Ma. The ensuing degradational regime continued until <14.5 Ma, when aggradation of the Fence Lake Formation began, and continued until about 7 Ma, although cutting and filling of local insets within the Fence Lake may have continued until about 5 Ma. Local late Miocene basaltic volcanism accompanied the waning stages of Fence Lake sedimentation. Post-Fence Lake erosion continued until 3-4 Ma, when aggradation of the Quemado Formation began. The end of Quemado deposition appears to have been diachronous, possibly ending about 4 Ma near Springerville, Arizona, about 2.5 Ma near Cow Springs Draw in western New Mexico, and <1 Ma near Red Hill, New Mexico. Late Pliocene-Pleistocene volcanism occurred during and after Quemado sedimentation. A regime of fluvial degradation has prevailed in most areas since the end of Quemado deposition.

INTRODUCTION

The Red Hill-Quemado volcanic field of western New Mexico (Fig. 1) is one of a series of late Cenozoic volcanic fields that define the Jemez lineament (Mayo, 1958; Aldrich, 1986). The Red Hill-Quemado field is probably the least studied of the volcanic fields along the Jemez lineament. Previous investigations of the late Cenozoic geology and volcanology were largely reconnaissance (Willard and Weber, 1958; Crumpler and Aubele, 1990; Chamberlin et al., 1994). The only detailed studies are those of Hoffer and Corbitt (1989), who described the volcanology of the Quemado crater, and Baldridge et al. (1989) who reported several K-Ar ages for Tertiary volcanic rocks in the Red Hill-Quemado area. Geothermal data and some additional K-Ar results were reported by Minier et al. (1988). No systematic investigation of the petrology or geochemistry of the volcanic rocks has yet been attempted. This preliminary report integrates $^{40}Ar/^{39}Ar$ ages for late Cenozoic volcanic rocks of the Red Hill-Quemado field with the sedimentologic history and landscape development recently elucidated by reconnaissance mapping by Chamberlin et al. (1994) and by geological reconnaissance in eastern Arizona by the junior author.

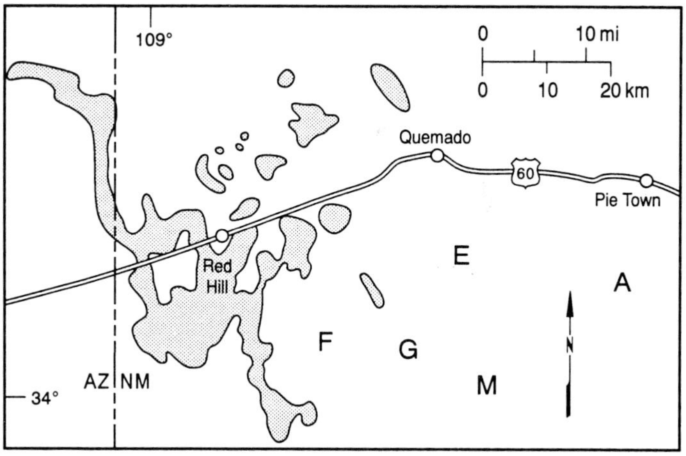

FIGURE 1. Map showing generalized distribution of late Miocene to Pleistocene volcanic rocks in the Red Hill-Quemado volcanic field and selected landforms, after New Mexico Geological Society (1982) and Reynolds (1988). **A**, Alegres Mountain; **E**, Escondido Mountain; **M**, Mangas Mountains; **G**, Gallo Mountains; **F**, Fox Mountain.

DATING METHODS

Hand samples, typically 1-2 kg of unweathered lava, were collected from 18 localities. Wherever possible, samples were collected from holocrystalline flow interiors, avoiding glassy, vesicular, or brecciated flow margins or areas of hydrothermal alteration. Groundmass concentrates of samples were prepared by crushing, then sieving to 250-500 μm. After ultrasonic cleaning in dilute nitric acid and removal of phenocrysts by Frantz magnetic separator and hand picking, 40-60 mg aliquots of concentrate were packaged in tinfoil and sealed in quartz irradiation tubes with alternating flux monitors of Fish Canyon Tuff sanidine (27.84 Ma). The sample-filled quartz tubes were then irradiated for one to five hours in the L67 position of the Ford reactor at the University of Michigan.

Analyses were performed at the New Mexico Geochronology Research Laboratory at New Mexico Tech in Socorro, New Mexico. This facility includes an MAP 215-50 mass spectrometer attached to a computer-automated all-metal argon extraction line. Monitor sanidines were fused by CO_2 laser for 15 seconds, and reactive gases were removed using a SAES GP-50 getter prior to expansion into the mass spectrometer. Extraction line blanks during these laser analyses ranged from 5×10^{-17} to 2×10^{-16} moles ^{40}Ar and 5×10^{-19} to 2×10^{-18} moles ^{36}Ar. The pooled results of four three-to-four crystal subsamples (approximately 1 mg) of each monitor allowed the neutron flux (J-values) along the irradiation tubes to be determined to a precision of ±0.25%. A double-vacuum resistance furnace was used to heat the unknown samples. Samples were preheated to 500°C to degas the tin foil package, and then samples were incrementally heated in eight to ten steps from 650°C to 1650°C. Heating times of eight to ten minutes were followed by five to ten minutes of cleanup with SAES AP-10 and GP-50 getters. Data from each heating step were corrected for extraction line blank, which for furnace analyses ranged from 3 x10^{-16} to 3 x 10^{-15} moles ^{40}Ar and from 2 x 10^{-18} to 9 x 10^{-18} moles ^{36}Ar.

RESULTS

Results of step-heating analyses from each sample are displayed as age spectra in Appendix 1, are detailed in Appendix 2, and ages are summarized in Table 1. Age spectra are accompanied by plots of apparent K/Ca (a function of measured $^{37}Ar/^{39}Ar$) and radiogenic yield (the percent of measured ^{40}Ar that cannot be attributed to trapped argon of atmospheric composition, i.e. $^{40}Ar/^{36}Ar$=295.5).

Of the 18 samples analyzed (Table 1, Appendix 1), ten yielded age spectra that satisfy age "plateau" criteria of three or more contiguous steps, totaling more than 50% of ^{39}Ar released, that agree within analytical error at the 95% confidence level. With the exception of some low-temperature or high-temperature steps with large uncertainties related to low radiogenic yield and low signal-to-blank ratios (e.g. Appendix 1: 2b), most individual steps have analytical uncertainties of approximately ±10-50 ka (1σ). Plateau ages were calculated for each of these ten samples by weighting steps by the inverse of their variance (Appendix 1, 2).

Seven samples have age spectra with relatively flat segments which approach, but do not meet plateau criteria. Of these samples, two (nm1158, Appendix 1:d; nm1108, Appendix 1:h) gave spectra with flat segments constituting less than 50% of the ^{39}Ar released and the rest yielded spectra with central segments that are slightly climbing (nm1160, Appendix 1:k) or falling (nm1111, Appendix 1:j; nm1159, Appendix 1:n; SMC-1, Appendix 1:p; SMC-5, Appendix 1:q). These slightly disturbed spectra may be related to alteration, interstitial glass, or minor recoil of ^{39}Ar from fine-grained groundmass minerals during irradiation. For each of the seven samples that have slightly disturbed spectra, mean ages were calculated as the inverse-variance-weighted average of the heating steps that comprise the nearly flat parts of the age spectra. Sample nm1155 (Appendix 1:e) yielded a highly disturbed spectrum that fell far short of meeting plateau criteria; this sample is rejected from further consideration other than as indicating an approximate age near 1.0 Ma.

The K/Ca and radiogenic-yield plots show common behavior. Typically, radiogenic yields start low, increase to maxima at heating steps of 800-1000°C, then decrease at higher temperatures (Appendix 1). K/Ca ratios generally decline steadily with increasing step-heating temperatures. These behaviors are probably related to initial degassing of atmospheric contamination from interstitial glass and intergrain boundaries, followed by degassing dominated by groundmass plagioclase, and ending with increasing degassing of retentive, low-K groundmass pyroxene. This behavior of radiogenic yield and K/Ca ratios during step heating, as well as the general form of the age spectra, are similar to results reported from young alkali basalts elsewhere (e.g. Laughlin et al., 1993).

The age spectra in Appendix 1 are accompanied by inverse isochron diagrams, which offer an alternative means of calculating ages from step-heating data, as well as a means of testing the assumption, implicit in the apparent ages presented in each age spectrum, that trapped, non-radiogenic argon components have atmospheric composition ($^{40}Ar/^{36}Ar$ = 295.5). The isochron ages for all samples are within 1 σ of the plateau age or the mean age of near-plateau steps. Mean standard weighted deviation (MSWD) values for the isochrons range from 1.2 to 11.4 for the samples that satisfied plateau criteria and from 2.8 to 44.7 for the samples which nearly satisfy plateau criteria (Appendix 1, 2). These relatively high MSWD values indicate that much of the scatter cannot be attributed solely to analytical error, and may reflect minor alteration or heterogeneous trapped initial argon compositions. $^{40}Ar/^{36}Ar$ intercept values range from 290 to 300, but most are not significantly different (±2 σ) from atmospheric argon. These isochron results indicate that excess argon is not a significant problem, and support the interpretation that the plateau ages, mean ages of near-plateau steps, or isochron ages represent the time of crystallization of the sampled units.

Our $^{40}Ar/^{39}Ar$ results are in reasonable accord with K-Ar data reported by Baldridge et al. (1989) and Dethier et al. (1986), which include late Miocene 6.4-6.7-Ma ages for mesa-capping basaltic lavas in the Quemado-Red Hill volcanic field. Our results are also in accord with the K-Ar results for older basalts of Minier et al. (1988), who similarly indicated late Miocene 5.7-6.7-Ma ages for some of the mesa-capping lavas. Minier et al. (1988), however, indicated mid-Pleistocene (1.03 ± 0.27 to 1.23 ± 0.23 Ma) ages for Red Hill and Blaines Lake lavas, in

TABLE 1. Summary of $^{40}Ar/^{39}Ar$ ages for Red Hill-Quemado volcanic rocks.

Unit	Sample #	Age Type	Age (Ma)	± Err(1s)
lava, Red Hill cinder cone	nm1161	plateau	0.071	0.012
ring intrusive, Zuni Salt Lake	nm1107	plateau	0.086	0.031
bomb, Zuni Salt Lake	nm1106	plateau	0.114	0.038
lava, hornito fissure, NE of Blaines Lake	nm1158	isochron	0.206	0.034
lava, overlying Quemado Fm., Ponderosa Tank quad	nm1155	disturbed spectrum	ca. 1.0	
lava, interbedded with Quemado Fm., E of Blaines Lake	nm1157	plateau	0.970	0.014
lava, interbedded with Quemado Fm., Rt. 60, SE of Blaines Lk.	nm1156	plateau	1.51	0.01
lava, wall of Quemado Crater	nm1108	isochron	1.55	0.01
lava, S of Cerro la Mula	nm1110	plateau	1.95	0.05
lava, N of Mangitas Flat, Black Peak quad	nm1111	isochron	2.05	0.02
lava, overlying Bidahochi Fm., N of Mesa Parada, AZ	nm1160	isochron	2.37	0.03
lava, overlying Quemado Fm., US 60, E of NM/AZ border	SMC-3	plateau	2.46	0.04
lava, interbedded in Fence Lake Fm., US 60, W of Red Hill	SMC-2	plateau	5.20	0.03
lava, mesa capping, N of Blaines Lake	nm1159	isochron	6.08	0.04
lava, capping SW edge of Cimarron Mesa	SMC-4	plateau	6.05	0.02
lava, overlying Fence Lake Fm., E of Cow Springs Draw	SMC-1	isochron	6.03	0.02
lava, Flat Top, S of Springerville, AZ (White Mountains field)	SMC-5	isochron	6.80	0.02
intermediate intrusive body, El Portocito	nm1104	plateau	7.92	0.20

sharp contrast to our ages at 0.07 ± 0.01 and 0.206 ± 0.03 Ma, respectively. We cannot offer a specific explanation for why these K-Ar data differ so much from the $^{40}Ar/^{39}Ar$ results for the young basalts, but note that the $^{40}Ar/^{39}Ar$ method is inherently more precise than K-Ar, offers means of testing whether Ar systematics have been disturbed (e.g. excess argon and argon loss), and has in numerous studies been shown to be more accurate than K-Ar for dating of volcanic rocks (e.g. McIntosh et al., 1990). Based on the observed discordance with our $^{40}Ar/^{39}Ar$ results, we regard the younger (Plio-Pleistocene) ages reported by Minier et al. (1988) as incorrect.

CONSTRAINTS ON STRATIGRAPHY AND LANDSCAPE DEVELOPMENT

Table 1 and Figure 2 summarize results for the 17 Red Hill-Quemado samples discussed above; the plateau ages are considered best estimates of eruptive age for the ten samples that satisfy plateau criteria, and isochron ages are used for the seven samples that closely approach plateau criteria. These ages range from 7.92 Ma to 71 ka and fall into two distinct eruptive intervals: 7.9-5.2 Ma and 2.5 Ma–71 ka. The volcanic rocks within these two eruptive intervals occupy distinct landscape positions within the study area and constrain the timing of episodes of erosion and sedimentation.

In the area south of Quemado, the topographically highest landforms (Alegres Mountain, Escondido Mountain, Mangas Mountain, Gallo Mountains, Fox Mountain; Fig. 1) consist of an aggradational sequence of mid-Tertiary volcanic and volcaniclastic rocks in the northwest part of the Mogollon-Datil volcanic field that are locally as young as 24-26 Ma (Bearwallow Mountain Andesite). In the topographically lower areas to the north, near US-60 in western New Mexico and eastern Arizona, dated basalts of late Miocene age (7.9-5.2 Ma) cap many of the topographically most prominent landforms. These landforms include Flat Top, Cimarron Mesa, Tejana Mesa, and the unnamed mesa east of Cow Springs Draw (Figs. 3, 4). Basalts capping other high mesas in the area (Mesa Redonda, unnamed mesas west of Tejana Mesa) have not yet been dated, but appear to be correlative with the dated upper Miocene basalts on the basis of concordance of elevation, flow thickness and hand-specimen petrology. The upper Miocene basalts generally overlie the Fence Lake Formation, and thus constrain its upper age limit (about 5-7 Ma). The lower age limit of the Fence Lake is about 14.5 Ma or younger, based on vertebrate remains (Lucas and Anderson, this volume). Possible inset relations within the Fence Lake are indicated by the landscape position of a 5.2-Ma basalt within the Fence Lake Formation west of Red Hill Draw (Fig. 3).

The Plio-Pleistocene basalts (2.5 Ma-71 Ka) occupy landscape positions inset below the late Miocene basalts (Fig. 4), but above the level of the modern drainages. Two basalt flows in eastern Arizona near Nutrioso Creek (Fig. 4) were K-Ar dated by Laughlin et al. (1979, 1980) at 3.87 and 3.67 Ma. These flows are topographically

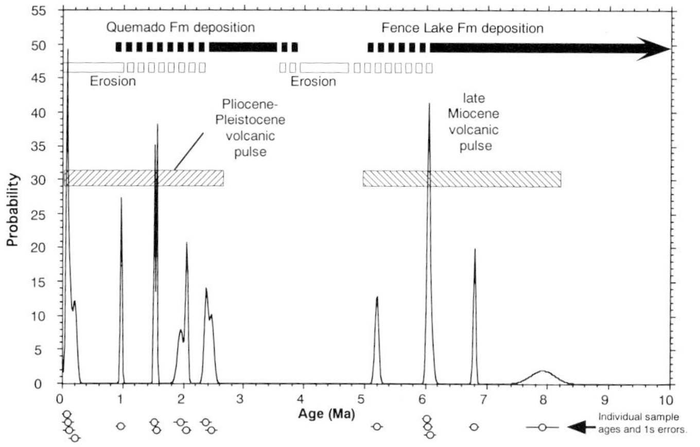

FIGURE 2. Ideogram showing age distribution of dated samples and inferences about timing of volcanic activity, aggradation and erosion.

lower than, and apparently inset against, the 6.80-Ma basalt that caps nearby Flat Top.

Plio-Pleistocene basalts locally interfinger with and cap the Quemado Formation (Cather and McIntosh, this volume). At the type section in the Cow Springs Basin, the Quemado is capped by Pliocene basalt (about 2.5 Ma). In the Nutrioso Basin of eastern Arizona, thin gravelly sequences possibly correlative with the Quemado are capped by lower Pliocene basalt (3.87-3.67 Ma, Fig. 4). In the Red Hill Basin east of Cimarron Mesa, the Quemado Formation is interbedded with basalts as young as 0.970 Ma (Fig. 4), significantly younger than in basins to the west. In the Largo Basin, the Quemado can only be constrained to be younger than the 6.73 ± 0.18 Ma basalt that caps Tejana Mesa (K-Ar date of Dethier et al., 1986). The age of the Quemado in the Omega Basin to the east is poorly constrained due to lack of fossils or interbedded lavas.

Based on the relationships summarized above, it is apparent that a regime of net degradation, punctuated by at least two episodes of fluvial aggradation, has prevailed in western New Mexico and eastern Arizona since the late Oligocene. Initial degradation began after eruption of the Bearwallow Mountain Andesite (24-26 Ma), and continued until aggradation of the Fence Lake Formation began sometime after 14.5 Ma. Culminal aggradation of the Fence Lake occurred about 7 Ma, although local cutting and filling of insets within the Fence Lake may have continued until about 5 Ma (Fig. 4), when regional degradation continued following the cessation of late Miocene volcanism. Aggradation again ensued at 3-4 Ma, resulting in deposition of the Quemado Formation within incised paleovalley/basin systems. Quemado deposition ceased diachronously from west to east, ending about 4 Ma for possible Quemado correlatives near Springerville, about 2.5 Ma for the Quemado in the Cow Springs Basin, and <1 Ma for Quemado deposits in the Red Hill Basin. Age constraints for the cessation of Quemado deposition in the Largo Basin and the Omega Basin are lacking. The apparent easterly decrease in the age of Quemado deposition may be due to the proximity of the western outcrops to the trunk stream of the Little Colorado River, where headward migration of erosion and degradation may have been manifested earlier than in the more distal basins to the east. Alternatively, disruption of drainage by voluminous late Pliocene-Pleistocene volcanism in the Cow Springs and Red Hill Basins may have prolonged deposition in these areas. A degradational regime continues today in most areas; the youngest dated flows in the area (Red Hill flow, 71 ka; unnamed flow north of Blaines Lake, 206 ka) are topographically a few meters higher than modern drainages.

SPACE-TIME VARIATIONS IN MAGMATISM

In their reconnaissance mapping of the Quemado 30'x 60' quadrangle, Chamberlin et al. (1994) recognized over 40 volcanic vents of late Cenozoic age. These vents are restricted to the Red Hill-Quemado field in the western part of the quadrangle, between Quemado and the New Mexico-Arizona state line (Fig 5). We divide these vents into two groups (late Miocene and Plio-Pleistocene) on the basis of radioisotopic constraints on vents or associated lavas, or by topographic and petrologic correlation with adjacent, dated flows. The late Miocene vents are restricted to a strikingly narrow and linear zone, which begins at Tejana Mesa and Cerro Prieto (a volcanic neck about 13 km north of Tejana Mesa, dated 6.0 ± 0.5 Ma [K-Ar, Baldridge et al., 1989] and 5.98 ± 0.14 [K-Ar, Minier et al., 1988]) and extends southwestward toward the late Miocene eruptive centers in the White Mountains of Arizona (Merrill and Péwé, 1977). About half of the late Miocene vents are closely associated with known faults or fracture zones. No age trends are apparent within the late Miocene vent zone.

Plio-Pleistocene vents occupy a broad zone that trends southward from an apex near Zuni Salt Lake (Fig. 5) to an ill-defined terminus in the area south of Black Peak. The Plio-Pleistocene vent zone is much broader than, and trends athwart, the late Miocene vent zone. Although several Plio-Pleistocene vents are closely associated with fracture systems, the Plio-Pleistocene vents are not as closely associated with known fractures as are the late Miocene vents. In contrast to the late Miocene eruptive centers, Plio-Pleistocene vents become distinctly

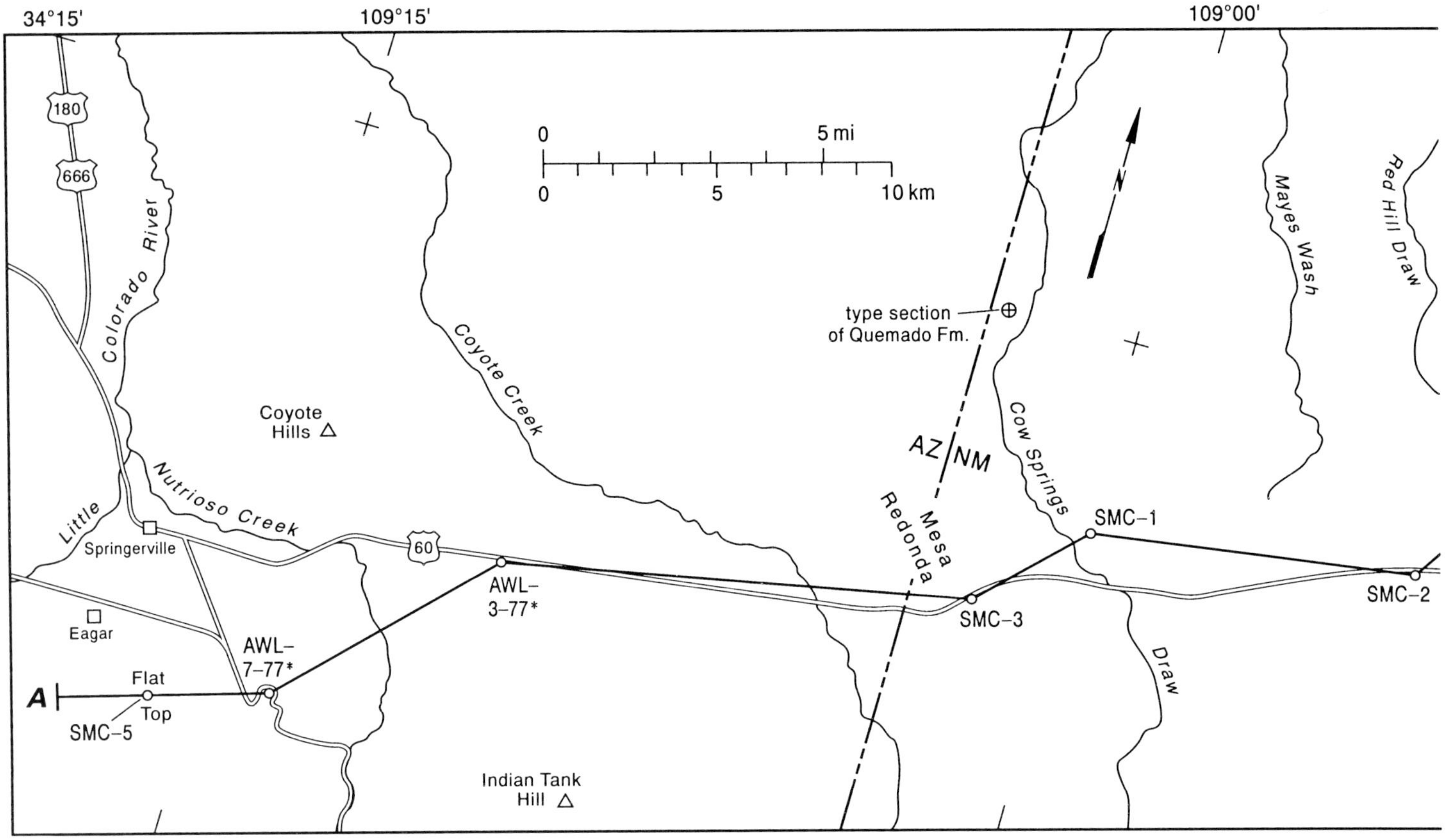

FIGURE 3. Map showing selected drainages and landforms in west-central New Mexico and eastern Arizona, sample numbers and locations, and location of cross section A-A' (Fig. 4). All sample localities are from this study except * = Laughlin et al. (1979), 1980; ** = Dethier et al. (1986).

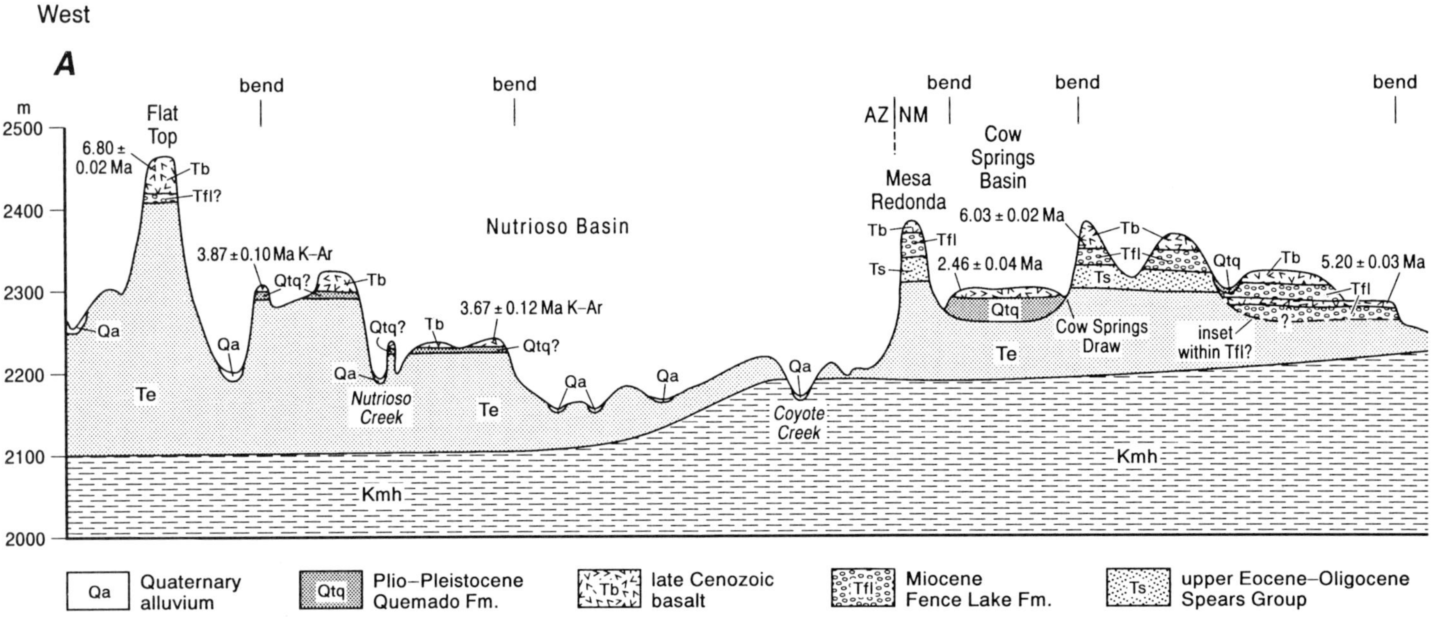

FIGURE 4. Cross section A-A' showing stratigraphy and landscape positions of dated basalts. Location of cross section shown in Figure 3. Geology based on Chamberlin et al. (1994) and on reconnaissance in Arizona by junior author. All depicted dates are from this study except the 3.87 Ma and 3.67 Ma dates (K-Ar) near Nutrioso Creek (Laughlin et al., 1979, 1980) and the 6.73 Ma date (K-Ar) on Tejana Mesa (Dethier et al., 1986).

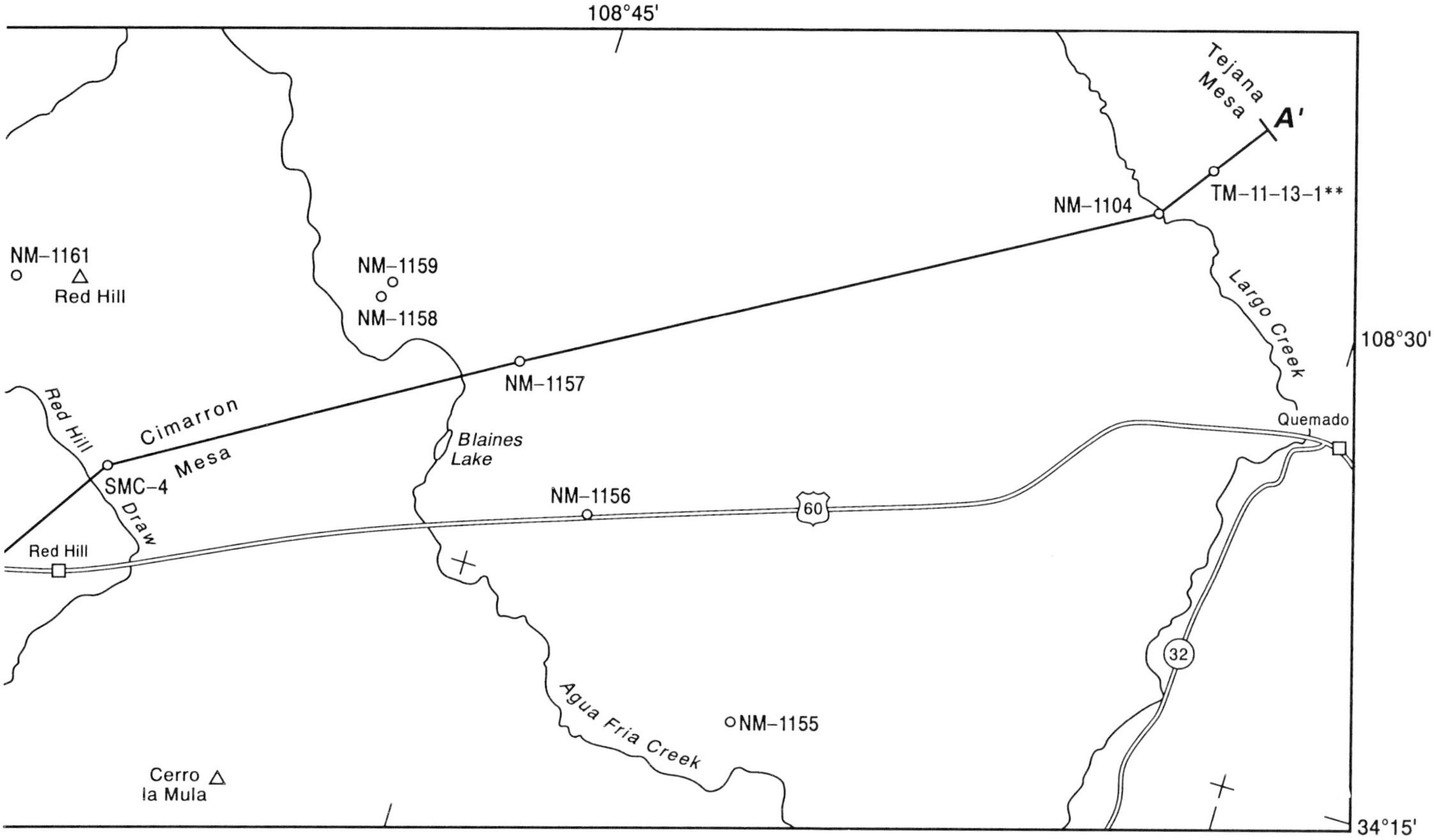
108°45'
Tejana Mesa
A'
TM-11-13-1**
NM-1104
NM-1161
Red Hill
NM-1159
NM-1158
Largo Creek
108°30'
NM-1157
Red Hill
Cimarron
Mesa
SMC-4
Draw
Blaines Lake
Quemado
NM-1156
60
Red Hill
32
Agua Fria Creek
NM-1155
Cerro la Mula
34°15'

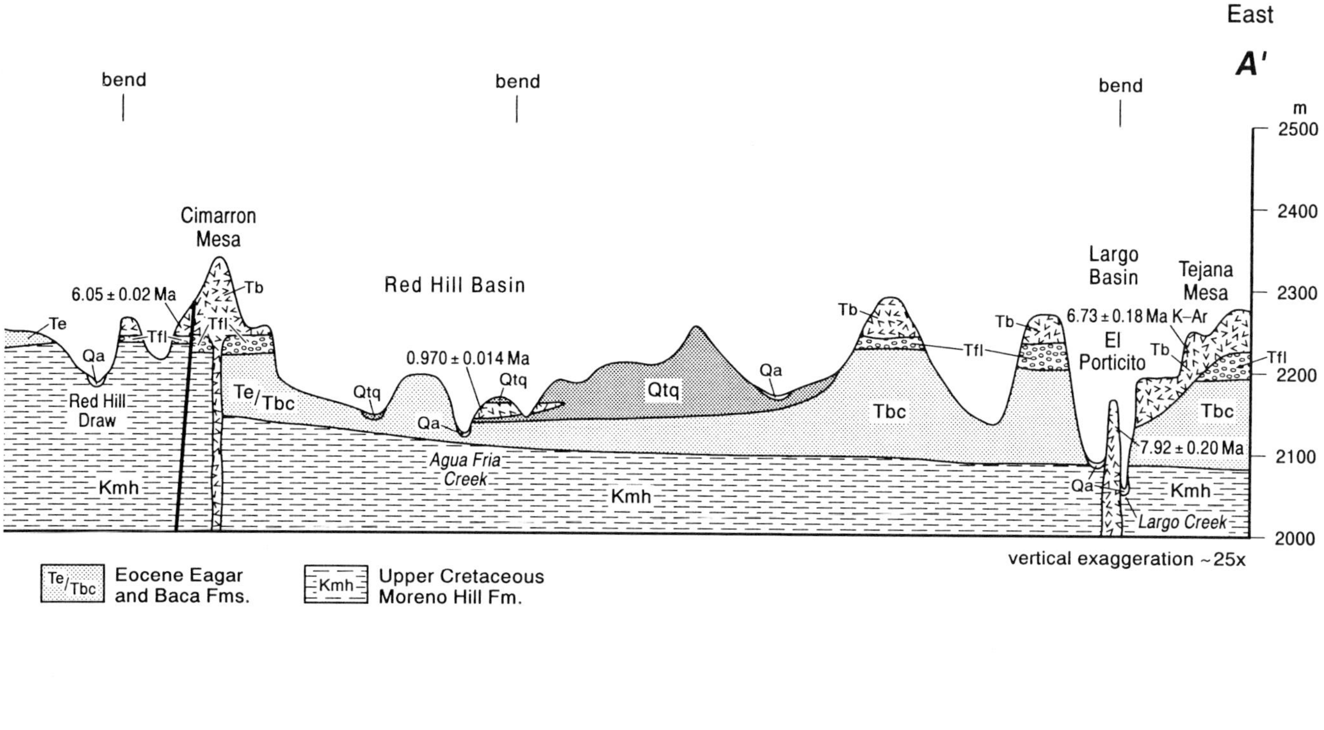
East
A'
bend
bend
bend
m
2500
2400
2300
2200
2100
2000
Cimarron Mesa
6.05 ± 0.02 Ma
Tb
Te
Tfl
Qa
Red Hill Draw
Te/Tbc
Kmh
Red Hill Basin
0.970 ± 0.014 Ma
Qtq
Qa
Agua Fria Creek
Tbc
Largo Basin
Tejana Mesa
6.73 ± 0.18 Ma K-Ar
El Porticito
7.92 ± 0.20 Ma
Largo Creek
vertical exaggeration ~25x
Te/Tbc Eocene Eagar and Baca Fms.
Kmh Upper Cretaceous Moreno Hill Fm.

younger to the north. In the southern part of Figure 5 at the latitude of Black Peak, dates range from 1.9 to 2.5 Ma (our sample nm1111; samples 420 and 429 of Baldridge et al., 1989). Plio-Pleistocene basalts at the latitude of US-60 range from 0.97 to 1.95 Ma (our samples nm1155, nm1157, nm1156, nm1108, nm1110). In the northern part of the Plio-Pleistocene vent zone, samples range from 71 ka to 206 ka (our samples nm1161, nm1107, nm1106, nm1158). These age trends and the youthfulness of the northern vents suggest that locus of volcanism is migrating northward and that future eruptions may be expected in this area.

TECTONIC IMPLICATIONS

The late Cenozoic tectonic history of the Red Hill-Quemado area is characterized by extension; all of the faults depicted in Figure 5 are normal. For the purpose of the following analysis, all faults are assumed to be dip-slip. Although no systematic kinematic analysis has been attempted, observations by R. M. Chamberlin (oral commun., 1994) during mapping of the eastern part of the Quemado 30' x 60' quadrangle indicates a dominance of dip-slip on late Cenozoic faults, based on slickenlines and fault offsets on dikes.

Most of the late Cenozoic normal faults in the Red Hill-Quemado area strike northeast (average about N35°E). In contrast, the narrow zone of late Miocene vents trends east-northeast (about N60°E), subparallel to the average trend of the Jemez lineament (N52°E, Laughlin et al., 1982). At least two northeast-striking faults controlled the location of eruptive centers within the late Miocene vent zone; these faults are exposed on Cimarron and Tejana Mesas (Fig. 5). Assuming dip-slip behavior, the en echelon arrangement of faults within the vent zone implies left-oblique transtension across the zone in the late Miocene. The obliquity between the trend of the late Miocene vent zone and the strike of the normal faults within it is similar to the obliquity between orientations of the Jemez lineament and faults in the Zuni-Bandera field (Laughlin et al., 1982).

Obliquity also exists between the northerly trend of the Plio-Pleistocene vent zone and the northeast strike of the faults within it (Fig. 5). If dip-slip is assumed on these en echelon faults, then the Plio-Pleistocene vent zone comprised an area of right-oblique transtension. Although volcanic rocks within the vent zone become younger to the north, a similar trend for ages of faulting is not apparent. Indeed, the most youthful-appearing fault scarps occur in the central part of the Plio-Pleistocene vent area, near Quemado crater and Cerro la Mula.

ACKNOWLEDGMENTS

The paper was improved by reviews from Matt Heizler and Dave Love. We thank Matt Heizler for assistance in the lab. The manuscript was typed by Lynne Hemenway and figures were drafted by Becky Titus.

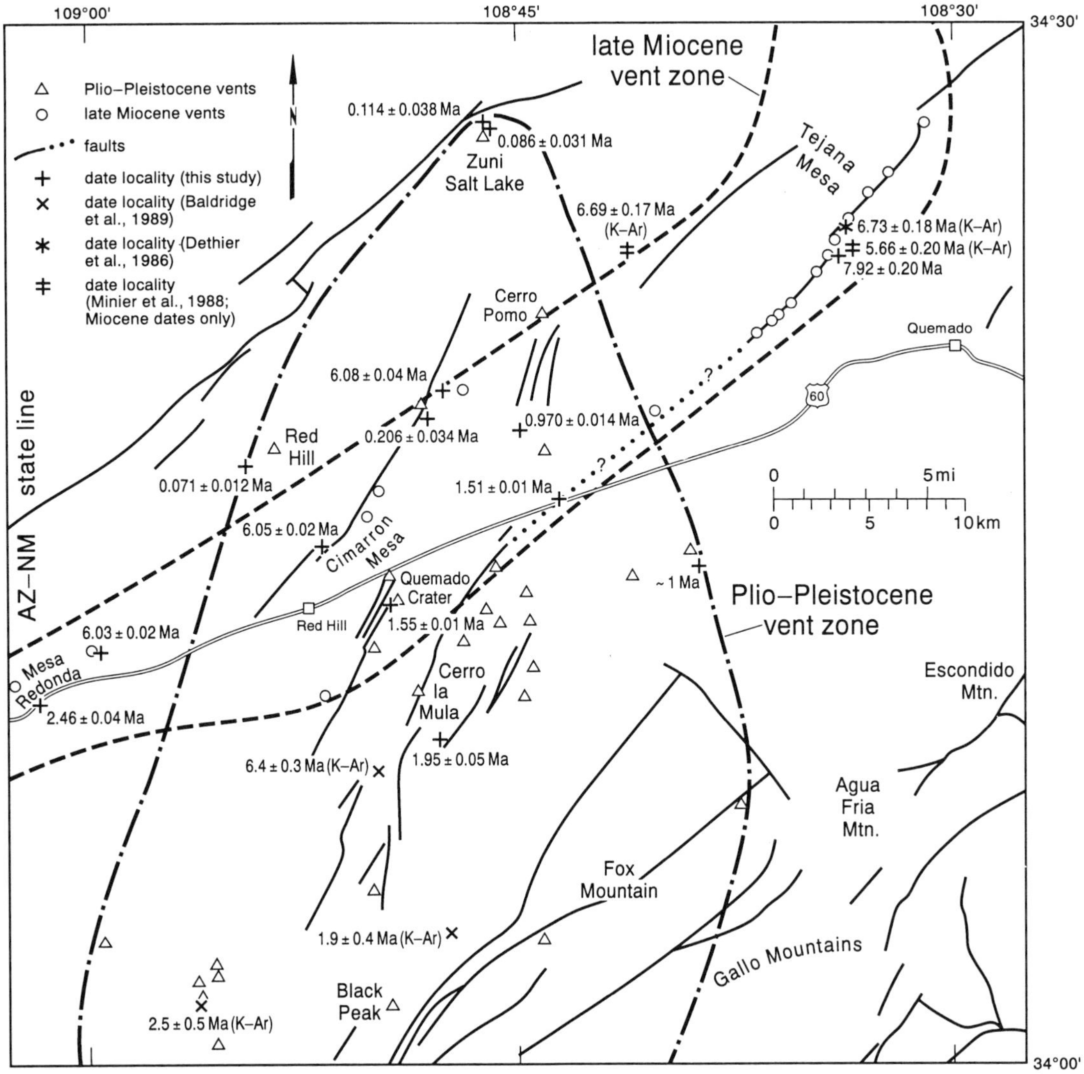

FIGURE 5. Map showing distribution of late Miocene vents and Plio-Pleistocene vents, faults, date localities, and selected landforms in the Red Hill-Quemado area (after Chamberlin et al., 1994).

REFERENCES

Aldrich, M. J. Jr., 1986, Tectonics of the Jemez lineament in the Jemez Mountains and Rio Grande rift: Journal of Geophysical Research, v. 91, p. 1753-1762.

Baldridge, W. S., Perry, F. V., Vaniman, D. T., Nealey, L. D., Leavy, B. D., Laughlin, A. W., Kyle, P., Bartov, Y., Steinitz, G. and Gladney, E. S., 1989, Magmatism associated with lithospheric extension: middle to late Cenozoic magmatism of the southeastern Colorado Plateau and central Rio Grande rift, New Mexico and Arizona: New Mexico Bureau of Mines and Mineral Resources, Memoir 46, p. 187-230.

Cather, S. M. and McIntosh, W. C., 1994, The Plio-Pleistocene Quemado Formation of west-central New Mexico: New Mexico Geological Society, Guidebook 45.

Chamberlin, R. M., Cather, S. M., Anderson, O. J. and Jones, G. E., 1994, Reconnaissance geologic map of the Quemado 30′x 60′ quadrangle, Catron County, New Mexico: New Mexico Bureau of Mines and Mineral Resources, Open-file Report 406, scale 1:100,000.

Crumpler, L. S. and Aubele, J. C., 1990, Red Hill volcanic field, New Mexico; in Wood, C. A. and Kienle, J., eds., Volcanoes of North America: Cambridge University Press, p. 306-308.

Dethier, D. P., Aldrich, M. J., Jr. and Shafiqullah, M., 1986, New K-Ar ages for Miocene volcanic rocks from the northeastern Jemez Mountains and Tejana Mesa, New Mexico: Isochron/West, no. 47, p. 12-14.

Hoffer, J. M. and Corbitt, L. L., 1989, A preliminary note on some unusual explosion-collapse craters, Quemado basalt field, Catron County, New Mexico: New Mexico Geological Society, Guidebook 40, p. 157-158.

Laughlin, A. W., Brookins, D. G., Damon, P. E. and Shafiqullah, M., 1979, Late-Cenozoic volcanism of the central Jemez zone, Arizona-New Mexico: Isochron/West, no. 25, p. 5-8.

Laughlin, A. W., Damon, P. E. and Shafiqullah, M., 1980, New K-Ar dates from the Springerville volcanic field, central Jemez zone, Apache County, Arizona: Isochron/West, no. 29, p. 3-4.

Laughlin, A. W., Aldrich, M. J., Jr., Ander, M. E., Heiken, G. H. and Vaniman, D. T., 1982, Tectonic setting and history of late-Cenozoic volcanism in west-central New Mexico: New Mexico Geological Society, Guidebook 33, p. 279-284.

Laughlin, A. W., Perry, F. V., Damon, P. E., Shafiqullah, M., WoldeGabriel, G., McIntosh, W., Harrington, C. D., Wells, S. G. and Drake, P. G., 1993, Geochronology of Mount Taylor, Cebollita Mesa, and Zuni-Bandera volcanic fields, Cibola County, New Mexico: New Mexico Geology, v. 15, p. 81-92.

Lucas, S. G. and Anderson, O. J., 1994, Miocene Proboscidean from the Fence Lake Formation, Catron County, New Mexico: New Mexico Geological Society, Guidebook 45.

McIntosh, W. C., Sutter, J. F., Chapin, C. E. and Kedzie, L. L., 1990, High-precision $^{40}Ar/^{39}Ar$ sanidine geochronology of ignimbrites in the Mogollon-Datil volcanic field, southwestern New Mexico: Bulletin of Volcanology, v. 52, p. 584-601.

Mayo, E. B., 1958, Lineament tectonics and some ore districts of the Southwest: American Institute of Mining Engineers Transactions, v. 10, p. 1169-1175.

Merrill, R. K. and Péwé, T. L., 1977, Late Cenozoic geology of the White Mountains, Arizona, Arizona Bureau of Geology and Mineral Technology, Special Paper 1, 65 p.

Minier, J., Reiter, M., Shafiqullah, M. and Damon, P. E., 1988, Geothermal studies in the Quemado area, New Mexico: Geothermics, v. 17, p. 735-755.

New Mexico Geological Society, 1982, New Mexico highway geologic map: scale 1:1,000,000.

Reynolds, S. J., 1988, Geologic map of Arizona: Arizona Geological Survey, Map 26, scale 1:1,000,000.

Willard, M. E. and Weber, R. H., 1958, Reconnaissance geologic map of Cañon Largo thirty-minute quadrangle: New Mexico Bureau of Mines and Mineral Resources, Geologic Map 6, scale 1:126,720.

APPENDIX 1

Age spectra and isochron diagrams for samples dated by the $^{40}Ar/^{39}Ar$ incremental heating method. Letters a to r correspond to sample number 1 to 18 in Table 1.

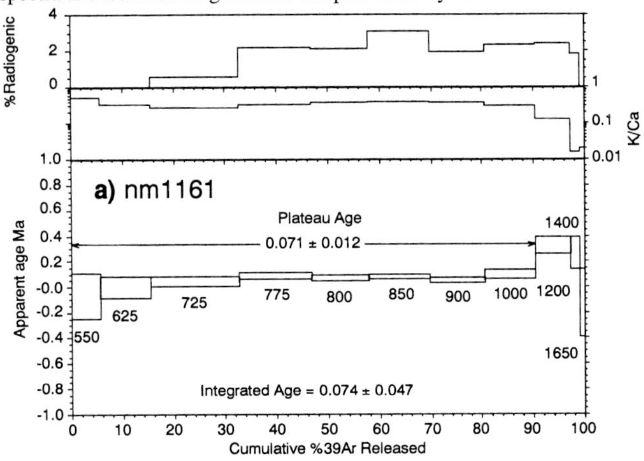

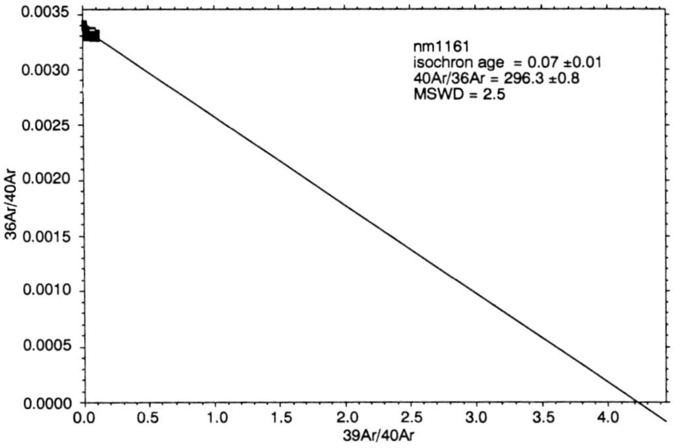

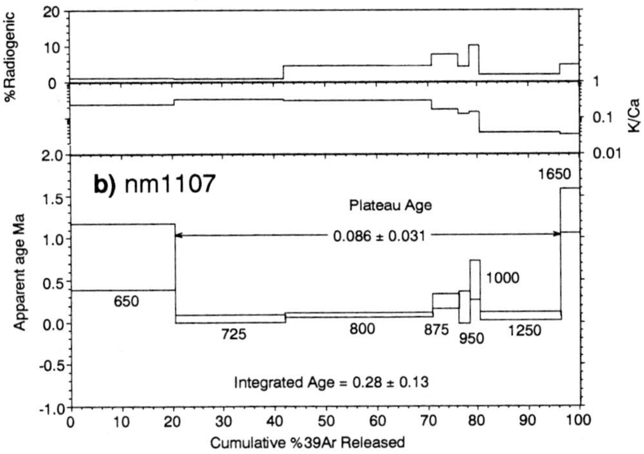

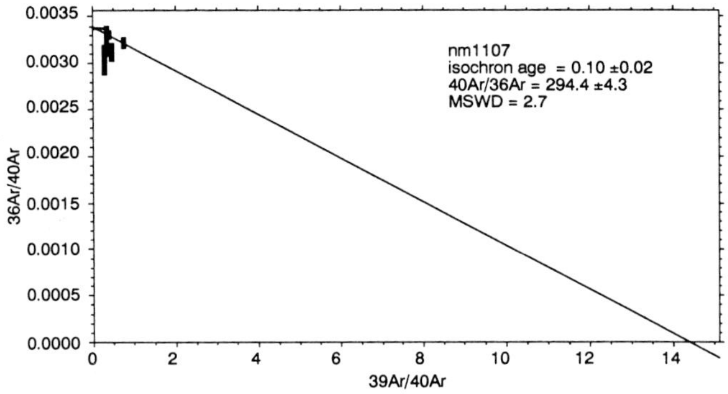

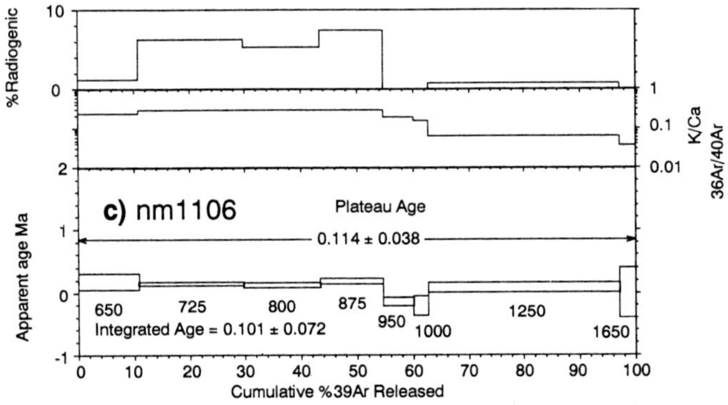

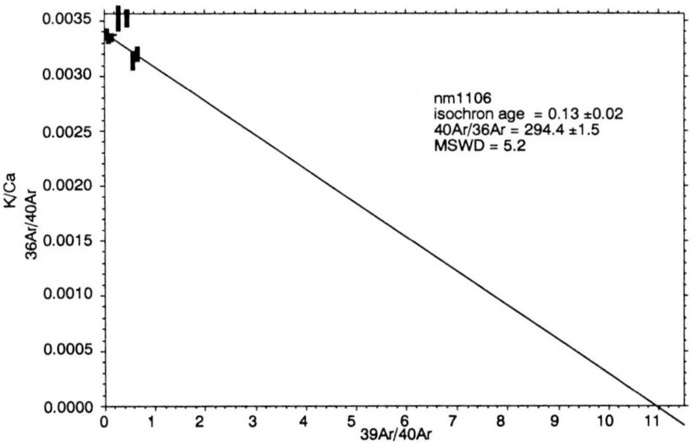

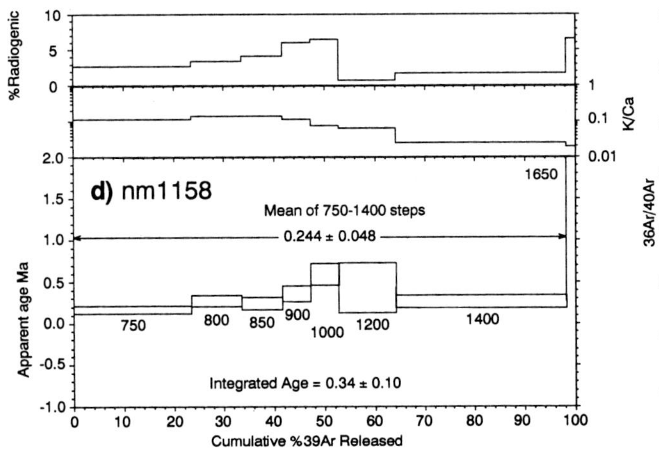

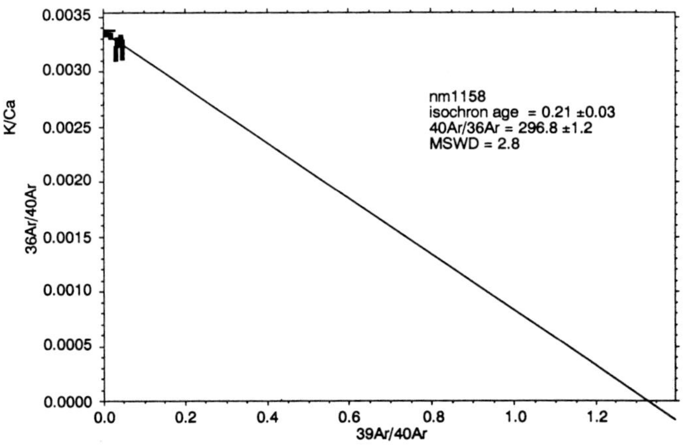

APPENDIX 1 continued

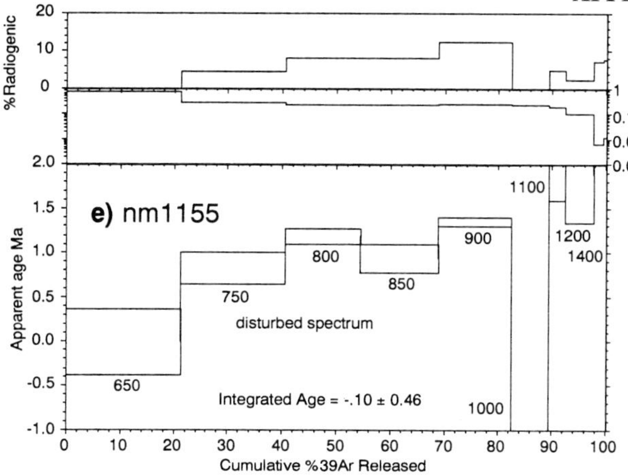

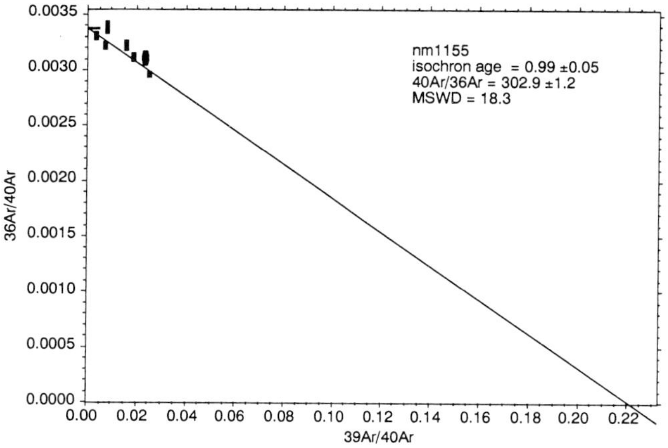

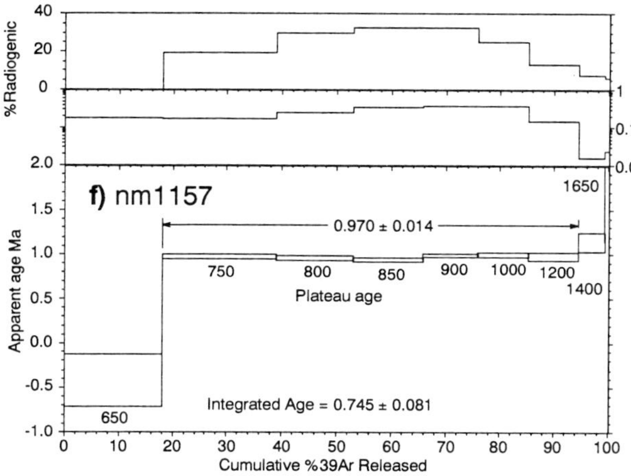

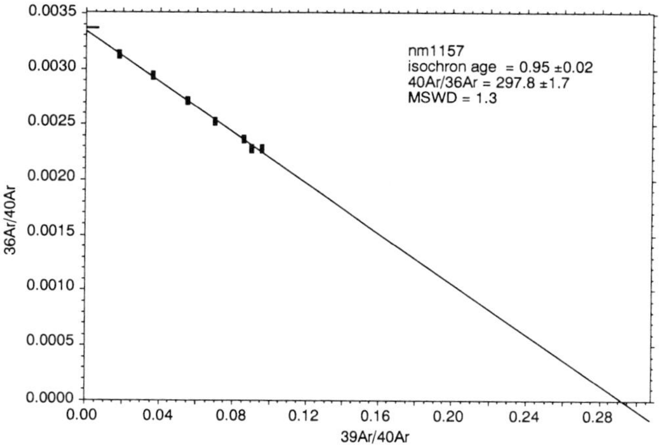

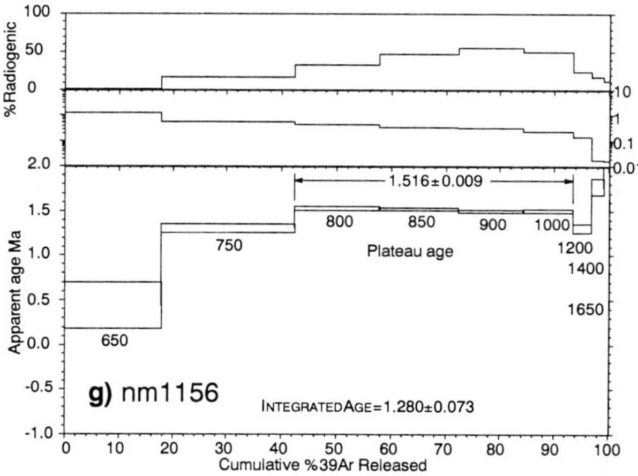

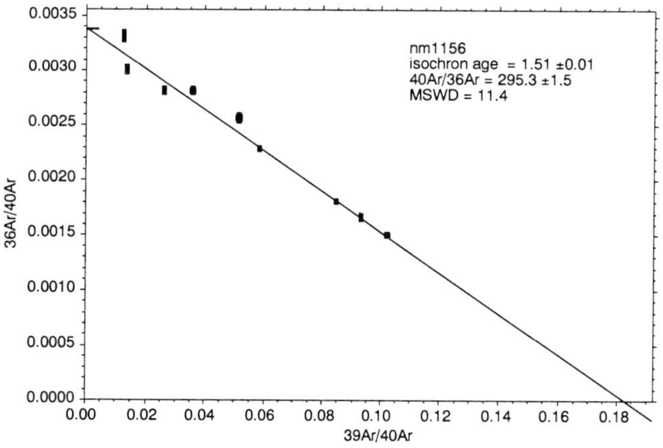

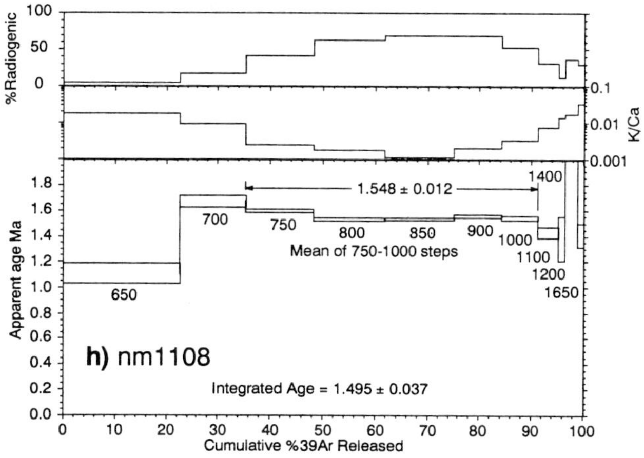

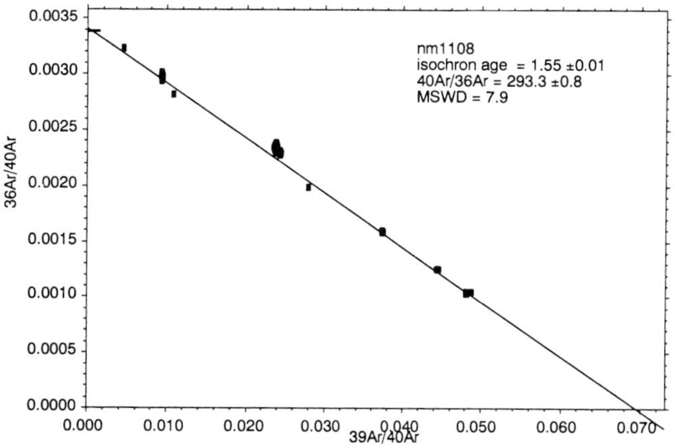

APPENDIX 1 continued

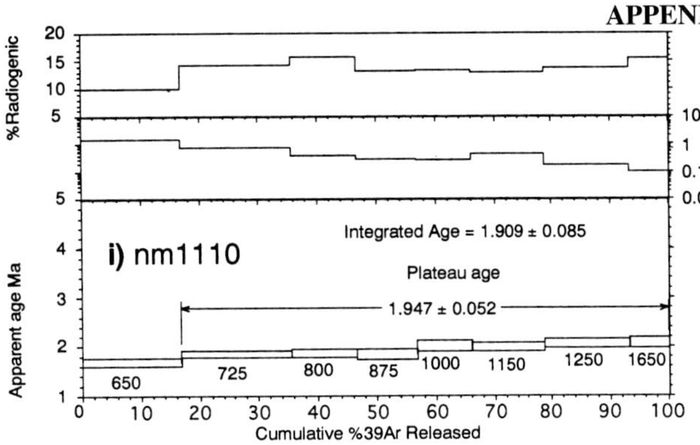

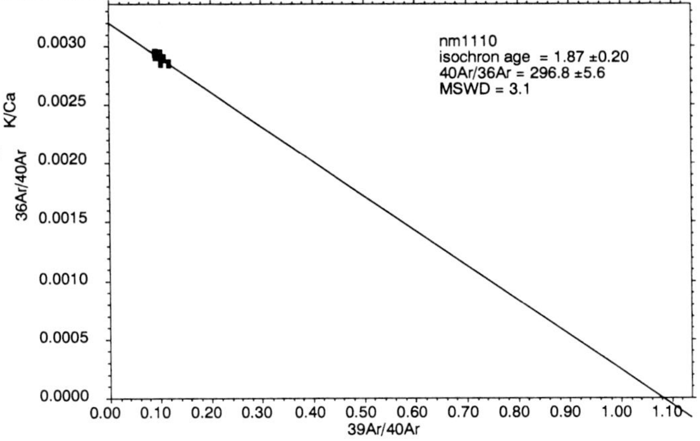

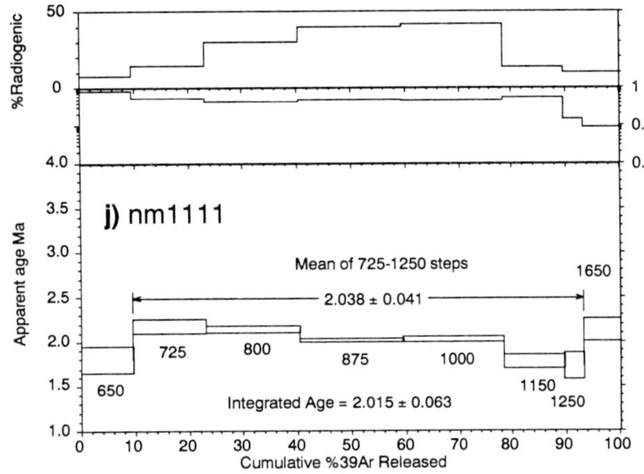

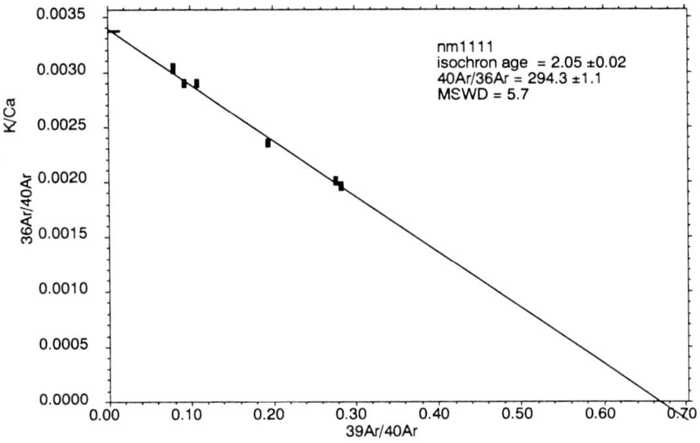

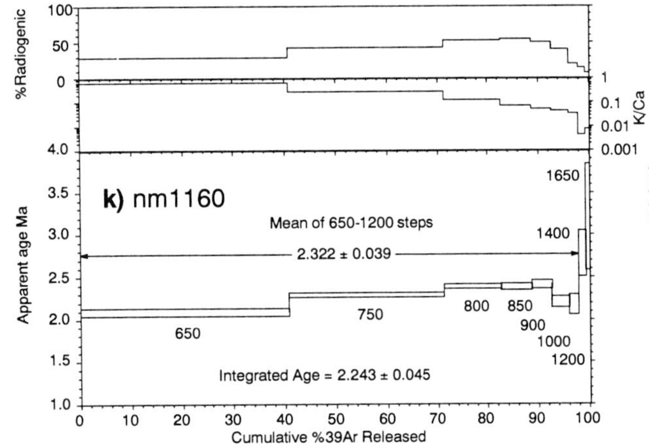

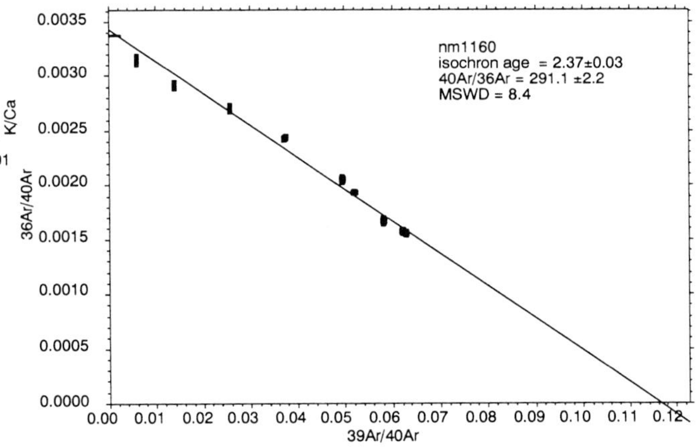

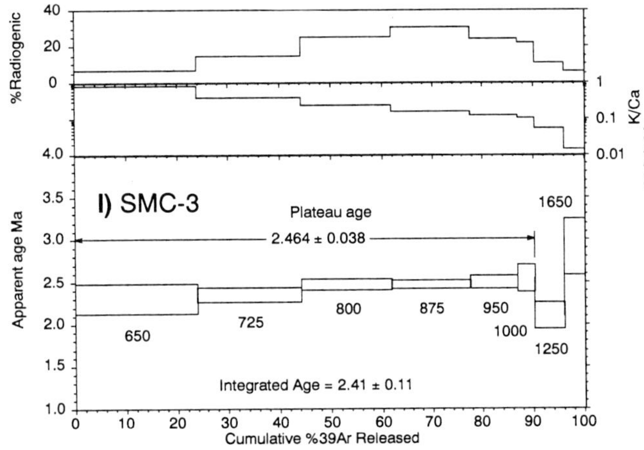

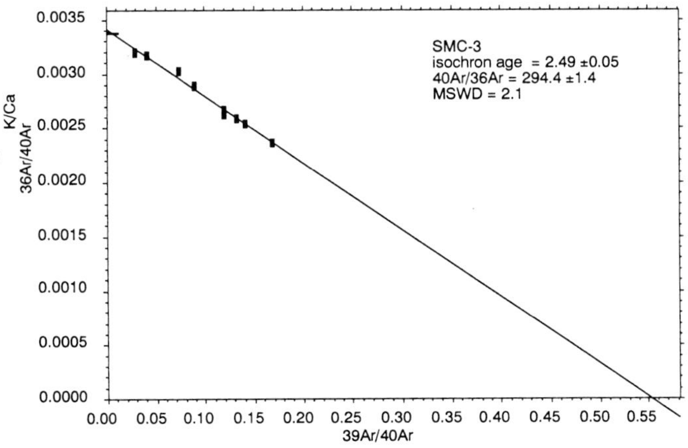

APPENDIX 1 continued

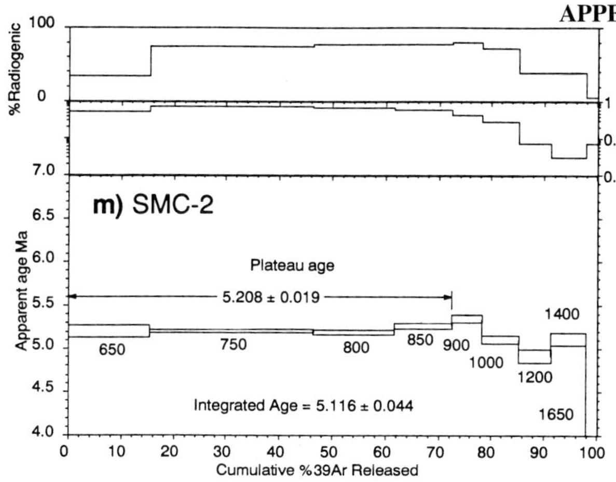

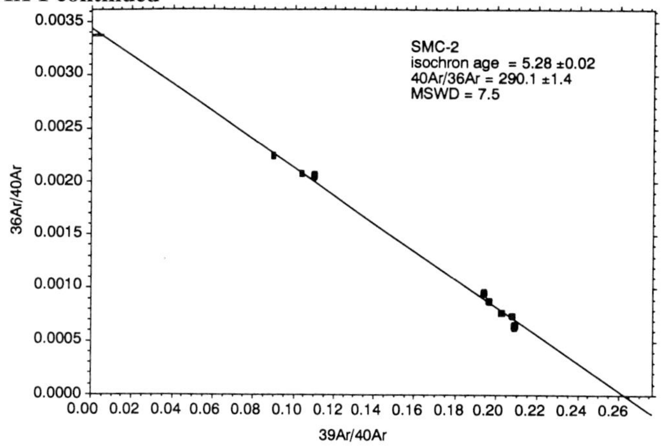

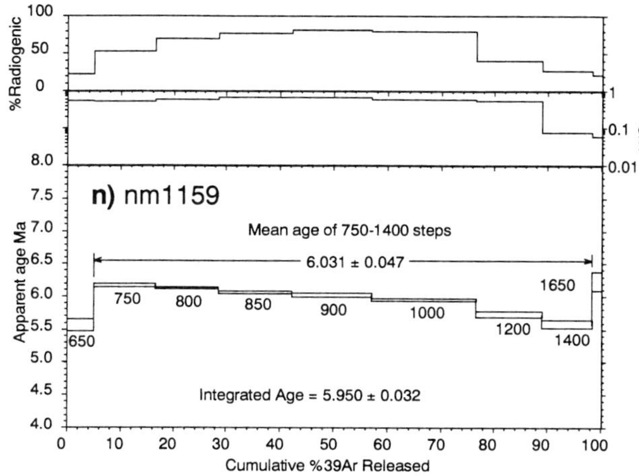

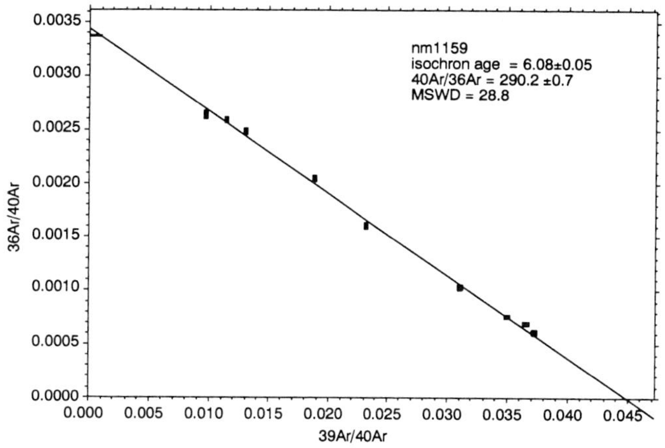

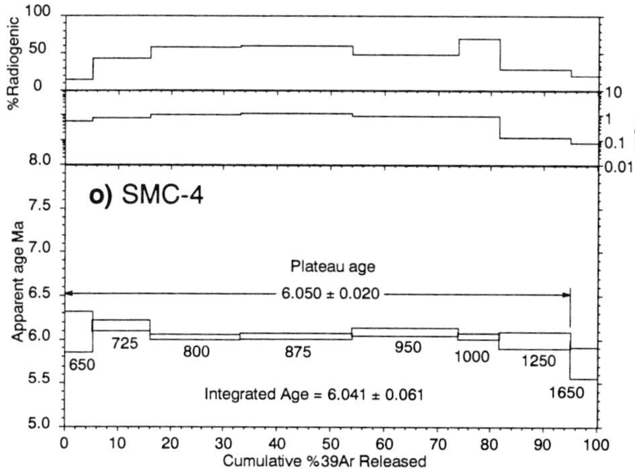

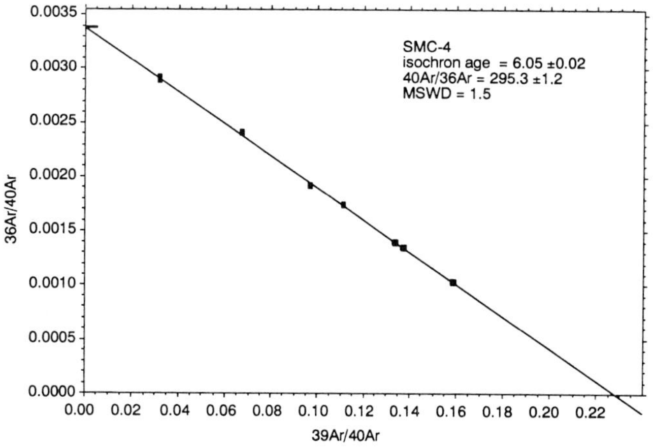

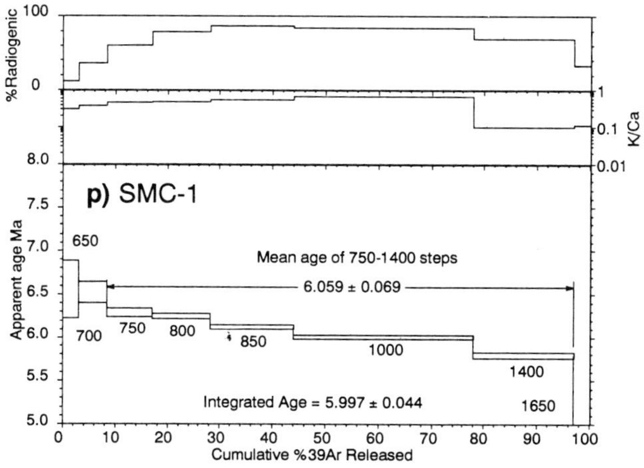

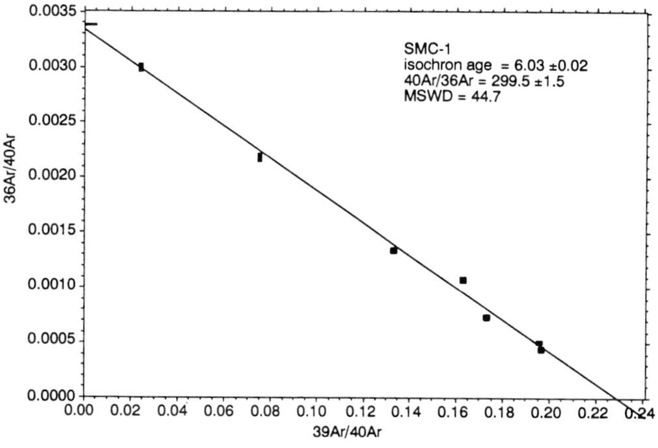

APPENDIX 1 continued

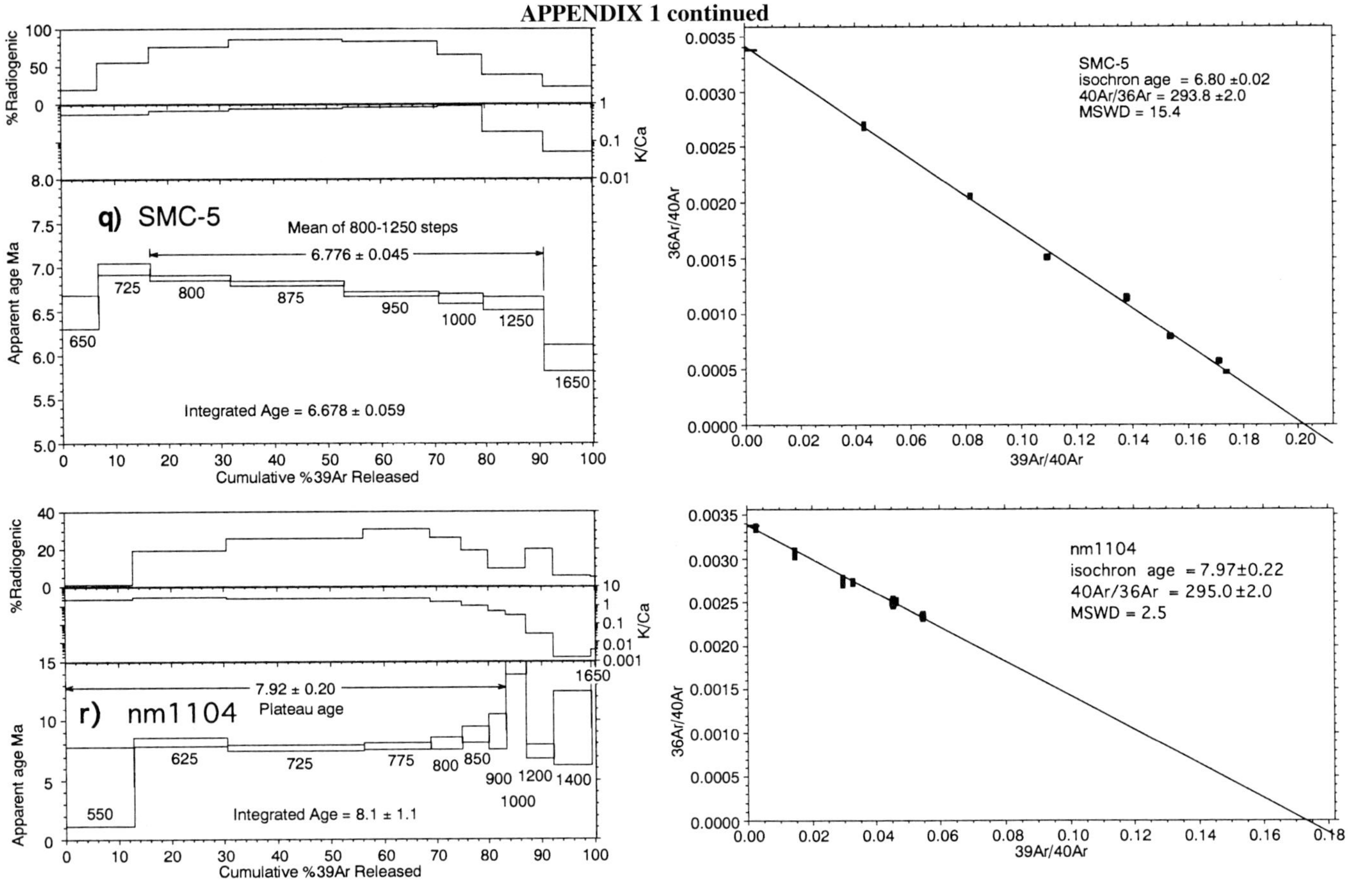

APPENDIX 2

$^{40}Ar/^{39}Ar$ step-heating results for Red Hill-Quemado volcanic rocks.

Lab#	Temp	40/39	37/39	36/39	39K moles	K/Ca	Cl/K	%40*	%39Ar	Age	± Err
nm1161, basaltic lava, Red Hill cinder cone, J=0.0001543 ± 0.0000009, Goat Spring quad., 34.288N, 108.913W.											
1005-01A	550	1.18e+02	1.01e+00	4.00e-01	6.9e-16	0.5	3.9e-02	-0.2	5.7	-0.071	0.174
1005-01B	625	6.60e+01	1.58e+00	2.24e-01	1.2e-15	0.3	2.8e-02	0.0	15.4	0.003	0.081
1005-01C	725	2.93e+01	1.87e+00	9.89e-02	2.1e-15	0.3	1.0e-02	0.5	32.8	0.044	0.037
1005-01D	775	1.47e+01	1.56e+00	4.91e-02	1.7e-15	0.3	2.6e-03	2.2	46.8	0.089	0.022
1005-01E	800	1.15e+01	1.33e+00	3.84e-02	1.4e-15	0.4	1.8e-03	2.1	57.9	0.068	0.020
1005-01F	850	9.52e+00	1.30e+00	3.15e-02	1.5e-15	0.4	1.2e-03	3.1	69.8	0.082	0.018
1005-01G	900	9.38e+00	1.37e+00	3.14e-02	1.3e-15	0.4	1.5e-03	1.9	80.6	0.051	0.019
1005-01H	1000	1.48e+01	1.69e+00	4.94e-02	1.2e-15	0.3	2.9e-03	2.4	90.5	0.098	0.033
1005-01I	1200	4.84e+01	4.01e+00	1.61e-01	8.2e-16	0.1	7.8e-03	2.4	97.2	0.326	0.066
1005-01J	1400	5.13e+01	3.17e+01	1.79e-01	2.2e-16	0.0	1.2e-02	1.8	98.9	0.263	0.127
1005-01K	1625	9.49e+01	2.41e+01	3.29e-01	1.3e-16	0.0	1.3e-02	-0.5	100.0	-0.124	0.266
total gas age					1.2e-14				n=11	0.074	0.046
plateau age									A-H	**0.071**	**0.012**
isochron age					40/36=296.3±0.8	MSWD=2.5			A-H	0.066	0.011
nm1107, ring intrusive, Zuni Salt Lake, J=0.000768 ± 0.000001, Zuni Salt Lake quad., 34.456N, 108.769W.											
849-01A	650	5.25e+01	2.23e+00	1.76e-01	3.6e-15	0.2	1.4e-03	1.1	20.6	0.784	0.392
849-01B	653	3.02e+00	1.61e+00	1.05e-02	3.7e-15	0.3	1.7e-04	0.9	42.0	0.040	0.043
849-01C	786	1.34e+00	1.71e+00	4.70e-03	5.1e-15	0.3	4.8e-04	4.6	71.1	0.086	0.025
849-01D	665	2.29e+00	2.96e+00	7.84e-03	9.1e-16	0.2	1.4e-03	7.6	76.3	0.242	0.087
849-01E	950	2.97e+00	3.96e+00	1.06e-02	3.7e-16	0.1	2.1e-03	4.3	78.4	0.177	0.195
849-01F	1000	3.54e+00	3.54e+00	1.16e-02	3.4e-16	0.1	2.3e-03	10.2	80.4	0.500	0.234
849-01G	1250	2.64e+00	1.28e+01	1.20e-02	2.8e-15	0.0	2.9e-03	2.0	96.2	0.075	0.050
849-01H	1614	1.98e+01	1.42e+01	6.75e-02	6.7e-16	0.0	4.6e-03	4.8	100.0	1.322	0.265
total gas age					1.7e-14				n=8	0.283	0.128
plateau age									B-G	**0.086**	**0.031**
isochron age					40/36=294.4±4.3	MSWD=2.7			B-G	0.095	0.024
nm1106, bomb in hydrovolcanic tuff, Zuni Salt Lake, J=0.0007665 ± 0.000001, Zuni Salt Lake quad., 34.546N, 108.768W.											
850-01A	650	1.13e+01	2.17e+00	3.84e-02	2.4e-15	0.2	4.0e-03	1.1	10.9	0.179	0.128
850-01B	725	1.75e+00	1.79e+00	5.93e-03	4.2e-15	0.3	1.4e-03	6.2	29.7	0.150	0.031
850-01C	800	1.55e+00	1.80e+00	5.37e-03	3.1e-15	0.3	2.0e-03	5.3	43.5	0.115	0.033
850-01D	875	1.79e+00	1.83e+00	6.02e-03	2.5e-15	0.3	3.1e-03	7.4	54.7	0.185	0.043
850-01E	950	2.28e+00	2.64e+00	8.66e-03	1.2e-15	0.2	5.8e-03	-4.3	60.0	-0.136	0.065
850-01F	1000	3.54e+00	3.16e+00	1.32e-02	5.8e-16	0.2	8.0e-03	-4.2	62.6	-0.207	0.153
850-01G	1250	7.97e+00	8.22e+00	2.88e-02	7.7e-15	0.1	1.0e-02	0.8	97.1	0.084	0.069
850-01H	1615	2.87e+01	1.40e+01	1.01e-01	6.4e-16	0.0	9.1e-03	0.0	100.0	-0.003	0.402
nm1106 (cont'd)											
total gas age					2.2e-14				n=8	0.101	0.072
plateau age									A-H	**0.114**	**0.038**
isochron age					40/36=294.4±1.5	MSWD=5.2			A-H	0.126	0.016
nm1158, basaltic lava, hornito fissure, N of Blaines Lake, J=0.0001522 ± 0.0000009, Blaines Lake quad., 34.311N, 108.801W.											
1000-01B	750	2.23e+01	4.39e+00	7.45e-02	9.7e-16	0.1	2.5e-03	2.6	23.6	0.162	0.040
1000-01C	800	2.98e+01	3.59e+00	9.82e-02	4.1e-16	0.1	2.2e-03	3.4	33.6	0.277	0.070
1000-01D	850	2.15e+01	3.56e+00	7.05e-02	3.4e-16	0.1	2.5e-03	4.1	41.8	0.243	0.072
1000-01E	900	2.17e+01	4.52e+00	6.99e-02	2.3e-16	0.1	2.3e-03	6.0	47.5	0.360	0.087
1000-01F	1000	3.37e+01	6.68e+00	1.08e-01	2.3e-16	0.1	5.1e-03	6.4	53.0	0.593	0.129
1000-01G	1200	2.45e+02	8.29e+00	8.24e-01	4.6e-16	0.1	1.4e-02	0.6	64.3	0.434	0.298
1000-01H	1400	5.81e+01	2.07e+01	1.99e-01	1.4e-15	0.0	2.2e-02	1.7	98.3	0.268	0.070
1000-01I	1623	1.77e+02	2.58e+01	5.67e-01	7.0e-17	0.0	7.5e-02	6.6	100.0	3.254	0.640
total gas age					4.1e-15				n=8	0.335	0.103
mean age of near-plateau steps								A-H	0.244	0.048	
isochron age					40/36=296.8±1.2	MSWD=2.8			A-H	**0.206**	**0.034**
nm1155, basalt lava, overlying Quemado Fm., J=0.000153 ± 0.0000009, Ponderosa Tank quad., 34.230N, 108.657W.											
1002-01A	650	1.26e+02	6.79e-01	4.26e-01	2.1e-15	0.8	1.9e-02	0.0	21.3	-0.006	0.376
1002-01B	750	6.37e+01	1.79e+00	2.06e-01	1.9e-15	0.3	1.3e-02	4.7	40.6	0.828	0.176
1002-01C	800	5.31e+01	2.33e+00	1.66e-01	1.4e-15	0.2	4.6e-03	8.1	54.4	1.182	0.082
1002-01D	850	4.21e+01	2.23e+00	1.31e-01	1.4e-15	0.2	2.7e-03	8.1	69.0	0.941	0.159
1002-01E	900	3.95e+01	1.98e+00	1.18e-01	1.3e-15	0.3	1.1e-03	12.4	82.5	1.352	0.048

APPENDIX 2 continued

Lab#	Temp	40/39	37/39	36/39	39K moles	K/Ca	Cl/K	%40*	%39Ar	Age	± Err
1002-01F	1000	1.03e+03	2.15e+00	3.68e+00	6.8e-16	0.2	4.9e-03	-5.3	89.5	-15.139	3.600
1002-01G	1000	1.35e+02	2.66e+00	4.36e-01	2.9e-16	0.2	2.5e-03	4.9	92.4	1.818	0.226
1002-01H	1200	2.96e+02	4.86e+00	9.79e-01	5.2e-16	0.1	6.8e-03	2.3	97.7	1.885	0.546
1002-01I	1400	2.83e+02	7.67e+01	9.08e-01	1.6e-16	0.0	1.3e-02	7.2	99.4	5.904	0.485
1002-01J	1627	3.34e+02	4.23e+01	1.05e+00	6.4e-17	0.0	3.1e-02	7.8	100.0	7.396	1.136
total gas age					9.8e-15				n=10	-0.109	0.456
disturbed spectrum - no mean age calculated											
isochron age				40/36=302.9±1.2		MSWD=18.3			A-J	0.990	0.050
nm1157, basaltic lava, interbedded with Quemado Fm., NE of Blaines Lake, J=0.0001525 ± 0.0000009, Blaines Lake quad., 34.305N, 108.751W.											
1001-01A	650	1.88e+02	2.75e+00	6.41e-01	1.5e-15	0.2	3.7e-02	-0.8	18.1	-0.418	0.292
1001-01B	750	1.81e+01	3.08e+00	5.01e-02	1.7e-15	0.2	1.2e-02	19.6	39.0	0.980	0.024
1001-01C	800	1.17e+01	2.03e+00	2.82e-02	1.2e-15	0.3	5.8e-03	29.9	53.0	0.962	0.020
1001-01D	850	1.05e+01	1.51e+00	2.42e-02	1.1e-15	0.3	4.2e-03	32.6	65.8	0.940	0.018
1001-01E	900	1.11e+01	1.36e+00	2.57e-02	8.3e-16	0.4	4.4e-03	32.5	75.9	0.992	0.021
1001-01F	1000	1.44e+01	1.35e+00	3.68e-02	7.6e-16	0.4	8.9e-03	25.2	85.1	1.001	0.028
1001-01G	1200	2.76e+01	3.41e+00	8.23e-02	7.7e-16	0.1	1.2e-02	12.9	94.5	0.982	0.044
1001-01H	1400	5.49e+01	3.19e+01	1.80e-01	3.9e-16	0.0	1.3e-02	7.4	99.3	1.145	0.103
1001-01I	1623	1.74e+02	2.15e+01	5.56e-01	5.8e-17	0.0	5.6e-02	6.3	100.0	3.049	0.557
total gas age					8.2e-15				n=9	0.745	0.081
plateau age									B-G	**0.970**	**0.014**
isochron age				40/36=297.8±2.4		MSWD=1.3			B-G	0.949	0.024
nm1156, basaltic lava, interbedded with Quemado Fm., US 60, SE of Blaines Lake, J=0.0001534 ± 0.0000009, Tejana Mesa SW quad., 34.270N, 108.719W.											
1003-01A	650	8.17e+01	3.98e-01	2.71e-01	2.5e-15	1.3	2.0e-02	1.9	17.9	0.44	0.26
1003-01B	750	2.81e+01	8.61e-01	7.92e-02	3.5e-15	0.6	1.5e-02	16.8	42.4	1.30	0.05
1003-01C	800	1.71e+01	1.23e+00	3.96e-02	2.2e-15	0.4	8.9e-03	32.2	57.9	1.53	0.02
1003-01D	850	1.19e+01	1.46e+00	2.18e-02	2.1e-15	0.3	4.6e-03	46.4	72.4	1.53	0.01
1003-01E	900	9.81e+00	1.66e+00	1.51e-02	1.7e-15	0.3	2.8e-03	55.4	84.3	1.51	0.01
1003-01F	1000	1.08e+01	2.14e+00	1.85e-02	1.3e-15	0.2	4.2e-03	50.6	93.4	1.51	0.02
1003-01G	1200	1.95e+01	3.52e+00	5.11e-02	4.9e-16	0.1	1.3e-02	24.1	96.8	1.31	0.05
1003-01H	1400	3.77e+01	2.90e+01	1.14e-01	3.0e-16	0.0	1.3e-02	16.7	98.9	1.77	0.09
1003-01I	1625	7.36e+01	3.05e+01	2.29e-01	1.5e-16	0.0	1.6e-02	11.2	100.0	2.32	0.22
total gas age					1.4e-14				n=9	1.28	0.07
plateau age									C-F	**1.51**	**0.01**
isochron age				40/36=295.3±1.5		MSWD=11.4			A-I	1.51	0.01
nm1108, basaltic lava, wall of Quemado crater, J=0.00006 ± 0.0000002, Red Hill quad., 34.232N, 108.828W.											
1310-01B	650	2.27e+02	1.26e+00	7.34e-01	1.2e-15	0.4	1.8e-02	4.5	22.8	1.11	0.08
1310-01C	700	9.28e+01	1.76e+00	2.62e-01	6.8e-16	0.3	9.6e-03	16.6	35.3	1.67	0.04
1310-01D	750	3.58e+01	1.69e+00	7.16e-02	7.1e-16	0.3	2.6e-03	41.2	48.2	1.60	0.02
1310-01E	800	2.25e+01	1.47e+00	2.87e-02	7.4e-16	0.3	1.8e-03	62.8	61.8	1.53	0.01
1310-01F	850	2.06e+01	1.45e+00	2.20e-02	7.3e-16	0.4	1.1e-03	68.9	75.1	1.54	0.01
1310-01G	900	2.08e+01	1.63e+00	2.20e-02	5.0e-16	0.3	2.0e-03	69.2	84.2	1.56	0.01
1310-01H	1000	2.68e+01	1.65e+00	4.30e-02	3.9e-16	0.3	3.3e-03	53.0	91.3	1.54	0.02
1310-01I	1100	4.13e+01	1.89e+00	9.54e-02	2.2e-16	0.3	7.7e-03	32.0	95.2	1.43	0.04
1310-01J	1200	1.05e+02	3.25e+01	3.20e-01	6.7e-17	0.0	1.4e-02	11.9	96.5	1.38	0.17
1310-01K	1400	7.46e+01	2.96e+01	1.67e-01	1.3e-16	0.0	1.7e-02	36.8	98.9	3.03	0.06
1310-01L	1570	4.16e+01	2.39e+01	1.04e-01	6.1e-17	0.0	3.4e-02	30.7	100.0	1.41	0.09
total gas age					5.5e-15				n=11	1.49	0.04
mean age of near-plateau steps								D-H	1.55	0.01	
isochron age				40/36=293.3±0.8		MSWD=7.9			A-L	**1.55**	**0.01**
nm1110, basaltic lava, S side of Cerro la Mula, J=0.0007672 ± 0.000001, Red Hill quad., 34.160N, 108.832W											
851-01A	650	1.21e+01	3.40e-01	3.70e-02	6.8e-15	1.5	8.2e-03	10.0	16.9	1.68	0.09
851-01B	725	9.41e+00	6.73e-01	2.74e-02	7.6e-15	0.8	7.8e-03	14.2	35.7	1.85	0.07
851-01C	800	8.60e+00	1.32e+00	2.48e-02	4.5e-15	0.4	7.0e-03	15.7	46.7	1.87	0.07
851-01D	875	1.00e+01	1.82e+00	2.98e-02	4.1e-15	0.3	7.0e-03	13.2	56.8	1.83	0.10
851-01E	1000	1.10e+01	2.00e+00	3.26e-02	3.8e-15	0.3	9.0e-03	13.3	66.2	2.01	0.11
851-01F	1150	1.12e+01	1.25e+00	3.31e-02	5.1e-15	0.4	9.2e-03	13.0	78.8	2.01	0.08
851-01G	1250	1.10e+01	2.84e+00	3.27e-02	5.8e-15	0.2	8.9e-03	13.7	93.2	2.08	0.08
851-01H	1603	9.77e+00	5.44e+00	2.93e-02	2.8e-15	0.1	8.4e-03	15.4	100.0	2.09	0.10
total gas age					4.0e-14				n=8	1.91	0.09

APPENDIX 2 continued

Lab#	Temp	40/39	37/39	36/39	39K moles	K/Ca	Cl/K	%40*	%39Ar	Age	± Err
plateau age									B-H	**1.95**	**0.05**
isochron age					40/36=296.8±5.6	MSWD=3.1			A-H	1.87	0.20
nm1111, basaltic lava, N of Mangitas Flat, J=0.000769 ± 0.000001, Black Peak quad., 34.104N, 108.853W.											
852-01A	650	1.69e+01	6.24e-01	5.30e-02	5.5e-15	0.8	2.0e-02	7.7	9.7	1.81	0.14
852-01B	725	1.10e+01	9.66e-01	3.19e-02	7.6e-15	0.5	1.2e-02	14.3	23.0	2.18	0.08
852-01C	800	5.15e+00	1.19e+00	1.24e-02	9.9e-15	0.4	4.4e-03	30.1	40.3	2.15	0.04
852-01D	875	3.62e+00	1.07e+00	7.54e-03	1.1e-14	0.5	1.9e-03	40.2	59.4	2.02	0.02
852-01E	1000	3.54e+00	1.10e+00	7.22e-03	1.1e-14	0.5	2.0e-03	41.4	78.4	2.03	0.03
852-01F	1150	9.33e+00	9.39e-01	2.74e-02	6.4e-15	0.5	6.8e-03	13.7	89.7	1.78	0.07
852-01G	1250	1.25e+01	3.54e+00	3.91e-02	2.0e-15	0.1	1.2e-02	10.0	93.1	1.74	0.15
852-01H	1616	1.49e+01	5.95e+00	4.68e-02	3.9e-15	0.1	1.1e-02	10.3	100.0	2.14	0.12
total gas age					5.7e-14				n=8	2.02	0.06
mean age of near-plateau steps								B-G	2.04	0.04	
isochron age					40/36=294.3±1.1	MSWD=5.7			A-G	**2.05**	**0.02**
nm1160, basaltic lava, overlying Bidahochi Fm., N of Mesa Parada, AZ, J=0.0001537 ± 0.0000009, Mesa Parada quad., 34.460N, 109.091W.											
1004-01A	650	2.68e+01	7.26e-01	6.54e-02	3.1e-15	0.7	1.2e-02	28.1	40.9	2.09	0.05
1004-01B	750	1.93e+01	1.81e+00	3.77e-02	2.3e-15	0.3	1.0e-02	42.9	71.5	2.29	0.02
1004-01C	800	1.61e+01	4.08e+00	2.63e-02	8.4e-16	0.1	8.0e-03	53.5	82.7	2.39	0.02
1004-01D	850	1.59e+01	7.19e+00	2.64e-02	4.6e-16	0.1	5.0e-03	54.1	88.8	2.39	0.04
1004-01E	900	1.71e+01	1.02e+01	3.11e-02	3.0e-16	0.1	3.6e-03	50.7	92.7	2.42	0.05
1004-01F	1000	2.00e+01	1.25e+01	4.42e-02	2.5e-16	0.0	3.6e-03	39.5	96.1	2.21	0.06
1004-01G	1200	3.86e+01	1.39e+01	1.08e-01	1.4e-16	0.0	9.8e-03	20.2	98.0	2.18	0.11
1004-01H	1400	6.69e+01	1.16e+02	2.25e-01	9.7e-17	0.0	1.0e-02	13.9	99.2	2.78	0.27
1004-01I	1637	1.63e+02	6.22e+01	5.30e-01	5.8e-17	0.0	1.9e-02	6.8	100.0	3.22	0.63
total gas age					7.5e-15				n=9	2.24	0.04
mean age of near-plateau steps								A-G	2.32	0.04	
isochron age					40/36=291.1±2.2	MSWD=8.4			A-I	**2.37**	**0.03**
SMC-3, basaltic lava, overlying Quemado Fm., US 60, NM-AZ border, J=0.0007629 ± 0.000001, Nelson Reservoir NE quad., 34.025N, 109.028W.											
848-01A	650	2.45e+01	6.03e-01	7.74e-02	6.6e-15	0.8	2.2e-02	6.8	23.9	2.31	0.18
848-01B	725	1.13e+01	1.25e+00	3.28e-02	5.6e-15	0.4	1.6e-02	15.1	44.1	2.35	0.08
848-01C	800	7.12e+00	2.16e+00	1.85e-02	4.9e-15	0.2	7.7e-03	25.2	61.9	2.48	0.06
848-01D	875	5.95e+00	3.12e+00	1.48e-02	4.3e-15	0.2	3.4e-03	30.2	77.5	2.48	0.05
848-01E	950	7.59e+00	4.31e+00	2.06e-02	2.6e-15	0.1	3.1e-03	23.9	86.8	2.50	0.07
848-01F	1000	8.38e+00	4.84e+00	2.33e-02	9.6e-16	0.1	6.3e-03	22.0	90.3	2.54	0.16
848-01G	1250	1.38e+01	9.03e+00	4.37e-02	1.6e-15	0.1	1.1e-02	11.0	96.0	2.10	0.15
848-01H	1612	3.42e+01	3.39e+01	1.18e-01	1.1e-15	0.0	1.4e-02	6.0	100.0	2.91	0.33
total gas age					2.8e-14				n=8	2.41	0.11
plateau age									A-F	**2.46**	**0.04**
isochron age					40/36=294.4±1.4	MSWD=2.1			A-H	2.49	0.05
SMC-2, basaltic lava, interbedded in Fence Lake Fm., US 60, W of Red Hill, J=0.0007649 ± 0.000001, Cow Springs quad., 34.213N, 108.896W.											
844-01A	650	1.12e+01	9.87e-01	2.52e-02	8.7e-15	0.5	1.0e-02	33.7	15.4	5.20	0.07
844-01B	750	5.11e+00	6.99e-01	4.64e-03	1.8e-14	0.7	5.5e-03	73.8	46.4	5.20	0.02
844-01C	800	4.84e+00	7.34e-01	3.78e-03	8.7e-15	0.7	4.1e-03	77.7	61.7	5.18	0.02
844-01D	850	4.96e+00	8.76e-01	4.04e-03	6.2e-15	0.6	3.5e-03	76.8	72.6	5.25	0.03
844-01E	900	4.81e+00	1.22e+00	3.41e-03	3.2e-15	0.4	4.3e-03	80.5	78.3	5.34	0.04
844-01F	1000	5.17e+00	1.80e+00	5.41e-03	3.9e-15	0.3	5.7e-03	71.4	85.3	5.09	0.04
844-01G	1200	9.08e+00	6.91e+00	2.05e-02	3.4e-15	0.1	6.8e-03	39.0	91.2	4.91	0.07
844-01H	1400	9.53e+00	1.62e+01	2.40e-02	3.7e-15	0.0	1.3e-02	38.4	97.8	5.10	0.07
844-01I	1621	2.92e+01	7.25e+00	9.57e-02	1.2e-15	0.1	6.3e-03	5.1	100.0	2.05	0.31
total gas age					5.7e-14				n=9	5.11	0.04
plateau age									A-D	**5.20**	**0.03**
isochron age					40/36=290.1±1.4	MSWD=7.5			A-H	5.23	0.02
nm1159, basaltic lava, caps mesa N of Blaines Lake, J=0.0001518 ± 0.0000009, Blaines Lake quad., 34.315N, 108798W.											
999-01A	650	8.78e+01	9.07e-01	2.28e-01	9.1e-16	0.6	2.7e-02	23.2	5.1	5.58	0.09
999-01B	750	4.31e+01	9.80e-01	6.97e-02	2.1e-15	0.5	8.1e-03	52.3	16.7	6.17	0.03
999-01C	800	3.22e+01	8.39e-01	3.34e-02	2.1e-15	0.6	3.4e-03	69.5	28.6	6.13	0.02
999-01D	850	2.86e+01	7.51e-01	2.19e-02	2.5e-15	0.7	2.0e-03	77.6	42.2	6.07	0.02

APPENDIX 2 continued

Lab#	Temp	40/39	37/39	36/39	39K moles	K/Ca	Cl/K	%40*	%39Ar	Age	± Err
999-01E	900	2.69e+01	7.46e-01	1.67e-02	2.7e-15	0.7	1.6e-03	81.8	57.1	6.02	0.02
999-01F	1000	2.74e+01	8.77e-01	1.92e-02	3.5e-15	0.6	1.9e-03	79.5	76.4	5.96	0.02
999-01G	1200	5.31e+01	9.45e-01	1.09e-01	2.2e-15	0.5	4.8e-03	39.4	88.9	5.73	0.04
999-01H	1400	7.65e+01	6.74e+00	1.92e-01	1.7e-15	0.1	1.3e-02	26.6	98.2	5.58	0.06
999-01I	1622	1.03e+02	8.52e+00	2.75e-01	3.3e-16	0.1	5.1e-02	21.9	100.0	6.24	0.14
total gas age					1.8e-14				n=9	5.95	0.03
mean age of near-plateau steps								B-H	**6.03**	**0.05**	
isochron age					40/36=290.2±0.7		MSWD=28.8		A-I	6.08	0.04
SMC-4, basaltic lava, capping SW edge Cimarron Mesa, J=0.0007624 ± 0.000001, Red Hill quad., 34.250N, 108.869W.											
847-01A	650	3.13e+01	8.60e-01	9.12e-02	3.7e-15	0.6	3.4e-03	14.1	5.1	6.08	0.23
847-01B	725	1.04e+01	6.34e-01	2.01e-02	7.7e-15	0.8	9.2e-04	43.1	16.0	6.15	0.06
847-01C	800	7.49e+00	5.05e-01	1.06e-02	1.2e-14	1.0	3.2e-04	58.6	33.0	6.03	0.03
847-01D	875	7.31e+00	4.42e-01	9.91e-03	1.5e-14	1.2	3.4e-04	60.1	53.9	6.03	0.03
847-01E	950	9.07e+00	5.31e-01	1.58e-02	1.4e-14	1.0	7.7e-04	48.8	73.8	6.08	0.04
847-01F	1000	6.33e+00	5.19e-01	6.63e-03	5.5e-15	1.0	1.6e-03	69.4	81.5	6.03	0.03
847-01G	1250	1.49e+01	3.77e+00	3.67e-02	9.6e-15	0.1	7.4e-03	29.1	95.0	5.99	0.09
847-01H	1624	2.00e+01	6.24e+00	5.53e-02	3.6e-15	0.1	8.9e-03	20.7	100.0	5.73	0.18
total gas age					7.1e-14				n=8	6.03	0.06
plateau age									A-G	**6.05**	**0.02**
isochron age					40/36=295.3±1.2		MSWD=1.5		A-H	6.05	0.03
SMC-1, basaltic lava, overlying Fence Lake Fm., E of Cow Springs Draw, J=0.0007694 ± 0.000001, Cow Springs quad., 34.198N, 108.996W.											
845-01A	650	4.16e+01	1.68e+00	1.25e-01	1.3e-15	0.3	4.1e-03	11.4	3.1	6.58	0.33
845-01B	700	1.33e+01	1.34e+00	2.94e-02	2.2e-15	0.4	1.8e-03	35.4	8.5	6.54	0.12
845-01C	750	7.54e+00	1.10e+00	1.04e-02	3.6e-15	0.5	9.2e-04	60.3	17.2	6.30	0.05
845-01D	800	5.80e+00	1.04e+00	4.52e-03	4.5e-15	0.5	6.9e-04	77.9	28.1	6.26	0.03
845-01E	850	5.11e+00	9.26e-01	2.49e-03	6.6e-15	0.6	7.1e-04	86.6	44.1	6.14	0.02
845-01F	1000	5.13e+00	7.48e-01	2.79e-03	1.4e-14	0.7	1.9e-03	84.6	78.0	6.02	0.02
845-01G	1400	6.15e+00	5.15e+00	7.95e-03	7.8e-15	0.1	1.1e-02	67.9	97.0	5.81	0.03
845-01H	1603	6.89e+00	4.45e+00	1.68e-02	1.2e-15	0.1	7.4e-03	32.5	100.0	3.11	0.11
total gas age					4.1e-14				n=8	6.01	0.04
mean age of near-plateau steps								C-G	6.06	0.07	
isochron age					40/36=299.5±1.5		MSWD=44.7		A-G	**6.03**	**0.02**
SMC-5, basaltic lava, Flat Top, S of Springerville, AZ, White Mtns field, J=0.0007643 ± 0.000001, Springerville quad., 34.354N, 109.271W.											
846-01A	650	2.31e+01	9.43e-01	6.24e-02	3.0e-15	0.5	1.8e-02	20.4	6.9	6.48	0.18
846-01B	726	9.16e+00	9.11e-01	1.40e-02	4.3e-15	0.6	6.8e-03	55.4	16.6	6.98	0.06
846-01C	800	6.54e+00	7.39e-01	5.35e-03	6.7e-15	0.7	3.1e-03	76.3	31.8	6.87	0.03
846-01D	875	5.77e+00	6.55e-01	2.90e-03	9.4e-15	0.8	1.9e-03	85.6	53.0	6.81	0.02
846-01E	950	5.86e+00	6.17e-01	3.49e-03	7.9e-15	0.8	2.0e-03	82.8	70.9	6.68	0.03
846-01F	1000	7.28e+00	5.66e-01	8.41e-03	3.8e-15	0.9	3.8e-03	66.2	79.5	6.63	0.05
846-01G	1250	1.22e+01	2.88e+00	2.59e-02	5.0e-15	0.2	7.0e-03	39.0	90.7	6.58	0.07
846-01H	1612	1.82e+01	9.70e+00	4.97e-02	4.1e-15	0.1	1.0e-02	23.6	100.0	5.96	0.15
total gas age					4.4e-14				n=8	6.67	0.06
mean age of near-plateau steps								C-G	6.78	0.05	
isochron age					40/36=293.8±2.0		MSWD=15.4		B-G	**6.80**	**0.02**
nm1104, intermediate intrusive body, El Porticito, J=0.0007662 ± 0.000002, Tejana Mesa quad., 34.389N, 108.569W.											
1006-01A	550	3.54e+02	2.42e-01	1.19e+00	2.4e-16	2.1	1.4e-02	0.9	13.1	4.48	3.34
1006-01B	625	3.06e+01	1.93e-01	8.35e-02	3.3e-16	2.6	8.4e-03	19.4	31.1	8.20	0.35
1006-01C	725	2.17e+01	2.04e-01	5.46e-02	4.8e-16	2.5	8.2e-03	25.8	57.1	7.72	0.23
1006-01D	775	1.83e+01	2.07e-01	4.26e-02	2.4e-16	2.5	7.0e-03	31.1	69.9	7.85	0.28
1006-01E	800	2.22e+01	3.21e-01	5.53e-02	1.1e-16	1.6	7.4e-03	26.3	76.2	8.06	0.48
1006-01F	850	3.34e+01	5.53e-01	9.15e-02	9.2e-17	0.9	7.4e-03	19.2	81.2	8.84	0.68
1006-01G	900	6.83e+01	1.02e+00	2.09e-01	6.0e-17	0.5	1.1e-02	9.6	84.5	9.09	1.48
1006-01H	1000	1.22e+02	1.82e+00	3.75e-01	7.0e-17	0.3	2.9e-02	9.5	88.3	16.07	2.18
1006-01I	1200	2.66e+01	1.63e+01	7.63e-02	9.6e-17	0.0	9.7e-02	19.9	93.5	7.37	0.57
1006-01J	1400	9.31e+01	2.99e+02	3.75e-01	1.0e-16	0.0	6.5e-01	5.7	99.2	9.32	3.13
1006-01K	1634	4.18e+02	1.13e+02	1.37e+00	1.5e-17	0.0	2.2e-01	5.2	100.0	32.61	11.64
total gas age					1.8e-15				n=11	8.11	1.09
plateau age									A-G	**7.92**	**0.20**
isochron age					40/36=295.0±2.0		MSWD=2.5		A-G	7.97	0.22

TRACE ELEMENT AND MINERALOGICAL ALTERATION ASSOCIATED WITH MODERATE AND ADVANCED DEGREES OF K-METASOMATISM IN A RIFT BASIN AT SOCORRO, NEW MEXICO

N.W. DUNBAR[1,2], C.E. CHAPIN[1], D.J. ENNIS[2] and A.R. CAMPBELL[2]

[1]New Mexico Bureau of Mines and Mineral Resources, Socorro, NM 87501; [2]Geoscience Department, New Mexico Institute of Mining and Technology, Socorro, NM 87801

Abstract—Silicic ignimbrites (ash-flow tuffs) and interbedded mafic lavas of Oligocene age were altered by alkaline, saline brines beneath the Popotosa basin of the Rio Grande rift near Socorro, New Mexico. This alteration caused a variety of chemical and mineralogical changes in the affected rocks. The characteristic alteration mineralogy in highly metasomatized rock is adularia + hematite ± quartz ± clay minerals (Lindley, 1985), but we have determined, based on detailed mineralogical analysis of altered plagioclase, that the ratio of phases is highly variable, possibly as a function of degree of metasomatism. The relative proportion of adularia to clay minerals, and the composition of clay minerals is variable within the metasomatized rocks, possibly resulting from variations in fluid composition and temperature. During metasomatism, elements such as K, As, Rb, Sb, Ag, Cs, Au, Pb, Th and U were enriched in the altered rocks, whereas Na, Mg, Ca and Sr were depleted. Some trace elements, rather than varying smoothly with degree of rock alteration (estimated by K/Na ratio), change dramatically during either low or high degrees of alteration. These data suggest that the trace-element signature of alteration may be dependent on the mineralogical alteration assemblage, which may, in turn, depend on the composition of fluids that cause alteration. Dating of the metasomatic event by $^{40}Ar/^{39}Ar$ suggests that the alteration was active between 7.4 and 8.9 Ma. The estimated age range for deposition of playa clays suggests that metasomatism may have begun as early as 15 Ma, hence the metasomatism may have been active for many millions of years. The long duration of the metasomatic event and implied long time available for elements to be progressively scavenged from surrounding rocks may explain the lack of a distinctly depleted source area for elements enriched by the metasomatic alteration.

INTRODUCTION

Some areas of the western US, particularly in regions of Basin-and-Range extension, are characterized by surface enrichments of potassium (K) (Duval, 1990). In many areas, particularly one around Socorro, NM, this enrichment is thought to be caused by secondary enrichment of K by the process of K-metasomatism (Chapin and Lindley, 1986), in which K and other elements are concentrated in altered rocks by a K-rich fluid, while another set of elements are leached (depleted) from the altered rock. This type of fluid interaction with rocks has been associated with ore deposition in some areas (Roddy et al., 1988; Smith et al., 1991) and enrichment of some elements of economic interest has been observed in the Socorro area. However, the chemical nature of the altering fluid, the mineralogical reactions that occur during alteration, and the timing of alteration in the Socorro area are poorly understood. In this paper, we discuss results of ongoing research on the problem of K-metasomatism, particularly the chemical and mineralogical alteration that occurred in several rhyolitic ignimbrites and basaltic andesites as a result of intermediate and advanced degrees of K-metasomatism. We also present preliminary $^{40}Ar/^{39}Ar$ dates for minerals that were produced during the alteration process.

BACKGROUND

Chemical and mineralogical alteration of rocks by interaction with fluids is a common process, especially associated with ore deposition (e.g., Barnes, 1979), but also common in diagenesis (Awwillter, 1993; Wintsch and Kvale, 1994). Enrichment of K by alteration fluids is a chemical signature observed in many places, associated both with hydrothermal/geothermal systems and lower temperature alkaline-saline brines (e.g., Saunders and Tuach, 1991; Turner and Fishman, 1991). A large area around Socorro shows evidence of K enrichment (Fig. 1; D'Andrea, 1981, Lindley, 1985 and Chapin and Lindley, 1986). This area of enrichment covers more than 1800 km^2 and is 1.5 km

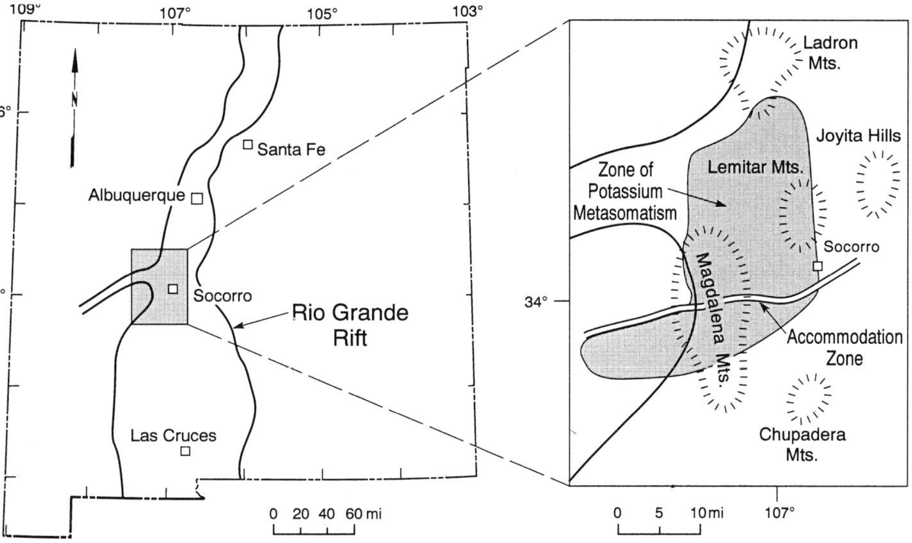

FIGURE 1. Map of the Socorro area, showing the extent of K–metasomatism. (after Lindley, 1985).

thick. The K_2O content of altered rocks can be as high as 13.5 wgt % in rock that initially contained 4-5 wgt %. Other components, such as Na_2O, CaO, and MgO, are removed during the alteration process (Chapin and Lindley, 1986).

Although chemical changes in metasomatized rocks are dramatic, the mineralogical effect of alteration is subtle (Lindley, 1985). The original rock textures, particularly in extrusive volcanic rocks, are preserved, even in rocks that have been highly altered. High-K phenocrysts in volcanic rocks, such as sanidine and biotite, are apparently unaffected by the metasomatic process, and remain fresh-looking even in a highly altered rock. However, low-K phases, such as plagioclase, and reactive phases, such as fine-grained groundmass, Fe-oxides and glass are strongly affected by the metasomatic process (Lindley, 1985). The alteration mineralogy typically consists of a mixture of K-feldspar (adularia) + hematite ± quartz ± clay minerals. Plagioclase phenocrysts in altered volcanic rocks typically are replaced by a fine-grained mixture of quartz + adularia + clay minerals (Lindley, 1985). Glass is also altered to a similar crystalline mixture. Pyroxene phenocrysts are hematized, as is groundmass magnetite (Lindley, 1985).

The fluids responsible for K-metasomatism could be derived from several geological environments. The chemical and mineralogical characteristics of alteration have been observed in intrusion-related hydrothermal systems (e.g., Saunders and Tuach, 1991), in rocks that have been altered by seawater at high water:rock ratios (Munhá et al., 1980), as well as in rocks altered by alkaline-saline brines (Sheppard and Gude, 1973; Duffin, 1989). The Socorro K anomaly was originally interpreted as an ancient geothermal system (D'Andrea et al., 1983), but was reinterpreted as widespread alteration by alkaline-saline brines derived from a large playa lake (Chapin and Lindley, 1986). This reinterpretation is based on three factors. First, the presence of thick playa mudstones in the Socorro area suggests that long-lived playas were present. Second, the size of the anomaly is large enough that if a hydrothermal system were responsible, it would have to be very large. The only magmatic systems in the area, which may have been large enough to cause a hydrothermal system of this magnitude, were active between about 27 and 32 Ma (McIntosh et al., 1992), and because rocks as young as 15 Ma are metasomatized (S. Cather, personal commun., 1994), the older, large magma chambers are ruled out as hydrothermal sources. Magmatism in the Socorro area younger than 15 Ma is very small-volume. Finally, altered rocks are enriched in $\delta^{18}O$ compared to fresh rocks, in contrast to the pattern expected in a hydrothermal system (Lindley, 1985). Basin-and-Range topography around the Socorro area, and a dry climate at the time of metasomatism are consistent with the presence of alkaline-saline playa brines; other K-metasomatized areas in the Western United States (Duval, 1990) may have formed by similar mechanisms.

Although ore deposits are associated with areas of K-metasomatism, in some cases associated with alkaline-saline brines (Roddy et al., 1988), the mobility of elements of economic interest in these systems is poorly understood. The objective of this research is to study the mobility of all elements, including elements of economic interest, in the Socorro K anomaly. This includes investigation of the mineralogy and chemistry that resulted from the alteration process, in order to better understand the fluid composition that was responsible for the alteration, and the location in the alteration assemblage in which different elements reside. The Socorro area is well suited to this type of study because of the presence of numerous regional ignimbrites, as well as basaltic-andesite lava flows, which are crosscut by the K anomaly. Therefore, it is possible to compare the chemical composition and mineralogy of altered rocks with their unaltered parent compositions, and obtain a clear picture of the changes caused by the metasomatic fluid.

ANALYTICAL METHODS

Whole rock geochemistry

Samples were taken of rhyolitic ignimbrites and basaltic-andesite flows at a wide range of locations in the K-metasomatized area. These samples have undergone varying degrees of metasomatism. Fresh samples from each of the ignimbrites and the basaltic-andesite lavas were also collected, so that fresh and altered compositions could be directly compared. The samples were ground and sent to XRAL analytical services for analysis of major elements and a wide range of trace elements. Analyses were made by x-ray fluorescence, neutron activation analysis, and inductively-coupled mass spectrometry.

Mineralogy

In addition to whole-rock geochemical analysis, some detailed mineralogical analysis were made using x-ray diffraction. The instrument used was a Rigaku DMAX-I x-ray diffractometer with JADE software. In order to best characterize the mineralogical processes that took place during alteration, material was hand-picked from the interior of former plagioclase crystals, still recognizable in the samples of basaltic-andesite lavas or rhyolitic ignimbrites, but partially to completely replaced by an alteration assemblage. Samples were also taken of material that was likely to have been originally composed of glass. By using the technique of sampling a starting material of a known simple composition, the mineralogical changes as a function of degree of alteration could be better characterized. Also, the confusion of trying to differentiate between remnant unaltered mineral phases and alteration-produced phases was avoided (Lindley, 1985).

X-ray diffraction scans were run between 2 and 70° 2θ using a Cu anode x-ray tube. The step width was 0.05° using a 0.5 second count time at 25 mA and 40 kV. Mineral phases were identified using standard peak search techniques, and approximate relative abundance of different phases was estimated using full-width, half-maximum measurements of peak size. Although it is not possible to exactly determine the ratio of mineral phases in a sample using XRD without a wide range of standards, the relative importance of different mineral phases within a group of mineralogically similar samples can be estimated based on XRD relative peak heights.

Age determinations

The age of K-metasomatic alteration in the Socorro area was determined by $^{40}Ar/^{39}Ar$ geochronology, using the New Mexico Geochonological Research Laboratory at the New Mexico Institute of Mining and Technology. The material dated was a mixture of adularia and quartz that replaced plagioclase feldspar during the alteration process. The samples were analysed with XRD prior to dating, in order to ascertain that no residual plagioclase, with an age signature representative of eruption age, rather than alteration age, remained in the hand-picked sample. The samples were repeatedly rinsed in deionized water, in order to remove as much secondary clay as possible. Although the clay would have formed during the alteration process, and would hence yield a useful age, the small crystal size of clay leads to recoil problems during irradiation, and is therefore undesirable (McDougal and Harrison, 1988). The presence of quartz mixed with adualaria in the sample was not a problem, because the K and Ar content of quartz is negligible.

The $^{40}Ar/^{39}Ar$ dating facility at New Mexico Tech offers state-of-the-art equipment similar to that now operating in several other $^{40}Ar/^{39}Ar$ laboratories. The mass spectrometer is a MAP-215-50 instrument equipped with Faraday and electron multiplier collectors. The electrostatic filter and adjustable collector slit allow for a routine mass resolution of 600 at mass 40. The ^{40}Ar sensitivity is 7×10^{-4} A/Torr at normal source operating conditions (Trap Current = 200 μA, total emission 725 μA and 4 kV accelerating voltage). The instrument background at mass 36 is presently $\sim 7 \times 10^{-19}$ moles and the ^{40}Ar static rate of rise is $\sim 5 \times 10^{-13}$ cc/min.

RESULTS AND DISCUSSION

Mineralogical investigation

Other investigations have shown that the alteration mineralogy produced by K-metasomatism consists of adularia + quartz + hematite ± clay minerals. However, in some studies, the mineralogy was studied using a bulk sample that contained original unaltered minerals as well as those produced by the alteration process (Lindley, 1985). Using this approach, it is difficult to quantify the relative abundances of phases produced by alteration because of the number of interfering XRD peaks caused by primary minerals in the samples. However, by selectively sampling phases within an altered rock, such as a low-K phases (plagioclase) or reactive phases such as glass, the mineralogy caused by alteration can be better understood. Because the altered phases did not contain magnetite, this technique neglects analysis of the hematite phase of the alteration assemblage, but we feel it nevertheless expands our understanding of the mineralogical processes that occur during K-metasomatism.

Although previous studies have not reported major mineralogical variations between different K-metasomatized samples, our more detailed analysis suggests that wide mineralogical variations are present (Figs. 2 A-D). Some altered samples consist of a combination of quartz, adularia and clay minerals (Fig. 2A), but the proportions of these phases vary strongly between samples. In addition, some samples consist almost entirely of a single alteration phase such as clay minerals (Fig. 2B), or adularia + quartz (Fig. 2C). Several clay minerals have been observed in the samples, including mixed-layer illite-smectite (I/S) and kaolinite. Although some samples contain only one type of clay (either I/S or kaolinite) (Figs. 2A, D), others contain multiple clay mineral types in the same sample. Residual plagioclase (Fig. 2D, see peak at ~22°) is observed in samples that have presumably undergone a lesser degree of alteration. Plagioclase coexists with adularia, clay minerals and quartz. In the samples that we analysed, residual plagioclase always co-exists with both clay minerals and adularia. Clay minerals in samples with residual plagioclase are dominantly kaolinitic, suggesting that plagioclase first alters to quartz+kaolinite and then a progressive reaction occurs to form adularia with an increased addition of K^+. This is consistent with mineralogical changes that can occur as a function of variable fluid composition, particularly the K^+/H^+ ratio (ratio of potassium ions to hydrogen ions), suggesting that adularia should form as the altering fluid becomes more concentrated with respect to K (Meyer and Hemley, 1967). Alternatively, the transition from clay minerals to adularia can be caused by increasing alteration temperatures at a constant fluid composition (Meyer and Hemley, 1967).

The starting material that undergoes alteration may also have an effect on the alteration mineralogy. Two samples, collected in close proximity to one another, one which appears to have been glassy before the alteration occurred (based on faintly preserved flow banding) and the other from which altered plagioclase was sampled, have very different alteration mineralogies. The sample that was originally glassy altered to dominantly adularia + quartz, whereas the other sample contains dominantly clay minerals, mostly of illitic composition. The variation in alteration products may be due to the low-K, high-Na and -Ca nature of the plagioclase crystal versus the higher K composition of the glass. A starting material higher in K might tend to alter to a more K-rich alteration product (adularia) whereas a lower-K mineral might tend to alter to a lower-K product, such as illite, smectite or kaolinite clay.

Geochemical analyses

Geochemical analyses of altered and unaltered samples of rhyolitic ignimbrites and basaltic andesite lavas show a wide range in composition.

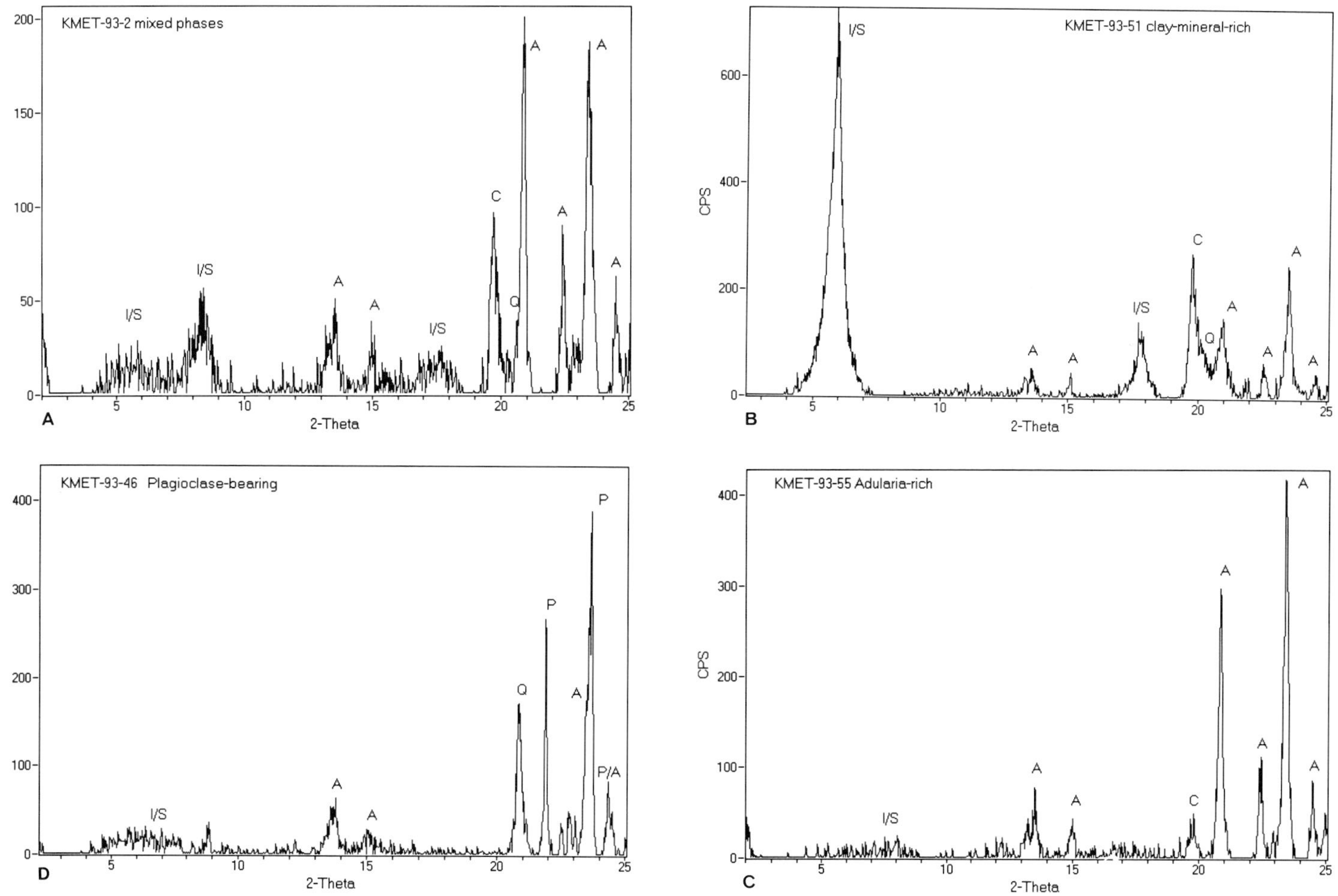

FIGURE 2. X-ray diffraction scans of alteration products of plagioclase showing peaks representative of different minerals. A, sample with mixed alteration phases, B, clay-mineral-rich sample, C, adularia–rich sample, and D, sample containing residual plagioclase. The letters marked on the scans identify main peaks for the different mineral phases. A = adularia; I/S = illite– smectite mixed layer clay minerals; C = clay minerals for which composition cannot be determined; Q = quartz; K = kaolinite; P = plagioclase. Quartz peaks occur as shoulders on adularia peaks, and adularia occurs as a shoulder on a plagioclase peak, hence the letters associated with these shoulders.

Five mafic altered-unaltered pairs of basaltic andesite samples were analysed, including the basaltic andesite of Silver Creek and the LaJara Peak basaltic andesite. Seven altered-unaltered pairs of rhyolitic ignimbrite samples were analysed, including the Hell's Mesa, LaJencia, Upper and Lower Lemitar and Upper and Lower South Canyon tuffs. In order to understand the processes of geochemical variations that occur as a result of K-metasomatism, enrichment or depletion factors (composition of altered rock/ composition of fresh rock) were calculated for elements in each altered/fresh rock pair. These enrichment/depletion factors were then averaged for the rhyolitic ignimbrite and the basaltic andesite samples (Fig. 3 A-C). Although there are some variations in the enrichment/depletion values for individual elements between samples, as should be expected in any geochemical alteration process by a fluid, we feel that averaging the behavior of elements in a number of samples of the same rock type yields the most representative chemical information about the alteration process.

Alteration of a rock by a fluid can lead to mass loss or gain in the original rock, by preferential removal of some phases or deposition of others. Mass gain as a result of K-metasomatism was reported by Glazner (1988), who noted a 14% increase in rock volume as a function of metasomatism. However, the volume of rocks in the Socorro area does not appear to have been changed by the metasomatic process. As can be seen from Fig. 3A, the abundance of Al, an element typically considered immobile during chemical alteration processes, is unchanged by K-metasomatism, in both the rhyolitic ignimbrite samples and the mafic flows. Si, which can be immobile in some types of alteration, is also unchanged. Finally, Ti, which is typically immobile in alteration processes involving fluids, is invariant in the rhyolitic ignimbrites but is slightly depleted in the more mafic flows. The depletion is probably due to alteration of magnetite to hematite, and removal of Ti in the magnetite mineral structure. Considered together, however, the lack of chemical variation of the three immobile elements, Al, Si and Ti, strongly suggest that no net mass gain occurred during the process of K-metasomatism.

Chemical enrichments and depletions

Although there was no net mass change during K-metasomatism, several elements underwent significant enrichments or depletions during the alteration process. Notably K and Rb were strongly enriched (Fig. 3A, 4), and Mg, Ca, Mn, Na and Sr were depleted (Fig. 3A), as was reported by Chapin and Lindley (1986), Glazner (1988) and Brooks (1986). The enrichment of K and depletion of Na and Ca are consistent with the alteration mineralogy observed. As plagioclase and glass in the samples are altered to some combination of adularia, quartz, hematite and clay minerals, Na and Ca are lost, and K is incorporated into the adularia structure, thereby increasing the K content of the sample. Rb will substitute for K in the feldspar structure (Deer, Howie and Zusmann, 1966) and will thereby also be enriched, whereas Sr, which tends to be concentrated in plagioclase and glass relative to potassic feldspar, will be progressively depleted as the plagioclase crystal structure or glass is replaced by potassic feldspar and clay minerals. Mg is probably lost largely through alteration of mafic silicate phases, but the mineralogical details of this alteration process are not well known because our investigation concentrated on alteration of plagioclase and glass. However, the relatively larger depletion of Mg from the basaltic-andesite as compared to rhyolitic ignimbrites favor this interpretation.

Elements of economic interest

In addition to the elements mentioned above, some elements of economic interest (Cu, Ag, Pt, Au, Sn, Pb), or elements that are indicator elements for elements of economic interest (As, Sb, Hg), are enriched in the K-metasomatized rocks (Fig. 3B). Although enrichments of some of these elements have been reported in other areas of K-metasomatism (Roddy et al., 1988, Brooks, 1986), they have not been measured before in the Socorro area. The absolute magnitude of enrichment is small in some cases (Fig. 4), but these elements are consistently enriched in the metasomatized rocks, suggesting that the fluid that caused K-alteration could produce economic ore deposits, if highly concentrated.

The location of these elements in the alteration assemblage (associated with, or incorporated into the structure of, which minerals) has not

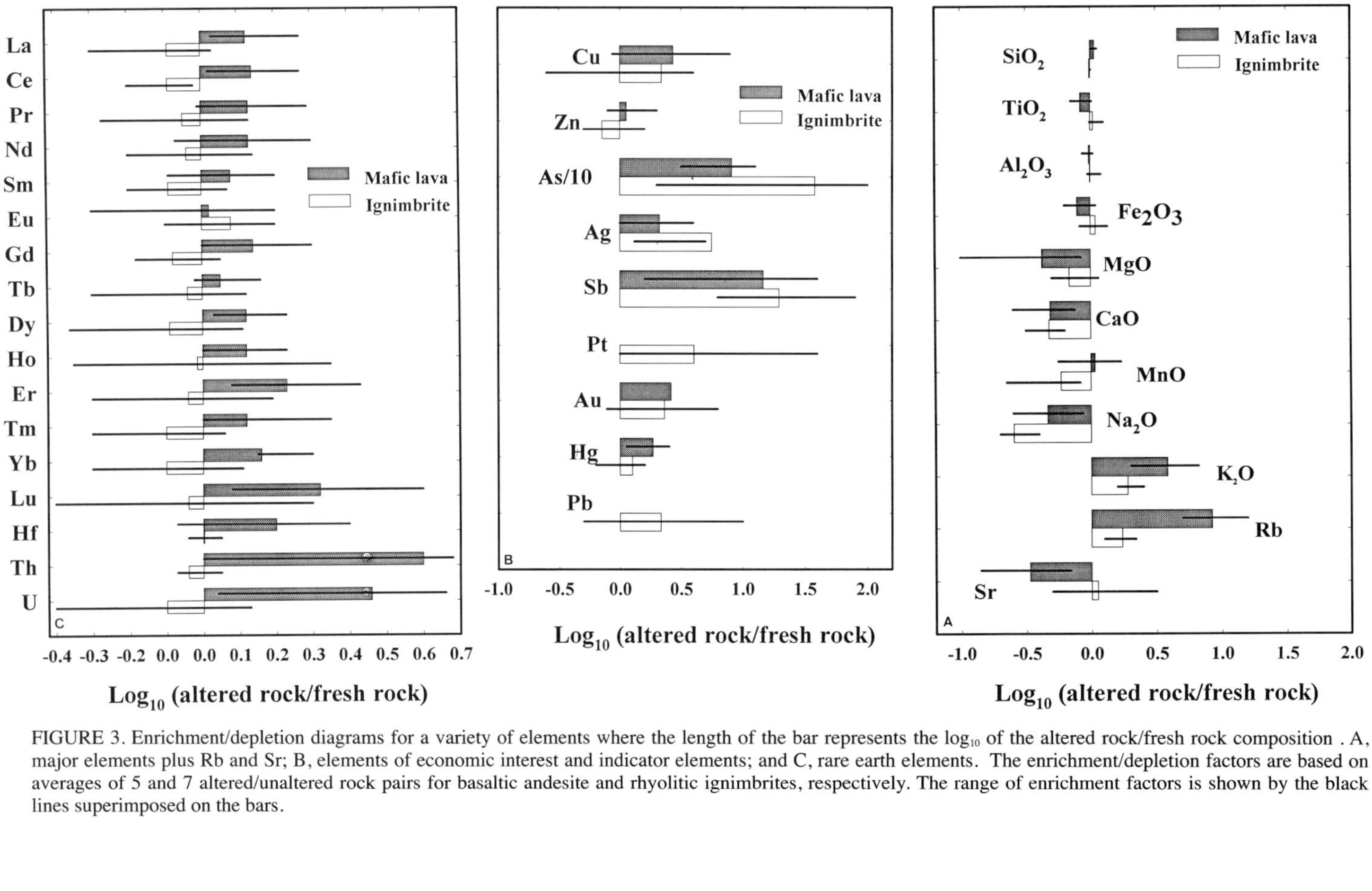

FIGURE 3. Enrichment/depletion diagrams for a variety of elements where the length of the bar represents the $\log_{10}$ of the altered rock/fresh rock composition . A, major elements plus Rb and Sr; B, elements of economic interest and indicator elements; and C, rare earth elements. The enrichment/depletion factors are based on averages of 5 and 7 altered/unaltered rock pairs for basaltic andesite and rhyolitic ignimbrites, respectively. The range of enrichment factors is shown by the black lines superimposed on the bars.

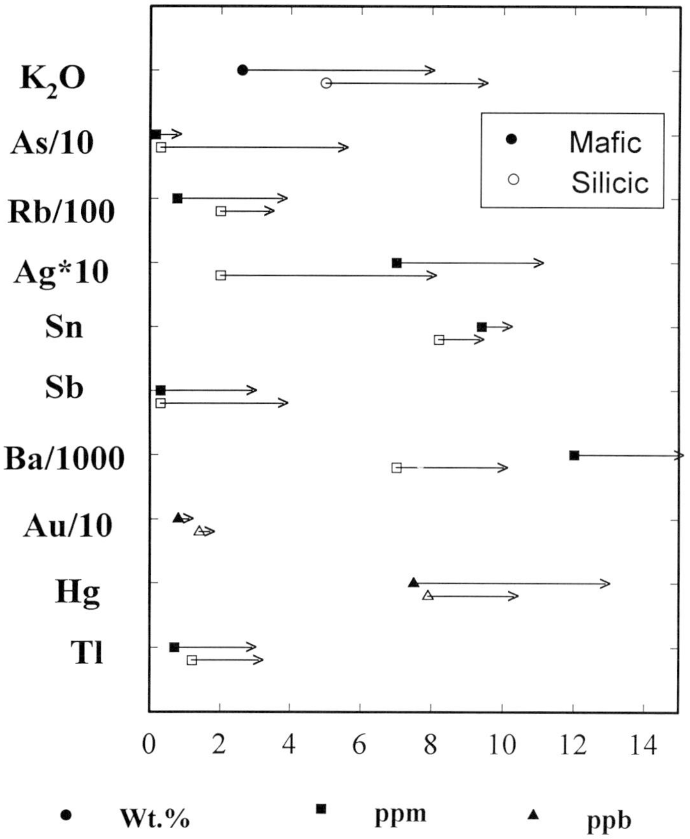

FIGURE 4. Average absolute enrichments of elements in silicic and mafic rocks. The symbols (circle for weight percent, square for ppm, and triangle for ppb) represent the unaltered rock composition. The length of the line shows the average absolute enrichment.

yet been determined, but studies to elucidate this problem are underway. The location and stability in the alteration assemblage for some elements, particularly As, are of interest because of the possible impact on the potability of local water supplies.

Rare-earth elements

Rare-earth elements (REE) have not been studied previously in K-metasomatized rocks. These were analysed in the rocks from the Socorro area, and were found to behave differently in the rhyolitic ignimbrites compared to the more mafic lavas. The REE are consistently enriched in the basaltic-andesite lavas, but are unchanged to slightly depleted in the rhyolitic ignimbrites (Fig. 3C). Although REE have not been determined for most areas of K-metasomatism, Glazner (1988) reported a lack of change in REE by the process of K-metasomatism in rhyolitic rocks from the Mojave Desert. However, subtle variations of REE may have been difficult to detect due to the interpreted mass increase of those rocks.

The strikingly regular variations of REE in the rhyolitic and mafic rocks in the Socorro area may be a function of the alteration mineralogy produced by the metasomatic process. Another possible explanation is that the metasomatic fluid composition plays a role in controlling the REE enrichments and depletions in the mafic and silicic rocks, respectively. However, this would imply that the fluid that altered the basaltic andesite rocks is chemically different than that which altered the rhyolitic rocks, and the close geographical proximity of these two rock types suggests that this is unlikely. Rare-earth elements are not compatible in quartz or adularia minerals produced by metasomatism, but are compatible to varying degrees in clay minerals, and these minerals can be important in controlling the REE composition of certain rocks (e.g., Cullers et al., 1977; Condie, 1991; Elliot, 1993). This leads to the suggestion that the enrichment or depletion of REE in K-metasomatized rocks may be a function of the composition and abundance of clay minerals produced during metasomatism. The clay minerals produced by alteration of a rhyolitic rock are likely to be different than those produced by a mafic rock, due to the different abundances of Na, Ca and K in the two rock types. This interpretation is speculative and more detailed chemical and mineralogical investigations are needed.

Chemical variations as a function of intermediate or advanced metasomatism

In addition to investigating the chemical composition of fresh and highly K-metasomatized mafic and rhyolitic rocks, several samples of rhyolitic rocks believed to be intermediately altered were also analysed. The rocks characterized by an intermediate degree of alteration contain K_2O contents intermediate between fresh rhyolite (~4-5 weight percent K_2O) and highly altered rhyolite (~13 weight percent K_2O). Other elements were analysed in the intermediately altered rocks: not all were found to vary in direct relationship to K_2O (Fig. 5). Some elements, such as Rb and Sr, vary strongly in the first alteration step (from fresh rock to intermediately altered), whereas others, such as Ba, As and Sb, are relatively invariant in the first step of alteration, but are strongly increased in the second step (intermediate to highly altered). The several possible explanations for this chemical behavior all relate to the alteration mineralogy. First, the marked decrease in Sr is likely due to progressive destruction of plagioclase in the host rock. As mentioned earlier, remnant plagioclase is found in some samples that have not undergone extensive alteration, and Sr is a compatible element in the plagioclase crystalline structure. Rb is apparently highly compatible in the phases that form in the early stages of alteration, and less so in the later phases. Our working model of mineralogical alteration involves clay minerals being formed early in the alteration process, and adularia being formed later (e.g., Meyer and Hemley, 1967), and although Rb is known to be compatible in the adularia structure, where it substitutes for K, it is probably more compatible in clay mineral structures and some is actually excluded during the clay mineral-adularia transition. Ba, As, Sb and La are apparently more compatible in the later-formed mineralogical assemblage. In contrast, Cu is highly enriched in the early alteration stages, but depleted later. Cu, and other metals, can be adsorbed onto the outside of mineral grains, rather than being incorporated into the actual chemical structure (Maynard,1991). The larger surface area of clay minerals relative to other phases (Aagaard, 1974) may allow a greater amount of Cu to be adsorbed onto a clay mineral assemblage relative to an assemblage of adularia + quartz. Alternatively, the slightly different fluid composition or temperature responsible for deposition of adularia+quartz relative to clay may cause adsorbed Cu to be mobilized.

Age determinations and implications for source of enriched elements

$^{40}Ar/^{39}Ar$ age determinations of 7.4 ± 0.1 and 8.7 ± 0.1 Ma have been

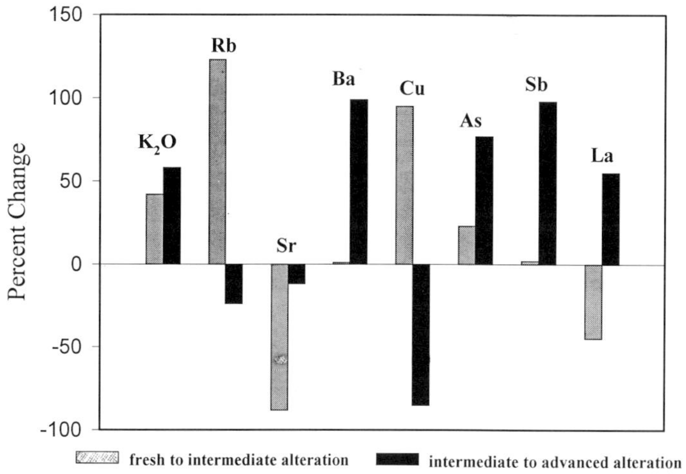

FIGURE 5. Percent change of elements as a function of intermediate and advanced degrees of alteration for a single rhyolitic ignimbrite. Length of the bar represents the percent change during each alteration step.

made on hand-picked adularia from two highly K-metasomatized samples (Fig. 6). The errors quoted are analytical errors to 1σ. Although we attempted to separate as much clay material as possible from the samples, the K/Ca ratio of between 10 and 100, determined as part of the $^{40}Ar/^{39}Ar$ analysis in the younger of the two samples, indicates that some clay minerals may have remained in the sample, and x-ray diffraction analyses confirms this idea. Although the clay minerals mixed in the sample are likely to have also formed by the metasomatic process, $^{40}Ar/^{39}Ar$ of clay minerals is hampered by recoil of ^{39}Ar out of the sample during irradiation, yielding an apparently old age. Thus, the younger of the two ages may be slightly older than the age of formation of adularia in that sample.

Even given that the younger age may be slightly too old, these dates, in conjunction with geological evidence, allow us to speculate about the timing and duration of K-metasomatism. In the northern part of the metasomatized area, playa clays are interbedded with 15 Ma volcanic ash (S. Cather, personal commun., 1994). The youngest known ash bed interbedded with playa clays is 7.85 Ma, but playa clays may have continued to be deposited until as recently as 4.1 Ma (R. Chamberlin, personal commun., 1994). Assuming that K-metasomatism could have occurred during the time when alkaline-saline brines were present, and that the alkaline-saline lake systems were the source of the playa clays, the above estimates of timing of playa clay deposition approximately bracket the time during which metasomatism may have occurred. The $^{40}Ar/^{39}Ar$ ages mentioned above are from one of the most highly metasomatized areas near Socorro and may have been among the latest material metasomatized at a time when the playa lake was smallest and the fluids were the most concentrated with respect to dissolved ions. Playa clay deposits are found near the sampling site for the two dated samples. However, metasomatism may have begun as early as 15 Ma, and may therefore have been an ongoing process for at least as long as 8 Ma. During this time, the playa lake in the enclosed basin or basins may have varied in size and alkalinity as a function of local climate, but nevertheless would have had continuous water input for a long time. The water fed into the basin would be derived from surrounding areas, and through chemical leaching of the rock through and over which the water travelled, would contain a small, but measurable quantity of a wide range of elements. These elements would be concentrated in the playa brines by evaporation, producing a dense fluid that could circulate through underlying rocks (Tyler et al., 1993; Leising et al., 1993). The possible long duration of the metasomatic event, and scavenging of elements from a wide geographical area over this long period could explain the lack of any discernable depleted source area for the elements that were enriched during metasomatism. The lack of a depleted source area has been puzzling, particularly for the element K, for which 6.4 x 10^{10} tons have been added to the rocks in the Socorro area during metasomatism (Chapin and Lindley, 1986).

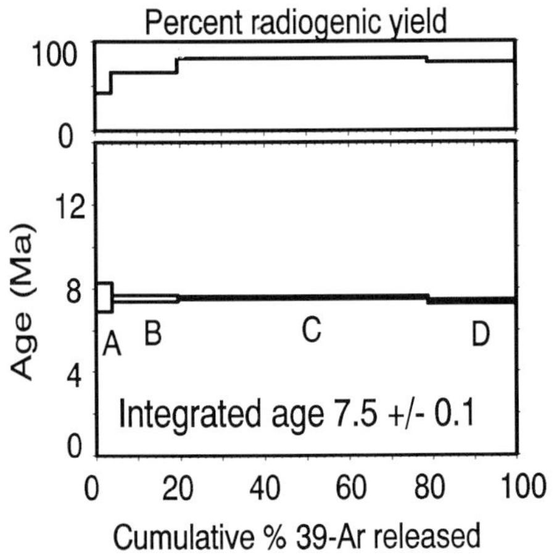

FIGURE 6. $^{40}Ar/^{39}Ar$ age spectrum and percent radiogenic gas release for a step–heated adularia sample produced by K–metasomatism. The A, B, C and D steps represent heating steps during analysis.

CONCLUSIONS

K-metasomatism within an area of central New Mexico, a process caused by alkaline, saline brines in a playa lake environment, caused reactive phases, such as plagioclase phenocrysts and glass, to be altered to a range of secondary minerals, including quartz, adularia, and a range of clay mineral types. The mineralogy produced by alteration is complex, and may have been a function of fluid composition and fluid temperature, as well as composition of the altered rock. The metasomatism caused local rocks to be enriched in K and Rb, as well as other ore-forming or indicator elements, such as Au, Ag, Pt, Hg, As and Sb, but did not cause any net mass change in the altered rocks. During the alteration process, other elements, such as Mg , Ca, Na, Mn and Sr, were depleted. Although alteration of rhyolitic and basaltic-andesite rocks is chemically similar in many respects, some elements, particularly REE, behave differently, suggesting that the presence and composition of clay minerals in the alteration products may play a role in controlling alteration chemistry. The chemical enrichments and depletions of rocks appear to be a multi-step process, probably controlled by progressive mineralogical changes. Based on $^{40}Ar/^{39}Ar$ age determinations on alteration phases, K-metasomatism appears to have been occurring between 7.4 and 8.7 Ma, but may have begun as early as 15 Ma, based on independent geological evidence. The long duration of the playa lake system may account for the lack of a discernable, depleted source area for elements enriched during metasomatism.

ACKNOWLEDGMENTS

This research was supported by the New Mexico Bureau of Mines and Mineral Resources and by NSF EAR-9219758. We would like to thank Chris McKee for assistance with the XRD analyses and Bill McIntosh and Matt Heizler for assistance with the $^{40}Ar/^{39}Ar$ analyses. Thanks to George Austin, Richard Chamberlin and Peter Mozley for helpful reviews of this paper.

REFERENCES

Aagaard, P., 1974, Rare earth elements adsorption on clay minerals: Bulletin Groupe Francais Argiles, v. 26, p. 193–199.

Awwiller, D. N., 1993, Illite/smectite formation and potassium mass transfer during burial diagenesis of mudrocks: a study from the Texas Gulf Coast Paleocene–Eocene: Journal of Sedimentary Petrology, v. 63, p. 501–512.

Barnes, H. L., 1979, Geochemistry of Hydrothermal Ore Deposits [2nd ed.]: New York, John Wiley and Sons.

Brooks, W. E., 1986, Distribution of anomalously high K_2O volcanic rocks in Arizona: Metasomatism at the Picacho Peak detachment fault: Geology, v. 14, p. 339–342.

Chapin, C. E. and Lindley, J. I., 1986, Potassium metasomatism of igneous and sedimentary rocks in detachment terranes and other sedimentary basins: Economic implications: Arizona Geological Society Digest, v. 16, p. 118–126.

Condie, K. C., 1991, Another look at rare earth elements in shales: Geochimica et Cosmochimica Acta, v. 55, p. 2527–2531.

Cullers, R. L., Chaudhuri, S., Arnold, B., Lee, M. and Wolf, C. W., 1975, Rare earth distributions in clay minerals and in the clay–sized fraction of the Lower Permian Haversville and Eskridge shales of Kansas and Oklahoma: Geochimica et Cosmochimica Acta, v. 39, p. 1691–1703.

D'Andrea, J. F., 1981, The geochemistry of the Socorro K2O anomaly, New Mexico [MS thesis]: Tallahassee, Florida State University, 243 p.

D'Andrea–Dinkelman, J. F., Lindley, J. I., Chapin, C. E. and Osburn, G. R., 1983, The Socorro K2O anomaly: a fossil geothermal system in the Rio Grande Rift: New Mexico Geological Society, Guidebook 34, p. 77–78.

Deer, W. A., Howie, R. A. and Zussman, J., 1966, An introduction to the rock forming minerals [1st ed.]: London, Longman Group Limited, v. 1, 528 p.

Duffin, M. E., 1989, Nature and origin of authigenic K–feldspar in Precambrian basement rocks of the North American midcontinent: Geology, v. 17, p. 765–768.

Duval, J. S., 1990, Modern aerial gamma–ray spectrometry and regional potassium map of the conterminеous United States: Journal of Geochemical Exploration, v. 39, p. 249–253.

Elliott, W. C., 1993, Origin of the Mg–smectite at the Cretaceous/Tertiary (K/T) boundary at Stevns Klint, Denmark: Clays and Clay Minerals, v. 41, p. 442–452.

Glazner, A. F., 1988, Stratigraphy, structure, and potassic alteration of Miocene volcanic rocks in the Sleeping Beauty area, central Mojave Desert, California: Geological Society of America Bulletin, v. 100, p. 424–435.

Leising, J. F., Tyler, S. W., Chapin, C. E. and Miller, W. W., 1993, Convection of saline brines in enclosed lacustrian sedimentary basins as a mechanism for potassium metasomatism: Geological Society of America, Abstracts with Program, v. 25, p. A–244.

Lindley, J. I., 1985, Potassium metasomatism of cenozoic volcanic rocks near Socorro, New Mexico [Ph.D. dissertation]: Chapel Hill, University of North Carolina, 563 p.

Maynard, J. B., 1991, Copper: product of diagenesis in rifted basins: Reviews in Economic Geology, p. 199–207.

McDougal, I. and Harrison, T. M., 1988, Geochronology and thermochronology by the $^{40}Ar/^{39}Ar$ argon method [1st ed.]: Oxford, U.K., Oxford Press, 212 p.

McIntosh, W. C., Chapin, C. E., Ratte, J. C. and Sutter, J. F., 1992, Time–stratigraphic framework for the Eocene–Oligocene Mogollon–Datil volcanic field, southwestern New Mexico: Geological Society of America Bulletin, v. 104, p. 851–871.

Meyer, C. and Hemley, J. J., 1967, Wall rock alteration; in Barnes, H. L., ed., Geochemistry of hydrothermal ore deposits [1st ed.]: New York, Holt, Rinehart, and Winston, Inc, v. 1, p. 166–232.

Munhá H., Fyfe, W. S. and Kerrich, R., 1980, Adularia, the characteristic mineral of felsic spilites: Contributions to Mineralogy and Petrology, v. 75, p. 15–19.

Roddy, M. S., Reynolds, S. J., Smith, B. M. and Ruiz, J., 1988, K–metasomatism and detachment–related mineralization, Harcuvar Mountains, Arizona: Geological Society of America Bulletin, v. 100, p. 1627–1639.

Saunders, C. M. and Tuach, J., 1991, Potassic and sodic alteration accompanying gold mineralization in the Rattling Brook deposit, Western White Bay, Newfoundland Appalachians: Economic Geology, v. 86, p. 555–569.

Sheppard, R. A. and Gude, A. J., 1973, Zeolites and associated authigenic silicate minerals in tuffaceous rocks of the Big Sandy Formation, Mohave County, Arizona: U. S. Geological Survey, Professional Paper 830, p. 36.

Smith, B. M., Reynolds, S. J., Day, H. W. and Bodnar, R. J., 1991, Deep–seated fluid involvement in ductile–brittle deformation and mineralization, South Mountains metamorphic core complex, Arizona: Geological Society of America Bulletin, v. 103, p. 559–569.

Turner, C. E. and Fishman, N. S., 1991, Jurassic lake T'oo'dichi': a large alkaline, saline lake, Morrison Formation, eastern Colorado Plateau: Geological Society of America Bulletin, v. 103, p. 538–558.

Tyler, S. W., Ferguson, J., Lyons, W. B. and Wooding, R. A., 1993, Evidence of brine fingering in ground water beneath saline lakes: Geological Society of America, Abstracts with Program, v. 25, p. A–182.

Wintsch, R. P., and Kvale, C. M., 1994, Differential mobility of elements in burial diagenesis of siliciclastic rocks: Journal of Sedimentary Research, v. A64, p. 349–361.

Day 1 Stop 3
Jordan 1993

ALPINE 1/FEDERAL COREHOLE — SUBSURFACE STRATIGRAPHY OF THE EASTERN WHITE MOUNTAINS, APACHE COUNTY, ARIZONA

JAMES C. WITCHER[1], W. RICHARD HAHMAN[2] and CHANDLER A. SWANBERG[3]

[1]New Mexico State University, Southwest Technology Development Institute, Las Cruces, NM 88003; [2]Consultant, 780 Suzanne Ave., Las Cruces, NM 88005; [3]Consultant, 5478 E. Cholla St., Scottsdale, AZ 85254

Abstract—The Alpine 1/Federal corehole provides the first observations of Permian, Upper Cretaceous, and lower Tertiary rocks in the eastern White Mountains of Arizona. Due to erosion beneath a pre-Late Cretaceous unconformity, no Triassic to Lower Cretaceous rocks exist. Only 74 ft of Permian San Andres Formation is preserved. The Permian Glorieta Sandstone, 203 ft thick, is intruded by basaltic rocks that also truncate the top of the underlying Corduroy Member of the Supai Formation. An additional basaltic intrusion separates the Corduroy Member, 515 ft thick, from the underlying Fort Apache Limestone, 48 ft thick. Only 49 ft of Big A Butte Member was cored before encountering another basaltic intrusion. Permian carbonate units represent restricted marine and coastal sabkha environments and hydrocarbon shows indicate source potential. Carbonaceous Upper Cretaceous Dakota (?) sandstone, 116 ft thick, is preserved beneath a Laramide unconformity. Homoclinal downwarping, less than one degree of dip to the south and with possible faults, has formed a basin with at least 1600 ft of structural relief on the Laramide erosional surface. An unnamed early (?) Laramide unit, 107 ft thick, unconformably rests on reddened Cretaceous sandstone and underlies the basal conglomerate of the Eocene Eagar Formation, 1087 ft thick. Volcaniclastic rocks of the Eocene to Oligocene Spears Group, 2032 ft thick, conformably overlie the Eagar Formation. The only distinct volcanic flow that was cored is a 20-ft-thick Datil (?) Group rhyolite ash-flow tuff. Composite thickness of the Tertiary sediments in the corehole is 3226 ft. Basaltic intrusions in Paleozoic rocks may range in age from mid-Tertiary to Quaternary and could represent several magmatic events.

INTRODUCTION

The primary mission of the Alpine 1/Federal drilling project was to assess the hot dry rock (HDR) geothermal energy potential of the eastern White Mountains region. However, the purpose of this paper is to present an overview of field geologic logs of core for the Alpine 1/Federal hole at Alpine Divide. Our discussion attempts to correlate units within a regional stratigraphic framework. Also, special attention is given to the details of stratigraphy and depositional environments of the Permian Supai units because of potential oil and gas implications.

No definite pre-Tertiary stratigraphic information exists for the eastern White Mountains area. There are a lack of outcrops as well as deep borehole data beneath Tertiary volcanic and sedimentary rocks in the region for at least 40 to 70 mi to the east, south and west, and 20 to 40 mi to the northwest, north and northeast. Alpine Divide is located on the Colorado Plateau adjacent to the boundary with the transition zone of the Basin and Range province, near the southwest margin of the Paleozoic Defiance-Zuni positive area, and along the Jemez volcanic zone or lineament (Fig. 1). All of these features are related to regional crustal structures that have had a major influence on the nature of Paleozoic through Tertiary deposition and erosion. As a result, projection of regional Paleozoic isopachs, structure, and facies into the subsurface of the area, based on the nearest well and outcrop control to the northwest, north and northeast, has been uncertain. The Alpine 1/Federal corehole provides the first observations of the lower Tertiary, Upper Cretaceous and Permian rocks beneath the eastern White Mountains region.

LOCATION AND CORE

The Tonto Alpine 1/Federal (state permit 878) is located 15 mi from the New Mexico border in Arizona in the SW1/4 sec. 23 T6N R30E, 1153 ft FWL and 2122 ft FSL, at an elevation of 8556 ft at Alpine Divide on the east side of U. S. Highway 180/190/666. Alpine Divide is in the Apache-Sitgreaves National Forest, 5 mi north of Alpine, in Apache County in the eastern White Mountains area (Fig. 1).

Core recovery was over 99% for the two sizes of core that were obtained. The HQ (2.50 in. diameter) core from 500 to 3265.5 ft, which comprises the Tertiary stratigraphic section, is stored at the New Mexico Bureau of Mines and Mineral Resources, in Socorro, New Mexico. The HQ and NQ (1.875 in. diameter) core from 3265.5 to 4505 ft, which comprises the Mesozoic and Paleozoic stratigraphic section, is stored at the Arizona Geological Survey, in Tucson, Arizona. Table 1 summarizes formation footages.

STRATIGRAPHY

Previous studies

Reconnaissance geologic studies are the only available information specific to the Alpine region (Sirrine, 1956; Weber and Willard, 1959; Wrucke, 1961). Weber and Willard (1959) and Wrucke (1961) mapped Pennsylvanian carbonate rocks in isolated outcrops on the northeast side of Escudilla Mountain and just across the New Mexico-Arizona border. Sirrine (1956) and Wrucke (1961) described the lower Tertiary Eagar Formation and the middle Tertiary Datil Formation (now Spears Group) rocks in the area. Gerrard (1966) and Winters (1963) described Supai Formation stratigraphy in the eastern Mogollon Rim region; Foster (1964) discussed the stratigraphy of deep oil and gas tests; Ratte et al. (1969) and Ratte (1989) mapped the Tertiary geology of areas south and southeast of Alpine Divide; and Merrill and Pewe (1977) covered the late Cenozoic volcanism in the White Mountains.

Permian Supai Formation

Nomenclature

Sandstones, evaporites and carbonate rocks below the Permian Coconino and Glorieta sandstones are the subject of debate over regional correlation and use of nomenclature. Eastward in New Mexico, rocks below the Glorieta Sandstone are assigned to the Yeso Formation. West of the Alpine Divide site, these rocks are assigned to the Schenebly Hill Formation (Blakey, 1980), the Supai Formation (Winters, 1963), and Supai Group (Peirce, 1989). The nomenclature of Peirce (1989) and Blakey (1980) is formulated upon regional interpretations of stratigraphy far removed from the Alpine area. The measured sections used by Winters (1963) are within 50 to 70 mi of Alpine Divide and contain a limestone unit, the Fort Apache Limestone Member of the Supai Formation, that is easily recognized in the Alpine 1/Federal. From top to bottom, the Winters nomenclature for Supai units are the Corduroy Member, Fort Apache Limestone Member, Big A Butte Member, and Amos Wash Member.

Gerrard (1966) redefined the Fort Apache Limestone Member of Winters (1963) as the Fort Apache Member by including all contiguous limestone, dolomite and evaporites at the base of the Corduroy Member. This redefinition correlates subsurface well data in the Holbrook basin with the Mogollon Rim outcrops of Winters (1963). Gerrard's (1966) correlations assume that the lowest and thickest dolomite/evaporite sequence in the Corduroy Member is the same

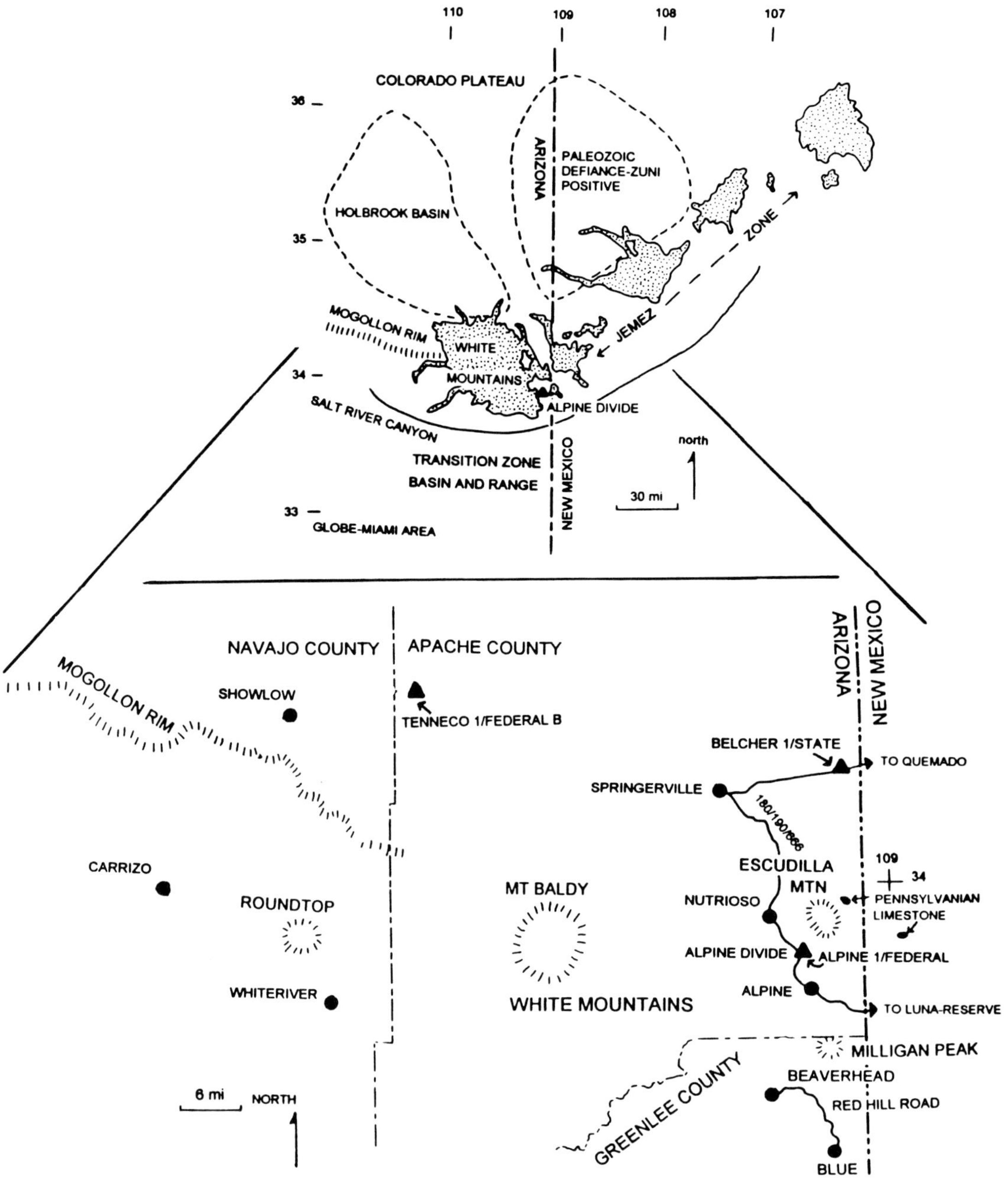

FIGURE 1. Location maps for the Alpine 1/Federal corehole. Stippled areas are Cenozoic (<15 Ma) volcanic rocks of the Jemez zone.

laterally continuous unit as the Fort Apache Limestone in outcrop.

Blakey (1980) retained Gerrard's (1966) Corduroy and Fort Apache Member designations in his Schenebly Hill Formation, whereas Peirce (1989) upgraded Winters' (1963) Corduroy Member to formation status in the Supai Group. Peirce's (1989) Corduroy Formation includes the Fort Apache Limestone as a member. Because of potential differences in depositional environment implied by the dolomite/evaporite (intertidal-to-supratidal coastal sabkha) and the limestone (possible restricted shallow marine), we choose to use Winters' (1963) nomenclature.

Big A Butte Member

The base of the cored 49-ft-thick section of the Big A Butte Member, representing only the upper part of the unit, has an intrusive contact with basaltic rocks at 4454 ft (Fig. 2). The majority of the Big A Butte Member from 4405 to 4454 ft is a brown, very fine-grained sandstone and siltstone. Between 4421 and 4428 ft a fine-to-medium crystalline, nodular-to-mosaic anhydrite interval overlies 1 ft of gray dolomite with black shale laminations.

Fort Apache Limestone

The Fort Apache Limestone Member, a dark gray to black, fetid, fossiliferous micrite and biomicrite with vuggy and moldic porosity, is identified between 4322 and 4327 ft and 4362 and 4405 ft (Fig. 2). Hydrocarbon films and staining occurs between 4399 and 4402 ft. The Fort Apache Limestone is conformable with the Big A Butte Member. A basaltic dike or sill intrudes the Fort Apache Limestone from 4327 to 4362 ft. If the basalt is a sill, the Fort Apache Limestone is thinned to about 50 ft in the Alpine subsurface compared to a thickness of about 95 to 120 ft in the region around Whiteriver, west of the borehole (Winters, 1963). However, if the basalt is a dike, the Fort Apache at Alpine Divide has similar thickness to outcrops around Whiteriver.

Corduroy Member

Basaltic rocks intrude the contact between the Corduroy Member and the Fort Apache Limestone from 4260 to 4422 ft. For descriptive purposes, the Corduroy Member is divided into six major subdivisions (Fig. 2). Mostly brown microcrystalline dolomite and dolomitic limestone, with interbedded nodular and mosaic blue-gray anhydrite, forms the basal unit 1, from 4226 to 4260 ft. Some of the dolomite horizons in this interval are fetid and have wavy laminations and black shale laminae, whereas other dolomite layers exhibit scattered blue-gray anhydrite grains and nodules.

From 4158 to 4226 ft, unit 2 forms a sandy section of light brown to dark brown, dolomitic, fine-grained sandstone and muddy siltstone with wavy laminations, some cross laminations, bioturbation and minor anhydrite at 4203 ft.

A mostly carbonate section, unit 3, is observed from 4038 to 4158 ft. Lost circulation occurred while coring unit 3. Finely crystalline dolomite, silty and sandy dolomite, and dolomitic limestone are interbedded with minor, fetid and fossiliferous, laminated limestone. Gray-brown and brown colors predominate and wavy laminations and black shale laminae are common. A few dolomite horizons have small (less than 1.0 in.) blue-gray anhydrite nodules. From 4128 to 4145 ft, numerous calcite-filled fractures occur.

The carbonate strata in unit 4, between 3906 and 4038 ft, are essentially the same as in unit 3, except more sandy and silty dolomite is present. Unit 4 also has two interbeds of 10-to-12 ft-thick, brown to red brown siltstone and fine-grained sandstone with flaser, wavy laminations, scour and rip-up clasts. Two 8 to 10 ft intervals of anhydrite were

TABLE 1. Summary of formation footages.

lower Oligocene (?) upper Spears Group		
	0 to 854 ft	thickness 854 ft
upper Eocene (?) andesite breccia (possible clastic unit associated with andesite of Dry Leggett Canyon)		
	854 to 942 ft	thickness 88 ft
upper Eocene (?) middle Spears Group - conglomerate and sandstone (volcanic-clast conglomerate and sandstone)		
	942 to 1018 ft	thickness 76 ft
upper Eocene (?) Datil Group ignimbrite (possibly the tuff of Bishop Peak)		
	1018 to 1038 ft	thickness 20 ft
upper Eocene (?) middle Spears Group - conglomerate and sandstone (volcanic-clast conglomerate and mixed-clast conglomerate with volcanic and Precambrian granite and Paleozoic limestone clasts)		
	1038 to 1358 ft	thickness 320 ft
upper Eocene (?) middle Spears Group - pumiceous sandstone		
	1358 to 1574 ft	thickness 216 ft
upper Eocene lower Spears Group - andesitic sandstone		
	1574 to 2052 ft	thickness 478 ft
Eocene Eagar Formation		
	2052 to 3139 ft	thickness 1087 ft
Upper Cretaceous(?)/ Paleocene (?) /lower Eocene (?) unnamed beds		
	3139 to 3246 ft	thickness 107 ft
Cretaceous Dakota (?) Sandstone		
	3246 to 3362 ft	thickness 116 ft
Permian San Andres Formation		
	3362 to 3436 ft	thickness 74 ft
Permian Glorieta Sandstone		
	3436 to 3639 ft	thickness 203 ft
Tertiary (?)/Quaternary(?) basaltic intrusion		
	3639 to 3751 ft	thickness 112 ft
Permian Corduroy Member Supai Formation		
	3751 to 4266 ft	thickness 515 ft
Tertiary (?)/Quaternary(?) basaltic intrusion		
	4260 to 4322 ft	thickness 62 ft
Permian Fort Apache Limestone Member Supai Formation		
	4322 to 4327 ft	thickness 5 ft
Tertiary (?)/Quaternary(?) basaltic intrusion		
	4327 to 4362 ft	thickness 35 ft
Permian Fort Apache Limestone Member Supai Formation		
	4362 to 4405 ft	thickness 43 ft
Permian Big A Butte Member Supai Formation		
	4405 to 4454 ft	thickness 49 ft
Tertiary (?)/Quaternary(?) basaltic intrusion		
	4454 to 4505 ft	thickness 51 ft

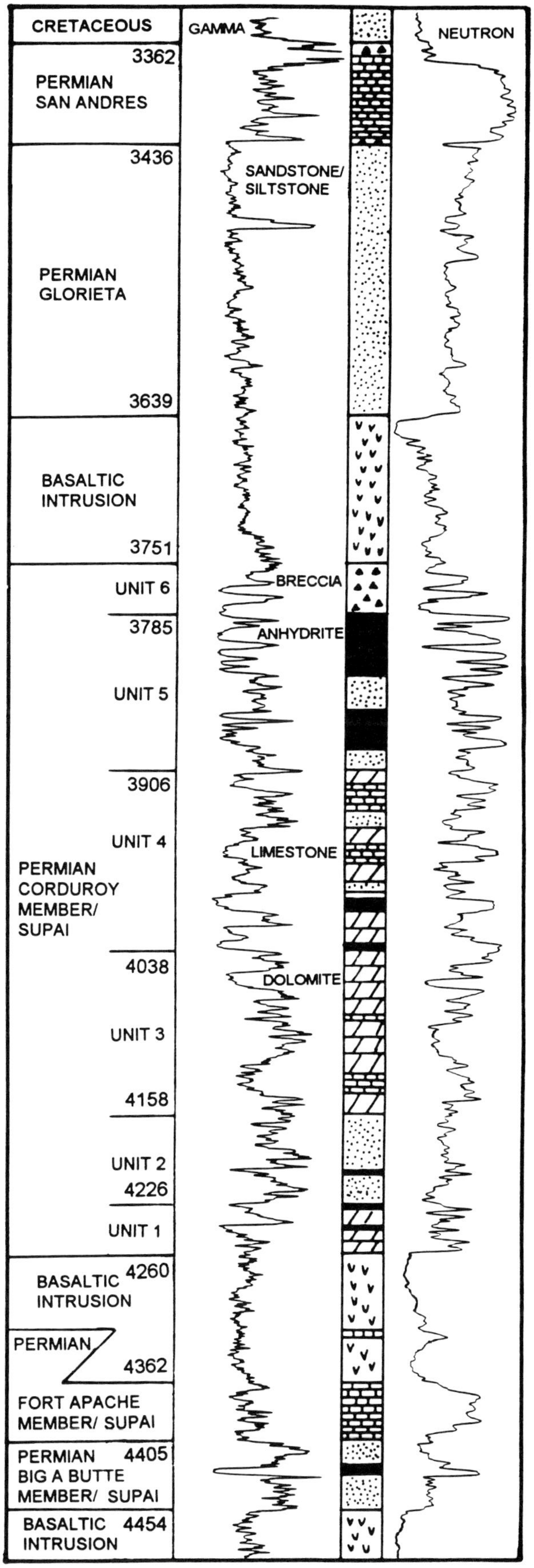

FIGURE 2. Summary stratigraphic column of the Paleozoic rocks cored in the Alpine 1/Federal hole.

also encountered in the lower one-third of unit 4. Rauzi (1994) observed hydrocarbon stains and leakage from around vertical fractures in a fetid limestone from 4027 to 4031 ft after the core had been placed in storage in Tucson.

Unit 5, from 3906 to 3785 ft, consists of interbedded light gray to gray anhydrite and mostly brown to gray-brown siltstone and fine sandstone. The anhydrite intervals ranged in thickness from 20 to 5 ft. Mosaic textures predominate. The anhydrite intervals are generally subdivided into several 1-to-3-ft thick beds separated by scour, very thin siltstone beds, sandy dolomite or black wavy shale laminae. The sandstone and siltstone beds contain scour, rip-up clasts, soft-sediment deformation, and anhydrite nodules.

The upper 35 ft, unit 6, of the Corduroy Member is characterized by a probable solution-collapse of rubble breccia. This zone was characterized by major lost circulation during drilling. The lower breccia, starting at 3785 ft, is vuggy sucrosic dolomite that contains "saddle" dolomite crystals. Dolomite breccia at the top of the zone is black with abundant microfractures. Scattered anhydrite in the breccia shows mosaic textures. Basaltic rocks intruded the top of the Corduroy Member breccia unit at 3751 ft.

Discussion and interpretation of Supai rocks

A restricted, shallow marine or subtidal lagoon depositional environment is interpreted for the Fort Apache Limestone. Overall, a coastal sabkha depositional environment is indicated for the Big A Butte Member and Corduroy Members. A supratidal facies apparently dominates unit 5 of the Corduroy. Units 3 and 4 of the Corduroy probably indicate a range of sabkha facies from subtidal, intertidal to supratidal. Unit 2 may also represent a supratidal environment with abundant terrestrial clastic influx. A detailed study is required to delineate sequence stratigraphy and define the depositional environments in detail. Also, source terrains and provenance studies of the clastic units in the Corduroy and Big A Butte Member may significantly augment regional stratigraphic correlations and paleogeographic interpretations.

A total of 515 ft of the Corduroy Member was cored in the Alpine 1/Federal. If basalt intrusions at the base and top of the Corduroy are dikes, the Corduroy Member beneath Alpine Divide may have a thickness approaching 680 ft. Winters (1963) reported approximatelty 330 ft of the Corduroy Member west of Whiteriver. The Corduroy Member in the Alpine area also contains considerably more carbonates and evaporites than measured sections to the west near Whiteriver.

Permian Glorieta Sandstone

The Glorieta is a fluvial equivalent of the Coconino eolian sandstones to the west (Peirce, 1989). Contact between the Glorieta Sandstone and the underlying Corduroy Member is obscured by a basalt intrusion (Fig. 2). The Glorieta Sandstone, from 3436 to 3639 ft, consists of medium- to fine-grained, well-sorted, light-gray and white sandstone with wavy and parallel laminations, ripple cross-laminations, cross laminations, and cross bedding. Calcite, quartz and dolomite (?) cement is present. Isolated zones, generally less than 0.5 in. diameter, contain marcasite or pyrite and carbonaceous material and represent local reducing conditions ("reduction spots"). Marcasite or pyrite and black carbonaceous material is common along wavy laminations. Healed and open high-angle fractures are present. Open fractures contain calcite crystals and contributed to major lost circulation during drilling, especially between 3450 and 3520 ft. Intergranular porosity is indicated by drill mud sieving (mud buildups) on outer core surfaces, except on darker gray "reduction spots." A total of 203 ft of Glorieta Sandstone was cored.

Permian San Andres Formation

The Permian San Andres Formation is distinguished from equivalent strata (Kaibab Formation) to the west by a more restricted marine depositional environment (Peirce, 1989). The lower 3 ft of the San Andres Formation, just above the Glorieta Sandstone, is a probable solution-collapse breccia with a black shale and dark micrite matrix, possibly rich in carbonaceous material (Fig. 2). The San Andres limestone, between 3362 and 3436 ft, is fetid, finely crystalline, medium to dark gray, and brown. The San Andres micrites and biomicrites have black shale laminae and black, parallel and wavy laminations. Dark sutured stylolites are common. Hydrocarbon films are present along stylolite and fracture surfaces. Vertical fractures and minor small-scale vugs, partially filled with calcite crystals, are common. Several fractures probably contributed to lost circulation during coring.

The top of the San Andres Formation was picked at a silicified and brecciated zone at 3362 ft. Only 74 ft of San Andres is preserved. The silicified zone represents a post-Triassic to pre- Late Cretaceous regional unconformity.

Cretaceous Dakota (?) Formation

Unconformably above the San Andres Formation is a carbonaceous, medium- to coarse-grained sandstone with calcite cement. This unit is tentatively correlated with the Cretaceous Dakota Formation (Nations, 1989). In the Alpine 1/Federal, the Dakota Formation is cross laminated and has abundant ripple cross-laminated zones with much carbonaceous laminae. This unit is a light gray, dark gray and black, moderately well-sorted sandstone. Marcasite or pyrite is abundant, especially in association with carbonaceous zones. Where carbonate cements are lacking in ripple troughs or in high-porosity coarse-sand laminae, the carbonaceous material appears to be diagenetic. A 4 in. bentonite bed was observed at 3309 ft, together with laminated coal and carbonaceous shale at 3339 feet depth. The upper 10 ft, from 3246 to 3256 ft, of the Cretaceous sandstone is yellow to orange brown, but shows the same depositional structures and textures as underlying dark-gray sandstone (Fig. 3). We interpret this upper 10 ft zone to be oxidized by weathering.

The depositional environment of the Cretaceous sandstone is uncertain. A marine barrier island system, delta, or terrestrial fluvial environment are all possible, but further study of this unit is needed.

Cretaceous or Tertiary unnamed beds

An unnamed stratigraphic unit, from 3139 to 3246 ft, was encountered unconformably above the Cretaceous sandstone in the Alpine 1/Federal corehole (Fig. 3). A limestone-pebble conglomerate marks the base of the unit. These beds, 107 ft thick, are orange-red and orange, mostly siltstone and fine- to medium-grained silty sandstone with calcite cement. A similar orange and orange-red unit of limited lateral extent, beneath the Eocene Mogollon Rim basal conglomerate, unconformably overlies Cretaceous sediments near Round Top, about 55 mi to the west of Alpine Divide (Potochnik, 1989). The depositional setting of this unit is different than the overlying Eagar Formation and minor granule-to-cobble conglomerate beds suggest a different provenance. Clasts are predominantly limestone, with minor intermediate volcanic clasts, in contrast with gravels in the overlying Eagar Formation that are rich in Precambrian plutonic and metamorphic

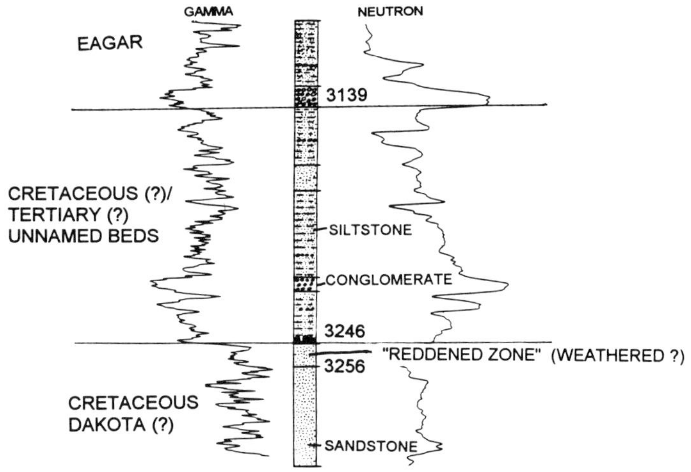

FIGURE 3. Stratigraphic relationships of the Cretaceous or Tertiary unnamed beds.

clasts. The beds show both parallel and cross laminations and some ripple cross-laminated zones. The occurrence of intermediate volcanic clasts may indicate a Late Cretaceous to early Tertiary (early Laramide) age. It is possible that the unit is a southern Colorado Plateau/transition zone equivalent to early Laramide units, namely, the Fort Crittenden and Ringbone formations of the Basin and Range of southeastern Arizona and southwestern New Mexico (Dickinson et al., 1989; Lawton et al., 1993). Alternatively, the beds may record the initial deposition of the overlying Eagar Formation. The depositional structures and the fine-grained and regionally discontinuous nature of the unit suggests low-energy fluvial deposits that filled isolated, low-relief valleys cut into the Cretaceous sandstones.

Tertiary Eagar Formation

Unconformably overlying the unnamed Cretaceous or Tertiary unit at 3139 ft are conglomerate beds containing abundant plutonic and metamorphic clasts with Precambrian affinity. Lower Tertiary sediments that have Precambrian basement and Paleozoic provenance are variously designated in the region as the Baca Formation (Cather and Johnson, 1984), Eagar Formation (Sirrine, 1956), Mogollon Rim Formation (Potochnik, 1989), and Rim Gravels (Cooley and Davidson, 1963). Potochnik's (1989) and Sirrine's (1956) sections also contain intermediate volcanic clasts in the basal conglomerate, which is not a characteristic of the Baca Formation to the east in New Mexico. Precambrian-clast conglomerate strata in the Alpine 1/Federal also contain andesite at the base, and are therefore, correlated more closely with the Eagar and Mogollon Rim formations. The Baca, Eagar and Mogollon Rim formations are Eocene age (Cather and Johnson, 1984; Potochnik, 1989). Ash-fall tuffs, 125 and 220 ft below the top of the Mogollon Rim Formation, range from 37.6 to 37.5 Ma, respectively (Potochnik, 1989). Peirce et al. (1979) reported that three volcanic clasts in the Mogollon Rim conglomerate range in age from 54.6 to 66.7 Ma.

The Eagar Formation at Alpine Divide is divided into two major units. The lower unit, from 2580 to 3139 ft, is composed of matrix-supported conglomerate and red-brown mudstone (Fig. 4). Clasts within the conglomerate beds are composed mostly of Precambrian lithologies, ranging from granite and metamorphic rocks to younger Precambrian limestone and sandstone of the Apache Group. The conglomerate beds also contain intermediate volcanic porphyry clasts, possibly derived from the Globe-Miami area of Arizona (Potochnik, 1989). Matrix in the conglomerate beds consists of poorly consolidated, sandy silt and clay. This interval presented drilling problems due to washouts, caving and maintenance of proper drilling mud properties. The gravels rich in Precambrian lithologies at the base of the Eagar and Mogollon Rim formations represent the uplift and unroofing of basement-cored uplifts, associated with the later phases of the Laramide Orogeny in the Salt River Canyon region of east-central Arizona (Potochnik, 1989).

From 2052 to 2574 ft, the Eagar Formation consists of mostly siltstone and fine- to medium-grained sandstone with minor interbedded conglomerate and red-brown mudstone. The upper unit of the Eagar Formation is predominantly a brown and brown-gray arkosic litharenite, showing bioturbation, parallel laminations, ripple cross-lamination, flaser, soft-sediment deformation, very thin to medium bedding, weak pedogenic calcite, and root casts.

A detailed study of the core is required to correlate the Alpine Divide section with facies of the Mogollon Rim Formation (Potochnik, 1989) and the Baca Formation (Cather and Johnson, 1984). However, the depositional environment of the Eagar Formation in the Alpine 1/Federal appears to be dominated by fluvial channel deposits in the lower unit and floodplain deposits in the upper unit.

Cather and Johnson (1984) defined the top of the Baca Formation at a level where volcanic debris, characteristic of the overlying Spears Formation (now Spears Group), becomes abundant. Potochnik (1989) defined the top of the Mogollon Rim Formation at the horizon where volcanic clasts roughly equal Paleozoic and Precambrian clasts. With Potochnik's (1989) criteria, the top of the Eagar Formation is approximately 2052 depth in the Alpine Divide corehole. This top occurs where coarse, quartz-rich sandstones of the Eagar Formation give way to fine- to medium-grained, quartz-poor, andesitic sandstones of the lower Spears Group (S. M. Cather, personal commun., 1994).

Tertiary stratigraphic nomenclature

Outcrops of Tertiary volcaniclastic sedimentary rocks and volcanic rocks in the Alpine area have been designated as the Tertiary Datil Formation or Tertiary Datil Group in most previous studies in the area. However, Osburn and Chapin (1983) restricted the use of Datil nomenclature in its type locality in the eastern Mogollon-Datil region in New Mexico to volcanic rocks and related sediments beneath the 32 Ma Hells Mesa Tuff. The Hells Mesa Tuff is not known to extend into the Alpine region or into much of the western Mogollon-Datil region of New Mexico and Arizona (Marvin et al., 1987; Ratte, 1989).

In the Alpine area, the 900 ft of basaltic lavas that cap Escudilla Mountain may represent the Bearwallow Mountain Andesite (R. M. Chamberlin, personal commun., 1994). Marvin et al. (1987) bracketed the Bearwallow Mountain andesite at 23 to 27 Ma. Also, a sanidine-rich ash-flow tuff crops out a short distance below the Escudilla Mountain basaltic rocks and near the top of the Tertiary volcaniclastic sedimentary section (Wrucke, 1961). This ash-flow tuff may correlate with the regionally extensive Bloodgood Canyon Tuff that McIntosh et al. (1991) showed is 28.1 Ma. Volcaniclastic and minor volcanic rocks in the Alpine area overlie the Eagar Formation and lie below the

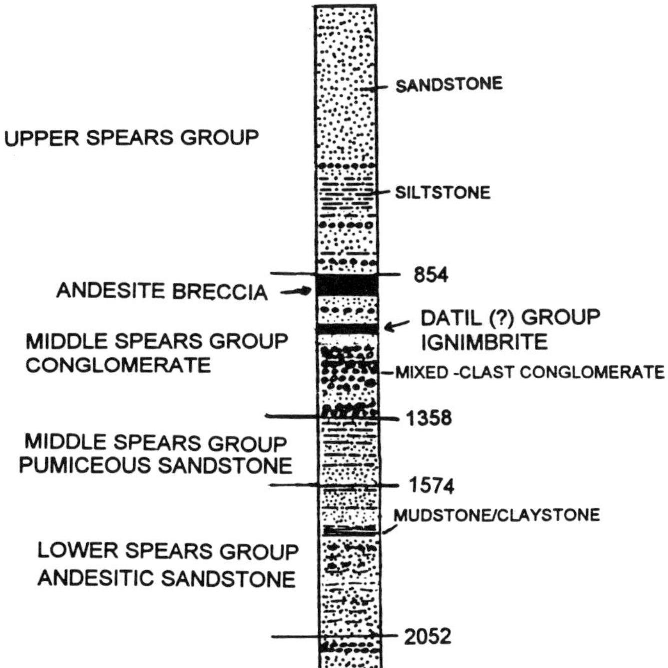

FIGURE 4. Summary stratigraphic column of the Tertiary rocks in the Alpine 1/Federal hole.

presumed Bearwallow Mountain Andesite. Therefore, volcaniclastic rocks in the Alpine area may be younger, at least in part, than the Hells Mesa Tuff; especially, the 650 ft of Tertiary sediments exposed on the surface at Alpine Divide upward to the base of basaltic rocks on Escudilla Mountain.

Cather et al. (1994, this volume) redefine Tertiary nomenclature in the Mogollon-Datil volcanic field of western New Mexico and eastern Arizona. Volcanic rocks older than 31.4 Ma are assigned to the Datil Group, whereas younger Oligocene volcanic rocks are placed in the Mogollon Group. More importantly to this paper, all sedimentary rocks above the Baca Formation (Eagar Formation) and below the Santa Fe Group (Gila Group), but interbedded with the Mogollon and Datil Group volcanic sequences, are referred to as the Spears Group. Since nearly all of the post-Eagar Formation rocks in the Alpine Divide corehole are volcaniclastic sedimentary rocks, the Spears Group nomenclature is most appropriate.

Lower Spears Group - andesite sandstone unit

The lower Spears Group, from 2052 to 1574 ft, grades downward into the Eagar Formation. Generally, both units have similar depositional structures and lithology except that the lower Spears contains more andesite volcanic detritus, less quartz, and fewer beds of red-brown mudstone, and it coarsens upward. This unit is mostly gray, clayey siltstone and clayey, fine- to medium-grained, andesitic sandstone that shows possible bioturbation, parallel laminations, ripple cross-lamination, flaser, soft-sediment deformation and very thin to medium bedding. The lower Spears Group sediments probably represent fluvial plain deposits and distal alluvial apron deposits of andesitic volcanoes to the south and southeast.

Middle Spears Group - pumiceous sandstone unit

Light gray, gray-green, and pink muddy sandstone and mudstone with abundant felsic pumice lapilli occur from 1358 to 1574 ft. This unit is assigned to the middle Spears Group because it may record the initiation of silicic pyroclastic volcanism in the Mogollon-Datil volcanic field to the south and east of Alpine Divide. At 1438 ft, a 2-to-3-in.-thick, ash-fall tuff was encountered. An isotopic-age determination of the tuff may assist to constrain the volcanotectonic transition from Laramide compression to mid-Tertiary extension at about 35 to 37 Ma (Cather, 1989; Cather and Chapin, 1989). This pumiceous middle Spears unit was deposited on an alluvial apron, spreading outward from volcanoes to the southeast and south.

Middle Spears Group - conglomerate unit

A middle Spears Group conglomerate unit in the Alpine 1/Federal is separated into two major lithologic types, mixed-clast conglomerate beds and mostly volcanic-clast sandstone and conglomerate beds (Fig. 4). The first distinctive mixed-clast conglomerate bed was observed at 1093 ft. This conglomerate unit from 1093 to 1266 ft, contains the first down-hole occurrence of red granite (Precambrian). The unit is a sandy, granule-to-cobble conglomerate, interbedded with coarse- to medium-grained sandstone and pebbly sandstone. The mixed-clast conglomerate beds are moderately indurated, matrix-supported, and contain well rounded to subrounded clasts of silicic and intermediate volcanics, limestone and red granite. From 1266 to 1325 ft, the middle Spears Group consists of siltstone that coarsens downward into fine- to medium-grained sandstone and pebbly sandstone and pumiceous sandstone. Between 1325 and 1358 ft, the middle Spears Group again consists of interbedded, mixed-clast matrix-supported granule-to-cobble conglomerate and medium- to coarse-grained sandstone.

Between 854 and 942 ft, a dark blue-green to gray-green andesite porphyry breccia is present in the middle Spears Group sequence. This unit may be a toe-breccia that is peripheral to an andesite flow. The andesite flow mapped at the surface along the northern and southeastern flanks of Escudilla Mountain is about 500 to 600 ft higher in elevation than the andesite in the corehole (Wrucke, 1961). Wrucke's (1961) flow, a medium-crystalline andesite porphyry lava with seriate texture, is tentatively correlated with the andesite of Dry Leggett Canyon of Ratté (1989) (S. M. Cather, personal commun., 1994). Further study is needed to determine the relationship, if any, of the cored andesite breccia and the andesite lavas mapped in the area.

Datil Group (?) ignimbrite

A partially- to densely-welded, orange-brown to pale red rhyolite ash-flow tuff between 1018 and 1038 ft is the only distinctly volcanic unit intercalated in the Spears Group. This moderately crystal-rich unit contains brown and white pumice fragments and medium-grained phenocrysts of clear sanidine and black biotite, but little or no quartz. This ash-flow tuff is tentatively correlated with the tuff of Bishop Peak, one of the Datil Group ignimbrites. McIntosh et al. (1991) gave a $^{40}Ar/^{39}Ar$ age of 34.7 Ma for the tuff of Bishop Peak and an apparently equivalent unit, the tuff of Lebya Well, in the Horse Springs, New Mexico area about 65 mi east of Alpine Divide. If this tentative correlation is correct, this ignimbrite out-flow sheet probably originated from a New Mexico source to the east.

Discussion of mixed-clast conglomerate beds in middle Spears Group

Ratté (1989) mapped a conglomerate with limestone, granite, gneiss and volcanic clasts near the base (?) of the Pueblo Creek Formation (middle Spears Group) and stratigraphically below the tuff of Bishop Peak in the San Francisco Mountains about 30 mi southeast of Alpine Divide.

Ratté et al. (1969) and Landis et al. (1971) described other mixed-clast conglomerate beds near the base of the exposed Tertiary volcanic and volcaniclastic section (Spears Group and Datil (?) Group) along the Red Hill Road between Beaverhead and Blue about 20 mi south of Alpine and at Milligan Peak, 5 mi northeast of the Red Hill Road. Many conglomerate strata in the Red Hill Road section have over seventy percent non-volcanic clast composition. Landis et al. (1971) indicated that virtually all of the fossiliferous limestone clasts are of Middle and Late Pennsylvanian age and may be derived from strata correlative with the Horquilla Limestone of southeastern Arizona. They discussed the evidence for a source within 10 mi or less of the Red Hill Road section. Limestone boulders, 2 ft in length, provide the most compelling evidence for a possible nearby Eocene uplift.

In the Nutrioso area, about 5 mi north of the Alpine Divide, Wrucke (1961) described additional gravels with volcanic, granite and limestone clasts. Mixed-clast conglomerate beds intercalated in the Spears Group are clearly not localized to the Alpine 1/Federal corehole and appear to have regional extent and importance. It is not clear whether the conglomerates observed in these widely separated localities represent the same unit or several discontinous conglomerate lenses with similar stratigraphic position.

The mixed-clast conglomerate beds in the middle Spears Group (1093 to 1358 ft) may indicate drainage incision into pre-Tertiary rocks due to regional thermal uplift associated with Tertiary magmatism, an episode of very late-stage Laramide uplift of basement-involved structures, or reworking of lower Eagar and Mogollon Rim gravel units. If late-stage Laramide compressional uplift to the south and west in Arizona was concurrent with volcanism in the Mogollon-Datil volcanic field to the east and southeast, these conglomerate beds may represent southwesterly-derived upper tongues of the Eagar Formation that are intercalated with southerly- and southeasterly-derived volcaniclastic sediments of the Spears Group. Further study is needed to detail provenance, timing, and tectonic setting of the mixed-composition conglomerate.

Upper Spears Group

The upper Spears Group was encountered from the surface to 854 ft depth. The cored portion of this unit consists mostly of light brown and brown fluvial and alluvial volcanic arenites and conglomerates. These sediments were deposited on the alluvial aprons of volcanic complexes to the south and southeast. Surface outcrops in the vicinity of Alpine Divide also contain white and light cream-colored cross-bedded and moderately well-sorted eolian (?) volcaniclastic sandstones.

TERTIARY-QUATERNARY BASALTIC INTRUSIONS

Several basaltic intrusions were cored. Age of the magmatism is unknown, but a range from mid-Tertiary to Quaternary is reasonable. Whether the intrusions represent dike or sill geometry is unknown, but the two upper intrusions are between formation contacts, suggesting concordance (sills).

The first intrusion, a porphyritic basalt or basaltic andesite, separates the Glorieta Sandstone and the Corduroy Member of the Supai Formation and is 112 ft thick. The upper 35 ft of the Corduroy Member is a solution-collapse breccia, which would allow ingress of magma.

The second intrusion at 4260 to 4322 ft is between the Corduroy Member and the Fort Apache Limestone and is 62 ft thick. A third intrusion, 5 ft below the second, in the Fort Apache Limestone is 35 ft thick. The second and third intrusions are both aphanitic olivine basalts with a composite thickness of 97 ft.

A fourth intrusion, a pyroxene basalt with minor olivine, intrudes the Big A Butte Member at 4454 ft. Because the hole reached total depth at 4505 ft in this intrusion, its total thickness is unknown.

Three different intrusive events may be indicated and each intrusion may have cooled to ambient temperature before the next intrusive event. Modeling of transient thermal fronts associated with the intrusions may constrain the role the intrusions play in hydrocarbon maturation of source rocks in the Permian section.

STRUCTURAL IMPLICATIONS

Winters (1963) reported that the Big A Butte and Amos Wash Members of the Supai Formation total about 805 ft thickness in the White River region. Therefore, the bottom of the Alpine 1/Federal corehole may be at least 750 ft above Precambrian basement. If Pennsylvanian and older rocks are present, the remaining depth would approach 2,000 ft. The Tenneco 1/Federal B well, 45 mi northwest of Alpine Divide, has 1167 ft of Pennsylvanian and older rocks (Peirce et al., 1977) On the other hand, the Belcher 1/State well, 20 mi northeast of the Alpine drill site, has 1670 ft of section between Precambrian basement and the Coconino/Glorieta Sandstone (Foster, 1964). The Fort Apache Limestone is not present in the Belcher well. The difference in the cored Supai thickness at Alpine Divide and pre-Glorieta rocks in the Belcher well suggests that the Alpine 1/Federal bottom is about 800 ft above Precambrian basement.

Weber and Willard (1959) and Wrucke (1961) mapped allochthonous (?) outcrops of Pennsylvanian carbonate rocks in Tertiary andesitic rocks on the northeast side of Escudilla Mountain and just across the New Mexico-Arizona border east of the Alpine drill site (Fig. 1). Harrison (1989) described similar allochthonous Pennsylvanian rocks in andesitic Tertiary rocks in New Mexico more than 100 mi southeast of the Alpine area. Harrison (1989) postulated that the Pennsylvanian limestone outcrops are gravity-slide deposits associated with wrench faulting. Individual slide blocks may be tens of miles from their origin as a result of wrench faulting and slide detachment. Also, Cather and Chapin (1989) described additional allochthonous Paleozoic limestone blocks in Tertiary volcaniclastic rocks about 70 mi east of the Alpine area. Major strike-slip faulting in the region east and south of Alpine was hypothesized by Chamberlin and Anderson (1989). Confirmation of the occurrence of Pennsylvanian and older rocks in the subsurface of the Alpine area will require deeper drilling.

An inferred and unknown thickness of Jurassic rocks, all of the Triassic Chinle Formation, and most of the Permian San Andres Formation, were removed by erosion beneath a sub- Late Cretaceous unconformity and most of the Cretaceous Dakota Sandstone was removed by erosion beneath a Laramide unconformity. Both unconformities resulted from upwarp along the southern margin of the Colorado Plateau (Fig. 5). The Laramide unconformity is approximately 1600 ft lower in elevation in the Alpine 1/Federal than in the Belcher 1/State. Up to 1600 ft of structural displacement, down to the south, has occurred since erosion of the low-relief Laramide surface that bevels the Dakota Sandstone. Determination of the exact location and nature of major subsurface structure between Alpine Divide and Springerville requires detailed geologic mapping. However, several structural geometries are possible. One or more buried and unmapped early Tertiary faults may create a basin. On the other hand, a gently-dipping homoclinal slope with less than one degree of dip to the south is all that is required to displace the Laramide surface. Wrucke (1961)

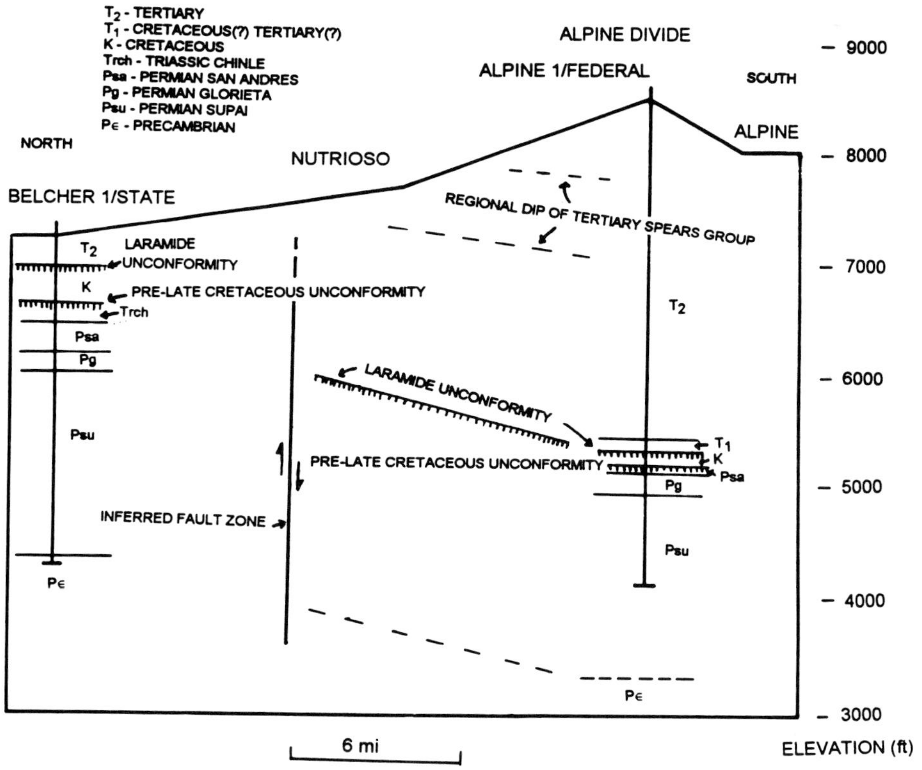

FIGURE 5. Diagrammatic structure cross section between the Alpine 1/Federal and the Belcher 1/State boreholes.

indicated that Tertiary sediments in the area have a gentle southerly dip. In conclusion, the subsurface of the eastern White Mountains is typical of the Mogollon Slope to the east in New Mexico. Subsurface structure in the New Mexico segment of the Mogollon Slope shows a post- Late Cretaceous south-facing downwarp toward the southern Colorado Plateau margin. The subsurface in the region north of the Mogollon Rim, to the west in Arizona, generally rises upward toward the southern Colorado Plateau margin.

SUMMARY

The Alpine 1/Federal drilling provides valuable new information on the geology of the region. The eastern White Mountains region of Arizona is tectonically situated over a south-facing downwarp of the Mogollon Slope on the southern Colorado Plateau. Unnamed Upper Cretaceous or lower Tertiary beds were cored. Several basalt sills or dikes of unknown age intrude the Paleozoic section. Cretaceous rocks rest unconformably on the Permian San Andres Formation. The Corduroy Member of the Supai Formation is thicker and has a greater percentage of carbonate and evaporite rocks than in sections east of Springerville and around Whiteriver.

Minimum depth to the Precambrian is estimated to be 5300 ft at Alpine Divide. If Pennsylvanian or older rocks are present, the estimated depth to Precambrian will approach 6500 ft.

The organic-rich Dakota Sandstone and fetid and petroliferous dolomites and limestones in the San Andres Formation, and the Corduroy and Fort Apache Limestone Members of the Supai indicate potential as source rocks for petroleum maturation. Detailed petrographic analysis and facies studies of core, hydrocarbon maturation studies, delineation of the burial and thermal history of the Cretaceous and Permian rocks, assessment of the hydrodynamic history of the area, and analysis of potential structural and stratigraphic traps and potential reservoir rocks is required to detail the oil and gas potential of the area. The Alpine 1/Federal core and logs provide key information for such an assessment.

ACKNOWLEDGMENTS

The Alpine drilling was administered by the State of Arizona Department of Commerce, Energy Office, Phoenix. Funding was a grant from the U. S. Department of Energy, Geothermal Division, Washington, D. C to the State of Arizona with a matching contribution from Arizona's share of the Petroleum Violation Escrow Funds. Tonto Drilling Services, Inc. of Salt Lake City was the operator and drilling contractor. Southwest Geophysical Services, Inc. of Farmington performed the geophysical logs. Discussions with Richard Chamberlin and Steve Cather greatly enhanced our understanding of the region. Katherine Giles and William Seager are thanked for helpful and constructive reviews.

REFERENCES

Blakey, R. C., 1980, Pennsylvanian and early Permian paleogeography, southern Colorado Plateau and vicinity; *in* Fouch, T. D. and Magathan, E. R., eds., Paleozoic paleogeography of west-central United States: Society of Economic Paleontologists and Mineralogists, Denver, p. 239-257.

Cather, S. M., 1989, Post-Laramide tectonic and volcanic transition in west-central New Mexico: New Mexico Geological Society, Guidebook 40, p. 91-97.

Cather, S. M. and Chapin, C. E., 1989, Day 2: Field guide to upper Eocene and lower Oligocene volcaniclastic rocks of the northern Mogollon-Datil volcanic field: New Mexico Bureau of Mines and Mineral Resources, Memoir 46, p. 60-68.

Cather, S. M. and Johnson, B. D., 1984, Eocene tectonics and depositional setting of west-central New Mexico and eastern Arizona: New Mexico Bureau of Mines and Mineral Resources, Circular 192, 33 p.

Cather, S. M., Ratté, J. C., and Chamberlin, R. M., 1994, Tertiary stratigraphy and nomenclature for western New Mexico and eastern Arizona: New Mexico Geological Society, Guidebook 45.

Chamberlin, R. M. and Anderson, O. J., 1989, The Laramide Zuni uplift, southeastern Colorado Plateau: a microcosm of Eurasian-style indentation-extrusion tectonics?: New Mexico Geological Society, Guidebook 40, p. 81-90.

Cooley M. E. and Davidson, E. S., 1963, The Mogollon highlands - their influence on Mesozoic and Cenozoic erosion and sedimentation: Arizona Geological Society, Digest 6, p. 7-35.

Dickinson, W. R., Fiorillo, A. R., Hall, D. L., Monreal, R., Potochnik, A. R. and Swift, P. N., 1989, Cretaceous strata of southern Arizona: Arizona Geological Society, Digest 17, p. 447-461.

Foster, R. W., 1964, Stratigraphy and petroleum possibilities of Catron County, New Mexico: New Mexico Bureau of Mines and Mineral Resources, Bulletin 85, 55 p.

Gerrard, T. A. , 1966, Environmental studies of Fort Apache Member, Supai Formation, east-central Arizona: Amerian Association of Petroleum Geologists Bulletin, v. 50, p. 2434-2463.

Harrison, R. W., 1989, Exotic blocks with the early Tertiary Rubio Peak Formation in the north-central Black Range, New Mexico: occurence, insights into post-emplacement tectonic activity, economic implications and emplacement hypothesis: New Mexico Geological Society, Guidebook 40, p. 99 -106.

Landis, E. R., Ratté, J. C. and Myers, D. A., 1971, Carbonate clasts in epiclastic volcanic rocks and their paleotectonic implications, Blue Range Primitive Area, Arizona and New Mexico: U. S. Geological Survey, Professional Paper 750-C, p. C58-C64.

Lawton, T. F., Basabilvazo, G. T., Hodgson, S. A., Wilson, D. A., Mack, G. H., Mcintosh, W. C., Lucas, S. G. and Kietzke, K. K., 1993, Laramide stratigraphy of the Little Hatchet Mountains, southwestern New Mexico: New Mexico Geology, v. 15, p. 9-15.

Marvin, R. F., Naeser, C. W., Bikerman, M., Mehnert, H. H. and Ratté, J. C., 1987, Isotopic ages of post-Paleocene igneous rocks within and bordering the Clifton 1°x2° quadrangle, Arizona-New Mexico: New Mexico Bureau of Mines and Mineral Resources, Bulletin 118, 63 p.

McIntosh, W. C., Kedzie, L. L. and Sutter, J. F., Paleomagnetism and $^{40}Ar/^{39}Ar$ ages of ignimbrites, Mogollon-Datil volcanic field, southwestern New Mexico: New Mexico Bureau of Mines and Mineral Resources, Bulletin 135, 79 p.

Merrill, R. K. and Pewé, T. L., 1977, Late Cenozoic geology of the White Mountains, Arizona: Arizona Bureau of Geology and Mineral Technology, Special Paper 1, 65 p.

Nations, J. D., 1989, Cretaceous history of northeastern and east-central Arizona: Arizona Geological Society, Digest 17, p. 435-446.

Osburn, G. R. and Chapin, C. E., 1983, Nomenclature for Cenozoic rocks of northeast Mogollon-Datil volcanic field, New Mexico: New Mexico Bureau of Mines and Mineral Resources, Stratigraphic Chart 1.

Peirce, H. W., 1989, Correlation problems of Pennsylvanian-Permian strata of the Colorado Plateau of Arizona: Arizona Geological Society, Digest 17, p. 349-368.

Peirce, H. W., Damon, P. E. and Shafiqullah, M., 1979, An Oligocene (?) Colorado Plateau edge in Arizona: Tectonophysics, v. 61, p. 1-24.

Peirce, H. W., Jones, N. and Rogers, R., 1977, A survey of uranium favorability of Paleozoic rocks in the Mogollon rim and slope region-east central Arizona: Arizona Bureau of Geology and Mineral Technology, Circular 19, 60 p.

Potochnik, A. R., 1989, Depositional style and tectonic implications of the Mogollon Rim formation (Eocene), east-central Arizona: New Mexico Geological Society, Guidebook 40, p. 107-118.

Ratté, J. C., 1989, Geologic map of the Bull Basin Quadrangle, Catron County, New Mexico: U. S. Geological Survey, Geologic Quadrangle Map GQ-1651, scale 1:24,000.

Ratté, J. C., Landis, E. R., Gaskill, D. L., Raabe, R. G. and Eaton, G. P., 1969, Mineral resources of the Blue Range primitive area, Greenlee County, Arizona, and Catron County, New Mexico with a section on aeromagnetic interpretation: U. S. Geological Survey, Bulletin 1261-E, 91 p.

Rauzi, S. L., 1994, Implications of live oil shows in eastern Arizona geothermal test: Arizona Geological Survey, Open-file Report 94-1, 16 p.

Sirrine, G. K., 1956, Geology of the Springerville-St Johns area, Apache County, Arizona [PhD dissertation]: Austin, University of Texas, 248 p.

Weber, R. H. and Willard, M. E., 1959, Reconnaissance geologic map of Reserve thirty-minute quadrangle: New Mexico Bureau of Mines and Mineral Resources, Geologic Map 12, scale 1:126,720.

Winters, S. S., 1963, Supai Formation (Permian) of eastern Arizona: Geological Society of America, Memoir 89, 99 p.

Wrucke, C. T., 1961, Paleozoic and Cenozoic rocks in the Alpine-Nutrioso area Apache County, Arizona: U. S. Geological Survey, Bulletin 1121-H, 26 p.

TRIASSIC STRATIGRAPHY IN THE LUCERO UPLIFT, CIBOLA, VALENCIA AND SOCORRO COUNTIES, NEW MEXICO

SPENCER G. LUCAS[1] and ANDREW B. HECKERT[2]
[1]New Mexico Museum of Natural History and Science, 1801 Mountain Road N.W., Albuquerque, New Mexico 87104;
[2]Department of Earth and Planetary Sciences, University of New Mexico, Albuquerque, New Mexico 87131-1116

Abstract—Triassic strata exposed in the Lucero uplift of Cibola, Valencia and Socorro Counties are nonmarine red beds of the Moenkopi Formation and Chinle Group. Moenkopi Formation strata disconformably overlie Middle Permian (Guadalupian) limestones and dolomites of the San Andres Formation and are mostly grayish red, trough-crossbedded and ripple-laminated micaceous litharenites intercalated with beds of intraformational conglomerate and grayish red, micaceous siltstone and mudstone. Charophytes, ostracods and capitosauroid amphibians indicate the Moenkopi Formation in the Lucero uplift is of Middle Triassic (early Anisian) age. Overlying Chinle Group strata consist of (ascending) the Shinarump, Bluewater Creek, Petrified Forest, Owl Rock and Rock Point Formations. In the Lucero uplift, strata of the San Pedro Arroyo Formation laterally replace Bluewater Creek Formation strata south of the Rio Salado. Shinarump Formation strata are extrabasinal conglomerates (mostly Paleozoic limestone clasts) and sandstones as much as 17 m thick that disconformably overlie Moenkopi strata. Locally, pedogenically modified sandstones, conglomerates and siltstones - the "mottled strata" - laterally replace Shinarump conglomerates and sandstones. Bluewater Creek Formation strata are at least 70 m thick and are mostly reddish brown siltstones, mudstones and ripple-laminated sandstones The distinctive McGaffey Member - 5-6 m of mostly ripple-laminated sandstone - is present in the upper half of the Bluewater Creek Formation in the northern part of the Lucero uplift. Below the McGaffey Member, the Bluewater Creek Formation contains an Adamanian (late Carnian) tetrapod fauna that includes cf. Buettneria, indeterminate phytosaurs and Stagonolepis. The base of the Petrified Forest Formation in the Lucero uplift is the Sonsela Member; evidently the Blue Mesa Member has been removed by erosion at the pre-Sonsela unconformity. Sonsela strata are as much as 15 m thick and are mostly trough-crossbedded micaceous litharenite and intrabasinal calcrete-pebble conglomerate. Overlying Painted Desert Member strata are mostly reddish-brown bentonitic mudstone at least 100 m thick. Owl Rock and Rock Point strata are only exposed in the Petaca Pinta area (SW 1/4 T6N, R6W and SE 1/4 T6N, R7W). The Owl Rock is mostly grayish-red and very pale green, calcareous mudstone and grayish-red purple bioturbated sandy siltstone and is up to 37 m thick. Rock Point strata are up to 70 m thick and consist of laterally extensive beds of massive, moderate-reddish brown sandy siltstone and laminar to trough-crossbedded moderate reddish brown quartzose sandstone. Middle Jurassic (Entrada) to Upper Cretaceous (Dakota) strata disconformably overlie Upper Triassic Chinle Group rocks in the Lucero uplift.

INTRODUCTION

Nonmarine Triassic red beds exposed in the Lucero uplift of Cibola, Valencia and Socorro Counties, New Mexico, represent one of the largest single outcrop belts of Triassic strata in the state (Fig. 1). Yet, beyond basic mapping by Kelley and Wood (1946), Tonking (1957) and Jicha (1958), little study has been undertaken of these rocks. Here, we present a comprehensive lithostratigraphy of Triassic strata in the Lucero uplift and correlate them to nearby Triassic rocks.

PREVIOUS STUDIES

Early geological fieldworkers in central New Mexico recognized that Triassic strata are exposed throughout the Lucero uplift. Thus, Darton (1910, p. 49-50) summarized earlier observations by J.S. Newberry, C.L. Herrick and H.N. Herrick. Darton (1910, pl. 1) mapped the Triassic strata in the Lucero uplift as "near 500 feet" of "undifferentiated Triassic." Darton (1928, p. 124, pl. 30) subsequently identified these strata as "Triassic red shales" (Fig. 2).

Kelley and Wood (1946) first attempted to subdivide and apply formal nomenclature to the Triassic strata exposed in the Lucero uplift (Fig. 2). They assigned them to the Shinarump Conglomerate overlain by the Chinle Formation. Kelley and Wood (1946) inferred a maximum thickness of about 400 ft and stated that "little or no conglomerate is present in the Lucero area, and the formation more nearly resembles the possibly equivalent Santa Rosa sandstone east of the Rio Grande Valley than the Shinarump conglomerate of western New Mexico." They described the Shinarump conglomerate as "red-brown shale, siltstone, and sandstone, locally contains limestone-conglomerate lenses, 300 ± feet thick".

FIGURE 1. Distribution of Triassic strata in New Mexico, showing location of Lucero uplift Triassic outcrops.

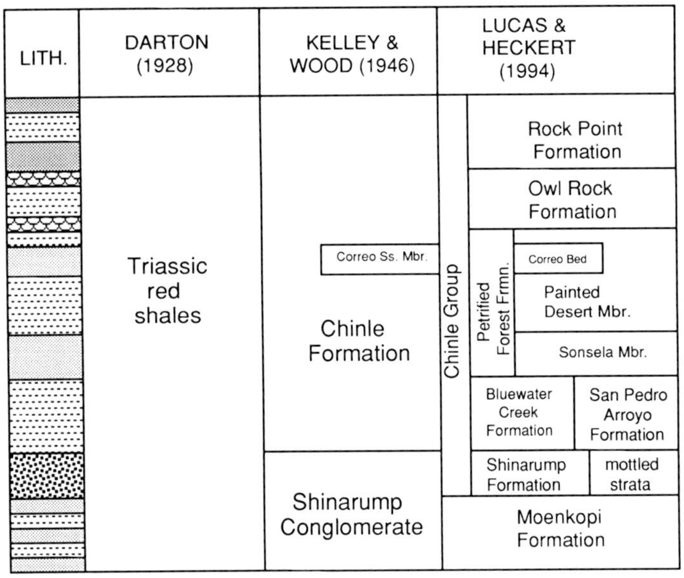

FIGURE 2. Summary of development of stratigraphic nomenclature applied to Triassic strata in the Lucero uplift.

Kelley and Wood (1946) divided strata they termed Chinle Formation in the Lucero uplift into two members, a red shale member overlain by the Correo Sandstone Member. They described the red shale member as "red-brown soft shale, 1000 ± feet thick; the basal part is not exposed." The Correo Sandstone Member was described as "dark-brown and buff, medium to massive, cross-bedded sandstone, 90-120 feet thick" that only crops out "along the lower bench at the south edge of Mesa Gigante."

Subsequent geologic mapping and study of the southern part of the Lucero uplift by Tonking (1957) used the Triassic nomenclature of Kelley and Wood (1946). Jicha (1958), mapping in the northern part of the uplift, essentially followed suit, but did not separate the Shinarump from the Chinle.

Apparently no work was undertaken on Triassic strata in the Lucero uplift for about 30 years. Mapping along the western edge of the uplift by Maxwell (1988a, b) identified the Triassic strata as Chinle Formation overlain by the Rock Point Member of the Wingate Sandstone. Hunt et al. (1989), Lucas and Hayden (1989), and Hayden and Lucas (1989) presented some observations on the Triassic strata exposed in the Lucero uplift. They assigned these rocks to the Moenkopi Formation overlain by the Chinle Formation, consisting of the Shinarump and Petrified Forest Members. Lucas and Hunt (1990) identified the Owl Rock and Rock Point Members of the Chinle Formation at Petaca Pinta on the northwestern edge of the uplift. Recently, Lucas (1993) revised the nomenclature of the Upper Triassic strata in the American Southwest, elevating the Chinle to group status and recognizing the following units in west-central New Mexico: Shinarump Formation, Bluewater Creek Formation, Petrified Forest Formation, (including Blue Mesa, Sonsela and Painted Desert Members) and the Owl Rock and Rock Point Formations.

STRATIGRAPHY

Our study of the Triassic stratigraphy of the Lucero uplift is based on several measured stratigraphic sections and a few fossil localities (Fig. 3).

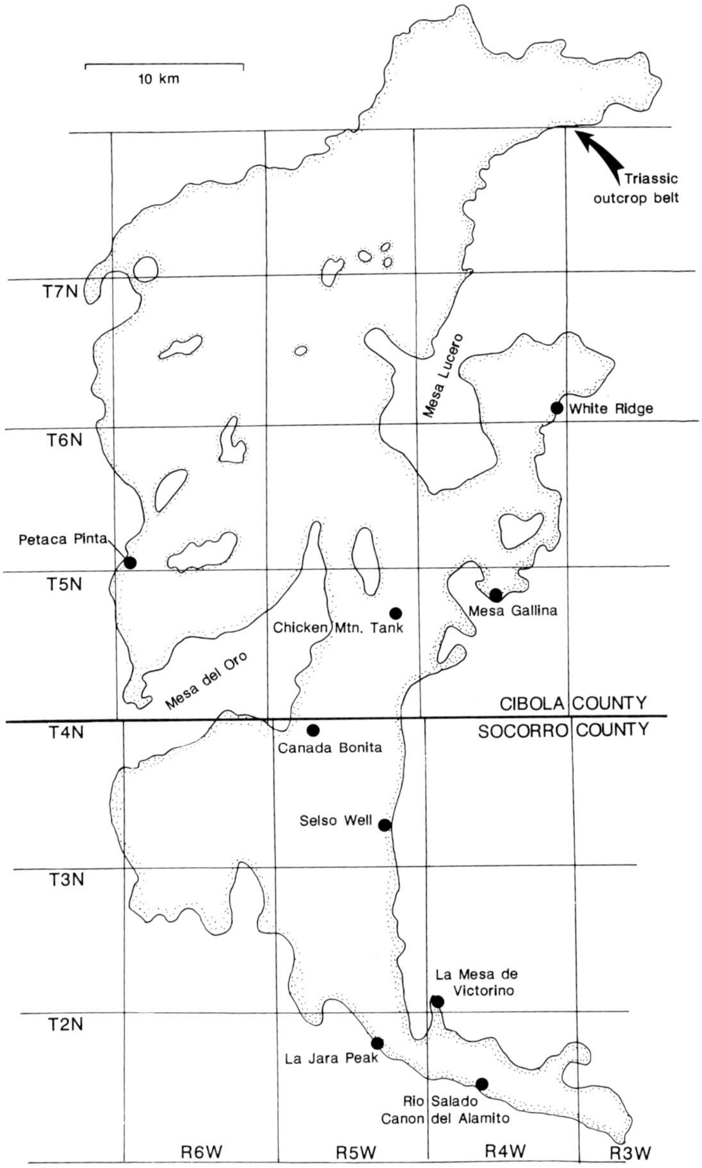

FIGURE 3. Distribution of Triassic strata in the Lucero uplift (after Dane and Bachman, 1965) and location of measured stratigraphic sections and fossil localities of this study.

Moenkopi Formation

Strata of the Moenkopi Formation (Anton Chico Member) in the Lucero uplift consist of 20 to 70 m of interbedded conglomerates, sandstones, siltstones and mudstones that are disconformably bounded by the San Andres Formation (below) and the Chinle Group (above) (Fig. 4). The Moenkopi Formation is exposed throughout the Lucero uplift both west and east of the San Andres Formation axis. Basal Moenkopi strata rest on dramatically eroded surfaces of the San Andres Formation, including both incised channels (Lucas and Hayden, 1989) and paleokarst topography (Stewart et al., 1972). The uppermost Moenkopi is covered by either "mottled strata" or sandstones and conglomerates of the Shinarump Formation (Fig. 4). In addition to the measured sections reported by Lucas and Hayden, we document two complete sections at La Mesa de Victorino and Selso Well (Fig. 3).

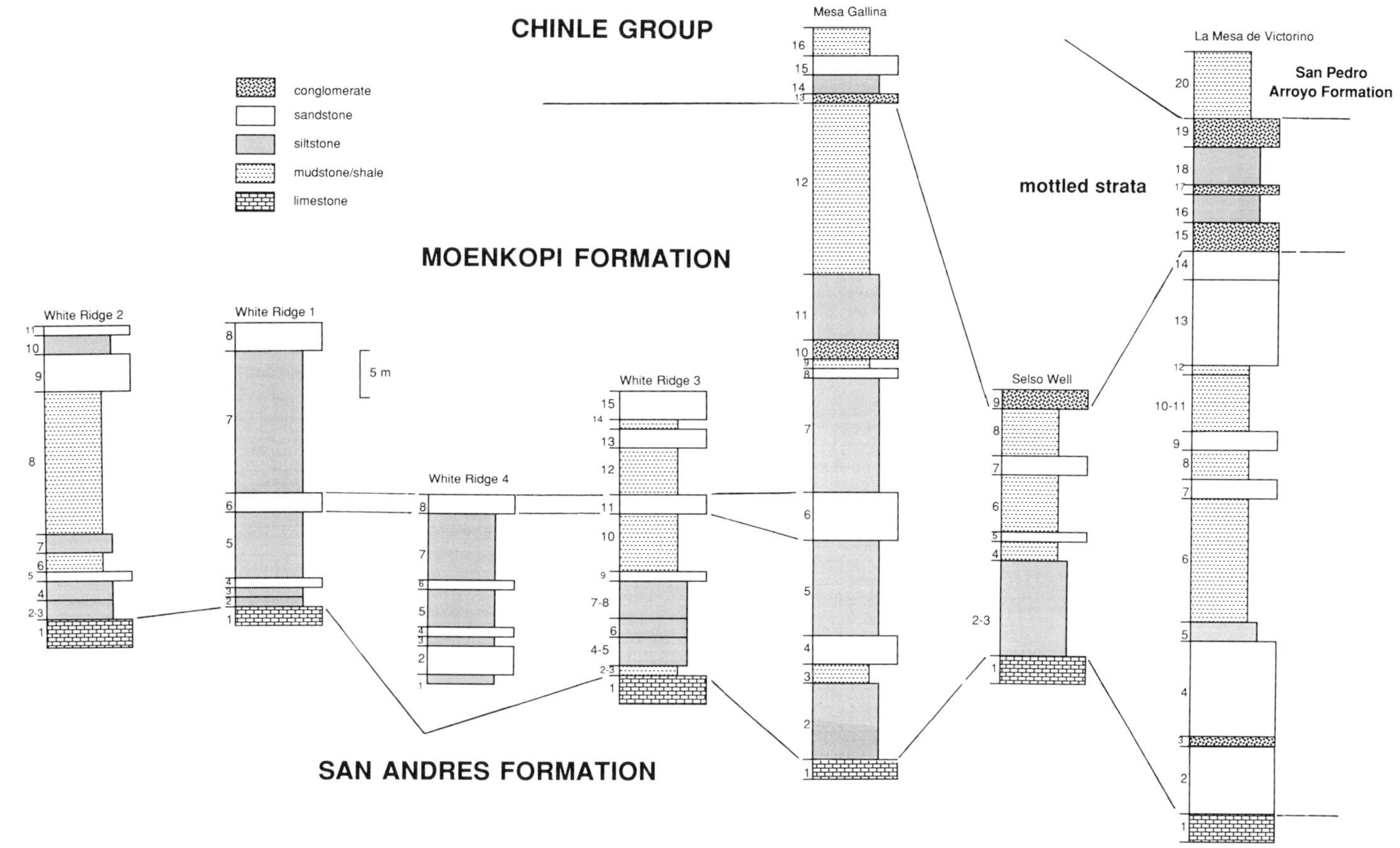

FIGURE 4. Measured stratigraphic sections of the Moenkopi Formation in the Lucero uplift.

The basal contact of the Moenkopi Formation is typically an interbedded series of white to yellowish gray mudstones, siltstones and sandstones. Lucas and Hayden (1989) referred to this lowermost interval of the Moenkopi in the Lucero uplift as a bleached zone of mudstones and siltstones. At La Mesa de Victorino (the section measured by Tonking, 1957) we place the San Andres-Moenkopi contact at the change from yellowish gray micritic dolomite of the San Andres Formation to yellowish gray to grayish orange sandstones and siltstones of the Moenkopi Formation (Fig. 4). This is approximately 12.4 m below the base of the Triassic described by Tonking (1957). Likewise, at Selso Well north of Puertecito, yellow to yellowish gray siltstones mark this lowermost bleached horizon in the Moenkopi.

Above these strata, the Moenkopi Formation is dominated by poorly sorted reddish brown to reddish gray mudstones, siltstones, sandstones and conglomerates. Mudstones are typically silty with some interbeds of siltstone, whereas siltstones are typically muddy and calcareous. Sandstones are poorly sorted litharenites typically displaying either ripple laminae or, more frequently, trough crossbeds (Fig. 5B). Some sandstones are conglomeratic, with clasts consisting of intraformational mudstone rip-ups. Conglomerates in the Moenkopi are also trough crossbedded and characterized by intraformational clasts of siltstone, mudstone and calcrete up to 2 cm in diameter.

Thickness of the Anton Chico Member of the Moenkopi Formation varies considerably due to unconformities that bound the unit. In the extreme eastern side of the uplift, closely spaced measured sections show that Moenkopi thicknesses average approximately 30 m (Lucas and Hayden, 1989). Farther west the section varies from a low of approximately 25 m at our measured section at Selso Well to almost 70 m at Mesa Gallina (Lucas and Hayden, 1989). In the southern portion of the uplift we measured 51 m of Moenkopi at La Mesa de Victorino. Sections capped by mottled strata generally appear to be thicker than those capped by the Shinarump, indicating a greater amount of erosion associated with pre- and syn-Shinarump deposition.

Chinle Group

The Chinle Group in the Lucero uplift consists of (in ascending order) the Shinarump, Bluewater Creek, San Pedro Arroyo, Petrified Forest, Owl Rock and Rock Point Formations (Fig. 6). Mottled strata, as originally described by Stewart et al. (1972) are also locally present throughout the Lucero uplift at the base of the Chinle Group. Near the Rio Salado, strata of the Bluewater Creek Formation grade into those of the San Pedro Arroyo Formation. Thus, rocks identified by us as Bluewater Creek Formation are present throughout most of the uplift, whereas the San Pedro Arroyo Formation is the equivalent unit south of the Rio Salado. These rocks are disconformably overlain by the Sonsela Member of the Petrified Forest Formation, which in turn is overlain by the Painted Desert Member. There are no strata of the Blue Mesa Member of the Petrified Forest Formation in the Lucero uplift; evidently these rocks were eroded during the pre-Sonsela unconformity. In the area of Petaca Pinta the Chinle succession is capped by the Owl Rock and Rock Point Formations.

Shinarump Formation/"mottled strata"

Both the Shinarump Formation and the correlative "mottled strata" are present in the Lucero uplift to varying extents. We identify conglomerates (25%) and conglomeratic sandstones (75%) of the Shinarump Formation in the north near Chicken Mountain, and at Selso Well farther south (Fig. 6). Mottled strata were first reported from the Lucero uplift by Stewart et al. (1972) and are evident at Mesa Gallina (Lucas and Hayden, 1989), at La Mesa de Victorino, and just south of the Rio Salado (Fig. 6).

Strata identified by us as Shinarump consist of light olive-gray to brownish gray extrabasinal conglomerate and conglomeratic sandstones (Figs. 7A-B). Bedding is trough- to wedge-planar crossbedded,

FIGURE 5. Photographs of selected Moenkopi outcrops in the Lucero uplift. A, Overview of La Mesa de Victorino section showing Moenkopi Formation (1) overlain by San Pedro Arroyo Formation (2) followed by Quaternary travertine (3). B, Typical cross-bedded Moenkopi litharenite, unit 9 of La Mesa de Victorino section.

with much scouring of underlying strata (Fig. 7B). At Selso Well, conglomerate clasts consist primarily of Paleozoic limestones with a few chert and jasper pebbles. Similar lithologies predominate at Chicken Mountain Tank, where the Shinarump also crops out (Fig. 7A). Due to structural complexities the entire unit could not be measured at either locality, so known thickness is a minimum of 10 m.

"Mottled strata" are identified from numerous localities in the Lucero uplift. These pedogenically altered siltstones, mudstones and conglomerates have been inferred to represent a paleoweathering surface that developed either during the Tr-3 unconformity of Pipiringos and O'Sullivan (1978) and/or contemporaneously with Shinarump deposition (e.g. Lucas and Hayden, 1989; Lucas, 1991). In the southern portion of the Lucero uplift "mottled strata" are widely distributed, with particularly informative outcrops at La Mesa de Victorino, in the section originally measured by Tonking (1957) (Fig. 7C), and at our Rio Salado section to the south and east of Tonking's section (Fig. 7D).

At La Mesa de Victorino, "mottled strata" consist of nearly 15 m of mud pellet conglomerates (70%) and muddy sandy siltstones (30%) (Fig. 7C). Mud-pellet conglomerates are pale to grayish red with mottles of reddish orange, reddish brown, and dark gray. Clasts are rip-ups of Moenkopi mudstone. Siltstones are pale to grayish red, calcareous and extensively pedogenically altered. In contrast, within sight of this section, at Rio Salado "mottled strata" consist of a thin mixture of calcrete and silcrete less than a meter thick (Fig. 7D). Mottles are dark gray, grayish purple, light gray and pale red. Conglomeratic clasts include quartzite and jasper pebbles up to 2.0 cm in diameter.

Bluewater Creek Formation

The Bluewater Creek Formation in the Lucero uplift is typified by at least 70 m of ripple-laminated sandstones, muddy siltstones, and silty mudstones (Fig. 6). Strata of the Bluewater Creek Formation crop out in extensive, structurally complex badlands in the vicinity of Chicken Mountain Tank, where they rest with apparent disconformity on strata of the Shinarump Formation. In this region the McGaffey Member crops out as 5 m of ripple-laminated litharenite near the top of the measured section. Elsewhere in the uplift, Bluewater Creek strata are often preserved as several meters of bentonitic mudstone red beds underlying hogbacks of the Sonsela Member of the Petrified Forest Formation. In the southernmost portion of the uplift, rocks of the Bluewater Creek Formation grade into strata assigned by us to the San Pedro Arroyo Formation.

At Chicken Mountain, basal Bluewater Creek strata consist of approximately 20 m of olive-gray, ripple-laminated micaceous quartzarenites. Some minor trough crossbedding is evident. Sandstones are typically fine to very fine grained. Above these strata the Bluewater Creek Formation is typified by bentonitic mudstones (63%) with minor crossbedded sands and clays (37%) (Fig. 7E). Approximately 30% of the mudstones display varying degrees of calcrete nodule development.

Due to structural deformation and cover in the Lucero uplift, we were not able to locate a single, complete section of the Bluewater Creek Formation. Above the mudstones at Chicken Mountain Tank, the McGaffey Member of Anderson and Lucas (1993) is present. In other localities, including Cañada Bonita and numerous other localities, trough crossbedded sandstones of the Sonsela Member of the Petrified Forest Formation rest disconformably on the Bluewater Creek Formation. These uppermost Bluewater Creek sediments are "red beds" of trough-crossbedded quartzarenites and bentonitic mudstones.

McGaffey Member

To the north and west of Chicken Mountain, sandstones of the McGaffey Member crop out in a prominent bench near the road (Fig. 6). These strata are hummocky-bedded and ripple laminated sandstones at least 5 m thick. We identify this ledgy sand as the McGaffey Member on the basis of lithology, thickness, outcrop pattern and nature of the contact with underlying strata. These sandstones are micaceous, lithic arenites that are regionally persistent and form a distinct bench. As in the type section of the McGaffey Member (Anderson and Lucas, 1993), the unit is approximately 5-7 m thick. The presence of "redbeds" of fine sandstones and siltstones beneath these rocks indicates that they are not basal Bluewater Creek Formation sandstones. For all of these reasons we believe this sandy unit is the first identified outcrop of the McGaffey Member in the Lucero uplift. Hunt et al. (1989) earlier misidentified this outcrop as Sonsela.

San Pedro Arroyo Formation

Near the southern terminus of the Lucero uplift, strata pertaining to the San Pedro Arroyo Formation (Lucas, 1991) grade into muddier Bluewater Creek Formation sediments (Fig. 6). We identify the San Pedro Arroyo Formation in the Lucero uplift as a series of mudstones (89%) interrupted by numerous thin, persistent ledges of sandstone (7%), conglomerate (1%), and very minor siltstone and calcrete (Fig. 7F). Mudstones are dominantly bentonitic, calcareous and grayish red purple with light gray mottles. Sandstones and conglomerates typically crop out as thin, 0.5 to 2.0 m thick, laterally extensive sheets. The sandstones are laminated, micaceous litharenites, whereas conglomerates are typified by intraformational calcrete and mudstone clasts. This sequence of strata is approximately 90 m thick in the vicinity of our Rio Salado section, and represents a complete section from the underlying mottled strata to the base of the Sonsela Member (Fig. 6).

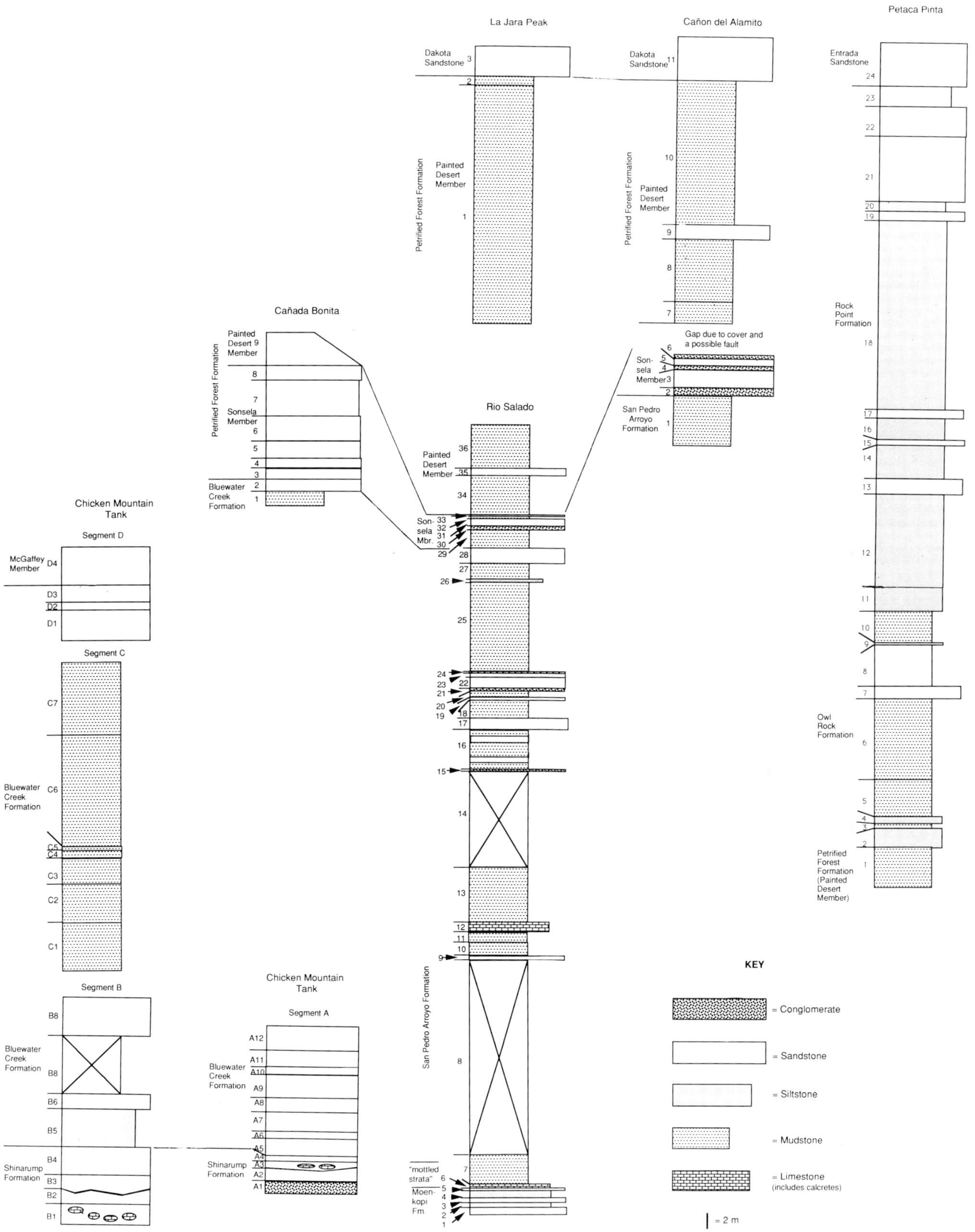

FIGURE 6. Measured stratigraphic sections of Chinle Group strata in the Lucero Uplift.

FIGURE 7. Photographs of selected Chinle Group outcrops in the Lucero uplift. A, Paleozoic limestone cobble conglomerate of Shinarump Formation, unit A1 of Chicken Mountain Tank section. B, Planar and crossbedded sandstones and conglomeratic sandstones of the Shinarump Formation near the Chicken Mountain Tank section. C, Base of the Chinle Group at the La Mesa de Victorino section shows mottled strata (1=unit 19 of measured section) overlain by mudstone (2=unit 20 of measured section) at the base of the San Pedro Arroyo Formation. D, Calcrete and silcrete nodules of the mottled strata at the Rio Salado section (unit 6). E, Mudstones and heterolithic cross strata of the Bluewater Creek Formation in the Chicken Mountain Tank section (unit C6). F, Typical mudstones, sandstones and conglomerates of the San Pedro Arroyo Formation, units 20-28 of the Rio Salado section.

Petrified Forest Formation

Lucas (1993) subdivided the Petrified Forest Formation into the Blue Mesa, Sonsela and Painted Desert members. We recognize only strata pertaining to the Sonsela and Painted Desert Members in the Lucero uplift. The Blue Mesa Member appears to have been eroded during the pre-Sonsela unconformity. The Sonsela Member thus rests disconformably on the Bluewater Creek Formation or laterally equivalent sediments of the San Pedro Arroyo Formation.

Sonsela Member

The lowest unit of the Petrified Forest Formation in the Lucero uplift north of the Rio Salado is the Sonsela Member. It rests directly on the Bluewater Creek Formation and is overlain by the Painted Desert Member (Fig. 6). South of the Rio Salado, the Sonsela Member separates the San Pedro Arroyo Formation (below) from the Painted Desert Member (above). Excellent outcrops of the Sonsela Member are present throughout the Lucero uplift, especially: (1) around Cañada Bonita and Cerro Pelon in secs. 5,6,8,20 and 29, T4N, R5W; (2) near Paint Tank in secs. 4 and 5, T3N, R5W; (3) in Cañada Bonita near Bill Dobbs Well in sec. 8, T3N, R5W; and (4) south of the Rio Salado near Cañon del Alamito in secs. 16 and 17, T2N, R4W.

Two measured sections, one to the north, the other to the south, well encompass the lithologic variation of the Sonsela Member in the Lucero uplift. To the north, at Cañada Bonita, the Sonsela Member is 15.3 m thick. It is almost entirely sandstone with a conglomeratic sandstone at its base (Figs. 6, 8A-C). This conglomerate mostly consists of mudstone clasts, but a few siliceous and Paleozoic limestone clasts are present. The sandstones are feldspathic litharenites that are either laminated or trough crossbedded. Grayish red and grayish pink color banding is characteristic (Fig. 8C). At Cañada Bonita, the Sonsela forms a resistant bench between redbed mudstone-dominated slopes, the Bluewater Creek Formation below and Painted Desert Member above (Fig. 8A).

To the south, along Cañon del Alamito near Riley, the Sonsela Member is thinner, about 5 m thick. Conglomerates are much more prominent in the Sonsela here and their clasts are almost exclusively calcrete rip-ups. Laminar and trough crossbedded sandstones are also litharenites and grayish purple and olive gray color banding is common.

Painted Desert Member

The Painted Desert Member conformably overlies the Sonsela Member throughout the Lucero uplift (Fig. 6). In the Petaca Pinta area, the Owl Rock Formation of the Chinle Group conformably overlies the Painted Desert Member (Fig. 8D). At Mesa Gigante, Middle Jurassic strata (Entrada Sandstone) disconformably overlie the Painted Desert Member. Throughout the rest of the Lucero uplift, the Cretaceous Dakota Group rests disconformably on the Painted Desert Member. This Cretaceous-Triassic disconformity is particularly well exposed from the unnamed mesa just northwest of Puertecito in secs. 18, 19 and 30 of T3N, R5W to the southeast at Cañon del Alamito in secs. 16, 20, and 21, T2N, R4W.

In the Lucero uplift, as is the case elsewhere on the Colorado Plateau (Lucas, 1993), the Painted Desert Member is mostly reddish brown and grayish red bentonitic mudstones (Fig. 8D). Typical lithologies are described here (see La Jara Peak and Cañon del Alamito sections).

Because it is not a resistant unit, the Painted Desert Member is not well exposed in the Lucero uplift. It typically forms broad valleys or slopes much mantled by soil or covered with colluvial debris. Because of this, we were not able to measure a single, completely exposed section of the member in the Lucero uplift. The closest we came to this is Cañon del Alamito where the measured 34.7 m is only a few meters less than the total thickness of the Painted Desert Member, which we estimate as about 40 m thick here. Just northwest of Puertecito, between the Sonsela Member outcrops in sec. 8, T3N, R5W at the road crossing of Cañada Bonita, the Painted Desert Member is nearly flat lying to its contact with the Dakota in sec. 18, T3N, R5W over an elevation gain of about 73 m. This suggests a thickening of the Painted Desert Member from south to north across the Lucero uplift.

In the northern part of the Lucero uplift, at Mesa Gigante, the upper part of the Painted Desert Member contains the Correo Bed, about 26 m of cliff-forming sandstone and conglomerate (Kelley and Wood, 1946; Stewart et al., 1972; Lucas et al., 1987). Correo Bed conglomerates are intrabasinal, being composed of siltstone and limestone clasts, and the sandstones are quartzose. In the Lucero uplift, the Correo Bed crops out only in its type area along the southern flank of Mesa Gigante.

Owl Rock Formation

The Owl Rock Formation rests conformably on the Painted Desert Member in the vicinity of Petaca Pinta (Fig. 6). In the Lucero uplift, its outcrops are limited to the Petaca Pinta area in T6N, R6W. To the south and southeast, where Cretaceous Dakota Group strata rest disconformably on the Chinle Group, no Owl Rock (or overlying Rock Point) strata are preserved. At the northern edge of the Lucero uplift, at Mesa Gigante, the Entrada Sandstone rests disconformably on the Painted Desert Member, so here too Owl Rock and Rock Point strata are missing.

At Petaca Pinta, we measured a 35.4-m-thick section of the Owl Rock Formation. Here, the Owl Rock is mostly mudstone (78% of the measured section) with much less siltstone (17% of the section) and minor silty sandstone (5% of the section). The mudstones are variegated hues of red, purple and pale green. They are bentonitic and typically contain numerous calcrete nodules (Fig. 8E). The siltstones are brown and purple and very bioturbated, as is the one ledge-forming sandstone near the middle of the formation. At Petaca Pinta, as elsewhere on the Colorado Plateau (Lucas, 1993), the Owl Rock Formation conformably overlies the Painted Desert Member of the Petrified Forest Formation and is disconformably overlain by the Rock Point Formation.

Rock Point Formation

Like the Owl Rock Formation, the Rock Point Formation is only present on the western fringe of the Lucero uplift at Petaca Pinta (Fig. 6), and its distribution in the Lucero uplift is identical to that of the Owl Rock Formation. We measured 69.1 m of Rock Point Formation at Petaca Pinta.

At Petaca Pinta the Rock Point Formation is repetitively interbedded, laterally persistent sandy siltstone (71% of the measured section) and silty sandstone (29% of the section) (Fig. 8F). These strata are mostly reddish brown. The siltstones are massive, ledgy, and not bentonitic. The sandstones are mostly laminated, ripple laminated or massive micaceous quartzarenites. The Middle Jurassic Entrada Sandstone overlies the Rock Point Formation disconformably at Petaca Pinta.

PALEONTOLOGY

No systematic effort has been undertaken to recover fossils from the extensive Triassic exposures in the Lucero uplift, but a few biochronologically significant fossils are known. These are mostly from the Moenkopi Formation and lower part of the Chinle Group.

Kietzke (1988, 1989) reported charophytes and ostracods from the lower part of the Moenkopi Formation at White Ridge, in the northern part of the Lucero uplift. These fossils indicate an Early or Middle Triassic age. From Mesa Gallina, also in the northern part of the Lucero uplift, Lucas and Hayden (1989) described bones of a capitosauroid amphibian and an indeterminate archosaur from the Moenkopi Formation. Capitosauroids (*Eocyclotosaurus*) from elsewhere in the Anton Chico Member of the Moenkopi Formation indicate an early Middle Triassic (early Anisian) age (Lucas and Morales, 1985; Lucas and Hunt, 1987; Morales, 1987; Ochev and Shishkin, 1989; Hunt and Lucas, 1993).

Hunt et al. (1989) reported a vertebrate fossil assemblage from the Bluewater Creek Formation near Chicken Mountain Tank in the northern Lucero uplift. Taxa represented are cf. *Buettneria* sp. (= *Metoposaurus* sp. of Hunt et al., 1989, here revised after Hunt, 1993), *Apachesaurus* sp. (="*Anaschisma* n. sp." of Hunt et al., 1989, here revised after Hunt, 1993), *Stagonolepis* sp. (=*Calyptosuchus* sp. of Hunt et al, 1989, here revised after Hunt and Lucas, 1993) and

FIGURE 8. Photographs of selected Chinle Group outcrops in the Lucero uplift. A, Sonsela Member forming bench between red-bed mudstones of the Bluewater Creek Formation (below) and the Painted Desert Member (above) in the Cañada Bonita section. B, Close-up of basal conglomeratic sandstone of Sonsela Member, unit 3 of Cañada Bonita section. C, Typical color-banded crossbeds of Sonsela Member, unit 4 of Cañada Bonita section. D, Overview of La Jara Peak section, showing red-bed mudstone slope of Painted Desert Member of Petrified Forest Formation (1) overlain by bench-forming sandstone and conglomerate of Dakota Sandstone (2). E, Pedogenically altered, variegated mudstone of Owl Rock Formation, unit 5 of Petaca Pinta section. F, Overview of upper part of Petaca Pinta section, showing evenly bedded and intercalated slope-forming sandstones and siltstones of the Rock Point Formation (1=top of formation) overlain by Middle Jurassic eolianites of the Entrada Sandstone (2).

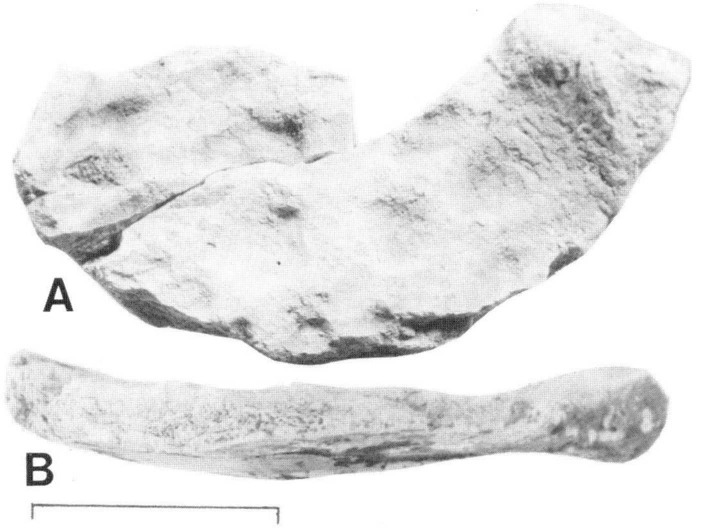

FIGURE 9. NMMNH P-22297, part of the paramedian scute of *Desmatosuchus* sp. from NMMNH locality 2810 in the San Pedro Arroyo Formation. A, Dorsal view. B, Edge view. Bar scale = 2 cm.

Phytosauridae. Unionid bivalves also are present with the vertebrate fossils (Hunt et al., 1989). The presence of *Stagonolepis* indicates an Adamanian (late Carnian) age for the Bluewater Creek Formation at Chicken Mountain Tank (Lucas, 1993; Lucas and Hunt, 1993).

We document here another Adamanian vertebrate fossil occurrence in the Lucero uplift. This is a partial paramedian scute of *Desmatosuchus* sp. (Fig. 9) from NMMNH locality 2810 in the San Pedro Arroyo Formation near our Rio Salado section (Fig. 6). The ornamentation of this scute fragment consists of large, deep oblong pits that are arranged randomly and lack a radial pattern. A raised marginal lamina is evident along the preserved margins of this scute. These features strongly support identification of NMMNH P-22297 as *Desmatosuchus* (Case, 1922; Long and Ballew, 1985). This indicates an Adamanian age (Lucas and Hunt, 1993).

The youngest Triassic fossils reported from the Lucero uplift are petrified wood, water-worn fragments of phytosaur bone and nonmarine gastropods. Lucas et al., (1987, p. 88, 93) reported these fossils from the upper part of the Painted Desert member of the Petrified Forest Formation, including the Correo Bed, at the Correo type section on the south flank of Mesa Gigante. These fossils indicate a Late Triassic age, but are not a more precise age indicator.

CORRELATION

A fence diagram (Fig. 10) depicts our interpretation of stratigraphic correlation of Triassic strata across the Lucero uplift and with nearby Triassic outcrops. The diagram correlates four composite sections: (1) Bluewater Creek - Thoreau, T12-13-14 N, R 11-12-13 W, Cibola County, based on data in Stewart et al. (1972) and Lucas and Hayden (1989); (2) Northern Lucero uplift, T6-7-8-9, R 4-5-6, Cibola County, based on data presented here; (3) southern Lucero uplift, T 2-3 N, R 3-4-5 W, Valencia and Socorro Counties, based on data presented here; and (4) Carthage area, T4-5 S, R 2 E, Socorro County based on data in Lucas (1991).

Correlation of the Bluewater Creek-Thoreau composite section with the section in the northern Lucero uplift is very direct because all Triassic stratigraphic units in west-central New Mexico, except the

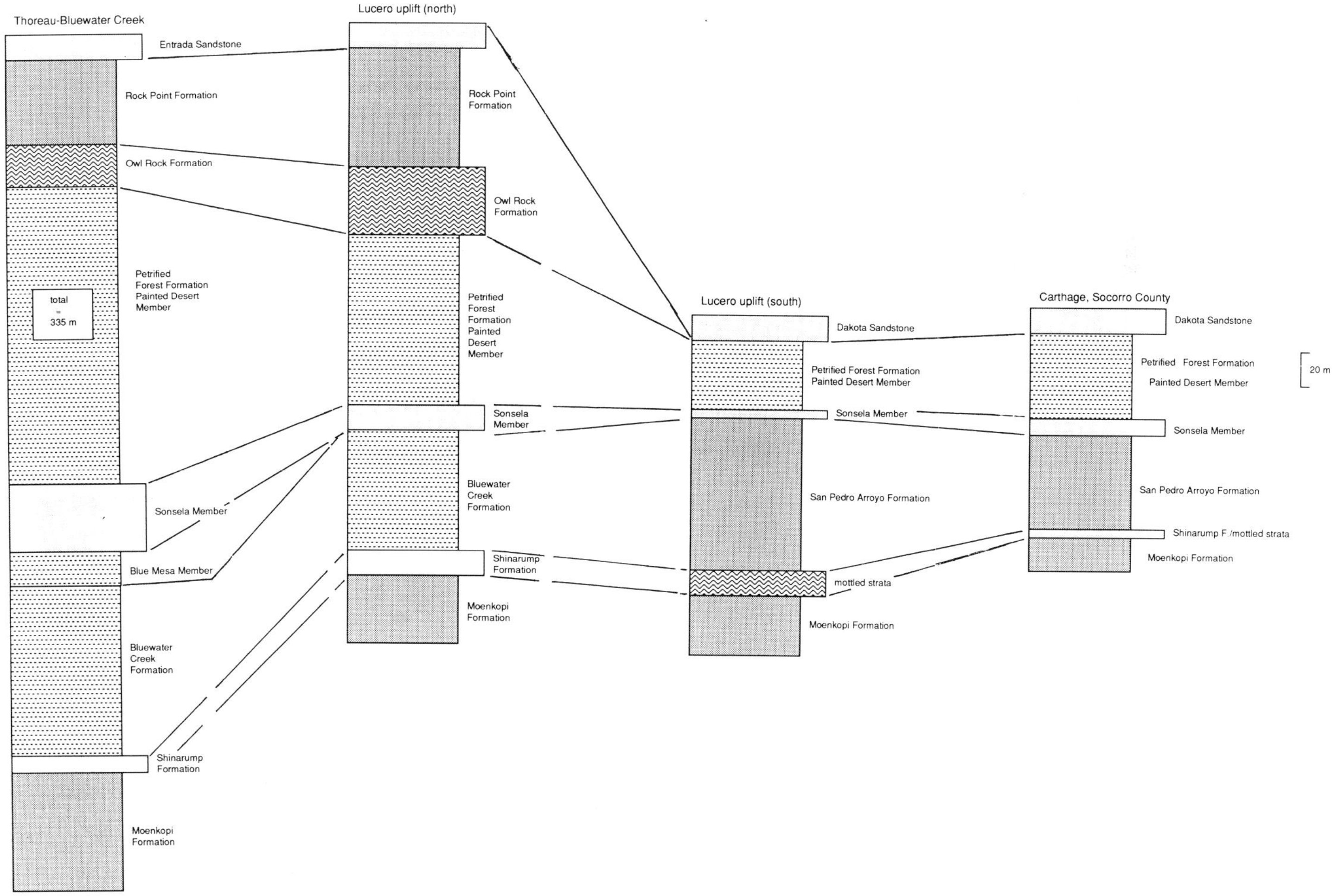

FIGURE 10. Correlation of Triassic strata in the Lucero uplift and nearby areas.

Blue Mesa Member of the Petrified Forest Formation are present in the Lucero uplift. The Blue Mesa Member shows a regional trend of thinning from west to east; in the Petrified Forest National Park of Apache County, Arizona, it is at least 83 m thick but in McKinley County, west-central New Mexico it is 20-40 m thick and in the Lucero uplift it is absent. Over a similar traverse, the Bluewater Creek Formation thickens from about 50 m in McKinley County to more than 70 m in the northern Lucero uplift. This thickening is not, however, as great as the thinning of the Blue Mesa member, so it cannot be argued that the Blue Mesa Member is simply grading eastward into red beds of the upper Bluewater Creek Formation. Clearly, erosion associated with the pre-Sonsela unconformity (Tr-4 unconformity of Lucas, 1993) must account for some (or perhaps all) of the eastward thinning and disappearance of the Blue Mesa Member across the southern Colorado Plateau.

Correlation of the composite sections in the northern and southern Lucero uplifts relies on the persistence of the Moenkopi Formation, Shinarump Formation/mottled strata, Sonsela Member and Painted Desert Member across the Rio Salado (Fig. 10). Strata below the Sonsela Member south of the Rio Salado differ strikingly from the Bluewater Creek Formation in being sandier, having beds of conglomerate and including significant thicknesses of grayish purple mudstones. These strata south of the Rio Salado are lithologically identical to the lower part of the San Pedro Arroyo Formation to the southeast across the Rio Grande at Carthage (Lucas, 1991). For this reason we assign strata between the mottled strata and Sonsela Member in the southern Lucero uplift to the San Pedro Arroyo Formation. These strata occupy the same stratigraphic position as the Bluewater Creek Formation to the north and Adamanian-age aetosaurs also support correlation of the Bluewater Creek and San Pedro Arroyo Formation across the Lucero uplift. Strata above the Sonsela Member in the southern Lucero uplift are red-bed mudstones of the Painted Desert Member identical to Painted Desert Member strata to the north.

Identification and correlation of the San Pedro Arroyo Formation resolves the problems encountered when correlating this unit in its type area at Carthage, Socorro County, discussed by Lucas (1991). Resolution of these problems and a redefinition of the San Pedro Arroyo Formation in its type area are embodied in our fence diagram (Fig. 10). We correlate the pre-Sonsela San Pedro Arroyo Formation strata of the southern Lucero uplift with only the lower part (actually units 2-9) of the San Pedro Arroyo Formation type section. In other words, we limit the San Pedro Arroyo Formation to units 2-9 of Lucas' (1991) original type section, and we identify his unit 10 as Sonsela Member and his units 11-18 as Painted Desert Member. This redefinition recognizes the southeastward continuity of the Sonsela and Painted Desert members from the southern Lucero uplift eastward across the Rio Grande rift to Carthage.

Limited biostratigraphic data and magnetostratigraphy further support these correlations. The Bluewater Creek and lower part of the original San Pedro Arroyo Formation produce some Adamanian-age vertebrates, as does the Santa Rosa and Garita Creek Formation of east-central New Mexico (Lucas, 1993; Hunt and Lucas, 1993). Magnetostratigraphically, the lower San Pedro Arroyo Formation also has the same sequence of magnetic polarity as the Santa Rosa Formation, normal-reversed-normal (Molina-Garza et al., 1993),and this supports correlation of the two units.

What emerges is a much clearer correlation of Chinle Group strata in central New Mexico than was available to previous workers. The redefined San Pedro Arroyo Formation is an Adamanian (upper Carnian) unit that extends from the Lucero uplift across the Rio Grande to as far east as Lincoln County (Lucas, 1991). In Socorro County, the Sonsela Member, followed by the Painted Desert Member (both Petrified Forest Formation) overlie the San Pedro Formation. What the precise equivalents of these upper Chinle Group strata are to the east, in Lincoln County, will have to be resolved with data not now available.

ACKNOWLEDGMENTS

The New Mexico Museum of Natural History and Science supported this research. Adrian Hunt provided some field assistance. Orin Anderson and John Lorenz reviewed an earlier version of this manuscript.

REFERENCES

Anderson, O. J. and Lucas, S. G., 1993, McGaffey Member of Upper Triassic Bluewater Creek Formation, west-central New Mexico: *in* Lucas and Morales, eds., The nonmarine Triassic: New Mexico Museum of Natural History and Science, Bulletin 3, p. G30-G31.

Case, E.C, 1922, New reptiles and stegocephalians from the Upper Triassic of Western Texas: Washington, 84 p.

Dane, C.H. and Bachman, G.O., 1965, Geologic map of New Mexico. Denver, U.S. Geological Survey, scale 1:500,000.

Darton, N.H., 1910, A reconnaissance of parts of northwestern New Mexico and northern Arizona: U.S. Geological Survey, Bulletin 435, 88 p.

Darton, N.H., 1928, "Red beds" and associated formations in New Mexico: U.S. Geological Survey, Bulletin 794, 356 p.

Goddard, E.N., Trask, P.D., DeFord, R.K., Rove, O.N., Singewald, G.T., Jr. and Overbeck, R.M., 1984, Rock color chart. Boulder, Geological Society of America.

Hayden, S.N. and Lucas, S.G., 1988a, Stratigraphy of the Middle Triassic Moenkopi Formation, Lucero uplift, west-central New Mexico: New Mexico Geology, v. 10, p. 67.

Hayden, S.N. and Lucas S.G., 1988b, Stratigraphy of the Permo-Triassic boundary in northern New Mexico: Abstracts of the Symposium on Southwestern Geology and Paleontology 1988 [Museum of Northern Arizona, Flagstaff]. p. 5.

Hayden, S.N. and Lucas, S.G., 1989, Triassic stratigraphy of west-central New Mexico: New Mexico Geological Society, Guidebook 40, p. 191-211.

Hunt, A.P., 1993, Revision of the Metoposauridae (Amphibia: Temnospondyli) and description of a new genus from western North America: Museum of Northern Arizona Bulletin 59, p. 27-50.

Hunt, A.P. and Lucas, S.G., 1993, Triassic vertebrate paleontology and biochronology of New Mexico: New Mexico Museum of Natural History and Science, Bulletin 2, p. 49-60.

Hunt, A.P. Lucas, S.G., Martini, K. and Martini, T., 1989, Triassic stratigraphy and paleontology, Mesa del Oro, Valencia County, New Mexico: New Mexico Geological Society, Guidebook 40, p. 8-9.

Jicha, H.L. Jr., 1958, Geology and mineral resources of Mesa del Oro quadrangle, Socorro and Valencia counties, New Mexico: New Mexico Bureau of Mines and Mineral Resources, Bulletin 56, 67 p.

Kelley, V.C. and Wood, G.H., 1946, Lucero uplift, Valencia, Socorro, and Bernalillo counties, New Mexico: U.S. Geological Survey, Oil and Gas Investigations Preliminary Map OM-47.

Kietzke, K.K, 1988, The calcareous microfauna of the Moenkopi Formation (Triassic, Scythian or Anisian) of central New Mexico: New Mexico Geology, v. 10, p. 64-65.

Kietzke, K.K., 1989, Calcareous microfossils from the Moenkopi Formation (Triassic, Scythian or Anisian) of central New Mexico: New Mexico Geological Society, Guidebook 40, p. 181-190.

Long, R.A. and Ballew, K.L., 1985, Aetosaur dermal armor from the Late Triassic of southwestern North America, with special reference to material from the Chinle Formation of Petrified Forest National Park: Museum of Northern Arizona Bulletin, v. 54, p. 45-68.

Lucas, S.G., 1991, Triassic stratigraphy, paleontology and correlation, south-central New Mexico: New Mexico Geological Society, Guidebook 42, p. 243-259.

Lucas, S.G., 1993, The Chinle Group: revised stratigraphy and biochronology of Upper Triassic nonmarine strata in the western United States: Museum of Northern Arizona Bulletin 59, p. 27-50.

Lucas, S.G. Allen, B.D. and Hayden, S.N., 1987, Type section of the Triassic Correo Sandstone Bed, Chinle Formation, Cibola County, New Mexico: New Mexico Journal of Science, v. 27, p. 87-93.

Lucas, S.G. and Hayden, S.N., 1989, Triassic stratigraphy of west-central New Mexico: New Mexico Geological Society, Guidebook 40, p. 191-211.

Lucas, S.G. and Hunt, A.P., 1987, Stratigraphy of the Anton Chico and Santa Rosa Formations, Triassic of east-central New Mexico: Journal of the Arizona-Nevada Academy of Science, v. 25, p. 21-33.

Lucas, S.G. and Hunt, A.P., 1990, Upper Triassic Owl Rock and Rock Point Members, Chinle Formation, Petaca Pinta, Cibola County, New Mexico: New Mexico Geology, v. 12, p. 92.

Lucas, S.G. and Hunt, A. P., 1993, Tetrapod biochronology of the Chinle Group (Upper Triassic), western United States: *in* Lucas and Morales, eds., The nonmarine Triassic: New Mexico Museum of Natural History and Science, Bulletin 3, p. 327-329.

Lucas, S.G. Martini, K. and Martini, T., 1988, Upper Triassic Correo Sandstone Bed, Petrified Forest Member, Chinle Formation, Hagan basin, Sandoval County, New Mexico: New Mexico Geology, v. 10, p. 65.

Lucas, S.G. and Morales, M., 1985, Middle Triassic amphibian from basal Santa Rosa Formation, east-central New Mexico: New Mexico Geological Society, Guidebook 36, p. 56-58.

Maxwell, C.H., 1982, Mesozoic stratigraphy of the Laguna-Grants region: New Mexico Geological Society, Guidebook 33, p. 261-266.

Maxwell, C.H., 1988a, Geologic map of the Cerro del Oro quadrangle, Cibola County, New Mexico: U.S. Geological Survey, Miscellaneous Field Studies Map MF-2033, scale 1:24,000.

Maxwell, C.H., 1988b, Geologic map of the Marmon Ranch quadrangle, Cibola County, New Mexico: U.S. Geological Survey, Miscellaneous Field Studies Map MF-2049, scale 1:24,000.

Molina-Garza, R. S., Geissman, J. W. and Lucas, S. G., 1993, Late-Carnian Early-Norian magnetostratigraphy from nonmarine strata, Chinle Group, New Mexico: contributions to the Triassic magnetic polarity time scale and the correlation of nonmarine and marine Triassic faunas: *in* Lucas and Morales, eds., The nonmarine Triassic: New Mexico Museum of Natural History and Science, Bulletin 3, p. 345-352.

Ochev, V.G. and Shishkin, M.A., 1989, On the principles of global correlation of the continental Triassic on the tetrapods: Acta Paleontologica Polonica, v. 34, p. 149-173.

Pipiringos, G.N. and O'Sullivan, 1978, Principal unconformities in Triassic and Jurassic Rocks, western interior United States-a preliminary survey: U.S. Geological Survey, Professional Paper 1035-A, 29 p..

Stewart, J.H., Poole, F.G. and Wilson, R.F., 1972, Stratigraphy and origin of the Chinle Formation and related Upper Triassic strata in the Colorado Plateau region: U.S. Geological Survey, Professional Paper 690, 336 p.

Tonking, W.H., 1957, Geology of the Puertecito quadrangle, Socorro County, New Mexico: New Mexico Bureau of Mines and Mineral Resources, Bulletin 41, 67 p.

APPENDIX–STRATIGRAPHIC SECTIONS

Petaca Pinta

Section measured in the SW 1/4 SW1/4 NW 1/4 sec. 31, T6N, R6W, Cibola County by S.G. Lucas and A.P. Hunt, 14 October 1989. Strata are flat-lying

unit	lithology	thickness (m)
	Entrada Sandstone:	
24.	Silty sandstone; grayish yellow (5Y8/4) and moderate reddish brown (10R4/6); quartzarenite; very fine grained; subrounded; moderately sorted; moderately calcareous; trough crossbedded; forms a cliff.	not measured
	unconformity (J-2 unconformity of Pipiringos and O'Sullivan, 1978).	
	Chinle Group:	
	Rock Point Formation:	
23.	Sandy siltstone; grayish red (10R4/2); very calcareous.	2.7
22.	Silty sandstone; moderate reddish brown (10R4/6); quartzarenite; very fine grained; subangular-subrounded; well sorted; slightly micaceous; very calcareous; trough crossbedded and ripple laminated.	5.2
21.	Silty sandstone; moderate reddish brown (10R4/6) and pale reddish brown (10R5/4); micaceous quartzarenite; very fine to fine grained; subrounded; moderately sorted; very calcareous; lower 3 m massive, next 1.5 m trough crossbedded, remainder laminated	9.1
20.	Sandy siltstone; moderate reddish brown (10R4/6); moderately calcareous; climbing ripple laminations.	1.3
19.	Silty sandstone; moderate reddish brown (10R4/6); micaceous quartzarenite; very fine grained; subangular-subrounded; well sorted; very calcareous; massive; forms a persistent ledge.	1.5
18.	Sandy siltstone; pale reddish brown (10R5/4); a few resistant ledges 0.3-0.6 m thick every 3-5 m.	25.1
17.	Silty sandstone; yellowish gray (5Y7/2); quartzarenite; very fine grained; subrounded; well sorted; moderately calcareous; massive; forms a prominent ledge.	1.2
16.	Sandy siltstone, same color and lithology as units 12 and 14.	3.0
15.	Silty sandstone; very pale green (10G8/2) with moderate reddish brown (10R4/6) spots; same lithology as unit 19; massive and blocky; forms a prominent ledge.	0.8
14.	Sandy siltstone; same color and lithology as unit 12.	4.7
13.	Silty sandstone; same color and lithology as unit 15; massive and blocky; top of unit bioturbated; forms a prominent ledge.	2.1
12.	Sandy siltstone; moderate reddish brown (10R4/6) with yellowish gray (5Y7/2) mottles; very calcareous; forms a slope with some thin ledges	13.1
	Total thickness of Rock Point Formation:	69.8
	unconformity (Tr-5 unconformity of Lucas, 1991, 1993).	
	Owl Rock Formation:	
11.	Sandy siltstone; pale reddish brown (10R5/4); very calcareous; much more poorly sorted than overlying units; very bioturbated and blocky	3.3
10.	Silty mudstone; same color and lithology as unit 8.	4.3
9.	Siltstone; grayish red purple (5RP5/2); very calcareous; blocky and mottled ledge.	0.3
8.	Silty mudstone; variegated grayish red purple (5RP4/2) and pale greenish yellow (10Y8/2); calcite veins; approximately in middle of unit is 1.3 m-thick nodular calcrete zone.	5.6
7.	Silty sandstone; mottled grayish red purple (5RP4/2) and very pale green (10G8/2); quartzarenite; very fine to medium grained; subrounded; poorly sorted; very calcareous; bioturbated; forms a blocky ledge.	1.7
6.	Mudstone; variegated very pale green (10G8/2) and grayish red (10R4/2); very calcareous	11.1
5.	Mudstone; variegated grayish red (10R4/2) and very pale green (10G8/2); very calcareous; very bentonitic; contains grayish red (10R4/2) calcrete nodules.	6.1
4.	Sandy siltstone and sandstone; Grayish red purple (5RP4/2), mottles of very pale green (10G8/2); sandstone is very fine to fine grained; poorly sorted; subangular.	1.1
3.	Mudstone; grayish purple (5P4/2); slightly calcareous.	0.6
2.	Sandy siltstone and clay-pebble conglomerate; grayish red purple (5RP4/2) and pale yellowish green (10GY7/2); very calcareous, fills fissures and scours in underlying unit.	2.4
	Thickness of Owl Rock Formation:	36.5
	Petrified Forest Formation:	
	Painted Desert Member:	
1.	Mudstone; dark reddish brown (10R3/4) with moderate yellowish green (5GY7/4) spots; not calcareous; blocky.	not measured

Chicken Mountain Tank

Composite section consisting of four partial sections. These are segments A-D, in ascending order.

Segment D measured at UTM 3830751N, 13285105E. Strata dip 12° to the NW.

unit	lithology	thickness (m)
	Chinle Group:	
	Bluewater Creek Formation:	
	McGaffey Member:	
D4	Fine-grained sandstone; hummocky bedded and ripple laminated; fresh material is yellow gray (5Y7/2) weathering to grayish brown (5YR3/2); some mostly yellow gray (5Y7/2) iron- or magnesium-rich beds; forms persistent, extensive benches 0.5m between scours. Top is a stripped surface.	5.2
	Bluewater Creek Formation:	
D3	Same color and lithology as unit D1.	2.6
D2	Sandstone; hummocky bedded, ripple laminated, bench-forming litharenite; grayish orange pink (5YR7/2) fresh; weathering light brownish gray (5YR6/1); well-sorted, subrounded to subangular very fine grained; calcareous.	1.0
D1	Silty sandstone; very fine light olive gray (5Y6/1) ; well-sorted, subrounded-subangular micaceous litharenite; calcareous.	4.5 +

Segment C measured from UTM 3838867N, 13286397E to 3828809N, 13286065E, where basaltic rubble caps section. Apparent dip of strata is 10° to N 70°W.

	Chinle Group:	
	Bluewater Creek Formation:	
C7	Same color and lithology as unit C5.	10.0 +
C6	Variegated blue, purple and reddish-gray clay with some trough-bedded sandstones in heterolithic crossbeds; sandstones are very fine-grained grayish red (5R4/2) sublitharenite; micaceous; calcareous; weathers to yellow gray (5Y8/1); NMMNH locality 248 produced fossils of the unionid bivalve *Antediplodon* sp. near the middle of this unit.	15.5

unit	lithology	thickness (m)
C5	Mudstone; waxy grayish red (10R4/2); bentonitic.	0.5
C4	Mudstone; grayish purple (5P4/2); bentonitic; calcareous; mottled very light gray (N8).	1.1
C3	Sandy mudstone; grayish red purple (5RP4/2); bentonitic.	3.8
C2	Bentonitic mudstone; grayish purple (5P4/2); calcareous; large calcrete bed with nodules up to 10 cm in diameter in first 0.3 m of this unit.	5.0
C1	Slightly calcareous grayish purple (5P4/2) bentonitic mudstones; calcrete nodules up to 10 cm in diameter.	6.5 +

Segment B was measured WSW down Horse Mountain Draw on the west limb of the anticline measured for segment A. Strata dip 15° to due west. Section was measured from UTM 3838396N, 13286359E to 3838540N, 13286752E.

Chinle Group:
Bluewater Creek Formation:

unit	lithology	thickness (m)
B8	Same color, lithology, and weathering profile as unit A12.	5.3
B7	Slope; much covered by debris from B8.	8.2
B6	Sandstone; yellow gray (5Y7/2) to light olive gray (5Y5/2); lithologically identical to B2; calcareous.	2.2
B5	Sandy siltstone and silty sandstone; light olive gray (5Y6/1); micaceous.	5.0

Shinarump Formation:

unit	lithology	thickness (m)
B4	Same lithology as unit B; more hematized; weathers brownish gray (5YR4/1); fresh color is light to medium gray (N7-N6).	3.9
B3	Trough scour on B2; sandstone identical to unit B2; trough has relief of .5 m.	1.8
B2	Very fine-grained micaceous quartz arenite; well sorted; subrounded-subangular; calcareous; grayish yellow green (5GY7/2) weathering to pale red (5R6/2); planar to wedge planar bedded.	2.3
B1	Trough crossbedded fine- to medium-grained quartzarenite; some beds of limestone cobble conglomerate; sands are well sorted, subrounded, and slightly calcareous; very light gray (N8), weathering pale red (5R6/2); conglomerate clasts are Paleozoic limestones up to 5 cm in diameter; moderate yellow brown (5YR5/4)to pale reddish brown (10R5/4).	2.8

Segment A measured on the east limb of an anticline in the Chicken Mountain Draw/Horse Canyon area, UTM13287136E, 3838429N. Strata dip 15° to S 70°E.

Chinle Group:
Bluewater Creek Formation:

unit	lithology	thickness (m)
A12	Very fine-grained micaceous quartz arenite; yellow gray (5Y8/1); weathers pale olive (10Y6/2) well-sorted; subrounded; calcareous.	3.2
A11	Same color and lithology as unit A7.	2.5
A10	Same color and lithology as unit A6.	1.0
A9	Much covered, same color and lithology as unit A7.	2.5
A8	Ledge-forming ripple-laminated sandstone, same lithology as unit A6.	1.8
A7	Fine-grained micaceous litharenite and grayish red (10R4/2) silty mud; sandstone is pale olive (10Y6/2); calcareous; sandstone is subangular.	2.6
A6	Very fine-grained micaceous litharenite; well-sorted; subangular; ripple laminated; interbeds of siltstone; sandstone is light olive gray (5Y6/1); calcareous.	1.0
A5	Flaggy ripple-bedded thin sandstones like A4 with very micaceous sandy mud; sandstone is light olive gray (5Y5/2); mudstone is pale olive (10Y6/2).	2.5

Shinarump Formation:

unit	lithology	thickness (m)
A4	Planar bedded flaggy micaceous quartzarenite with some mudballs and Paleozoic limestone clasts; light brownish gray (5YR6/1) and light olive gray (5Y6/1); hematitic; beds are .25m or thinner.	0.9
A3	Scour; conglomerate at base, trough crossbedded sandstone above; same lithology and color as unit A2.	0.8
A2	Very fine- to fine-grained quartzarenite; trough cross-bedded with some pebbly lenses like A1; olive gray (5Y3/2); calcareous.	1.5
A1	Massive, hummocky-bedded conglomerate with interbedded fine- to medium-grained quartzarenite; sandstone is pinkish gray (5YR8/1), subrounded, and calcareous; conglomerate clasts are up to 1.5 cm diameter and consist of greenish gray (5GY6/1) Paleozoic limestone pebbles and grayish red (10R4/2) Moenkopi mudchips.	1.0 +

Cañada Bonita

Section begins at UTM 3830910N, 13280990E and ends at 3830897N, 13281040E. Strata dip 20° to due north.

unit	lithology	thickness (m)
	Chinle Group:	
	Petrified Forest Formation:	
	Painted Desert Member:	
9.	Muddy sandstone; grayish red (5R4/2) with very light gray (N8) spots; very fine- to fine-grained quartzarenite; subrounded; well sorted; not calcareous; forms a slope much covered by terrace gravels.	not measured
	Sonsela Member:	
8.	Sandstone; banded grayish pink (5R8/2) and pale red (5R6/2); feldspathic litharenite; very fine grained; subrounded; well sorted; not calcareous; laminated; forms a resistant bench.	2.6
7.	Sandstone; mottled grayish red (10R4/2) and very light gray (N8); feldspathic litharenite; medium to coarse grained; subangular; moderately sorted; very calcareous; laminated; friable, forms a slope.	5.0
6.	Sandstone; very light gray (N8); some very dusky red purple bands (5RP2/2); micaceous litharenite; medium grained; subrounded; moderately sorted; slightly calcareous; trough cross-bedded.	3.2
5.	Sandstone; grayish pink (5RP 8/2); feldspathic quartzarenite; subangular; fine to medium grained; moderately sorted; not calcareous; some isolated clasts of Paleozoic limestone up to 4 cm in diameter; trough crossbedded.	2.2
4.	Sandstone; pinkish gray (5YR8/1) with bands of very dusky red purple (5RP2/2); quartzarenite; fine grained; subrounded; well sorted; not calcareous; trough crossbedded.	1.2
3.	Conglomeratic sandstone; very light gray (N8), weathers pale red (5R6/2) to grayish red (5R4/2); sandstone is same lithology as unit 2; clasts are very dusky red (10R2/2) mudstone chips, a few cherts and Paleozoic limestones; trough crossbedded.	1.3
	Thickness of Sonsela Member:	15.3
	unconformity (Tr-4 unconformity of Lucas, 1991, 1993).	
	Bluewater Creek Formation:	
2.	Sandstone; pale pink (5RP8/2); lithic quartzarenite; fine to medium grained; subrounded; moderately sorted; calcareous; some biotite; trough cross-bedded; some thin lenses of mudstone of unit 1 color and lithology	1.5
1.	Mudstone; grayish red (5R4/2) and grayish purple (5P2/3); bentonitic; not calcareous.	not measured

Selso Well

Section measured at UTM 3824229N, 13285390E. Strata dip 10° to N 80°E.

unit	lithology	thickness (m)
	Chinle Group:	
	Shinarump Formation:	
9.	Conglomerate and conglomeratic sandstone; light olive gray (5Y6/1) to brownish gray (5YR4/1); conglomerate clasts are mostly gray Paleozoic limestone up to 5 cm in diameter, but a few clasts of yellow and red chert and jasper are present; matrix is fine- to coarse-grained sandstone that is a poorly sorted, subrounded, calcareous litharenite; trough crossbedded; forms a bench.	1.2 +
	unconformity (Tr-3 unconformity of Pipiringos and O'Sullivan, 1978).	
	Moenkopi Formation:	
	Anton Chico Member:	
8.	Silty mudstone; same colors and lithology as unit 6; forms a slope.	4.8
7.	Sandstone and conglomerate; sandstone is grayish red (5R4/2) to pale reddish brown (10R5/4); very fine grained, well sorted, subrounded; non-calcareous micaceous litharenite; conglomerate is grayish red (10R4/2), matrix-supported,	

	calcareous mudstone pellets up to 1 cm in diameter; trough crossbedded and ripple laminated with conglomerate in trough bases; forms the only prominent bench in the Moenkopi slope.	2.0
6.	Silty mudstone; moderate reddish brown (10R4/6); calcareous; contains lenses of greenish gray (5GY6/1), calcareous, bioturbated micaceous silt-stone.	6.3
5.	Sandstone; light olive gray (5Y6/1); very fine grained; well sorted; subrounded; calcareous; micaceous litharenite; trough crossbedded; forms a ledge; has some lenses of conglomerate of unit 7 lithology with unidentifiable bone fragments.	0.5
4.	Silty mudstone and muddy siltstone; moderate reddish brown (10R4/6).	2.0
3.	Siltstone and sandy siltstone; yellowish gray (5Y7/2) but weathers pale reddish brown (10R5/4); calcareous; ripple laminated.	0.8
2.	Covered slope; probably yellow siltstone.	8.5
	Thickness of Moenkopi Formation:	24.9

unconformity (Tr-1 unconformity of Pipiringos and O'Sullivan, 1978).

San Andres Formation (Permian)

1.	Limestone; pale yellowish orange (10YR8/6); weathers grayish orange (10YR7/4); forms a bench.	not measured

La Mesa de Victorino

Measured in the NE 1/4 SW 1/4 sec. 31, T3N, R4W, Socorro County. Section begins at UTM 3813209N, 13288841 E and ends at 3813427N, 13288823E. Strata dip 40° to N 30° W. This is the section described by Tonking (1957, p. 58-59).

unit	lithology	thickness (m)
	Chinle Group:	
	San Pedro Arroyo Formation:	
20.	Silty and sandy mudstone; pale reddish brown (10R5/4); calcareous; forms a slope.	10.5 +
	"mottled strata"	
19.	Mud-pellet conglomerate; pale red (5R6/2), weathers grayish red (10R4/2); calcareous; mottled and bioturbated; contains pale red (10R6/2) calcrete nodules; ledgy and blocky.	3.5
18.	Muddy, sandy siltstone; grayish red (10R4/2) with light gray (N7) reduction spots; calcareous.	3.8
17.	Mud-pellet conglomerate; pale red (5R6/2) with color mottling like unit 15; calcareous; some crude trough crossbeds.	1.3
16.	Muddy sandy siltstone; pale red (5R6/2); calcareous; heavily modified pedogenically.	2.8
15.	Mud-pellet conglomerate; grayish red (10R4/2) with color mottling of moderate reddish orange (10R6/6), moderate reddish brown (10R4/6), pale red (5YR6/2) and medium dark gray (N4); not calcareous; pellets up to 0.5 cm in diameter; rip-ups of Moenkopi mud-stone.	3.0
	Total thickness of "mottled strata":	14.4
	unconformity (Tr-3 unconformity of Pipiringos and O'Sullivan, 1978)	
	Moenkopi Formation:	
	Anton Chico Member:	
14.	Sandstone; same color and lithology as unit 7.	3.5
13.	Sandstone; grayish red (5R4/2); micaceous litharenite; very fine grained; subrounded; well sorted; calcareous; trough-crossbedded; fines upward, grades into overlying unit.	8.8
12	Silty mudstone; same color and lithology as unit 6.	1.1
11.	Sandstone; grayish red (5R4/2); very fine grained; micaceous litharenite; subangular-subrounded; well sorted; has iron-rich (goethitic) laminae that are brownish black (5YR2/1) to black (N1); wavy laminated and shallow trough crossbeds.	0.3
10.	Silty mudstone; same colors and lithology as unit 6; lower 0.5 m of unit is a calcareous paleosol.	6.0
9.	Sandstone; pale red (10R6/2); same lithology as unit 7; trough crossbedded; contains some pebbly beds like those of unit 7; forms a prominent bench; section is offset at top of this unit ~ 100 m to NW.	2.5
8.	Silty mudstone; same colors and lithology as unit 6; top of unit is a paleosol.	3.5
7.	Sandstone; grayish red (10R4/2); micaceous litharenite; fine grained; subangular-subrounded; well sorted; calcareous; trough crossbedded; has some pebbly beds of grayish red mudstone/siltstone; forms a bench.	1.8
6.	Silty mudstone; grayish red (10R4/2); calcareous; has some lenses of mud-pellet conglomerate that are pale red (10R6/2) to light olive gray (5Y6/1); much pedogenic modification; forms a hackly slope; base of this unit=unit 1 of Tonking's section.	12.5
5.	Muddy siltstone; yellowish gray (5Y8/1); calcareous; ripple laminated and bioturbated.	2.3
4.	Sandstone; light brownish gray (5YR6/1); weathers light gray (N7); micaceous litharenite; medium grained; subrounded; well sorted; calcareous; trough crossbedded; some pebbly lenses (of unit 3 lithology) at bases of trough axes.	10.0
3.	Conglomerate and conglomeratic sandstone; conglomerate is pale red (5R6/2) and light olive gray (5Y6/1) with clasts of San Andres Formation limestone up to 2 cm in diameter; sandstone is moderate reddish brown (10R4/6) and very light gray (N8) micaceous litharenite that is fine-medium grained, subangular, moderately sorted and hematitic; trough crossbedded.	0.8
2.	Sandstone and siltstone; sandstone is yellowish gray (5Y8/1) and light greenish gray (5GY8/1), very fine-medium grained, poorly sorted, subangular, calcareous, micaceous litharenite; siltstone is grayish orange (10YR7/4), muddy and very calcareous; sandstone is hummocky laminated and ripple laminated.	6.5
	Total thickness of Moenkopi Formation:	51.6
	unconformity (Tr-1 unconformity of Pipiringos and O'Sullivan, 1978)	
	San Andres Formation:	
1.	Dolomitic limestone; yellowish gray (5Y7/2); micritic.	not measured

La Jara Peak

Measured in the SE 1/4 sec. 10, T2N, R5W, Socorro County. Section begins at UTM 3810274N, 13284528E in arroyo bottom and ends at 3810365N, 13284483 E. Strata are flat-lying.

unit	lithology	thickness (m)
	Dakota Sandstone:	
3.	Sandstone and conglomeratic sandstone; pale red (5R6/2), weathers dark yellowish orange (10YR6/6); quartzarenite, very fine to fine grained; subangular-subrounded; poorly sorted; hematitic; not calcareous; siliceous pebbles; trough crossbedded; forms a prominent cliff.	not measured
	unconformity	
	Chinle Group:	
	Petrified Forest Formation:	
	Painted Desert Member:	
2.	Mudstone; grayish orange (10YR7/4); not calcareous.	1.1
1.	Silty and sandy mudstone; grayish red (5R4/2) with light greenish gray (5GY8/1) reduction spots; not calcareous; forms a slope much covered by colluvium.	32.8 +

Rio Salado

Section starts at UTM 2809765N, 13291284E and ends at 3809108N, 13290989E. Strata dip 7° due South.

unit	lithology	thickness (m)
	Chinle Group:	
	Petrified Forest Formation:	
	Painted Desert Member:	
36.	Mudstone; dark reddish brown (10R3/4); bentonitic; calcareous.	not measured
35.	Sandstone; light brownish gray (5YR6/1); very fine grained; subrounded; well sorted; litharenite; not calcareous; trough crossbedded; forms a cuesta.	1.0

unit	lithology	thickness (m)
34.	Mudstone; grayish red (5R8/2) and grayish pink (5R8/2); bentonitic; calcareous.	5.2
	Sonsela Member:	
33.	Sandstone; same colors and lithology as unit 31; laminated in shallow troughs; a porphyritic dike cuts the top of this unit locally.	0.2
32.	Mudstone; grayish red purple (5RP4/2); bentonitic; calcareous.	0.8
31.	Sandstone; very light gray (N8); very fine grained; well sorted; subrounded; litharenite; not calcareous; trough crossbedded.	1.0
30.	Conglomerate; grayish green (5G5/2); weathers grayish red purple (5RP4/2); clasts are rounded calcrete, mudstone and siltstone pebbles up to 0.5 cm in diameter; clast supported.	0.5
	Total thickness of Sonsela Member:	2.5
	unconformity (Tr-4 unconformity of Lucas, 1993)	
	San Pedro Arroyo Formation:	
29.	Mudstone; same color and lithology as unit 25.	2.5
28.	Sandstone; grayish red purple(5RP4/2); very fine grained; micaceous litharenite; subrounded; well sorted; trough crossbedded; some pebbly beds; forms a ledge.	2.2
27.	Mudstone; same color and lithology as unit 25.	3.3
26.	Siltstone; light gray (N7); not calcareous; massive; very well indurated; forms a ledge.	0.5
25.	Mudstone; grayish red (5R4/2); bentonitic; calcareous; contains numerous medium light gray (N6) calcrete nodules.	12.3
24.	Conglomerate; pale yellowish brown; clast supported; clasts are rounded calcrete and mudstone pebbles; massive.	0.1
23.	Sandstone; light gray (N7); same lithology as unit 17; planar crossbedded; forms a cuesta.	0.5
22.	Sandstone; light brownish gray (5YR6/1); weathers brownish gray (5YR 4/1); very fine-fine grained; micaceous litharenite; subangular, poorly sorted; laminated.	1.5
21.	Conglomerate; medium gray (N5) and bluish white (5B9/1); clasts are rounded calcrete pebbles up to 2 cm in diameter; contains *Desmatosuchus* sp. at NMMNH locality 94-18.	0.9
20.	Mudstone; same color and lithology as unit 11.	0.7
19.	Sandstone; same colors and lithology as unit 17 except rock is mostly light gray (N7) and weathers brownish gray (5YR4/1).	0.3
18.	Mudstone; same colors and lithology as unit 11.	2.5
17.	Sandstone; pale red (5R6/2); very fine grained; litharenite; well sorted; subrounded; not calcareous; laminated in shallow trough crossbeds; crest of ridge.	1.5
16.	Mudstone and sandstone, interbedded; mudstone same colors and lithology as unit 11; sandstone is banded light olive gray (5Y6/1) and grayish purplè(5R4/2), fine grained micaceous litharenite, subangular, moderately sorted, not calcareous, laminated.	5.5
15.	Conglomerate; brownish gray (5YR4/1); clast supported; clasts are rounded calcrete pebbles up to 1 cm in diameter; some trough crossbeds.	0.3
14	Covered slope; probably same lithology as units 13 and 11.	13.2
13.	Mudstone; same colors and lithology as unit 11.	7.5
12.	Calcrete (limestone); massive to nodular; olive gray (5Y4/1).	1.3
11.	Mudstone; grayish red purple (5RP4/2) with very light gray (N8) mottles; bentonitic; calcareous; forms a powdery slope.	1.5
10.	Silty mudstone; grayish red (10R4/2) with light greenish gray (5GY8/1) mottles; some sandy lenses; not calcareous.	1.5
9.	Sandstone; moderate red (5R5/4); weathers pale red (5R6/2) and moderate pink (5R7/4); very fine grained; micaceous litharenite; subangular; well sorted; calcareous; massive to ripple laminated; forms a thin ledge.	0.4
8.	Covered; gully cuts show red mudstone identical to unit 7.	26.0
7.	Mudstone; grayish red (10R4/2) with light greenish gray (5GY8/1) to light gray (N8) mottles; bentonitic; calcareous.	4.0
	Thickness of San Pedro Arroyo Formation:	90.0
	"mottled strata":	
6.	Calcrete (limestone) and silcrete; mottled medium dark gray (N4), grayish purple (5P4/2), light gray (N7) and pale red (10R6/2); some clasts of light olive gray quartzite up to 4 cm in diameter and moderate pink (5R7/4) to dark gray (N3) jasper up to 2 cm in diameter.	0.7
	unconformity (Tr-3 unconformity of Pipiringos and O'Sullivan, 1978)	
	Moenkopi Formation:	
	Anton Chico Member:	
5.	Sandstone; same color and lithology as unit 1.	0.3
4.	Muddy siltstone; grayish red (10R4/2); not calcareous.	1.0
3.	Sandstone; same color and lithology as unit 1.	0.8
2.	Muddy siltstone; grayish red (10R4/2); not calcareous.	0.5
1.	Sandstone; pale reddish brown (10R5/4); weathers grayish red (10R4/2); very fine grained; subrounded; well sorted; litharenite; not calcareous; ripple laminated with small trough crossbeds; forms a ledge.	0.8

Cañon del Alamito

Start at UTM 3810190N, 13290223E, end at 3908483 N, 13291592 E. Strata are nearly flat-lying.

unit	lithology	thickness (m)
	Dakota Group:	
11.	Sandstone; light olive gray(5Y6/1); weathers moderate orange pink (10R7/4); quartzarenite; very fine-fine grained; subrounded; poorly sorted; not calcareous; trough crossbedded; forms a bench capping the escarpment.	not measured
	unconformity	
	Chinle Group:	
	Petrified Forest Formation:	
	Painted Desert Member:	
10.	Mudstone; pale red (5R6/2); bentonitic; calcareous: forms a slope.	20.1
9.	Silty sandstone and mudstone interbeds; grayish red (10R4/2) with some white (N9) to light greenish gray (5GY8/1) mottles; very fine grained; litharenite; poorly sorted; subangular; not calcareous; trough crossbedded and ripple laminated; thin interbedding of sandstone and mudstone beds; tops of sandstone beds are bioturbated.	2.0
8.	Mudstone; dark reddish brown (10R3/4); calcareous; bentonitic; contains light gray (N7) calcrete nodules.	8.8
7.	Mudstone; grayish red (10R4/2); bentonitic; calcareous; contains medium gray (N5) calcrete nodules.	
	[There is a stratigraphic gap due to cover and a possible fault between units 6 and 7; this gap appears at most to be 3-4 m of the lowermost Painted Desert Member]	
	Sonsela Member:	
6.	Conglomerate; same colors and lithology as unit 2; top of the unit here is not its stratigraphic top but a stripped surface.	0.2 +
5.	Sandstone; same colors and lithology as unit 3; laminated and low angle trough crossbedded.	1.0
4.	Conglomerate; same colors and lithology as unit 2.	0.3
3.	Sandstone; pale red (5R6/2) with laminae of grayish red purple (5RP4/2); litharenite; fine grained; subangular; well sorted; not calcareous; trough and planar crossbedded; some laminated beds.	2.5
2.	Conglomerate; medium light gray (N6) with grayish red purple (5RP4/2) mottles; clasts are well rounded calcrete and mudstone pebbles up to 1 cm in diameter; some crude trough crossbeds.	0.8
	Thickness of Sonsela Member:	4.8
	unconformity (Tr-4 unconformity of Lucas, 1993)	
	San Pedro Arroyo Formation:	
1.	Mudstone; grayish red (5R4/2); bentonitic; calcareous.	6.8 +

WESTWARD EXTENSION OF SEBOYETA BAY, LATE CRETACEOUS (CENOMANIAN) OF WEST-CENTRAL NEW MEXICO

ORIN. J. ANDERSON[1] and SPENCER G. LUCAS[2]

[1]New Mexico Bureau of Mines and Mineral Resources, Socorro, NM 87801; [2]New Mexico Museum of Natural History and Science, 1801 Mountain Rd. NW, Albuquerque, NM 87104

Abstract—A molluscan fossil assemblage dominated by *Exogyra levis* Stephenson and *Granocardium enstromi* (Bergquist) is present in the upper part of the Dakota Sandstone along Cottonwood Canyon in extreme western Catron County. These fossils indicate correlation with the middle Cenomanian Paguate Tongue of the Dakota Sandstone. They thus extend the Seboyeta bay in which the Paguate was deposited 40 km to the west-southwest.

INTRODUCTION

The Upper Cretaceous section throughout large areas of west-central New Mexico is represented by an intertongued Dakota Sandstone-Mancos Shale sequence of Cenomanian age (Fig. 1). Landis et al. (1973) first recognized this intertonguing relationship and also named the various tongues. Several Dakota Sandstone tongues are very similar to each other lithologically, and without the entire sequence present stratigraphic assignment to a particular tongue can be problematic. Fortunately, an abundant molluscan fauna is present in these rocks, and is sufficiently varied to allow distinctions based on biostratigraphy (Cobban, 1977).

One Dakota unit, the Paguate Tongue, is of interest here for several reasons. It is recognized and may be traced biostratigraphically throughout much of the southern San Juan Basin. The ammonite *Acanthoceras amphibolum* is present together with the bivalves *Exogyra levis* and *Granocardium* sp. in the Paguate Tongue; *Acanthoceras amphibolum* is not present in the older Cubero Tongue of the Dakota or the younger Twowells Tongue (Cobban, 1977). The

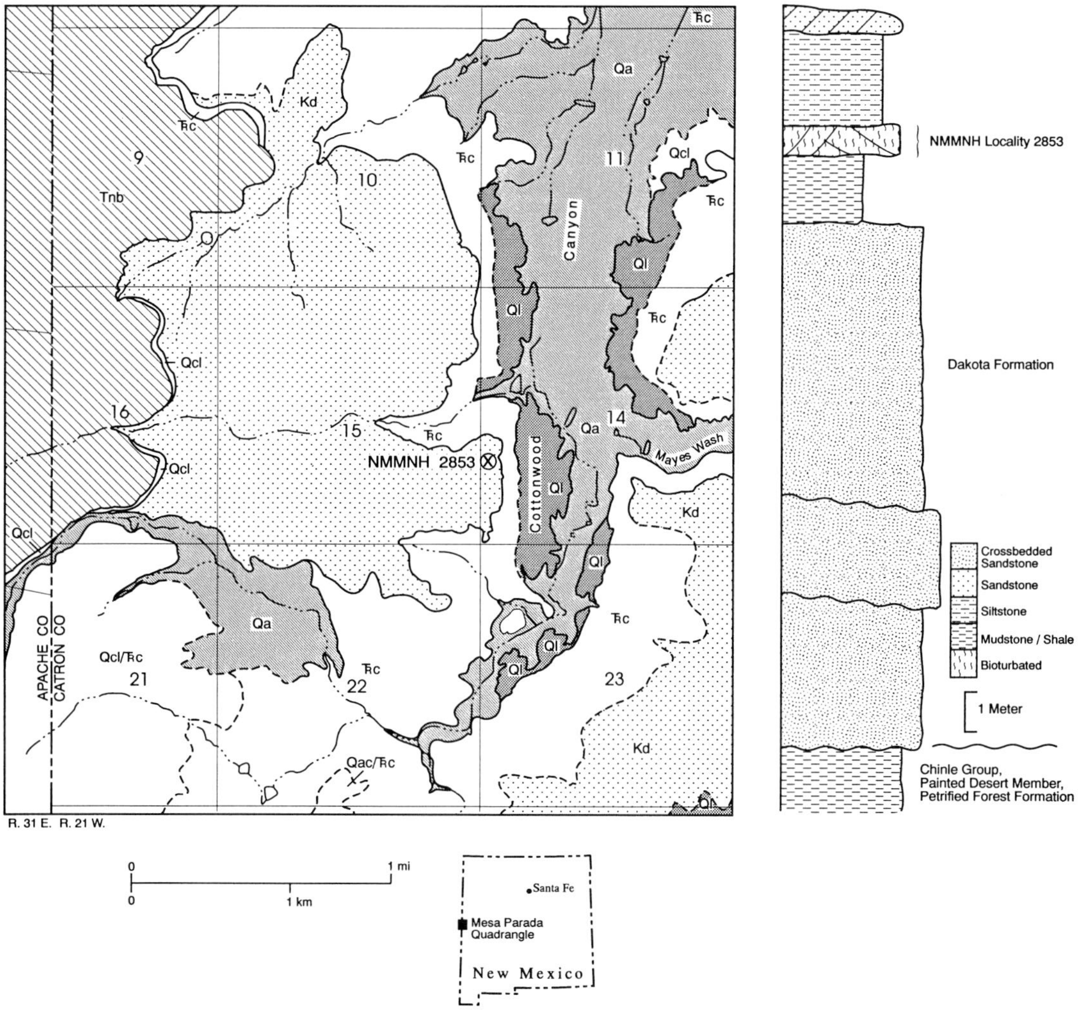

FIGURE 1. Map of a portion of the Mesa Parada quadrangle showing location of molluscan fossil assemblage (NMMNH 2583) in Dakota Sandstone on west side of Cottonwood Canyon. Stratigraphic section through the fossiliferous zone also shown.

Paguate Tongue is also associated with the concept of a Seboyeta bay, an embayment that reveals some of the details of the initial transgression of the Western Interior seaway in New Mexico.

The intertongued Dakota-Mancos sequence was recognized in the Zuni Basin of west-central New Mexico by Hook et al. (1980), who used the nomenclature of Landis et al. (1973) to describe and define the section. With the recognition of this intertongued sequence in the Zuni Basin many errors and uncertainties in the correlation of Upper Cretaceous rocks were addressed and corrected. Miscorrelations in this part of the section had appeared in Pike (1947), Dane (1959), Dane and Bachman (1965), and in numerous hydrologic reports.

The Paguate Tongue of the Dakota Sandstone was later recognized and mapped in the Atarque area by Anderson (1987) and in the Fence Lake-Moreno Hill area by Campbell (1989). Southward into Carrizo Wash, Tertiary erosion has stripped off most of the Cretaceous rocks, leaving Triassic rocks exposed. South of Carrizo Wash, however, recent reconnaissance mapping on the Quemado 30° x 60° quadrangle (Chamberlin et al., 1994) has revealed the presence of a discrete fossiliferous sandstone bed at the top of the Dakota Sandstone (Mesa Parada 7½' quadrangle, Fig. 1). The stratigraphic position and the faunal attributes of this sandstone are such that it may be correlated with the Paguate Tongue of the Dakota. Although the diagnostic ammonite *Acanthoceras amphibolum* was not found at this southern locality (NMMNH locality 2583, Fig. 1) the bivalve assemblage dominated by *Exogyra levis* Stephenson and *Granocardium enstromi* (Bergquist) identify it biostratigraphically as a middle Cenomanian Paguate equivalent.

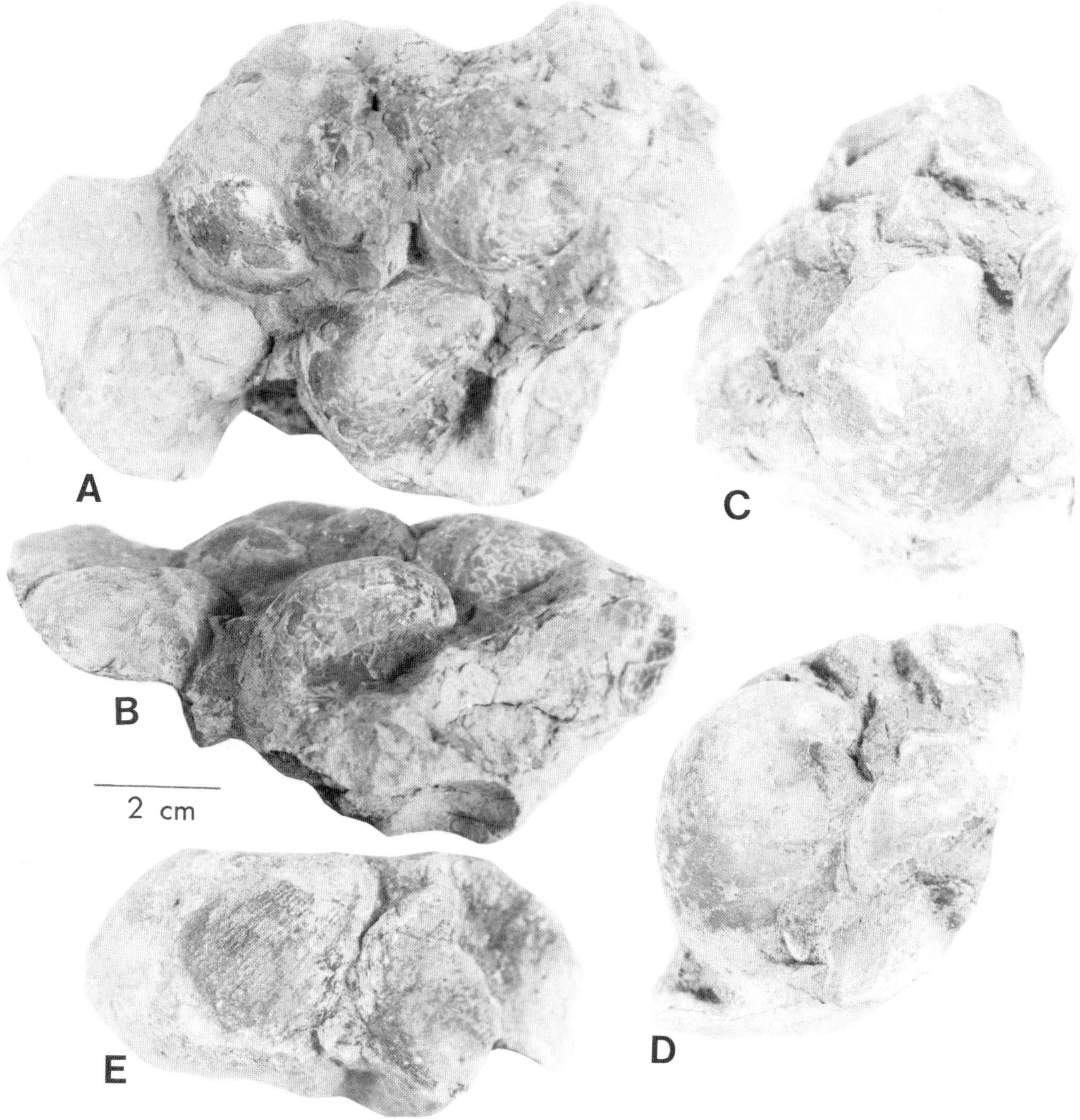

FIGURE 2. Selected fossils from NMMNH locality 2583. A-D, *Exogyra levis*, two views (A-B and C-D) of two samples of packstone, NMMNH P-22498; E, *Granocardium enstromi*, NMMNH P-22499.

The significance of this fossil assemblage is related to the concept of a Seboyeta bay (Hook et al., 1980). Seboyeta bay was an embayment along the western margin of the Interior seaway. Although embayments per se were not unusual, this one was large, as much as 120 km north-south, and in essence represents an arm of the sea that was transgressing at a faster rate than shoreline segments to the north or south. The molluscan assemblage we have found south of Carrizo Wash, along the tributary named Cottonwood Canyon (Fig. 1), is described below. This assemblage records the middle Cenomanian transgression of the Western Interior seaway and suggests that Seboyeta bay extended further westward than previously recognized.

DESCRIPTION OF LOCALITY AND FOSSIL ASSEMBLAGE

NMMNH locality 2583 is located near the crest of the escarpment just west of Cottonwood Canyon in the NW¼ NW¼ SW¼ sec. 14, T2N, R21W, Catron County (UTM 3807530N, 682350E, zone 12) (Fig. 1). The locality is in the Dakota Sandstone (undivided) 14.9 m above the base of the formation, which is its disconformable contact with underlying red-bed mudstones of the Chinle Group (cf. Lucas and Hayden, 1989). The fossiliferous unit is a 0.8-m-thick ledgy sandstone that is trough-crossbedded and bioturbated. The fossils occur as packstone lenses throughout the unit.

The packstones produce a low-diversity fossil assemblage of molluscs, many of which are poorly preserved as hematitic steinkerns and thus cannot be identified precisely. Most of the identifiable molluscs (about 90% of all macrofossils) belong to *Exogyra levis* Stephenson, and the next most common taxon is *Granocardium enstromi* (Bergquist), which is about 5% of all macrofossils. Representative specimens of *E. levis* (Fig. 2A-D) are catalogued as NMMNH P-22498, whereas those of *G. enstromi* (Fig. 2E) are designated NMMNH P-22499. Other macrofossils are poorly preserved unidentifiable bivalves and turitellid gastropods.

In west-central New Mexico, *Exogyra levis* and *Granocardium enstromi* are characteristic of the Paguate Tongue of the Dakota Sandstone (Cobban and Hook, 1989). The Paguate Tongue is of middle Cenomanian age and produces ammonites of the *Acanthoceras amphibolum* and *Pleisiacanthoceras wyomingense* zones. It was deposited in what Hook et al. (1980, p. 44) termed the Seboyeta bay, which was a conspicuous embayment of the Western Interior seaway that extended into west-central New Mexico during middle to Upper Cenomanian time (Fig. 3).

NMMNH locality 2853 is in strata that we believe are homotaxial to the Paguate Tongue of the Dakota Sandstone, although they are 40 km west-southwest of the nearest previously identified Paguate outcrops (McLellan et al., 1983). This locality plus another one on the Manuelito quadrangle immediately southwest of Gallup (Anderson, 1991) thus requires an extension of the western limit of the Seboyeta bay, 40 km further to the west-southwest (Fig. 3).

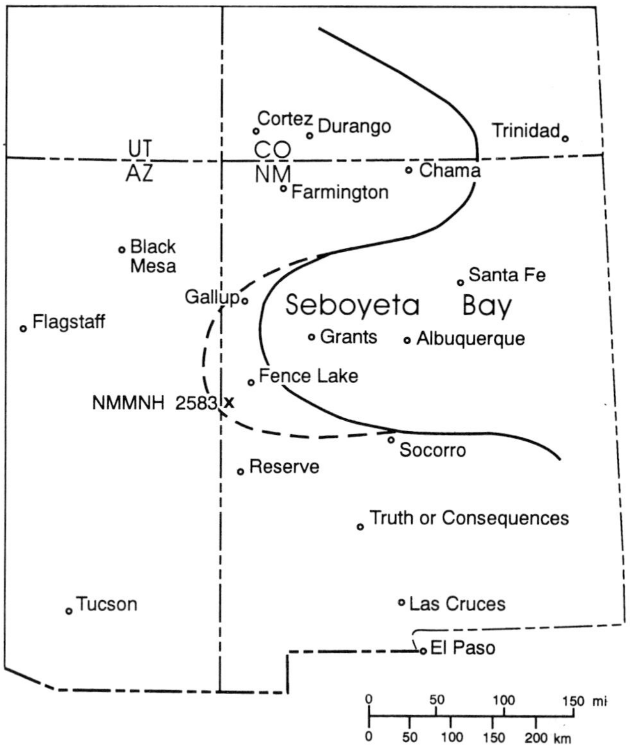

FIGURE 3. Seboyeta bay in west-central New Mexico, illustrating the extent to which the bay was extended based on the molluscan fossil assemblage at Cottonwood Canyon.

REFERENCES

Anderson, O. J., 1987, Geology and coal resources of the Atarque Lake 1:50,000 quadrangle, New Mexico (NW quadrant of Fence Lake 1:100,000): New Mexico Bureau of Mines and Mineral Resources, GM-61, 2 sheets, scale 1:50,000.

Anderson, O. J., 1991, Geology and mineral resources of Manuelito quadrangle, McKinley county, New Mexico: New Mexico Bureau of Mines and Mineral Resources, GM-66, 1 sheet with text, scale 1:24,000.

Campbell, F., 1989, Geology and coal resources of the Fence Lake 1:50,000 quadrangle, New Mexico (SW quadrant of Fence Lake 1:100,000): New Mexico Bureau of Mines and Mineral Resources, GM-62, 2 sheets, scale 1:50,000.

Chamberlin, R. M., Cather, S. M., Anderson, O. J. and Jones, G. E., 1994, Geology of the Quemado 30 x 60 degree quadrangle: New Mexico Bureau of Mines and Mineral Resources, Open-file Report 406.

Cobban, W. A., 1977, Characteristic marine molluscan fossils from the Dakota Sandstone and intertongued Mancos Shale, west-central New Mexico: U.S. Geological Survey, Professional Paper 1009, 30 p.

Dane, C. H., 1959, Historical background of the type locality of Tres Hermanos Sandstone Member of Mancos Shale: New Mexico Geological Society, Guidebook 10, p. 85–91.

Dane, C. H. and Bachman, G. O., 1965, Geologic map of New Mexico: U.S. Geological Survey and New Mexico Bureau of Mines and Mineral Resources, scale 1:500,000.

Hook, S. C., Cobban, W. A. and Landis, E. R., 1980, Extension of the intertongued Dakota Sandstone-Mancos Shale terminology into southern Zuni basin: New Mexico Geology, v. 2, p. 42–46.

Landis, E. R., Dane, C. H., and Cobban, W. A., 1973, Stratigraphic terminology of the Dakota Sandstone and Mancos Shale, west-central New Mexico: U.S. Geological Survey, Bulletin 1372-J, 44 p.

Lucas, S. G. and Hayden, S. M., 1989, Triassic stratigraphy of west-central New Mexico: New Mexico Geological Society, Guidebook 40, p. 191–212.

Pike, W. S., 1947, Intertonguing marine and nonmarine upper Cretaceous deposits of New Mexico, Arizona, and southwestern Colorado: Geological Society of America, Memoir 24, 103 p.

Pie Town Community Center. View is N89°E. Handbill on the door announces "The Country Coins" musical group. The community center is south of US-60 and east of dirt road that leads south out of Pie Town, New Mexico. Wayne Lambert photograph No. 93L57. August 12, 1993, 3:35 p.m. MDT.

TERTIARY STRATIGRAPHY AND NOMENCLATURE FOR WESTERN NEW MEXICO AND EASTERN ARIZONA

S. M. CATHER[1], R. M. CHAMBERLIN[1], and J. C. RATTÉ[2]
[1]New Mexico Bureau of Mines and Mineral Resources, Socorro, New Mexico 87801; [2]P. O. Box 382, Black Hawk, Colorado 80422

Abstract—This report delineates new lateral boundaries between lithostratigraphic units of Tertiary age in western New Mexico and eastern Arizona, including the Eocene Baca, Eagar and Mogollon Rim Formations and the Mio-Pliocene Santa Fe Group, Gila Group, Fence Lake Formation, Bidahochi Formation, Quemado Formation and Sheep Crossing Formation. New group-rank units for the upper Eocene to Oligocene rocks of the Mogollon-Datil volcanic field are also proposed. In the proposed nomenclature, all sedimentary rocks of the volcaniclastic apron surrounding the field are subsumed into the Spears Group. Volcanic rocks below a locally prominent disconformity representing a previously recognized 31.4 to 29.0 Ma hiatus in ignimbrite volcanism are assigned to the Datil Group (redefined); volcanic rocks above the disconformity are assigned to the Mogollon Group.

INTRODUCTION

Throughout western New Mexico and eastern Arizona, nomenclature for Tertiary stratigraphic units has typically been conceived in isolated and disparate areas of local study. Through further work, nomenclature within correlative rock units then commonly spreads amoeba-like until it collides with an adjoining nomenclatural domain. Several things may then occur. Due to precedence or the preferred use by geologists in the region, one of the competing terminologies may be abandoned and replaced by the neighboring nomenclature. Alternatively, adjoining stratigraphic terminologies may be delimited by the perceived paleogeography, such as paleodrainage divides between basins, or by lithologic differences (i.e., provenance or paleoenvironmental factors) between sediments within the same basin. In some cases, the interface between adjacent nomenclatures has been defined arbitrarily by modern geographic or political boundaries, such as drainage divides or state lines.

Western New Mexico and eastern Arizona are areas where collisions between competing nomenclatures for Tertiary rocks are presently occurring, and this paper is an attempt to resolve some of these conflicts and their resulting confusion. We divide our discussion into three parts: (1) middle to upper Eocene rocks (late Laramide synorogenic deposits of the Baca, Eagar and Mogollon Rim Formations), (2) upper Eocene-Oligocene rocks (volcanic and sedimentary rocks of the northern Mogollon-Datil volcanic field); and (3) Miocene-Pliocene rocks (extension-related deposits of the Santa Fe Group, Gila Group, Fence Lake Formation, Bidahochi Formation, Quemado Formation, and correlative units).

MIDDLE TO UPPER EOCENE

Late Laramide synorogenic sediments were deposited in two structurally related basins in west-central New Mexico and eastern Arizona. The largest and westernmost of these basins, the Baca basin, extended westward from Socorro, New Mexico, to near Show Low, Arizona (Fig. 1). Correlative middle to upper Eocene deposits within the Baca basin have been assigned, from east to west, to the Baca Formation (Wilpolt et al., 1946; Snyder, 1971; Cather and Johnson, 1984; 1986), the Eagar Formation (Sirrine, 1956), and the Mogollon Rim Formation (Hunt, 1956; Cooley and Davidson, 1963; Peirce et al., 1979; Potochnik, 1989). Although the term Baca Formation should have precedence due to first usage, the terms Eagar Formation and Mogollon Rim Formation (or gravels) are firmly entrenched in the geologic literature for the western Baca basin. We advocate that all three of these formational terms be retained, as extension of the term Baca to eastern Arizona would result in needless confusion. Instead, we propose to define lateral boundaries between the three correlative formations.

The Eagar and Mogollon Rim Formations are compositionally similar, both consisting of fluviatile conglomerate, sandstone and mudstone derived largely from Paleozoic and Mesozoic sedimentary strata and Precambrian quartzites, metavolcanics and granitoids. Much of the Mogollon Rim Formation, however, is significantly coarser and more conglomeratic than the Eagar. Intermediate-composition, porphyritic volcanic clasts are also present throughout both formations. These clasts increase in abundance upsection and, in the Mogollon Rim Formation, have yielded early Tertiary K-Ar ages (Peirce et al., 1979). Potochnik (1989) defined the top of the Mogollon Rim Formation at the first up-section occurrence of greater than 50% volcanic clasts within a pebbly sandstone or conglomerate. A similar top to the Eagar Formation may also be defined, although conglomeratic horizons are much more sparse and future workers may have to resort to sandstone compositions to precisely place the contact. In some areas, the lower contact of the Eagar Formation has been incorrectly placed within the Upper Cretaceous section by previous workers (Sirrine, 1956; Akers, 1964; Reynolds, 1988; see minipaper by Cather in Day Two road log, this guidebook) due to the local, strongly reddened nature of Cretaceous strata beneath the Tertiary unconformity within the Baca basin (e. g., Chamberlin, 1981).

Prior to the recognition that the Mogollon Rim and Eagar Formations are elements within a tract of contiguous depositional systems (e. g. Johnson, 1978, Cather and Johnson, 1984; 1986; Potochnik, 1989), mapping utilized the large volcanic-covered interval of the Eocene outcrop belt north of the White Mountains as the divide between the outcrop areas of the two formations (Fig. 1). We advocate the continuation of this practice, although future drilling beneath the interval covered by the Springerville and White Mountains volcanic fields may require a more rigorous definition of the units based on composition or grain size.

The contact between the laterally equivalent Eagar and Baca Formations has been arbitrarily placed at or near the state line by previous workers. Recent reconnaissance mapping (Chamberlin et al., 1994), however, has revealed a mappable lithologic contact between the two units. As noted above, the Eagar and Mogollon Rim Formations contain significant amounts of Tertiary volcanic detritus. In contrast, the Baca Formation is essentially free of volcanogenic material. The lateral transition between volcanic-bearing and volcanic-free lithologies within the Baca basin occurs about 20 km east of the Arizona-New Mexico state line, near Cimarron Mesa (Fig. 1). In this area, an uncommonly coarse boulder gravel intervenes between the volcanic-bearing Eagar Formation to the west and the Baca Formation to the east. The boulder gravel itself contains volcanic clasts so we also assign it to the Eagar Formation. The anomalously coarse-grained deposits of the Eagar Formation appear to overlie, and possibly intertongue with, the typical volcanic-free deposits of the Baca Formation near Cimarron Mesa. The origin of the lithologic and textural variations in this area is in need of further study.

Because of the near absence of volcanic materials in the Baca

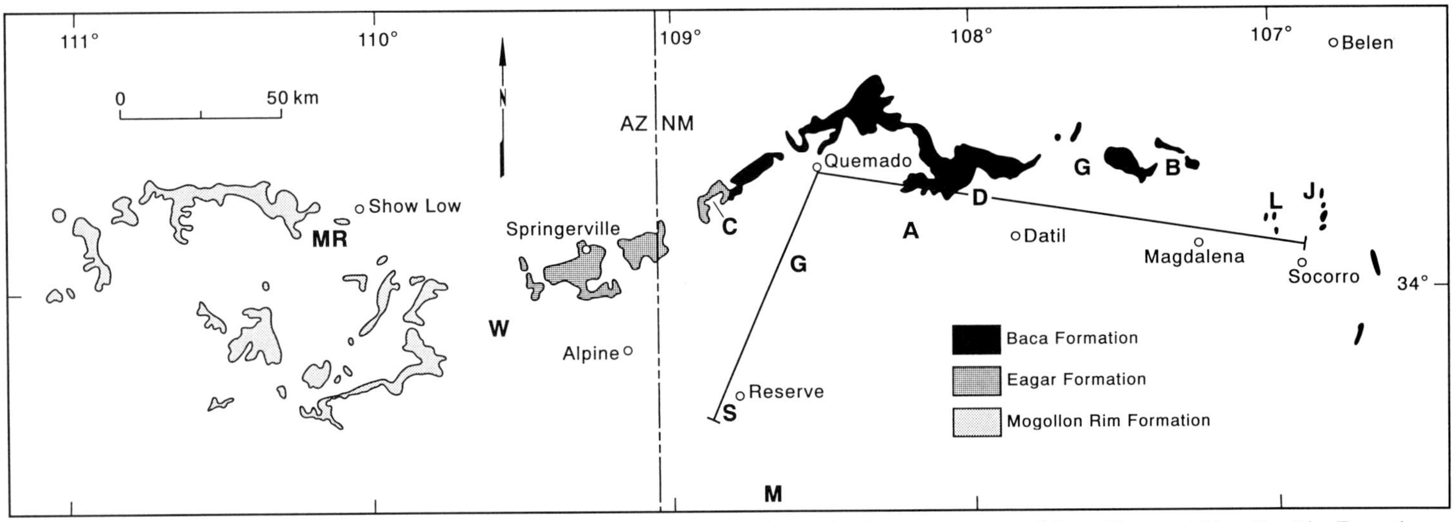

FIGURE 1. Outcrop map for Eocene deposits in western New Mexico and eastern Arizona, showing outcrop areas of Baca, Eagar and Mogollon Rim Formations. Line connecting Reserve, Quemado, and Socorro areas is location of fence diagram in Fig. 2. Physiographic features are MR, Mogollon Rim; W, White Mountains; C, Cimarron Mesa; G, Gallo Mountains; S, Saliz Mountains; M, Mogollon Mountains; A, Allegres Mountain; D, Datil Mountains; G, Gallinas Mountains; B, Bear Mountains; L, Lemitar Mountains; and J, Joyita Hills. Modified from Reynolds (1988) and New Mexico Geological Society (1982).

Formation, the upper contact typically has been defined at the first occurrence of megascopically visible volcanic detritus (e. g. Cather, 1980; 1986). This contrasts with equivalent rocks to the west, where the top of the Mogollon Rim Formation is defined at the fifty-percentile abundance of volcanic detritus (Potochnik, 1989). For the sake of continuity, we recommend that in future studies the 50-percentile criterion also be applied to the upper contact of the Baca Formation. This would require only slight revision of existing maps and measured thicknesses for the Baca Formation, as the stratigraphic transition between the Baca and the overlying volcaniclastics of the Spears Formation (now Spears Group; see below) typically occurs within a few meters.

The inception of late Eocene volcaniclastic sedimentation may have been somewhat older in west-central New Mexico than in eastern Arizona. Ash-fall deposits in the upper Mogollon Rim Formation have yielded K-Ar ages of 37.6 ± 0.8 Ma and 37.5 ± 0.8 Ma (Potochnik, 1989). In contrast, rocks *above* the Baca Formation yield radioisotopic ages (mostly K-Ar) between 39.6 and 37.02 Ma (Cather et al., 1987) in west-central New Mexico, suggesting that the lower part of the Spears Group is chronostratigraphically equivalent to the upper Mogollon Rim Formation. Such time-transgressive behavior of the inception of mid-Tertiary volcaniclastic deposition may be an expectable consequence of the diachronous onlap and burial of proximal portions of the Baca basin in eastern Arizona. Alternatively, the apparent time-transgressive nature of the contact may simply be a statistical aberration due to the low precision of the K-Ar technique. More $^{40}Ar/^{39}Ar$ ages are needed to further test these hypotheses.

Lucas and Williamson (1993) recently proposed the term Hart Mine Formation (formerly Baca Formation) for the non-volcanic deposits of the Eocene Carthage-La Joya basin east of Socorro. They claimed (p. 148) that the deposits of the Carthage-La Joya basin are readily distinguished from those of the Baca basin by clast content. However, the granite- and limestone-bearing conglomerates that characterize the deposits of Carthage-La Joya basin near Socorro are also present in the eastern Baca basin, in the Bear and Lemitar Mountains areas. These conglomerates reflect derivation from the Laramide Sierra uplift that intervened between the two basins (Cather, 1983; Cather and Johnson, 1984; 1986), and are not solely restricted to the Carthage-La Joya basin. Because of the effects of later volcanism and rift-related tectonism, the boundaries of the Carthage-La Joya basin are highly interpretive. For example, Cather (1980) and Cather and Johnson (1984, 1986) interpreted the northern part of the basin to include a thick (1146 m) red-bed sequence in the subsurface of the modern southern Albuquerque Basin (Foster, 1978) and a conglomerate consisting entirely of clasts of Abo Formation on Hubbell bench east of Belen (Kelley, 1977). Lozinsky (1988) subsequently reinterpreted the subsurface red beds to be Triassic and regarded the conglomerates on the Hubbell bench to possibly pertain to the Santa Fe Group (see summary in Cather, 1992). As such, any definition of the Hart Mine Formation that depends on the paleogeographic reconstruction of the Carthage-La Joya basin would seem inadvisable at present. Furthermore, because no workable lithologic criteria has yet been devised to distinguish the rocks of the Carthage-La Joya basin from those of the eastern Baca basin, we recommended abandonment of the term Hart Mine Formation and a return to usage of Baca Formation for these rocks. Indeed, although named for Baca Canyon in the Bear Mountains, the term "Baca Formation" was first used by Wilpolt et al. (1946) to describe rocks in the Joyita Hills, in what is now considered to be part of the Carthage-La Joya basin.

UPPER EOCENE TO OLIGOCENE

Introduction

The evolution of stratigraphic nomenclature for the volcanic and volcaniclastic rocks (~ 40 to 24 Ma) of the northern Mogollon-Datil volcanic field has been extraordinarily convoluted (see summaries in Elston, 1976; Osburn and Chapin, 1983; and Cather, 1986). Lateral changes in the lithology of pyroclastic and sedimentary units between areas of local study and the relative imprecision of early K-Ar ages resulted in many ambiguities and errors in correlation. Recently, paleomagnetic correlations and precise $^{40}Ar/^{39}Ar$ geochronology of ignimbrites by McIntosh et al. (1991; 1992), have enabled the conceptualization of a regional stratigraphic framework spanning most or all of the Mogollon-Datil field. McIntosh et al. (1991, 1992) recognized four episodes of eruptive activity within the 36.2 to 24.3 Ma history of ignimbrite volcanism in the Mogollon-Datil field. These eruptive episodes are separated by hiatuses of 1.5 to 3.0 Ma duration that are of regional extent. Using the stratigraphic and geochronologic framework proposed by McIntosh et al. (1991, 1992) and our own mapping and stratigraphic analysis in the northern and western parts of the field, a new group-rank nomenclature for the upper Eocene-Oligocene rocks of the Mogollon-Datil field is proposed below.

Prior to this paper, the Datil Group was the only group-rank unit commonly recognized in the Mogollon-Datil volcanic field. Winchester (1920) coined the term Datil Formation, and applied it to a thick sequence of tuff, rhyolite, conglomerate and sandstone that forms the Datil Mountains. His type section, however, was measured in the Bear Mountains, and consists of a 480-m-thick section of rocks that begins at the Cretaceous-Tertiary unconformity and ends within or at the top of a thick quartz rhyolite (Hells Mesa Tuff, 32.0 Ma). Wilpolt et al. (1946) separated the lower, non-volcanic deposits from Winchester's Datil Formation and renamed them the Baca Formation.

Tonking (1957) then extended the Datil Formation upward to include the basaltic and basaltic andesite rocks above Winchester's section, thereby lumping the entire 1,500+-m-thick section of mid-Tertiary volcanic rocks of the northern Datil-Mogollon field into a single formation. Tonking also divided the Datil Formation into three members which, in ascending stratigraphic order, are the Spears, Hells Mesa and La Jara Peak Members. Willard (1959) removed the La Jara Peak Member from Tonking's Datil Formation, and Weber (1971) then raised the Datil Formation to group status (but not synonymous with the Datil Group of this report). Chapin (1971) raised the Spears, Hells Mesa, and La Jara Peak to formation rank. Elston (1976) noted that the term Datil had been extended so far geographically and stratigraphically as to virtually mean "mid-Tertiary volcanic rocks of the Mogollon-Datil province, undifferentiated". He recommended its abandonment.

Osburn and Chapin (1983) subsequently resurrected the Datil Group, but restricted usage to the volcanic and volcaniclastic rocks which occur *below* the Hells Mesa Tuff in the northeast part of the field. In this sense, the Datil Group is equivalent to the Spears Formation (or member) of previous studies. The main advantage of Osburn and Chapin's nomenclature is that it allows the regionally persistent and mappable tuffs below the Hells Mesa Tuff to be assigned formation status. The term Spears Formation was retained for the volcaniclastic sediments that underlie and interfinger with these tuffs.

The concept of the Datil Group as proposed by Osburn and Chapin (1983) is tenable and useful in the northeastern part of the field, where the Hells Mesa Tuff effectively divides the largely sedimentary strata below from the dominantly extrusive and pyroclastic rocks above. The usefulness of their concept, however, breaks down in other parts of the field because (1) The Hells Mesa Tuff occurs only in the northeastern and northern parts of the volcanic field (McIntosh et al., 1992) and thus cannot be used to define the top of the Datil Group in other parts of the field; and (2) The post-Hells Mesa (post-32 Ma) section in the western and northwestern parts of the field contains voluminous sedimentary rocks which, because the Hells Mesa Tuff is absent, cannot be readily distinguished from the underlying strata of Osburn and Chapin's Datil Group.

Based on the stratigraphic framework of McIntosh et al. (1991, 1992) and recent mapping and stratigraphic analysis by the authors, we believe a unifying, group-rank nomenclature may now be proposed that has applicability to most or all of the Mogollon-Datil volcanic field. To achieve this, we utilize locally prominent disconformities (e. g., Ratté, 1989; Osburn and Chapin, 1983) that represent a regional hiatus in ignimbrite volcanism between 31.4 and 29.0 Ma (McIntosh et al., 1992) to divide the *volcanic* rocks (tuffs, lavas, domes, etc.) of the field into upper (Mogollon Group) and lower (Datil Group, redefined) units. This disconformity appears to be manifested throughout the field, except possibly in the south-central part (see discussion below). We assign all of the *sedimentary apron deposits* in the field to the Spears Group, which interfingers with both of the aforementioned volcanic groups (Fig. 2). A similar subdivision of volcanic and sedimentary rocks within the Mogollon-Datil field will be utilized in the upcoming geologic map of New Mexico (Anderson and Jones, in preparation). Each of the three proposed groups consists of numerous formation-rank units, although some of the formations within the Spears Group are redefined or raised from member rank in this report (Fig. 3).

Datil Group

As redefined here, the Datil Group is restricted to the ignimbrites and lavas that were erupted prior to development of the regional disconformity that represents the 31.4 to 29.0 Ma hiatus in volcanism delineated by McIntosh et al. (1991; 1992); intercalated sedimentary apron deposits are assigned to the Spears Group. In the northeast part of the field, the stratigraphic range of the proposed Datil Group is similar to that of Winchester's (1920) Datil Formation, except that sedimentary rocks are omitted. We retain Winchester's original type section in the Bear Mountains but broaden the geographic range of the Datil by designating the type sections of constituent formations (Fig. 3) as reference sections (North American Commission on Stratigraphic Nomenclature, 1983, Article 22c). The Datil Group consists largely of intermediate to mafic lavas and numerous formation-rank ignimbrites of mostly low-silica rhyolite composition (Fig. 3). The contact between the Datil Group and the overlying Mogollon Group is typically marked by a disconformity, except in distal areas where sediments of the Spears Group intervene and serve to divide the volcanic groups. (Fig. 2; see discussion below). Our definition of the Datil Group differs from that of Osburn and Chapin in that we *exclude* all sedimentary rocks but *include* the Hells Mesa Tuff.

Mogollon Group

The name of this group derives from the Mogollon Mountains area (e.g. Ratté, 1981; Ratté and Gaskill, 1975), which contains widespread exposures of the upper part of the mid-Tertiary volcanic sequence. Herein, we reinstate and revise the term Mogollon, which was first applied by Ferguson (1927) to an andesite in the Mogollon mining district but has subsequently been abandoned (Ratté and Finnell, 1978; Ratté 1981). The base of the Mogollon Group marks the culmination of the great ignimbrite flare-up (pulse 3 of McIntosh et al., 1992) which terminated the post-Datil hiatus in ignimbrite volcanism. During early Mogollon Group volcanism, nine major regional ignimbrites and at least 11 subregional to local units were erupted within a period of 1.6 Ma. (McIntosh et al., 1991). Most of these ignimbrites are high-silica rhyolites. Mafic to intermediate lavas are widespread in the Mogollon Group, particularly in the upper part where thick sequences of flows comprise the Bearwallow Mountain Andesite (western part of field), La Jara Peak Basaltic Andesite (northeast), Uvas Basaltic Andesite (southeast) and correlative local units. The top of the Mogollon Group is typically unconformably overlain by Miocene rift-basin or rift-shoulder deposits (Santa Fe Group, Gila Group, Fence Lake Formation, Bidahochi Formation, and correlative units), although in some areas the upper contact of the Mogollon Group may be gradational with the younger units.

Spears Group

The Spears Member of the Datil Formation was originally defined by Tonking (1957) to include all volcanic and volcaniclastic rocks below his Hells Mesa Member (the basal part of which is now termed the Hells Mesa Tuff) in the Bear Mountains. Chapin (1971) raised the Spears to formation rank, and Osburn and Chapin (1983) restricted the Spears Formation (within their Datil Group) to include only the *volcaniclastic sedimentary rocks* beneath the Hells Mesa Tuff. We propose to raise the Spears to group rank and to broaden its scope to include *all* upper Eocene to Oligocene volcaniclastic sedimentary apron deposits in the field. In so doing, we retain the original type section of the Spears (Tonking, 1957), except that we exclude all volcanic rocks. We designate the type sections (or type areas) of the component sedimentary units of the proposed Spears Group (see Fig. 3) as supplementary reference sections (North American Commission on Stratigraphic Nomenclature, 1983, Article 22c). Further, we specifically exclude all sedimentary deposits not related to the sedimentary apron of the Mogollon-Datil field from the Spears Group, such as caldera-fill deposits. Our proposal causes only minor nomenclatural changes in the northeastern part of the field, as only minor volumes of sedimentary apron deposits are present above the 32.0 Ma Hells Mesa Tuff in this area. In other parts of the field, however, the Spears Group includes sedimentary rocks as young as 26-27 Ma (Fig. 2), considerably younger than can be encompassed by Osburn and Chapin's nomenclature.

We also recommend that the formal members of the Spears Formation of Osburn and Chapin (1983) be raised to formation rank. These units (Dog Springs Formation, Chavez Canyon Formation, and Rincon Windmill Formation) crop out in the northern part of the field and are readily mappable at the present scale of investigation (Harrison, 1980; Coffin, 1981; Robinson, 1981; Broulliard, 1984).

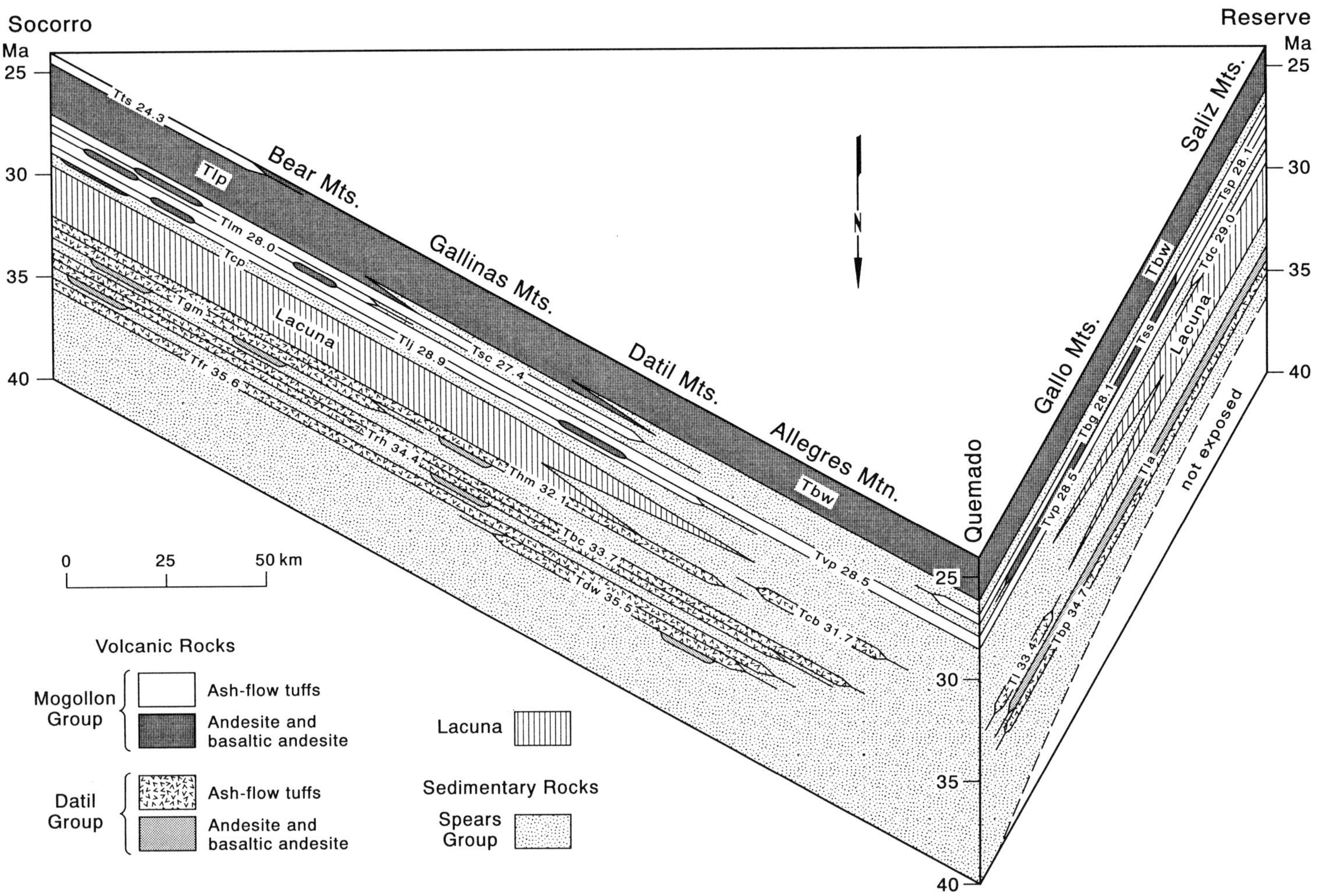

FIGURE 2. Schematic chronostratigraphic fence diagram for 40 to 24 Ma rocks of the northern Mogollon-Datil field, showing maximum inferred extent of units within the Datil, Mogollon, and Spears Groups. Note that the Spears Group interfingers with both the Datil Group and Mogollon Group and that the Datil and Mogollon Groups are separated in their proximal areas by a lacuna (i.e., a time-stratigraphic gap representing a period of non-deposition and erosion). Line of section for fence diagram is shown in Fig. 1. Numerals adjacent to unit abbreviations are $^{40}Ar/^{39}Ar$ ages (Ma) from McIntosh et al. (1991). Unit abbreviations are Tts, tuff of Turkey Springs; Tlp, La Jara Peak Basaltic Andesite; Tsc, South Canyon Tuff; Tlm, Lemitar Tuff; Tvp, Vicks Peak Tuff; Tlj, La Jencia Tuff; Tcp, South Crosby Peak Formation; Tcb, Caballo Blanco Tuff; Thm, Hells Mesa Tuff; Tgm, Tuff of Granite Mountain; Tbc, Blue Canyon Tuff; Trh, Rock House Canyon Tuff; Tfr, tuff of Farr Ranch; Tdw, Datil Well Tuff; Tbw, Bearwallow Mountain Andesite; Tbg, Bloodgood Canyon Tuff; Tsp, Shelley Peak Tuff; Tss, Squirrel Springs Canyon Andesite; Tdc, Davis Canyon Tuff; Tl, tuff of Luna; Tla, andesite of Dry Leggett Canyon; Tbp, tuff of Bishop Peak. See text for discussion.

Several previously defined formations within the Spears Group are present elsewhere in the Mogollon-Datil field (Fig. 3). Reconnaissance by the authors in the northwest part of the field has revealed several informal formations within the Spears Group (sandstone of Escondido Mountain, volcaniclastic unit of Cañon del Leon, volcaniclastic unit of Largo Creek; Chamberlin and Harris, this volume).

Discussion

Because of the complex lateral gradations between volcanic and sedimentary rocks in the Mogollon-Datil field, it seems doubtful that any lithologic unit of regional extent is present that would enable the stratigraphic subdivision of the entire volcanic field. Even the major 31.4 to 29.0 Ma hiatus in ignimbrite volcanism recognized by McIntosh et al. (1991) has no apparent correlative within the distal part of volcaniclastic aprons, such as that exposed near Quemado (Fig. 2). Based on these observations, we believe the best approach to the regional stratigraphy is for development of *separate* nomenclatures for volcanic versus sedimentary rocks in the field. In the proposed nomenclature, the sedimentary Spears Group interfingers with the volcanic rocks of the superposed Datil and Mogollon Groups in much the same manner as the Upper Cretaceous Mancos Shale interfingers with both the Dakota Group and the Mesaverde Group throughout New Mexico.

The principal difficulty created by the proposed nomenclature is that several existing formation-rank units in the Mogollon-Datil field contain *both* sedimentary and volcanic rocks [e. g., Rubio Peak Formation (Jicha, 1954; Elston, 1957; Jones et al., 1967), Palm Park Formation (Kelley and Silver, 1952; Seager et al., 1971), Pueblo Creek Formation (Ratté, 1989)]. In many outcrop areas of these hybrid formations, sedimentary strata have already been mapped separately from volcanic facies; we suggest that, in future studies, these sedimentary units be accorded separate nomenclatures as well. In areas where sedimentary and volcanic rocks are too finely intercalated to be separated at a given scale of mapping, we recommend group assignment based on the dominant lithology (e. g. 50 percentile abundance of volcanic versus sedimentary rocks). In such areas, the map segregation of volcanic from sedimentary rocks will increase as more detailed mapping is attempted. Alternatively, in near-vent areas where delineation of volcanic, plutonic, and sedimentary rocks is not feasible at a given scale of mapping, the lithodemic term "volcanic complex" may be employed (North American Commission on Stratigraphic Nomenclature, 1983, Article 37b).

We use the regional disconformity that resulted from the 31.4 to 29.0

	Volcanic Rocks		Sedimentary Rocks
Mogollon Group	tuff of Turkey Springs	Spears Group	South Crosby Peak Formation**
	Bearwallow Mountain Andesite		Rincon Windmill Formation*
	Uvas Basaltic Andesite		Chavez Canyon Formation*
	La Jara Peak Basaltic Andesite		Dog Springs Formation*
	South Canyon Tuff		Rubio Peak Formation**
	Lemitar Tuff		Palm Park Formation**
	Apache Springs Tuff		Pueblo Creek Formation**
	Bloodgood Canyon Tuff		Bell Top Formation**
	Squirrel Springs Andesite		Rock Springs Formation**
	Shelley Peak Tuff		Red Rock Ranch Formation**
	Vicks Peak Tuff		Piloncillo Sediments
	La Jencia Tuff		unit of East Red Canyon
	Davis Canyon Tuff		sandstone of Monument Park
	Sawmill Canyon Formation		sandstone of Escondido Mountain
	Luis Lopez Formation		volcaniclastic unit of Cañon del Leon
	basaltic andesite of Poverty Creek		volcaniclastic unit of Largo Creek
	lacuna (31.4–29.0 Ma)		
Datil Group	Tadpole Ridge Tuff		
	Caballo Blanco Tuff		
	Hells Mesa Tuff		
	Box Canyon Tuff		
	Blue Canyon Tuff		
	Cooney Tuff		
	andesite of Dry Leggett Canyon		
	Rock House Canyon Tuff		
	tuff of Bishop Peak (tuff of Lebya Well)		
	Kneeling Nun Tuff		
	Bell Top Formation tuffs		
	Sugarlump Tuff		
	tuff of Farr Ranch		
	Datil Well Tuff		
	andesite of White House Canyon		
	Doña Ana Tuff		
	Squaw Mountain Tuff		
	Achenback Park Tuff		
	Cueva Tuff		

*formerly member
**includes sedimentary part only

FIGURE 3. Chart showing selected major formation-rank units within proposed Mogollon, Datil and Spears Groups. Data sources are Elston (1976), Osburn and Chapin (1983), and McIntosh et al. (1992).

FIGURE 4. Erosional unconformity between the Davis Canyon Tuff, Tdc, at the base of the Mogollon Group and laharic breccias of the Spears Group, Ts, on the west side of Yam Canyon in the Alma Mesa 7½ minute quadrangle, Greenlee County, Arizona, about 15 km northwest of Alma, New Mexico. View is west toward Horse Mountain, a dacitic eruptive center, Tad, of Bearwallow Mountain Andesite age. Cliffs of Bloodgood Canyon Tuff, Tbc, overlie Shelley Peak Tuff (not shown) and Davis Canyon Tuff in center of photo (Ratté et al., 1969).

hiatus in major ignimbrite volcanism (McIntosh et al., 1991) to define the boundary between the Datil and Mogollon Groups, except in distal areas where sediments of the Spears Group intervene and serve to divide the volcanic groups. In some areas of the Mogollon-Datil field, this hiatus was slightly narrowed by intervening episodes of volcanism. Except possibly for the south-central part of the field, however, there is no evidence that volcanism entirely spanned the hiatus between the two volcanic groups.

The contact between the Datil and Mogollon Groups is not exposed in the central part of the field. Contact relations around the periphery of the field are as follows [except where noted, all age data are from McIntosh et al., 1992]:

Southwest. In the Summit Mountains area (Steeple Rock mining district; McLemore, 1993) about 60 km north-northwest of Lordsburg, New Mexico, the base of the Mogollon Group consists locally of either the Bloodgood Canyon Tuff (28.05 ± 0.04 Ma) or the Davis Canyon Tuff (29.01 ± 0.11 Ma). These tuffs unconformably overlie the 31.3 Ma Summit Mountain Formation (McLemore, 1993), which composes the upper part of the Datil Group in this area.

West. In the Saliz Mountains area near Reserve, New Mexico (Ratté, 1981), and in adjacent Greenlee County, Arizona (Fig. 4), a moderate-relief unconformity at the base of the Davis Canyon Tuff (29.01 ± 0.11 Ma) marks the contact between the Mogollon Group and the underlying Spears Group. Tongues of Spears Group sediments are intercalated with underlying volcanic rocks of the Datil Group (~32 to 34.7 Ma; see Ratté, 1981 and Fig. 2) and separate the Datil Group from the Mogollon Group.

Northwest. In the Quemado area (Fig. 2), a 120-240-m-thick sequence of upper Spears Group strata intervene between the basal Mogollon Group (Vicks Peak Tuff, 28.56 ± 0.04 Ma) and the upper part of the Datil Group (Caballo Blanco Tuff, 31.65 ± 0.06 Ma; tuff of Luna, 33.39 ± 0.12 Ma; andesite of Dry Leggett Canyon, ± 34 Ma; Fig. 2).

North and Northeast. The basal ignimbrite of the Mogollon Group (La Jencia Tuff, 28.85 ± 0.04 Ma) is separated from the Datil Group by volcanic and volcaniclastic strata of the South Crosby Peak and the Luis Lopez Formations. The uppermost tuff of the Datil Group (Hells Mesa Tuff, 32.06 ± 0.10 Ma) is locally thin or missing beneath the basal disconformity of the South Crosby Peak Formation (Osburn and Chapin, 1983; Coffin, 1981), which we regard as part of the Mogollon Group. Recent $^{40}Ar/^{39}Ar$ dating of volcanic rocks from the Luis Lopez Formation (29.85 ± 0.31 Ma, 28.61 ± 0.17 Ma; W. C. McIntosh, unpublished dates) and the lower South Crosby Peak Formation (29.67 ± 0.86 Ma; McIntosh and Chamberlin, this volume) indicate these units are close in age to the overlying La Jencia Tuff. These new ages narrow, but do not span, the hiatus between major ignimbrites of Mogollon and Datil Groups and support the assignment of volcanic rocks within the South Crosby Peak and the Luis Lopez Formations to the Mogollon Group.

East. In the northern Black Range, a volcanic sequence about 200 m thick is assigned to the basaltic andesite of Poverty Creek (Eggleston, 1987; Harrison, 1990). The flows of the Poverty Creek intervene between the basal ignimbrite of the Mogollon Group (Tuff of Little Mineral Creek, 29.01 ± 0.10 Ma) and the upper ignimbrite of the Datil Group (Caballo Blanco Tuff, 31.65 ± 0.06 Ma). On the basis of the presence of a pronounced, regional disconformity with as much as 100 m of paleorelief at its base (Eggleston, 1987, p. 37), we assign the Poverty Creek to the Mogollon Group. This assignment is supported by the only age determination for the Poverty Creek (28.8 ± 0.6 Ma; K/Ar age in Woodard, 1982), which is slightly younger than (but within analytical error of) the age for the overlying tuff of Little Mineral Creek.

Southeast. In the Sierra de las Uvas area northwest of Las Cruces, New Mexico, a tongue of the Spears Group (the upper sedimentary member of the Bell Top Formation; Clemons, 1976) intervenes between the upper ignimbrite of the Datil Group (Bell Top 6 or Box Canyon, 33.51 ± 0.13 Ma) and the lower ignimbrite of the Mogollon Group (Bell Top 7 or Vicks Peak Tuff, 28.56 ± 0.04 Ma). No volcanic rocks intervene between the two groups in this area (W. R. Seager, oral commun., 1994).

South-central. In the southeastern Mogollon Mountains-Gila Cliff Dwellings-Mimbres Valley-southern Black Range area, the Alum Mountain Formation (or Group; Elston, 1968; Ratté and Gaskill, 1975; Krier, 1980; Farris, 1981) underlies or interfingers with the basal ignimbrites of the Mogollon Group and overlies with local unconformity the upper ignimbrites of Datil Group (Caballo Blanco Tuff, tuff of Terry Canyon, tuff of Monument Canyon). The Alum Mountain consists of a thick sequence of lavas, breccias and volcaniclastic sediments that has yielded K-Ar ages of 30.5 ± 1.0 Ma and 30.1 ± 1.0 Ma (Ratté and Gaskill, 1975). The great thickness and the existing K-Ar ages of the Alum Mountain Formation suggest that it represents an episode of volcanism in the southern Mogollon-Datil field that may have spanned the hiatus manifested between the Datil and Mogollon Groups elsewhere in the field. On the other hand, the local pronounced basal unconformity of the Alum Mountain Formation (Krier, 1980; Farris, 1981) is similar to that beneath the correlative basaltic andesite of Poverty Creek and may argue for inclusion of the Alum Mountain in the Mogollon Group. It is clear, however, that the hiatus between the Datil and Mogollon Groups must disappear at some point to the south, as the corresponding period in Chihuahua, Mexico was a time of voluminous volcanism (McDowell and Mauger, 1994). Whether or not the proposed volcanic nomenclature is applicable to the south-central part of the Mogollon-Datil field will require further study of the Alum Mountain Formation.

MIOCENE AND PLIOCENE

Extension-related deposits of late Tertiary age in western New Mexico and eastern Arizona have been assigned variously to the Santa Fe Group, Gila Group (or Formation or Conglomerate), Fence Lake Formation, Bidahochi Formation, and correlative local units. The Santa Fe and Gila Groups were mostly deposited in extensional basins, whereas the Fence Lake, Quemado (Cather and McIntosh, this volume; McIntosh and Cather, this volume), and Bidahochi Formations typically fill paleovalleys and broad, erosional basins on the Colorado Plateau (Chapin and Cather, in press). Most of the Gila and Santa Fe are middle Miocene to Pliocene in age, although in some areas beds may be as old as late Oligocene or early Miocene (Leopoldt, 1981; Osburn and Chapin, 1983; Marvin et al., 1987; Chapin, 1988; Chapin and Cather, in press; Cather et al., in press; Cross, in preparation). The Bidahochi Formation ranges from middle or late Miocene to early Pliocene in age (see summary by Love, 1989). The Fence Lake Formation appears to be mostly middle to late Miocene (McLellan et al., 1982, Lucas and Anderson, this volume) and may be correlative with the lower part of the Bidahochi Formation (Love, 1989). The Quemado Formation is early Pliocene to Pleistocene in age, and may be in part equivalent to the upper Bidahochi Formation (Cather and McIntosh, this volume).

Dane and Bachman (1965) utilized modern drainage divides to arbitrarily delimit the Santa Fe and Gila Groups. Their usage was summarized by Elston and Netelbeek (1965, p. 36):

> "The Santa Fe Group [is] restricted to all regions where the surface drainage is to the Rio Grande and to closed basins east of the Rio Grande. The Gila Conglomerate or Gila Formation [is] restricted to use in all areas where the surface drainage is to the Colorado River and its tributaries *or to closed basins west of the Rio Grande above Hatch*, and west of the Sierra de las Uvas and the West Potrillo Mountains" [italics added].

The above definition is unwieldy with regard to the assignment of deposits within "closed basins west of the Rio Grande above Hatch" to the Gila Group. By this criteria, deposits within the San Agustin Basin and the southern part of the La Jencia Basin, both areas of internal drainage, are Gila Group, but deposits within the intervening Abbe Spring Basin and in the northern part of the La Jencia Basin are Santa Fe Group. This is inelegant, at best.

Instead, we recommend usage of the continental divide as the arbitrary boundary between Gila and Santa Fe throughout the northern Mogollon-Datil field (Fig. 5). This avoids the problems inherent with the previous definition, and is conceptually simpler as the continental divide is well marked on all topographic maps. The main drawback of our proposed boundary is that it crosses an outcrop area of possible late Tertiary sediments (QToa of Richter, 1987; Richter and Lawrence, 1989) southwest of the San Agustin Basin (Fig. 5). We have arbitrarily and tentatively assigned these rocks to the Gila Group, although further

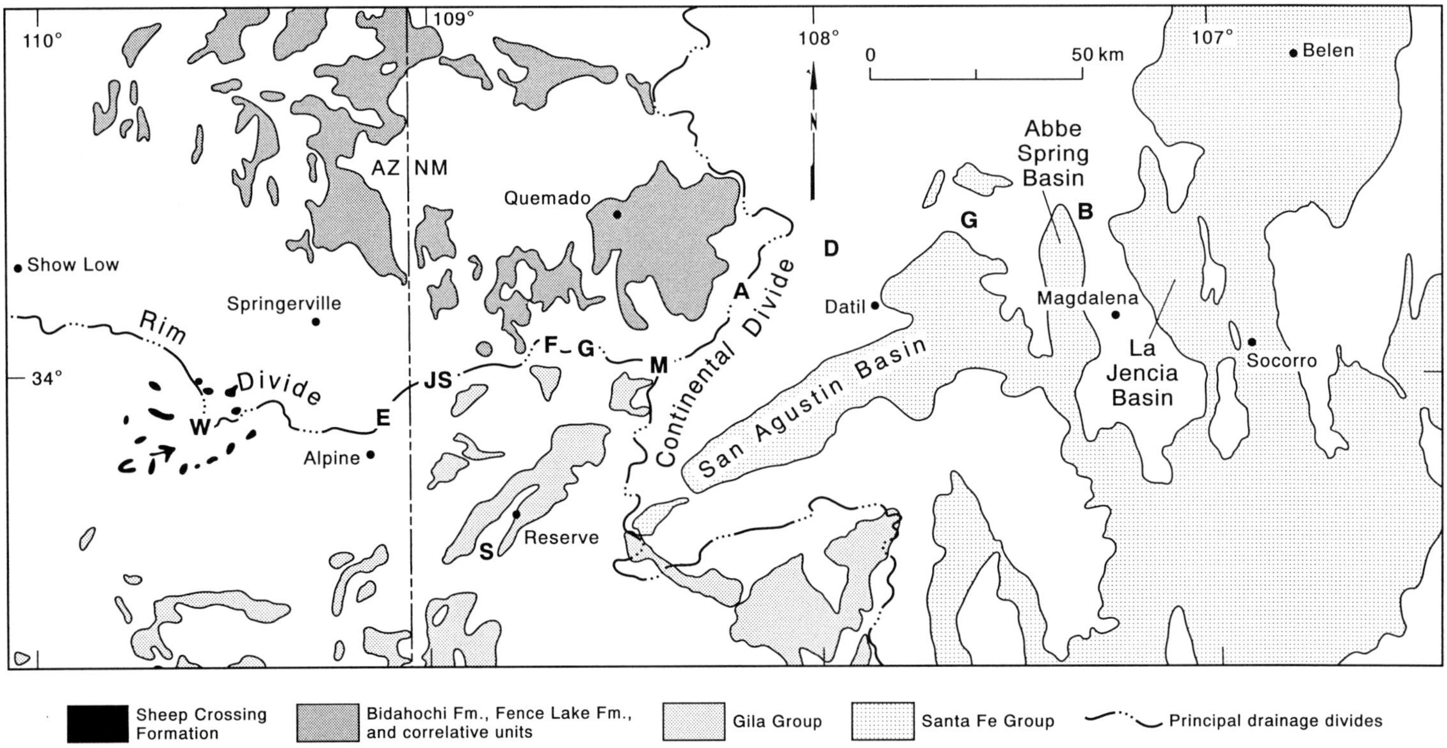

FIGURE 5. Generalized map of Miocene and Pliocene deposits in eastern Arizona and western New Mexico showing proposed nomenclature subdivision. See text for discussion. Abbreviations are W, White Mountains; E, Escudilla Mountain; JS, Jim Smith Peak; S, Saliz Mountains; F, Fox Mountain; G, Gallo Mountains; M, Mangas Mountains; A, Allegres Mountain; D, Datil Mountains; G, Gallinas Mountains; B, Bear Mountains.

study of the lithology and paleodrainage affinity of these deposits may argue for an alternative correlation.

We further recommend that the term Gila be restricted to the area *south* of the Rim Divide (Fig. 5; the Rim Divide separates the watersheds of the San Francisco and Salt Rivers to the south from the Little Colorado drainage to the north). This follows the usage of most workers (e.g., Repenning and Irwin, 1954; Sirrine, 1956; Marr, 1956; Repenning et al., 1958; Akers, 1964; McLellan et al., 1982; Love, 1989), who assigned Mio-Pliocene beds north of the Rim Divide to the Fence Lake or Bidahochi Formations. Reconnaissance mapping by Chamberlin et al. (1994) has delineated several previously unrecognized outcrop areas of the Fence Lake and Quemado Formations in west-central New Mexico, in the regions to the west, south and east of Quemado (Fig. 4).

Miocene and Pliocene rocks were not deposited, or have been removed by erosion, from most areas along the Rim Divide, except peripheral to the White Mountains where upper Miocene mafic lavas have buried and preserved the Miocene Sheep Crossing Formation (Fig. 5; Merrill and Péwé, 1971; 1977). Because the Sheep Crossing Formation straddles the Rim Divide, it is not amenable to the nomenclature proposed above. Due to its proximal location high on the Rim Divide and its partly pyroclastic nature, the Sheep Crossing Formation differs lithologically from correlative deposits to the north and south. We suggest that the term Sheep Crossing Formation be retained for such proximal deposits in the White Mountains area, as they are significantly more coarse and less sorted than their probable lateral equivalents (such as the Bidahochi outcrops north of Springerville). More work will be necessary to constrain the stratigraphic relations among Mio-Pliocene deposits in the Springerville-Quemado region.

ACKNOWLEDGMENTS

The manuscript was improved by reviews from C. E. Chapin, J. W. Hawley, D. W. Love, S. G. Lucas, W. R. Seager and G. A. Smith. Lengthy discussions with W. E. Elston were of value in modifying and sharpening our views on the stratigraphy of the Mogollon-Datil volcanic field, although he did not agree with several aspects of our methods and conclusions. The manuscript was typed by Terry Telles and figures were drafted by Becky Titus.

REFERENCES

Akers, J. P., 1964, Geology and ground water in the central part of Apache County, Arizona: U.S. Geological Survey, Water-supply Paper 1771, 107 p.

Anderson, O. J. and Jones, G., in preparation, Geologic map of New Mexico: New Mexico Bureau of Mines and Mineral Resources and U. S. Geological Survey, scale 1:500,000.

Broulliard, L. A., 1984, Geology of the northeastern Gallinas Mountains, Socorro County, New Mexico [M. S. thesis]: Socorro, New Mexico Institute of Mining and Technology, 161 p.

Cather, S. M., 1980, Petrology, diagenesis, and genetic stratigraphy of the Eocene Baca Formation, Alamo Navajo Reservation, Socorro County, New Mexico [M. A. thesis]: Austin, University of Texas, 243 p.

Cather, S. M., 1983, Laramide Sierra uplift: Evidence for major pre-rift uplift in central and southern New Mexico: New Mexico Geological Society, Guidebook 34, p. 99-101.

Cather, S. M., 1986, Volcano-sedimentary evolution and tectonic implications of the Datil Group (latest Eocene-early Oligocene), west-central New Mexico [Ph. D. dissertation]: Austin, University of Texas, 482 p.

Cather, S. M., 1992, Suggested revisions to the Tertiary tectonic history of north-central New Mexico: New Mexico Geological Society, Guidebook 43, p. 109-122.

Cather, S. M., Chamberlin, R. M., Chapin, C. E. and McIntosh, W. C., in press, Stratigraphic consequences of episodic extension in the Lemitar Mountains, central Rio Grande rift: Geological Society of America, Special Paper 291.

Cather, S. M. and Johnson, B. D., 1984, Eocene tectonics and depositional setting of west-central New Mexico and eastern Arizona: New Mexico Bureau of Mines and Mineral Resources, Circular 192, 32 p.

Cather, S. M. and Johnson, B. D., 1986, Eocene depositional systems and tectonics in west-central New Mexico and eastern Arizona: American Association of Petroleum Geologists, Memoir 41, p. 623-652.

Cather, S. M. and McIntosh, W. C., 1994, The Plio-Pleistocene Quemado Formation of west-central New Mexico: New Mexico Geological Society, Guidebook 45.

Cather, S. M., McIntosh, W. C. and Chapin, C. E., 1987, Stratigraphy, age, and rates of deposition of the Datil Group (upper Eocene-lower Oligocene), west-central New Mexico: New Mexico Geology, v. 9, p. 50-54.

Chamberlin, R. M., 1981, Uranium potential of the Datil Mountains-Pie Town area, Catron County, New Mexico: New Mexico Bureau of Mines and Mineral Resources, Open-file Report 138, 51 p.

Chamberlin, R. M., Cather, S. M., Anderson, O. J. and Jones, G. E., 1994, Reconnaissance geologic map of the Quemado 60´x30´ quadrangle: New Mexico Bureau of Mines and Mineral Resources, Open-file Report, scale 1:100,000.

Chamberlin, R. M. and Harris, J. S., 1994, Upper Eocene and Oligocene volcaniclastic sedimentary stratigraphy of the Quemado-Escondido Mountain area, Catron County, New Mexico: New Mexico Geological Society, Guidebook 45.

Chapin, C. E., 1971, K-Ar age of the La Jara Peak Andesite and its possible significance to exploration in the Magdalena mining district, New Mexico: Isochron/West, no. 2, p. 43-44.

Chapin, C. E., 1988, Axial basins of the northern and central Rio Grande rift; a summary; *in* Riecker, R. E., ed., Rio Grande rift: Tectonics and magmatism: Washington, D. C., American Geophysical Union, p. 1-5.

Chapin, C. E. and Cather, S. M., in press, Tectonic setting of the axial basins of the northern and central Rio Grande rift: Geological Society of America, Special Paper 291.

Clemons, R. E., 1976, Sierra de las Uvas ash-flow field, south-central New Mexico: New Mexico Geological Society, Special Publication 6, p. 115-121.

Coffin, G. C., 1981, Geology of the northwestern Gallinas Mountains, Socorro County, New Mexico [M.S. thesis]: Socorro, New Mexico Institute of Mining and Technology, 202 p.

Cooley, M. E. and Davidson, E. S., 1963, The Mogollon highlands - their influence on Mesozoic and Cenozoic erosion and sedimentation: Arizona Geological Society Digest, v. 6, p. 7-35.

Cross, S. G., in preparation, Reconnaissance survey of Neogene clastic sediments in the southeastern Mogollon-Datil volcanic province, New Mexico [M. S. thesis]: Socorro, New Mexico Institute of Mining and Technology.

Dane, C. H. and Bachman, G. O., 1965, Geologic map of New Mexico: U. S. Geological Survey, 2 sheets, scale 1:500,000.

Eggleston, T. L., 1987, The Taylor Creek district, New Mexico: Geology, petrology, and tin deposits [Ph.D. dissertation]: Socorro, New Mexico Institute of Mining and Technology, 473 p.

Elston, W. E., 1957, Geology and mineral resources of Dwyer quadrangle, Grant, Luna, and Sierra Counties, New Mexico: New Mexico Bureau of Mines and Mineral Resources, Bulletin 38, 86 p.

Elston, W. E., 1968, Terminology and distribution of ash flows of the Mogollon-Silver City-Lordsburg region, New Mexico: Arizona Geological Society, Guidebook 3 to southern Arizona, p. 231-240.

Elston, W. E., 1976, Glossary of stratigraphic terms of the Mogollon-Datil volcanic province, New Mexico: New Mexico Geological Society, Special Publication 5, p. 131-144.

Elston, W. E. and Netelbeek, T. A., 1965, Road log from Mimbres Valley to Silver City: New Mexico Geological Society, Guidebook 16, p. 36-43.

Farris, S. R., 1981, Geology and mid-Tertiary volcanism of the McKnight Canyon area, Black Range, Grant County, New Mexico [M.S. thesis]: Albuquerque, University of New Mexico, 87 p.

Ferguson, H. G., 1927, Geology and ore deposits of the Mogollon mining district: U. S. Geological Survey, Bulletin 787, 100 p.

Foster, R. W., 1978, Selected data for deep drill holes along the Rio Grande rift in New Mexico: New Mexico Bureau of Mines and Mineral Resources, Circular 163, p. 236-237.

Harrison, R. W., 1980, Geology of the northeastern Datil Mountains, Socorro and Catron Counties, New Mexico [M. S. thesis]: Socorro, New Mexico Institute of Mining and Technology, 137 p.

Harrison, R. W., 1990, Cenozoic stratigraphy, structure, and epithermal mineralization of the north-central Black Range, New Mexico, in the regional geologic framework of south-central New Mexico [Ph.D. dissertation]: Socorro, New Mexico Institute of Mining and Technology, 402 p.

Hunt, C. B., 1956, Cenozoic geology of the Colorado Plateau: U.S. Geological Survey, Professional Paper 279, 99 p.

Jicha, H. L. Jr., 1954, Geology and mineral deposits of Lake Valley quadrangle, Grant, Luna, and Sierra Counties, New Mexico: New Mexico Bureau of Mines and Mineral Resources, Bulletin 37, 93 p.

Johnson, B. D., 1978, Genetic stratigraphy and provenance of the Baca Formation, New Mexico, and the Eagar Formation, Arizona [M. A. thesis]: Austin, University of Texas, 150 p.

Jones, W. R., Hernon, R. M. and Moore, S. L., 1967, General geology of Santa Rita quadrangle, Grant County, New Mexico: U.S. Geological Survey, Professional Paper 555, 144 p.

Kelley, V. C., 1977, Geology of the Albuquerque Basin, New Mexico: New Mexico Bureau of Mines and Mineral Resources, Memoir 33, 59 p.

Kelley, V. C. and Silver, C., 1952, Geology of the Caballo Mountains, with special reference to regional stratigraphy and structure, and to mineral resources, including oil and gas: University of New Mexico Publications in Geology No. 5, 286 p.

Krier, D. J., 1980, Geology of the southern part of the Gila Primitive Area, Grant County, New Mexico [M.S. thesis]: Albuquerque, University of New Mexico, 113 p.

Leopodt, W., 1981, Neogene geology of the central Mangas graben, Cliff-Gila area, Grant County, New Mexico [M. S. thesis]: Albuquerque, University of New Mexico, 159 p.

Love, D. W., 1989, Bidahochi Formation: an interpretive summary: New Mexico Geological Society, Guidebook 40, p. 273-280.

Lozinsky, R. P., 1988, Stratigraphy, sedimentology, and sand petrology of the Santa Fe group and pre-Santa Fe Tertiary deposits in the Albuquerque Basin, central New Mexico [Ph.D. dissertation]: Socorro, New Mexico Institute of Mining and Technology, 298 p.

Lucas, S. G. and Williamson, T. E., 1993, Eocene vertebrates and late Laramide stratigraphy of New Mexico: New Mexico Museum of Natural History and Science, Bulletin 2, p. 145-158.

Marr, R. J., 1956, Geology of the Lynch ranches, Catron and Valencia Counties, New Mexico [M.A. thesis]: Austin, The University of Texas, 113 p.

Marvin, R. F., Naeser, C. W., Bikerman, M., Mehnert, H. H. and Ratté, J. C., 1987, Isotopic ages of post-Paleocene igneous rocks within and bordering the Clifton 1°x2° quadrangle, Arizona-New Mexico: New Mexico Bureau of Mines and Mineral Resources, Bulletin 118, 63 p.

McDowell, F. W. and Mauger, R. L., 1994, K-Ar and U-Pb zircon chronology of Late Cretaceous and Tertiary magmatism in central Chihuahua State, Mexico: Geological Society of America Bulletin, v. 106, p. 118-132.

McIntosh, W. C., Kedzie, L. L. and Sutter, J. F., 1991, Paleomagnetism and $^{40}Ar/^{39}Ar$ ages of ignimbrites, Mogollon-Datil field, southwestern New Mexico: New Mexico Bureau of Mines and Mineral Resources, Bulletin 135, 79 p.

McIntosh, W. C., Chapin, C. E., Ratté, J. C. and Sutter, J. F., 1992, Time-stratigraphic framework for the Eocene-Oligocene Mogollon-Datil volcanic field, southwest New Mexico: Geological Society of America Bulletin, v. 104, p. 851-871.

McIntosh, W. C. and Cather, S. M., 1994, $^{40}Ar/^{39}Ar$ geochronology of basaltic rocks and constraints on late Cenozoic stratigraphy and landscape development in the Red Hill-Quemado area, New Mexico: New Mexico Geological Society, Guidebook 45.

McLellan, M., Robinson, L., Haschke, L., Carter, M. D. and Medlin, A., 1982, Fence Lake Formation (Tertiary), west-central New Mexico: New Mexico Geology, v. 4, p. 53-55.

McLemore, V. T., 1993, Geology and geochemistry of the Steeple Rock district, Grant County, New Mexico and Greenlee County, Arizona [Ph.D. dissertation]: El Paso, University of Texas at El Paso, 499 p.

Merrill, R. K. and Péwé, T. L., 1971, The Sheep Crossing Formation: a new late Cenozoic epiclastic formation in east-central Arizona: Arizona Academy of Science Journal, v. 6, p. 226-229.

Merrill, R. K. and Péwé, T. L., 1977, Late Cenozoic geology of the White Mountains, Arizona: Arizona Bureau of Geology and Mineral Technology, Special Paper 1, 65 p.

New Mexico Geological Society, 1982, New Mexico Highway Geologic Map, scale 1:1,000,000, 1 sheet.

North American Commission on Stratigraphic Nomenclature, 1983, North American Stratigraphic Code: American Association of Petroleum Geologists Bulletin, v. 67, p. 841-875.

Osburn, G. R. and Chapin, C. E., 1983, Nomenclature for Cenozoic rocks of northeast Mogollon-Datil volcanic field, New Mexico: New Mexico Bureau of Mines and Mineral Resources, Stratigraphic Chart 1, 2 sheets.

Peirce, H. W., Damon, P. E. and Shafiqullah, M., 1979, An Oligocene (?) Colorado Plateau edge in Arizona: Tectonophysics, v. 61, p. 1-24.

Potochnik, A. R., 1989, Depositional style and tectonic implications of the Mogollon Rim Formation (Eocene), east-central Arizona: New Mexico Geological Society, Guidebook 40, p. 107-118.

Ratté, J. C., 1981, Geologic map of the Mogollon quadrangle, Catron County, New Mexico: U.S. Geological Survey, Geological Quadrangle Map GQ-1557, scale 1:24,000.

Ratté, J. C., 1989, Geologic map of the Bull Basin quadrangle, Catron County, New Mexico: U.S. Geological Survey, Geological Quadrangle Map GQ-1651, scale 1:24,000.

Ratté, J. C. and Finnell, T. L., 1978, Third day road log from Silver City to Reserve via Glenwood and the Mogollon mining district: New Mexico Geological Society, Special Publication 7, p. 49-63.

Ratté, J. C. and Gaskill, D. L., 1975, Reconnaissance geologic map of the Gila Wilderness study area, southwestern New Mexico: U. S. Geologic Survey, Miscellaneous Investigations Map I-886, scale 1:50,000.

Repenning, C. A. and Irwin, J. H., 1954, Bidahochi Formation of Arizona and New Mexico: American Association of Petroleum Geologists Bulletin, v. 38, p. 1821-1826.

Repenning, C. A., Lance, J. F. and Irwin, J. H., 1958, Tertiary stratigraphy of the Navajo country: New Mexico Geological Society, Guidebook 9, p. 123-129.

Reynolds, S. J., 1988, Geologic map of Arizona: Arizona Geological Survey, Map 26, scale 1:1,000,000.

Richter, D. H., 1987, Geologic map of the O Bar O Canyon East quadrangle, Catron County, New Mexico: U.S. Geological Survey, Miscellaneous Field Studies Map MF-1976, scale 1:24,000.

Richter, D. H. and Lawrence, V. A., 1989, Geologic map of the O Bar O Canyon West quadrangle, Catron County, New Mexico: U.S. Geological Survey, Miscellaneous Field Studies Map MF-2075, scale 1:24,000.

Robinson, B. R., 1981, Geology of the D Cross Mountain quadrangle, Socorro and Catron Counties, New Mexico [Ph.D. dissertation]: El Paso, University of Texas at El Paso, 213 p.

Seager, W. R., Hawley, J. W. and Clemons, R. E., 1971, Geology of the San Diego Mountain area, Doña Ana County, New Mexico: New Mexico Bureau of Mines and Mineral Resources, Bulletin 97, 38 p.

Sirrine, G. K., 1956, Geology of the Springerville-St. Johns area, Apache County, Arizona [Ph.D. dissertation]: Austin, University of Texas, 248 p.

Snyder, D. O., 1971, Stratigraphic analysis of the Baca Formation, west-central New Mexico [Ph.D. dissertation]: Albuquerque, University of New Mexico, 160 p.

Tonking, W. H., 1957, Geology of the Puertecito quadrangle, Socorro County, New Mexico: New Mexico Bureau of Mines and Mineral Resources, Bulletin 41, 67 p.

Weber, R. H., 1971, K-Ar ages of Tertiary igneous rocks in central and western New Mexico: Isochron/West, no. 1, p. 33-45.

Willard, M. E., 1959, Tertiary stratigraphy of northern Catron County, New Mexico: New Mexico Geological Society, Guidebook 10, p. 92-99.

Wilpolt, R. H., MacAlpin, A. J., Bates, R. L. and Vorbe, G., 1946, Geologic map and stratigraphic sections of Paleozic rocks of Joyita Hills, Los Pinos Mountains, and northern Chupadera Mesa, Valencia, Torrance, and Socorro Counties, New Mexico: United States Geological Survey, Oil and Gas Investigations, Preliminary Map 61.

Winchester, D. E., 1920, Geology of Alamosa Creek Valley, Valley, Socorro County, New Mexico; with special reference to the occurrence of oil and gas: U. S. Geological Survey, Bulletin 716-A, p. 1-15.

Woodard, T. W., 1982, Geology of the Lookout Mountain area, northern Black Range, Sierra County, New Mexico [M.S. thesis]: Albuquerque, University of New Mexico, 95 p.

STRATIGRAPHIC NOMENCLATURE OF THE TERTIARY OF WESTERN NEW MEXICO AND EASTERN ARIZONA: AN ALTERNATIVE PROPOSAL

WOLFGANG E. ELSTON

Department of Earth and Planetary Sciences, University of New Mexico, Albuquerque, NM 87131

I am very grateful to Steve Cather, Jim Ratté and Dick Chamberlin for permitting me to review the previous paper (Cather et al., this volume) and to present some alternatives. The six-fold Tertiary classification (T1 through T6) of NMGS (1982) is here proposed as a chronostratigraphic framework (Fig. 1) to which lithostratigraphic units can be attached, in accordance with the stratigraphic code of NACSN (1983). In the Mogollon-Datil volcanic field, boundaries between T2, T3, and T4 are based on dated ignimbrite sheets that do not transgress time lines; their chronostratigraphic boundaries can also serve as lithostratigraphic boundaries. It is further proposed that NMGS sponsor a committee to review the Tertiary terminology of New Mexico and make recommendations on the following issues: (1) The adequacy and validity of proposed chronostratigraphic boundaries, (2) Boundary problems in interfingering volcanic rocks from diverse sources, (3) The placement of arbitrary boundaries between laterally equivalent units with different traditional names, e.g. Mogollon Rim-Baca Formations, Gila-Santa Fe Groups, and (4) The nomenclature of rock bodies, keeping in mind the NACSN (1983) rules regarding establishment and maintenance of type localities, naming of units (with special regard to the rule of priority), and abandonment of obsolete or redundant terms.

Specifically, it is recommended that the term *Spears* be retained as a formation with the type section and stratigraphic limits originally defined by Tonking (1957). On the north side of the Mogollon-Datil field, a predominantly volcaniclastic distal apron takes the place of the entire T2-T4 volcanic section. It should be given a unit name other than Spears. The term *Datil* should be dropped as a stratigraphic term on the grounds of "widespread misuse in diverse ways which compound confusion" (NACSN, 1983), as illustrated by the variety of stratigraphic ranges assigned to it by different authors (Fig. 1). It survives as part of the name for the entire field. The concept of two major units, equivalent to the proposed Datil and Mogollon Groups of Cather et al. (this volume), is workable if a mappable boundary can be demonstrated. Over much of the southern part of the field, the two proposed groups are separated by a thick section of basaltic to andesitic lavas (T3a of NMGS, 1982). The disconformity (local

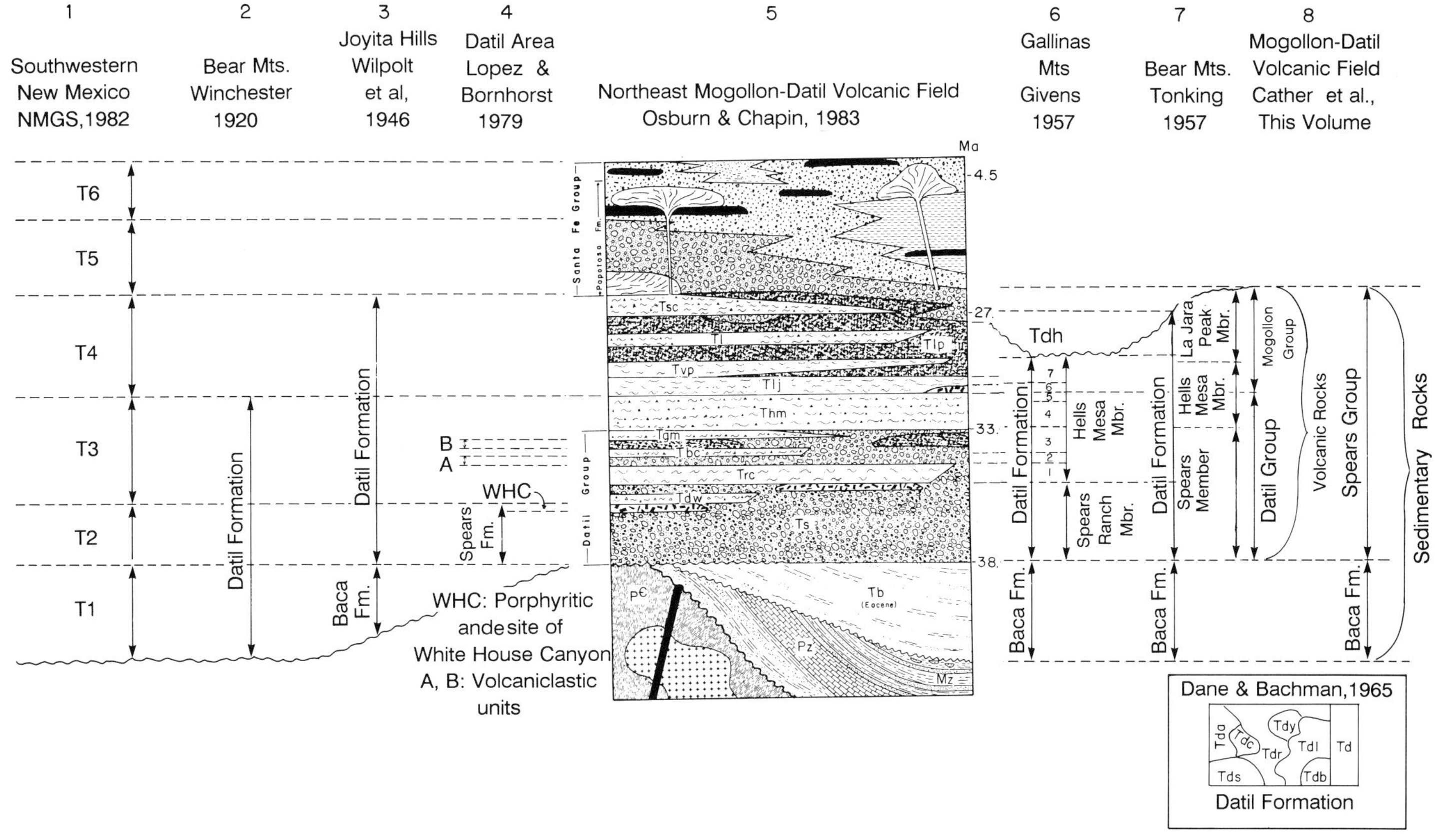

FIGURE 1. Columns showing the stratigraphic positions and ranges assigned by various authors to units mentioned in the text, northeastern part of Mogollon-Datil volcanic field. Inset: The "facies" of the Datil Formation according to Dane and Bachman (1965). Td-Datil Formation, undivided; Tdy-rhyolitic flows; Tda-andesite and basaltic andesite; Tds-volcaniclastic rocks; Tdc-boulder beds; Tdr-rhyolite tuffs; Tdl-latite; Tdb-pumiceous breccia. Symbols for Column 5: Tb-Baca Formation; Tbc-Blue Canyon Tuff; Tdw-Datil Well Tuff; Tgm-tuff of the Granite Mountain; Thm-Hells Mesa Tuff; Tl-Lemitar tuff; Tlj-La Jencia Tuff; Tlp-La Jara Peak Basaltic Andesite; Trc-Rock House Canyon Tuff; Ts-Spears Formation; Tsc-South Canyon Tuff; Tvp-Vicks Peak Tuff (nomenclature from Osburn and Chapin, 1983).

unconformity) and petrologic discontinuity (Davis et al., 1993) at the base of the Bear Springs and Poverty Creek basaltic andesites (T3) may be part of such a boundary. If so, it is recommended that the two groups proposed by Cather et al. (this volume) be raised to supergroups. The new name *Cobre Supergroup* (after the Cobre Mountains, Grant Co., east of Hurley, NM) is here recommended in lieu of Datil Group; it comprises T2 and T3 without Bear Springs Basalt. The term *Mogollon Supergroup* is suitable for T4, including Bear Springs Basalt. New group names are needed for T2a, T3r and T4r of NMGS (1982); the existing names *Alum Mountain Group* and *Bearwallow Mountain Group* are proposed for T3a and T4a, respectively. Other existing formations, members, and tongues would remain unchanged.

This alternative proposal will be described in detail in a forthcoming issue of *New Mexico Geology*. I hope that it will stimulate further discussions and lead to a consensus.

SUPPLEMENT TO REFERENCES CITED BY CATHER ET AL. (THIS VOLUME)

Dane, C.H. and Bachman, G.O., 1965, Geologic map of New Mexico: U.S. Geological Survey, 2 sheets, scale 1:500,000.

Davis, J.M., Elston, W.E. and Hawkesworth, C.J., 1993, Basic and intermediate volcanism of Mogollon-Datil volcanic field: implications for mid-Tertiary tectonic transitions in southwestern New Mexico, U.S.A.; *in* Prichard, H.M., Alabaster, T., Harris, N.B.W., and Neary, C.R., eds., Magmatic processes and plate tectonics: London, The Geological Society, Special Publication 76, p. 469-488.

Lopez, D.A. and Bornhorst, T.J., 1979, Geologic map of the Datil area, Catron County, New Mexico: U.S. Geological Survey, Miscellaneous Investigations Map I-1098, 1:50,000.

NACSN (North American Commission on Stratigraphic Nomenclature), 1983, North American stratigraphic code: American Association of Petroleum Geologists Bulletin, v. 67, p. 841-875.

UPPER EOCENE AND OLIGOCENE VOLCANICLASTIC SEDIMENTARY STRATIGRAPHY OF THE QUEMADO-ESCONDIDO MOUNTAIN AREA, CATRON COUNTY, NEW MEXICO

RICHARD M. CHAMBERLIN[1] and JAMES S. HARRIS[2]

[1]New Mexico Bureau of Mines and Mineral Resources, Socorro, NM 87801; [2]New Mexico Institute of Mining and Technology, Socorro, NM 87801

Abstract—An upper Eocene to Oligocene (ca. 40-26 Ma) volcaniclastic sedimentary apron derived from the Mogollon-Datil volcanic field can be divided into three mappable lithostratigraphic units in the Quemado-Escondido Mountain area. From oldest to youngest we propose the new names volcaniclastic unit of Largo Creek (VLC), volcaniclastic unit of Cañon del Leon (VCDL), and sandstone of Escondido Mountain (SEM) as informal divisions of the gently south-southeast-dipping Spears Group. A geologic cross section of the area indicates that VLC, VCDL and SEM are approximately 600, 180 and 240 m thick, respectively. Gray, quartz-poor andesitic sandstones and interbedded red mudstones of the main body of the VLC (lower 540 m) conformably overlie the middle to upper Eocene Baca Formation. The main body of the VLC represents the distal fluvial equivalent of andesitic debris-flow deposits of the Dog Springs Formation (ca. 39.6-36.9 Ma), which is well exposed in the Datil region southeast of Quemado. An upward coarsening interval (~60 m) of gray to light brown, quartz-poor to moderately quartz-rich, andesitic conglomeratic sandstones forms a local (unmapped) transition zone in the upper VLC. A unique quartz-microcline-rich basement-derived sandstone near the top of the VLC may be correlative with mixed-clast conglomerates (basement and volcanic derived) recently observed in the middle Spears Group in the Alpine No. 1 Federal Well near Alpine, Arizona. A cliff-forming marker bed composed of light gray tuffaceous sandstones and thin pumiceous mudstones as much as 18 m thick defines the base of the VCDL and conformably overlies the upper VLC. Numerous gray tuffaceous sandstone beds intercalated with brownish gray to bluish-green andesitic conglomeratic sandstones comprise the VCDL at its type section in upper Cañon del Leon. Sanidines from pumice in syneruptive tuffaceous sandstones and plagioclase from a rhyolitic ash bed in the VCDL yield $^{40}Ar/^{39}Ar$ ages that imply aggradation on a distal alluvial apron from approximately 35.3 to 34.2 Ma. Tuffaceous sandstones of the VCDL thicken to the southwest and appear to form part of the andesitic volcaniclastic apron of the Pueblo Creek Formation, which was shed northward from a late Eocene eruptive center near Reserve. Pale yellowish-brown, medium-grained, planar to high-angle cross-bedded, volcaniclastic sandstones of the SEM are locally well exposed where they paraconformably overlie VCDL in uppermost Cañon del Leon. These eolian sandstones of the SEM are discontinuously exposed on the flanks of several ranges capped by Oligocene basaltic andesites, from Alegres Mountain westward to Escudilla Mountain in eastern Arizona. The SEM appears to represent a southeasterly tapering erg that was blown over the north-facing epiclastic apron of the middle Spears Group (e.g. VCDL) by prevailing westerly winds off the Colorado Plateau in Oligocene time (ca. 32-26 Ma).

INTRODUCTION

Reconnaissance geologic mapping of the Quemado 30´x60´ quadrangle (Chamberlin et al., 1994) has shown that a thick interval of volcaniclastic sedimentary rocks, locally well exposed along the drainage areas of Largo Creek and Cañon del Leon between Quemado and Escondido Mountain, can be informally subdivided into three mappable lithostratigraphic units. These informal subdivisions of the Spears Group (Cather et al., this volume) conformably overlie the middle to upper Eocene Baca Formation and are conformably overlain by middle Oligocene basaltic-andesite lavas and ignimbrites of the Mogollon Group (Chamberlin et al., 1994; Cather et al., this volume). Numerous new $^{40}Ar/^{39}Ar$ ages for volcanic and volcaniclastic rocks in the Quemado region (McIntosh and Chamberlin, this volume) provide a geochronologic framework for these volcaniclastic sedimentary units of the Spears Group. A petrographic study of volcaniclastic sandstones in the Quemado-Escondido Mountain area (J. S. Harris, unpubl. report, NMIMT, March 1994) provides most of the compositional data summarized in this article.

As shown on the reconnaissance geologic map of the Quemado-Escondido Mountain area (Fig. 1) we propose to informally name these three divisions of the Spears Group the volcaniclastic unit of Largo Creek, volcaniclastic unit of Cañon del Leon and sandstone of Escondido Mountain. Here we summarize the lithology, compositional variations, age constraints, correlation, thickness and origin for each of these units. Previous reconnaissance geologic maps of the Quemado region (Willard, 1957; Willard and Weber, 1958) grouped these Eocene to Oligocene units and the unconformably overlying Miocene Fence Lake Formation together as the volcanic sediment facies of the Datil Formation.

At this time we purposely choose not to identify these mappable volcaniclastic sedimentary units as formations or members of formations. Additional mapping is necessary to relate properly these informal map units to formally designated formations in the region (e.g. Chapin and Osburn, 1983; Ratté, 1989; Cather et al., this volume; also see stratigraphic nomenclature chart in back of this guidebook). For a description of the regional geologic setting of the Quemado-Escondido Mountain area see McIntosh and Chamberlin (this volume). Descriptions of these Spears Group units near Quemado can also be found in the First- and Second-Day road logs of this guidebook.

Samples of volcaniclastic sandstones were collected at several of the better exposures along the east flank of Largo Creek and upper Cañon del Leon. Petrographic thin sections were prepared from 20 samples from within the study area and six additional samples of correlative units from outside the area (Fig. 1, Table 1). Three samples of pumiceous muddy siltstone and one primary ash bed were collected from the volcaniclastic unit of Cañon del Leon in order to evaluate the age of this unit by $^{40}Ar/Ar^{39}$ geochronology. Plagioclase and sanidine phenocrysts were hand picked from individual pumice clasts, in order to avoid older detrital components. Feldspar separates from these pumice clasts and the ash bed were then dated by $^{40}Ar/^{39}Ar$ geochronologic methods (see McIntosh and Chamberlin, this volume).

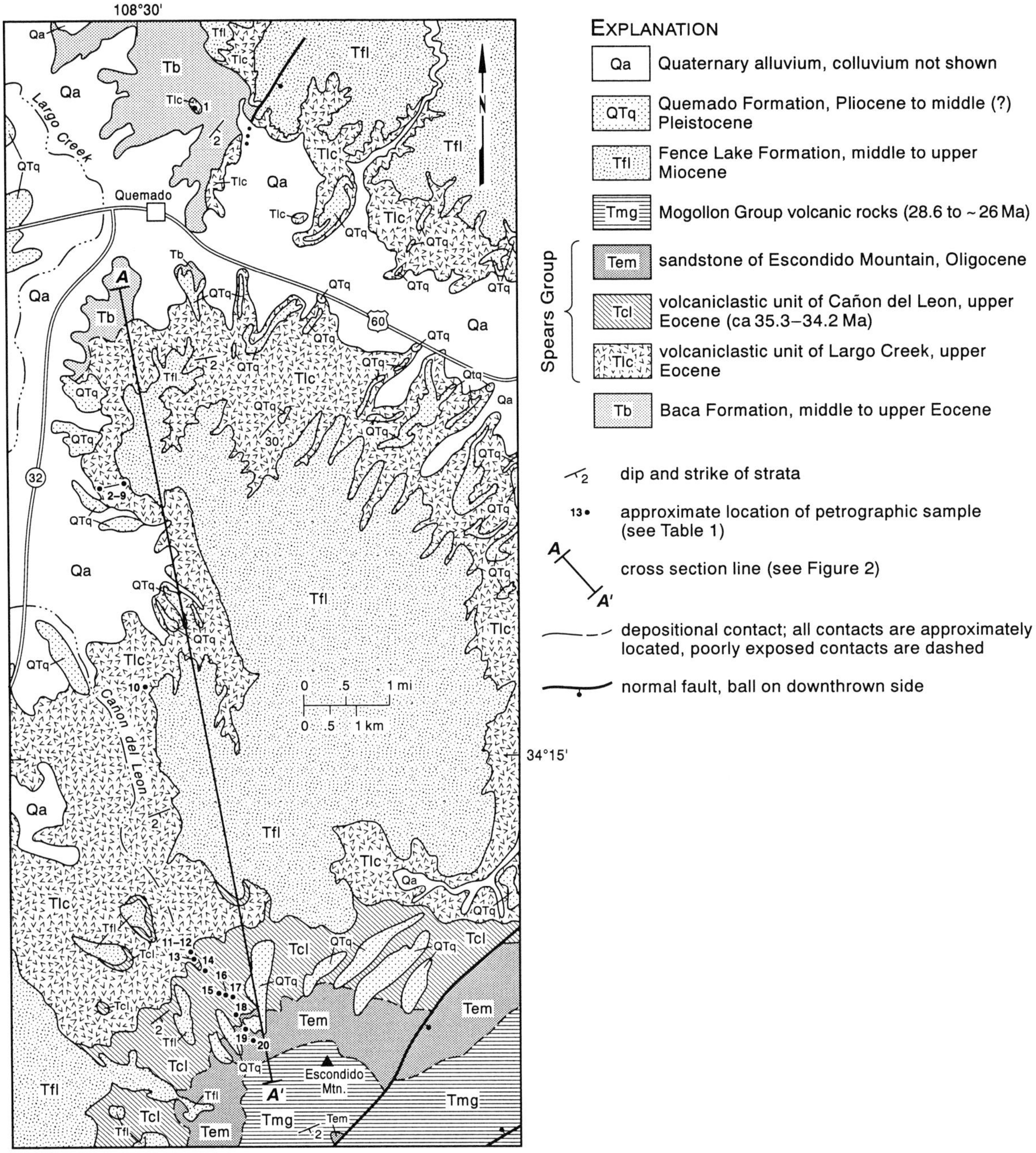

FIGURE 1. Geologic map of the Quemado-Escondido Mountain area, Catron County New Mexico; modified after Chamberlin et al. (1994). Some smaller areas of Quaternary alluvium are not shown.

VOLCANICLASTIC UNIT OF LARGO CREEK

The new informal stratigraphic term "volcaniclastic unit of Largo Creek" (VLC) is proposed here for a thick sequence of mostly gray andesitic sandstones, reddish mudstones, and minor andesitic conglomeratic sandstones that conformably overlie the Baca Formation in the Quemado-Escondido Mountain area (Fig. 1). We designate the eastern side of the valley of Largo Creek from the Quemado Community Cemetery (SW¼ sec. 26, T2N, R16W) southward to the cliffs of upper Cañon del Leon (SW¼ sec. 15, T1S, R16W) as the type area for this unit. It is locally well exposed along the west side of a large mesa capped by middle to upper Miocene Fence Lake Formation (Fig. 1). This mesa defines the upper Miocene piedmont slope that descends northward from its source at Escondido Mountain (Fig. 2). Plio-Pleistocene gravels of the Quemado Formation (Cather and McIntosh, this volume) are locally inset below the Fence Lake Formation.

Contacts and lithology

The basal contact of the VLC is defined by the upsection change from buff quartzofeldspathic sandstones and bright red mudstones of the Baca Formation to the gray andesitic sandstones and dull red mudstones of the VLC. This contact is conformable, gradational and may locally intertongue over an interval of approximately 10-15 m, in the Quemado area (Guilinger, 1981; Chamberlin et al., 1994). We propose the unnamed hill northeast of the Quemado Community Cemetery (NW¼ sec. 35, T2N, R16W) as the basal boundary stratotype for the VLC. The contact is well exposed here at an elevation of approximately 2202 m. The VLC is also locally well exposed in landslide scars below the cliff-forming Fence Lake Formation from Quemado south to Cañon del Leon; each individual exposure typically presents 30 to 60 m of gray slope-forming andesitic sandstones and red mudstones.

In most VLC exposures, the fine- to medium-grained sandstone beds do not stand out as distinct ledges separated by mudstones, but rather the VLC typically weathers to a soft mud- covered slope. Bedding is commonly planar and horizontal, but trough cross beds occur locally. Upward-fining conglomeratic sandstone beds are locally present. Sandstone beds are generally 0.3 to 3 m thick and reddish mudstones may occur as thin drapes to massive looking beds as much as 5-7 m thick.

The upper part of the VLC, approximately 60 m thick, is formed by light gray to pale brown andesitic sandstones and conglomeratic sandstones that define an upward coarsening sequence. Bluish-green andesitic sandstones and conglomerates are common near the top of the VLC. The bluish-green color is attributed to celadonite cement, which in part may be associated with diagenetic hydration of glass shards in the overlying tuffaceous sandstones. This upper subunit of the VLC appears to be transitional into the overlying volcaniclastic unit of Cañon del Leon, defined below. Rare, thin, pumiceous sandstones, and a general increase in quartz content (from near zero to approximately 30%) also help distinguish the upper transition zone from the main body of the VLC. The vague boundary between the main body of the VLC and upper transition zone has not been mapped. The approximate position of the transition zone (Tlct) is shown only in cross section (Fig. 2), and its regional significance is uncertain.

Petrographic data for the main body and transition zone of the VLC (Table 1; Fig. 3) show that VLC sandstones are predominantly quartz-poor volcanic litharenites of andesitic composition. Large grains of oscillatory zoned plagioclase (phenocryst derived) and microlitic andesite rock fragments comprise most of the framework grains in these sandstones (Fig. 3B). Hornblende, biotite and Fe-Ti oxides commonly form 2-5% of these sandstones; placer-type concentrations of these heavy minerals, however, locally form laminations with 10-20% heavy minerals (Table 1). Quartz is typically sparse to absent in the main body of the VLC.

The most quartz-rich sample (28% quartz; No. 13, Fig. 1, Table 1) in the entire Eocene-Oligocene section was collected about 10 m below the top of the VLC, near the mouth of upper Cañon del Leon. This gray, medium- to coarse-grained sandstone bed is compositionally unique and contains a large component (~80%) of basement-derived grains. Most of the quartz shows undulose extinction; cross-hatched microcline represents at least half of the alkali feldspar grains, and sericitic microgranular felsite is the most common type of rock fragment. These are not typical rhyolitic rock fragments with shard outlines, or banded and spherulitic textures. The sericitic fragments are texturally similar to hydrothermally altered (silicified) lavas, but their association with basement-type quartz and feldspar suggests that these microgranular rock fragments represent basement-derived "metarhyolite". The possible significance of this basement-derived sandstone near the middle of the Spears Group is discussed later.

Age and correlation

The VLC has not been directly dated in the Quemado area. Its age, however, is bracketed by the underlying Baca Formation (ca. 46-40 Ma; Cather et al., 1987) and the overlying volcaniclastic unit of Cañon del Leon (ca. 35.3-34.2 Ma; McIntosh and Chamberlin, this volume). Regional mapping (Chamberlin, et al., 1994) indicates that the VLC is temporally equivalent to andesitic debris-flow deposits of the Dog

TABLE 1. Point count data[1] for volcaniclastic sandstones[2] in the Quemado-Escondido Mountain area and other localities in the Quemado region.

Sample No.[3] or Location	Field Number	Quartz	Plagioclase	Alkali Feldspar	Andesitic Rock Fragments	Rhyolitic Rock Fragments	Fe-Mg Minerals and Opaques	Grain Size ϕ
sandstone of Escondido Mountain (SEM)								
20	CDL-1	2.4	11.8	0.0	82.2	0.0	3.6	2.0
19	CDL-3	16.4	17.6	2.8	2.4	59.2	1.6	2.0
Killion Can.	KC-1	24.5	22.5	14.0[4]	3.2	35.8	0.0	2.0
Killion Can.	KC-2	8.0	10.4	3.6	3.6	74.4	0.0	2.0
volcaniclastic unit of Cañon del Leon (VCDL)								
18	CDL-2	7.3	8.5	0.0	4.0	80.2	0.0	3.0
17	CDL-5	3.6	16.8	2.8	16.8	53.5	6.4	2.5
16	CDL-6	3.2	18.3	4.5	7.1	66.9	0.0	3.5
15	CDL-7	5.6	19.3	15.7	12.0	44.2	3.2	2.5
14	CDL-12	0.7	0.2	0.0	0.0	98.8	0.1	3.5
Castle Rock	CR-1	0.6	0.3	0.3	0.0	97.6	1.2	3.0
volcaniclastic unit of Largo Creek, (upper transition zone, VLCT)								
13	CDL-11A	27.7	8.8	20.0[4]	4.8	37.0[5]	1.6	1.5
12	CDL-10A	4.4	7.8	1.6	74.1	9.4	2.4	1.5
11	CDL-10B	8.0	34.8	4.8	24.4	17.6[5]	10.4	1.5
Castle Rock	CR-2	2.0	4.4	0.0	81.0	11.3	1.2	2.5
volcaniclastic unit of Largo Creek, (main body, VLCM)								
10	WR-9	2.4	51.2	0.0	35.6	0.0	9.6	2.5
9	WR-8	0.0	51.6	0.0	41.2	0.0	5.2	1.0
8	WR-7	6.0	49.6	2.0	38.0	0.0	4.0	3.0
7	WR-6	6.4	38.0	2.0	49.2	0.0	3.6	2.5
6	WR-5	1.6	64.0	0.0	9.2	1.2	23.6	2.0
5	WR-4	0.0	4.0	0.0	74.4	0.0	21.2	0.5
4	WR-3	0.0	7.2	0.0	88.4	0.0	3.6	2.5
3	WR-2	11.6	36.8	8.4	38.0	0.0	4.8	2.5
2	WR-1	4.4	49.6	0.0	43.6	0.0	1.6	1.5
1	QUE-1	4.4	23.3	0.0	62.0	3.2	1.6	2.0
Pie Town	MP-51.8	2.8	74.4	0.0	14.4	0.0	8.8	2.5
Pie Town	MP-62.4	2.6	41.4	0.0	40.8	0.0	15.2	2.5

[1]Point count data by J. S. Harris; percent of total framework grains. Total Count = 250 grains per sample.
[2]Unit correlations from Chamberlin et al. (1994).
[3]Sample No. refers to Fig. 1.
[4]Includes microcline.
[5]Includes minor chert and siltstone.

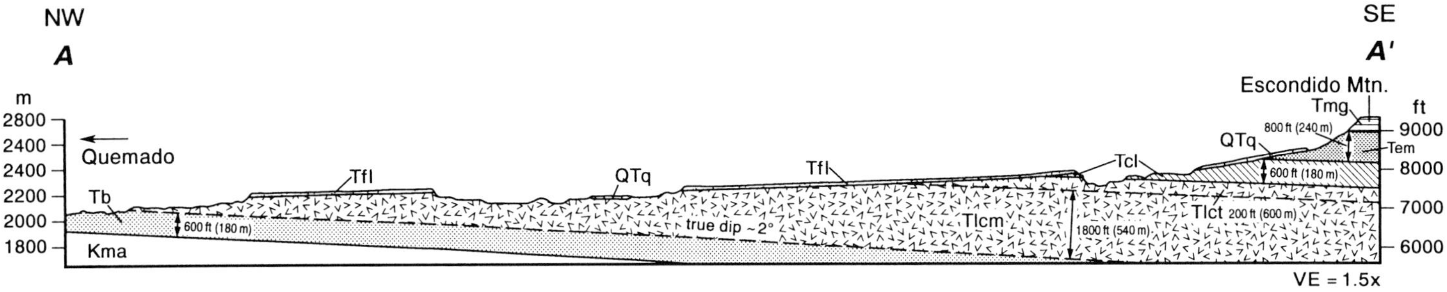

FIGURE 2. Geologic cross section of the Quemado-Escondido Mountain area; original cross section used to estimate thickness of units was drawn to true scale at 1:24,000. See Figure 1 for explanation of symbols. Tlcm=main body of the volcaniclastic unit of Largo Creek. Tlct=upper transition zone of the volcaniclastic unit of Largo Creek. Kma=Cretaceous Moreno Hill-Atarque Formations. Thickness of Baca Formation from Guilinger (1981).

Springs Formation exposed in the Datil region southeast of Quemado. Isotopic ages of Dog Springs volcaniclastic rocks range from 39.6 ± 1.5 Ma (Osburn and Chapin, 1983; Cather et al., 1987) to 36.94 ± 0.07 Ma (McIntosh and Chamberlin, this volume). The above age constraints support the correlation of the volcaniclastic unit of Largo Creek at Quemado with the Dog Springs Formation at Datil. The VLC at Quemado is here considered to be of late Eocene age (ca. 40-35 Ma).

Thickness and origin

The thickness of the VLC south of Quemado was estimated by constructing a geologic cross section (at true scale) along the regional south-southeast dip. This cross section is vertically exaggerated here for clarity (Fig. 2). Assuming a uniform 2° dip, this yields a 600 m thickness for the VLC. This estimate is a maximum, since the VLC is locally folded and faulted (see First-Day road log, mile 149.7-151.4). Intercepts and surface observations at the Hunt No. 1-16 State Well, 43 km southeast of Quemado (Broadhead and Chamberlin, this volume) indicate a total thickness of 462 m for the quartz-poor andesitic lower Spears Group (VLC equivalent). Quartz-poor, gray, andesitic sandstones similar to the VLC at Quemado were intersected in the Alpine No. 1 Federal Well (62 km SW of Quemado) between 480 m and 625 m, a total intercept of 145 m. These sandstones overlie 363 m of basement-derived Eagar Formation and lie below 66 m of pumiceous sandstones and 97 m of mixed-clast conglomerate, the latter derived from both volcanic and basement terranes. We suggest that the basement-derived sandstone bed near the top of the VLC may be in part correlative with the mixed-clast conglomerate interval of the Alpine No. 1 Federal Well. The above observations suggest that the actual thickness of the VLC near Quemado is approximately 300 to 400 m.

The dominance of fine- to medium-grained, well-sorted, andesitic sandstones and interbedded mudstones, with minor pebbly sandstones, suggest that the main body of the VLC in the Quemado area is a distal basin-floor fluvial deposit. These fluvial andesitic sandstones appear to grade easterly or southeasterly into argillaceous andesitic sandstones (also assigned to the VLC, Chamberlin et al., 1994) in the Pie Town area. These better-indurated argillaceous sandstones may primarily represent hyperconcentrated flood deposits derived from downstream dilution of andesitic debris-flow deposits in the Datil-Crosby Mountains area. These volcanogenic facies transitions or transformations (Fisher and Smith, 1991) are consistent with a source for the VLC (main body) somewhere under the south-central San Agustin plains (see First-Day road log, mile 71.4 and 101.4).

A few observations of pebble imbrications in the upper transitional VLC suggest northeasterly paleoflow. In conjunction with the quartz-rich, basement-derived sandstone observed in this zone, the available data suggest that the upper VLC represents a northeast prograding alluvial apron derived from an andesitic volcanic center in the Reserve area. This younger volcanic center is presumably the same one documented by Ratté (1989) as the source area for volcaniclastic alluvial deposits of the Pueblo Creek Formation and the andesite of Dry Leggett Canyon.

VOLCANICLASTIC UNIT OF CAÑON DEL LEON

We propose the new informal name "volcaniclastic unit of Cañon del Leon" for approximately 180 m of interbedded rhyolitic tuffaceous sandstones and andesitic to rhyolitic conglomeratic sandstones that conformably overlie the VLC in upper Cañon del Leon (Figs. 1, 2). We designate the eastern side of upper Cañon del Leon from the SE¼SE¼ sec. 15, T1S, R16W to the SE¼NW¼ sec. 23, T1S, R16W as the type section. Access to this type section within the Apache National Forest may be gained by Forest Service roads along Escondido Creek. The most advantageous entry to upper Cañon del Leon, which avoids large cliffs, lies approximately 1.7 km south-southwest of Escondido Tank (SE¼ sec. 11, T1S, R16W, Escondido Mountain 7.5′ quadrangle).

Contacts and lithology

The basal contact of the volcaniclastic unit of Cañon del Leon (VCDL) is well exposed at the type section on the east wall of Cañon del Leon (Fig. 4). The basal stratotype here consists of a thin bedded, cliff-forming, light gray, angular-jointed tuffaceous sandstone and pumiceous mudstone that conformably overlies medium-bedded, bluish-green to brownish-gray and rare reddish-brown conglomeratic sandstone of the upper VLC. The base of this cliff-forming marker bed (basal VCDL) is present at an approximate elevation of 2378 m (34.212°N, 108.487°W, Escondido Mountain 7.5′ quadrangle). An alternative, more accessible, boundary stratotype is at the east end of Castle Rock at an elevation of 2430 m in the Largo Mesa 7.5′ quadrangle (34.158°N, 108.548°W). The latter locality is described in the First-Day road log of this guidebook (see mile 156.5). The top of the VCDL is locally exposed in upper Cañon del Leon (east fork) at an elevation of 2500 m, where it appears to be paraconformably overlain by well-sorted sandstone of the sandstone of Escondido Mountain (new name), as described in a following section.

The basal VCDL is a cliff-forming marker bed, as much as 18 m thick, that is composed of light gray tuffaceous sandstones, 0.1-1 m thick, with thin (~ 1 cm) pale red muddy siltstone beds. The thin siltstone beds appear to be slackwater drape deposits and preferentially contain larger (0.5-2 cm) clasts of white fibrous pumice. Pumice at the type section and Castle Rock consists of a moderately phenocryst-rich (biotite, sanidine, plagioclase and quartz) pumice, and a phenocryst-poor pumice with dominantly plagioclase crystals. The former is distinctly rhyolitic and the latter may be of andesitic to dacitic composition. Sandstones in the basal cliff-forming marker bed consists almost entirely of rhyolitic glass shards mixed with minor reworked phenocrysts of quartz, plagioclase and sanidine (Table 1, CR-1 and CDL-12). A 5-cm-thick white rhyolitic ash-bed occurs in the narrow box of upper Cañon del Leon within the cliff-forming marker bed. This ash bed contains approximately 2% of phenocrystic quartz and plagioclase, 10% small angular devitrified felsite clasts and about 88% flattened pumice clasts and shards (McIntosh and Chamberlin, this volume, sample CDL-10/2-9B).

Interbedded andesitic conglomeratic sandstone and thin tuffaceous to pumiceous sandstone comprise most of the VCDL at the type section. Andesitic sandstones consist of upward fining, coarse- to medium-

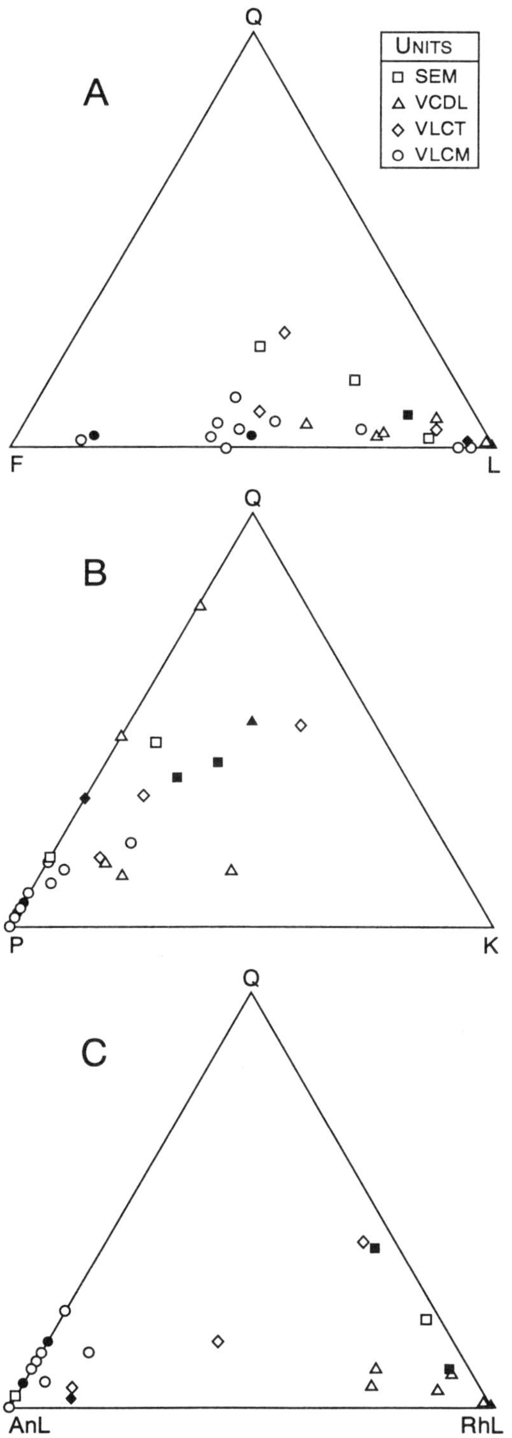

FIGURE 3. Ternary diagrams showing compositional variations in the main body of the volcaniclastic unit of Largo Creek (VLCM), in the upper transition zone of the volcaniclastic unit of Largo Creek (VLCT), in the volcaniclastic unit of Cañon del Leon (VCDL), and in the sandstone of Escondido Mountain (SEM). All data are from Table 1; open symbols represent samples from the Quemado-Escondido Mountain area (Fig. 1). Solid symbols represent samples collected outside the Quemado-Escondido Mountain area (Table 1). 3A, variation in total quartz, Q; total feldspar, F and total lithic fragments, L. 3B, variation in total quartz, Q; total plagioclase feldspar, P; and total alkali feldspar, K. 3C, variation in total quartz, Q; andesitic lithic fragments, AnL; and rhyolitic lithic fragments, RhL.

grained beds generally 0.3 to 2 m thick, with common clasts of dark gray to reddish-brown andesite porphyry and rare clasts of light gray rhyolite. The conglomeratic sandstone beds are mostly light brownish

FIGURE 4. View to west-southwest of upper Cañon del Leon. Basal contact of volcaniclastic unit of Cañon del Leon (Tcl) is indicated by dashed line. Unconformity (unc.) at base of Fence Lake Formation (Tfl) is indicated by solid line. Basal 18 m of the volcaniclastic unit of Cañon de Leon is a cliff-forming tuffaceous sandstone and pumiceous mudstone marker bed, just above white tuffaceous sandstone bed. Transitional sandstones and conglomeratic sandstones of upper part of the volcaniclastic unit of Largo Creek (Tlct) locally form badlands topography.

gray and moderately indurated. A few bluish-green conglomeratic sandstones, presumably cemented with celadonite, are also present. Tuffaceous sandstones are mostly light gray, thin bedded (0.1-1 m) and are commonly associated with pale red pumiceous mudstone.

Preliminary petrographic study of the VCDL sandstones (Table 1, Fig. 3) indicate that they generally represent a mixture of rhyolitic and andesitic detritus similar to the underlying transition zone in the upper VLC and much more variable than the andesitic main body of the VLC (Fig. 3). In thin section (Table 1, Fig. 3) sandstones of the VCDL are classified as volcanic lithic arenites of rhyolitic to andesitic composition (Fig. 3C). Relatively iron-rich microlitic grains (with finely disseminated opaques) and plagioclase porphyry clasts are the most common types of andesitic lithic fragments. Glass shards, sand-sized pumice fragments and minor microgranular felsites (iron poor) are the most common type of rhyolitic rock fragments. Rare siliceous siltstone clasts and chert were counted with rhyolitic fragments in some VCDL sandstones. Low birefringent clay minerals or zeolites are the most common cement; sparry calcite may be present in some conglomeratic sandstones. Bluish-green sandstones are apparently cemented with fibrous, moderately birefringent celadonite. Sparse grains of embayed quartz and sanidine form a minor rhyolitic component of VCDL sandstones; these were not observed in the main body of the VLC.

Age and correlation

A rhyolitic ash bed in the lower VCDL, approximately 12 m above the base at the type section, yielded an imprecise $^{40}Ar/^{39}Ar$ depositional age of 35.19 ± 0.42 Ma (McIntosh and Chamberlin, this volume). Sanidine phenocrysts hand-picked from pumice clasts at the base of VCDL at Castle Rock produced a precise $^{40}Ar/^{39}Ar$ age of 35.33 ± 0.06 Ma (McIntosh and Chamberlin, this volume). Also, a single sanidine crystal from pumice at the top of the VCDL at the type section, apparently derived from the Rock House Canyon Tuff, yielded an $^{40}Ar/^{39}Ar$ age of 34.17 ± 0.13 Ma (McIntosh and Chamberlin, this volume). One additional plagioclase separate from pumice at the base of the type section gave an imprecise $^{40}Ar/^{39}Ar$ age of 35.48 ± 0.42 Ma (McIntosh and Chamberlin, this volume), which is within analytical error of the more precise age determination at Castle Rock.

The above ages from pumice clasts would normally be interpreted as maximum ages of deposition. However, the field characteristics of these pumice samples, and the fact that their maximum ages are in agreement with the stratigraphic sequence suggest that these pumice-clast ages are also the approximate age of deposition of the host beds. We interpret the VCDL at the type section to range in age from approximately 35.3 to 34.2 Ma. This age range includes several regional igimbrites, including the 35.0-Ma Datil Well Tuff, the 34.9-Ma tuff of Bishop Peak, the 34.9-Ma Kneeling Nun Tuff, and the 34.2-Ma Rock House Canyon Tuff (McIntosh et al., 1991; McIntosh and Chamberlin, this volume). Older pumice clasts near the base of the VCDL and the rhyolitic ash bed may have been derived from smaller pyroclastic eruptions associated with silicic or intermediate lava domes, although such lava domes have not been recognized in the Quemado region.

The tuffaceous cliff-forming marker bed at the base of the VCDL can be traced across the north flank of Escondido Mountain westward to Castle Rock and the northern toe of Agua Fria Mountain. West of Agua Fria Mountain the marker bed is locally truncated by a splay of the Spur Lake fault, but it reappears to the west at Cañon Blanco (SW¼ sec. 25, T2S, R18W) and Steel Canyon (SW¼ sec. 8, T3S, R18W), where it underlies the upper Eocene andesite of Dry Leggett Canyon. Light brown andesitic conglomeratic sandstone beds that overlie the andesite of Dry Leggett Canyon west of Agua Fria Mountain also contain the 33.4-Ma tuff of Luna (McIntosh and Chamberlin, this volume), which is apparently absent at the type section in Cañon del Leon. These map relationships (Chamberlin et al., 1994) suggest a local unconformity may be present at the top of VCDL at the type section.

Pumiceous sandstones and bluish-green andesitic sandstones, similar to the VCDL, are known to underlie the andesite of Dry Leggett Canyon at Deep Canyon (30 km to the SW of the type section, see Third-Day road log, mile 65.95). Pumiceous sandstones also occur in the middle Spears Group of the Alpine Federal No. 1 Well (Witcher et al., this volume, fig. 4), and above the Rock House Canyon Tuff at Saulsberry Ranch (see First-Day road log, mile 84.1). The VCDL is in part correlative with the upper Eocene Chavez Canyon Formation and the Rincon Windmill Formation (Osburn and Chapin, 1983; Cather et al., this volume) as mapped in the Datil region. Also, the VCDL appears to be in part correlative with the Pueblo Creek Formation as mapped in the Reserve area (e.g., Ratté, 1989).

Thickness and origin

At its type section the VCDL is approximately 180 m thick, as determined in cross section (Fig. 2). This unit may thicken to the southwest where 250 m of pumiceous and tuffaceous sandstones have been observed at Deep Canyon (Third-Day road log, mile 65.95). Only 20 m of pumiceous sandstones are present in the roughly equivalent stratigraphic interval at Saulsberry Ranch (First-Day road log, mile 84.1). A few observations of pebble imbrications in andesitic conglomeratic sandstones at the type section suggest northeasterly paleoflow for the fluvial system that delivered both the andesitic sandstones and the rhyolitic tuffaceous sandstones. The thin-bedded character of tuffaceous sandstones and preferential concentration of pumice in thin, muddy siltstones (drape deposits) suggest that this stream system was periodically oversaturated with pyroclastic debris. These clogged stream systems then formed rapidly-aggraded, sheet-like, tuffaceous deposits shortly after the explosive eruptions. The cliff-forming basal marker bed and similar tuffaceous sandstones higher in the VCDL are classified as syneruptive volcaniclastic sedimentary rocks (Smith, 1991). Andesitic conglomeratic sandstones within the VCDL probably represent medial to distal alluvial-apron deposits derived from a late Eocene volcanic center, to the south near Reserve (e.g. Ratté, 1989).

SANDSTONE OF ESCONDIDO MOUNTAIN

The new and informal name "sandstone of Escondido Mountain" is here proposed for well-sorted, fine- to medium-grained sandstone, as much as 240 m thick (Fig. 2) that paraconformably overlies the VCDL at Cañon del Leon. The base of this pale yellowish-brown sandstone is locally well exposed in the upper reach of Cañon del Leon at an elevation of approximately 2500 m (SE¼ NW¼ sec. 23, T1S, R16W). The sandstone of Escondido Mountain (SEM) is commonly but discontinuously exposed on the upper flanks of Escondido Mountain, for which it is named. The base of the SEM locally exposed on the east rim of Cañon del Leon (cited above) is proposed as the boundary stratotype. Here the well-sorted sandstone is planar bedded, but approximately 150 m up-canyon the medial SEM shows large high-angle cross beds more typical of the main body of this unit. The upper third of the SEM in Cañon del Leon is covered with colluvial blocks of andesite porphyry derived from overlying cliffs of 28.4-Ma Squirrel Springs Canyon Andesite (Chamberlin, et al., 1994; McIntosh and Chamberlin, this volume). The base of these cliff-forming lavas on the north face of Escondido Mountain at an elevation of about 2737 m is mapped as the approximately located top of the SEM (Fig. 1). We suggest the lower north flank of El Caso Peak (sec. 18, T2S, R16W), upper Cañon del Macho (sec. 27, T1S, R15W) and the east flank of Canon del Lolo (sec. 32, T1S, R15W) as partial reference sections for the SEM. All proposed reference sections are on the flanks of Escondido Mountain, which is suggested as the type area for the SEM. Cross-bedded SEM is also locally well exposed in roadcuts of Forest Road 13, NE of Quemado Lake.

Contacts and lithology

Regional mapping suggests that the base of the SEM at Cañon del Leon is a paraconformity. However, regional mapping also implies that the SEM once intertongued with or graded into epiclastic conglomeratic sandstones of the South Crosby Peak Formation between Alegres Mountain and the Crosby Mountains. The base of the SEM is typically poorly exposed and thus its regional character is not well known. Southwest of Escondido Mountain the SEM is generally in fault contact with the middle Spears Group and Datil Group along the Spur Lake fault zone (Chamberlin et al., 1994). The upper contact of the SEM is also poorly exposed in most areas and is generally defined by mafic lavas of the Oligocene Mogollon Group. including the 28.4-Ma Squirrel Springs Canyon Andesite and the 26.1-Ma Bearwallow Mountain Andesite (McIntosh and Chamberlin, this guidebook).

Where well exposed, the SEM typically consists of well-sorted, pale yellowish-brown, planar to high-angle cross-bedded, medium-grained, volcaniclastic sandstone. Pebbles and mudstones typical of underlying fluviatile deposits are normally absent. Cross bed sets as much as 10 m high have locally been observed. The steepest cross beds (25-30°) most commonly face to the east (from northeast to southeast). Weathered outcrops are often littered with small (~1 cm) yellowish-brown sandstone concretions.

Limited petrographic study (Table 1) indicates that SEM sandstones are primarily volcaniclastic, and composed of a mixture of andesitic to rhyolitic lithic fragments and associated phenocryst-derived minerals (plagioclase, sanidine and quartz). Assuming samples from Killion Canyon (Table 1; NE¼ sec. 3, T3S, R14W) are representative of the main body of SEM, the unit generally appears to contain more quartz than the underlying Eocene volcaniclastic units. Minor amounts of recycled or basement-derived microcline are also present in the samples from Killion Canyon. A very low birefringent zeolite (possibly heulandite) appears to be the most common type of cement. Diagenetic calcite may also be present.

Age and correlation

The age of the SEM is established by thin distal ignimbrites that are intercalated with this unit in the Quemado region (Chamberlin et al., 1994; McIntosh and Chamberlin, this volume). The upper SEM is locally interbedded with the 28.1-Ma Bloodgood Canyon Tuff near the head of Cañon del Macho (sec. 27, T1S, R15W) and with the 28.6-Ma Vicks Peak Tuff at Killion Canyon. The lower SEM also contains the 31.6-Ma Caballo Blanco Tuff in the conical hills east of Mangas (sec. 35, T1S, R14W). A possible correlative high-angle, cross-bedded, volcaniclastic sandstone is reported to underlie the 32.1-Ma Hells Mesa Tuff on Sugarloaf Mountain, 14 km southwest of Datil (Lopez and Bornhorst, 1979; Osburn and Chapin, 1983). The Hells Mesa Tuff is the oldest volcanic unit known to overlie cross-bedded SEM-type sandstone; the youngest volcanic unit that typically caps the SEM in the Quemado-Datil region is the Bearwallow Mountain Andesite, which has yielded an $^{40}Ar/^{39}Ar$ isochron age of 26.1 ± 0.1 Ma from a basal lava flow (McIntosh and Chamberlin, this volume). Available age data indicate the SEM is of Oligocene age and ranges from about 32 to 26 Ma.

Epiclastic conglomeratic sandstones in the South Crosby Peak Formation, southwest of Datil are known to occupy the same age range (ca. 32-27 Ma; McIntosh and Chamberlin, this volume) as the SEM near Quemado. Presumably these northwesterly transported (Osburn and Chapin, 1983) South Crosby Peak conglomerates once intertongued with the SEM in the erosional gap between the Crosby Mountains and Alegres Mountain. Jones (1980) mapped correlative eolian sandstones in the upper part of his "sandstone of Alegros Mountain". High-angle, cross-bedded, volcaniclastic sandstones in the Alpine, Arizona region mapped by Wrucke (1961) as the "upper sedimentary formation" are probably correlative to the SEM. Similar cross-bedded sandstones in the upper Rincon Windmill Formation (Osburn and Chapin, 1983) are also probably correlative with the SEM.

Thickness and origin

Based on cross section (Fig. 2), the estimated thickness of the SEM on the north face of Escondido Mountain is 240 m. Wrucke (1961) reported that a high-angle cross-bedded sandstone of probable eolian origin in the Alpine, Arizona area is as much as 360 m thick. Other regional observations suggest that SEM thins to the south (James C. Ratté, oral commun., 1994) and to the east (Lopez and Bornhorst, 1979; Osburn and Chapin, 1983; Chamberlin et al., 1994). The SEM is approximately 120-180 m thick in the Alegres Mountain-Mangas area (Chamberlin et al., 1994).

The common occurrence of east-facing high-angle cross beds, uniform fine- to medium-grain size and absence of fluviatile features (e.g. pebbles) indicates that the SEM is primarily an eolian deposit derived from the distal alluvial apron of the Mogollon-Datil volcanic pile. This volcaniclastic sandstone was probably blown in from the west and northwest off the Colorado Plateau. It generally overlapped the toe of epiclastic alluvial-apron deposits derived from volcanic highlands to the south. The SEM presumably formed as a widespread southeast tapering wedge of dune sands, and most likely represents a dry climatic period on the Colorado Plateau in Oligocene time.

ACKNOWLEDGMENTS

We thank S. M. Cather and J. M. Barker for helpful reviews of this article. The manuscript was typed by Terry Telles. Rebecca Titus and Jesse Dengate made the maps and line drawings. RMC wishes to thank his family, Louise, Robert and Chris for their patience and support during the preparation of this article and this volume.

REFERENCES

Broadhead, R. F. and Chamberlin, R. M., 1994, Hunt Oil Co. No. 1-16 State Well, wireline log responses, Tertiary and Cretaceous sections: New Mexico Geological Society, Guidebook 45.

Cather, S. M. and Johnson, B. D., 1984, Eocene tectonics and depositional setting of west-central New Mexico and eastern Arizona: New Mexico Bureau of Mines and Mineral Resources, Circular 192, 33 p.

Cather, S. M. and McIntosh, W. C., 1994, The Plio-Pleistocene Quemado Formation of west-central New Mexico: New Mexico Geological Society, Guidebook 45.

Cather, S. M., Chamberlin, R. M. and Ratté, J. C., 1994, Tertiary stratigraphy and nomenclature for western New Mexico and eastern Arizona: New Mexico Geological Society, Guidebook 45.

Cather, S. M., McIntosh, W. C. and Chapin, C. E., 1987, Stratigraphy, age and rates of deposition of the Datil Group (Upper Eocene-lower Oligocene), west-central New Mexico: New Mexico Geology, v. 9, p. 50-54.

Chamberlin, R. M., Cather, S. M., Anderson, O. J. and Jones, G. E., 1994, Reconnaissance geologic map of the Quemado 30′x60′ quadrangle, Catron County, New Mexico: New Mexico Bureau of Mines and Mineral Resources, Open-file Report 406, scale 1:100,000.

Fisher, R. V. and Smith, G. A., 1991, Volcanism, tectonics and sedimentation: SEPM Special Publication no. 45, p. 1-5.

Guilinger, D. R., 1982, Geology and uranium potential of the Tejana Mesa-Hubbell Draw area, Catron County, New Mexico: New Mexico Bureau of Mines and Mineral Resources, Open-file Report 176, 137 p.

Jones, D. P., 1980, Volcanic geology of the Alegros Mountain area, Catron County, New Mexico [M.S. thesis]: Albuquerque, University of New Mexico, 76 p.

Lopez, D. A. and Bornhorst, T. J., 1979, Geologic map of the Datil Area, Catron County, New Mexico: U.S. Geological Survey, Miscellaneous Investigations Map I-1098, scale 1:50,000.

McIntosh, W. C. and Chamberlin, R. M., 1994, $^{40}Ar/^{39}Ar$ geochronology of middle to late Cenozoic ignimbrites, mafic lavas, and volcaniclastic rocks in the Quemado-Datil region, New Mexico: New Mexico Geological Society, Guidebook 45.

Osburn, G. R. and Chapin, C. E., 1983, Nomenclature for Cenozoic rocks of the northeast Mogollon-Datil volcanic field, New Mexico: New Mexico Bureau of Mines and Mineral Resources, Stratigraphic Chart 1.

Ratté, J. C., 1989, Geologic map of the Bull Basin quadrangle, Catron County, New Mexico: U.S. Geological Survey, Geologic Quadrangle Map GQ-1651, scale 1:24,000.

Smith, G. A., 1991, Facies sequences and geometries in continental volcaniclastic sediments: SEPM Special Publication no. 45, p. 109-121.

Willard, M. E., 1957, Reconnaissance geologic map of the Piñonville thirty minute quadrangle: New Mexico Bureau of Mines and Mineral Resources, Geologic Map 3; scale 1:126,720.

Willard, M. E. and Weber, R. H., 1958, Reconnaissance geologic map of Cañon Largo thirty-minute quadrangle: New Mexico Bureau of Mines and Mineral Resources, Geologic Map 6; scale 1:126,720.

Witcher, J. C., Hahman, W. R. and Swanberg, C. A., 1994, Alpine 1/Federal—Subsurface stratigraphy of the eastern White Mountains, Apache County, Arizona: New Mexico Geological Society, Guidebook 45.

Wrucke, C. T., 1961, Paleozoic and Cenozoic rocks in the Alpine-Nutrioso area, Apache County, Arizona: U.S. Geological Survey, Bulletin 1121-H, 26 p.

Lower slopes of San Francisco Mountains approximately 17 km west of Reserve, New Mexico. View is N74°E. Layered volcaniclastic conglomerates in foreground (Heifer Basin) are assigned to the Pueblo Creek Formation (upper Eocene to lower Oligocene) of the middle Spears Group. Tularosa Mountains are on the distant horizon east of the Reserve graben. John Kerr Peak, at center on horizon (touching clouds), is an upper Miocene rhyolite lava dome. Camera station is along side Highway 180 in SW ¼ sec. 32, T6S, R20W. Altitude is about 2375 m. Wayne Lambert photograph No. 93L61. August 14, 1993, 12 noon MDT.

MIOCENE PROBOSCIDEAN FROM THE FENCE LAKE FORMATION, CATRON COUNTY, NEW MEXICO

SPENCER G. LUCAS[1] and ORIN J. ANDERSON[2]

[1]New Mexico Museum of Natural History and Science, 1801 Mountain Road N.W., Albuquerque, New Mexico 87104;
[2]New Mexico Bureau of Mines and Mineral Resources, Socorro, New Mexico 87801

Abstract—An incomplete left innominate of a proboscidean was collected from the lower Fence Lake Formation southwest of Quemado in Catron County. This fossil establishes a maximum age for the base of the Fence Lake Formation of 14.5 Ma (beginning of late Barstovian).

INTRODUCTION

McLellan et al. (1982) named the Fence Lake Formation for exposures 3 mi (4.8 km) south of Fence Lake, New Mexico (Fig. 1). They postulated a Miocene age for the unit based on nothing more than prevailing thought and the observation that the source area was the "Oligocene volcanics of the Datil Mountain region." While this was an acceptable age assignment, it did leave open to question the exact age of the basal Fence Lake Formation.

We report here a fossil mammal bone from the lower part of the Fence Lake Formation southwest of Quemado in Catron County (Fig. 1). The junior author and his assistant, Glen E. Jones, discovered this fossil during reconnaissance work in 1992 and it was subsequently collected by Adrian Hunt, Peter Reser and the senior author. The fossil locality is NMMNH (New Mexico Museum of Natural History and Science) locality 2808 located at UTM 3779380N, 716820E, zone 12 (NE¼, sec. 18, T2S, R17W; Ponderosa Tank U.S.G.S. 7.5-minute quadrangle). The fossil bone, NMMNH P-22299, was in a block of grayish brown sandstone of the Fence Lake Formation approximately 1 ft (0.3 m) above its unconformable contact with the underlying Spears Formation (Fig. 2A).

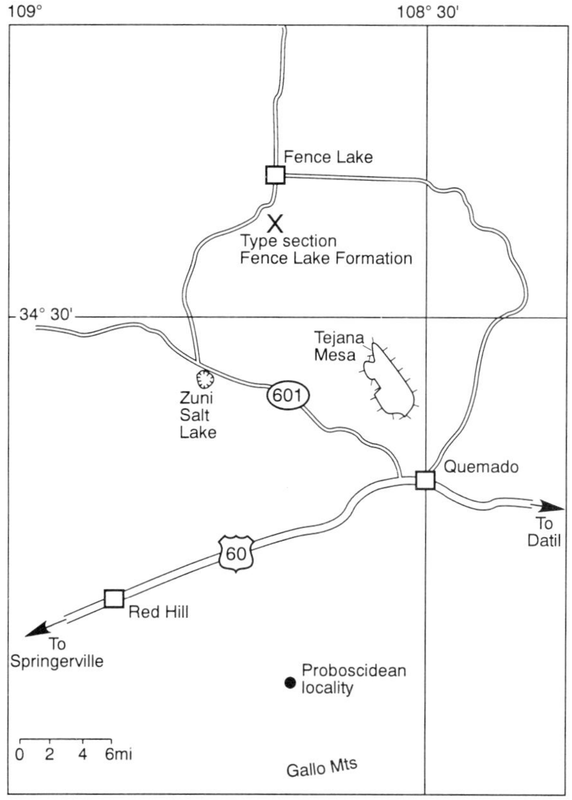

FIGURE 1. Index map of Quemado-Fence Lake area showing proboscidean locality, and location of type section of Fence Lake Formation.

DESCRIPTION AND DISCUSSION

NMMNH P-22299 (Fig. 2B-C) is an incomplete left innominate preserving the acetabulum and the acetabular roots of the ilium, ischium and pubis. The posterior portion of the acetabulum together with the acetabular root of the ischium have been fractured and displaced anteriorly, so as to greatly distort the shape of the preserved portion of the obturator foreman so it is more rectangular than ovoid. The acetabulum is shallow, eroded in the area of the acetabular fossa and has a maximum diameter of 115 mm. The pubic border of the acetabulum has a low blunt ridge that runs dorso-ventrally.

The very large size and morphology of NMMNH P-22299 indicate it belongs to a proboscidean. Its relatively small size, shallow acetabulum, poorly formed acetabular fossa and/or lack of a pronounced tuberosity on the pubic border of the acetabulum preclude assignment of NMMNH P-22299 to either *Mammuthus*, the mammoth, or *Mammut*, the American Mastodon (cf., Olsen, 1979, figs. 24-28). Instead, these features suggest assignment to an older, more primitive proboscidean, a gomphothere (Osborn, 1936). Full fusion of all pelvic bones suggests that NMMNH P-22299 is an adult individual, so it is not a juvenile of one of the larger, post-Miocene proboscideans.

The oldest record of proboscideans in North America is the time of their immigration from Eurasia at the beginning of the late Barstovian, about 14.5 Ma (Tedford et al., 1987). NMMNH P-22299 thus indicates the Fence Lake Formation outcrop from which it was collected is no older than 14.5 Ma. The minimum age for the Fence Lake Formation of 6.9 Ma established by the basalt flow at Tejana Mesa is not contradicted by the gomphothere fossil reported here. Indeed, this fossil cannot be any younger than late Miocene or Pliocene. It thus seems that the gomphothere fossil reported here, besides being the first fossil described from the Fence Lake Formation, refines estimates of the initiation of Fence Lake deposition to no older than middle Miocene, 14.5 Ma.

ACKNOWLEDGMENTS

Glen Jones, Adrian Hunt and Pete Reser provided field assistance. D. Love and R. Chamberlin reviewed an earlier version of this article.

REFERENCES

Dethier, D. P., Aldrich, M. J., Jr. and Shafiqullah, M., 1986, New K-Ar ages for Miocene volcanic rocks from the northeastern Jemez Mountains and Tejana Mesa, New Mexico: Isochron/West, no. 47, p. 12-14.

McLellan, M., Robinson, L., Haschke, L., Carter, M. D. and Medlen, A., 1982, Fence Lake Formation (Tertiary), west-central New Mexico: New Mexico Geology, v. 4, p. 53-55.

Olsen, S. J., 1979, Osteology for the archaeologist. No. 3. The American mastodon and the woolly mammoth: Papers of the Peabody Museum of Archaeology and Ethnology, Harvard University, v. 56, no. 3, 47 p.

Osborn, H. F., 1936, Proboscidea Vol. I: Moeritherioidea, Deinotherioidea and Mastodontoidea: New York, American Museum of Natural History, 802 p.

Tedford, R. H., Skinner, M. F., Fields, R. W., Rensberger, J. M., Whistler, D. P., Galusha, T., Taylor, B. E., Macdonald, J. R. and Webb, S. D., 1987, Faunal succession and biochronology of the Arikareean through Hemphillian interval (late Oligocene through earliest Pliocene epochs) in North America; *in* Woodburne, M. O., ed., Cenozoic mammals of North American geochronology and biostratigraphy: Berkeley, University of California Press, p. 153-210.

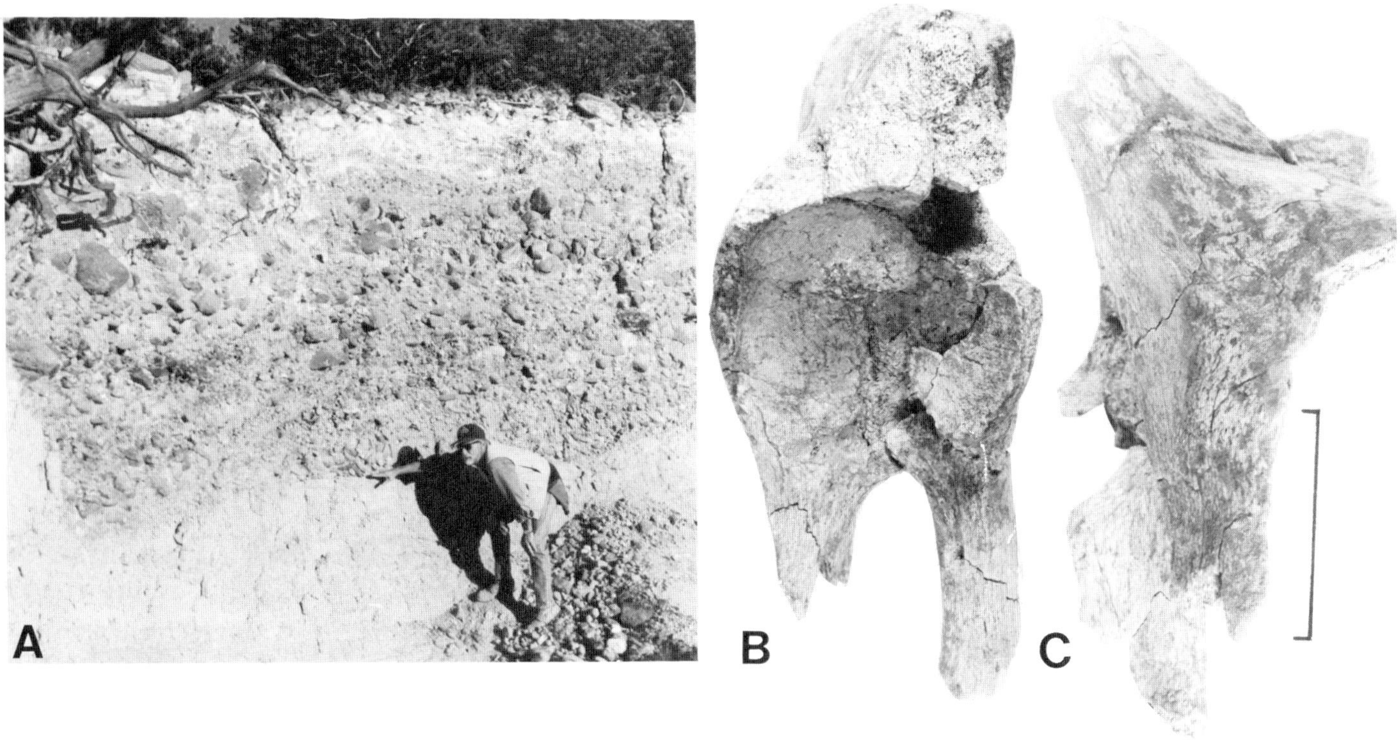

FIGURE 2. A, Fence Lake Formation boulder conglomerate overlying Spears Formation (Man's hand marks contact) at NMMNH locality 2808. B-C, NMMNH P-22299, incomplete left innominate of a gomphothere, ventral (B) and dorsal (C) views. Bar scale = 10 cm.

THE PLIO-PLEISTOCENE QUEMADO FORMATION OF WEST-CENTRAL NEW MEXICO

STEVEN M. CATHER and WILLIAM C. MCINTOSH
New Mexico Bureau of Mines and Mineral Resources, Socorro, New Mexico 87801

Abstract—The Plio-Pleistocene Quemado Formation (new name) consists of 29.5 m of volcaniclastic and siliciclastic sandstone and conglomerate at its type section in northwestern Catron County, New Mexico. The Quemado ranges in age from ~4 to <1 Ma, and consists of fluvial deposits that accumulated in several basin/paleovalley systems in upper reaches of tributaries to the ancestral Little Colorado River. The Quemado Formation is, at least in part, correlative with the upper part of the Bidahochi Formation.

INTRODUCTION

Reconnaissance geologic mapping of the Quemado 30′x 60′ quadrangle (Chamberlin et al., 1994) has led to improved understanding of the stratigraphy, structure and geomorphologic evolution of a large area of west-central New Mexico. Numerous $^{40}Ar/^{39}Ar$ ages from volcanic units within the map area (McIntosh and Chamberlin, McIntosh and Cather, this volume) have created a geochronologic framework for the area. Based on the geologic and geochronologic relations, it is apparent that this area of west-central New Mexico has experienced a regime of general fluvial degradation since the late Oligocene, interrupted by at least two periods of fluvial aggradation (McIntosh and Cather, this volume). The earliest episode of aggradation is represented by the Fence Lake Formation (Marr, 1956; McLellan et al., 1982), which accumulated <14.5 Ma to ~7 Ma. The Fence Lake Formation and associated late Miocene basalts cap the high mesas in the vicinity of US-60 in western New Mexico (e.g. Tejana Mesa, Largo Mesa, Cimarron Mesa, Mesa Redonda, and numerous unnamed mesas, Fig. 1).

The younger aggradational sequence (mapped as Quaternary alluvium by Willard [1957] and Willard and Weber [1958]) is topographically lower than, and inset against, the Fence Lake Formation that typically caps interfluves between the younger aggradational basins. The younger aggradational deposits range in age from ~4 Ma to <1 Ma, and accumulated in a series of erosional basins or paleovalleys in the upper reaches of the tributaries to the ancestral Little Colorado River. We propose the new name Quemado Formation for these Plio-Pleistocene deposits.

The Quemado Formation, which is nowhere exposed in its entirety, crops out discontinuously within at least four basins in west-central New Mexico (Fig. 1). These are, from east to west, the Omega Basin, the Largo Basin, the Red Hill Basin and the Cow Springs Basin. Thin gravelly sequences beneath early Pliocene basalts in eastern Arizona (McIntosh and Cather, this volume; see also Day Two road log) may represent remnants of deposits within a fifth basin in eastern Arizona (Nutrioso Basin), although correlation of these deposits is uncertain and will require further work.

LITHOLOGY AND TYPE SECTION

The type section of the Quemado Formation is designated in W/2 sec. 4, T1S, R21W, Catron County, New Mexico. The type section (Fig. 2) is exposed in roadcuts where an unimproved road ascends a

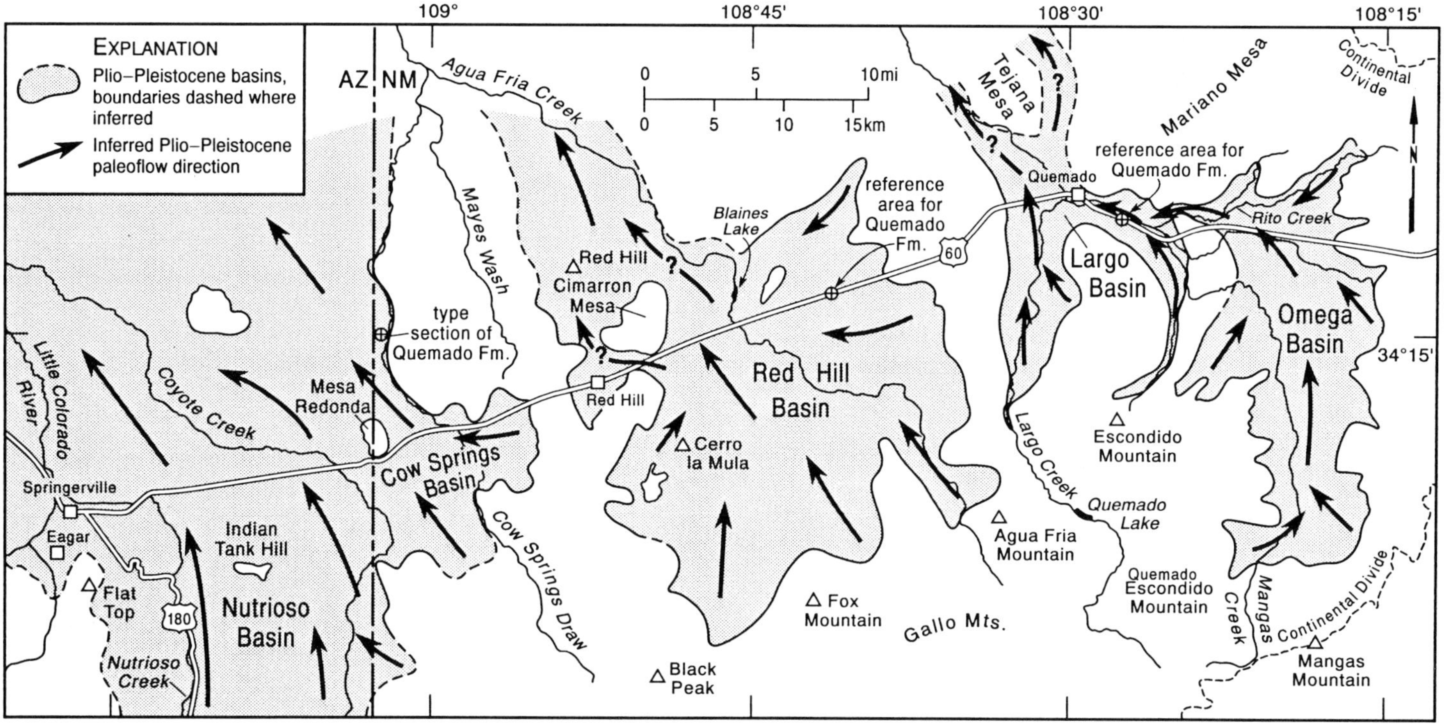

FIGURE 1. Map showing basins that contain sediments of Quemado Formation, inferred paleoflow directions, selected geomorphic features, and locations of type section and reference areas of Quemado Formation.

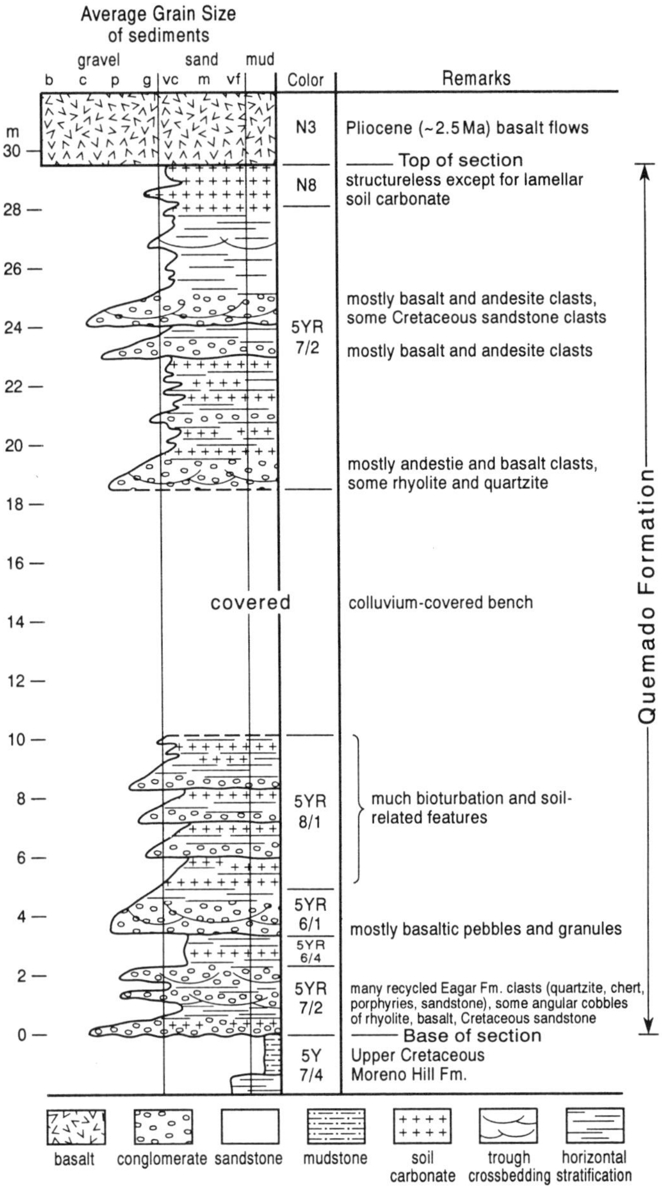

FIGURE 2. Type section of the Quemado Formation in W/2 sec. 4, T1S, R21W, Catron County, New Mexico. Section was measured with Jacob staff and Brunton compass by S. M. Cather.

basalt-capped mesa about 9 km north of US-60 and 0.5 km east of the New Mexico-Arizona state line.

At the type section the Quemado Formation consists of 29.5 m of mostly light brown to buff sandstone and conglomerate. The unit contains a mixture of volcanic detritus reworked from the Fence Lake Formation and Spears and Datil Groups as well as siliciclastic materials reworked from Upper Cretaceous sandstones and from Eocene Baca/Eagar strata. Cements appear to be largely calcite. The base of the Quemado is conglomeratic and rests unconformably on grayish-yellow mudstone and sandstone of the Upper Cretaceous Moreno Hill Formation. The middle part of the Quemado Formation is not exposed at the type section due to colluvial cover. The upper part of the Quemado is disconformably(?) overlain by late Pliocene (~2.5 Ma) basalt. Calcareous paleosols are common in the type section.

The lithology of the Quemado Formation at the type section reflects its location near the margin of the Cow Springs Basin. The proximal deposits at the type section are considerably coarser than is typical in the centers of most Quemado-age basins, and paleosols are typically more abundant than in distal, basin-center deposits. In order to represent the range of Quemado lithologies present in the region, we designate two additional reference areas, one each in the central parts of the Largo and Red Hill Basins. These reference areas consist of roadcut exposures of the Quemado which reflect the typical lithology (dominantly sandstone and siltstone) of the basin-center areas. The reference areas are located along US-60 3.4 km east and 19.9 km west, respectively, from the Quemado, New Mexico, post office (see Fig. 1 and Day Two road log).

AGE AND CORRELATION

At the type section in the Cow Springs Basin, the Quemado is inset against an unnamed mesa that is capped by basalts ranging in age from 6.03 Ma to 5.20 Ma (McIntosh and Cather, this volume). The basalt that caps the type section is one of a series of related flows that cap both the Quemado Formation to the south near US-60 (there dated 2.46 ± 0.04 Ma) and the Bidahochi Formation in eastern Arizona north of Mesa Parada (there dated 2.37 ± 0.03 Ma). These relationships argue for an age range for the Quemado type section of <5.2 to ~2.5 Ma, and suggest that the Quemado is, at least in part, temporally equivalent to the upper Bidahochi Formation.

West of the type section, in the Nutrioso Basin, possible Quemado-equivalent gravels are younger than and apparently inset against a 6.80-Ma basalt that caps Flat Top near Springerville, Arizona. These gravels are capped by 3.87-3.67-Ma basalts (see fig. 5 of McIntosh and Cather, this volume). East of the type section, in the Red Hill Basin, the Quemado Formation is inset against a 5.20-Ma basalt and interfingers with basalts as young as 0.970 Ma. In the Largo Basin, the Quemado can only be constrained to be younger than the 6.73-Ma basalt that caps Tejana Mesa (Dethier et al., 1986). Quantitative constraints on the age of the Quemado Formation in the Omega Basin are lacking.

The Quemado Formation is lithologically similar to parts of the upper Bidahochi Formation (Reagan, 1924, 1932; Repenning and Irwin, 1954; Love, 1989). The Quemado is also, at least in part, temporally equivalent to the upper Bidahochi (Love, 1989), although the upper part of the Quemado in the Red Hill Basin apparently post-dates the Bidahochi. It is possible that both the Quemado Formation and the Fence Lake Formation may represent proximal correlatives of the Bidahochi and that the inset nature of the Quemado in proximal areas may give way to normal stratigraphic superposition in the type area of the Bidahochi to the northwest. Further research will be necessary to evaluate this hypothesis. The Quemado Formation is temporally equivalent to parts of the upper Santa Fe Group to the east and upper Gila Group to the south, although it is clear that the paleodrainage net within the Quemado basins was not interconnective with those of the Gila or Santa Fe.

THICKNESS AND GENESIS

In the basins depicted in Figure 1, the Quemado ranges in thickness from a feather edge to as much as about 60 m in the Omega Basin. The Quemado is anomalously thick in the Omega Basin due to the effects of subsidence along the Hickman fault zone (Chamberlin et al., 1994). Other basins to the west are largely erosional paleovalleys which probably contain substantially thinner (<50 m?) accumulations of Quemado sediments, although thickness constraints are lacking in many areas.

The Quemado Formation appears to be entirely alluvial in origin. Although no paleocurrent data were collected, basin configurations indicate paleoflow was generally northerly (Fig. 1). The Quemado was deposited within paleovalley/basin systems that developed in the upper reaches of tributaries to the ancestral Little Colorado River. The paleovalley/basin systems depicted in Figure 1 become substantially narrower where they cross a series of late Miocene eruptive centers that compose a narrow, linear zone through Mesa Redonda, Cimarron Mesa and Tejana Mesa. The basalts that erupted from these vents just prior to incision and filling of the Quemado paleovalleys were relatively resistant to erosion and thus constricted the width of the paleovalleys where they crossed the late Miocene vent zone. South of the vent zone, the Quemado paleovalleys became broader where they were incised into the poorly indurated sediments of the Spears Group.

ACKNOWLEDGMENTS

The paper was improved by reviews from Matt Heizler and Dave Love. The manuscript was typed by Terry Telles and the figures were drafted by Becky Titus.

REFERENCES

Dethier, D. P., Aldrich, M. J., Jr. and Shafiqullah, M., 1986, New K-Ar ages for Miocene volcanic rocks from the northeastern Jemez Mountains and Tejana Mesa, New Mexico: Isochron/West, no. 47, p. 12-14.

Love, D. W., 1989, Bidahochi Formation: an interpretive summary: New Mexico Geological Society, Guidebook 40, p. 273-280.

Marr, R. J., 1956, Geology of the Lynch Ranches, Catron and Valencia Counties, New Mexico [M.A. thesis]: Austin, The University of Texas.

McIntosh, W. C. and Cather, S. M., $^{40}Ar/^{39}Ar$ geochronology of basaltic rocks and constraints on late Cenozoic stratigraphy and landscape development in the Red Hill-Quemado area, New Mexico: New Mexico Geological Society, Guidebook 45.

McIntosh, W. C. and Chamberlin, R. M., 1994, $^{40}Ar/^{39}Ar$ geochronology of middle to late Cenozoic ignimbrites, mafic lavas, and volcaniclastic rocks in the Quemado-Datil region, New Mexico: New Mexico Geological Society, Guidebook 45.

McLellan, M., Robinson, L., Haschke, L., Carter, M. D. and Medlin, A., 1982, Fence Lake Formation (Tertiary), west-central New Mexico: New Mexico Geology, v. 4, p. 53-55.

Reagan, A. B., 1924, Stratigraphy of the Hopi Buttes volcanic field, Arizona: Pan-American Geologist, v. 41, p. 355-366.

Reagan, A. B., 1932, The Tertiary-Pleistocene of the Navajo country in Arizona, with a description of some of its fossils: Kansas Academy of Science Transactions, v. 35, p. 253-259.

Repenning, C. A. and Irwin, J. H., 1954, Bidahochi Formation of Arizona and New Mexico: American Association of Petroleum Geologists Bulletin, v. 38, p. 1821-1826.

Willard, M. E., 1957, Reconnaissance geologic map of the Piñonville thirty-minute quadrangle: New Mexico Bureau of Mines and Mineral Resources, Geologic Map 3, scale 1:126,720.

Willard, M. E. and Weber, R. H., 1958, Reconnaissance geologic map of Cañon Largo thirty-minute quadrangle: New Mexico Bureau of Mines and Mineral Resources, Geologic Map 6, scale 1:126,720.

Day 2 Stop 4

COAL GEOLOGY OF THE LOWER MORENO HILL FORMATION, SALT LAKE FIELD, WEST-CENTRAL, NEW MEXICO

GRETCHEN K. HOFFMAN
New Mexico Bureau of Mines and Mineral Resources, Socorro, NM 87801

Abstract—The Salt Lake coal field in west–central New Mexico is defined by the arcuate outcrops of the Moreno Hill Formation of Late Cretaceous, Turonian age. This formation is divided into three informal units, the lower, middle and upper members. The lower member contains the Antelope, Cerro Prieto and Rabbit informal coal zones, in ascending order. The Antelope zone, just above the underlying marine Atarque Sandstone, has two relatively thin, high sulfur (1.2% avg.) coals in a fine-grained sequence deposited in a back–barrier swamp environment. As many as five coal beds are in the medial Cerro Prieto zone, although only two of these coastal–plain coals have lateral continuity. These coals range from 1 to 14 ft thick and often contain persistent claystone (tonstein) partings. The low sulfur (0.7%) coastal–plain Cerro Prieto coals developed landward of the marine Fite Ranch Member (Tres Hermanos Formation) pinchout and represent the majority of the economic coal resource in the Salt Lake field. The Rabbit zone contains up to four coal beds of fluvial swamp origin. The second bed above the base attains economic thickness (7.5 ft). All of the Moreno Hill lower member coals are relatively high in ash (18%) and combustion analyses indicate an average apparent rank of subbituminous C, although vitrinite reflectance data show a slightly higher rank, high volatile bituminous C. These coals tend to have a higher percentage of inertinite macerals than other coals from northwestern New Mexico, probably a result of greater oxidation or charring of the swamps before burial.

INTRODUCTION

The Salt Lake coal field covers 750 mi^2 in west–central New Mexico. It is separated from the Zuni coal field by the Atarque monocline, which trends southeast near Ojo Caliente, and by a tongue of the Bandera lava flows. Outcrops of the coal-bearing Moreno Hill Formation (McLellan et al., 1983) define the Salt Lake field, forming a west-facing arcuate belt centered around the Zuni Salt Lake and are traversed by NM-601 and NM-36 (Fig. 1). The field extends from the western border of New Mexico east to R15W and from T1N to T8N. The coal-bearing Moreno Hill strata are relatively flat lying, with dips of 3–5° to the southeast, have undergone minor displacement along faults, and show minor flexures in different parts of the field due to Tertiary volcanism (Campbell, 1989).

STRATIGRAPHY

General Features

The Moreno Hill Formation was named by McLellan et al. (1983) in the northwestern Salt Lake coal field. It consists of three informal members of continental origin, which overlie the marine Atarque Sandstone deposited during the first major retreat of the Late Cretaceous shoreline (R-1; Molenaar, 1983; Fig. 2) from the San Juan Basin, northwest New Mexico. Molenaar (1983) placed a pivot point near Gallup for this oldest, regressive Late Cretaceous shoreline (Fig. 2). The shoreline trend northwest of this point was N60°W and N30°W southeast. The pivot point appears to be a result of increased sediment supply to the southeast (Anderson, personal commun., 1993), resulting in braided stream deposits and short-lived flood-plain coal deposits. The Moreno Hill Formation at the type section (sec. 7 T4N R18W) is 520 ft thick and laterally equivalent to the Tres Hermanos, Gallup and Crevasse Canyon formations in the southwestern San Juan Basin (Fig. 3; Hook et al., 1983).

The lower member of the Moreno Hill consists of channel sandstones and crevasse splays that grade laterally into siltstone, mudstone and coal. Sandstone in this unit tends to be fine grained, well sorted, well rounded, predominantly quartzose, and is generally cemented by siliceous material. Carbonaceous mudstone containing coalified plant fragments is common. The lower member varies in thickness from 100 to 400 ft, thinning towards the west. Roybal (now Hoffman) and Campbell (1981) recognized the Antelope, Cerro Prieto, and Rabbit coal zones in the lower member of the Moreno Hill (Fig. 4). The lower member of the Moreno Hill is considered in part equivalent to the Carthage and Fite Ranch Members of the Tres Hermanos Formation (Anderson and Stricker, 1987).

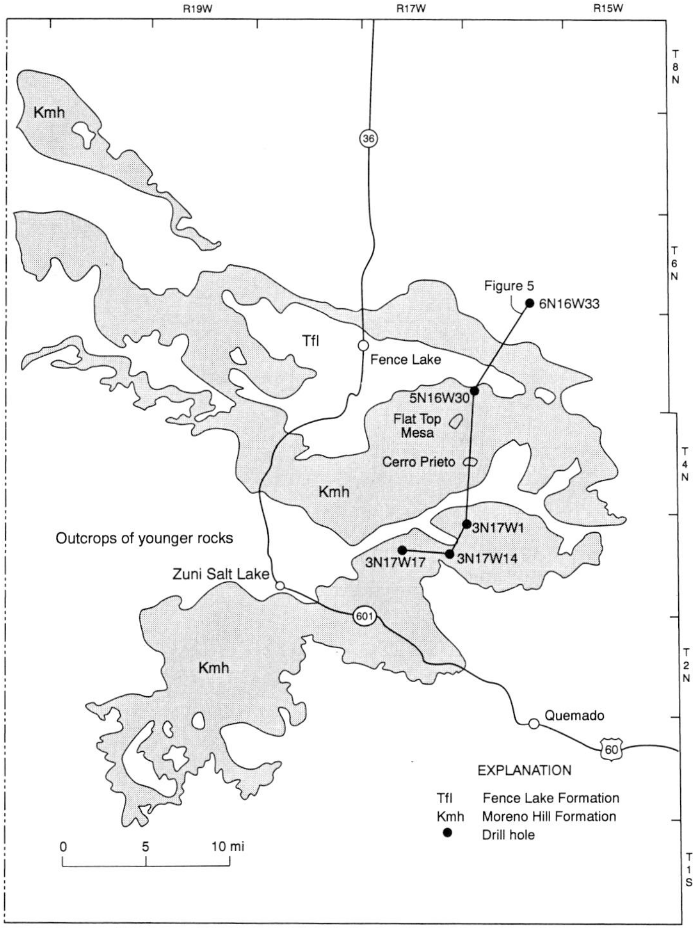

FIGURE 1. Salt Lake coal field, Moreno Hill Formation with drill site locations from NMRDI coal quality study. Kmh= Moreno Hill Formation, Tfl= Fence Lake Formation. Base map modified from Dane and Bachman, 1965.

The middle member of the Moreno Hill is 60 ft of medium– to coarse-grained, poorly–sorted, subangular to subrounded, grain-

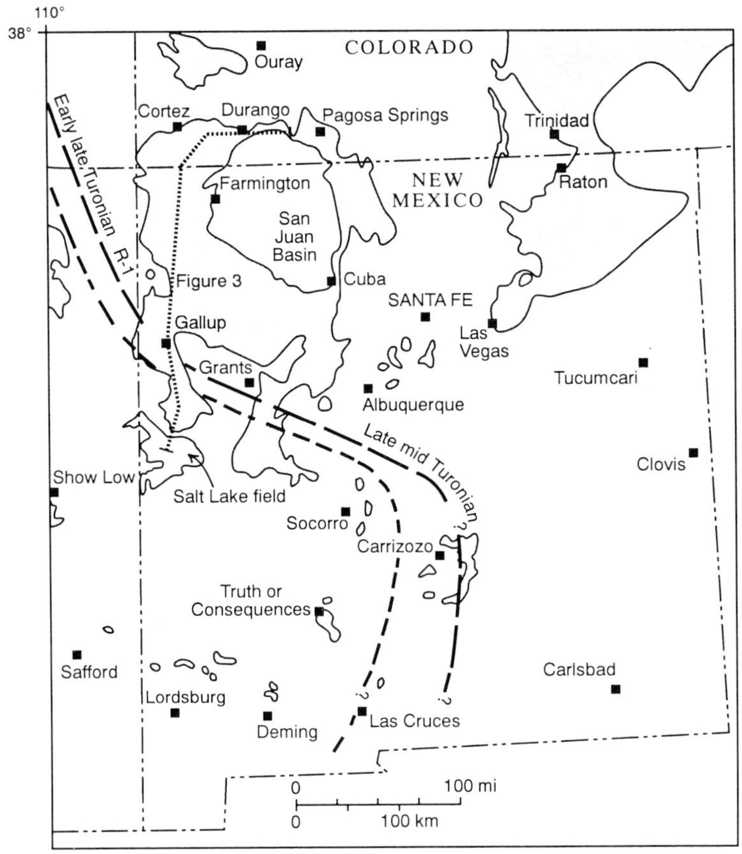

FIGURE 2. Restored position of the Turonian shoreline during maximum regression (R-1). Line of short dashes is seaward extent of subaerial–plain deposits of the regression (Moreno Hill and Tres Hermanos formations; see Fig. 3). Line of long dashes is seaward extent of shoreface sandstones of the regression (Atarque Sandstone). Modified from Molenaar (1983).

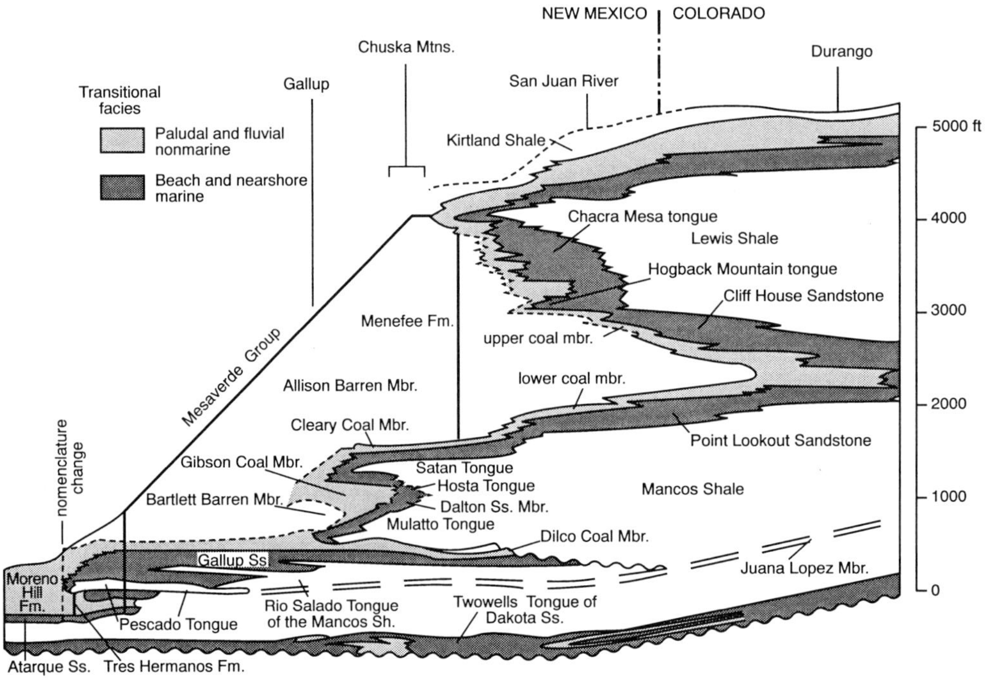

FIGURE 3. Stratigraphic diagram showing sequence, thickness and nomenclature of Cretaceous rocks in the San Juan Basin and Salt Lake coal field, New Mexico and Colorado. Modified from Beaumont (1982, fig. 2).

supported quartzose sandstone with a significant percentage of feldspars. Iron-stained concretions and a pinkish–orange color are common features of the cross-bedded middle member sandstone exposures. This unit represents a braided stream environment with coalescing channel sandstones that filled scours in the lower member. The middle member is equivalent to the Torrivo Member of Gallup Sandstone in the San Juan Basin (Campbell, 1981).

The upper member of the Moreno Hill has a similar lithologic sequence to the lower member, although this unit has a greater percentage of silty sandstone, siltstone and claystone. The total thickness of this unit varies from 30 to 250 ft. A few channel sandstones are within the first 100 ft and the Twilight coal zone (Fig. 4; Roybal and Campbell, 1981) is within this interval. The upper member was deposited in a low energy continental environment and is equivalent to the lower Crevasse Canyon Formation in the San Juan Basin.

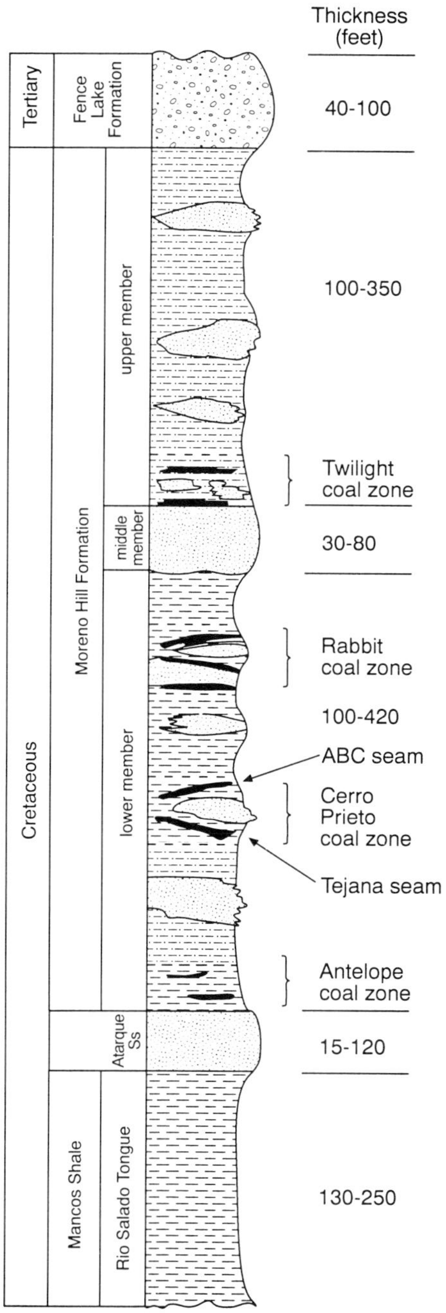

FIGURE 4. Generalized stratigraphic column of the Moreno Hill Formation, Salt Lake coal field with coal zones. Modified from Campbell (1989).

Coal zones of the lower member, Moreno Hill Formation

The Antelope zone described by Campbell (1987, 1989) is within the 50 ft interval above the Atarque Sandstone. A maximum of two relatively thin coals (1 to 3 ft) are in this zone in the southeastern part of the Salt Lake field.

Campbell (1989) characterized the Cerro Prieto zone as being approximately 150 ft below the middle member of the Moreno Hill Formation. In general, the Cerro Prieto zone contains some of the thickest coals in the Salt Lake field, ranging from 1 to 14 ft thick. A major coal bed, the Double T of Campbell (1987, 1989), in this zone is 6–14 ft thick in the vicinity of Cerro Prieto (Fig. 1), a Tertiary volcanic neck. The Double T has two laterally continuous kaolinitic partings, tonsteins, possibly of volcanic origin (0.5 and 0.25 in. thick, lower and upper, respectively). In some areas this bed splits into two coal beds with up to 20 ft of mudstone and sandstone between them. The upper split of the Double T is the A seam and the lower split is the BC seam (Fig. 4) of the Salt River Project (SRP; Ken Rodgers, personal commun., 1994). Three other thin coal beds are in the Cerro Prieto zone, although these tend to be discontinuous, including the Tejana seam named by the Salt River Project (SRP).

The uppermost coal zone of the lower member is the Rabbit zone, approximately 60 ft below the base of the middle member. A maximum of four beds are within this zone; the average total combined thickness is 4 ft and the average individual bed thickness is 3 ft. The second bed from the bottom of this zone has a kaolinitic parting and can be up to 7.5 ft thick (Campbell, 1989).

RECENT STUDIES

The New Mexico Bureau of Mines and Mineral Resources (NMBMMR) and the U.S. Geological Survey (USGS) mapped most of the Salt Lake coal field during the 1980s, studying the geology and coal resources of the area. Much of this work is summarized in a 1:100,000 geologic map (Anderson, 1986) and 1:50,000 geologic maps by Anderson (1987) and Campbell (1989). The coal geology of the Salt Lake and adjacent Zuni Basin were discussed in articles by Campbell (1987) and Anderson and Stricker (1987).

As part of a 1988 coal quality study at the NMBMMR, funded in part by the New Mexico Research and Development Institute (NMRDI), five drill sites were completed in the Salt Lake field along a southwest-northeast trend (Fig. 1, Fig. 5), perpendicular to the general shoreline trend during the Late Cretaceous. Drill sites were at the base or just above exposures of the middle member of the Moreno Hill Formation and most were completed in the Rio Salado Tongue of the Mancos Shale, below the marine Atarque Sandstone. A total of 220 to 300 ft of the lower member of the Moreno Hill was penetrated in these drill holes (Fig. 5).

All of the drill sites had coal beds greater than 1.5 ft thick in the lower member. Eleven coal core samples were taken from the five sites. At these locations, two to five lower member coal beds are within 260 ft of the surface. A majority of the coals in this sequence are thin, varying in thickness from 1.2 to 2.5 ft in the drilled sections. Due to the scale, these coals do not appear as individual beds on the cross section (Fig. 5). Lower member Moreno Hill coals have moderate banding and abundant vitrain, a lithotype in coal derived from woody tissue that is vitreous, and has high luster, some pyrite, and displays good cleat. The thicker coals (5.4 and 7.4 ft) often contain a thin kaolinitic parting or tonstein, characteristic of the Double T seam (Campbell, 1987, 1989).

The greatest frequency of coal in the drilled sections is 50 to 180 ft above the marine Atarque Sandstone. Thin coals just above the Atarque Sandstone are in the Antelope zone (Campbell, 1987, 1989). The thickest coals (5.4 and 7.4 ft) are 100 to 170 ft above the base of the lower member of the Moreno Hill Formation in the two holes near the geographic center of the field (3N17W1, 5N16W30). These thicker coal beds are assigned to the second coal zone above the Atarque Sandstone, the Cerro Prieto zone (3N17W14, 3N17W1, 5N16W30; Campbell, 1989). Some of the thicker coal beds are the lower split of the Double T seam. Cerro Prieto zone coals attain significant thickness (avg. 4.2 ft, max. 13.9 ft) and extent (12 to 15 mi) and have economic

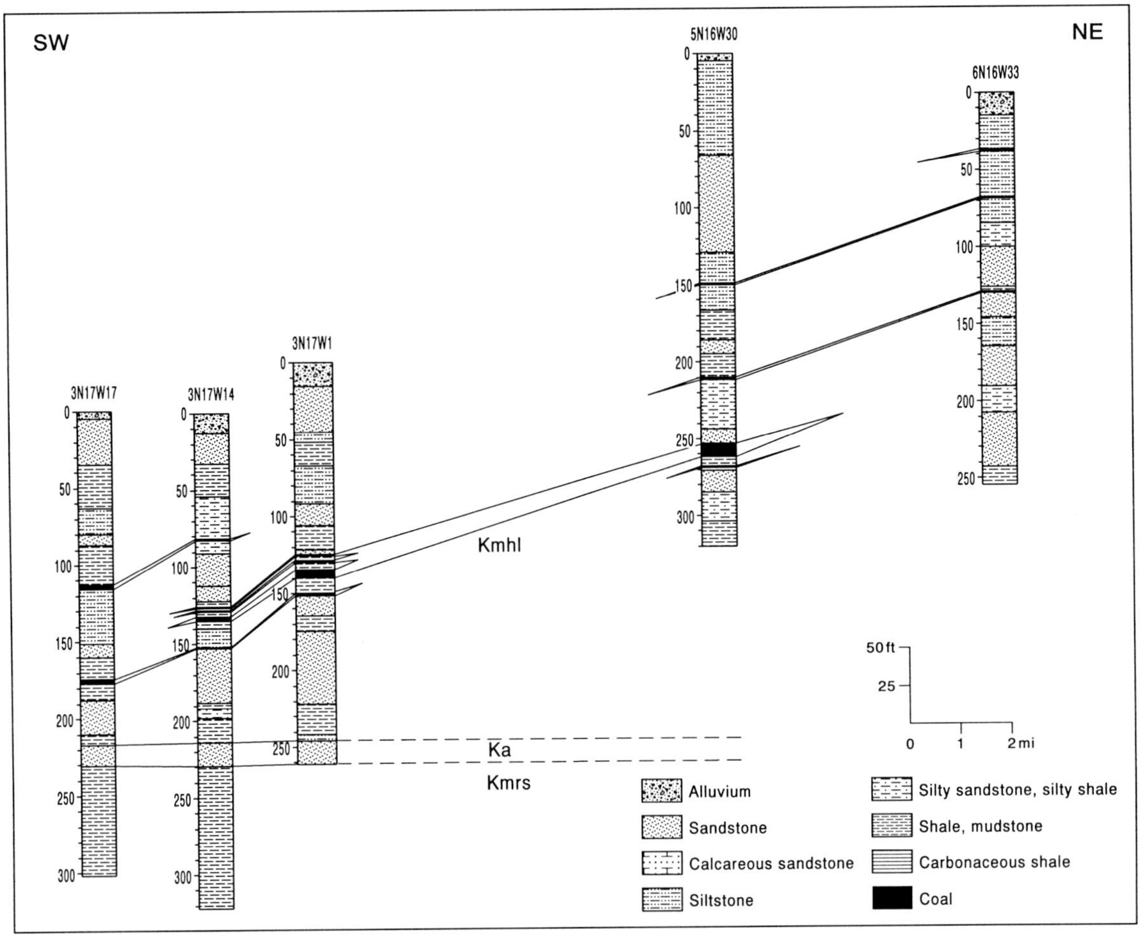

FIGURE 5. Correlation of drill holes in the Salt Lake field from NMRDI coal quality project. Kmrs=Rio Salado Tongue of the Mancos Shale; Ka=Atarque Sandstone; Kmhl=lower member of the Moreno Hill Formation. Modified from Hoffman, Campbell, and Beaumont (1993).

significance in this field. The Rabbit zone coals in the upper section of the lower Moreno Hill Formation are thinner (avg. 2.4 ft, max. 7.5 ft; Campbell, 1987).

Coal beds in the Antelope and Cerro Prieto zones are associated with shales. The Antelope zone coals, within close proximity of the marine Atarque Sandstone, are back barrier swamp deposits (Anderson and Stricker, 1987), and the coals in the Cerro Prieto zone are upper coastal plain deposits. However, the coals higher in the section (Rabbit zone) are commonly associated with siltstone, and the entire sequence becomes more siltstone- and sandstone-dominated upward, indicating greater influx of sediment. The Rabbit zone coals are fluvial in origin (Campbell, 1987). Comparison of the Cerro Prieto and Rabbit zone

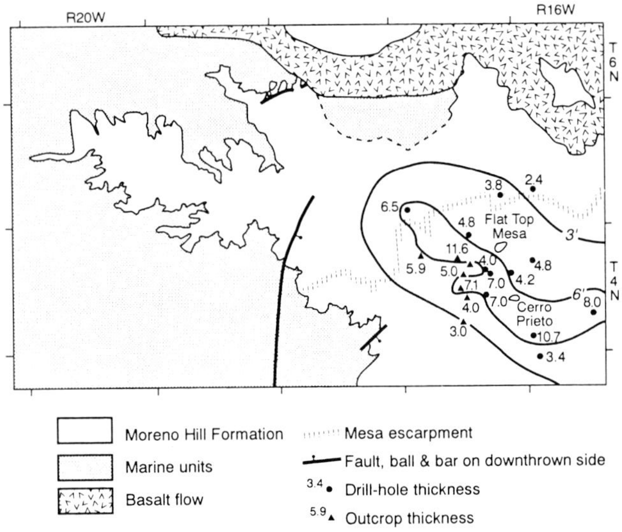

FIGURE 6. Isopach of total coal thickness, Cerro Prieto zone, eastern Salt Lake coal field (from Campbell, 1989).

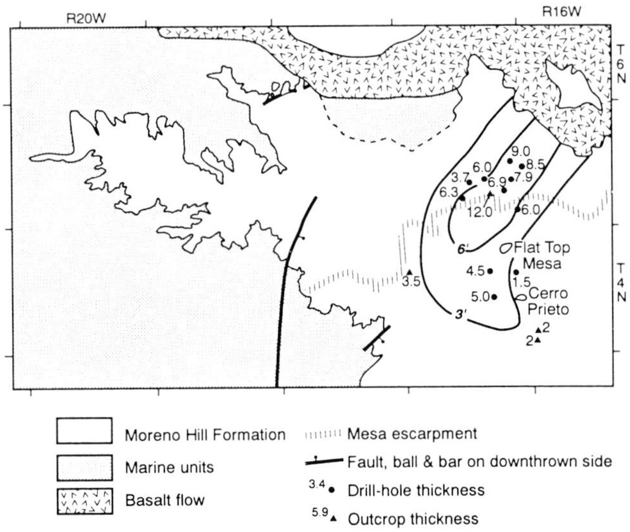

FIGURE 7. Isopach of total coal thickness, Rabbit zone, eastern Salt Lake coal field (from Campbell, 1989).

total coal isopachs (Figs. 6, 7) show a change from a northwest trend to a northeast trend, indicating a change from coastal plain swamps that tend to parallel the shoreline to fluvial–dominated swamps that parallel stream drainage patterns. Anderson and Stricker (1987) also noted the trend of the Cerro Prieto coal isopach and the relatively thick coals in this zone. They related these factors to the close proximity (6 to 8 mi) of the landward pinchout of the laterally equivalent Fite Ranch Member of the Tres Hermanos Formation to the northeast. It is common for thicker coals to have developed landward of marine sandstone pinchouts that represent reversals in the direction of shoreline migration; nearly stable conditions are created in coastal swamps during these periods of reversal. On a smaller scale, the Cerro Prieto coal zone is similar to the Gibson-Cleary coal members of the Menefee-Crevasse Canyon formations in the Gallup area (Fig. 3), southwest of the Point Lookout Sandstone pinchout.

COMPARISON OF ANALYTICAL DATA

The three major coal-bearing zones (Antelope, Cerro Prieto and Rabbit) were sampled during the NMRDI coal quality study. The population sizes were insufficient for comparison by coal zones. Table 1 is a statistical summary of the weighted average analyses (as–received basis) of the Salt Lake field coals, comparing the NMRDI coal quality study samples to previous studies samples (see footnote, Table 1). The sample analyses were weighted by the total individual bed thickness and all analyses in the comparison are from drill cores.

The weighted–average t-test comparison of these two data sets show statistical differences in the hydrogen and sulfur values, except for the sulfate, a consequence of oxidation. A progression from higher to lower sulfur with increased stratigraphic distance above the Atarque Sandstone is typical in both data sets, but a majority of the analyses (96%) in the data set from other studies are from the Cerro Prieto and Rabbit coal zones, resulting in a lower average sulfur value. Diversity in the sulfur values may result from 40% of the NMRDI study samples taken from the Antelope coals and lower seams in the Cerro Prieto zone. The Antelope coals of back barrier swamp origin were influenced by seawater invasion during shoreline oscillations or major storms and consequently tend to have greater percentage of total sulfur (avg. of all analyses 1.20%) than the Cerro Prieto and Rabbit zones coals (avg. of all analyses 0.85%), deposited in coastal–plain

TABLE 1. Weighted average analyses of the Salt Lake field, Moreno Hill Formation coals*.

	NMRDI data			Other studies[1]			t-test values	
	No. of samples	Mean	Std. dev.	No. of samples	Mean	Std. dev.	t-values	Degrees of freedom
Eq. moisture	5	14.72	6.64	0				
Moisture	5	14.56	0.36	53	14.73	3.34		
Ash	5	15.94	1.71	53	17.18	4.17		
Volatile matter	5	33.53	1.32	53	31.50	2.65		
Fixed carbon	5	35.96	1.19	53	36.21	3.88		
Calorific value	5	9070	252	51	9175	865		
Carbon	5	55.04	1.25	43	50.92	5.85		
Hydrogen	5	4.29	0.12	43	3.70	0.47		
Nitrogen	5	1.05	0.03	43	0.93	0.17		
Oxygen	5	7.95	0.16	43	9.73	1.98		
Total sulfur	5	1.17	0.25	22	0.68	0.13	6.39	25
Sulfide	5	0.51	0.28	22	0.15	0.13	4.47	25
Sulfate	5	0.01	0.01	22	0.06	0.06		
Organic sulfur	5	0.65	0.09	22	0.47	0.10	3.71	25
Ashed oxides								
SiO_2	4	56.83	4.59	29	56.84	2.79		
Al_2O_3	4	25.33	2.42	29	28.66	3.71		
TiO_2	4	1.62	0.20	29	1.44	0.26		
Fe_2O_3	4	7.43	3.73	29	3.66	0.90	4.89	31
MgO	4	1.09	0.31	29	0.58	0.19	4.48	31
CaO	4	4.41	0.86	29	4.08	1.24		
K_2O	4	0.43	0.05	29	0.19	0.18		
Na_2O	4	0.77	0.38	29	0.54	0.43		
Total oxides		97.91			95.99			

*Analyses on as-received basis. All values in percent except calorific value (Btu/lb).
[1]Sources of data from other studies: Campbell, 1981; Roybal and Campbell, 1981; and confidential data.

and alluvial–plain environments. Overall, the Moreno Hill lower member coals have a low total sulfur content (0.70%), predominantly organic sulfur. Average apparent rank is subbituminous A, although some individual coal beds have a high volatile bituminous C rank. These coals are moderately high ash (17%), similar to that of the Dilco Coal Member coals (Crevasse Canyon Formation) in the Gallup field (Fig. 3).

Ash in the lower coal member consists principally of silica and aluminum oxides (Table 1). Comparison of major oxide (weighted average) analyses of ash for the Salt Lake field between the NMRDI coal quality study and other studies indicate statistically significant differences between the iron and magnesium oxides; remaining major oxides are within the same population. The NMRDI samples are higher in Fe_2O_3, and MgO; the higher iron is a consequence of the higher percentage of sulfur in these samples, particularly in those analyses from the Antelope coal zone. Campbell (1989) also found the iron from whole coal oxide analyses to be greater in the Antelope coals compared to the Cerro Prieto or Rabbit zone coals. Iron oxides can influence the behavior of coal in boilers. Depending on the state of the iron, the ash fusion temperature may increase or decrease.

PETROGRAPHIC ANALYSES

During the NMRDI coal quality study petrographic analyses for vitrinite reflectance and maceral content were done on 11 coal samples from the Moreno Hill Formation. These analyses were compared to 64 Fruitland, 28 Menefee, and 20 Crevasse Canyon petrographic analyses from the same study.

Maceral content

Coal organic constituents are divided into three major maceral groups; vitrinite, liptinite and inertinite. Maceral analyses measure the proportion by volume of these organically derived components. Vitrinite, derived from wood, leaf or root tissue, is the dominant maceral (70–89%) in the Crevasse Canyon, Menefee and Fruitland coals in the San Juan Basin. The Moreno Hill Formation coals have a much lower and statistically different vitrinite percentage (avg. 63%) than the other coal–bearing formations (Fig. 8) within the San Juan Basin.

The source material of inertinite is cellulose and lignin from the cell walls of the plant, the same as that of vitrinite. This plant material has been oxidized either biochemically or by fire or desiccation early in the coalification process. The Moreno Hill Formation coals contain a significantly greater percentage of inertinite (27%) than the Crevasse Canyon, Menefee and Fruitland coals. Most of the inertinite maceral in the Moreno Hill coals is semifusinite resulting from charring.

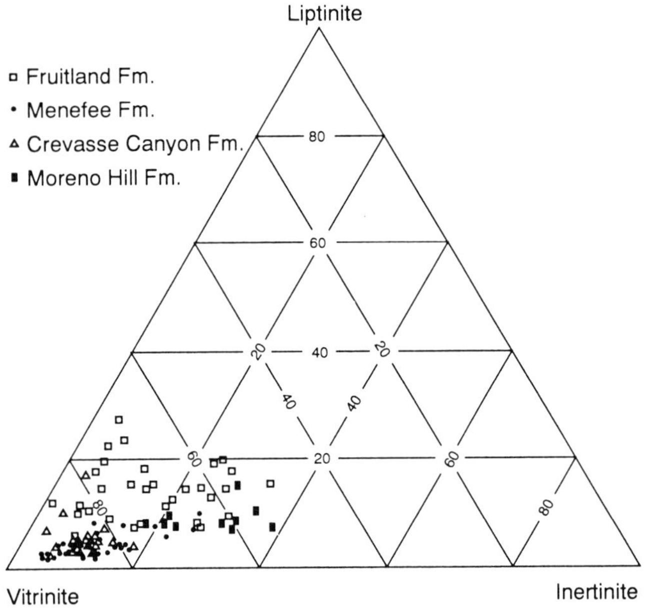

FIGURE 8. Maceral–group content (%) by formation of NMRDI coal quality samples (from Bellis, 1993.)

Liptinite is composed of spore and resin material. In the coal–bearing sequences analyzed during the NMRDI study, 2 to 27% of the total macerals were in this group. The Fruitland coals have the greatest percentage of liptinite, specifically resinite (13–17%; Hoffman, Campbell and Beaumont, 1993). The Moreno Hill Formation coals averaged 7% liptinitic material, a greater percentage than the Menefee and Crevasse Canyon coals but lower than the Fruitland Formation coals.

Maceral composition is affected by the nature of the plant material forming the peat, as well as by its oxidation. Statistical comparison of the coal–bearing formations in northwestern New Mexico indicates the Fruitland Formation and the Moreno Hill Formation are different from the other formations sampled for vitrinite, liptinite and inertinite content (Fig. 8). The Fruitland coals contain a greater percentage of liptinite macerals, in particular resinite, suggesting a swamp rich in trees rather than shrubs and grasses. The Moreno Hill has a higher percentage of inertinite, indicating that the swamps in which these coals formed were subjected to greater oxidation, perhaps the result of more numerous swamp fires before burial. The Menefee and Crevasse Canyon coals are statistically within the same population for all three major maceral groups and are vitrinite rich.

Vitrinite reflectance

Reflectance of the macerals increases with the degree of coalification. The petrographic properties of vitrinite change uniformly with increasing coalification, making it a good indicator of rank. The vitrinite reflectance methods used in the NMRDI study are described in Roybal et al. (1986, p.80). Vitrinite is used for reflectance studies because it is the most abundant maceral and is assumed to be representative of the reflectance of the coal. Moreno Hill coals have a slightly higher vitrinite reflectance (R_O= 0.49) than Crevasse Canyon coals (R_O= 0.47). The Moreno Hill coals are older than the Crevasse Canyon and Menefee coals (Turonian vs Conacian through Campanian) and may have experienced some thermal influence from Tertiary and Quaternary volcanics in the area. These factors lead to greater coalification and slightly higher vitrinite reflectance values for the Moreno Hill coals. The average vitrinite reflectance values of the Moreno Hill, Crevasse Canyon, and Menefee formations (0.49, 0.46, 0.50, respectively) indicate an apparent rank of high volatile bituminous C.

SUMMARY AND CONCLUSIONS

The lower member of the Moreno Hill Formation in west–central New Mexico contains moderately high–ash, low sulfur, subbituminous A to high–volatile C bituminous coals in three zones. The decreasing sulfur content and change in isopach trends of the coals in these zones demonstrate a progression from a back barrier environment (Antelope zone) to an alluvial plain environment (Rabbit zone). Maceral analyses indicates some oxidation or charring of the plant material in these swamps took place before burial. Tonsteins within several of the coal beds indicate swamp development was interrupted by deposition of fine-grained material, possibly volcanic ash. If these tonsteins are of volcanic origin, they could be the cause of charring or oxidation in the swamps, resulting in greater amounts of semifusinite in the Moreno Hill coals.

Economic coal beds in the lower Moreno Hill are in the medial Cerro Prieto zone. Coal in this zone averages 4.2 ft thick and beds show some lateral continuity (12 to 15 mi). Recent resource evaluations at the NMBMMR (Campbell, 1981, 1989; Roybal and Campbell, 1981; NMBMMR coal data base) indicate 341 million st of demonstrated coal resources (minimum 2.5 ft thick) with less than 250 ft of overburden in the Moreno Hill Formation, Salt Lake field. The majority of these resources are in the east–central part of the coal field (T17N to T15N, R3W to R4W).

ACKNOWLEDGMENTS

All of the drill sites for the NMRDI study were on private surface land. Thanks to the SRP, Mr. Odell, and Mr. and Mrs. Nardine for permission to drill on their land. NMRDI analyses were done at the NMBMMR coal lab by Frank Campbell, Jeanne Verploegh, Lynn Brandvold, Cecilia McCord and Barbara Popp. Vitrinite reflectance analyses for the NMRDI study were done by Frederick J. Kuellmer, New Mexico Institute of Mining and Technology (NMIMT); Arthur D. Cohen in conjunction with the Earth Sciences and Resource Institute (ECRI) of the University of South Carolina; Frederick Rich, formerly with Environmental and Coal Associates (ECA); and Deborah W. Kuehn. Maceral content for this study was analyzed by Kuellmer, Rich (ECA) and Kuehn. Data from other studies were collected for data entry into the National Coal Resource Data System through a cooperative grant with the USGS. Thanks to Orin Anderson for reviewing this paper and previous insightful discussions on the Salt Lake coal field. I appreciate Les Presmyk and Ken Rodgers of SRP for taking time from their hectic schedules to review this article and for their cooperation in past and present projects. The thorough review of this paper by Frank Kottlowski and his constant encouragement is greatly appreciated.

This article is dedicated to the memory of Frank Campbell (1947–1988) who spent several years doing geologic mapping, and coal quality and resource evaluations in the Salt Lake coal field. His initial mapping and that of others from the NMBMMR and USGS helped stimulate interest in the economic potential of coals in this field. Frank's open–file reports (1981) and final compilation of the coal geology in the southeast Salt Lake field (1989) are significant contributions to the geologic knowledge of this area. His work continues to be referenced by investigators in academia and industry.

REFERENCES

Anderson, O.J., 1986, Geologic map of Fence Lake, New Mexico 1:100,000 metric sheet: New Mexico Bureau of Mines and Mineral Resources, Open-file Report 220, 4 p.

Anderson, O.J., 1987, Geology and coal resources of Atarque Lake 1:50,000 quadrangle, New Mexico (NW quadrant of Fence Lake 1:100,000 sheet): New Mexico Bureau of Mines and Mineral Resources Geologic Map GM-61, scale 1:50,000.

Anderson, O.J. and Stricker, G.D., 1987, Stratigraphy and coal occurrences of the Tres Hermanos Formation and Gallup Sandstone (Upper Cretaceous), Zuni Basin, west-central New Mexico: New Mexico Bureau of Mines and Mineral Resources, Bulletin 121, p. 59-63.

Beaumont, E.C., 1982, Geology of New Mexico coal deposits and geological setting for field trips; *in* Coal–bearing sequences— modern geological concepts for exploration and development: American Association of Petroleum Geologists, Short Course Notes, March 1982, fig. 2.

Bellis, D., 1993, Petrographic analyses: assessment of methods and results: New Mexico Bureau of Mines and Mineral Resources, Bulletin 141, p. 71-79.

Campbell, F.W., 1981, Geology and coal resources of Cerro Prieto and The Dyke quadrangles: New Mexico Bureau of Mines and Mineral Resources, Open-file Report 144, 68 p.

Campbell, F.W., 1987, Coal geology of the Salt Lake coal field: New Mexico Bureau of Mines and Mineral Resources, Bulletin 121, p. 65-73.

Campbell, F.W. 1989, Geology and coal resources of Fence Lake 1:50,000 quadrangle, New Mexico: New Mexico Bureau of Mines and Mineral Resources, Geologic Map 62, scale 1:50,000.

Hoffman, G.K., Campbell, F.W. and Beaumont, E.C., 1993, Quality assessment of strippable coals in northwestern New Mexico: Fruitland, Menefee, and Crevasse Canyon Formation coals in San Juan Basin, and Moreno Hill Formation coals in Salt Lake field: New Mexico Bureau of Mines and Mineral Resources, Bulletin 141, 84 p.

Hook, S.C., Molenaar, C.M. and Cobban, W.A., 1983, Stratigraphy and revision of nomenclature of upper Cenomanian to Turonian (Upper Cretaceous) rocks of west-central New Mexico: New Mexico Bureau of Mines and Mineral Resources, Circular 185, p. 7-28.

McLellan, M., Haschke, L., Carter, M.D. and Medlin, A., 1984, Middle Turonian and younger Cretaceous rocks, northern Salt Lake coal field, Cibola and Catron Counties, New Mexico: New Mexico Bureau of Mines and Mineral Resources, Circular 185, p. 41-47.

Molenaar, C.M., 1983, Major depositional cycles and regional correlations of Upper Cretaceous rocks, southern Colorado Plateau and adjacent areas; *in* Mesozoic paleogeography of the west-central United States: Rocky Mountain Paleogeography Symposium 2, Rocky Mountain Section, Society of Economic Paleontologists and Mineralogists, p. 201-224.

Roybal, G.H. and Campbell, F.W., 1981, Stratigraphic sequence in drilling data from Fence Lake area: New Mexico Bureau of Mines and Mineral Resources, Open-file Report 145, 63 p.

Roybal, G.H., Campbell, F.W.,Beaumont, E.C., Cohen, A.D., Kuellmer, F.J. and Kottlowski, F. E., 1986, Quality assessment of strippable coals in New Mexico, Year 1, Phase II, Fruitland and Cleary coals in the San Juan Basin of northwestern New Mexico: New Mexico Research and Development Institute Report 2-73-4304, 90 p.

Valley of Apache Creek about 12 km north of Apache Creek, New Mexico. View is S47°E, downstream. Looking down Hardcastle Canyon; rim rocks consist of cliff-forming andesite of Dry Leggett Canyon overlying volcaniclastic conglomerates and sandstones of the Pueblo Creek Formation. Several low terraces are present, but the grassy valley floor is largely undissected. Camera station is alongside Highway 32 in NW ¼ sec. 25, T4S, R18W. Altitude is about 2100 m. Wayne Lambert photograph 93L74. August 13, 1993, 9:53 a.m. MDT.

SALT RIVER PROJECT'S PROPOSED FENCE LAKE MINE

KENNETH W. RODGERS
Salt River Project, P. O. Box 52025, Phoenix, AZ 85072-2025

Abstract—Salt River Project's (SRP's) proposed Fence Lake Mine is situated in the Salt Lake coal field 14 mi northwest of Quemado, New Mexico. The coal will serve as fuel primarily for SRP's coal- fired Coronado Generating Station, located approximately 6 mi northeast of St. Johns, Arizona. Coal delivery will be provided by rail, necessitating the construction of approximately 45 mi of new line. A Surface Coal Mine Permit Application was submitted to the New Mexico Mining and Minerals Division on November 1, 1993. Initial coal production is scheduled for the end of 1997. Projected mine life is 40 years with an estimated total coal production of 81.3 million tons. The Cerro Prieto coal zone of the Moreno Hill Formation (Upper Cretaceous) will be mined. From the exploration drilling data, two previously unmapped faults have been detected.

INTRODUCTION

The proposed Fence Lake Mine is situated in the Salt Lake coal field approximately 14 mi northwest of Quemado, New Mexico, mostly in Catron County (Fig. 1). The north edge of the permit area extends slightly into Cibola County (New Mexico), and the proposed rail corridor extends from Catron County, New Mexico, into Apache County, Arizona, to the Coronado Generating Station (CGS), located approximately 6 mi northeast of St. Johns, Arizona.

SRP's objective for the Fence Lake Mine is to provide a self- owned source of coal capable of meeting the fuel demand at CGS, as a low-cost viable alternative to all other sources of coal in the regional market. CGS is wholly owned and operated by SRP, and consists of two 350 megawatt units. This coal-fired generating station began operation in 1979 and is presently supplied by Pittsburg and Midway's McKinley Mine, located near Gallup, New Mexico. Due to increased demand at CGS, coal from the Fence Lake Mine will initially serve as a supplement to the coal supply from the McKinley Mine. By the year 2006, the Fence Lake Mine may become the sole supplier of coal to CGS.

SRP's earliest exploration drilling in the Fence Lake area was in 1975. Land acquisition for the project began in 1981. By 1993, SRP had obtained full right-of-entry for a potential Logical Mining Unit (LMU) of nearly 18,000 acres (federal, state and fee holdings) and for the entire 45-mi rail corridor to CGS.

Federal Coal Leases were awarded to SRP in 1991 after a Lease by Application effort was initiated in 1988. The leasing process required an Environmental Impact Statement (EIS), which assessed the impacts of the Federal action of leasing the Federal coal and the subsequent mining of the LMU. The EIS was based on exhaustive environmental baseline studies, which have been on-going for the project since as early as 1982.

Bulk testing of the Fence Lake coal was initiated in 1985, when 12 tons of coal were extracted under an exploration permit for a pilot-scale test burn. In 1986 a surface mining permit was issued for the Fence Lake No. 1 Mine, which was opened during the first half of 1987 for the extraction of 100,000 tons of coal, for a full-scale test burn at CGS (Greenberg, 1987). The mine has been reclaimed since 1987 and the last seeding effort for establishing revegetation was in 1990. The Fence Lake No. 1 Mine is still under active permit but is closed under a temporary cessation of operations status.

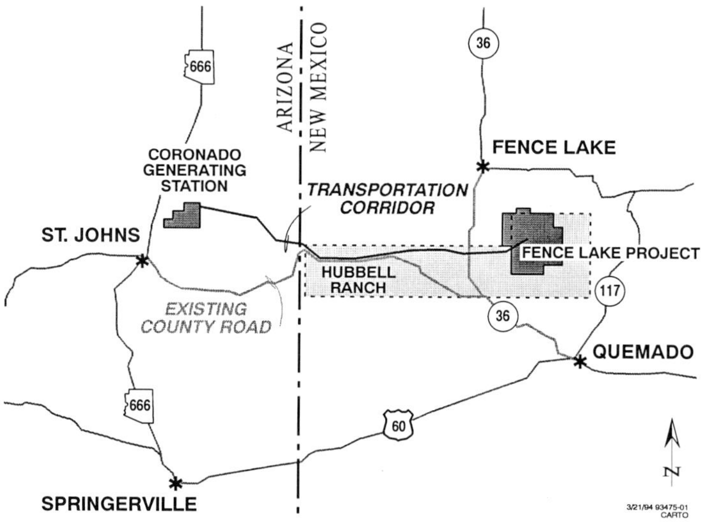

FIGURE 1. General location map of the proposed Fence Lake Mine.

TENTATIVE MINE DEVELOPMENT SCHEDULE

For the proposed Fence Lake Mine, a Surface Coal Mine Permit Application was submitted to the New Mexico Mining and Minerals Division on November 1, 1993. Approximately two years have been blocked out in the mine development schedule for securing all of the necessary permits and environmental clearances. Construction (preceded by archaeological mitigation) for both the mine infrastructure and the railroad, is scheduled for 1996 and 1997, and the first coal production is scheduled for the end of 1997.

DESCRIPTION OF PROPOSED OPERATION

The estimated mine life is 40 years, with production rates ranging from 1.8 to 3.2 million tons of coal per year, for a total estimated production of 81.3 million tons. Actual production rates may vary depending on electrical demands at CGS.

The surface mining method will incorporate a combination of area mining and contour mining, extracting coal from an area up to a maximum highwall height of approximately 200 feet. Most of the overburden and some of the interburden will be drilled and blasted for initial fracturing, and subsequently removed with an electric-powered dragline, with possible truck and shovel support where greater overburden thickness is encountered; further assistance will be provided by dozers, scrapers and castblasting. Coal removal will chiefly utilize trucks and shovels or loaders.

From the central loadout in the mine area, coal haulage will be by rail, requiring the construction of approximately 45 mi of new line to complete a rail link to the existing rail spur at CGS. The proposed rail corridor basically parallels the natural drainage pattern in the area, following the general course of Nations Draw, Largo Creek and Carrizo Wash, mostly involving an overall descent from the mine area to the station.

TRIBAL CONSIDERATIONS

SRP has closely coordinated the development of the mining project with five major southwestern tribal groups, including the Acoma, Hopi, Navajo, Ramah Navajo and Zuni. Although none of the land in the project area is tribal in ownership, extensive research has been conducted to establish the ethnohistory and the ancestral interests of each of the tribes in the area.

GEOLOGIC SETTING

The coal-bearing strata in the mine area are of the Moreno Hill Formation (Upper Cretaceous; formerly mapped as the Mesaverde Group). The particular coal seams of economic interest which occur in the lower member of the formation belong to the Cerro Prieto coal zone (Fig. 2). The primary seam, referred to as the ABC, splits variably into as many as three coal beds. This seam typically contains a white tonstein parting approximately 3 in. thick throughout much of the reserve. Where the seams have coalesced into one body, the bed attains a maximum thickness of 13 ft. The next coal horizon lower in the stratigraphic section is the Tejana seam, which in some parts of the reserve area is too thin to mine. This seam attains a maximum thickness of 6.5 ft and is separated from the coal bed(s) of the ABC seam by interburden varying in thickness from about 10 ft to as much as 50 ft.

Two significant faults or fault zones have been detected in the mine area and are well displayed in the drilling data. The dominant fault (which forms the southeastern boundary for the surface coal reserve; see Fig. 3) displays a northeasterly trend with a maximum vertical displacement of approximately 110 ft. The second fault zone displays a northwesterly trend coincident with Tejana Draw, with a maximum vertical displacement of approximately 70 feet. Lesser faults have been noted in the mine area and likely more will be revealed as the mine development continues.

REFERENCES

Campbell, F., 1989, Geology and coal resources of Fence Lake 1:50,000 quadrangle, New Mexico: New Mexico Bureau of Mines and Mineral Resources, Geologic Map 62.

Greenberg, M. A., 1987, Large-scale combustion test of Fence Lake coal at the Coronado Generating Station: New Mexico Bureau of Mines and Mineral Resources, Bulletin 121, p. 79-83.

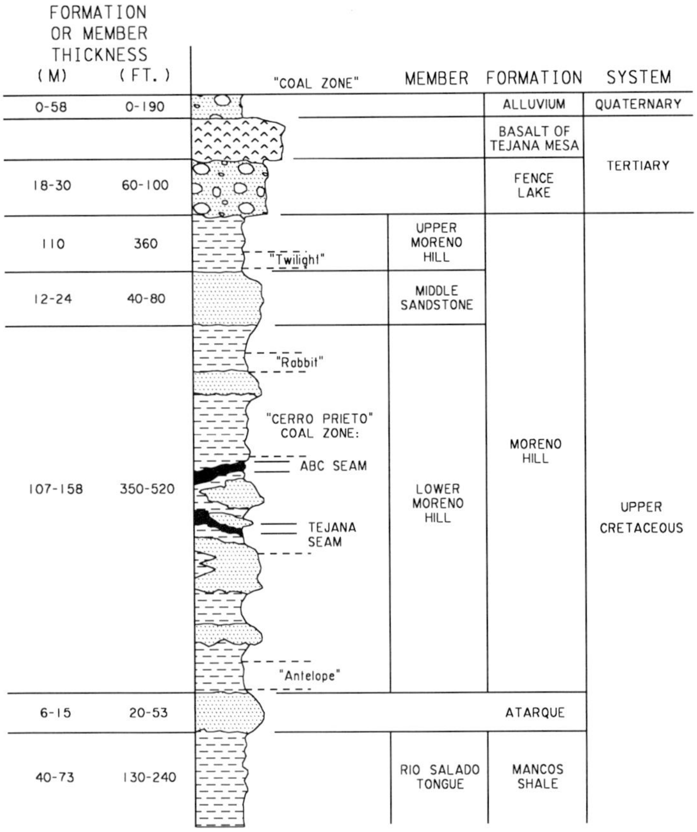

FIGURE 2. Generalized coal stratigraphy of the Fence Lake region.

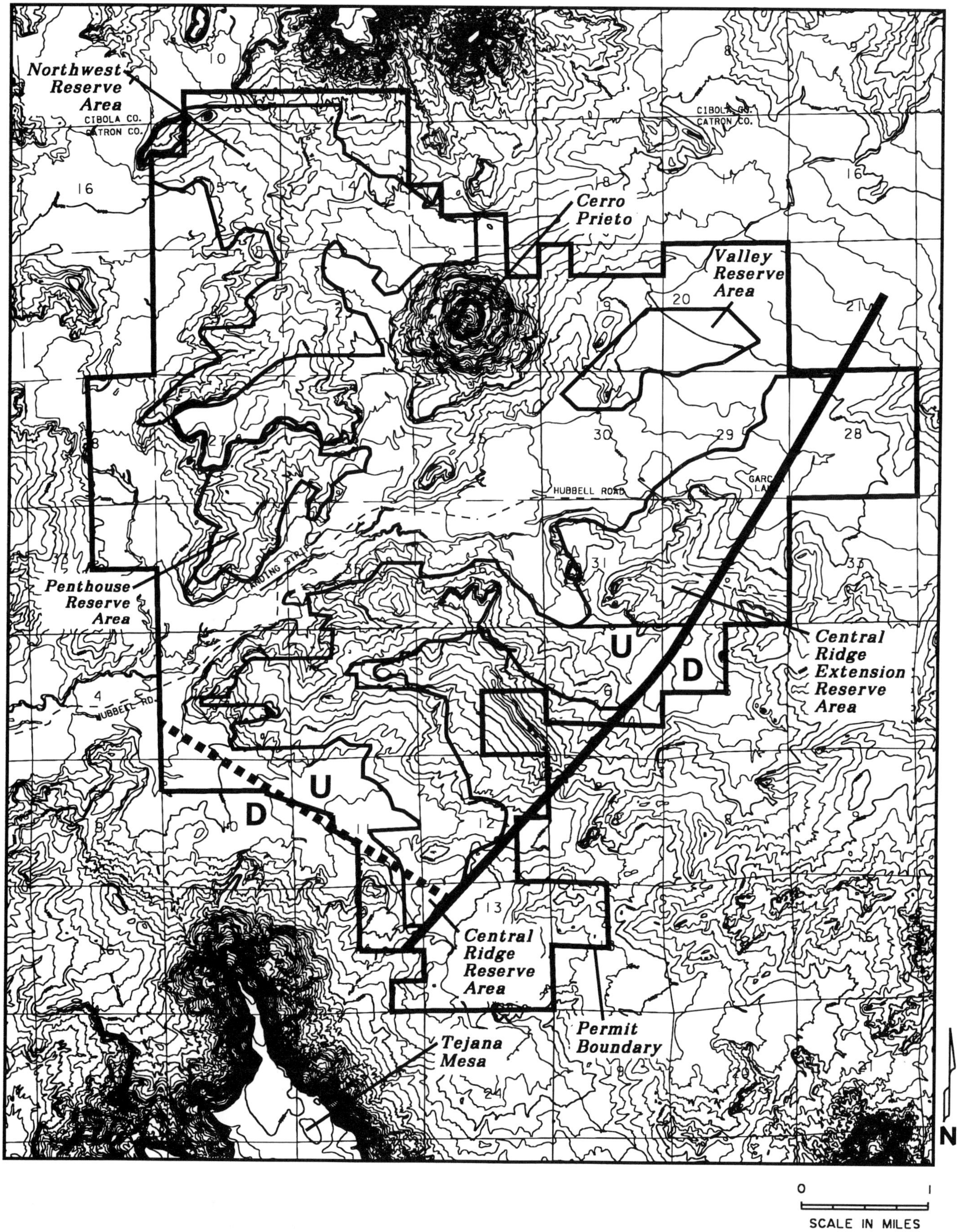

FIGURE 3. Major reserve areas and fault locations.

Day 2 Stop 1
Jordan 1993

GEOLOGY AND PETROLEUM SOURCE ROCKS IN THE HUNT NO. 1-16 STATE WELL

RONALD F. BROADHEAD
New Mexico Bureau of Mines and Mineral Resources, Socorro, NM 87801

Abstract—The Hunt Oil Company No. 1-16 State well was drilled for petroleum exploration to a total depth of 6890 ft in north-central Catron County, New Mexico during late 1989. The well drilled through the top of the Spears Group (Tertiary) near the surface, the Baca Formation (Tertiary) at 1370 ft, the Upper Cretaceous at 3290 ft, the Triassic at 3630 ft, the San Andres Formation (Permian) at 3890 ft, the Glorieta Sandstone (Permian) at 4290 ft, the Yeso Formation (Permian) at 4560 ft and the Abo Formation (Permian) at 6253 ft. Precambrian metarhyolites were encountered from 6792 ft to total depth. An intrusive diorite of probable Tertiary age is present within the Yeso from 5180 to 5580 ft. Drill cuttings were analyzed for petroleum source-rock potential. Dark-gray Cretaceous shales contain sufficient total organic carbon (TOC) to be petroleum source rocks but are thermally immature. Yellowish to brown San Andres limestones also contain sufficient TOC to be source rocks and are moderately mature. Kerogens in the San Andres are dominantly amorphous and herbaceous; a mixture of oil and gas is likely to have been generated. Hydrocarbon shows were encountered in the San Andres. Yellowish to brown Yeso limestones also contain sufficient TOC to be source rocks. The Yeso is thermally mature and yielded several hydrocarbon shows. Maximum thermal maturation has been attained by carbonates proximal to the diorite intrusive, indicating that heat emitted from the intrusive enhanced thermal maturation.

INTRODUCTION

The Hunt Oil Company No. 1-16 State Well was drilled to a total depth of 6890 ft in sec. 16, T3S, R13W, Catron County, New Mexico (Figs. 1, 2; Table 1). It was completed as an unsuccessful wildcat well on November 7, 1989. It antedated the three wells that Shell drilled as part of the "Magic play" (Fig. 1). A full set of drill cuttings was donated to the New Mexico Bureau of Mines and Mineral Resources (NMBMMR) at the conclusion of drilling operations. Selected drill cuttings were analyzed as part of the NMBMMR Petroleum Source-Rock Project. Those analyses are the basis for this paper.

GEOLOGY OF THE WELL

The Hunt No. 1-16 State well penetrated rocks from Cenozoic to Precambrian age (Fig. 3). The well was spudded in Quaternary alluvi-

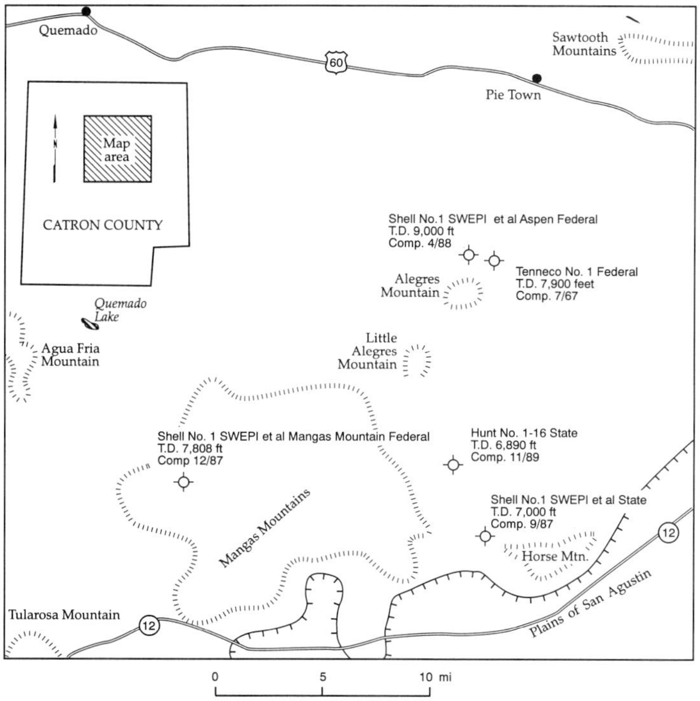

FIGURE 1. Hunt Oil Company No. 1-16 State well, the three exploratory wells drilled by Shell as part of the "Magic play", and the Tenneco No. 1 Federal well.

FIGURE 2. Hunt Oil Company No. 1-16 State drilling near Mangas Mountains, November 1989.

TABLE 1. Engineering and geologic data for the Hunt Oil Co. No. 1-16 State well.

Operator:	Hunt Oil Company
Well name & number:	State No. 1-16
Location:	1090 ft FSL, 660 ft FEL Sec. 16, T3S, R13W Catron County, New Mexico
Surface Elevation:	7573 ft, ground level
Total Depth:	6,890 ft (plugged back to 1,825 ft)
Spud Date:	October 6, 1989
Completion Date:	November 9, 1989, Plugged & Abandoned
Casing:	20 inch at 40 ft 13⅜ inch at 454 ft 9⅝ inch at 4,563 ft

um and drilled through 1370 ft of gray andesitic sandstones of the lower Spears group (Tertiary) and 1920 ft of siliceous sandstones, conglomerates and mudstones of the Tertiary Baca Formation. The probable top of Cretaceous strata was encountered at 3290 ft. The Mancos Shale was drilled from 3290 to 3590 ft and consists of dark-gray shales interbedded with minor light-gray, fine-grained sandstones. The Twowells Sandstone Tongue of the Dakota Sandstone (Cretaceous) is present from 3526 to 3550 ft. The main body of the Dakota Sandstone (Cretaceous) (3590 to 3630 ft) consists of white, fine- to medium-grained sandstone. No Jurassic section was identified in the well; the Cretaceous rests unconformably on the Triassic.

Triassic strata, encountered from 3630 to 3890 ft, consist of red to brown shales and minor interbedded siltstones. Prominent fine- to coarse-grained sandstones are present from 3630 to 3660 ft and 3880 to 3890 ft.

The unconformity at the top of Paleozoic strata was drilled at 3890 ft where the Triassic red beds rest on the San Andres Formation (Permian). The San Andres, present from 3890 to 4290 ft, consists of dark-gray to brown, finely crystalline dolostones and limestones. Circulation was lost from 3970 to 4150 ft and was only regained after cement was pumped into the wellbore. The dipmeter log indicates chaotic dip orientations in the San Andres from 3890 to 4100 ft. Lost

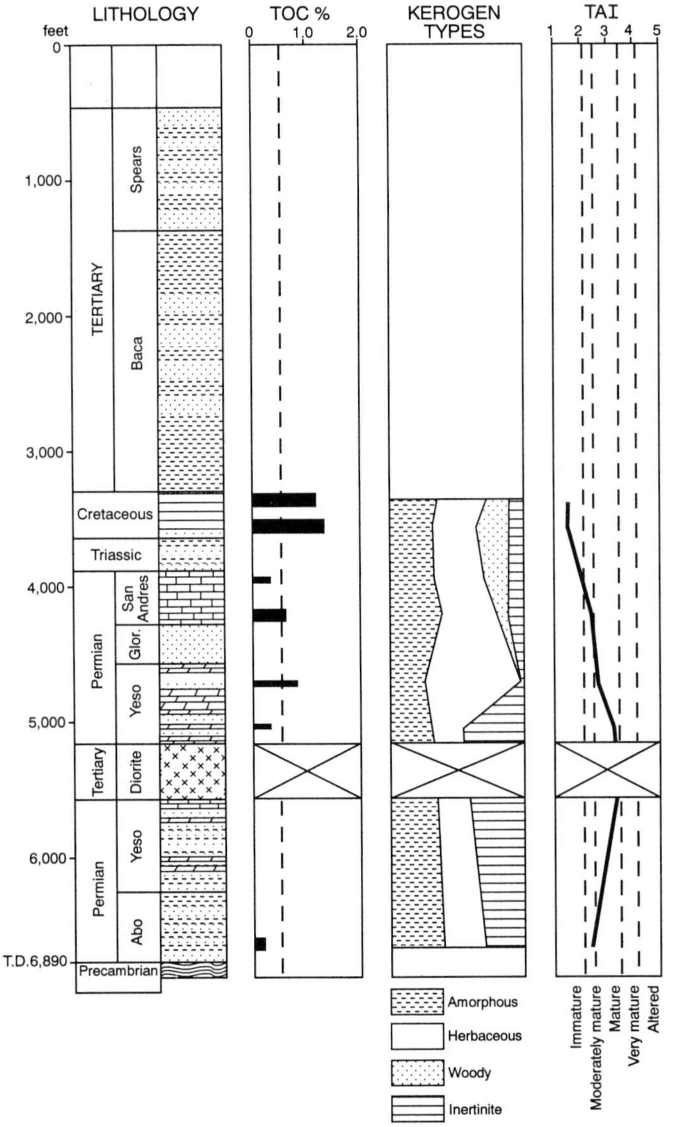

FIGURE 3. Stratigraphic section and profiles of source-rock data for Hunt Oil Company No. 1-16 State well. See Table 2 and Bayliss (1990) for source-rock data.

circulation was probably caused by penetration of an extremely permeable zone of cavernous porosity. Cavernous porosity is widespread in the San Andres of west-central New Mexico and developed during two stages of karstification (White and Kelly, 1989). The first stage occurred during formation of the Permo-Triassic unconformity. The second stage occurred after the San Andres was exposed on the flanks of the Zuni uplift; this exposure may have occurred as early as Laramide time. The chaotic patterns recorded on the dipmeter log may result from poor resolution of caverns within the carbonate sequence or these patterns may reflect collapse of overlying strata into caverns within the San Andres. Chaotic dips in the lower 30 ft of the Triassic tend to confirm the latter possibility.

Numerous hydrocarbon shows were encountered in the San Andres. The mud log indicates numerous occurrences of hydrocarbon fluorescence and hydrocarbon cut in chloroethane throughout the San Andres; stains of oil and asphalt were also common.

The Glorieta Sandstone was drilled from 4290 to 4560 ft. It consists mostly of white, very fine- to medium-grained sandstone. Interbedded brown, finely crystalline dolostones are present in the upper 80 ft. The sandstones and dolostones are visibly porous. Circulation was lost from 4470 to 4550 ft and was reestablished only after an intermediate string of 9 ⅝ inch diameter casing was set at 4563 ft.

The Yeso Formation (Permian) is present from 4560 to 6253 ft. A diorite sill has intruded the Yeso from 5180 to 5580 ft. The Yeso consists of interbedded anhydrites, dolostones, sandstones and shales. The dolostones are light gray and light to dark brown and are microcrystalline to finely macrocrystalline; visible porosity is common in the dolostone cuttings. The sandstones are white and very fine to fine grained; some sandstones have visible intergranular porosity and other sandstones have no visible porosity and are well cemented with anhydrite, calcite and dolomite. Yeso shales are orange to brown in color.

Several hydrocarbon shows were reported in the Yeso. Most shows occurred above the diorite sill. The mudlog indicates that dolostones at 4600 ft and at 4720 ft exhibited hydrocarbon fluorescence and yielded a yellowish-blue cut in chloroethane. Black asphaltic and oily stains were described in dolostones above the sill. Numerous shows of hydrocarbon gases were recorded by the gas detector on the mudlogging unit. Gas concentrations of 5000 ppm in the drilling mud occurred in 10-200 ft thick zones; although composed mostly of methane, these gases also contained ethane and propane and appear to have emanated from dolostones.

Intrusive diorite of probable Tertiary age was drilled from 5180 to 5580 ft. Thin-section and x-ray diffraction analyses indicate it is composed dominantly of plagioclase (albite-oligoclase) with minor quartz and potassium feldspar. Crystal size averages 0.08 mm.

The Abo Formation (Permian), drilled from 6253 to 6792 ft, consists of orange, very fine- to coarse-grained, arkosic sandstone interbedded with red shale, orange siltstone, and minor brown microcrystalline to cryptocrystalline dolostone. No hydrocarbon shows were encountered in the Abo.

Precambrian basement was drilled from 6792 ft to total depth of 6890 ft. The Precambrian is a foliated igneous rock composed of phenocrysts of strained quartz and potassium feldspar in a microcrystalline matrix of silica and mica. It appears to be of the same composition and texture as the Precambrian basement in the Shell No. 1 Mangus Mountains Federal well (Fig. 1) where samples were recovered in sidewall cores and the foliation was observed to be horizontal. Although lack of a large fragment of the Precambrian makes identification problematic, the composition and foliated porphyritic-aphanitic texture indicate it may be a metarhyolite.

PETROLEUM SOURCE ROCKS

Drill cuttings from the Hunt No. 1-16 State well were analyzed for source-rock potential as part of the NMBMMR Petroleum Source-Rock Evaluation Project (Bayliss, 1990). Seven samples of drill cuttings were collected for analysis (Fig. 3; Table 2). Each sample consists of cuttings picked from several 10 ft depth intervals. Individual cuttings were selected carefully under the microscope by the author so that only lithologies likely to be indigenous source rocks were included in the samples (i.e. cavings and non-source materials were removed from the sample). Samples were analyzed by Geoff Bayliss of Geochem Laboratories, Inc. for (1) Total organic carbon (TOC); (2) Kerogen type by visual kerogen analysis; (3) Thermal Alteration Index (TAI) by visual kerogen analysis and (4) Rock-eval pyrolysis to determine maturity and other source-rock parameters. The TOC and visual kerogen analysis of kerogen type and thermal maturity are stressed in this report. These analyses are generally considered more accurate than pyrolysis techniques, which are best used to screen samples for more rigorous analytical procedures (Philp and Galvez-Sinibaldi, 1991).

Total organic carbon (TOC) was measured to determine which stratigraphic units contain sufficient organic carbon to be source rocks. Jarvie (1991) summarized the minimum amount of TOC required to form a petroleum source rock (Table 3). Shales in the Mancos and Dakota (Cretaceous) and limestones in the San Andres and Yeso (Permian) contain sufficient TOC to be petroleum source rocks. The red Abo shales contain insufficient TOC to be source rocks. Because the analyzed samples are composites of cuttings from several 10-ft drill intervals, the TOC values are inexact estimates of the average TOC content in the sampled interval; some parts of the analyzed interval are likely to exceed this average value and other parts of the same interval are likely to be less than this average value.

The types of kerogens that constitute the organic fraction of a source rock determine if it will generate oil, gas, or both oil and gas (Hunt, 1979; Jarvie, 1991). Generally, algal and amorphous kerogens produce paraffinic oils and gas upon maturation; they have the highest percentage of convertible carbon (>70% by weight) and will produce more hydrocarbons per unit weight than other kerogen types. Herbaceous kerogens produce a mixture of liquid hydrocarbons and gases; they have an intermediate percentage of convertible carbon (30-70%) by weight. Woody kerogens produce mostly gas; they have only a small percentage of convertible carbon (<30% by weight) and will produce fewer hydrocarbons than amorphous, algal, or herbaceous kerogens per unit weight. Inertinite is carbonized kerogen that is capable of producing only small amounts of gas.

In the Hunt No. 1-16 State well, both oil-prone and gas-prone kerogens are present in possible source rocks (Fig. 3; Table 2). The gray shales of the Mancos and Dakota contain predominantly mixed amorphous and herbaceous kerogens, but significant amounts of woody kerogen and inertinite are also present. San Andres and Yeso limestones with more than 0.5% TOC contain predominantly herbaceous and amorphous kerogens and contain only minor woody kerogen and inertinite; these rocks should have generated oil upon maturation.

Visual analysis of kerogen was also used to assess thermal maturation of source rocks. The color of kerogen in transmitted light varies from yellow to orange to brown to black as thermal maturation increases (Staplin, 1969). The kerogen is assigned a numerical value of thermal maturity (TAI or Thermal Alteration Index) based upon its color; TAI values range from 1.0 for yellow-colored immature kerogen to 5.0 for black metamorphosed kerogen. TAI has been correlated with vitrinite reflectance and other maturation indices by several workers, including Dow (1977), Geochem Laboratories (1980), and Sentfle and Landis (1991).

TAI values in the Hunt No. 1-16 State well vary from 1.5 (immature) in the Cretaceous section to 3.2 (mature) in the Yeso Formation (Fig. 3; Table 2). Thermal maturity increases with depth from immature in the Cretaceous to moderately mature in the San Andres Formation (Permian) to mature in the Yeso Formation (Permian). This is expected because of increasing temperatures associated with increasing burial depths. However, maturity then decreases to moderately mature with deeper burial in the Abo Formation (Permian).

The maturation maximum at 5010 to 5030 ft within the Yeso may have been caused by heat emitted from the diorite sill that intruded the Yeso from 5180 to 5580 ft. Heat exuded by this sill during its intrusion and subsequent cooling elevated temperatures so that the Yeso entered the mature stage of oil generation but was not matured past the stage of oil preservation (TAI ≈ 3.4). The sill, rather than destroying any previously generated

TABLE 2. Hydrocarbon source-rock analyses, Hunt Oil Company No. 1-16 well. Data from Bayliss (1990)

Sample depth (ft)	Stratigraphic Unit	Lithology	Total Organic carbon*		Source type		Maturation	
			%	source potential	kerogen types	character	TAI	Maturity
3,300 - 3,400	Mancos (Cretaceous)	shale, dark grey	1.17	adequate	36% amorphous 36% herbaceous 19% woody 9% inertinite	oil + gas	1.5	immature
3,500 - 3,590	Dakota (Cretaceous)	shale, dark grey	1.28	adequate	33% amorphous 33% herbaceous 25% woody 12% inertinite	oil + gas	1.5	immature
3,940 - 3,970	San Andres (Permian)	limestone, brownish grey	0.27	inadequate	36% amorphous 36% herbaceous 19% woody 9% inertinite	nonsource	2.0	moderately immature
4,180 - 4,250	San Andres (Permian)	limestone, yellowish brown	0.57	marginal	40% amorphous 40% herbaceous 10% woody 10% inertinite	oil + gas	2.3	moderately mature
4,700 - 4,730	Yeso (Permian)	limestone, yellowish brown	0.80	marginal	28% amorphous 72% herbaceous 0% woody 10% inertinite	oil + gas	2.6	mature
5,010 - 5,030	Yeso (Permian)	dolostone, brownish grey	0.30	inadequate	33% amorphous 22% herbaceous 0% woody 45% inertinite	nonsource	3.2	mature
6,600 - 6,700	Abo (Permian)	shale, dark reddish brown	0.14	inadequate	42% amorphous 29% herbaceous 0% woody 29% inertinite	nonsource	2.4	moderately mature

TABLE 3. Source-rock potential based on total organic carbon (TOC) content. From Jarvie (1991).

TOC (Wt. %)	Source-rock potential
0.0 -0.5	inadequate
0.5 - 1.0	marginal
>1.0	adequate

liquid hydrocarbons, elevated temperatures in the Yeso so that the Yeso reached optimum oil generation. It is plausible that evaporite and shale seals within the Yeso prevented migration of heated formation waters to strata above and below (San Andres, Glorieta and Abo formations), resulting in thermal maturation stratigraphically confined to the Yeso.

Petroleum source-rock analyses are also available for the Tenneco Oil Co. No. 1 Federal well (Fig. 1; Bayliss and Schwarzer, 1987). Stratigraphic tops for this well are: Cretaceous – 3730 ft; Triassic – 4870 ft; San Andres – 5390 ft; Glorieta – 5710 ft; Yeso – 6050 ft; Abo- 7260 ft; Pennsylvanian – 7690 ft; and Precambrian basement – 7860 ft. Stratigraphic units below the Cretaceous are 1200 - 1400 ft deeper than in the Hunt well. Most of this difference is accounted for by the thicker Cretaceous section (1140 ft) present in the Tenneco well. Thickness of the Tertiary section is approximately the same in both wells (3290 ft in Hunt well and 3775 ft in Tenneco well). In the Cretaceous, San Andres and Yeso of the Tenneco well, TOC values and kerogen types are similar to those in the Hunt well. TAI values in the Tenneco well are 2.2 (moderately mature) in the Mancos Shale, 2.3 (moderately mature) in the San Andres and upper Yeso, and 2.4 (moderately mature) in the lower Yeso. Source-rock analyses were not performed on stratigraphic units below the Yeso. The Mancos is more mature in the Tenneco well than in the Hunt well. However, maturation levels in the San Andres and Yeso are similar to those in the Hunt well, except for the samples within 1000 ft of the intrusion in the Hunt well, where the highest levels of maturity are attained. This similarity supports the hypothesis that maximum kerogen maturation in the Hunt well is related to the diorite intrusion.

IMPLICATIONS FOR PETROLEUM EXPLORATION

The source-rock analyses of samples from the Hunt No. 1-16 State well have favorable implications for petroleum exploration in west-central New Mexico and east-central Arizona. Dark-gray Mancos shales (Cretaceous) and limestones in the San Andres and Yeso Formations (Permian) contain sufficient TOC to be petroleum source rocks. Significant hydrocarbon shows were encountered in both the San Andres and the Yeso. Generated hydrocarbons are likely to be both oil and gas, although oil may predominate in the San Andres and Yeso because of high percentages of oil-prone kerogens. The Cretaceous may not be sufficiently mature to have generated significant amounts of hydrocarbons. The San Andres and Yeso are likely to be moderately mature to mature throughout much of the region. Permian source rocks will have reached optimum maturity either where they have a history of deeper burial (in Paleozoic or Laramide basins) or where they are proximal to igneous heat sources.

Observations of a recently drilled geothermal test well core near Alpine, Arizona (see Witcher et al., this volume) confirm that significant hydrocarbons have been generated within the Yeso. Rauzi (1994 a, b) described oil shows in limestones of the Fort Apache Limestone Member of the Corduroy Formation (Yeso equivalent). Limestones may be widespread source rocks in the Yeso.

ACKNOWLEDGMENTS

Hunt Oil Company and Lee Amoroso of Rocky Mountain Geo-Engineering provided the drill cuttings and access to the rig while the Hunt No. 1-16 State well was being drilled, and Roy Foster helped the author correlate and identify stratigraphic units. Geoff Bayliss of Geochem Laboratories, Inc. provided the source-rock analyses as part of the New Mexico Petroleum Source Rock Evaluation Project. Richard Chamberlin (of Guidebook fame) provided the stratigraphic correlations for the Tertiary sediments. Chris McKee provided x-ray diffraction analyses of the diorite sill. Terry Telles typed the manuscript and Kathy Campbell drafted the illustrations. Frank Kottlowski and David Schoderbek reviewed the manuscript and provided helpful suggestions and criticisms.

REFERENCES

Bayliss, G. S., 1990, Hydrocarbon source-rock evaluation, Hunt Oil Co. No. 1-16 State well, sec. 16, T3S, R13W, Catron County, New Mexico: New Mexico Bureau of Mines and Mineral Resources, Open-file report 365, 13 p.

Bayliss, G. S. and Schwarzer, R. R., 1987, Hydrocarbon source-rock evaluation of Tenneco Oil Co. No. 1 Federal, sec. 35, T1S, R13W, Catron County, New Mexico: New Mexico Bureau of Mines and Mineral Resources, Open-file report 291, 15 p.

Dow, W. G., 1977, Kerogen studies and geological interpretations: Journal of Geochemical Exploration, v. 7, p. 79-99.

Geochem Laboratories, 1980, Source rock evaluation reference manual: Geochem Laboratories, Inc., Houston, 107 p.

Hunt, J. M., 1979, Petroleum geochemistry and geology: W. H. Freeman and Co., San Francisco, 617 p.

Jarvie, J. M., 1991, Total organic carbon (TOC) analysis; *in* R. K. Merrill, ed., Source and migration processes and evaluation techniques: American Association of Petroleum Geologists, Treatise of Petroleum Geology, p. 113-118.

Philp, R. P., and Galvez-Sinibaldi, A., 1991, Characterization of organic matter by various pyrolysis techniques; *in* R. K. Merrill, ed., Source and migration processes and evaluation techniques: American Association of Petroleum Geologists, Treatise of Petroleum Geology, p. 107-118.

Rauzi, S. L., 1994a, Geothermal test hints at oil potential in eastern Arizona volcanic field: Oil and Gas Journal, v. 92, no. 1, p. 52-54.

Rauzi, S. L., 1994b, Implications of live oil shows in eastern Arizona geothermal test: Arizona Geological Survey, Open-file report 94-1, 16 p.

Sentfle, J. T. and Landis, C. R., 1991, Vitrinite reflectance as a tool to assess thermal maturity; *in* R. K. Merrill, ed., Source and migration processes and evaluation techniques: American Association of Petroleum Geologists, Treatise of Petroleum Geology, p. 119-125.

Staplin, F. L., 1969, Sedimentary organic matter, organic metamorphism, and oil and gas occurrence: Bulletin of Canadian Petroleum Geology, v. 17, p. 47-66.

White, W. D. and Kelly, T. E., 1989, The San Andres-Glorieta aquifer in west-central New Mexico: New Mexico Geological Society, Guidebook 40, p. 331-335.

VOLCANIC-EPITHERMAL DEPOSITS IN THE MOGOLLON-DATIL VOLCANIC FIELD, WEST-CENTRAL NEW MEXICO

VIRGINIA T. McLEMORE
New Mexico Bureau of Mines and Mineral Resources, Socorro, NM 87801

Abstract—Most volcanic-epithermal precious-metal deposits in New Mexico are restricted to Oligocene to Miocene volcanic rocks and to areas immediately adjacent to volcanic centers in the Mogollon-Datil volcanic field in west-central New Mexico. These deposits occur in areas of complex regional and local structures that appear to control emplacement of magmas and subsequent flow of hydrothermal fluids. Although the mechanism of ground preparation varies, most districts occur along the ring-fracture zones of calderas, in the vicinity of rhyolitic domes, and/or along faults associated with regional crustal lineaments. Two types of volcanic-epithermal deposits occur in the Mogollon-Datil volcanic field. Most deposits are of the low-sulfidation (quartz-adularia) class and formed by multiple cycles of low salinity (<5 eq. wt.% NaCl), slightly acidic to neutral pH fluids, at temperatures between 150° and 300°C, at relatively shallow depths (<1500 m) and low pressures (<150 bars). The mineral deposits at Alum Mountain are high-sulfidation (acid-sulfate) deposits. In addition, acid-sulfate alteration similar to that found in high-sulfidation deposits is present in several districts and may indicate that additional high-sulfidation precious-metal deposits may occur in these areas as well. Limited geologic and geochemical data are consistent with formation of acid-sulfate alteration by acidic fluids at temperatures below 340°C at relatively shallow depths (<1.5 km) in a magmatic-hydrothermal environment. The acid-sulfate alteration characteristic of the high-sulfidation deposits appears to be restricted to 27-33 Ma in the Mogollon-Datil volcanic field, whereas the low-sulfidation, volcanic-epithermal vein deposits are younger than 27 Ma. The economic resource potential for low-sulfidation, volcanic-epithermal vein deposits in many districts is good because many favorable structures have not been adequately explored. Also, there is good potential for the discovery of high-sulfidation precious-metal deposits in areas of acid-sulfate alteration.

INTRODUCTION

Epithermal deposits have been recognized as an important class of mineral deposits for many years. Some of the largest gold and silver deposits in the world are epithermal deposits, including the "bonanza" gold deposits of the Comstock Lode (286 tons contained gold) and Sleeper (83.6 tons contained gold) deposits in Nevada; Cripple Creek, Colorado (830.5 tons contained gold); and El Indio, Chile (154 tons contained gold; Sillitoe, 1993). Lindgren (1922, 1933) defined the term "epithermal" to include a broad range of deposits that formed by ascending waters at shallow to moderate depths (<1500 m), low to moderate temperatures (100-200°C), and which are typically associated with intrusive and/or volcanic rocks. From fluid inclusion and isotopic data, it is now generally accepted that epithermal deposits were formed at slightly higher temperatures (50-300°C) and relatively low pressures (few hundred bars). Not all epithermal deposits occur in igneous rocks; for example, the McLaughlin deposit in California is hosted in mafic and ultramafic metamorphic and sedimentary rocks (Lehrman, 1986). Many Carlin-type deposits appear to be epithermal and are found in sedimentary rocks (Bagby and Berger, 1985; Berger and Henley, 1989).

Most precious-metal epithermal deposits in New Mexico are restricted to volcanic rocks and areas immediately adjacent to volcanic fields and are called volcanic-epithermal deposits (North and McLemore, 1986, 1988). The majority of the volcanic-epithermal deposits in New Mexico are precious-metal deposits, hosted by Oligocene to Miocene volcanic and intrusive rocks, located within the Mogollon-Datil volcanic field in west-central New Mexico (Table 1, Fig. 1; North and McLemore, 1986, 1988). Not all epithermal deposits consist of precious metals. The Taylor Creek tin deposit in Sierra County is an epithermal deposit without any precious metals (Eggleston and Norman, 1986). Some fluorite and manganese deposits are also epithermal deposits but do not contain any precious metals (such as the Bishops Cap fluorite and Rincon manganese deposits in Doña Ana County; Dunham, 1935).

This paper compares and contrasts geologic, mineralogic and geochemical characteristics of volcanic-epithermal deposits in the Mogollon-Datil volcanic field and offers some theories concerning structural controls, genesis and resource potential of these deposits. The deposits in the Mogollon-Datil volcanic field are briefly described in Table 2 and located in Figure 2. Production is in Table 3. This work is part of ongoing studies of mineral deposits in New Mexico and includes updates and revisions of prior work by North (1983) and North and McLemore (1986, 1988).

TYPES OF VOLCANIC-EPITHERMAL DEPOSITS

An important advance in knowledge of epithermal-mineral deposits was Hayba et al's. (1985) two-fold classification (adularia-sericite and acid-sulfate) of volcanic-hosted epithermal deposits based on mineralogy and associated alteration. In order to define both classes according to mineralogy, Berger and Henley (1989) refined this classification to adularia-sericite (illite) and alunite-kaolinite (±pyrophyllite). Earlier models referred to these deposits as quartz-adularia and quartz-alunite (Cox and Singer, 1986, model #25). However, one problem with applying these classifications in the field is that some of the minerals, such as alunite, diaspore and adularia, are too fine grained to be identified easily (Berger and Henley, 1989). Thus White and Hedenquist (1990) refined this classification even further to low-sulfidation and high-sulfidation deposits (Table 4). Bonham (1988) added alkalic-related, epithermal precious-metal deposits to this scheme. The alkalic-related epithermal deposits in New Mexico are included with the Great Plains Margin deposits (North and McLemore, 1986, 1988).

Most volcanic-epithermal deposits in New Mexico, including those in the Mogollon-Datil volcanic field, are of the low-sulfidation (quartz-adularia) class (Tables 1, 2, 4). The mineral deposits at Alum Mountain (#20) are high-sulfidation deposits based on the alteration mineral assemblage. In addition, acid-sulfate alteration is present in several districts (Tables 1, 2) and if these areas are formed by magmatic-hydrothermal fluids they may contain high-sulfidation precious-metal deposits.

LOW-SULFIDATION DEPOSITS

Most of the largest gold and silver districts in New Mexico are low-sulfidation, volcanic-epithermal deposits in the Mogollon-Datil volcanic field (Tables 1, 3; North and McLemore, 1986). However, none of the precious-metal deposits in New Mexico are considered giant (>220 tons of contained gold) or bonanza (>33 tons of gold in ore grading 1 oz/ton gold) deposits as defined by Sillitoe (1993). Nearly 600,000 oz of gold and 26.6 million oz of silver have been produced from volcanic-epithermal districts in the Mogollon-Datil volcanic province (Table 3). Both the Mogollon and Steeple Rock (#2) districts

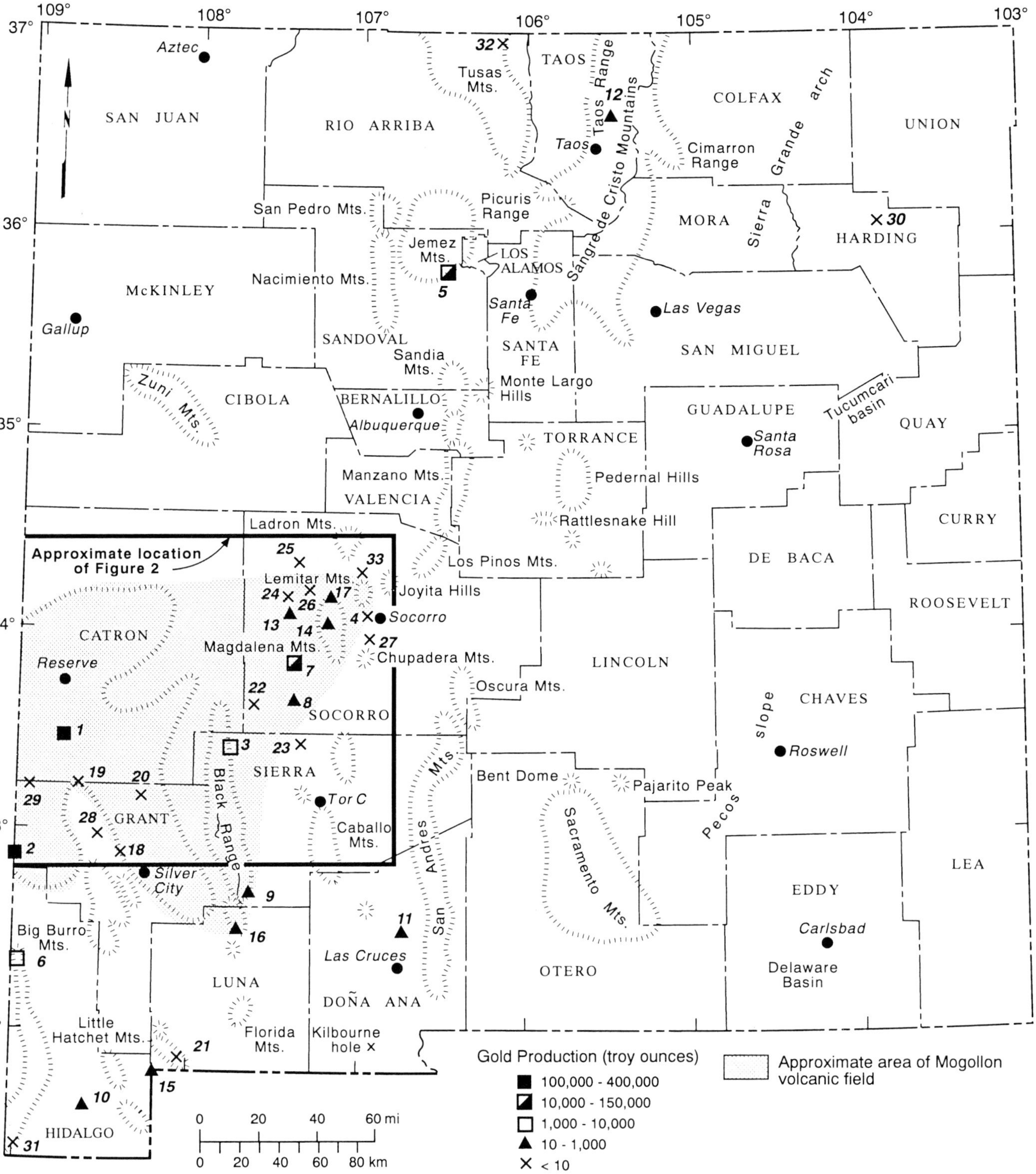

FIGURE 1. Volcanic-epithermal deposits in New Mexico, showing estimated gold production in troy ounces. Numbers refer to Tables 1 and 2. Modified from North and McLemore (1986, 1988), McIntosh et al. (1992a, b), and New Mexico Geological Society (1982).

rank in the top ten of gold-producing districts in the state. The Mogollon district (#1) has produced more silver than any other district in New Mexico. In addition, 5.8 million lbs of copper, 7.5 million lbs of lead, and 4 million lbs of zinc have also been produced from these deposits. Locally, barite, fluorite, manganese, clay, tellurium and tin have been produced from some districts as well (Table 2). In recent years, production from the Mogollon and Steeple Rock districts has been for use as silica flux.

Most volcanic-epithermal deposits in the Mogollon-Datil volcanic province contain predominantly silver with lesser amounts of gold, unlike giant and bonanza gold deposits elsewhere in the world (Fig. 3; Sillitoe, 1993). Grades vary, but typically deposits contain 0.15-0.20 oz/ton gold and 7-10 oz/ton silver (Table 5). The Rosedale district (#7) is one of the few volcanic-epithermal districts in the state that contains predominantly gold with low silver. The Socorro Peak district (#4) contains predominantly silver with little gold (Lasky, 1932). Most volcanic-epithermal deposits in New Mexico contain minor amounts of base metals, although base-metal content typically increases with depth

Map number (Fig. 1)	District	County	Gold (oz)	Silver (oz)
1.	Mogollon*	Catron	365,000	>20,000,000
2.	Steeple Rock*,‡	Grant	151,000	3,400,000
3.	Chloride*	Sierra	9,000	2,000,000
4.	Socorro Peak*	Socorro	--	750,000
5.	Cochiti‡	Sandoval	42,000	210,000
6.	Kimball	Hidalgo	1,500	400,000
7.	Rosedale*	Socorro	28,000	10,000
8.	San Jose*,‡	Socorro	<1,000	13,000
9.	Macho	Sierra	100	20,000
10.	Gillespie	Hidalgo	5	16,000
11.	Doña Ana Mountains	Doña Ana	100	5,000
12.	Red River‡	Taos	65	1,103
13.	Cat Mountain*	Socorro	53	1,302
14.	Hop Canyon*	Socorro	82	1,137
15.	Fremont	Hidalgo	10	10,000
16.	Old Hadley	Luna	150	533
17.	North Magdalena*	Socorro	35	149
18.	Cora Miller*	Grant	W	W
19.	Wilcox*,‡	Catron	some	17
20.	Alum Mountain*,‡	Grant	1	21
21.	Carrizalillo	Luna	some	some
22.	Taylor*	Socorro	--	some
23.	Goldsboro*	Sierra	some	some
24.	Council Rock*	Socorro	--	some
25.	Abbe Spring*	Socorro	--	some
26.	Bear Mountains*	Socorro	--	--
27.	Luis Lopez*	Socorro	--	--
28.	Gila Fluorspar*,‡	Grant	--	--
29.	San Francisco*	Grant	--	--
30.	Bueyeros	Harding	--	--
31.	Silver Tip	Hidalgo	--	--
32.	Cruces Basin	Rio Arriba	--	--
33.	San Lorenzo*	Socorro	--	--

TABLE 1. Volcanic-epithermal precious-metal deposits in New Mexico. *Deposits in Mogollon-Datil volcanic field. All deposits are low-sulfidation deposits (see Table 4) except for Alum Mountain (#20) which is of the high-sulfidation class. Alteration in some districts, designated by ‡, indicate that high-sulfidation deposits may occur in these districts.

in many districts (Table 2). The Steeple Rock district has yielded more lead and zinc than any other volcanic-epithermal district in New Mexico (Table 3). The Carlisle and Center mines in the Steeple Rock district contain significant base metals together with silver and gold; these veins consist of 5-20% sulfides as galena, sphalerite, chalcopyrite and pyrite (McLemore, 1993).

Nearly all volcanic-epithermal deposits in the Mogollon-Datil volcanic field are hosted by Oligocene to Miocene volcanic and intrusive rocks (Table 2). Epithermal veins are hosted locally by Paleozoic and Mesozoic sedimentary rocks in a few districts, such as Chloride (#3), Cat Mountain (#13) and Abbe Spring (#25), but these districts are near Oligocene-Miocene volcanic centers. Most low-sulfidation, volcanic-epithermal veins are Oligocene or younger in age and are significantly younger than the host rocks (Fig. 4, Table 2). In districts where age determinations and/or stratigraphic and structural relationships can be used to date mineralization, mineralization is typically 5-15 Ma younger than the host rocks (Table 2).

The low-sulfidation, volcanic-epithermal vein deposits are similar with respect to vein emplacement, mineralogy, chemistry, alteration and textures (Table 2). They are structurally controlled and fill open spaces along faults, fractures, and locally bedding planes. The veins commonly form prominent outcrops and are bifurcating, sinuous and pinch and swell along strike. Complex vein textures, especially brecciation, banded layering and cementation by quartz, are common to most districts and locally are associated with higher concentrations of metals (Buchanan, 1981;McLemore, 1993). Gangue minerals generally include quartz, calcite, sericite, pyrite, chlorite, iron and manganese oxides, and locally fluorite, barite and adularia. Gold and silver occur as native species, electrum, various silver minerals, and as solid inclusions within pyrite or galena. Base-metal sulfides, predominantly galena, sphalerite, chalcopyrite and pyrite, are found in most districts in at least trace amounts. Silicification and argillic, propylitic, sericitic and chloritic alteration are common in most districts. Older acid-sulfate alteration is also found in a few districts (Tables 1, 2); these areas are discussed separately in this report. Textures, fluid inclusion data and isotopic data indicate that boiling occurred in most districts (Table 2). Zoning of metals varies through a district. In some districts (e.g., Steeple Rock), base-metal content increases and calcite decreases with increasing depth. In the Mogollon district, copper and lead are more abundant in the northwestern and northern portions of the district, whereas silver and gold are predominant elsewhere (Ferguson, 1927; Kamilli, 1993). Geologic, geochemical and mineralogic data indicate that mineralization occurred in multiple pulses and was produced from low salinity (<5 eq. wt.% NaCl), slightly acidic to neutral pH fluids, at temperatures between 150° and 300°C, at relatively shallow depths (<1500 m) and low pressures (<150 bars; Buchanan, 1981; Kent, 1983; Eggleston et al., 1983; Forureu, 1984; Harrison, 1990; McLemore, 1993; Kamilli, 1994).

Volcanic-epithermal deposits occur in areas of complex regional and local structures that appear to control emplacement of magmas and subsequent flow of hydrothermal fluids. Many districts occur along or near the ring-fracture zone of calderas (i.e. Mogollon; Fig. 2; Table 2), but not all calderas are mineralized (Rytuba, 1981). A few individual volcanic-epithermal deposits occur in the vicinity of rhyolitic domes (i.e. Chloride; Harrison, 1988, 1990). Regional crustal lineaments appear to control mineralized faults in other volcanic-epithermal districts (i.e. the

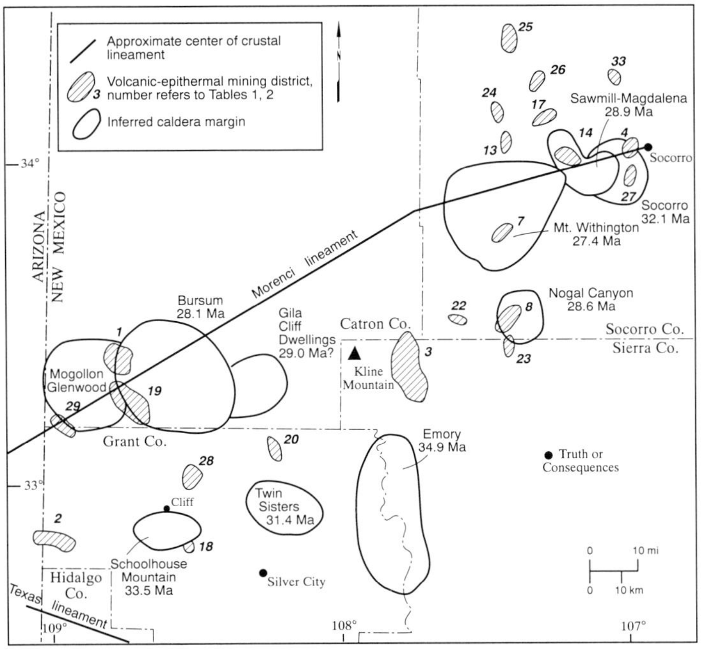

FIGURE 2. Volcanic-epithermal districts in the Mogollon-Datil volcanic field, west-central New Mexico. Numbers refer to Tables 1 and 2 and to text. Modified from McIntosh et al. (1992a, b) and Ratté (1989).

Texas lineament near Steeple Rock, Figs. 1, 2; McLemore, 1993). Although the mechanism of ground preparation varies from district to district, local faulting, fracturing and brecciation of the host rocks are critical in forming volcanic-epithermal vein deposits.

HIGH-SULFIDATION DEPOSITS

Only one deposit in New Mexico is classified as a high-sulfidation deposit, Alum Mountain (#20). Alum Mountain consists of intensely altered lava flows and pyroclastic and volcaniclastic rocks that are cut by rhyolite and quartz monzonite dikes and sills (Ratté et al., 1979). Alteration minerals include quartz, opal, clays, alunite, halotrichite and alunogen (Hayes, 1907; Ratté et al., 1979). Disseminated veinlets, thin stockwork veins and disseminations of quartz and calcite contain precious metals (Ratté et al., 1979) and were mined for gold and silver in 1945 (USBM Mineral Yearbooks). Drilling results are not known. However, areas of intense acid-sulfate alteration are found elsewhere in the Mogollon-Datil volcanic field, which may have potential for additional undiscovered high-sulfidation deposits.

High-sulfidation epithermal deposits are characterized by areas of acid-sulfate alteration. Acid-sulfate alteration, a subclass of advanced-argillic alteration, is spatially associated with many types of mineral deposits, including volcanic-epithermal deposits. Acid-sulfate alteration is characterized by the mineral assemblage containing quartz, kaolinite, alunite, and locally pyrite and pyrophyllite, and forms as a result of base leaching by acidic fluids containing H_2SO_4. Original textures and mineralogy are commonly destroyed by this process. Recognition and understanding the genesis of acid-sulfate alteration are important in mineral exploration and the understanding of the formation of mineral deposits, because specific alteration types are associated with specific mineral deposit types.

Four types of environments can produce acid-sulfate alteration in an epithermal system (Rye et al., 1992): (1) atmospheric oxidation of sulfides (supergene); (2) atmospheric oxidation at the water table in a steam-heated environment (steam-heated); (3) disproportionation of magmatic sulfur to H_2S and H_2SO_4 (magmatic-hydrothermal); and (4) magmatic steam produced by rapid release of fluids from a magma (magmatic-steam). Specific types of mineral deposits are associated locally with each type of acid-sulfate alteration (Rye et al., 1992). Supergene acid-sulfate alteration may overlie a sulfide zone containing precious metals or it may overprint older sulfide mineralization. Steam-heated acid-sulfate alteration also may overlie sulfide mineralization. Gold mineralization occurs within acid-sulfate altered areas in a magmatic-hydrothermal environment. Magmatic-steam environments may overlie porphyry mineralization or are adjacent to low-sulfidation deposits.

Acid-sulfate alteration occurs in several volcanic-epithermal districts in the Mogollon-Datil volcanic field (Tables 1, 2): Steeple Rock, San Jose, Wilcox, Alum Mountain, and Gila Fluorspar. Acid-sulfate alteration also occurs at Kline Mountain near the Taylor Creek epithermal tin deposit in northwestern Sierra County (A, Fig. 2; Hall, 1978; Eggleston and Norman, 1986; Eggleston, 1987). Most areas of acid-sulfate alteration in the Mogollon-Datil volcanic field have been examined only briefly and the genesis of the alteration is not well established. Areas of acid-sulfate alteration occur in the vicinity of porphyry copper deposits at Chino and Tyrone. Stable-isotopic data on alunite from the Chino porphyry-copper deposit confirm a supergene origin (Field, 1966; Rye et al., 1992). K/Ar age determinations on alunite from the Tyrone porphyry-copper deposit indicate an age of 19.5 Ma, and other geologic data confirm a supergene origin for acid-sulfate alteration at Tyrone (Cook, 1993).

Several areas of acid-sulfate alteration in the Steeple Rock district were studied extensively by McLemore (1993). Three of these altered areas occur adjacent to younger low-sulfidation, epithermal-vein deposits, whereas other areas are in the vicinity of vein deposits. The acid-sulfate altered areas are mineralogically, chemically, texturally and probably temperature zoned, including a clay zone (outermost, lowest temperature), silicified zone (intermediate), and silica/chert zone (innermost, highest temperature). Alunite occurs in all three zones but enargite has not been identified from the district (McLemore, 1993). Some areas contain anomalously high gold assays. The acid-sulfate alteration grades outward to argillic and then to prophylitic alteration. The acid-sulfate alteration is of a similar age as the host rocks, as indicated by a K/Ar age determination on quartz-alunite whole rock from Saddleback Mountain (at Steeple Rock) of 31.3±1.1 Ma (Marvin et al., 1987, #175; Hedlund, 1993). The host andesite of Summit Mountain (at Steeple Rock) is dated as 31.3±2.2 Ma (K/Ar on hornblende, Hedlund, 1993). Some areas of acid-sulfate alteration in the Steeple Rock district are as young as 27 Ma, as determined from stratigraphic relationships (McLemore, 1993). Limited stable-isotopic data from Bitter Creek in the northern part of the Steeple Rock district (Field, 1966), together with mineral and chemical zonations, preserved textures, multiple horizons, stratigraphic relationships and age determinations (McLemore, 1993), are consistent with a magmatic-hydrothermal origin by acidic fluids at temperatures less than 340°C in a relatively shallow environment (<1.5 km) as defined by Rye et al. (1992). The acidic fluids are produced by disproportionation of magmatic sulfur with decreasing temperature as the fluids rose toward the surface. Gold is commonly carried in these magmatic-hydrothermal fluids and gold assays in acid-sulfate altered areas indicate good potential for undiscovered high-sulfidation gold deposits (Table 4; McLemore, 1993). The epithermal veins are younger than the acid-sulfate alteration, as shown by crosscutting relationships, stratigraphic position and age determinations (McLemore, 1993).

Elsewhere in the Mogollon-Datil volcanic field, known areas of acid-sulfate alteration have not been studied in the same detail as in the Steeple Rock district. However, evidence (such as gold assays and similar ages of the alteration and host rocks) suggests some of these areas may have been formed in a magmatic-hydrothermal environment (Table 4; White and Hedenquist, 1990; Rye et al., 1992). Anomalous gold and silver assays were reported in altered rocks at Alum Mountain (up to 0.28 oz/ton Au and 0.36 oz/ton Ag; Ratté et al., 1979, #528) and in the Gila Fluorspar district (0.02 oz/ton Au and 0.22 oz/ton Ag; Ratté et al., 1979, #476). Recent work by McOwen (1993) suggests that the quartz-sericite/illite-pyrite alteration in the Gila Fluorspar district is older than the regional chloritic (or argillic) alteration. The quartz-sericite/illite-pyrite alteration was produced by acidic, sulfur-rich fluids but alunite and kaolinite have not been found. K/Ar age determinations

TABLE 2. Descriptions of volcanic-epithermal deposits in the Mogollon-Datil volcanic field, New Mexico. [1]Names and ages of formations updated from most recent reference and Osburn and Chapin (1983).[2]Interpretations from cited references and/or this author. [3]Includes element associations identified by Laughlin (1985) from NURE data. [4]From fluid inclusion data. [*]Indicates districts examined by the author.

Map no.	District	Host rocks[1]	Mineralization Age	Mineralization Style (Table 4)	Commodities produced (present) (Table 3)[3]	Predominant mineralogy of veins	Structural setting	Alteration	Evidence of boiling[2]	Temperature range °C[4]	Salinities eq. wt.% NaCl[4]	Comments	Selected references
1	Mogollon* (Coney)	Oligocene andesite flows, ash-flow tuffs, and rhyolite intrusives (33-23 Ma)	Miocene 16-18 Ma on adularia (Marvin et al., 1987, #78; Kamilli, 1994)	low sulfidation; quartz-calcite veins, hydrothermal brecciation, stockworks	Au,Ag,Cu,Pb, Zn,U (Ba,Mn,F)	quartz, calcite, chlorite, adularia, pyrite, electrum, fluorite, galena, sphalerite, bornite, chalcopyrite, gold,	along normal faults associated with the Bursum caldera, veins trend N60W 75-85N, N10E 70S	argillic, silicification, chloritization, sericitic, propylitic	very fine-grained quartz, fluid inclusion evidence, ore shoots with flat bottoms, loss of CO_2, bladed calcite	150-300	<5	copper and lead are more abundant in the north	Jones (1904), Ferguson (1927), Buchanan (1981), Kent (1983), Bornhorst et al. (1984), Wolfe (1985), Marvin et al. (1987), Kamilli (1993, 1994), Ratté (1981), Bornhorst and Kent (1985), Avery (1988)
2	Steeple Rock* (Carlisle, Duncan, Twin Peaks, Hells Hole)	Summit Mt. and Dark Thunder Canyon Fm (31-28 Ma; Oligocene)	Miocene veins 28-18 Ma; acid sulfate alteration 27-31 Ma (McLemore, 1993)	low sulfidation veins, high sulfidation alteration older than veins; quartz veins, hydrothermal brecciation	Au,Ag,Cu,Pb, Zn,Mn,F (Mo, alunite, clay)	quartz, calcite, pyrite, fluorite, galena, sphalerite, chalcopyrite, malachite, gold, silver, Mn and Fe oxides, sericite, alunite, jarosite, chlorite, epidote	on northern flank of Texas lineament, normal faults trend NE to E-W with steep dips to the W or S	older acid-sulfate grading laterally to argillic alteration; silicification, chloritization and propylitic adjacent to veins	very fine-grained quartz, loss of CO_2, bladed calcite and quartz pseudomorphs, hydrothermal brecciation	240-325	<5	base-metals increase with depth, central zone of base metals with gold-silver grading to gold-silver to outermost zone of fluorite-manganese, quartz veins grade vertically upward to calcite veins, excellent potential for additional deposits in Hells Hole area	Griggs and Wagner (1966), Biggerstaff (1974), Hedlund (1990, 1993), McLemore and Clark (1993), McLemore (1993), Ratté et al. (1982a)
3	Chloride (Apache No. 1, Phillipsburg, Grafton, Black Range)	Rubio Peak Formation (Eocene-Oligocene), Magdalena Group (Pennsylvanian)	Miocene 25-27 Ma on adularia (Harrison, 1988, 1990; Behr, 1988)	low sulfidation; local jasperoids	Au,Ag,Cu,Pb, Zn,Sn (F,Mo,Ba)	quartz, bornite, sphalerite, galena, chalcocite, chalcopyrite, pyrite, calcite, adularia, epidote, sericite, fluorite, barite	southern part centered around Chloride Creek dome, northern veins trend N-S	propylitic, silicification, potassic, sericitic, argillic	very fine-grained quartz, hydrothermal brecciation, fluid inclusion evidence, gas analyses	220-300	<5 (Behr, 1988; Neal and Larson, 1986)	gold-copper veins toward the north grading southward to older silver-lead-zinc veins	Harley (1934), Maxwell and Heyl (1976, 1980), Willard (1973), Freeman and Harrison (1984) Harrison (1986, 1988, 1990), Lovering and Heyl (1989), Neal and Larson (1986), Behr (1988)
4	Socorro Peak*	Socorro Peak Rhyolite 12-7 Ma, Popotosa Formation	Miocene 10-12 Ma (Chamberlin, 1980)	low sulfidation; silver veins	Ag,Pb (Au, Ba,F,W,V, As,Br)	quartz, barite, clay, silver, malachite, calcite, Mn-Fe oxides, galena, fluorite, chlorapyrite, argentite	faults associated with Socorro caldera, veins strike N20W to N15E and N60-70E with steep dips	silicification, argillic, regional potassic alteration	hydrothermal brecciation	—	—	best ore associated with abundant fluorite, grades range up to 20 oz/ton Ag and 0.25 oz/ton Au (Lasky, 1932), areas of silicification may be older than 10-12 Ma veins	Lasky (1932), North (1983), Chamberlin (1980), Moats and Queen (1981)
7	Rosedale*	South Canyon Tuff 26.7 Ma, Lemitar Tuff 28.4 Ma	Miocene(?)	low sulfidation veins, hydrothermal brecciation	Au,Ag (U,Cu,Mn)	quartz, gold, Mn oxides, Fe oxides, fluorite chlorite	associated with Mt. Withington caldera ring-fracture zone, veins strike N20W 80E	argillic, silicification adjacent to veins	very fine-grained quartz, hydrothermal brecciation	—	—	veins 3-5 ft wide, grades up to 0.5 oz/ton Au, minor placer deposits	Jones (1904), Lasky (1932), Lindgren et al. (1910), North (1983), Ferguson (1990), Neubert (1983), Johnson (1972)
8	San Jose*	Spears Formation 33-39 Ma, Vicks Peak Tuff 28.6 Ma	Oligocene to Miocene(?) veins, acid-sulfate alteration less than 28.6 Ma	low sulfidation veins, hydrothermal brecciation, high sulfidation alteration	Au,Ag,Cu Pb (Mo)	quartz, gold, chlorargyrite, silver, malachite, stephanite(?), pyrite, adularia, sericite, cerargyrite, chlorite, alunite	associated with Nogal Canyon caldera ring-fracture zone, veins strike N10-30E	acid-sulfate alteration grading laterally to argillic and phyllic alteration; silicification near veins	bladed calcite and pseudomorphs of quartz, hydrothermal brecciation	—	—	veins 1-5 ft wide, central zone of acid-sulfate alteratin consists of massive silica cap	Lasky (1932), Lindgren et al. (1910), North (1983), Forurea (1984)
13	Cat Mountain*	Spears Formation 33-34 Ma, Abo Formation (Permian), Rock House Canyon Tuff	Oligocene to Miocene(?)	low sulfidation, veins, brecciation	Au,Ag,Cu (U,F,Ba,W)	quartz, barite, fluorite, calcite, pyrite, galena, malachite, cerussite, wulfenite, gold, chalcopyrite	associated with Morenci lineament veins strike N50-60E, 80-90NW and N30W, north of Mt. Withington caldera	silicification, argillic, propylitic, pyritic	—	180-230	<5	veins converge at depth, veins cut basaltic dikes, widths up to 8 ft, lead isotope data confirms an epithermal origin (modern leads)	Jones (1904), Wilkinson (1976), North (1983), Lasky (1932), Slawson and Austin (1962)
14	Hop Canyon (Hope Canyon)	Hells Mesa Tuff 32.1 Ma, Sawmill Canyon Formation 31-29 Ma, Lemitar Tuff 28.4 Ma	Oligocene to Miocene(?)	low sulfidation veins	Cu,Ag,Au, Pb (Ba,Zn)	quartz, gold, copper sulfides and carbonates	associated with Sawmill Canyon-Magdalena caldera, veins strike N30W N20E, N10W, 70SE	silicification, argillic	—	—	—	—	Lindgren et al. (1910), Lasky (1932), Allen (1979), North (1983)
17	North Magdalena (Pueblo, Silver Hill, St. Viciente)	La Jara Peak Basaltic Andesite 24-31 Ma	Oligocene to Miocene(?)	low sulfidation veins	Cu,Ag,Au,Pb (V,Zn,Ba)	quartz, calcite, copper minerals, galena, argentite sphalerite, barite, anglesite, hematite	associated with Socorro transverse shear zone, veins trend N70W to N35E, 60-90° dips north Sawmill Canyon-Magdalena caldera	silicification, argillic, propylitic, hematization	—	—	—	lead and vanadium high toward the east (Pueblo) and copper high westward (Silver Hill), veins up to 6 ft wide	North (1983), Lasky (1932) Simon (1973)
18	Cora Miller	Tertiary rhyolite, ash-flow tuff	Miocene to Oligocene(?)	low sulfidation veins	Au,Ag,Cu,Pb (Mn)	quartz, gold, malachite	veins strike N70-75E, associated with Schoolhouse Mountain caldera ring-fracture zone	silicification, argillic	—	—	—	1-5 ft wide vein	Gillerman (1964), Lindgren et al. (1910), Richter and Lawrence (1983)
19	Wilcox*	Oligocene andesite and rhyolite intrusives	Miocene veins Oligocene acid-sulfate alteration 31-33 Ma (Marvin et al., 1987, #53)	low sulfidation veins, older high sulfidation alteration	Cu,Au,Ag,Te, F, (Pb, Zn, Mo, Cd)	quartz, alunite, kaolinite, pyrite, chlorite, calcite, fluorite	veins associated with Bursum caldera ring-fracture zone	acid-sulfate, silicification, chloritization, argillic, sericitic, propylitic	hydrothermal brecciation	150-210	<5	5 tons of tellurium ore produced	Arnold (1974), Ratté et al. (1979), Ratté and Stotelmeyer (1984), Marvin et al. (1987), Bornhorst et al. (1984)

Map no.	District	Host rocks[1]	Mineralization Age	Mineralization Style (Table 4)	Commodities produced (present) (Table 3)[3]	Predominant mineralogy of veins	Structural setting	Alteration	Evidence of boiling[2]	Temperature range °C[4]	Salinities eq. wt.% NaCl[4]	Comments	Selected references
20	Alum Mountain (Gila River, Alunogen Copperas Creek)	volcanic complex of Alum Mountain 30 Ma (Ratté et al., 1979)	Oligocene acid-sulfate alteration 30 Ma (Ratté et al., 1979; Marvin et al., 1987, #98, 99)	high sulfidation disseminated and thin veins and stockwork veins	Au,Ag, alum (Cu,Pb,Zn, clay, Ga)	quartz, alum, alunogen, pyrite, gold, kaolinite, calcite, clay, sericite	veins associated with altered rhyolite domes, veins strike N50W, district is between Gila Cliff Dwellings and Twin Sisters calderas	acid-sulfate silicification	hydrothermal brecciation	—	—	one assay of 0.28 oz/ton Au, 0.36 oz/ton Ag (Ratté et al., 1979, #528)	Ratté and Gaskill (1975), Ratté et al. (1979), Ratté and Stotelmeyer (1984), Marvin et al. (1987), Hayes (1907)
22	Taylor (Ojo Caliente #2)	Tertiary rhyolite and andesite	Late Tertiary	low sulfidation veins	Cu,Pb,Ag (Au,Mn)	quartz, cerusite, chrysocolla, malachite, Mn and Fe oxides	veins strike N65-70E dip 75-80NW	silicification, prophylitic	—	—	—	mineralized zone up to 15 ft wide	Hillard (1969), Lasky (1932), North (1983), Griffitts and Alminas (1968)
23	Goldsboro (Goldsborough, Aragon Hill, Monticello)	Tertiary rhyolite intrusive and andesite flows	Oligocene to Miocene(?)	low sulfidation veins	Au,Ag (Sn, V)	quartz, fluorite, gold, pyrite, limonite	veins strike N10-15W 68-80SW	silicification	—	—	—	veins are 3-10 ft wide	Lasky (1932), Lovering and Heyl (1989), Harley (1934)
24	Council Rock* (Iron Mountain, Ten Mile)	Spears Formation 33-39 Ma, Hells Mesa Tuff 32.1 Ma	Oligocene to Miocene(?)	low sulfidation veins	Pb,Ag,Fe,Au, (Mn,Ba,F,Cu, U,Zn)	quartz, magnetite, barite, calcite, siderite, hematite, trace galena, Mn oxides, cerussite, gold, chlorargyrite	veins trend N10W to N45W and N30E to N60E	propylitic, argillic, pyritic, silicification	—	—	—	mines at intersections of veins forming pipe veins, veins are 2-5 ft wide, lead stable isotope data, confirms an epithermal origin (modern leads)	Lasky (1932), Chamberlin (1974), Jones (1904), North (1983), Slawson and Austin (1962)
25	Abbe Spring* (Springhill?)	faults cutting Baca and Chinle Formations and Dakota Sandstone	Late Tertiary	low sulfidation veins	Cu,Pb,Ag, (Ba,Zn)	chalcocite, quartz, clay, malachite, barite, bornite		argillic, silicification	—	—	—	—	Mayerson (1979), North (1983), Chapin et al. (1979)
26	Bear Mountains	La Jara Peak Basaltic Andesite 24-31 Ma	Oligocene to Miocene(?)	low sulfidation veins	(Cu,Ag,Sb,Zn)	quartz, pyrite hematite, calcite, chrysocolla	associated with Hells Mesa fault	argillic, silicification	—	—	—	veins up to 1 ft wide, assay reported by North (1985)—no Au, 0.75 oz/ton Ag	North (1985)
27	Luis Lopez*	Lemitar Tuff, South Canyon Tuff, Hells Mesa Tuff, 32.1 Ma, basaltic andesite	Miocene-Pliocene as young as 3-7 Ma (Eggleston et al., 1983)	low sulfidation Mn veins	Mn (Au,Ag,Zn, W,Pb)	Mn oxides, calcite, quartz	associated with Socorro caldera and Rio Grande rift structures, veins strike N15W to N30W with steep dips	regional potassic alteration with minor silicification adjacent to veins	fluid inclusion evidence, gas analyses	150-375	<1	veins fill vertical fractures, suggesting mineralization may be younger than 10-12 Ma (C. Chapin, personal commun., 1/24/94), oxidized epithermal system	Hewitt (1964), Norman et al. (1983), Eggleston et al. (1983), North and McLemore (1987)
28	Gila Fluorspar* (Brock Canyon)	Oligocene andesite flows and rhyolite intrusive	veins are Miocene, acid-sulfate alteration is older (Ratté et al., 1979; McOwen, 1993)	low sulfidation veins, high sulfidation alteration older than the low sulfidation veins	F (Au,Ag,U, Cu,Pb,Zn,Mo)	fluorite, quartz, Mn and Fe oxides, kaolinite, calcite, barite, epidote, chlorite, pyrite	normal faults strike N45E with steep dips, district is between Bursum and Schoolhouse Mountain calderas	acid-sulfate(?) and sericitic alteration with pyritization and silicification surrounds a core of argillic, propylitic and chloritic alteration	hydrothermal brecciation	—	—	—	Ratté et al. (1979), Gillerman (1964), Ratté and Stotelmeyer (1984), Rothrock et al. (1946), Backer (1974), McOwen (1993)
29	San Francisco*	Oligocene rhyolite and andesite flows and intrusives	Oligocene to Miocene(?)	low sulfidation veins	(Au,Ag,Cu, Mo,Sb,Mn)	quartz, calcite	veins associated with Mogollon caldera ring-fracture zone	argillic, silicification adjacent to veins	—	—	—	anomalous values as high as 5000 ppm Mo, 100 ppm Ag, 4 ppm Au, 500 ppm Cu, 100 ppm Sb, placer gold deposits reported	Ratté et al. (1982b), U.S. Geological Survey et al. (1969)
33	San Lorenzo* (Jerome)	Oligocene volcanic rocks	Oligocene to Miocene(?)	low sulfidation veins	Cu,Ag (U,Au)	malachite, chrysocolla, quartz, alunite, kaolinite, tridymite	north of Socorro caldera and in Rio Grande rift zone	argillic, silicification adjacent to veins	—	—	—	dump samples assay as high as 1.8 oz/ton Ag (North, 1983)	North (1983), Lasky (1932)
ACID-SULFATE ALTERED AREA (base and precious metals unknown)													
A	Kline Mountain	Oligocene volcanics	Oligocene 28.4 Ma (Eggleston and Norman, 1986)	high sulfidation alteration	clay (alunite)	quartz, alunite, kaolinite, tridymite	rhyolitic dome complex	silicification forms innermost and uppermost zone within acid-sulfate alteration	—	—	—	no precious or base metals indicated at surface, near Taylor Creek tin deposit, central zone of massive silica cap	Hall (1978), Eggleston and Norman (1986), Eggleston (1987)

on alunite and altered whole rock in the Wilcox district (33.8±1.2 Ma and 32.0±1.7 Ma; Marvin et al., 1987, #53) are essentially the same age as the host rocks. The alteration at Alum Mountain, dated as 30 Ma (K/Ar on feldspar and biotite, Marvin et al., 1987, #98, 99), is also of a similar age as the host rock. The acid-sulfate alteration at Kline Mountain, near the Taylor Creek tin deposit, is 28.4±1.2 Ma (K/Ar on alunite; Eggleston, 1987), similar to that of the host rock. The acid-sulfate alteration in the San Jose district occurs in the Vicks Peak Tuff, which is 28.6 Ma (McIntosh et al., 1992a). Reconnaissance examination by the author of acid-sulfate altered areas in some districts suggests chemical and mineralogical zonations may exist similar to those found in the Steeple Rock district. These limited data and observations are consistent, but not necessarily conclusive, with a magmatic-hydrothermal origin for these areas of acid-sulfate alteration.

The acid-sulfate alteration found in volcanic-epithermal districts in the Mogollon-Datil volcanic field appear to be restricted in time to 27-33 Ma (Table 2, Fig. 4). In addition, acid-sulfate alteration is typically older than the precious- and base-metal epithermal veins in these districts (Fig. 4), as indicated by crosscutting relationships and limited age determinations. All acid-sulfate altered areas should be examined in more detail to determine the origin of the acid-sulfate alteration and the resource potential for high-sulfidation gold and silver deposits.

TABLE 3. Estimated production of volcanic epithermal deposits in the Mogollon-Datil volcanic field (USBM Mineral Yearbooks).

Map No.	District	Year of discovery	Periods of production	Ore (short tons)	Copper (pounds)	Gold (ounces)	Silver (ounces)	Lead (pounds)	Zinc (pounds)	Comments	Additional references
1	Mogollon	1875	1902-1969	--	1,500,000	365,000	>20,000,000	1,200,000	some	1879-1957 over $24 million	Ferguson (1927), Anderson (1957), Jones (1965), North and McLemore (1986)
2	Steeple Rock	1860	1880-1991	365,000	1,200,000	151,000	3,400,000	5,000,000	4,000,000	Over $10 million produced	Griggs and Wagner (1966), Anderson (1957), McLemore (1993)
3	Chloride	1879	1934-1988	--	3,060,000	10,000	2,340,000	1,300,000	some	1880-1988 about $20 million produced	Harrison (1986, 1990)
4	Socorro Peak	1867	1867-1900	--	--	--	750,000	Some	--	<$1 million produced	North (1983)
7	Rosedale	1882	1882-1981	<100,000	--	27,000	10,000	--	--	Minor placer gold reported	North (1983), Johnson (1972), North and McLemore (1986)
8	San Jose	1900	Prior to 1946	5,000	250	900	13,000	100	--	--	North (1983), Neubert (1983)
13	Cat Mountain	1870	?	--	some	<100	1,300	--	--	--	North (1983)
14	Hop Canyon	1880	1913-1941	500	7,000	<100	1,100	200	--	--	Loughlin and Koshman (1942), North (1983)
17	N. Magdalena	1863	Prior to 1957	<500	some	35	<200	1,400	--	--	North (1983)
18	Cora Miller	1880	1940-1941	--	some	some	some	some	--	--	Gillerman (1964)
19	Wilcox	1879	1941	2	some	some	19	--	--	Additional ore produced about 1900	Jones (1904)
20	Alum Mountain		1893	1945	3	--	1	21	--	--	--
22	Taylor		?	--	some	--	some	some	--	--	North (1983)
23	Goldsboro		?	--	--	some	some	--	--	--	
24	Council Rock	1881	1880s	--	--	--	some	some	--	--	Jones (1904)
25	Abbe Spring	Prior to 1904	?	--	some	--	some	--	--	--	
26	Bear Mountain			None	--	--	--	--	--	--	-- --
27	Luis Lopez		None	--	--	--	--	--	--	Mn production	--
28	Gila Fluorspar	1880	None	--	--	--	--	--	--	--	
29	San Francisco		None	--	--	--	--	--	--	--	--
33	San Lorenzo	1901	None	--	--	--	--	--	--	--	
TOTAL PRODUCTION		--	--	--	5,767,250	554,135	26,515,640	7,501,700	--	--	

SUMMARY

Low-sulfidation, volcanic-epithermal vein deposits carrying precious metals are widespread in the Mogollon-Datil volcanic province. Mineralization is controlled by complex regional and local structures and proximity to centers of volcanic activity. Geologic and geochemical data suggest that mineralization occurred in multiple pulses in a shallow hydrothermal environment from both meteoric and magmatic fluids.

High-sulfidation (acid-sulfate), volcanic-epithermal vein deposits carrying precious metals are found only at Alum Mountain, although areas of acid-sulfate alteration are present in several areas. These areas of acid-sulfate alteration appear to be restricted in time to 27-33 Ma and are older than the low-sulfidation, volcanic-epithermal vein deposits. Geologic and geochemical data suggest that these altered areas are formed by acidic fluids in a shallow environment. Limited data (sulfur isotope data, mineral and chemical zonations, presence of pyrophyllite, gold assays) are consistent with a magmatic-hydrothermal origin, which is capable of carrying precious metals to form high-sulfidation deposits.

RESOURCE POTENTIAL

The potential for additional low-sulfidation, volcanic-epithermal vein deposits in most districts is good, because many favorable structures have not been adequately explored. Most districts have been examined only along known mineralized structures. Audio-magnetotelluric resistivity profiles, character of the veins, and probable offset by the Queen fault in the Mogollon district suggest that the area east of the Queen fault may contain significant mineral deposits (Kamilli, 1993, 1994). Two areas in the Steeple Rock district have been drilled and mineralization was encountered. Queenstake Resources Ltd. delineated 155,535 tons of ore averaging 0.11 oz/ton Au and 3.45 oz/ton Ag at the Jim Crow-Imperial mines (unpublished Queenstake Resources press release, 4/2/87; McLemore, 1993). Biron Bay Resources, Ltd. announced the discovery of a deposit containing 1.45 million tons of ore grading 0.179 oz/ton Au and 10.26 oz/ton Ag along the Summit vein (Petroleum and Mining Review, May 1992, p. 2; McLemore, 1993).

TABLE 4. Characteristics distinguishing the low-sulfidation (adularia-sericite) and high-sulfidation (alunite-kaolinite) deposits (from Hayba et al., 1985; Cox and Singer, 1986; Heald et al., 1987; Berger and Henley, 1989; White and Hedenquist, 1990).

	HIGH SULFIDATION (Alunite-kaolinite; acid-sulfate)	LOW SULFIDATION (Adularia-sericite)
Structural Setting	intrusive centers	structurally complex
Host rocks	rhyodacite common, acid to intermediate volcanics	various acid to intermediate volcanic and intrusive rocks
Timing of ore and host	typically similar ages of the host and ore	variable
Ore mineralogy	Higher sulfide mineral assemblage Enargite, pyrite, gold, electrum, base-metal sulfides including covellite no selenides Mn minerals rare Bismuthinite	Lower sulfide mineral assemblage Argentite, tetrahedrite, tennantite, silver, gold, base-metal sulfides selenides Mn minerals common No bismuthinite
Alteration/gangue mineralogy	Chlorite rare Chalcedony and adularia absent unless overprinting, abundant alunite and pyrophyllite	Chlorite present Chalcedony common, adularia, minor alunite and pyrophyllite
Production	Gold, silver, and copper	Gold, silver, variable base metals
Alteration	Extensive propylitic with advanced argillic to argillic with hypogene alunite and kaolinite, but no adularia	Extensive propylitic with sericitic to argillic with supergene alunite and some kaolinite and abundant adularia
Localizing controls	any faults or fracture zones related to volcanic centers	major regional faults or subvolcanics
General character of fluids	low to high salinity meteoric with magmatic waters pH acid from magmatic HCL and H_2S oxidized to reduced Total S typically high Base-metal content low to high	low salinity meteoric waters dominant pH near neutral reduced Total S low Base metal content variable

Areas of acid-sulfate alteration have been examined for the potential for porphyry-copper deposits without much success. At least three areas of acid-sulfate alteration in the Steeple Rock district were studied by companies for the potential of high-sulfidation epithermal deposits. Although intercepts containing gold assays were encountered, no reserves were developed (McLemore, 1993). Other areas of acid-sulfate alteration in the Mogollon-Datil volcanic province should be examined to determine the origin of acid-sulfate alteration and the potential for high-sulfidation precious-metal deposits. Acid-sulfate alteration in some cases is also found above porphyry copper and/or molybdenum deposits.

ACKNOWLEDGMENTS

This report is part of ongoing studies of mineral resources in New Mexico and, in part, related to a dissertation on the Steeple Rock mining district. The dissertation was sponsored by the New Mexico Bureau of Mines and Mineral Resources, the Department of Geological Sciences at the University of Texas at El Paso, and Mount Royal Mining Co. Many people have assisted with these studies over the years. I would particularly like to thank Charles Chapin, Richard Chamberlin, Robert Eveleth (NMBMMR), Miles Silberman, Robert Kamilli (U.S. Geological Survey), and Robert North (Phelps Dodge Corp.) for their enlightening discussions and access to unpublished reports. Thanks also to Tim Pease (NMIMT) for technical assistance. Robert Kamilli and Miles Silberman reviewed this manuscript and their comments are appreciated. Lynne Hemenway and Terry Telles typed the manuscript and Kathryn Campbell drafted the figures. I also thank Kenneth Clark (UTEP) for his advice, encouragement, patience and discussions throughout the time of the dissertation study. Finally, I would like to acknowledge the prior work of many geologists in the various districts, because without their detailed studies, this composite study could not have been completed at this time.

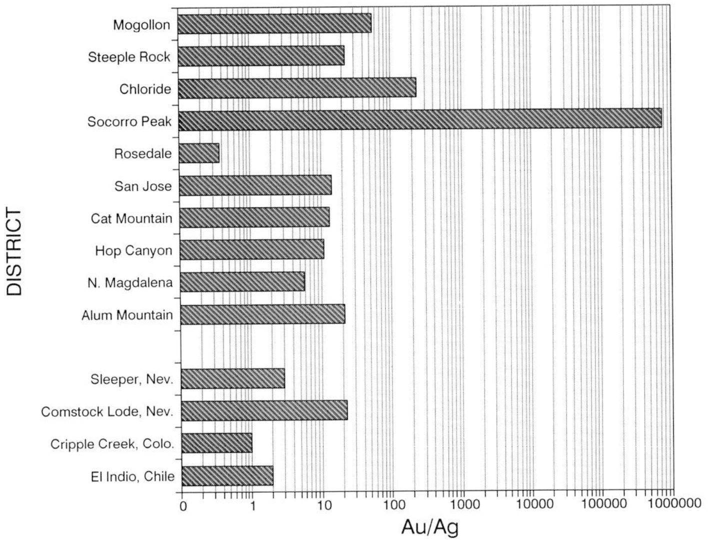

FIGURE 3. Approximate Ag/Au ratios for volcanic-epithermal deposits in New Mexico based on production (Table 3). Ag/Au ratios for Sleeper, Comstock Lode, Cripple Creek and El Indio deposits are from Sillitoe (1993).

REFERENCES

Allen, P., 1979, Geology of the west flank of the Magdalena Mountains south of the Kelly mining district, Socorro County, New Mexico [M. S. thesis]: Socorro, New Mexico Institute of Mining and Technology, 161 p. (New Mexico Bureau of Mines and Mineral Resources, Open-file Report 119.)

Anderson, E. C., 1957, The metal resources of New Mexico and their economic features through 1954: New Mexico Bureau of Mines and Mineral Resources, Bulletin 39, 183 p.

Arnold, R. I., 1974, Geology and mineral deposits of Little Whitewater Canyon, Holt gulch, and Goddard Canyon, Catron County, New Mexico [M. S. Thesis]: El Paso, University of Texas at El Paso, 63 p.

TABLE 5. Grade and tonnage from production of selected volcanic-epithermal deposits in the Mogollon-Datil volcanic field in New Mexico. *Estimated reserves of the St. Cloud mine as of 1982.

Map no.	District (period of production)	Ore (short tons)	Grade Au (oz/ton)	Ag (oz/ton)	Cu %	Pb %	Zn %	Source
1	Mogollon (1904-1925)	1,388,506	0.20	9.5	--	--	--	Ferguson (1927)
2	Steeple Rock (1904-1959)	199,095	0.19	7.7	0.3	1.3	1.1	Griggs and Wagner (1966)
3	Chloride* (1982)	375,217	0.04	8.9	1.4	1.0	1.3	Freeman and Harrison (1984)
7	Rosedale	81,828	0.16	0.12				North (1983)
8	San Jose	4,701	0.19	2.7	--	--	--	North (1983)

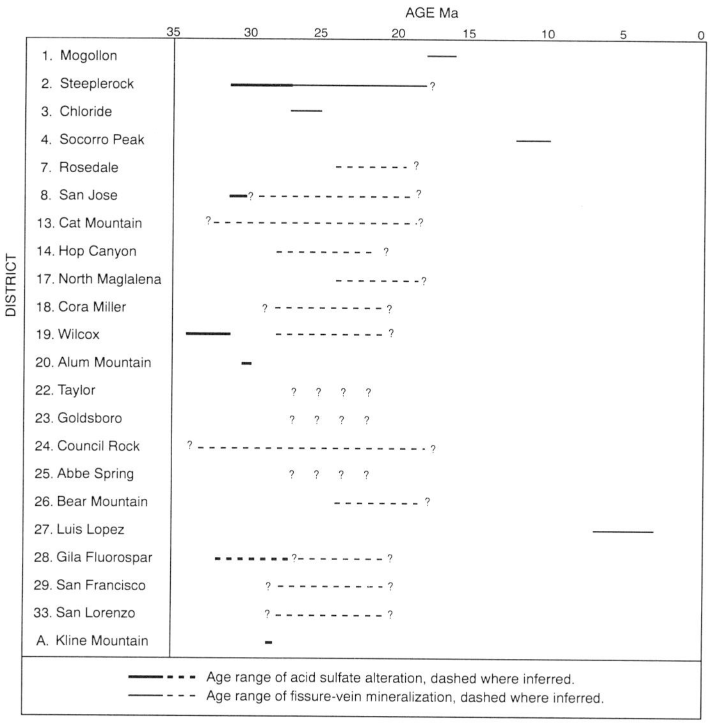

FIGURE 4. High-low graph showing age ranges of acid-sulfate alteration and mineralization in volcanic-epithermal districts in the Mogollon-Datil volcanic field. Dashed lines are inferred age ranges from stratigraphic data (Table 2). ? indicates age range may be greater or less than shown.

Avery, D. C., 1988, Flow structures and zonation of ring fracture domes and their control in the Mogollon mining district, Catron County, New Mexico [M. S. thesis]: Las Cruces, New Mexico State University.

Backer, H. A., 1974, Geology of the Gila fluorspar district (Brock Canyon volcanic complex), Grant County, New Mexico: New Mexico Geological Society, Guidebook 25, p. 377.

Bagby, W. C. and Berger, B. R., 1985, Geologic characteristics of sediment-hosted, disseminated precious-metal deposits in the western United States; *in* Berger, B. R. and Bethke, P. M., eds., Geology and geochemistry of epithermal systems: Reviews in Economic Geology, v. 2, p. 169-202.

Behr, C., 1988, Geochemical analyses of ore fluids from the St. Cloud-U.S. Treasury vein system, Chloride mining district, New Mexico[M. S. thesis]: Socorro, New Mexico Institute of Mining and Technology, 125 p.

Berger, B. R. and Henley, R. W., 1989, Advances in the understanding of epithermal gold-silver deposits, with special reference to the western United States; *in* Eays, R. R., Ramsay, W. R. H. and Groves, D. I., eds., The geology of gold deposits: the perspective in 1988: Economic Geology, Monograph 6, p. 405-423.

Biggerstaff, B. P., 1974, Geology and ore deposits of the Steeple Rock-Twin Peaks area, Grant County, New Mexico [M.S. thesis]: El Paso, University of Texas at El Paso, 102 p.

Bonham, H. F., Jr., 1988, Models for volcanic-hosted epithermal precious metal deposits; *in* Bulk mineable precious metal deposits of the western United States: Geological Society of Nevada, Symposium Proceedings, p. 259-271.

Bornhorst, T. J. and Kent, G. R., 1985, Geochemistry of host-rock alteration at the Eberle Mine, Mogollon mining district, southwestern New Mexico: New Mexico Bureau of Mines and Mineral Resources, Circular 199, p. 7-16.

Bornhorst, T. J., Kent, G. R., Mann, K. L. and Richey, S. R., 1984, Temperature of mineralizations in Mogollon mining district and vicinity, southwest New Mexico: New Mexico Geology, v. 6, p. 53-55.

Buchanan, L. J., 1981, Precious metal deposits associated with volcanic environments in the Southwest: Arizona Geological Society Digest, v. 14, p. 237-262.

Chamberlin, R. M., 1974, Geology of the Council Rock district, Socorro County, New Mexico [M.S. thesis]: Socorro, New Mexico Institute of Mining and Technology, 143 p. (New Mexico Bureau of Mines and Mineral Resources, Open-file Report 40).

Chamberlin, R. M., 1980, Cenozoic stratigraphy and structure of the Socorro Peak volcanic center, central New Mexico [Ph.D. dissertation]: Golden, Colorado School of Mines, 532 p. (New Mexico Bureau of Mines and Mineral Resources, Open-file Report 118).

Chapin, C. E., Osburn, G. R., Hook, S. C., Masingill, G. L. and Frost, S. J., 1979, Coal, uranium, oil, and gas potential of the Riley-Puertecito area, Socorro County, New Mexico: New Mexico Bureau of Mines and Mineral Resources, Open-file Report 103, 38 p.

Cook, S. S., 1993, Supergene copper mineralization at the Tyrone mine, Grant County, New Mexico: Society for Mining, Metallurgy and Exploration, Inc., 1993 Annual Meeting and Exhibit, Program with Abstracts, p. 139.

Cox, D. P., and Singer, D. A., editors, 1986, Mineral deposit models: U.S. Geological Survey, Bulletin 1693, 379 p.

Dunham, K. C., 1935, The geology of the Organ Mountains with an account of the geology and mineral resources of Doña Ana County, New Mexico: New Mexico Bureau of Mines and Mineral Resources, Bulletin 11, 272 p.

Eggleston, T. L., 1987, The Taylor Creek district, New Mexico: Geology, petrology, and tin deposits [Ph.D. dissertation]: Socorro, New Mexico Institute of Mining and Technology, 473 p.

Eggleston, T. L., Norman, D. I., Chapin, C. E. and Savin, S., 1983, Geology, alteration and genesis of the Luis Lopez manganese district, New Mexico: New Mexico Geological Society, Guidebook 34, p. 241-246.

Eggleston, T. L. and Norman, D. I., 1986, A summary of the geology, geochemistry, and tin occurrences in the Black Range, New Mexico: New Mexico Geological Society, Guidebook 37, p. 173-178.

Ferguson, C. A., 1990, Geology of the Grassy Lookout 7½ minute quadrangle, Socorro County, New Mexico: New Mexico Bureau of Mines and Mineral Resources, Open-file Report 366, 40 p.

Ferguson, H. G., 1927, Geology and ore deposits of the Mogollon mining district, New Mexico: U. S. Geological Survey, Bulletin 787, 100 p.

Field, C. W., 1966, Sulfur isotope method for discriminating between sulfates of hypogene and supergene origin: Economic Geology, v. 61, p. 1428-1435.

Forureu, J., 1984, Geology, alteration, and precious metal reconnaissance of the Nogal Canyon area, San Mateo Mountains, New Mexico [M.S. thesis]: Fort Collins, Colorado State University, 163 p.

Freeman, P. S. and Harrison, R. W., 1984, Geology and development of the St. Cloud silver deposit, Sierra County, New Mexico: Arizona Geological Society Digest, v. 15, p. 231-233.

Gillerman, E., 1964, Mineral deposits of western Grant County, New Mexico: New Mexico Bureau of Mines and Mineral Resources, Bulletin 83, 213 p.

Griffitts, W. R. and Alminas, H. V., 1968, Geochemical evidence for possible concealed mineral deposits near the Monticello Box, northern Sierra Cuchillo, Socorro County, New Mexico: U.S. Geological Survey, Circular 600, 13 p.

Griggs, R. L., and Wagner, H. C., 1966, Geology and ore deposits of the Steeple Rock mining district, Grant County, New Mexico: U.S. Geological Survey, Bulletin 1222-E, 29 p.

Hall, R. G., 1978, World nonbauxite aluminum resources-alunite: U.S. Geological Survey, Professional Paper 1076A, 35 p.

Harley, G. T., 1934, The geology and ore deposits of Sierra County, New Mexico: New Mexico Bureau of Mines and Mineral Resources, Bulletin 10, 220 p.

Harrison, R. H., 1986, General geology of the Chloride mining district, Sierra and Catron Counties, New Mexico: New Mexico Geological Society, Guidebook 37, p. 265-272.

Harrison, R. W., 1988, Mineral paragenesis structure, and "ore shoot" geometry at the U. S. Treasury Mine, Chloride mining district, New Mexico: New Mexico Geology, v. 10, p. 10-11, 15.

Harrison, R. W., 1990, Cenozoic stratigraphy, structure, and epithermal mineralization of the north-central Black Range, New Mexico, in the regional geologic framework of south-central New Mexico [Ph.D. dissertation]: Socorro, New Mexico Institute of Mining and Technology, 402 p.

Hayba, D. O., Bethke, P. M., Heald, P. and Foley, N. K., 1985, Geologic, mineralogic, and geochemical characteristics of volcanic-hosted epithermal precious-metal deposits; *in* Berger, B. R. and Bethke, P. M., eds., Geology and geochemistry of epithermal systems: Reviews in Economic Geology, v. 2, p. 129-168.

Hayes, C. W., 1907, The Gila River alum deposits, New Mexico: U.S. Geological Survey, Bulletin 315, p. 215-223.

Heald, P., Foley, N. K. and Hayba, D. O., 1987, Comparative anatomy of volcanic-hosted epithermal deposits: acid-sulfate and adularia-sericite types: Economic Geology, v. 82, p. 1-26.

Hedlund, D. C., 1990, Geology and mineral deposits of the Steeple Rock and Duncan mining districts, Grant and Hidalgo Counties, New Mexico and Greenlee County, Arizona: U.S. Geological Survey, Open-file Report 90-239, 27 p.

Hedlund, D. C., 1993, Geologic map of the Tillie Hall Peak quadrangle, Greenlee County, Arizona and Grant County, New Mexico: U. S. Geological Survey, Map GQ-1715, scale 1:24,000.

Hewitt, D. F., 1964, Veins of hypogene manganese oxide minerals in the southwestern United States: Economic Geology, v. 59, p. 1429-1472.

Hillard, P. D., 1969, Geology and beryllium mineralization near Apache Warm Springs, Socorro County, New Mexico: New Mexico Bureau of Mines and Mineral Resources, Circular 103, 16 p.

Johnson, M. G., 1972, Placer gold deposits of New Mexico: U. S. Geological Survey, Bulletin 1348, 46 p.

Jones, F. A., 1904, New Mexico mines and minerals: Santa Fe, New Mexican Printing Company, 349 p.

Jones, W. R., 1965, Copper: New Mexico Bureau of Mines and Mineral Resources, Bulletin 87, p. 160-176.

Kamilli, R. J., 1993, Data indicate possibility of additional deposits in Mogollon mining district: Mineral Resources Newsletter (U.S. Geological Survey), v. 4, no. 2, p. 4.

Kamilli, R. J., 1994, The Mogollon mining district: Results from mapping, paragenesis, and fluid inclusion studies: U.S. Geological Survey, Circular 1103A, p. 52–53.

Kent, J.R., 1983, Temperature and age of precious metal vein mineralization and geochemistry of host rock alteration at the Eberle Mine, Mogollon mining district, southwestern New Mexico [M.S. thesis]: Houghton, Michigan Tech University, 146 p.

Lasky, S. G., 1932, The ore deposits of Socorro County, New Mexico: New Mexico Bureau of Mines and Mineral Resources, Bulletin 8, 139 p.

Laughlin, A. W., 1985, A computer-assisted mineral resource assessment of Socorro and Catron Counties, New Mexico: Los Alamos Laboratory, Open-file Report LA-UR-85-375, 201 p.

Lehrman, N. J., 1986, The McLaughlin Mine, Napa and Yolo Counties, California; *in* Precious-metal mineralization in hot-springs systems, Nevada-California: Nevada Bureau of Mines and Geology, Report 41, p. 85-89.

Lindgren, W., 1922, A suggestion for the terminology of certain mineral deposits: Economic Geology, v. 17, p. 202-294.

Lindgren, W., 1933, Mineral deposits, 4th edition: New York, McGraw-Hill, 930 p.

Lindgren, W., Graton, L. C. and Gordon, C. H., 1910, The ore deposits of New Mexico: U.S. Geological Survey, Professional Paper 68, 361 p.

Loughlin, G. F., and Koschmann, A. H., 1942, Geology and ore deposits of the Magdalena mining district, New Mexico: U.S. Geological Survey, Professional Paper 200, 168 p.

Lovering, T. G. and Heyl, A. V., 1989, Mineral belts in western Sierra County, New Mexico, suggested by mining districts, geology, and geochemical anomalies: U.S. Geological Survey, Bulletin 1876, 49 p.

Marvin, R. F., Naeser, C. W., Bikerman, M., Mehnert, H. H. and Ratté, J. C., 1987, Isotopic ages of post-Paleocene igneous rocks within and bordering the Clifton 1 degree x 2 degree quadrangle, Arizona-New Mexico: New Mexico Bureau of Mines and Mineral Resources, Bulletin 118, 63 p.

Maxwell, C. H. and Heyl, A. U., 1976, Preliminary geologic map of the Winston quadrangle, Sierra County, New Mexico: U.S. Geological Survey, Open-file Report 76-858, scale 1:24,000.

Maxwell, C. H. and Heyl, A. V., 1980, Mineralization and structure of the Hermosa, Chloride, Philipsburg areas, New Mexico: Global Tectonics and Metallogery, v. 1, p. 129-133.

Mayerson, D. L., 1979, Geology of the Corkscrew Canyon-Abbe Springs area, Socorro County, New Mexico [M.S. thesis]: Socorro, New Mexico Institute of Mining and Technology, 133 p. (New Mexico Bureau of Mines and Mineral Resources, Open-file Report 112).

McIntosh, W. C., Chapin, C. E., Ratté, J. C. and Sutter, J. F., 1992a, Time-stratigraphic framework for the Eocene-Oligocene Mogollon-Datil volcanic field southwestern New Mexico: Geological Society of America Bulletin, v. 104, p. 851-871.

McIntosh, W. C., Geissman, J. W., Chapin, C. E., Kunk, M. J. and Henry, C. D., 1992b, Calibration of the latest Eocene-Oligocene geomagnetic polarity time scale using $^{40}Ar/^{39}Ar$ dated ignimbrites: Geology, v. 20, p. 459-463.

McLemore, V. T. 1993, Geology and geochemistry of the mineralization and alteration in the Steeple Rock district, Grant County, New Mexico and Greenlee County, Arizona [Ph.D. dissertation]: El Paso, University of Texas at El Paso, 525 p. (New Mexico Bureau of Mines and Mineral Resources, Open-file Report 397)

McLemore, V. T., and Clark, K. F., 1993, Alteration and epithermal mineralization in the Steeple Rock mining district, Grant County, New Mexico and Greenlee County, Arizona: Conference Program and Extended Abstracts, Integrated Methods in Exploration and discovery, Society of Economic Geologists, AB69-70.

McOwen, L. K., 1993, The Brock Canyon volcanic comples, Grant County, New Mexico: volcanic evolution, alteration, and mineralization [M.S. thesis]: Tucson, University of Arizona, 167 p.

Moats, W. P. and Queen, L. D., 1981, Minerals of the Socorro Peak district: New Mexico Geology, v. 3, p. 16.

Neal, W. S. and Larson, P. B., 1986, Mineral and fluid geochemistry of the Hoosier vein, Chloride mining district, New Mexico: Geological Society of America, Abstracts with Programs, v. 18, p. 703.

Neubert, J. T., 1983, Mineral investigation of the Apache Kid and Mt. Washington Wilderness Area, Socorro County, New Mexico: U.S. Bureau of Mines, MLA 72-83, 34 p.

New Mexico Geological Society (NMGS), 1982, Geologic highway map of New Mexico: New Mexico Geological Society, scale 1:1,000,000.

Norman, D. I., Bazafishan, K. and Eggleston, T. L., 1983, Mineralization of the Luis Lopez epithermal manganese deposits in light of fluid inclusion and geological studies: New Mexico Geological Society, Guidebook 34, p. 247-251.

North, R. M., 1983, History and geology of the precious metal occurrences in Socorro County, New Mexico: New Mexico Geological Society, Guidebook 34, p. 261-268.

North, R. M., 1985, Geology and mineralogy of the Bear Mountains mining district, Socorro County, New Mexico: New Mexico Geology, v. 7, p. 42.

North, R. M. and McLemore, V. T., 1986, Silver and gold occurrences in New Mexico: New Mexico Bureau of Mines and Mineral Resources, Resource Map 15, 32 p.

North, R. M. and McLemore, V. T., 1987, Mineralization in the Luis Lopez mining district, Socorro County, New Mexico—a summary; *in* Guidebook for the 1987 Conference, 24th Annual Meeting of the Clay Minerals Society and 36th Annual Clay Minerals Conference: New Mexico Bureau of Mines and Mineral Resources, p. 68–73.

North, R. M. and McLemore, V. T., 1988, A classification of the precious metal deposits of New Mexico: *in* Bulk minable precious metal deposits of the western United States: Geological Society of Nevada, Symposium Proceedings, p. 625-659.

Osburn, G. R. and Chapin, C. E., 1983, Nomenclature for Cenozoic rocks of northeast Mogollon-Datil volcanic field, New Mexico: New Mexico Bureau of Mines and Mineral Resources, Stratigraphic Chart 1, 1 sheet.

Ratté, J. C., 1981, Geologic map of the Mogollon quadrangle, Catron County, New Mexico: U.S. Geological Survey, Geologic Quadrangle Map GQ-1557, scale 1:24,000.

Ratté, J. C., 1989, Days 3 and 4: selected volcanic features of the western Mogollon-Datil volcanic field: New Mexico Bureau of Mines and Mineral Resources, Memoir 46, p. 68–85.

Ratté, J. C. and Gaskill, D. L., 1975, Reconnaissance geologic map of the Gila Wilderness study area, southwestern New Mexico: U.S. Geological Survey, Miscellaneous Investigations Map I-886, scale 1:62,500.

Ratté, J. C., Gaskill, D. L., Eaton, G. P., Peterson, D. L., Stotelmeyer, R. B. and Meeves, H. C., 1979, Mineral resources of the Gila Primitive Area and Gila Wilderness, New Mexico: U.S. Geological Survey, Bulletin 1451, 229 p.

Ratté, J. C., Hassemer, J. R., Martin, R. A. and Briggs, J. P., 1982a, Mineral resource potential map of the Hells Hole Further Planning Area, Greenlee County, Arizona, and Grant County, New Mexico: U.S. Geological Survey, Miscellaneous Field Studies Map MF-1344-E, scale 1:62,500, 7 p. of text.

Ratté, J. C., Hassemer, J. R., Martin, R. A. and Lane, M., 1982b, Mineral resources potential of the Lower San Francisco Wilderness Study Area and Contiguous Roadless Area, Greenlee County, Arizona, and Catron and Grant Counties, New Mexico: U. S. Geological Survey, Miscellaneous Field Studies Map MF-1463C, scale 1:62,500, 6 p. of text.

Ratté, J. C. and Stoelmeyer, R. B., 1984, Gila Wilderness, New Mexico: U.S. Geological Survey, Professional Paper 1300, v. 2, p. 811-813.

Richter, D. H. and Lawrence, V. A., 1983, Mineral deposit map of the Silver City 1°x2° quadrangle, New Mexico and Arizona: U. S. Geological Survey, Miscellaneous Investigations Series Map 1-1310-B, scale 1:250,000.

Rothrock, H. E., Johnson, C. H. and Hahn, A. D., 1946, Fluorspar resources in New Mexico: New Mexico Bureau of Mines and Mineral Resources, Bulletin 21, 245 p.

Rye, R. O., Bethke, P. M. and Wasserman, M. D., 1992, The stable isotope geochemistry of acid-sulfate alteration: Economic Geology, v. 87, p. 225-264.

Rytuba, J. J., 1981, Relation of calderas to ore deposits in the western United States: Arizona Geological Society Digest, v. 14, p. 227–236.

Sillitoe, R. H., 1993, Giant and Bonanza gold deposits in the epithermal environment: assessment of potential genetic factors; *in* Whiting, B. H., Hodgson, C. J. and Mason, R., eds, Giant ore deposits: Society of Economic Geologists, Special Publication No. 2, p. 125–156.

Simon, D. B., 1973, Geology of the Silver Hill area, Socorro County, New Mexico [M.S. thesis]: Socorro, New Mexico Institute of Mining and Technology, 107 p. (New Mexico Bureau of Mines and Mineral Resources, Open-file Report 41).

Slawson, W. F. and Austin, C. F., 1962, An isotope study defines a geologic structure: Economic Geology, v. 57, p. 21-29.

U.S. Geological Survey, Arizona Bureau of Mines, and U. S. Bureau of Reclamation, 1969, Mineral and water resources of Arizona: Arizona Bureau of Mines, Bulletin 180, 638 p.

White, N. C. and Hedenquist, J. W., 1990, Epithermal environments and styles of mineralization: variations and their causes, and guidelines for exploration: Journal of Geochemical Exploration, v. 36, p. 445-474.

Wilkinson, W. J., Jr., 1976, Geology of the Tres Montosas-Cat Mountain area, Socorro County, New Mexico [M.S. thesis]: Socorro, New Mexico Institute of Mining and Technology, 167 p. (New Mexico Bureau of Mines and Mineral Resources, Open-file Report 39.)

Willard, M. E., 1973, Geology of the Luis Lopez manganese district, New Mexico: New Mexico Bureau of Mines and Mineral Resources, Open-file Report OF-186, 81 p.

Wolfe, C. A., 1985, Geology of the Dripping Gold Prospect, Mogollon Mining district: New Mexico Bureau of Mines and Mineral Resources, Circular 199, p. 33-40.

Plains of San Agustin from Point of Rocks Canyon at the northwest base of the San Mateo Mountains, New Mexico. View is N75°W. From left to right, landmarks on the skyline include Horse Mountain, Mangas Mountain, Little Alegres Mountain, Alegres Mountain (bilevel), Sugarloaf Mountain, and Crosby Mountains. Alegres Mountain, elev. 3118 m, is the highest point on the Mogollon slope of west-central New Mexico. Horse Mountain is a dissected dacitic to rhyolitic volcano of late Miocene age. All the other landmarks are capped by Bearwallow Mountain Andesite of late Oligocene age. Flat-topped hills in middle ground, to left of center, define an intrabasin horst capped by an Oligocene ignimbrite of the Mogollon Group. C-N playa is at the floor of the basin below the flat-topped hills. Rocks in foreground are pink welded ash-flow tuff which underlies low hill northwest of Highway 52. Camera station is in SE ¼ sec. 16, T4S, R8W. Altitude about 2230 m. Wayne Lambert photograph No. 93L81. August 15, 1993, 3:58 p.m. MDT.

GEOLOGY AND ALTERATION OF THE KLINE MOUNTAIN KAOLIN DEPOSIT, SIERRA COUNTY, NEW MEXICO

ISKENDER ISIK[1], KENNETH F. CLARK[2], AND GEORGE S. AUSTIN[3]

[1]Dept. of Mining Engineering, Osmangazi University, Eskisehir, Turkey; [2]Dept. of Geological Sciences, University of Texas at El Paso, El Paso, TX 79968; [3]New Mexico Bureau of Mines and Mineral Resources, Socorro, New Mexico 87801

Abstract—The kaolinized tuff of the Kline Mountain area lies on the eastern margin of the Mogollon-Datil volcano-tectonic province, a major mid-Tertiary volcanic center. On the eastern margin of this center is the extensional environment of the Rio Grande rift. Faults in the study area display dominant northwest and northeast trends. The stratigraphy in the Kline Mountain clay deposit area consists of mid-Tertiary bimodal volcanic and volcaniclastic deposits. The kaolin deposit occurs within the advanced argillic zone of the tuff of Kline Mountain as a result of hydrothermal alteration. The deposit is a result of kaolinization, alunitization and silicification that is related to mid-Tertiary hydrothermal activity associated with a rhyolite porphyry intrusion. On the basis of calculated mineral compositions from chemical analyses and XRD data, there is an inverse relationship between kaolinite percentage and proximity to the intrusion. Conversely, there is a proportional relationship between alunite content and proximity to the intrusion. XRD data indicate that kaolinite is a principal mineralogical constituent in the deposit whereas smectite was identified locally. The other major constituent is fine-grained silica (cristobalite and/or tridymite?). Small amounts of opaque oxides, and hematite-cassiterite mineralization associated with fractures, have also been identified. That part of the deposit that was investigated in detail contains over 3 million tons of clay and the whole deposit is reported to contain over 200 million tons. In spite of high standard brightness of 94%, the silica content is considered to be a drawback for use in the paper industry. In 1969, 900 tons of the deposit were sold as an oil absorbent. Recently, experiments have shown the suitability of the deposit in the manufacture of white brick.

INTRODUCTION

The Kline Mountain (White Horse) kaolin deposit occurs within an advanced-argillic alteration zone in volcanic and volcaniclastic rocks of mid-Tertiary age (Patterson and Holmes, 1965; Ericsen et al., 1970; Coney, 1976; Eggleston, 1987; Eggleston and Norman, 1983; Harrison, 1986). The deposit is situated on the northwestern flank of Kline Mountain in the Black Range, Sierra County, New Mexico (Fig. 1). The elevation of the deposit is 2338 m above sea level, and it is located on the south side of NM-59 where it crosses the Continental Divide in Sierra County, and within the Gila National Forest. Nearby settlements are Winston and Chloride. The prevailing climate in the area is semi-arid, and precipitation varies from about 20 cm/yr to more than 50 cm/yr. Rainfall occurs predominantly in July and August.

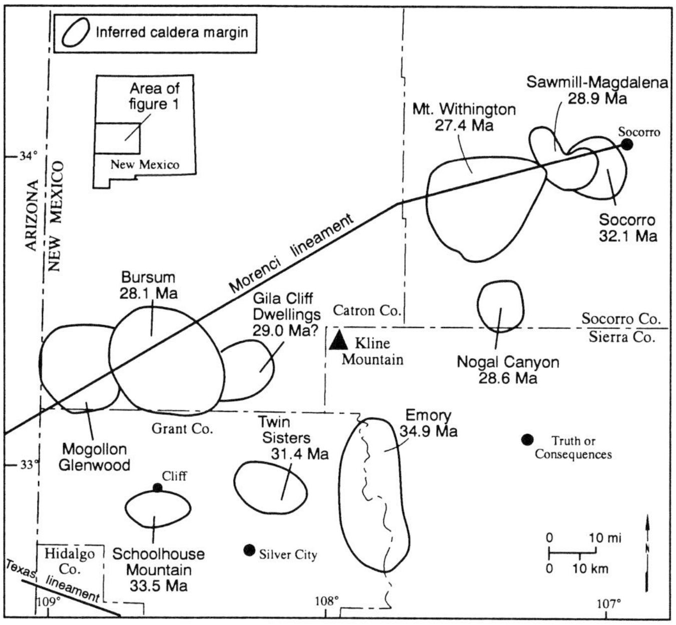

FIGURE 1. Location of Kline Mountain kaolinite deposit relative to major volcanic centers in the Mogollon-Datil volcanic field. Adapted from McLemore (this volume).

The deposit, investigated by F. L. Schneider in 1958, was used experimentally for its kaolin content in making ceramic tile. The deposit was sampled by the U.S. Bureau of Mines in 1959, and results of tests showed the kaolin to range from intermediate to a heavy-duty refractory material (Patterson and Holmes, 1965). Samples from the deposit were examined in some detail by DeVilliers (unpubl. reports for Inst. Paper Chemistry, Appleton, Wisconsin), Sallee in 1962 (unpubl. report for Inst. Paper Chemistry, Appleton, Wisconsin) and by Chowdhury in 1980 (unpubl. report for Dresser Industries). Evaluation for use of the kaolin as a paper coater was completed by the private sector. Fine-grained silica (cristobalite and/or tridymite?) within the kaolin has been reported as a major drawback for use in the paper industry. The whole deposit is estimated to have 200 million tons reserve (P. Roche, personal commun., 1991) with 38 to 39% Al_2O_3 and a very high standard brightness of 94%. About 900 tons of kaolin were sold in 1969 for use as an oil absorbent by Union Oil. The production of kaolin was by open-pit mining methods. After screening, crushed kaolin ore was shipped by trucks to Elephant Butte Lake for use in the Santa Barbara channel oil spill (P. Roche, personal commun., 1991). In 1993, firing experiments, conducted in El Paso, demonstrated the potential use of the Kline Mountain clay in white brick manufacture. This includes favorable characteristics of water absorption, plasticity and compressive strength. As of July, 1993, three different groups have shown interest in the Kline Mountain kaolin deposit: American Eagle Brick Company, El Paso, Texas; Internacional de Ceramica, Chihuahua, Mexico; and one unknown group that drilled the deposit between February and July, 1993 (Isik, 1993).

GEOLOGICAL SETTING

The kaolinized tuff of the Kline Mountain area lies on the eastern margin of the Mogollon-Datil volcano-tectonic province, a major mid-Tertiary volcanic center (Fig. 1). The eastern margin of this center is adjacent to the extensional environment of the Rio Grande rift. Local faults in the study area could have originated by the intrusion of the rhyolite porphyry of Kline Mountain dome, which may be part of the ring fracture intrusions related to the Gila Cliff Dwellings cauldron to the west (Rhodes, 1976). These faults have probably been reactivated by Basin-and-Range extension in the past 21 Ma (Elston, 1976). Faults in the study area display dominant northwest and northeast trends. The dominant structural style found in the study area is high-angle normal faulting. The stratigraphy in the Kline Mountain kaolin deposit area consists of mid-Tertiary bimodal volcanic and volcanoclastic deposits that consist of basaltic andesite lavas, high-silica rhyolite lavas and pyroclastic material. (Fig. 2). The lower units of the volcanic sequence were then intruded by the rhyolite porphyry of Kline Mountain (Coney, 1976; Eggleston, 1987). Accompanying hydrothermal alteration produced several mineralogical assemblages within and peripheral to the intrusion. Alunite in the advanced argillic assemblage was dated by the K-Ar method at 28.4 ± 1.2 Ma (Eggleston and Norman, 1987).

HYDROTHERMAL ALTERATION

Volcanic units and the rhyolite porphyry of Kline Mountain have undergone hydrothermal alteration. Alteration is most intense near the intrusive contact of the rhyolite porphyry and gradually diminishes distally. The altered zone has an elliptical shape that trends roughly northwest. Four distinct alteration assemblages are recognized in the study area (Isik, 1993): (1) weak propylitic alteration, (2) argillic alteration, (3) advanced argillic alteration, and (4) silicification. Alteration of the volcanic units occurs primarily as replacement, that is, a new mineral assemblage takes the place of the primary minerals. However, silicification is also due to primary introduction of silica by fluids and residue. The basaltic andesite of Poverty Creek displays weak propylitic alteration which is characterized by epidote, calcite and chlorite. It diminishes in intensity towards the northwest from the clay deposit (Eggleston, 1987).

The argillic alteration assemblage overlaps the advanced-argillic assemblage, being first noticeable about 2 km northeast of where highway NM-59 crosses the Continental Divide. It increases in intensity southward (Eggleston, 1987), and is characterized by a kaolinite and chalcedony assemblage that has replaced alkali feldspars in ignimbrites in this area. Alteration assemblages are distributed peripherally to the contact of the Kline Mountain intrusion, located approximately 1 km south-southeast of the open pit. The transition from argillic to advanced argillic assemblage occurs in a zone about 1 km wide and is characterized by the first occurrence of alunite.

The advanced argillic alteration zone contains commercial kaolin deposits that have been exploited intermittently. This alteration, characterized by a kaolinite-alunite-silica (chalcedony, tridymite[?], and quartz) mineral assemblage, is widespread in the study area. This pervasive alteration in the area developed in situ and is related to hydrothermal regimes developed concurrent with the emplacement of the Kline Mountain intrusion.

Kaolinite and alunite are the most diagnostic indicators of advanced-argillic alteration. On the basis of calculated mineral compositions, kaolinite is the dominant clay mineral in the northern part of this alteration zone where the abandoned mines are located, whereas alunite is the dominant mineral to the south near the contact of the intrusion (Fig. 3). The alunite-rich mineral assemblage is located around the intrusion and contains hematite-cassiterite(?) mineralization. Tin mineralization in this area, and associated with the Taylor Creek Rhyolite, has been documented by Eggleston and Norman (1986). On the basis of representative drill core fragments furnished by the New Mexico Bureau of Mines and Mineral Resources, the depth of kaolinization zone is at least 48 m (Fig. 4). According to Mr. P. Roche, the owner of some of the claims and agent for the others, the location of this drill core is estimated to be between the two clay mines in the study area. Between 0 and 48 m the tuff is completely altered to the kaolinite-alunite-silica mineral assemblage with a characteristic white color and softness. The texture of the tuff of Kline Mountain has been obliterated throughout most of this interval. However, drill core fragments display a relatively fresh breccia texture below 48 m, indicating the bottom of the commercially viable kaolin deposit at this location.

Silicification is considerable in the study area, but primarily located in three localities within the tuff of Kline Mountain and the rhyolite porphyry of Kline Mountain. They are (1) silicified tuffs at the top of the Kline Mountain tuff, which forms a cap up to 1000 m in lateral extent, (2) chalcedonic quartz within the kaolin deposit, and (3) widespread silicification at the top of the rhyolite porphyry of Kline Mountain, which forms a 500-m-wide cap. In the study area, silicification mainly occurs as fine-grained replacement of the groundmass and matrix of the tuff.

Hydrothermal activity, responsible for the silicification as well as kaolinization in the study area, is commonly associated with faults. Some of the silicified units exhibit fragments and chalcedonic nodules. Their origin is due to hydrothermal brecciation and/or redistribution by weathering and erosion. Many chalcedonic nodules and fragments are distributed on the surface to the north of the open pit location and are believed to be eroded silica fragments, possibly from the silica cap. These chalcedonic nodules and fragments developed a spongy porous texture, suggesting a possible ancient geothermal vent that produced acid leaching and resulting silica residue.

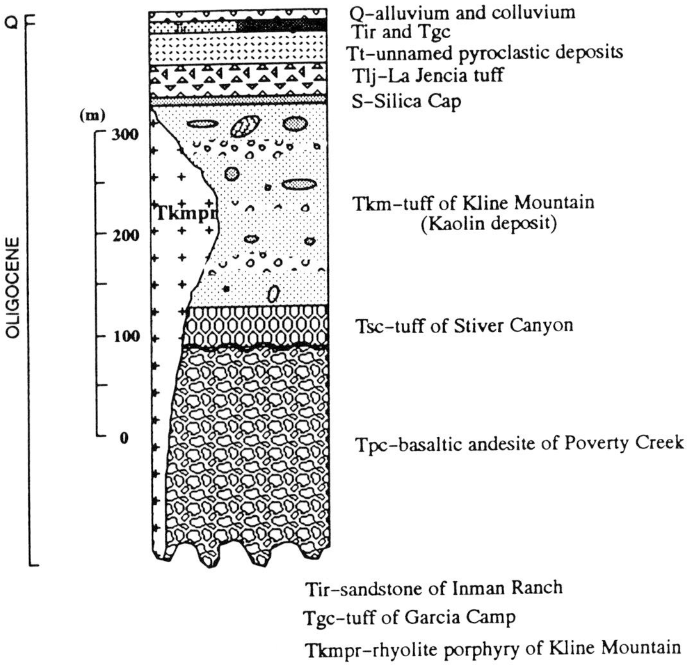

FIGURE 2. Stratigraphic column of the Kline Mountain kaolinite deposit (after Eggleston, 1987).

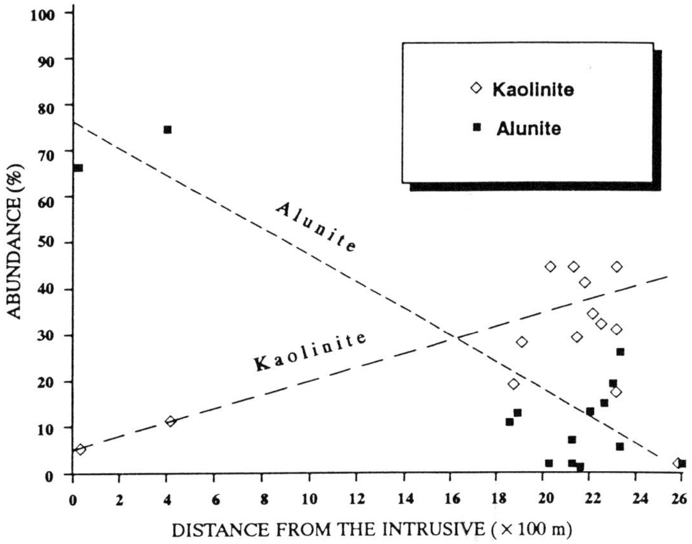

FIGURE 3. Graph showing kaolinite and alunite abundance in surface samples relative to the distance from the Kline Mountain intrusion (from Isik, 1993).

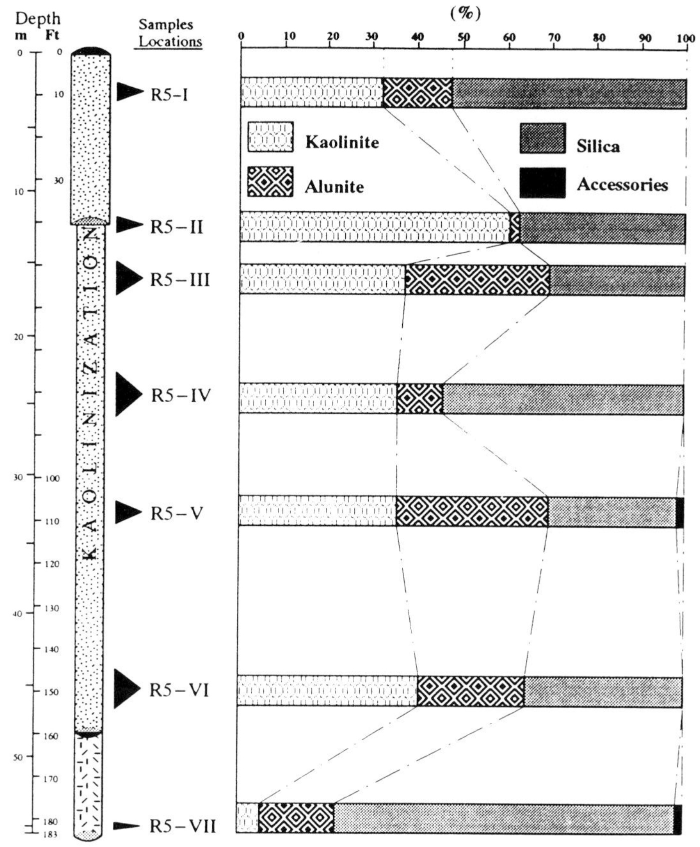

FIGURE 4. Kaolinization zone and mineral distribution in drill core R5 (from Isik, 1993).

Post-depositional weathering may have affected the hydrothermal minerals in the deposit to some extent. However, the extensive silica caps on the top of the deposit have protected the underlying clay at least in part, against the effects of post-depositional weathering, since they are highly resistant to both chemical and physical erosion.

MINERALOGY

Chemical analysis of 24 samples from the surface and core were accomplished by X-ray fluorescence at the New Mexico Bureau of Mines and Mineral Resources. Relative mineral proportions were then calculated from these analyses (Isik, 1993) and show that within the kaolinization zone, the kaolinite content varies from 31 to 58 weight percent. Also, XRD data from these 24 samples (Isik, 1993) indicate the widespread occurrence of kaolinite, while smectite was identified in two clay-size samples. Locally, alunite-rich samples show weak, small peaks of kaolinite reflection that would indicate a minor kaolinite content. Cristobalite is detected only in one of the bulk samples as a major constituent. Tridymite(?) is present in each sample at different concentrations except for alunitic tuff. As seen in Table 1, most of the tuff is composed of kaolinite, alunite, tridymite(?), quartz, cristobalite and accessory minerals. The clay minerals (<2 μm) of the tuff of Kline Mountain as determined by X-ray analysis, are listed in Table 2, in order of decreasing abundance. Scanning electron microscope images obtained by Isik (1993) show variable kaolinite textures in three surface samples from the advanced argillic assemblage. These textures are columnar and covered by silcretes, relatively poorly crystallized, and in stacks. Similarly, alunite crystals were imaged and reveal hexagonal and pseudo-rhombohedral crystals associated with silica lepispheres. Energy-dispersive spectroscopy clearly distinguished the chemistry of all these crystals.

CONCLUSIONS

The stratigraphy in the Kline Mountain kaolin deposit area consists of mid-Tertiary bimodal volcanic and volcaniclastic deposits that consist of basaltic-andesite lavas, high-silica rhyolite lavas and pyroclastic material. Faults in the study area display dominant northwest and northeast trends. They are silicified, altered and mineralized in some localities. Volcanic units in the field have undergone hydrothermal alteration. Alteration is most intense near the intrusive contact of the rhyolite porphyry of Kline Mountain and gradually diminishes distally. On the

TABLE 1. Whole rock mineralogy of the Kline Mountain kaolinite deposit, showing distribution of the minerals in the samples detected by X-ray diffraction (m: minor constituent). From Isik (1993).

Sample	Kaolinite	Alunite	Tridymite	Quartz	Cristobalite
R5-I	X	X	X	X	
R5-II	X	m	X	X	
R5-III	X	X	X	X	
R5-IV	X	X	X	X	
R5-V	X	X	X	X	
R5-VI	X	X	X	X	
R5-VII	m	X	X	X	
SM 200	X	X	X	m	
CS 200	X	X	X	m	
Tkm 1			X		X
Tkm 2	X	m	X		
Tkm 3	m	X	X	m	
2Tkm 2	X	X	X	X	
2Tkm 4	X	m	X	X	
2Tkm 5	X	X	X	X	
3Tkm 1	m	X	X		
3Tkm 3	X		X	X	
3Tkm 5	X		X	X	
Tkm 8	X		X	X	
S7	X	X	X	m	
S7.5		X	X	X	
S8	m	X	X	m	
Tkm 18	m	X	X		
Tkm 20	m	X		X	

TABLE 2. Clay-size fracture (<2μm) mineralogy, showing the distribution of minerals in samples of the tuff of Kline Mountain detected by X-ray diffraction analysis (the numerical order indicates decreasing order of abundance). From Isik (1993).

Sample	Kaolinite	Alunite	Tridymite	Quartz	Smectite
R5-I	1	3	2	4	
R5-II	1	2	3		
R5-III	1	2	4		3
R5-IV	1	3	2	4	
R5-V	1	2	3		
R5-VI	1	2	3		
R5-VII	2	1	3	4	
SM 200	1	2	4		3
CS 200	1	2	3		
Tkm 1	3		2		
Tkm 2	1	2	3		
Tkm 3	2	1			
2Tkm 2	3	1	2		
2Tkm 4	1		2	3	
2Tkm 5	2	1	3	4	
3Tkm 1	3	1	2		
3Tkm 5	1		2		
Tkm 8	1		2		
S7	4	2	1	3	
S8	2	1	3	4	
Tkm 18	?	1	2		
Tkm 20	?	1		2	

basis of X-ray diffractogram patterns, most of the whole-rock minerals are composed of kaolinite, alunite, tridymite(?), quartz, cristobalite and accessory minerals. Four distinct alteration assemblages are recognized in the study area. Alteration of the volcanic units and intrusion occur primarily by replacement, but also silicification has been produced by direct introduction of fluids and by formation of silica residue. The advanced argillic alteration zone contains commercial kaolin deposits that have been exploited intermittently. Advanced argillic alteration, characterized by a kaolinite-alunite-silica (chalcedony, tridymite[?], and quartz) mineral assemblage, is widespread in the study area. On the basis of chemical and XRD data and resultant calculation of mineral compositions, kaolinite is the dominant clay mineral in the northern part of this alteration zone where the abandoned clay mines are located, whereas alunite is the dominant mineral to the south near the contact of the intrusion. The resulting alteration mineral assemblages may best be described, form north to south, as a kaolinite zone with alunite grading to an alunite zone with kaolinite. On the basis of representative drill core fragments, the depth of the kaolinization zone is at least 48 m. Silicification is considerable in the study area, and is primarily located in three localities within the tuff of Kline Mountain and the rhyolite porphyry of Kline Mountain. Hydrothermal activity was responsible for silicification as well as kaolinization in the study area. Calculated mineral composition of outcrop samples from chemical analysis show an inverse relationship between kaolinite percentage and proximity to the intrusion. Conversely, there is a proportional relationship between alunite content and the proximity to the intrusion. Variations in the proportions of minerals found with respect to depth of the drill core show that kaolinite content varies from 31 to 58% but drops abruptly to about 5% at a depth of 48 m.

ACKNOWLEDGMENTS

Special thanks are due George S. Cudahy, President, American Eagle Brick Company, and the University of Texas at El Paso who supported the senior author during this investigation. Chemical analyses by XRF were performed by Christopher G. McKee, and sample preparation and XRD analysis were accomplished with the help of Ibrahim Gundiler and Murat Bengisa respectively, all of the New Mexico Bureau of Mines and Mineral Resources. Dennis J. Menaul assisted with the scanning electron microscopy at the University of Texas at El Paso, and Ron J. Clark performed firing experiments at El Paso Community College. Critical reviews of the manuscript by James M. Baker and Virginia T. McLemore are much appreciated. Palmira Hart typed the manuscript.

REFERENCES

Coney, P.J., 1976, Structure, volcanic stratigraphy and gravity across the Mogollon Plateau, New Mexico: New Mexico Geological Society, Special Publication 5, p. 29-41.

Eggleston, T.L. 1987, The Taylor Creek District, New Mexico [Ph. D. dissertation]: Socorro, New Mexico Tech, 437 p.

Eggleston, T.L. and Norman, D.I., 1983, Taylor Creek tin district – stratigraphy, structure and timing of mineralization: New Mexico Geology, v. 5, p. 1-4.

Eggleston, T.L. and Norman, D.I., 1986, A summary of the geology, geochemistry, and tin occurrences in the Black Range, New Mexico: New Mexico Geological Society, Guidebook 37, p. 173-177.

Eggleston, T.L. and Norman, D.I., 1987, New isotopic dates for the Taylor Creek rhyolite and associated rocks in the Black Range of New Mexico: Isochron/West, no. 48, p. 23-25.

Elston, W.E., 1976, Tectonic significance of Mid-Tertiary volcanism in the Basin and Range province: a critical review with special reference to New Mexico: New Mexico Geological Society, Special Publication 5, p. 93-102.

Ericsen, G.E., Wedow, H. Jr. and Eaton, G.P., 1970, Mineral resources of the Black Range Primitive area, Grant, Sierra, and Catron counties, New Mexico: U.S. Geological Survey, Bulletin 1319-E, 162 p.

Harrison, R.W., 1986, General geology of the Chloride mining district, Sierra and Catron countries, New Mexico: New Mexico Geological Society, Guidebook 37, p. 265-272.

Isik, I. 1993, White Brick manufacture utilizing Kline Mountain, New Mexico clay [Ph.D. dissertation]: El Paso, University of Texas at El Paso, 231 p.

McLemore, V.T., 1994, Volcanic-epithermal deposits in the Mogollon-Datil volcanic field, west-central New Mexico: New Mexico Geological Society, Guidebook 45.

Patterson, S.H. and Holmes, R.W., 1965, Clays: New Mexico Bureau of Mines and Mineral Resources, Bulletin 87, p. 312-322.

Rhodes, R.C., 1976, Petrologic framework of the Mogollon Plateau volcanic ring complex, New Mexico – surface expression of a major batholith: New Mexico Geological Society, Special Publication 5, p. 103-112.

PHOTOINTERPRETATION OF LATE PLEISTOCENE-HOLOCENE ROCK GLACIERS ON SACATON MOUNTAIN AND ESCUDILLA MOUNTAIN, DATIL-MOGOLLON UPLAND, WEST-CENTRAL NEW MEXICO AND EAST-CENTRAL ARIZONA

JOHN W. BLAGBROUGH
P.O. Box 8063, Albuquerque, NM 87198

Abstract—Rock glaciers on the flanks of Sacaton Mountain and Escudilla Mountain are below talus-mantled cliffs and slopes at an average elevation of about 2940-2980 m. They have surface features indicative of ice-cemented (permafrost) forms that moved by the flow of interstitial ice, and delineate ancient zones of alpine permafrost that probably existed in the late Wisconsinan and early Holocene. Within these zones the mean annual temperature was below freezing and snow cover was thin and of short duration. The rock glaciers probably formed during the late Wisconsinan under a periglacial climate characterized by much freezing and thawing, resulting in the generation of large volumes of talus. The rise in elevation of rock glacier occurrences from east to west across west-central New Mexico and east-central Arizona is attributed to greater snow fall in the Datil-Mogollon Upland, which reduced the depth and intensity of ground freezing near the late Wisconsinan 0°C air isotherm.

INTRODUCTION

Rock glaciers are periglacial mass movement deposits that commonly occur in alpine regions near the snow line. Potter (1972) defined a rock glacier as a tongue-like or lobate body composed of angular boulders that resembles a small glacier with ridges, furrows, and lobes on its surface and a steep front at the angle of repose. White (1981) distinguished two principal types of rock glaciers characterized by shape, topographic position and ratio of length to width. Tongue-shaped rock glaciers are elongated masses of rock debris that are longer than they are wide. They commonly occur in cirques or on valley floors. Lobate rock glaciers are broader than long and composed mainly of unsorted talus. They occur singly or in groups along valley walls and are usually extensions of talus.

Active rock glaciers occur either as an ice-cemented mass of debris that moved down slope due to the flow of interstitial ice (ice-cemented rock glacier) or as debris covering a small glacier that moved down slope due to the flow of glacial ice (ice-cored rock glacier) (Péwé, 1983a). Ice-cored rock glaciers are distinguished by saucer- or spoon-shaped depressions between the base of the cirque head wall and the rock glacier, longitudinal furrows along both sides, central meandering furrows, and conical or coalescing collapse pits (White, 1976). Ice-cemented rock glaciers lack depressions at their heads and have continuous talus and avalanche slopes feeding onto their heads (Luckman and Crockett, 1978). They are also defined by well-developed longitudinal and transverse ridges (White, 1976).

The formation of rock glaciers requires a periglacial climate that permits the dislodgement of blocks and slabs by intense freeze-thaw action and a mean annual temperature below 0°C that promotes the formation of permafrost (Corté 1987a). Active ice- cemented rock glaciers are good indicators of perennially frozen ground in alpine areas (Péwé, 1983b), and ancient forms imply the presence of former permafrost and a periglacial climate of sufficient intensity to permit ice to exist and deform under pressure (White, 1981). Permafrost can exist in alpine areas if the mean annual temperature at ground level is 0° to -1°C, especially when the snow cover is thin and of short duration (Péwé, 1983b). The lower limit of rock-glacier activity occurs in the zone of discontinuous permafrost and may be determined by mean annual air temperatures of 0° to -2°C (Barsch, 1978).

A comparison of the average elevation of rock glaciers at a specific locality to the elevation of the orographic snow line may suggest the regional climate (continental versus marine) under which the rock glaciers formed. Shumski (1964) pointed out that the equilibrium line of glaciers corresponds to the lower limit of permafrost in maritime climates with abundant snow precipitation. On the other hand, the lower limit of permafrost under a continental climate is far below the snow line and the preservation of glaciers is dependent mainly on low temperatures. Corté (1987b) noted that rock-glacier activity in the central Andes (latitude 33°S) is 1300 m below the equilibrium line of glaciers on the continental side in Argentina, whereas on the maritime side in Chile, rock-glacier activity is 700 m below the equilibrium line. He attributed this to the climatic differences between the two regions. Greater snow fall on the west side of the Andes reduces the intensity and depth of freezing and raises the elevation of the lower level of permafrost.

SACATON MOUNTAIN

Geographic and geologic setting

Sacaton Mountain, a prominent peak in the Mogollon Range, is located about 18 km southeast of Glenwood, New Mexico (Fig. 1). It is in the Gila Wilderness area (Gila National Forest) and is accessible by the Sacaton Mountain pack trail, which crosses the northern slope of the mountain (Fig. 2). The mountain has a summit elevation of 3249 m and is an isolated peak that extends about 600-760 m above the surrounding terrain. The upper slopes, above 2750 m, are less steep than the lower slopes and are dissected by canyons and gullies with maximum depths of about 200 m.

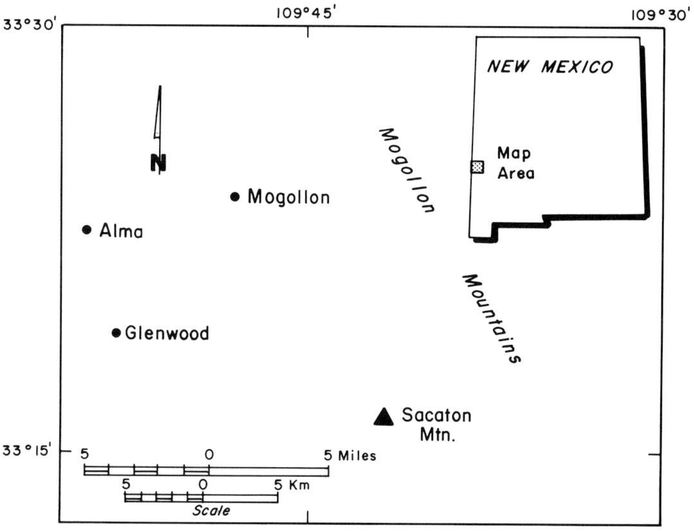

FIGURE 1. Location of Sacaton Mountain near Glenwood, New Mexico. Base map is U. S. Geological Survey topographic map, Clifton quadrangle, 1:250,000

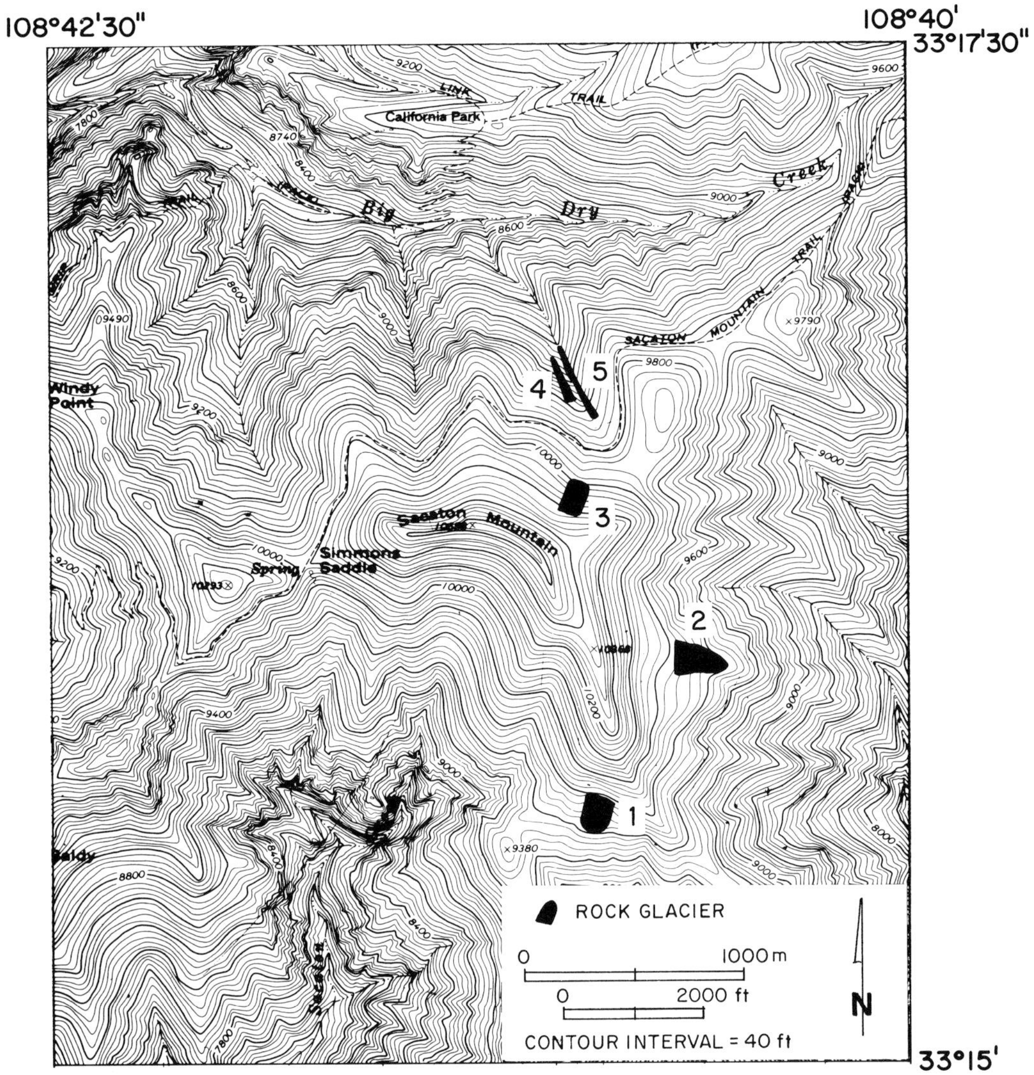

FIGURE 2. Topographic map showing the location of rock glaciers on Sacaton Mountain. Numbered localities indicate rock glaciers discussed in the text. Base map is the U. S. Geological Survey topographic map, Grouse Mountain, NM, 7.5 minute quadrangle.

The Mogollon Range is part of the Datil-Mogollon upland, which covers an extensive area of west-central New Mexico and east- central Arizona (Hawley, 1986). It includes several peaks above 3050 m and is eroded in lavas and tuffs of the Mogollon Plateau volcanic complex. The upper slopes of Sacaton Mountain, above 2600-2750 m, are formed by the Sacaton Mountain Quartz Latite, a porphyritic lava flow (Rhodes, 1970). On Sacaton Mountain this unit is a massive rock with blocky jointing. The Apache Springs Quartz Latite underlies the Sacaton Mountain Quartz Latite and forms the lower slopes of Sacaton Mountain. It is the major ash- flow tuff sheet in the Mogollon Range.

Talus sheets mantle extensive areas of the upper slopes of Sacaton Mountain. The debris is derived from the Sacaton Mountain Quartz Latite, a favorable source due to its massive outcrops and blocky jointing. Most deposits have smooth concave slopes; some have irregular or wavy surfaces that probably resulted from localized movement of the debris by solifluction or frost creep. Many of the talus sheets are devoid of vascular plants; on others, trees and shrubs project through the debris.

The town of Glenwood (elevation, 1441 m) is the closest weather station that maintains continuous long-term weather records. Thirty-six year normals for 1948-1983 indicate a mean annual temperature of 14.3°C (Kunkel, 1984). January is the coldest month (mean temperature, 4.9°C); July is the warmest (mean temperature 24.5°C). For the same period the average annual precipitation was 38 cm, with about 45% of the total occurring during the months of July, August and September.

Van Devender et al (1984) indicated a mean annual lapse rate in south-central New Mexico of 0.72°C/100 m. Utilizing this value, the following crude approximation of current temperature conditions were calculated for Sacaton Mountain: 5°C at about 2750 m, and 1°C for the crest (3249 m). Vertical precipitation gradients in the region suggest that the mean annual precipitation for the upper slopes and crest of Sacaton Peak may be about 75 cm. (Tuan et al, 1969). The cool temperatures and abundant precipitation result in a life zone characterized by spruce-fir on the upper slope and crest of the mountain.

Description

Five rock glaciers on Sacaton Mountain were identified on U. S. Forest Service aerial photographs (scale 1:15,840) and plotted on the U. S. Geological Survey topographic map of the area (Fig. 2). Their lengths and widths were determined and soil development and vascular plant cover were noted from the aerial photographs. The elevation of the fronts and heads, and their degree of slope were calculated from the topographic map.

Wahrhaftig and Cox (1959), in their studies in the Alaska Range, proposed a terminology of the gross features and surface relief of rock glaciers that is followed here with some modification. The front is the steep face that marks the down- valley end of the rock glacier. The sides are typically abrupt embankments that diminish in height up-valley, and the head is the place where the rock glacier terminates at the base of a cliff or steep slope. Lateral ridges are prominent embankments that border the sides.

Surface relief features include longitudinal, transverse and oblique ridges and furrows. Longitudinal ridges extend along the length of a rock glacier and parallel the apparent direction of flow. Transverse ridges and furrows are perpendicular to the apparent direction of movement and are generally convex down slope. Transverse ridges commonly die out against longitudinal or lateral ridges, or bend sharply to merge with them. Oblique ridges and furrows strike at about 45° to the apparent direction of movement.

The rock glaciers on Sacaton Mountain are below cliffs and in a ravine on the upper slopes. The elevation of the fronts average 2891 m and the heads 2987 m. They are tongue shaped, with lengths of 150 to 230 m and widths of 30 to 120 m. Heads terminate at talus which covers steep slopes or cliffs. Fronts are steep embankments with slopes of about 20° to 30°; the surfaces above the fronts slope valleyward at about 15°.

Rock glaciers 1, 2 and 3 are below talus-mantled cliffs on the south, east and north flanks of the mountain. Lateral ridges along their sides bend to form transverse ridges at the crest of the fronts. Surfaces above the fronts are hummocky due to many randomly oriented ridges and furrows. Rock glacier 2 is formed by two talus tongues with steep fronts (Fig. 3); the upper tongue appears to have overridden the head of the lower. Lateral ridges along the flanks of the upper tongue bend to form a transverse ridge at the crest of the front. A depressed area between the two lateral ridges extends from the base of the transverse ridge to the head.

Rock glaciers 4 and 5 are in a tributary ravine to Big Dry Creek on the north slope on the mountain (Fig. 4). The west flank of rock glacier 4 merges with talus on the west wall of the ravine and the east flank is delineated by an embankment that extends above the ravine floor. Many closely-spaced transverse ridges and furrows are on the front and the surface above the front. A small talus tongue with transverse ridges and furrows extends from the base of the front and may have been overridden by the front. The rock glacier appears to have developed from two talus lobes that accumulated along the west wall of the ravine. The lobes coalesced on the ravine floor and moved down slope as a single mass of debris.

The east flank of rock glacier 5 merges with talus on the east wall of the ravine and a well-defined lateral ridge along the west flank above the floor. Transverse ridges and furrows are on the front and many closely-spaced transverse and oblique ridges and furrows are on the surface above the front. A talus tongue extends from the base of the front and probably was overridden by the front.

The surface debris of the rock glaciers appears to be stable and is composed of fragments derived from outcrops of the Sacaton Mountain Quartz Latite. Soil development is minimal and may be confined to isolated pockets filling voids between the debris. Vascular plants cover from 1 to 10% of the rock glacier surfaces.

ESCUDILLA MOUNTAIN
Geographic and geologic setting

Escudilla Mountain is located about 8 km north of Alpine in Apache County, Arizona (Fig. 5). It is in the Escudilla Wilderness area within the Apache National Forest and its summit is accessible by a pack trail that ascends the south flank. The summit has an elevation of about 3260 m and three prominent peaks extend above 3280 m. The flanks are formed by steep cliffs about 240-370 m high and are dissected by gullies and ravines with maximum depths of about 250 m.

Escudilla Mountain is capped by a sequence of Tertiary basalt flows with a maximum thickness of about 275 m (Wrucke, 1961). A friable Tertiary sandstone of eolian origin underlies the basalt and crops out around the flanks. Talus derived from the basalt forms extensive sheets on the mountain slopes below about 3050 m. Many sheets have smooth concave slopes and some have irregular or wavy surfaces due to localized movement of the debris.

There are no long-term weather records for Escudilla Mountain. However, the natural vegetation suggests a humid climate for the upper slopes and crest with a mean annual precipitation greater than 75 cm. Regional lapse rates and temperature records at Alpine (Sellers, 1974) give an approximation of current temperature conditions on Escudilla Mountain of 3°C at 2900 m and 1°C for the summit.

Field observations on Escudilla Mountain recorded periglacial deposits of Wisconsinan age that suggest more severe climate. Barsch and Updike (1971) noted inactive rock glaciers and considered them to be ice-cemented (permafrost) forms because they occur well below the Wisconsinan snow line. Péwé (1983a) observed periglacial talus and thought it indicated a more rigorous climate but not necessarily requiring permafrost.

Description

Twelve rock glaciers on Escudilla Mountain were observed on U. S. Forest Service aerial photographs (scale 1:15,840) and plotted on U. S. Geological Survey topographic maps (Fig. 6). The rock glaciers are below talus in ravines and gullies on the north, east and west sides of the mountain. The elevation of the fronts average 2935 m and the heads

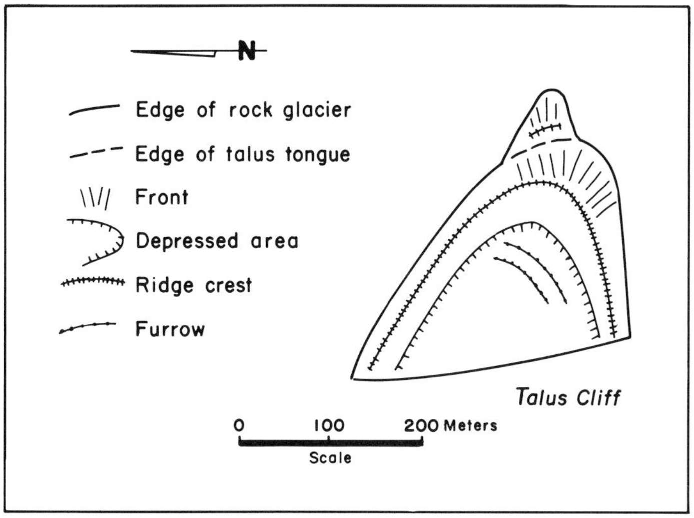

FIGURE 3. Sketch map of rock glacier 2 taken from U. S. Forest Service aerial photograph EDI-2-96. The rock glacier is below a talus-mantled cliff on the east side of Sacaton Mountain (Fig. 2).

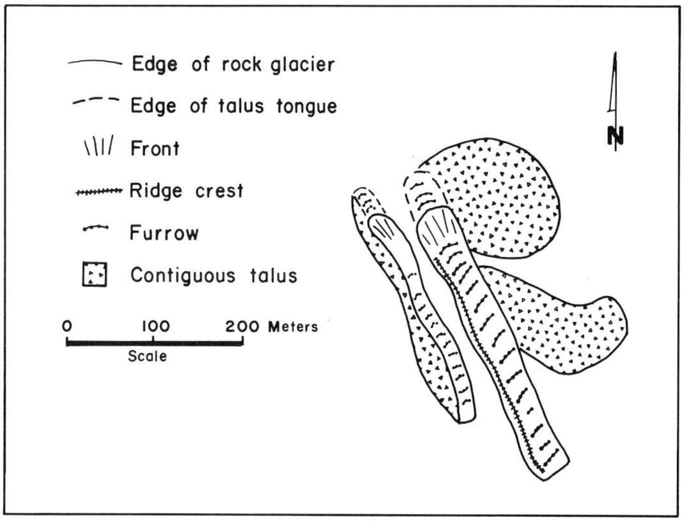

FIGURE 4. Sketch map of rock glaciers 4 and 5 taken from U. S. Forest Service aerial photograph EDI-295. The rock glaciers are in a ravine on the north side of Sacaton Mountain (Fig. 2).

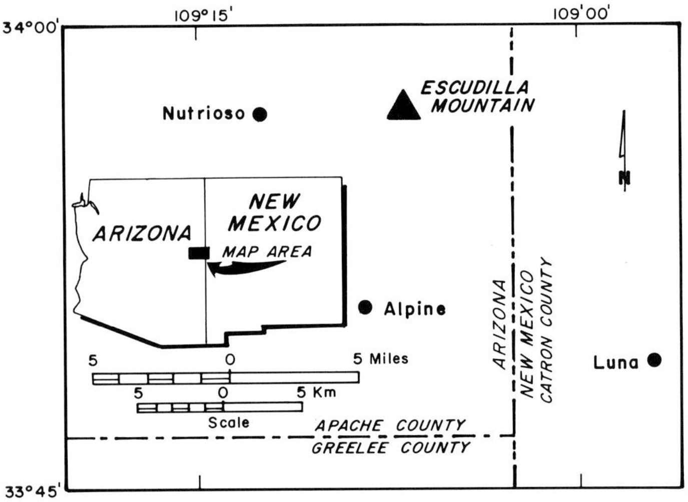

FIGURE 5. Location of Escudilla Mountain near Alpine, Arizona. Base map is U. S. Geological Survey topographic map, Clifton quadrangle, 1:250,000.

3030 m. They are tongue- shaped with average lengths of 130 m and widths of 55 m. Heads terminate at the base of talus sheets and fronts are steep embankments that slope at about 20 to 30°.

Most of these rock glaciers have lateral ridges along one or both sides that bend to form traverse ridges at the crests of the fronts. Transverse ridges and furrows occur on the fronts and the surfaces above the fronts. Some rock glaciers have depressed areas between the lateral ridges that extend from the crests of the fronts to the heads. Hummocky topography in the depressed areas is due to longitudinal, transverse and oblique ridges and furrows.

The debris on the surfaces of the rock glaciers appears to be stable and probably is composed predominately of fragments of basalt derived from the upper slopes of the mountain. Soil development is minimal and may be restricted to small pockets filling voids between the debris. Vascular plants cover from 0 to 10% of the rock glacier surfaces.

AGE

Glacial and periglacial deposits in the White Mountains in east-central Arizona provide a reference for estimating the age of the rock glaciers on Sacaton Mountain and Escudilla Mountain. Merrill and

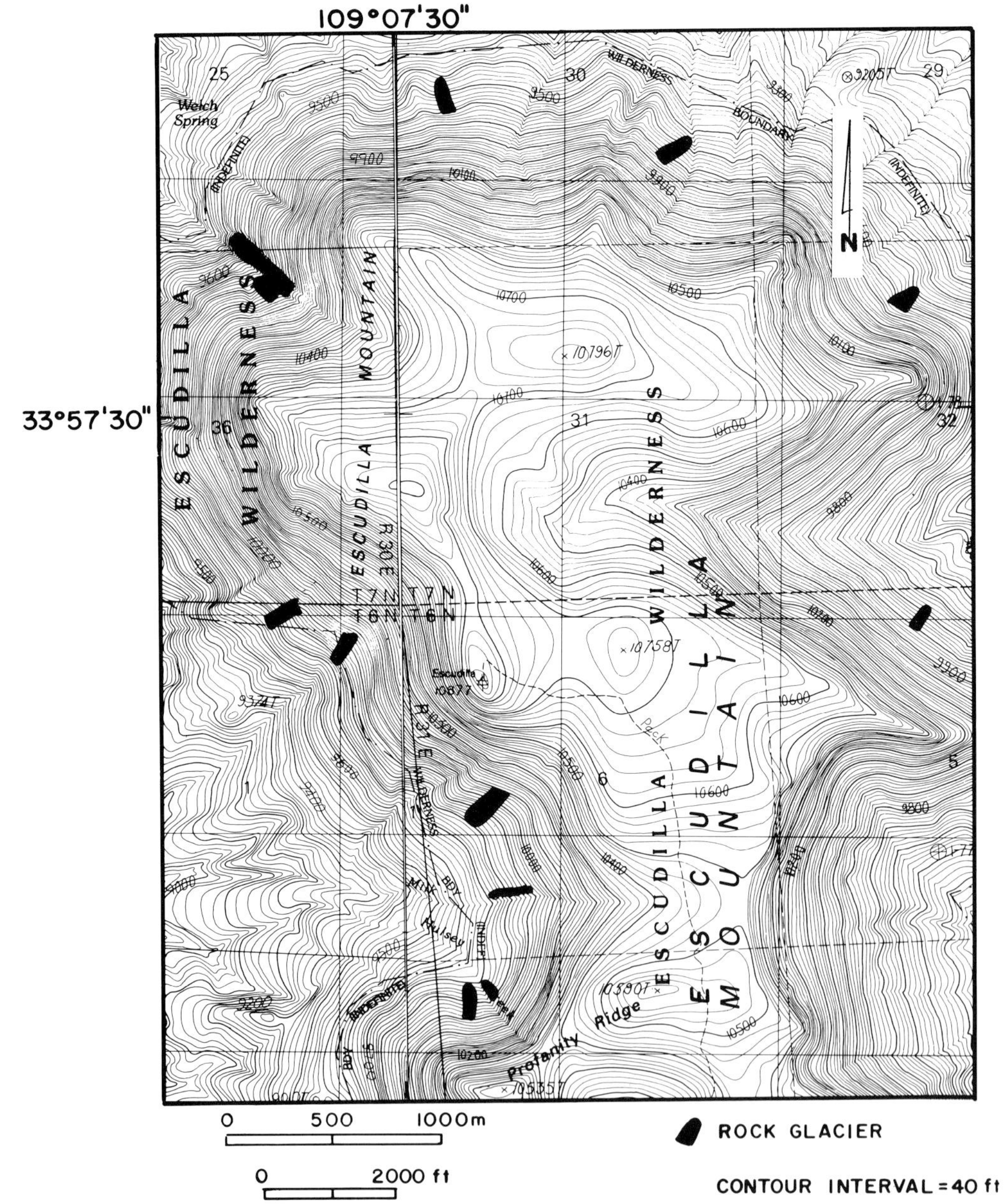

FIGURE 6. Topographic map showing the location of rock glaciers on Escudilla Mountain. Base maps are the U. S. Geological Survey topographic maps of the Escudilla Mountain, Arizona and New Mexico and the Nutrioso, Arizona, 7.5 minute quadrangles.

Péwé (1972, 1977) mapped moraines of pre-Wisconsinan, Wisconsinan and Holocene age in cirques and alpine valleys on Baldy Peak and Mount Ord. Moraines of the Purcell Glaciation of pre-Wisconsinan age are weathered to a depth of more than 50 cm and the surface boulders have deep weathering pits. The deposits are highly subdued and have a moderate to well-developed soil. Moraines of the Smith Cienega Glaciation (early Wisconsinan) bear a zonal soil 25-30 cm thick. Moraines of the Baldy Peak Glaciation (late Wisconsinan) are sharp crested and have a shallow azonal soil about 20 cm thick developed on their surfaces.

A moraine, protalus ramparts and periglacial talus are included in the Mount Ord Glaciation. The moraine is in a cirque on the northeast side of Mount Ord at about 3350 m. Its original surface is well-preserved and boulders on the surface are fresh. The moraine is forested and has a weak azonal soil about 12 cm thick developed on sediment between the boulders. Merrill and Péwé (1972) assigned the Mount Ord deposits an early Holocene or Neoglacial age. Richmond (1986) suggested that the Mount Ord moraine may be latest Wisconsinan in age because of its forest cover, a characteristic not reported on Holocene moraines elsewhere in the Rocky Mountain region.

In considering the age of rock glaciers it is important to be aware that the climate in which they formed may differ greatly from that in which they may have moved a little. Rock glaciers require a periglacial climate of sufficient intensity and duration to generate talus and to mobilize the talus into rock glaciers, whereas the creep of interstitial ice can take place in a climate with a mean annual temperature of 0° to -2°C of either long or short duration. Soil and vascular plants on rock glaciers probably date from the termination of the last episode of significant movement and do not necessarily indicate the age of formation.

The rock glaciers on Sacaton Mountain and Escudilla Mountain may have been active when the Mount Ord deposits were forming in the White Mountains because both have pronounced constructional relief. This suggests rock-glacier movement either during the late Wisconsinan or the Holocene. The sharply defined ridges and furrows, the restricted growth of vascular plants, and the scarcity of soil on the rock glaciers imply movement during the Holocene because well-preserved periglacial deposits in the southern Rocky Mountains generally are assigned a Holocene age (Richmond, 1986).

Recent studies in the Sacramento Mountains in south-central New Mexico (Van Devender et al, 1984, Van Devender, 1990) and in the San Agustin Plains in western New Mexico (Markgraf et al, 1984) indicate moderate cooling during the early Holocene and very little cooling during the middle and late Holocene. The early Holocene cooling may have resulted in the creep of interstitial ice in the rock glaciers on Sacaton Mountain and Escudilla Mountain and the accumulation of periglacial talus and protalus ramparts in the White Mountains. The late Wisconsinan glacial record in the White Mountains and an evaluation of the late Wisconsinan water balance of Lake San Agustin (Phillips et al., 1992) indicate a climate of sufficient intensity and duration to have favored the generation of rock glaciers on Sacaton Mountain and the Escudilla Mountain.

CLIMATE

The rock glaciers on Sacaton and Escudilla Mountains are permafrost forms because they have continuous talus extending onto their heads and well-defined longitudinal and transverse ridges and furrows on their surfaces. They indicate zones of alpine permafrost of probable late Wisconsinan and early Holocene age with a lower limit of about 2800 m on the north side of Sacaton Mountain. Within these zones the mean annual temperature was below freezing and the snow cover was thin and of short duration. During the late Wisconsinan, extended periods of diurnal freezing and thawing generated large volumes of talus which were mobilized into rock glaciers.

The altitudinal relationships between the rock glaciers on Sacaton Mountain and the modern and late Wisconsinan 0°C air isotherm, and the late Wisconsinan orographic snow line are shown in Fig. 7. The modern 0°C air isotherm is calculated at 3427 m using the lapse rate established by Van Devender et al., (1984) in south-central New Mexico, and climatic data from the weather station at Glenwood, New Mexico. The late Wisconsinan 0°C air isotherm is 1000 m below the modern 0°C air isotherm (Péwé, 1983, p. 175). The late Wisconsinan orographic snow line at 3260 m is based upon the average elevation of the cirque floors in the White Mountains in east-central Arizona (Melton, 1961).

The rock glaciers are about 290 m below the orographic snow line, which eliminates the possibility of movement due to glacial ice. They are about 500 m above the late Wisconsinan 0°C air isotherm, suggesting a climate of sufficient intensity to favor the formation of rock glaciers. The average elevation of the rock glacier fronts suggests that the early Holocene 0°C air isotherm was lowered a minimum of about 535 m. Utilizing this value and the lapse rate of Van Devender et al (1984) a mean annual temperature depression of about 4°C may be implied for the early Holocene.

The altitudinal ranges of the rock glaciers on Sacaton Mountain and Escudilla Mountain may be compared to the rock-glacier occurrences on South Baldy in the Magdalena Mountains (Blagbrough and Brown, 1983) and in the San Mateo Mountains (Blagbrough and Farkas, 1968) (Figs. 8, 9). The rise in altitude of the rock glaciers from east to west may be due to more abundant snow fall in the Datil-Mogollon Highland during the late Wisconsinan. A greater percentage of the total mean annual precipitation occurs during the winter months of December, January and February at weather stations in the Datil-Mogollon Highland than at those in the vicinity of the Magdalena Mountains and the San Mateo Mountains (Table 1).

These differences can be attributed to the location of the Datil-Mogollon Highland nearer to the Pacific Ocean, which is the principal source of winter moisture in the southwestern United States, and to orographic influences. The Datil-Mogollon Highland covers an extensive area in west-central New Mexico and east-central Arizona and has an important orographic control over the middle latitude Pacific-type winter storms moving from the southwest. This setting results in a mean annual precipitation of more than 100 cm in the White Mountains, the greatest in Arizona (Merrill and Péwé, 1972). The San Mateo Mountains and Mount Baldy are located on the leeward side of the Datil-Mogollon Highland which results in reduced winter precipitation.

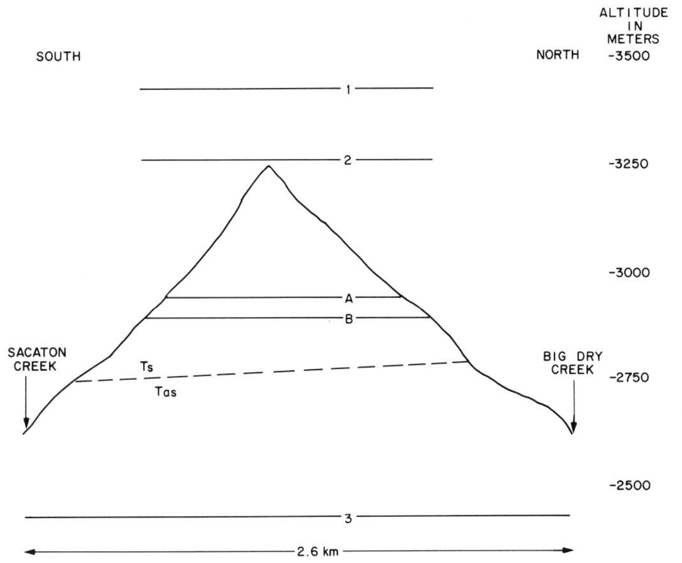

FIGURE 7. Diagrammatic cross section of Sacaton Mountain showing vertical distribution of glacial and periglacial features and the late Wisconsinan and modern 0°C air isotherm. 1, modern 0°C air isotherm; 2, late Wisconsinan orographic snow line; 3, late Wisconsinan 0°C air isotherm. A, average elevation of rock glacier heads; B, average elevation of rock glacier fronts. Tas, Apache Springs Quartz Latite; Ts, Sacaton Mountain Quartz Latite.

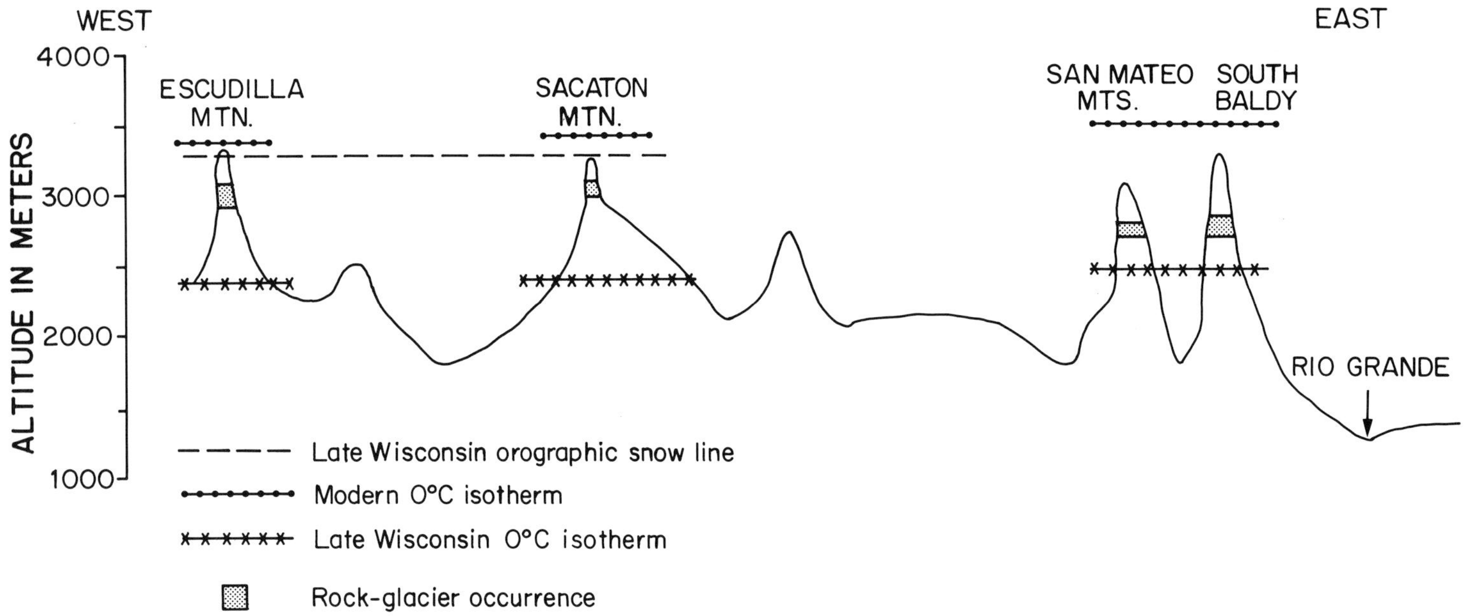

FIGURE 8. Generalized topographic profile across west-central New Mexico and east-central Arizona showing rock-glacier occurrences, the elevation of the modern and Late Wisconsin 0°C air isotherm, and the late Wisconsinan orographic snow line. Rock-glacier occurrences are determined by the average elevation of the fronts and heads at each site. The elevation of the modern 0°C air isotherm at each locality was calculated from lapse rates. The late Wisconsinan 0°C air isotherm is 1000 m below the modern 0°C air isotherm (Péwé, 1983a). The elevation of the late Wisconsinan orographic snow line is determined by the average elevation of the cirque floors in the White Mountains (Melton, 1961). The location of the rock- glacier sites are shown in Fig. 9.

The middle latitude Pacific-type winter storms probably were intensified and displaced southward during the late Wisconsinan (Spaulding et al., 1983), which would have enhanced winter precipitation in the Datil-Mogollon Highland. Greater snow fall on Sacaton Mountain and Escudilla Mountain apparently reduced the intensity and depth of ground freezing near the 0°C air isotherm and raised the elevation of the lower limit of permafrost to a higher elevation than in the San Mateo and Magdalena Mountains.

ACKNOWLEDGMENTS

The author expresses his appreciation to J. W. Hawley and S. E. White for their critical reviews of this paper and suggestions for its improvement. Appreciation also is extended to Michèle Brandwein for her help in editing the manuscript.

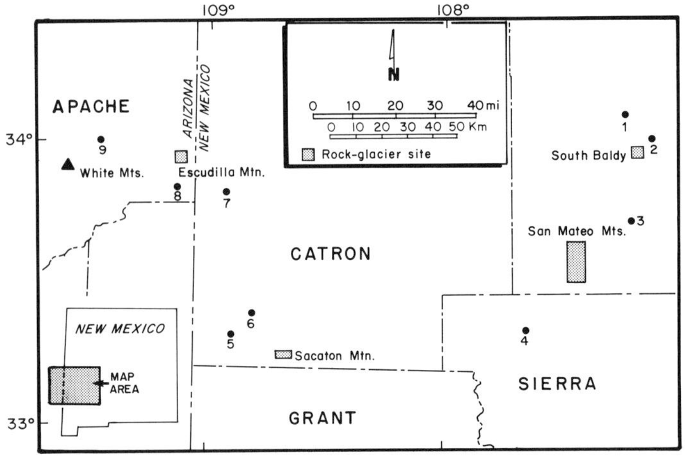

FIGURE 9. Location of rock-glacier sites in west-central New Mexico and east-central Arizona. Numbered localities identify weather stations in Table 1. Base maps are the U. S. Geological Survey's base maps of Arizona and New Mexico (scale 1:1,000,000).

REFERENCES

Barsch, D., 1978, Active rock glaciers as indicators for discontinuous alpine permafrost: an example from the Swiss Alps; in Third International Conference on Permafrost, Proceedings: Ottawa, National Research Council of Canada, v. 1, p. 349-352.

Barsch, D. and Updike, R. G., 1971, Late Pleistocene geomorphology (rock glaciers and blockfields) at Kendrick Peak, northern Arizona: Arizona Geological Society Digest, v. 9, p. 225-243.

Blagbrough, J. W. and Brown, H. G. III, 1983, Rock glaciers on the west slope of South Baldy, Magdalena Mountains, Socorro County, New Mexico: New Mexico Geological Society, Guidebook 34, p. 299-302.

Blagbrough, J. W. and Farkas, S. E., 1968, Rock glaciers in the San Mateo Mountains, south-central New Mexico: American Journal of Science, v. 266, p. 812-823.

Corté, A. E., 1987a, Central Andes rock glaciers: applied aspects; in Giardino, J. R., Shroder, J. F., Jr. and Vitek, J. D., eds., Rock Glaciers: Boston, Allen and Unwin, p. 289-303.

Corté, A. E., 1987b, Rock glacier taxonomy; *in* Giardino, J. R., Shroder, J. F. Jr. and Vitek J. D., eds., Rock Glaciers: Boston, Allen and Unwin, p. 27-39.

Gabin, V. L. and Lesperance, L. E., 1977, New Mexico climatological data: Socorro, W. K. Summers and Associates, 436 p.

Hawley, J. W., 1986, Physiographic provinces and land forms of New Mexico; *in* Williams, J. L., ed., New Mexico in maps: Albuquerque, University of New Mexico Press, p. 28-31.

Kunkel, K. E., 1984, Temperature and precipitation summaries for selected New Mexico locations: Las Cruces, New Mexico Department of Agriculture, 190 p.

Luckman, B. H. and Crockett, K. J., 1978, Distribution and characteristics of rock glaciers in the southern part of Jasper National Park, Alberta: Canadian Journal of Earth Science, v. 15, p. 540-550.

Markgraf, V., Bradbury, J. P., Forester, R. M., Singh, G. and Stemberg, R. S., 1984, San Agustin Plains, New Mexico, age and paleoenvironmental potential reassessed: Quaternary Research, v. 22, p. 336-343.

Melton, M. A., 1961, Multiple Pleistocene glaciation of the White Mountains, Apache Country, Arizona: Geological Society of America Bulletin, v. 72, p. 1279-1281.

Merrill, R. K. and Péwé, T. L., 1972, Late Quaternary glacial chronology of the White Mountains, east-central Arizona: Journal of Geology, v. 80, p. 493-501.

Merrill, R. K. and Péwé, T. L., 1977. Late Cenozoic geology of the White Mountains, Arizona: Arizona Bureau of Geology and Mineral Technology, Special Paper 1, 65 p.

Péwé, T. L., 1983a, The periglacial environment in North America during Wisconsin time; in Porter, S. C., ed., Late Quaternary environments of the United States, v.1, The late Pleistocene: Minneapolis, University of Minnesota Press, p. 157-189.

Péwé, T. L., 1983b, Alpine permafrost in the contiguous United States: a review: Arctic and Alpine Research, v. 15, p. 145- 156.

Phillips, F. M., Campbell, A. R., Kruger, C., Johnson, P., Roberts, R. and Keyes, E., 1992, A reconstruction of the response of the water balance in western United States lake basins to climatic change, v. 1, 1992: Las Cruces, New Mexico Water Resources Research Institute Report 269, 167 p.

Potter, N. Jr., 1972, Ice-cored rock glacier, Galena Creek, northern Absaroka Mountains, Wyoming: Geological Society of America Bulletin, v. 83, p. 3025-3057.

Rhodes, R. C., 1970, Volcanic rock associated with the western part of the Mogollon plateau volcano-tectonic complex, southwestern New Mexico [Ph. D. dissertation]: Albuquerque, University of New Mexico, 145 p.

Richmond, G. M., 1986, Stratigraphy and correlation of glacial deposits of the Rocky Mountains, the Colorado Plateau and the ranges of the Great Basin; *in* Sibrava, V., Bowen, D. Q. and Richmond, G. M., eds., Quaternary glaciation in the northern hemisphere: New York, Pergamon Press, p. 99-127.

Sellers, W. D. and Hill, R. H., 1974, Arizona climate: Tucson, University of Arizona Press, 616 p.

Shumskii, P. A., 1964, Principals of structural glaciology, the petrography of fresh water ice as a method of glaciological investigation: New York, Dover Publications, 497 p.

Spaulding, W. G., Leopold, E. B. and Van Devender, T. R., 1983, Late Wisconsin paleoecology of the American Southwest, in Porter, S. C., ed., Late Quaternary environments of the United States, v. 1, The late Pleistocene: Minneapolis, University of Minnesota Press, p. 259-293.

Tuan, Y. F., Everard, C. E. and Widdison, J. G., 1969, The climate of New Mexico: Santa Fe, New Mexico State Planning Office, 197 p.

Van Devender, T. R., 1990, Late Quaternary vegetation and climate of the Chihuahuan Desert, United States and Mexico; *in* Betancourt, J. L., Van Devender, T. R. and Martin, P. S., eds., Packrat middens: Tucson, University of Arizona Press, p. 104-133.

Van Devender, T. R., Betancourt, J. L. and Wimberly, M., 1984, Biogeographic implications of a packrat midden sequence from the Sacramento Mountains, south-central New Mexico: Quaternary Research, v. 22, p. 344-360.

Wahrhaftig, C. A. and Cox, A. W., 1959, Rock glaciers in the Alaska Range: Geological Society of America Bulletin, v. 70, p. 383-436.

White, S. E., 1976, Rock glaciers and block fields, review and new data: Quaternary Research, v. 6, p. 77-97.

White, S. E., 1981, Alpine mass movement forms (noncatastrophic): classification, description, and significance: Arctic and Alpine Research, v. 13, p. 127-137.

Wrucke, C. T., 1961, Paleozoic and Cenozoic rocks in the Alpine- Nutrioso area, Apache County, Arizona: U. S. Geological Survey, Bulletin 1121-H, 26 p.

TABLE 1. Mean annual precipitation and winter precipitation (December, January and February) at selected weather stations in west-central New Mexico and east-central Arizona. The locations of the weather stations are shown in Fig. 9. Climatic data is from Gabin and Lesperance (1977), Kunkel (1984), and Sellers and Hill (1974).

Station	Location	Map Locality	Elevation m	Years of Record	Mean Annual cm	Winter cm	Percent Winter Precipitation
Magdalena	Socorro County, New Mexico	1	1994	53	28.82	3.59	12.5
Kelly Ranch	Socorro County, New Mexico	2	2043	39	36.10	4.41	12.2
Rienhardt Ranch	Socorro County, New Mexico	3	1662	33	23.31	2.26	9.7
Winston	Sierra County, New Mexico	4	1890	35	31.48	3.08	10.0
Glenwood	Catron County, New Mexico	5	1441	36	37.95	9.21	24.3
Mogollon B	Catron County, New Mexico	6	2012	13	46.92	11.18	23.8
Luna	Catron County, New Mexico	7	2149	53	38.46	7.12	18.5
Alpine	Apache County, Arizona	8	2451	30	49.56	10.33	20.8
Greer	Apache County, Arizona	9	2591	20	61.54	16.54	26.8

Receding thunderstorm and pine trees in Gila National Forest southeast of Pelona Mountain, New Mexico. View is N30°E. Camera station is alongside Highway 163 in NW ¼ sec. 17, T8S, R11W. Wayne Lambert photograph No. 93L88. August 15, 1993, approximately 6:23 p.m. MDT.

HYDROGEOLOGY OF THE NATIONS DRAW AREA, WEST-CENTRAL NEW MEXICO

WILLIAM J. STONE
New Mexico Environment Department, Albuquerque, NM 87109

Abstract—Due to a lack of perennial streams, the main source of water in the Nations Draw area is groundwater. Most wells obtain water from valley-fill alluvium and/or Cretaceous sandstones at or near the surface. Direct groundwater recharge by precipitation is low, 0.02–0.08 inch/yr. Recharge also occurs through transmission loss along streams, but has not been quantified. Groundwater flows westerly, in response to topography, despite the southeasterly dip of strata. Water quality is generally poor, owing to the abundance of mudstone in the sequence, and the major use is stock watering. The discovery of good-quality artesian water in the Dakota Sandstone by the Salt River Project expands the known resources of the area, but this water may not be readily available to most potential users because of its great depth (1000 ft).

INTRODUCTION

Little hydrogeologic work has been done on many sparsely populated areas of New Mexico. Exceptions are irrigated agricultural areas or areas with economically significant mineral resources, such as around Nations Draw, a west-flowing ephemeral stream approximately 12 mi north of Quemado in Catron County. Leasing and exploration activity in the Salt Lake Coal Field during the 1980s prompted various hydrogeologic studies in the Nations Draw area. Several studies were done by the Salt River Project (SRP), Phoenix, AZ, the principal lease holder (Anonymous, 1983; Seifert and Greenberg, 1985). The Energy and Minerals Department of the New Mexico Mining and Minerals Division funded a study by the New Mexico Bureau of Mines and Mineral Resources (NMBMMR) of pre-mining groundwater recharge in major landscape settings in the Cerro Prieto 7.5' quadrangle (Stone, 1984). The Salt River Project contracted with the NMBMMR to formulate a conceptual hydrogeologic model of the region surrounding their leasehold in the Salt Lake Coal field. This involved a New Mexico Tech master's thesis covering a two-quadrangle (7.5') by four-quadrangle area centered on Nations Draw (McGurk and Stone, 1986). Implications of that study for coal mining in the area were presented by Stone and McGurk (1987). This paper briefly summarizes the findings of these previous studies and their bearing on groundwater resources of the area. For more detailed information, the reader is encouraged to consult the original references.

REGIONAL SETTING

The Nations Draw area is within the Zuni-Acoma section of the Colorado Plateau physiographic province. Topography is characterized by broad mesas with gently west-sloping tops, volcanic necks and broad stream valleys. Drainage is by west-flowing ephemeral streams that feed Carrizo Wash, a tributary of the Little Colorado River. The climate is arid to semiarid with annual precipitation averaging 9–15 in. across the area and potential evaporation amounting to two to three times precipitation. Vegetation consists mainly of piñon and juniper trees on the mesas and grasses in the valleys, but cacti, yucca and forbs (weeds) also occur.

Geologically, the Nations Draw area lies within the middle Eocene Baca basin, a narrow foreland basin at the southeastern margin of the Colorado Plateau. The geology and coal resources of the area were described by Campbell (1981, 1984). Coal-bearing rocks of the Cretaceous Moreno Hill Formation (Fig. 1) lie at the surface or beneath thin alluvium throughout most of the area. Tertiary strata of the Baca Formation, lower Spears Group, and Fence Lake Formation cap the Moreno Hill Formation at higher elevations. Underlying the Moreno Hill, but not cropping out in the immediate area, is the Atarque Sandstone. Little is known of the deeper units, but drilling by the Salt River Project (1983) penetrated the Dakota Sandstone at a depth of over 1000 ft on their lease. Regional dip is gentle, <3° to the southeast, except near volcanic features. The Tejana Mesa fault, a northeast-trending zone of high-angle normal faults crossing the southeastern corner of the area, causes up to 250 ft of stratigraphic throw in places (Guilinger, 1982).

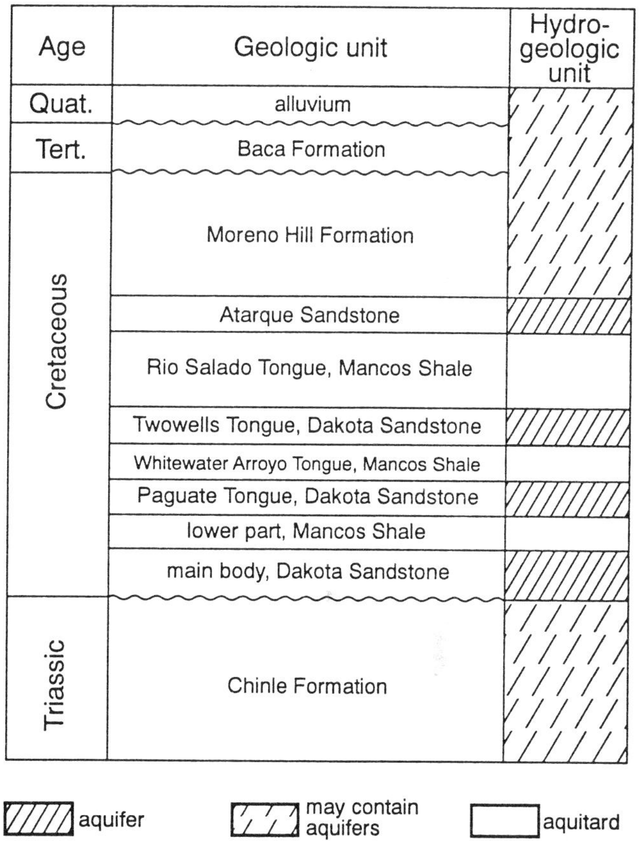

FIGURE 1. Hydrogeologic units in the Nations Draw area (McGurk and Stone, 1986). The lower Spears Group (Eocene) and Fence Lake Formation (Miocene), which are present near Nations Draw, are not shown because they generally cap mesas and lie above the regional zone of saturation. Diagram is not drawn to scale.

GROUND-WATER OCCURRENCE

Although water could be obtained from any of the Cretaceous sandstones, most wells in the area are screened in the shallower sandstones of the Moreno Hill Formation and/or the alluvium (Fig. 1). The depth to regional water table ranges from <20 ft in the major valleys in the west to >900 ft beneath higher elevations, such as Tejana Mesa. Perched water bodies may occur at shallower depths beneath mesas, especially where sandstone is underlain by mudstone.

Quaternary alluvium

The ephemeral stream valleys are underlain by unconsolidated deposits of gravel, fine- to coarse-grained sand, silt and clay. Two distinct units are generally present in this fill. The upper unit consists of

0–130+ ft of light-brown sand and sandy clay with gray sticky clay at the bottom. The lower unit consists of up to 80 ft of buried valley deposits of coarse sand and gravel. The gravel is composed of large, angular rock fragments near the base and volcanic cinders near the top. The lower unit is continuous within the buried valleys, but does not appear to extend beyond their boundaries. Between the lower unit and the underlying Moreno Hill Formation is typically a thin interval of weathered bedrock. The maximum known thickness of alluvium is 190 ft, observed near the confluence of Frenches and Nations Draw.

The alluvium is probably saturated in the center of major valleys, like Nations Draw and Tejana Draw. Saturated thickness in Nations Draw ranges from 0 to 79 ft, with the maximum occurring 1 mi southeast of Cerro Prieto windmill. In Tejana Draw saturated thickness ranges from 0 to 55 ft, with the maximum occurring 1 mi southeast of Tejana windmill.

SRP has completed a production well and several monitoring wells in the basal sand and gravel unit of the alluvium. A 15-day yield of 200 gpm was projected from test data for the production well (Anonymous, 1986). A specific capacity of 5.7 gpm/ft of drawdown was obtained by pumping at a rate of 350 gpm for 1 hr. Storativity determined at one well tested (0.000337) indicated the basal sand and gravel unit is at least partly confined by the overlying finer-grained materials of the upper unit.

Moreno Hill Formation

Nonmarine sandstones, mudstones and coal of the Moreno Hill Formation crop out everywhere in the area, except where overlain by alluvium or Tertiary deposits. These strata are believed to represent deposition in meandering to braided fluvial systems. The Moreno Hill may be divided into three members, based on the presence of a middle sandstone unit. The upper member (present only on higher mesas) consists of up to 600 ft of yellow and green siltstones and claystones with minor amounts of carbonaceous shale, coal and ledge-forming sandstones (Campbell, 1984). The middle member consists of up to 80 ft of pinkish-yellow, medium- to coarse-grained, crossbedded sandstone having little or no silt or clay matrix. The middle member pinches out in the subsurface along a northeast-southwest-trending line just east of the volcanic neck known as Cerro Prieto. East of this line, where the middle sandstone member is missing, the Moreno Hill is generally not subdivided. The lower member has a thickness of 0–600 ft and is lithologically similar to the upper member, but sandstone, carbonaceous shale and coal are more abundant.

The upper and middle members are above the regional water table and the saturated thickness of the Moreno Hill Formation (where the unit can be subdivided) is that of the lower member. Values range from zero in the northwestern part of the Nations Draw region to 1200 ft in the southwest, where regional dip takes the entire formation (undivided) below the water table. Hydraulic properties of the Moreno Hill Formation have not been extensively studied. Slug and pumping tests by SRP have provided some insight. Transmissivity (T) values from twelve tests of sandstone range from approximately 1 to 457 ft^2/day. Fractures were found to account for only 3% of the porosity of the coals but to have 20 times the conductivity of the unfractured matrix, based on response testing of piezometers. Transmissivity values obtained by SRP from five tests of coal range from 2 to 19 ft^2/day. A single measurement for shale gave a T value of approximately 4 ft^2/day.

Dakota Sandstone

In 1983, SRP penetrated the Dakota Sandstone at a depth of approximately 1000 ft in their leasehold. In this area the Dakota consists of three members separated by tongues of Mancos Shale: the Twowells, Paguate and main body, in descending stratigraphic order. The main body is 100 ft thick with coal measures at the top and fine- to medium-grained, white, quartzose fluvial sandstone at the bottom. Video logging showed the main body of the Dakota to be fractured and conglomeratic at the base. Water in the Dakota was found to be under artesian pressure and rose to a height of 191 ft above the ground surface. Pumping tests yielded T values of 709 ft^2/day and a 1-hr specific capacity of 0.86 gpm/ft (Anonymous, 1983).

GROUNDWATER MOVEMENT

Groundwater movement in any area has recharge, flow and discharge components. Recharge and discharge both include the area over which they take place, the process by which they are accomplished and the rate at which they occur. Flow includes both direction and rate of groundwater movement.

Recharge

Groundwater in the Nations Draw area is recharged in part by direct infiltration and percolation of precipitation on higher elevations and in part by transmission loss (seepage) and percolation along ephemeral stream channels. No attempt was made to measure recharge due to transmission loss: direct recharge was investigated using the chloride mass-balance method. The method gives a long-term average recharge rate and is described in various reports (e.g., Stone, 1990, 1992).

This recharge study involved sampling the unsaturated zone and determining the soil-water chloride content versus depth under five landscape settings believed to characterize the area (Stone, 1984). These settings were recognized on the basis of topography, geologic material and vegetation. They include thick alluvium, ephemeral lake, thin alluvium, bedrock/trees and bedrock/grass. The rates determined range from 0.02 inch/yr for the ephemeral lake to 0.08 inch/yr for the alluvium. Intermediate values of 0.05 inch/yr were obtained for the bedrock settings.

The volume of recharge occurring in an area is the sum of the volumes for each setting making up the area. The volumetric recharge for a given setting is calculated by multiplying its recharge rate by its area. The impact of coal mining on the hydrologic balance of an area may be determined by comparing the pre-mining and post-mining recharge volumes (Stone, 1990). This process in the Salt Lake Coal Field will be facilitated by the availability of pre-mining recharge data collected by Stone (1984).

Cumulative soil-water chloride data may be used to approximately date soil water (Allison et al., 1985). The application of this method to data from the Nations Draw area shows that the approximate age of soil water within 60 ft or so of the surface ranges from 5000 to 21,000 yrs. Recharge rates calculated for the older and deeper portions of the unsaturated profiles are generally higher than those for younger and shallower intervals. This agrees with the findings for the nearby San Agustin Plains (Markgraf et al., 1983), that the region was wetter between 8500 and 10,000 years ago, but started to dry out around 8000 yrs ago and that by 5000 yrs ago San Agustin Lake was desiccated.

Flow

Three scales and depths of groundwater flow seem likely in the Nations Draw area. Water in the Moreno Hill Formation and alluvium constitutes a local, shallow flow system. Movement is generally from upland recharge areas toward and then beneath stream valleys. Water in Cretaceous units below the Moreno Hill Formation presumably constitutes an intermediate flow system. There are too few data to determine its flow direction in any detail. However, there is an upward component, based on the artesian conditions associated with the Dakota Sandstone in SRP's deep well. Water in the underlying Permian and Triassic units probably constitutes a deep, regional flow system.

Flow rate depends on hydraulic conductivity (K) of the saturated material. Horizontal K values are greatest for the alluvium, ranging from 0.34 to 15 ft/day. The highest value is associated with the basal sand and gravel. Values for the Moreno Hill Formation range from 0.12 to 13 ft/day, with the high value corresponding to a sandstone.

Discharge

The ultimate discharge areas for all ground water in the Nations Draw area are the deeper stream valleys, especially those lying to the west. The shallow flow system discharges to stream beds not far from recharge areas. The intermediate and deep flow systems probably discharge much farther away. Production wells provide for artificial discharge of the hydrologic system. Although the rate or amount of discharge has not been quantified, it will essentially equal recharge.

GROUNDWATER QUALITY

Specific conductance, an electrical property related to the salinity of a water or the amount of dissolved solids in it, is a general measure of water quality. The larger the number, the poorer the quality. Values from the Nations Draw area show the quality of groundwater there to be rather poor. The major use is stock watering.

Specific conductance values for ground water from the Quaternary alluvium range from 211 to 1093 μmhos/cm. Analyses plot in the sodium-bicarbonate fields of a trilinear or Piper diagram. Water from the alluvium is suitable for livestock, marginally suitable for human consumption and unsuitable for irrigation.

Water from the Moreno Hill Formation is characterized by specific conductance values of 211–1250 μmhos/cm. Most analyses plot in the sodium-bicarbonate or calcium-bicarbonate fields. Some waters have subequal amounts of calcium, sodium, bicarbonate and sulfate. Shallower sandstones in the Moreno Hill provide water for stock and domestic use throughout the area. Some wells are open to both the alluvium and Moreno Hill.

Water in the Dakota Sandstone well is reported to be of good quality. Based on three readings over as many months, an average specific conductance value of 638 μmhos/cm and an average total dissolved solids content of 377 mg/L was obtained (Anonymous, 1983).

GEOLOGIC CONTROLS OF HYDROLOGY

The hydrology and water resources of the Nations Draw area are strongly controlled by its geologic setting. For example, major producible water is associated with the more porous and permeable materials such as alluvium and sandstones. Recharge is highest where the most porous and permeable of these lie at the surface. The westerly groundwater flow direction is contrary to the southwesterly dip of the strata and largely a consequence of topography/geomorphology. The poor water quality is attributable to the presence of fine-grained material in the section. Such material is characterized by a large particulate surface area and slow flow rates, both of which enhance dissolution of ions and thus salinity. The occurrence of fresher water in the Dakota Sandstone is due to its nonmarine origin here. This produced a cleaner (less fine material) aquifer than elsewhere in the region, where the Dakota is of marine origin and yields water with high total dissolved solids and elevated iron contents.

IMPLICATIONS FOR WATER RESOURCES

As there are no perennial streams in the Nations Draw area, the main source of water is groundwater in the alluvium along valleys and sandstones in the Cretaceous bedrock section. Such water is not overly abundant and care should be taken that pumping volumes do not exceed the low recharge volumes. Over-pumping the alluvium or Moreno Hill aquifers could cause water levels to drop, requiring the deepening of wells.

The discovery of good-quality artesian water in the Dakota Sandstone provides an alternative source of water for the area. However, its great depth will keep it out of reach of most potential users. When it is utilized, care should be taken so that pumping does not cause a reversal of gradient to the detriment of its water quality. Extensive pumping of the Dakota Sandstone could reduce artesian pressure such that gradients reverse, causing poor quality water from the overlying Cretaceous units to flow into the Dakota. Such a reversal occurred in the Grants region, where uranium-mine dewatering so lowered the head in the Morrison Formation that poor-quality water from the overlying Dakota Sandstone was being pumped from wells completed in the Morrison (Kelly et al., 1980).

ACKNOWLEDGMENTS

An early version of this article was improved through reviews by W. Haneberg and R. Chamberlin. L. Hemenway typed the manuscript and M. Wooldridge drafted the figure.

REFERENCES

Allison, G. B., Stone, W. J., and Hughes, M. W., 1985, Recharge in karst and dune elements of a semiarid landscape as indicated by natural isotopes and chloride: Journal of Hydrology, v. 76, p. 1–25.

Anonymous, 1983, Fence Lake coal leasehold water well drilling and testing: Salt River Project, Groundwater Planning Division, Water Resource Operations Report, 61 p.

Anonymous, 1983, Fence Lake no. 1 mine, permit application, 2 v.

Campbell, F. W., 1981, Geology and coal resources of Cerro Prieto and The Dyke quadrangles: New Mexico Bureau of Mines and Mineral Resources, Open-file Report 144, 60 p.

Campbell, F. W., 1984, Geology and coal resources of Cerro Prieto and The Dyke quadrangles, Cibola and Catron Counties, New Mexico: New Mexico Geology, v. 6, p. 6–19.

Guilinger, D. R., 1982, Geology and uranium potential of the Tejana Mesa–Hubbell Draw area, Catron County, New Mexico: New Mexico Bureau of Mines and Mineral Resources, Open-file Report 176, 137 p.

Kelly, T. E., Link, R. L. and Schipper, M. R., 1980, Effects of uranium mining on ground water in Ambrosia Lake area, New Mexico: New Mexico Bureau of Mines and Mineral Resources, Memoir 38, p. 313–319.

Markgraf, V., Bradbury, J. P., Forester, R. M., McCoy, W., Singh, G. and Sternberg, R., 1983, Paleoenvironmental reassessment of the 1.6-million-year-old record from San Agustin Basin, New Mexico: New Mexico Geological Society, Guidebook 34, p. 291–297.

McGurk, B. E. and Stone, W. J., 1986, Conceptual hydrogeologic model of the Nations Draw area, Catron and Cibola Counties, New Mexico: New Mexico Bureau of Mines and Mineral Resources report to Salt River Project, Phoenix, 104 p.

Seifert, G. G. and Greenberg, M. A., 1985, Fence Lake coal project, groundwater monitoring: Proceedings, Symposium on Groundwater Contamination and Reclamation, American Water Resources Association, Tucson, p. 19–25.

Stone, W. J., 1984, Recharge in the Salt Lake coal field based on chloride in the unsaturated zone: New Mexico Bureau of Mines and Mineral Resources, Open-file Report 214, 64 p.

Stone, W. J., 1990, Impact of mining on ground-water recharge, Navajo coal mine, New Mexico: Mining Engineering, v. 42, p. 1269–1272.

Stone, W. J., 1992, Paleohydrologic implications of some deep soil-water chloride profiles, Murray Basin, South Australia: Journal of Hydrology, v. 132, p. 201–223.

Stone, W. J. and McGurk, B. E., 1987, Hydrogeologic considerations in mining, Nations Draw area, Salt Lake coal field, New Mexico: New Mexico Bureau of Mines and Mineral Resources, Bulletin 121, p. 73–78.

Day 2 Stop 5
Jordan
93

LANDSLIDE REPAIRS ON US-180 BETWEEN MILEPOSTS 13 AND 19, NEAR LUNA, CATRON COUNTY, NEW MEXICO

SUSAN GALLAGHER
Materials Lab, New Mexico State Highway and Transportation Department, P.O. Box 1149, Santa Fe, NM 87504

Abstract—Three landslides within the Tertiary Pueblo Creek Formation were repaired on US-180 near Luna during 1993 in two separate projects. Data had been initially collected on one of the landslides as far back as 1979. All three had been continuing maintenance problems and regularly received maintenance asphalt patches before being repaired. One of the slide repairs involved excavating material above the slip surface and replacing it with select fill, which was keyed into bedrock for stability. The other two slides were repaired by excavating material above the slip surface and replacing it with local soils and reinforcing geogrid. Both repair methods had sufficiently high safety factors (1.25) when modeled on slope stability computer programs; each method of repair should perform adequately.

INTRODUCTION

In 1993 three landslide areas were repaired on US-180 near Luna, New Mexico at mileposts (MP) 13.5, 18.54 and 18.66 (Fig. 1). These repairs were made as part of two projects undertaken for major road improvements that were let to contract by the New Mexico State Highway and Transportation Department (NMSHTD). This portion of US-180 was originally constructed in the 1920s and 1930s by the Bureau of Public Roads and lies within the Apache and Gila National Forests. The highway alignment traverses the San Francisco Mountains and lesser volcanic hills of the Mogollon Slope.

The highway cut slopes are primarily in Tertiary volcanic and sedimentary rocks of the Pueblo Creek Formation (Ratté, 1989). These rocks are covered by Quaternary deposits. The volcanic rocks consist of hard andesites and rhyolites and less competent andesites and cobble-boulder tuff breccias laced with quartz stringers and some dikes. The sedimentary rocks consist of breccias, conglomerates, volcanic ashes and mud flows. The Quaternary material includes colluvium, rockslide debris and landslide debris.

The primary goal of the highway improvement projects was to create a wider roadway template to bring the roadway up to current safety standards. The roadway was increased from two 11-ft lanes with no shoulders to two 12-ft lanes with 6-ft shoulders. This required extensive presplitting and blasting of the rock sidewalls. Due to the weathered nature of the rock, an overrun of 10,000 yd^3 (20%) of blasted material was generated in the vicinity of milepost 17. The rock was more extensively fractured than expected and much of it fell onto the roadway. Traffic was halted and rerouted on a winding and bumpy dirt Forest Service road for 3 weeks.

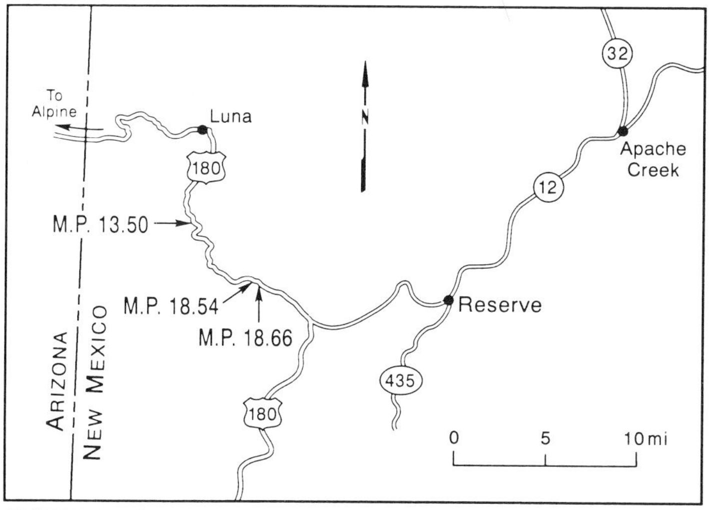

FIGURE 1. Index map of the Luna, New Mexico region, showing locations of landslides on US-180 repaired during 1993.

EARLY OBSERVATIONS

Roadway settlement along US-180 at MP 13.5 had been an ongoing maintenance problem, and was suspected to be the manifestation of a landslide. The settlement was investigated in 1979 using NMHSTD drill crews. Two exploratory holes were drilled into the slide mass and standard penetration tests (SPT) were performed at 5-ft intervals. Undisturbed soil samples were obtained for soil strength determination, and two slope inclinometers were installed in the exploratory holes. Number 1 was 8.3 ft right of centerline, and Number 2 was farther down the slide mass and 43.2 ft right of centerline.

The investigation indicated that the original ground surface had not been cleared and grubbed of its original vegetation and organic debris prior to placement of fill for the highway. The vegetation at the interface of the natural ground and overlying fill eventually decomposed to form a zone of weakness, which formed a shear surface when the overlying fill became saturated by infiltration of snowmelt and rain. The slip surface was located by the inclinometer readings, which showed movement at a 25-ft depth in inclinometer Number 1 and at a 20-ft depth in inclinometer Number 2. The investigation also revealed that the original natural ground was part of a dormant prehistoric landslide mass underlain by a fine- to medium-grained sandstone bedrock.

Slide repair recommendations were made to the NMSHTD by the Geotechnical Section after analyzing data from the site investigation. The recommendations were given in October of 1979, together with recommendations for controlling rock fall from cut slopes adjacent to the highway from the Arizona state line east for 15 mi.

RECENT OBSERVATIONS AND REPAIRS

Because of lack of funding and higher priorities, slope stabilization and highway widening projects for this stretch of roadway were not let to contract until 1991. Highway maintenance crews periodically patched the roadway through the landslide area during the 12 yr period since 1979 and regularly cleared fallen rock from the highway. After project construction began it was discovered that the slide correction had been left out of the plans.

During construction the slide mass began to move again. In early February 1992, the head of the slide dropped 10 in. in a 4-hr period. The ground was wet from recent rain and snow. A series of arcuate cracks and distresses had developed in the roadway over approximately a 200-ft length. The cracks extended from the shoulder of the east bound lane, across the centerline, and 8 ft into the west bound lane. A sag of up to 0.625 ft was measured along the centerline profile for a distance of 150 ft. The contractor cut a shoe fly detour into the adjacent hillside to accommodate traffic. By May 1992, the head of the slide had dropped a total of 4 ft. A typical landslide hummock had also developed on the side slope between the head of the slide and a creek located 50 ft vertically below the top of the moving mass. Based on the 1992 movement and information obtained in the 1979 investigation a corrective design was once again developed.

Remediation consisted of removing the mobile slide mass and organic shear plane and keying the new embankment material into bedrock. Approximately 12,000 yd^3 were excavated. A french drain, consisting of gravel wrapped in filter fabric, was placed along the backslope of the excavation in order to divert water away from the new embankment (Fig. 2). A perforated lateral drain collector pipe was placed at the bottom of the french drain and connected to a nonperforated pipe that discharged above the stream at the lowest elevation of excavation. Select fill was placed in 12 in. lifts and compacted to 95% of maximum obtainable density as determined by AASHTO T-99 (Proctor test). Select fill consisted of sandy, well-draining material with a high angle of internal friction (35°) and little or no cohesion. This material was obtained from cut slopes adjacent to the new Luna bridge 6 mi from the excavation. The replacement material was keyed into bedrock for a 20 ft width (see cross-section and profile, Figs. 3, 4). Construction took approximately 7 weeks to complete.

FIGURE 2. Landsle repair at MP 13.5, showing french drain being constructed on back slope of excavation, autumn 1993.

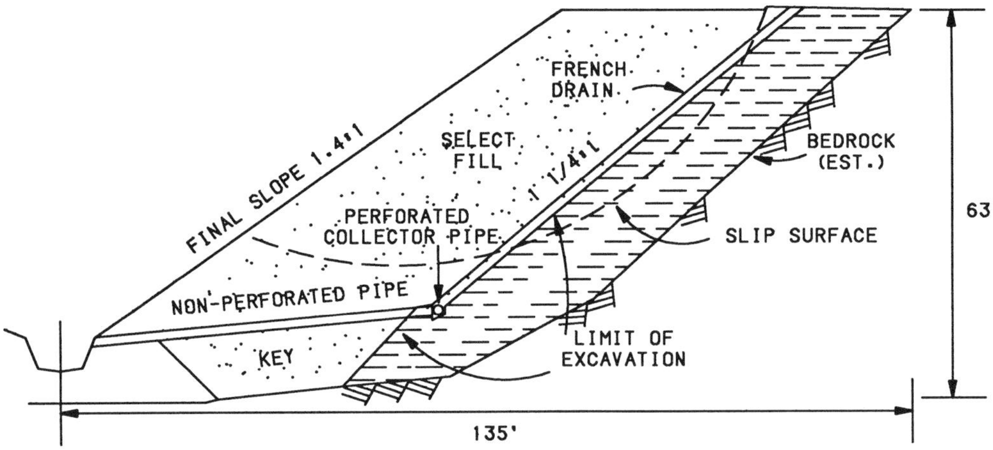

FIGURE 3. Cross-section at MP 13.5, detailing slide repair.

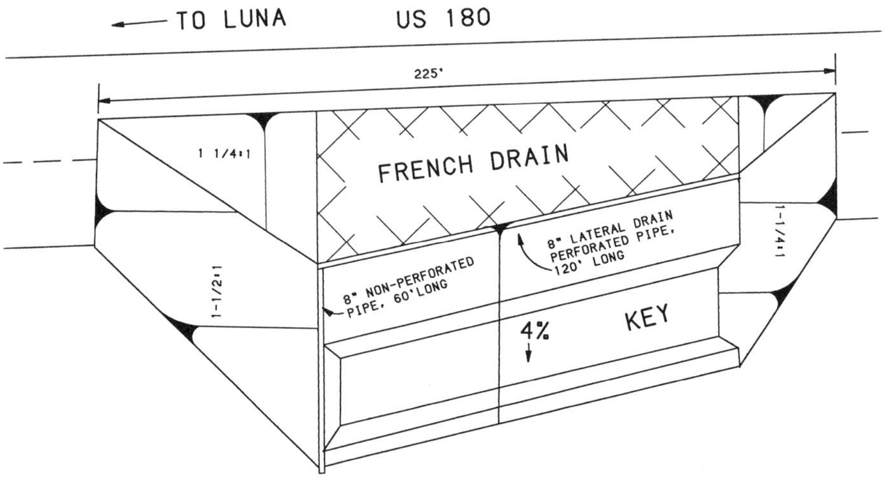

FIGURE 4. Plan view of excavated area at MP 13.5.

Two other potential slides were corrected in these projects at mileposts 18.54 and 18.66. Both were identified by the arcuate cracks in the road before any major mass movement occurred. Subsurface investigations were performed in January of 1990 and corrections were designed and included in the project plans. Organic material was found below the surface of these two areas, indicating the same type of failure mode as described earlier. These areas were also underlain by the same type of old landslide debris found in the slide at MP 13.5. Inclinometer readings at these two locations failed to identify a slip plane due to the less active state of these slide masses.

The problem of the slide movement at MP 18.66 was being compounded by a stream undercutting and removing material toward the base of the slide mass. During initial construction of this road, the stream was rechannelled away from the bank. After many years of eroding its way back to its natural channel, it began to cut away at the embankment built to support the road above. As erosion continued, the outlet section of a 48-in. culvert, 30 ft above the stream, lost support. This 8-ft section eventually fell off. The slope was gradually steepened to nearly vertical. MP 18.54 showed the same type of settlement of the pavement and shoulder as MP 18.66, but without loss of the side slope. Both areas had received as many as 14 in. of asphalt patches over the years.

These two areas were repaired using a soil reinforcing geogrid fabric. The first area (MP 18.54) was 250 ft in length and required an excavation of 2400 yd^3. It was reconstructed beginning at a depth of 12 ft with 10 layers of geogrid spaced 1 ft apart using backfill obtained from the opposite cut slope, after discarding large boulders and clumps of clay. The backfill was again obtained from the opposite cut slope. This area also contained a zone of unreinforced backfill on the outside of the reinforced area (Figs. 5, 6). Once begun, these two repairs took about 10 weeks to complete.

Two completely different approaches were taken in the repair of these slides, although their situations were similar. The two slides at MP 18.54 and 18.66 were drilled and analyzed with corrections designed using soil reinforcing geogrid and local soils with deleterious materials removed. The slide correction at MP 13.5 was also analyzed but an entirely different correction method was used. This involved replacing the slide mass with select borrow having the desired strength parameters and keying of the select borrow into the bedrock and providing for drainage.

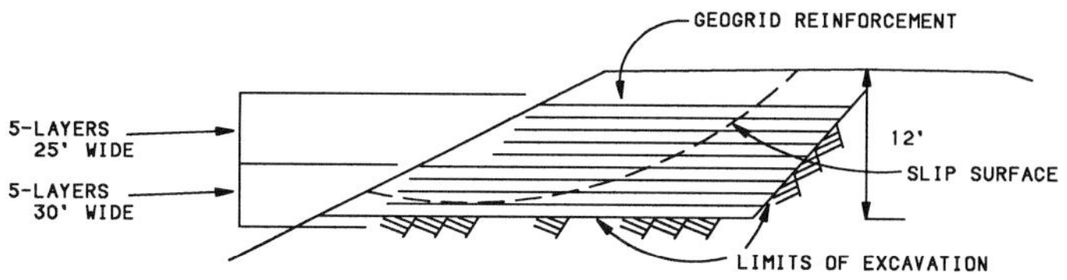

FIGURE 5. Cross-section at MP 18.54, showing geogrid placement in top 12 ft of roadway.

SAFETY FACTORS

For modeling purposes a safety factor of 1.0 is considered to represent failure while a 1.25 represents stability. Both repair methods described above were modeled with a safety factor of 1.25 based on slope stability computer programs, and each type of repair should perform adequately.

ACKNOWLEDGMENTS

Helpful reviews were provided by L. Garrabrant, G.D. Rawlings and W. Haneberg.

REFERENCES

Ratté, J. C., 1989, Geologic map of the Bull Basin Quadrangle, Catron County, New Mexico: U.S. Geological Survey, Geologic Quadrangle Map GQ-1651, scale: 1:24,000.

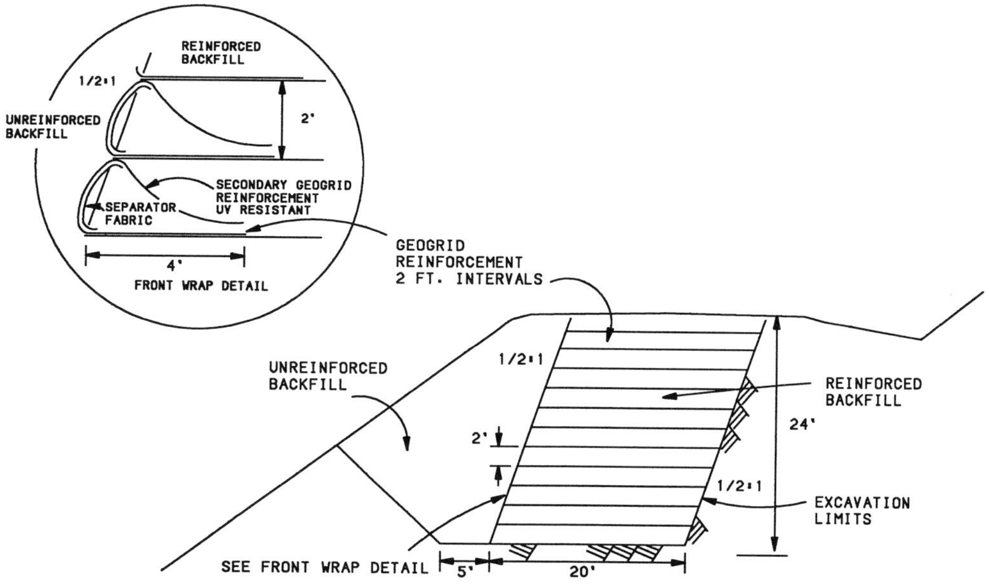

FIGURE 6. Cross-section at MP 18.66, showing geogrid reinforcement in top 24 ft of roadway.

COMMON PLANTS AND PLANT ASSOCIATIONS OF THE MOGOLLON SLOPE

JOSEPH C. CEPEDA[1] and PAMELA S. ALLISON[2]

[1]Department of Biology and Geosciences, West Texas A & M University, Canyon, TX 79016; [2]Battelle Pantex, P. O. Box 30020, Amarillo, TX 79177

Abstract—The landscape of the Mogollon Slope of west-central New Mexico and east-central Arizona supports a diverse flora of herbaceous wildflowers, shrubs, vines and trees. With an elevation range of 5500 to more than 10,000 ft above sea level, the following vegetational zones are represented: Desert and Plains-Mesa Grasslands (5000 to 6000 ft), Pinyon-Juniper Savanna and Woodlands (6000 to 7000 ft), Pine-Oak Forest (6500 to 8000 ft), and Fir-Aspen Forest (above 8000 ft). In addition, Chihuahuan Desert Scrub is well represented in the vicinity of Socorro, New Mexico. Riparian and wetland habitats support a variety of phreatophytes.

INTRODUCTION

The diverse landscapes of New Mexico harbor a varied flora that complements the many rock types that make this state a treasure of natural history. The present vegetation represents the response of plants to a variety of factors, including natural processes such as fires and natural grazing pressures, human disturbances such as overgrazing and the use of pesticides, and the introduction of opportunistic exotic species such as salt cedar and Russian thistle. The vegetation map of New Mexico (Dick-Peddie, 1993) lists 13 vegetation types, ranging from Alpine Tundra to Closed Basin Scrub vegetation. Most of the eastern part of the state is Plains-Mesa Grassland, the single most widespread vegetation type in the state. The western half of the state supports more numerous floral assemblages and greater diversity, consisting mostly of Desert Grassland but with large areas of Juniper Savanna, Coniferous and Mixed Woodland, Montane Coniferous Forest and Subalpine Coniferous Forest. The Mogollon Slope, lying at elevations ranging from 6000 to almost 10,000 ft, includes a broad slice of the vegetational diversity that typifies the western half of the state. U. S. Highway 60 between Socorro and Quemado traverses Chihuahuan Desert Scrub (in the vicinity of Socorro), Desert Grassland (between Socorro and Magdalena), Juniper Savanna (Magdalena to the edge of the Plains of San Agustin), Plains-Mesa Grassland (along the north margin of the Plains of San Agustin), and Coniferous and Mixed Woodland (west of Datil). Coniferous and Mixed Woodland and Montane Coniferous Forest are encountered in the vicinity of Alpine, Arizona.

Climatic extremes and the availability of water determine the types of plants that thrive at these elevations. The major factor for vegetation type at a particular latitude is elevation. However, plant associations may vary over a short distance depending on slope exposure or with distance from a stream or lake. This paper provides a short introduction to the common and abundant plant types in the vegetational zones traversed during the Mogollon Slope field conference.

METHODOLOGY

Field surveys along the scheduled route of the Field Conference were conducted in September and October 1993. Species in flower and/or fruit were identified and/or collected and pressed in the field. Most of the field identifications were verified by comparison with specimens in the West Texas A & M University Herbarium. These species (Table 1) will likely be conspicuous during the 1994 field trip. References consulted in identification included Correll and Johnston (1970), Elmore (1976), Ivey (1986), and Martin and Hutchins (1988). Additional field guides useful for this area include Patraw (1951), Dodge (1973), Spellenberg (1979), Arnberger (1982), and Whitson (1991).

CHIHUAHUAN DESERT SCRUB

The Chihuahuan desert of northern Mexico extends from near San Luis Potosi northward to southern New Mexico and Arizona, and western Texas. This desert, because it occupies generally higher elevations than other southwestern deserts, is cooler and has greater rainfall than other warm deserts (MacMahon, 1988). It is characterized by the presence of creosotebush (*Larrea tridentata*) and/or tarbush (*Fluorensia cernua*). The Chihuahuan Desert Scrub is dominated by the shrubby woody perennial plants listed above plus one or more of the following: honeybean mesquite (*Prosopis glandulosa*), lechugilla (*Agave lechugilla*), and whitethorn acacia (*Acacia constricta*), as well as a variety of cacti and grasses.

DESERT AND PLAINS-MESA GRASSLANDS

Desert Grassland spans the transition zone between the shortgrass prairie of the High Plains and the Chihuahuan Desert of northern Mexico. These grasslands are undergoing shrub invasion as grasses are gradually replaced by desert shrubs such as creosotebush, tarbush and mesquite. Probable reasons for this change include disturbance by overgrazing of domestic livestock producing erosion of the thin topsoil, fire suppression and a gradual warming of the climate in the southwest (Sims, 1988). The expansion of mesquite onto desert grasslands has been rapid where drought has thinned stands of grass, allowing the opportunistic mesquite to invade (Herbel et al., 1972). Dick-Peddie (1993) suggested that if desert grassland shrub species are typical of the Chihuahuan Desert, the desert community is probably a recent one. If the dominant forbs of a Desert Grassland are also major species of Montane Scrub vegetation, the community may represent an ecotone between Montane Scrub and Plains-Mesa Grassland.

The list of common plants in the Desert Grasslands of southern New Mexico is large because of the transitional nature of this habitat. Plants diagnostic of this vegetation type include (Dick-Peddie, 1993) lechugilla (*Agave lechugilla*), feather peavine (*Dalea formosa*), sotol (*Dasylirion wheeleri*), barrel cactus (*Ferocactus wislizenii*), snakeweed (*Gutierrezia sarothrae*), and tree cholla (*Opuntia imbricata*), as well as several species of grasses, including black grama (*Bouteloua eriopoda*), bush muhly (*Muhlenbergia porteri*) and burrograss (*Scleropogon brevifolius*).

Plains-Mesa Grassland is the most extensive vegetation type in New Mexico and includes the Plains of San Agustin (Fig. 1) in the west-central part of the state. This grassland type merges with Savanna or Juniper Woodlands (Fig. 2) at elevations of about 7000 feet, and with Desert Grassland at elevations below 6000 feet in southwestern New Mexico. Under optimum (climax) conditions, this vegetation type consists almost entirely of grasses, with less than 10% forbs and shrubs (Heerwagen, 1956). The dominant grass is blue grama (*Bouteloua gracilis*); other typical grasses include sideoats grama (*Bouteloua curtipendula*) and black grama (*Bouteloua eriopoda*). Typical forbs include red globemallow (*Sphaeralcea coccinea*), curly cup gumweed (*Grindelia squarrosa*) and zinnia (*Zinnia grandiflora*) (Fig. 3). In disturbed areas, rabbitbrush (*Chrysothamnus spp.*), woolly mullein (*Verbascum thapsus*) and snakeweed (*Gutierrezia sarothrae*) are common. Winter-fat (*Ceratoides lanata*) is a common shrub along the roadsides in this vegetation type, and apache plume (*Fallugia paradoxa*) is locally abundant (Fig. 4). In the vicinity of Quemado, grasslands have been invaded by rabbitbrush, a typical response to overgrazing. According to Dick-Peddie (1993, p.106), " It is not surprising that when stocking rates similar to those in Texas and Oklahoma were used on the relatively xeric and open grama grass (black grama), found on slopes with shallow soils, there was an extensive and rapid succession

TABLE 1. Plants identified in September and October 1993 along route of the 1994 New Mexico Geological Society Field Conference. Key to community types: 1 – Desert Scrub (less than 4800 ft elevation), 2 – Grassland (4800 to 6000 ft), 3 – Pinyon-Juniper Savanna and Woodland (6000 to 7000 ft), 4 – Pine-Oak Woodland (6500 to 8000 ft), 5 – Fir-Spruce-Aspen Forest (above 8000 ft, 6 – Riparian and Wetland communities.

Species	Common name	Community type	Species	Common name	Community type
Alnus tenuifolia	Thin-leaf alder	6	*Machaeranthera tanacetifolia*	Aster	1 2 3
Argemone sp.	Prickly poppy	2 3 4	*Machaeranthera tephrodes*	Gray aster	1 2 3
Aristida sp.	Three-awn	1	*Marubium vulgare*	Horehound	2 3 4
Artimisia filifolia	Sand Sage	1 2	*Mentzzelia rusbyi*	Rusby stickleaf	1 2 3 4
Artimisia frigida	Fringed Sage	2 3	*Mirabilis multiflora*	Wild four o'clock	3 4
Asclepias latifolia	Broadleaf milkweed	1 2 3	*Mirabilis oxybaphoides*	Wild four o'clock	3 4 5
Atriplex canescens	Four-winged Saltbush	1 2 3	*Monarda punctata*	Spotted horsemint	2 3 4
Berlanddiera lyrata	Green-eyes	1 2 3	*Muhlenbergia torreyi*	Ring muhly	1 2
Botriochloa laguroides	Silver bluestem	1 2 3	*Oenothera pallida*	Pale evening primrose	2 3 4
Bouteloua gracilis	Blue grama	1 2 3	*Opuntia imbricata*	Cholla	1 2
Bouteloua hirsuta	Hairy grama	3	*Oryzopsis bloomeri*	Indian rice grass	
Boutelous curtipendula	Sideoats Grama	1 2 3	*Parthenocissus vitacea*	Virginia creeper	4 5 6
Brickellia californica	Brickelbush	2 3	*Penstemon ambigues*	Penstemon	
Brickellia grandiflora	Brickellia	1 2 3 4	*Physalis hederaefolia*	Groundcherry	1 2 3
Castilleja integra	Indian paintbrush	2 3 4	*Picea engelmannii*	Engelmann spruce	5
Cercocarpus montanus	Mountain mahogany	4	*Picea pungens*	Blue spruce	5
Chrysothamnus nauseos	Rabbitbrush	1 2 3 4	*Pinus edulis*	Pinyon pine	3 4
Cirsium drummondi	Dwarf thistle	3 4 5	*Pinus monophylla*	Single-leaf pinon	3 4
Cirsium sp.	Yellow-spined thistle	1 2	*Pinus ponderosa*	Ponderosa pine	3 4 5
Clematis ligusticifolia	Goat's beard	2 3 4	*Populus angustifolia*	Narrowleaf cottonwood	5 6
Cleome serrulata	Rocky Mountain Bee plan	1 2 3 4	*Populus tremuloides*	Quaking aspen	5
Convolvulus incanus	Bindweed	1 2 3	*Proboscidea parviflora*	Devil's claw	2 3
Cosmos bipinnatus	Cosmos	2 3	*Prosopis glandulosa*	Honey Mesquite	1 2
Cowania mexicana	Mexican cliffrose	3 4	*Pseudotsuga menziesii*	Douglas fir	5
Cucurbita foetidessima	Buffalo gourd	2 3	*Psilostrophe tagetina*	Paper Daisy	2 3 4
Datura quercifolia	Oak-leaved thornapple	1 2 3	*Quercus arizonica*	Arizona oak	3 4
Datura stramonium	Jimsonweed	2 3	*Quercus gambellii*	Gambel's oak	3 4
Dithyrea wislizenii	Spectacle-pod	1 2 3	*Quercus grisea*	Gray oak	4
Dyssodia papposa	Fetid marigold	1 2 3 4	*Quercus turbinella*	Shrub live oak	1 2
Engelmannia pinnatifid	Engelmann daisy	2 3	*Rhamnus sp.*	Buckthorn	
Ephedra sp.	Mormon tea	1 2	*Rhus microphylla*	Little-leaf Sumac	1 2
Eragrostis curvula	Weeping lovegrass		*Rhus trilobata*	Skunkbush	1 2 3
Erigeron sp.	Fleabane	2 3 4	*Ribes wolfii*	Wolf currant	3 4
Eriogonum jamesii	Antelope sage	3 4	*Robinia neomexicana*	New Mexican locust	4
Eriogonum sp.	Whild buckwheat	2 3 4	*Rosa sp.*	Wild rose	3 4 5 6
Eurotia lanata	Winter fat	2 3	*Sambucus sp.*	Elderberry	5
Fallugia paradoxa	Apache plume	1 2 3	*Senecio douglasii*	Groundsel	1 2 3
Forestiera neomexicana	New Mexico forestiera	4	*Senecio multicapitatus*	Groundsel	2 3 4
Gaura pauciflora	Small-flowered gaura	1 2 3	*Sitanion hystrix*	Squirreltail	
Geranium fremontii	Fremont geranium	3 4 5	*Solanum heterodoxum*	Melonleaf nightshade	1 2 3
Grindelia squarrosa	Curlycup gumweed	2 3 4	*Solanum texanum*	Texas cotton	1 2 3
Gutierrezia sarothrae	Snakeweed	1 2 3 4	*Sphaeralcea grossulariaefo*	Globemallow	2 3
Helianthus sp.	Sunflower	1 2 3	*Sphaeralcea incana*	Globemallow	1 2 3 4
Ipomoea cristulata	Scarlet creeper	3	*Sporobolus contractus*	Spike dropseed	
Ipomopsis aggregata	Skyrocket or Scarlet gilia	2 3 4 5	*Sporobolus sp.*	Dropseed	
Ipomopsis longiflora	Throated trumpet	1 2 3	*Sporobolus texanus*	Texas dropseed	
Ipomopsis multiflora	Many-flowered ipomopsis	2 3	*Stephanomeria pauciflora*	Wire Lettuce	1 2 3
Juniperus deppeana	Alligator juniper	3 4	*Tamarix pentrandra*	Salt Cedar	6
Juniperus monosperma	One-seed juniper	3	*Tragopogon sp.*	Salsify	1 2
Juniperus osteosperma	Utah juniper	3	*Tribulus terrestris*	Goathead	1 2 3
Juniperus scopulorum	Rocky Mountain juniper	3	*Verbascum thapsus*	Mullein or Miner's Candle	1 2 3 4
Larrea tridentata	Creosotebush	1	*Verbena ciliata*	Creeping vervain	1 2 3
Lepidium montanum	Narrow-leaved peppergrass	1 2 3	*Vitis arizonica*	Canyon grape	3 4 5 6
Linum vernale	Flax	2	*Yucca elata*	Soaptree yucca	2 3
Machaeranthera bigelovii	Purple aster	1 2 3	*Zinnia grandiflora*	Zinnia	1 2 3

of these grasslands toward Desert Grassland or desert scrublands." The controversy over the roles that various factors play in the development and maintenance of grassland are admirably summarized by Anderson (1982, p. 298), " It is my opinion that the apparent confusion regarding the roles of fire and climate in grassland maintenance and distribution can be explained by understanding that fire and climate do not have separate, unrelated effects on grasslands, and that the response of grasslands to fire can be quite different under different climatic regimes. Furthermore, other factors, including grazing animals, edaphic patterns, and landscape topography, need to be considered to understand the mosaic pattern of forest, savannah, and grassland vegetation that can develop under a single climate."

JUNIPER SAVANNA

The Juniper Savanna plant association can be viewed as the result of the invasion of trees and shrubs onto grasslands (Anderson, 1982). In New Mexico, this ecotone between woodland and grassland is widespread and has been expanding during this century (Dick-Peddie, 1993). The Juniper Savanna in western New Mexico consists mostly of one-seed juniper (*Juniperus monosperma*) and Rocky Mountain

FIGURE 1. View southwest of the Plains of San Agustin and the VLA (Very Large Array) radiotelescope in the background. Low sand dunes in the foreground support a growth of sunflowers (*Helianthus sp.*), snakeweed and sparse grama grasses.

juniper (*Juniperus scopulorum*). In some areas pinyon pine is locally abundant. The dominant grasses are blue grama and sideoats grama. A large number of wildflowers, many still blooming in early fall, include the abundant zinnia and purple asters (*Machaeranthera sp.*).

CONIFEROUS AND MIXED WOODLAND

This vegetation type typically consists of a mixture of junipers, oaks and pinyon pine, forming a woodland carpeted with a layer of blue grama. It merges with pine forest (Fig. 5) at higher elevations and with juniper savanna and juniper woodland at lower elevations. In southwestern New Mexico the typical oak is either gambel oak (*Quercus gambelii*) or gray oak (*Quercus grisea*). Alligator juniper (*Juniperus deppeana*) may be locally abundant at elevations between 6000 and 8000 feet (Fig. 6). In the shade of the trees flowering plants can be found, including wild four-o'clock (*Mirabilis multiflora*) and zinnia (*Zinnia grandiflora*).

MONTANE AND SUBALPINE CONIFEROUS FOREST

In southwestern New Mexico the lowest elevation Montane Forest is dominated by ponderosa pine (*Pinus ponderosa*) and gambel oak, the so-called Pine-Oak Belt. At higher elevations aspen (*Populus tremuloides*) is locally abundant and readily recognized in the fall by the golden color of its leaves. At elevations above 8000 feet, in the vicinity of Alpine, Arizona, other conifers can be found, including Douglas fir (*Pseudotsuga menziesii*), blue spruce (*Picea pungens*), Engelmann spruce (*Picea engelmannii*) and limber pine (*Pinus flexilis*)—typical trees of the Fir-Aspen Belt (Table 1). Out of the carpet of pine and spruce needles wildflowers can be found, including the Fremont's geranium (*Geranium fremontii*).

RIPARIAN AND WETLAND HABITATS

Riparian and wetland habitats occur at all elevations, although because of reduced temperatures at the higher elevations, water there usually occurs throughout the growing season. At lower elevations of the Mogollon Slope, riparian habitats are characterized by narrowleaf cottonwood (*Populus angustifolia*), and a variety of shrubs and vines. Common shrubs and trees observed in the riparian zone include willows (*Salix spp.*), water birch (*Betula occidentalis*) and thinleaf alder (*Alnus tenuifolia*). Skunkbush (*Rhus trilobata*) and Arizona walnut

FIGURE 2. Pinyon-juniper savanna, 10 mi west of Magdalena.

FIGURE 3. The yellow-orange blooms of the native wild zinnia are an abundant component of the fall landscape in western New Mexico. The short (3 to 9 in. high) leafy stems support round clumps of yellowish-orange flowers 1-1.5 in. in diameter.

FIGURE 4. The feathery plumes of the apache plume, a shrub that grows to heights of 3 to 6 ft, are a striking and common sight along the roadsides of western New Mexico. The white rose-like blooms indicate that this plant is a member of the rose family.

(*Juglans major*) are common on the floodplains of major streams. Common vines include goatsbeard (*Clematis ligusticifolia*), Virginia creeper (*Parthenocissus vitaceae*), canyon grape (*Vitis arizonica*) and hops (*Humulus americanus*) (Fig. 7). The primitive horsetail (*Equisetum spp*.) is locally abundant in this habitat. In locations where standing water is present much of the year, cattails (*Typha latifolia*), smartweeds (*Polygonum spp*.) and sedges (*Scirpus spp)* are common.

FIGURE 5. Ponderosa pine forest with an understory of Gambel's oak, 19 mi south of Springerville, Arizona.

FIGURE 6. The thick and deeply furrowed bark give the alligator juniper its name. However, only mature trees exhibit this alligator skin texture. It may grow to a height of 40 ft (Elmore, 1976) and is common in the vicinity of Reserve.

FIGURE 7. The hop vine (*Humulus americanus*) occurs in the Rocky Mountain region (Kearney and Peebles, 1951), at elevations of 5500 to 9500 ft. The pistillate flowers of the closely related European species are used in the brewing of beer to impart a bitter flavor. According to Kearney and Peebles (1951), the southwestern plant was utilized by Native Americans, the Apache name for this plant meaning, "to make bread with it".

SUMMARY

The grasslands of western New Mexico merge at lower elevations with desert scrubland species typical of the Chihuahuan Desert and at higher elevations with coniferous trees ranging from juniper and piny-on pine to firs and spruce. The mosaic of vegetational types is the result of complex and incompletely understood interactions between natural and human disturbances. It hosts a variety of plant species for us to appreciate and admire.

ACKNOWLEDGMENTS

We thank Robert A. Wright and Larry C. Higgins for reviewing this manuscript. The editorial comments and revisions by Jan Pomeroy were helpful and are sincerely appreciated. The herbarium at the Department of Biology and Geosciences at West Texas A & M University was invaluable in comparison and identification of plant specimens collected for this project.

REFERENCES

Anderson, R. C., 1982, An evolutionary model summarizing the roles of fire, climate, and grazing animals in the origin and maintenance of grasslands: an end paper; *in* Ester, J. R., Tyrl, R. J. and Brunken, J. D., eds., Grasses and grasslands—systematics and ecology: University of Oklahoma Press, Norman, p. 297-308.

Arnberger, L. P., 1982, Flowers of the southwest mountains: Southwest Parks and Monuments Association, Tucson, Arizona, 139 p.

Correll, D. S. and Johnston, M. C., 1970, Manual of the vascular plants of Texas: Texas Research Foundation, Renner, Texas.

Dick-Peddie, W. A., 1993, New Mexico vegetation, past, present and future: University of New Mexico Press, Albuquerque, 244 p.

Dodge, N. N., 1973, Flowers of the Southwest deserts: Southwest Parks and Monuments Association, Globe, Arizona, 112 p.

Elmore, F. H., 1976, Shrubs and trees of the Southwest uplands: Southwest Parks and Monuments Association, Tucson, Arizona, 214 p.

Heerwagen, A., 1956, Mixed prairie in New Mexico; *in* Weaver, J. E. and Albertson, F. W., Grasslands of the Great Plains: Johnson Publishing Company, Lincoln, Nebraska, p. 284-300.

Herbel, C. H., Ares, F. N. and Wright, R. A., 1972, Drought effects on a semi-desert grassland range: Ecology, v. 53, p. 1084-1093.

Ivey, R. D., 1986, Flowering plants of New Mexico, second edition: Published by the Author, 371 p.

Kearney, T. H. and Peebles, R. H., 1951, Arizona flora: University of California Press, Berkeley, 1085 p.

MacMahon, J. A., 1988, Warm deserts; *in* Barbour, M. G. and Billings, W. D., eds., North American terrestrial vegetation: Cambridge University Press, Cambridge, p. 231-264.

Martin, W. C. and Hutchins, Charles R., 1988, Fall wildflowers of New Mexico: University of New Mexico Press, Albuquerque, 294 p.

Patraw, P. M., 1951, Flowers of the Southwest mesas: Southwest Parks and Monuments Association, Globe, Arizona, 112 p.

Sims, P. L., 1988, Grasslands; *in* Barbour, M. G. and Billings, W. D., eds., North American Terrestrial Vegetation: Cambridge University Press, Cambridge, p. 265-286.

Spellenberg, R., 1979, The Audubon Society field guide to North American wildflowers - western region: Alfred A. Knopf, New York, 862 p.

Whitson, T. D., ed., 1991, Weeds of the West: The Western Society of Weed Science, Western United States Land Grant Universities Cooperative Extension Service, and the University of Wyoming, 629 p.